ELECTRO

THEORY

BY

JULIUS ADAMS STRATTON

Professor of Physics

Massachusetts Institute of Technology

PREFACE

The pattern set nearly 70 years ago by Maxwell's *Treatise on Electricity and Magnetism* has had a dominant influence on almost every subsequent English and American text, persisting to the present day. The *Treatise* was undertaken with the intention of presenting a connected account of the entire known body of electric and magnetic phenomena from the single point of view of Faraday. Thus it contained little or no mention of the hypotheses put forward on the Continent in earlier years by Riemann, Weber, Kirchhoff, Helmholtz, and others. It is by no means clear that the complete abandonment of these older theories was fortunate for the later development of physics. So far as the purpose of the *Treatise* was to disseminate the ideas of Faraday, it was undoubtedly fulfilled; as an exposition of the author's own contributions, it proved less successful. By and large, the theories and doctrines peculiar to Maxwell—the concept of displacement current, the identity of light and electromagnetic vibrations—appeared there in scarcely greater completeness and perhaps in a less attractive form than in the original memoirs. We find that all of the first volume and a large part of the second deal with the stationary state. In fact only a dozen pages are devoted to the general equations of the electromagnetic field, 18 to the propagation of plane waves and the electromagnetic theory of light, and a score more to magnetooptics, all out of a total of 1,000. The mathematical completeness of potential theory and the practical utility of circuit theory have influenced English and American writers in very nearly the same proportion since that day. Only the original and solitary genius of Heaviside succeeded in breaking away from this course.

For an exploration of the fundamental content of Maxwell's equations one must turn again to the Continent. There the work of Hertz, Poincaré, Lorentz, Abraham, and Sommerfeld, together with their associates and successors, has led to a vastly deeper understanding of physical phenomena and to industrial developments of tremendous proportions.

The present volume attempts a more adequate treatment of variable electromagnetic fields and the theory of wave propagation. Some attention is given to the stationary state, but for the purpose of introducing fundamental concepts under simple conditions, and always with a view to later application in the general case. The reader must possess a general knowledge of electricity and magnetism such as may be acquired from an elementary course based on the experimental laws of Coulomb,

Ampère, and Faraday, followed by an intermediate course dealing with the more general properties of circuits, with thermionic and electronic devices, and with the elements of electromagnetic machinery, terminating in a formulation of Maxwell's equations. This book takes up at that point. The first chapter contains a general statement of the equations governing fields and potentials, a review of the theory of units, reference material on curvilinear coordinate systems and the elements of tensor analysis, concluding with a formulation of the field equations in a space-time continuum. The second chapter is also general in character, and much of it may be omitted on a first reading. Here one will find a discussion of fundamental field properties that may be deduced without reference to particular coordinate systems. A dimensional analysis of Maxwell's equations leads to basic definitions of the vectors **E** and **B**, and an investigation of the energy relations results in expressions for the mechanical force exerted on elements of charge, current, and neutral matter. In this way a direct connection is established between observable forces and the vectors employed to describe the structure of a field.

In Chaps. III and IV stationary fields are treated as particular cases of the dynamic field equations. The subject of wave propagation is taken up first in Chap. V, which deals with homogeneous plane waves. Particular attention is given to the methods of harmonic analysis, and the problem of dispersion is considered in some detail. Chapters VI and VII treat the propagation of cylindrical and spherical waves in unbounded spaces. A necessary amount of auxiliary material on Bessel functions and spherical harmonics is provided, and consideration is given to vector solutions of the wave equation. The relation of the field to its source, the general theory of radiation, and the outlines of the Kirchhoff-Huygens diffraction theory are discussed in Chap. VIII.

Finally, in Chap. IX, we investigate the effect of plane, cylindrical, and spherical surfaces on the propagation of electromagnetic fields. This chapter illustrates, in fact, the application of the general theory established earlier to problems of practical interest. The reader will find here the more important laws of physical optics, the basic theory governing the propagation of waves along cylindrical conductors, a discussion of cavity oscillations, and an outline of the theory of wave propagation over the earth's surface.

It is regrettable that numerical solutions of special examples could not be given more frequently and in greater detail. Unfortunately the demands on space in a book covering such a broad field made this impractical. The primary objective of the book is a sound exposition of electromagnetic theory, and examples have been chosen with a view to illustrating its principles. No pretense is made of an exhaustive treatment of antenna design, transmission-line characteristics, or similar topics of engineering importance. It is the author's hope that the present volume will provide the fundamental background necessary for a critical appreciation of original contributions in special fields and satisfy the needs of those who are unwilling to accept engineering formulas without knowledge of their origin and limitations.

Each chapter, with the exception of the first two, is followed by a set of problems. There is only one satisfactory way to study a theory, and that is by application to specific examples. The problems have been chosen with this in mind, but they cover also many topics which it was necessary to eliminate from the text. This is particularly true of the later chapters. Answers or references are provided in most cases.

This book deals solely with large-scale phenomena. It is a sore temptation to extend the discussion to that fruitful field which Frenkel terms the "quasi-microscopic state," and to deal with the many beautiful results of the classical electron theory of matter. In the light of contemporary developments, anyone attempting such a program must soon be overcome with misgivings. Although many laws of classical electrodynamics apply directly to submicroscopic domains, one has no basis of selection. The author is firmly convinced that the transition must be made from quantum electrodynamics toward classical theory, rather than in the reverse direction. Whatever form the equations of quantum electrodynamics ultimately assume, their statistical average over large numbers of atoms must lead to Maxwell's equations.

The m.k.s. system of units has been employed exclusively. There is still the feeling among many physicists that this system is being forced upon them by a subversive group of engineers. Perhaps it is, although it was Maxwell himself who first had the idea. At all events, it is a good system, easily learned, and one that avoids endless confusion in practical applications. At the moment there appears to be no doubt of its universal adoption in the near future. Help for the tories among us who hold to the Gaussian system is offered on page 241.

In contrast to the stand taken on the m.k.s. system, the author has no very strong convictions on the matter of rationalized units. Rationalized units have been employed because Maxwell's equations are taken as the starting point rather than Coulomb's law, and it seems reasonable to make the point of departure as simple as possible. As a result of this choice all equations dealing with energy or wave propagation are free from the factor 4π. Such relations are becoming of far greater practical importance than those expressing the potentials and field vectors in terms of their sources.

The use of the time factor $e^{-i\omega t}$ instead of $e^{+i\omega t}$ is another point of mild controversy. This has been done because the time factor is invariably discarded, and it is somewhat more convenient to retain the positive exponent e^{+ikR} for a positive traveling wave. To reconcile any formula with its engineering counterpart, one need only replace $-i$ by $+j$.

The author has drawn upon many sources for his material and is indebted to his colleagues in both the departments of physics and of electrical engineering at the Massachusetts Institute of Technology. Thanks are expressed particularly to Professor M. F. Gardner whose advice on the practical aspects of Laplace transform theory proved invaluable, and to Dr. S. Silver who read with great care a part of the manuscript. In conclusion the author takes this occasion to express his sincere gratitude to Catherine N. Stratton for her constant encouragement during the preparation of the manuscript and untiring aid in the revision of proof.

JULIUS ADAMS STRATTON.

CONTENTS

PAGE

CHAPTER II

STRESS AND ENERGY

CHAPTER III

THE ELECTROSTATIC FIELD

CHAPTER IV

THE MAGNETOSTATIC FIELD

CHAPTER V

PLANE WAVES IN UNBOUNDED, ISOTROPIC MEDIA

CHAPTER VI

CYLINDRICAL WAVES

CHAPTER VII

SPHERICAL WAVES

CHAPTER VIII

RADIATION

CHAPTER IX

BOUNDARY-VALUE PROBLEMS

APPENDIX I

APPENDIX II

APPENDIX III

APPENDIX IV

ELECTROMAGNETIC THEORY

CHAPTER I

THE FIELD EQUATIONS

A vast wealth of experimental evidence accumulated over the past century leads one to believe that large-scale electromagnetic phenomena are governed by Maxwell's equations. Coulomb's determination of the law of force between charges, the researches of Ampère on the interaction of current elements, and the observations of Faraday on variable fields can be woven into a plausible argument to support this view. The historical approach is recommended to the beginner, for it is the simplest and will afford him the most immediate satisfaction. In the present volume, however, we shall suppose the reader to have completed such a preliminary survey and shall credit him with a general knowledge of the experimental facts and their theoretical interpretation. Electromagnetic theory, according to the standpoint adopted in this book, is the theory of Maxwell's equations. Consequently, we shall postulate these equations at the outset and proceed to deduce the structure and properties of the field together with its relation to the source. No single experiment constitutes proof of a theory. The true test of our initial assumptions will appear in the persistent, uniform correspondence of deduction with observation.

In this first chapter we shall be occupied with the rather dry business of formulating equations and preparing the way for our investigation.

MAXWELL'S EQUATIONS

1.1. The Field Vectors.—By an electromagnetic field let us understand the domain of the four vectors **E** and **B**, **D** and **H**. These vectors are assumed to be finite throughout the entire field, and at all ordinary points to be continuous functions of position and time, with continuous derivatives. Discontinuities in the field vectors or their derivatives may occur, however, on surfaces which mark an abrupt change in the physical properties of the medium. According to the traditional usage, **E** and **H** are known as the intensities respectively of the electric and magnetic field, **D** is called the electric displacement and **B**, the magnetic induction. Eventually the field vectors must be defined in terms of the experiments by which they can be measured. Until these experiments are formulated, there is no reason to consider one vector more fundamental than another, and we shall apply the word intensity to mean indiscriminately the strength or magnitude of any of the four vectors at a point in space and time.

The source of an electromagnetic field is a distribution of electric charge and current. Since we are concerned only with its macroscopic effects, it may be assumed that this distribution is continuous rather than discrete, and specified as a function of space and time by the density of charge ρ, and by the vector current density $\mathbf{J}$.

We shall now *postulate* that at every ordinary point in space the field vectors are subject to the Maxwell equations:

$$\nabla \times \mathbf{E} + \frac{\partial \mathbf{B}}{\partial t} = 0, \tag{1}$$

$$\nabla \times \mathbf{H} - \frac{\partial \mathbf{D}}{\partial t} = \mathbf{J}. \tag{2}$$

By an ordinary point we shall mean one in whose neighborhood the physical properties of the medium are continuous. It has been noted that the transition of the field vectors and their derivatives across a surface bounding a material body may be discontinuous; such surfaces must, therefore, be excluded until the nature of these discontinuities can be investigated.

1.2. Charge and Current.—Although the corpuscular nature of electricity is well established, the size of the elementary quantum of charge is too minute to be taken into account as a distinct entity in a strictly macroscopic theory. Obviously the frontier that marks off the domain of large-scale phenomena from those which are microscopic is an arbitrary one. To be sure, a macroscopic element of volume must contain an enormous number of atoms; but that condition alone is an insufficient criterion, for many crystals, including the metals, exhibit frequently a microscopic "grain" or "mosaic" structure which will be excluded from our investigation. We are probably well on the safe side in imposing a limit of one-tenth of a millimeter as the smallest admissible element of length. There are many experiments, such as the scattering of light by particles no larger than 10^{-3} mm. in diameter, which indicate that the macroscopic theory may be pushed well beyond the limit suggested. Nonetheless, we are encroaching here on the proper domain of quantum theory, and it is the quantum theory which must eventually determine the validity of our assumptions in microscopic regions.

Let us suppose that the charge contained within a volume element Δv is Δq. The *charge density* at any point within Δv will be defined by the relation

$$\Delta q = \rho \, \Delta v. \tag{3}$$

Thus by the charge density at a point we mean the average charge per unit volume in the neighborhood of that point. In a strict sense (3) does not define a continuous function of position, for Δv cannot approach zero without limit. Nonetheless we shall assume that ρ can be represented by a function of the coordinates and the time which at ordinary points is continuous and has continuous derivatives. The value of the total charge obtained by integrating that function over a large-scale volume will then differ from the true charge contained therein by a microscopic quantity at most.

Any ordered motion of charge constitutes a current. A current distribution is characterized by a vector field which specifies at each point not only the intensity of the flow but also its direction. As in the study of fluid motion, it is convenient to imagine streamlines traced through the distribution and everywhere tangent to the direction of flow. Consider a surface which is orthogonal to a system of streamlines. The *current density* at any point on this surface is then defined as a vector $\mathbf{J}$ directed along the streamline through the point and equal in magnitude to the charge which in unit time crosses unit area of the surface in the vicinity of the point. On the other hand the current I across *any* surface S is equal to the rate at which charge crosses that surface. If $\mathbf{n}$ is the positive unit normal to an element Δa of S, we have

$$\Delta I = \mathbf{J} \cdot \mathbf{n} \, \Delta a. \tag{4}$$

Since Δa is a macroscopic element of area, Eq. (4) does not define the current density with mathematical rigor as a continuous function of position, but again one may represent the distribution by such a function without incurring an appreciable error. The total current through S is, therefore,

$$I = \int_S \mathbf{J} \cdot \mathbf{n} \, da. \tag{5}$$

Since electrical charge may be either positive or negative, a convention must be adopted as to what constitutes a positive current. If the flow through an element of area consists of positive charges whose velocity vectors form an angle of less than 90 deg. with the positive normal $\mathbf{n}$, the current is said to be positive. If the angle is greater than 90 deg., the current is negative. Likewise if the angle is less than 90 deg. but the charges are negative, the current through the element is negative. In the case of metallic conductors the carriers of electricity are presumably negative electrons, and the direction of the current density vector is therefore opposed to the direction of electron motion.

Let us suppose now that the surface S of Eq. (5) is closed. We shall adhere to the customary convention that *the positive normal to a closed surface is drawn outward.* In virtue of the definition of current as the flow of charge across a surface, it follows that the surface integral of the normal component of **J** over S must measure the loss of charge from the region within. There is no experimental evidence to indicate that under ordinary conditions charge may be either created or destroyed in macroscopic amounts. One may therefore write

$$\int_S \mathbf{J} \cdot \mathbf{n}\, da = -\frac{d}{dt}\int_V \rho\, dv, \tag{6}$$

where V is the volume enclosed by S, as a relation expressing the *conservation of charge.* The flow of charge across the surface can originate in two ways. The surface S may be fixed in space and the density ρ be some function of the time as well as of the coordinates; or the charge density may be invariable with time, while the surface moves in some prescribed manner. In this latter event the right-hand integral of (6) is a function of time in virtue of variable limits. If, however, the surface is fixed and the integral convergent, one may replace d/dt by a partial derivative under the sign of integration.

$$\int_S \mathbf{J} \cdot \mathbf{n}\, da = -\int_V \frac{\partial \rho}{\partial t}\, dv. \tag{7}$$

We shall have frequent occasion to make use of the divergence theorem of vector analysis. Let $\mathbf{A}(x, y, z)$ be any vector function of position which together with its first derivatives is continuous throughout a volume V and over the bounding surface S. The surface S is regular but otherwise arbitrary.[1] Then it can be shown that

$$\int_S \mathbf{A} \cdot \mathbf{n}\, da = \int_V \nabla \cdot \mathbf{A}\, dv. \tag{8}$$

As a matter of fact, this relation may be advantageously used as a definition of the divergence. To obtain the value of $\nabla \cdot \mathbf{A}$ at a point P within V, we allow the surface S to shrink about P. When the volume V has become sufficiently small, the integral on the right may be replaced by $V\nabla \cdot \mathbf{A}$, and we obtain

$$\nabla \cdot \mathbf{A} = \lim_{S \to 0} \frac{1}{V}\int_S \mathbf{A} \cdot \mathbf{n}\, da. \tag{9}$$

[1] A regular element of arc is represented in parametric form by the equations $x = x(t)$, $y = y(t)$, $z = z(t)$ such that in the interval $a \leq t \leq b$ x, y, z are continuous, single-valued functions of t with continuous derivatives of all orders unless otherwise restricted. A regular curve is constructed of a finite number of such arcs joined end to end but such that the curve does not cross itself. Thus a regular curve has no double points and is piecewise differentiable. A regular surface element is a portion of surface whose projection on a properly oriented plane is the interior of a regular closed curve. Hence it does not intersect itself. *Cf.* Kellogg, "Foundations of Potential Theory," p. 97, Springer, 1929.

The divergence of a vector at a point is, therefore, to be interpreted as the integral of its normal component over an infinitesimally small surface enclosing that point, divided by the enclosed volume. The flux of a vector through a closed surface is a measure of the sources within; hence the divergence determines their strength at a point. Since S has been shrunk close about P, the value of $\mathbf{A}$ at every point on the surface may be expressed analytically in terms of the values of $\mathbf{A}$ and its derivatives at P, and consequently the integral in (9) may be evaluated, leading in the case of rectangular coordinates to

$$(10) \qquad \nabla \cdot \mathbf{A} = \frac{\partial A_x}{\partial x} + \frac{\partial A_y}{\partial y} + \frac{\partial A_z}{\partial z}.$$

On applying this theorem to (7) the surface integral is transformed to the volume integral

$$(11) \qquad \int_V \left(\nabla \cdot \mathbf{J} + \frac{\partial \rho}{\partial t} \right) dv = 0.$$

Now the integrand of (11) is a continuous function of the coordinates and hence there must exist small regions within which the integrand does not change sign. If the integral is to vanish for arbitrary volumes V, it is necessary that the integrand be identically zero. The differential equation

$$(12) \qquad \nabla \cdot \mathbf{J} + \frac{\partial \rho}{\partial t} = 0$$

expresses the conservation of charge in the neighborhood of a point. By analogy with an equivalent relation in hydrodynamics, (12) is frequently referred to as the *equation of continuity.*

If at every point within a specified region the charge density is constant, the current passing into the region through the bounding surface must at all times equal the current passing outward. Over the bounding surface S we have

$$(13) \qquad \int_S \mathbf{J} \cdot \mathbf{n} \, da = 0,$$

and at every interior point

$$(14) \qquad \nabla \cdot \mathbf{J} = 0.$$

Any motion characterized by vector or scalar quantities which are independent of the time is said to be steady, or stationary. A steady-state flow of electricity is thus defined by a vector $\mathbf{J}$ which at every point within the region is constant in direction and magnitude. In virtue of the divergenceless character of such a current distribution, it follows that in the steady state all streamlines, or current filaments, close upon themselves. The field of the vector **J** is solenoidal.

1.3. Divergence of the Field Vectors.—Two further conditions satisfied by the vectors **B** and **D** may be deduced directly from Maxwell's equations by noting that the divergence of the curl of any vector vanishes identically. We take the divergence of Eq. (1) and obtain

$$\nabla \cdot \frac{\partial \mathbf{B}}{\partial t} = \frac{\partial}{\partial t} \nabla \cdot \mathbf{B} = 0. \tag{15}$$

The commutation of the operators ∇ and $\partial/\partial t$ is admissible, for at an ordinary point **B** and all its derivatives are assumed to be continuous. It follows from (15) that at every point in the field the divergence of **B** is constant. If ever in its past history the field has vanished, this constant must be zero and, since one may reasonably suppose that the initial generation of the field was at a time not infinitely remote, we conclude that

$$\nabla \cdot \mathbf{B} = 0, \tag{16}$$

and the field of **B** is therefore solenoidal.

Likewise the divergence of Eq. (2) leads to

$$\nabla \cdot \mathbf{J} + \frac{\partial}{\partial t} \nabla \cdot \mathbf{D} = 0, \tag{17}$$

or, in virtue of (12), to

$$\frac{\partial}{\partial t} (\nabla \cdot \mathbf{D} - \rho) = 0. \tag{18}$$

If again we admit that at some time in its past or future history the field may vanish, it is necessary that

$$\nabla \cdot \mathbf{D} = \rho. \tag{19}$$

The charges distributed with a density ρ constitute the sources of the vector **D**.

The divergence equations (16) and (19) are frequently included as part of Maxwell's system. It must be noted, however, that if one assumes the conservation of charge, these are not independent relations.

1.4. Integral Form of the Field Equations.—The properties of an electromagnetic field which have been specified by the differential equations (1), (2), (16), and (19) may also be expressed by an equivalent system of integral relations. To obtain this equivalent system, we apply a second fundamental theorem of vector analysis.

According to Stokes' theorem the line integral of a vector taken about a closed contour can be transformed into a surface integral extended over a surface bounded by the contour. The contour C must either be regular or be resolvable into a finite number of regular arcs, and it is assumed that the otherwise arbitrary surface S bounded by C is two-sided and may be resolved into a finite number of regular elements. The positive side of the surface S is related to the positive direction of circulation on the contour by the usual convention that an observer, moving in a positive sense along C, will have the positive side of S on his left. Then if $\mathbf{A}(x, y, z)$ is any vector function of position, which together with its first derivatives is continuous at all points of S and C, it may be shown that

$$\int_C \mathbf{A} \cdot d\mathbf{s} = \int_S (\nabla \times \mathbf{A}) \cdot \mathbf{n}\, da, \tag{20}$$

where $d\mathbf{s}$ is an element of length along C and $\mathbf{n}$ is a unit vector normal to the positive side of the element of area da. This transformation can also be looked upon as an equation defining the curl. To determine the value of $\nabla \times \mathbf{A}$ at a point P on S, we allow the contour to shrink about P until the enclosed area S is reduced to an infinitesimal element of a plane whose normal is in the direction specified by $\mathbf{n}$. The integral on the right is then equal to $(\nabla \times \mathbf{A}) \cdot \mathbf{n}S$, plus infinitesimals of higher order. The projection of the vector $\nabla \times \mathbf{A}$ in the direction of the normal is, therefore,

$$(\nabla \times \mathbf{A}) \cdot \mathbf{n} = \lim_{C \to 0} \frac{1}{S} \int_C \mathbf{A} \cdot d\mathbf{s}. \tag{21}$$

The curl of a vector at a point is to be interpreted as the line integral of that vector about an infinitesimal path on a surface containing the point, per unit of enclosed area. Since $\mathbf{A}$ has been assumed analytic in the neighborhood of P, its value at any point on C may be expressed in terms of the values of $\mathbf{A}$ and its derivatives at P, so that the evaluation of the line integral in (21) about the infinitesimal path can actually be carried out. In particular, if the element S is oriented parallel to the yz-coordinate plane, one finds for the x-component of the curl

$$(\nabla \times \mathbf{A})_x = \frac{\partial A_z}{\partial y} - \frac{\partial A_y}{\partial z}. \tag{22}$$

Proceeding likewise for the y- and z-components we obtain

$$\nabla \times \mathbf{A} = \mathbf{i}\left(\frac{\partial A_z}{\partial y} - \frac{\partial A_y}{\partial z}\right) + \mathbf{j}\left(\frac{\partial A_x}{\partial z} - \frac{\partial A_z}{\partial x}\right) + \mathbf{k}\left(\frac{\partial A_y}{\partial x} - \frac{\partial A_x}{\partial y}\right) = \begin{vmatrix} \mathbf{i} & \mathbf{j} & \mathbf{k} \\ \frac{\partial}{\partial x} & \frac{\partial}{\partial y} & \frac{\partial}{\partial z} \\ A_x & A_y & A_z \end{vmatrix} \tag{23}$$

Let us now integrate the normal component of the vector $\partial \mathbf{B}/\partial t$ over any regular surface S bounded by a closed contour C. From (1) and (20) it follows that

$$\int_C \mathbf{E} \cdot d\mathbf{s} + \int_S \frac{\partial \mathbf{B}}{\partial t} \cdot \mathbf{n}\, da = 0. \tag{24}$$

If the contour is fixed, the operator $\partial/\partial t$ may be brought out from under the sign of integration.

$$\int_C \mathbf{E} \cdot d\mathbf{s} = -\frac{\partial}{\partial t} \int_S \mathbf{B} \cdot \mathbf{n}\, da. \tag{25}$$

By definition, the quantity

$$\Phi = \int_S \mathbf{B} \cdot \mathbf{n}\, da \tag{26}$$

is the magnetic flux, or more specifically the flux of the vector **B** through the surface. According to (25) the line integral of the vector **E** about any closed, regular curve in the field is equal to the time rate of decrease of the magnetic flux through any surface spanning that curve. The relation between the direction of circulation about a contour and the positive normal to a surface bounded by it is illustrated in Fig. 1. A positive direction about C is chosen arbitrarily and the flux Φ is then positive or negative according to the direction of the lines of **B** with respect to the normal. The time rate of change of Φ is in turn positive or negative as the positive flux is increasing or decreasing.

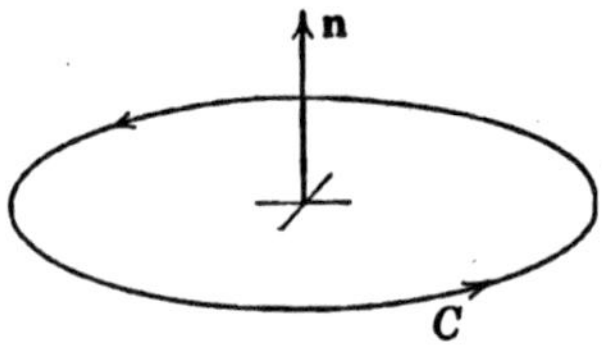

FIG. 1.—Convention relating direction of the positive normal **n** to the direction of circulation about a contour C.

We recall that the application of Stokes' theorem to Eq. (1) is valid only if the vector **E** and its derivatives are continuous at all points of S and C. Since discontinuities in both **E** and **B** occur across surfaces marking sudden changes in the physical properties of the medium, the question may be raised as to what extent (25) represents a general law of the electromagnetic field. One might suppose, for example, that the contour linked or pierced a closed iron transformer core. To obviate this difficulty it may be imagined that at the surface of every material body in the field the physical properties vary rapidly but *continuously* within a thin boundary layer from their values just inside to their values just outside the surface. In this manner all discontinuities are eliminated from the field and (25) may be applied to every closed contour.

The experiments of Faraday indicated that the relation (25) holds whatever the cause of flux variation. The partial derivative implies a variable flux density threading a fixed contour, but the total flux can likewise be changed by a deformation of the contour. To take this into account the Faraday law is written generally in the form

$$\int_C \mathbf{E} \cdot d\mathbf{s} = -\frac{d}{dt}\int_S \mathbf{B} \cdot \mathbf{n}\, da. \tag{27}$$

It can be shown that (27) is in fact a consequence of the differential field equations, but the proof must be based on the electrodynamics of moving bodies which will be touched upon in Sec. 1.22.

In like fashion Eq. (2) may be replaced by an equivalent integral relation,

$$\int_C \mathbf{H} \cdot d\mathbf{s} = I + \frac{d}{dt}\int_S \mathbf{D} \cdot \mathbf{n}\, da. \tag{28}$$

where I is the total current linking the contour as defined in (5). In the steady state, the integral on the right is zero and the conduction current I through any regular surface is equal to the line integral of the vector $\mathbf{H}$ about its contour. If, however, the field is variable, the vector $\partial\mathbf{D}/\partial t$ has associated with it a field $\mathbf{H}$ exactly equal to that which would be produced by a current distribution of density

$$\mathbf{J}' = \frac{\partial \mathbf{D}}{\partial t}. \tag{29}$$

To this quantity Maxwell gave the name "displacement current," a term which we shall occasionally employ without committing ourselves as yet to any particular interpretation of the vector $\mathbf{D}$.

The two remaining field equations (16) and (19) can be expressed in an equivalent integral form with the help of the divergence theorem. One obtains

$$\oint_S \mathbf{B} \cdot \mathbf{n}\, da = 0, \tag{30}$$

stating that the total flux of the vector $\mathbf{B}$ crossing any closed, regular surface is zero, and

$$\oint_S \mathbf{D} \cdot \mathbf{n}\, da = \int_V \rho\, dv = q, \tag{31}$$

according to which the flux of the vector $\mathbf{D}$ through a closed surface is equal to the total charge q contained within. The circle through the sign of integration is frequently employed to emphasize the fact that a contour or surface is closed.

MACROSCOPIC PROPERTIES OF MATTER

1.5. The Inductive Capacities ϵ and μ.—No other assumptions have been made thus far than that an electromagnetic field may be characterized by four vectors **E, B, D,** and **H,** which at ordinary points satisfy Maxwell's equations, and that the distribution of current which gives rise to this field is such as to ensure the conservation of charge. Between the five vectors **E, B, D, H, J** there are but two independent relations, the equations (1) and (2) of the preceding section, and we are therefore obliged to impose further conditions if the system is to be made determinate.

Let us begin with the assumption that at any given point in the field, whether in free space or within matter, the vector **D** may be represented as a function of **E** and the vector **H** as a function of **B**.

$$\mathbf{D} = D(\mathbf{E}), \qquad \mathbf{H} = H(\mathbf{B}). \tag{1}$$

The nature of these functional relations is to be determined solely by the physical properties of the medium in the immediate neighborhood of the specified point. Certain simple relations are of most common occurrence.

1. In *free space*, **D** differs from **E** only by a constant factor, as does **H** from **B**. Following the traditional usage, we shall write

$$\mathbf{D} = \epsilon_0 \mathbf{E}, \qquad \mathbf{H} = \frac{1}{\mu_0}\mathbf{B}. \tag{2}$$

The values and the dimensions of the constants ϵ_0 and μ_0 will depend upon the system of units adopted. In only one of many wholly arbitrary systems does **D** reduce to **E** and **H** to **B** in empty space.

2. If the physical properties of a body in the neighborhood of some interior point are the same in all directions, the body is said to be *isotropic*. At every point in an isotropic medium **D** is parallel to **E** and **H** is parallel to **B**. The relations between the vectors, moreover, are *linear* in almost all the soluble problems of electromagnetic theory. For the isotropic, linear case we put then

$$\mathbf{D} = \epsilon \mathbf{E}, \qquad \mathbf{H} = \frac{1}{\mu}\mathbf{B}. \tag{3}$$

The factors ϵ and μ will be called the inductive capacities of the medium. The dimensionless ratios

$$\kappa_e = \frac{\epsilon}{\epsilon_0}, \qquad \kappa_m = \frac{\mu}{\mu_0}, \tag{4}$$

are independent of the choice of units and will be referred to as the specific inductive capacities. The properties of a *homogeneous* medium are constant from point to point and in this case it is customary to refer to κ_e as the dielectric constant and to κ_m as the permeability. In general, however, one must look upon the inductive capacities as scalar functions of position which characterize the electromagnetic properties of matter in the large.

3. The properties of *anisotropic* matter vary in a different manner along different directions about a point. In this case the vectors **D** and **E**, **H** and **B** are parallel only along certain preferred axes. If it may be assumed that the relations are still linear, as is usually the case, one may express each rectangular component of **D** as a linear function of the three components of **E**.

$$D_x = \epsilon_{11}E_x + \epsilon_{12}E_y + \epsilon_{13}E_z,$$
$$D_y = \epsilon_{21}E_x + \epsilon_{22}E_y + \epsilon_{23}E_z, \tag{5}$$
$$D_z = \epsilon_{31}E_x + \epsilon_{32}E_y + \epsilon_{33}E_z.$$

The coefficients ϵ_{jk} of this linear transformation are the components of a symmetric tensor. An analogous relation may be set up between the vectors **H** and **B**, but the occurrence of such a linear anisotropy in what may properly be called macroscopic problems is rare.

The distinction between the microscopic and macroscopic viewpoints is nowhere sharper than in the interpretation of these parameters ϵ and μ, or their tensor equivalents. A microscopic theory must deduce the physical properties of matter from its atomic structure. It must enable one to calculate not only the average field that prevails within a body but also its local value in the neighborhood of a specific atom. It must tell us how the atom will be deformed under the influence of that local field, and how the aggregate effect of these atomic deformations may be represented in the large by such parameters as ϵ and μ.

We, on the other hand, are from the present standpoint sheer behaviorists. Our knowledge of matter is, to use a large word, purely phenomenological. Each substance is to be characterized electromagnetically in terms of a minimum number of parameters. The dependence of the parameters ϵ and μ on such physical variables as density, temperature, and frequency will be established by experiment. Information given by such measurements sheds much light on the internal structure of matter, but the internal structure is not our present concern.

1.6. Electric and Magnetic Polarization.—To describe the electromagnetic state of a sample of matter, it will prove convenient to introduce two additional vectors. We shall *define* the electric and magnetic polarization vectors by the equations

$$\mathbf{P} = \mathbf{D} - \epsilon_0\mathbf{E}, \qquad \mathbf{M} = \frac{1}{\mu_0}\mathbf{B} - \mathbf{H}. \tag{6}$$

The polarization vectors are thus definitely associated with matter and vanish in free space. By means of these relations let us now eliminate **D** and **H** from the field equations. There results the system

$$\nabla \times \mathbf{E} + \frac{\partial \mathbf{B}}{\partial t} = 0,$$

$$\text{(7)} \qquad \nabla \times \mathbf{B} - \epsilon_0\mu_0 \frac{\partial \mathbf{E}}{\partial t} = \mu_0 \left(\mathbf{J} + \frac{\partial \mathbf{P}}{\partial t} + \nabla \times \mathbf{M}\right),$$

$$\nabla \cdot \mathbf{B} = 0, \qquad \nabla \cdot \mathbf{E} = \frac{1}{\epsilon_0}(\rho - \nabla \cdot \mathbf{P}),$$

which we are free to interpret as follows: *the presence of rigid material bodies in an electromagnetic field may be completely accounted for by an equivalent distribution of charge of density* $-\nabla \cdot \mathbf{P}$, *and an equivalent distribution of current of density* $\frac{\partial \mathbf{P}}{\partial t} + \nabla \times \mathbf{M}$.

In isotropic media the polarization vectors are parallel to the corresponding field vectors, and are found experimentally to be proportional to them if ferromagnetic materials are excluded. The electric and magnetic susceptibilities χ_e and χ_m are defined by the relations

$$\text{(8)} \qquad \mathbf{P} = \chi_e\epsilon_0 \mathbf{E}, \qquad \mathbf{M} = \chi_m \mathbf{H}.$$

Logically the magnetic polarization **M** should be placed proportional to **B**. Long usage, however, has associated it with **H** and to avoid confusion on a matter which is really of no great importance we adhere to this convention. The susceptibilities χ_e and χ_m defined by (8) are dimensionless ratios whose values are independent of the system of units employed. In due course it will be shown that **E** and **B** are force vectors and in this sense are fundamental. **D** and **H** are derived vectors associated with the state of matter. The polarization vector **P** has the dimensions of **D**, not **E**, while **M** and **H** are dimensionally alike. From (3), (6), and (8) it follows at once that the susceptibilities are related to the specific inductive capacities by the equations

$$\text{(9)} \qquad \chi_e = \kappa_e - 1, \qquad \chi_m = \kappa_m - 1.$$

In anisotropic media the susceptibilities are represented by the components of a tensor.

It will be a part of our task in later chapters to formulate experiments by means of which the susceptibility of a substance may be accurately measured. Such measurements show that the electric susceptibility is always positive. In gases it is of the order of 0.0006 (air), but in liquids and solids it may attain values as large as 80 (water). An inherent difference in the nature of the vectors **P** and **M** is indicated by the fact that the magnetic susceptibility χ_m may be either positive or negative. Substances characterized by a positive susceptibility are said to be *paramagnetic*, whereas those whose susceptibility is negative are called *diamagnetic*. The metals of the ferromagnetic group, including iron, nickel, cobalt, and their alloys, constitute a particular class of substances of enormous positive susceptibility, the value of which may be of the order of many thousands. In view of the nonlinear relation of **M** to **H** peculiar to these materials, the susceptibility χ_m must now be interpreted as the slope of a tangent to the **M**-**H** curve at a point corresponding to a particular value of **H**. To include such cases the definition of susceptibility is generalized to

$$\chi_m = \frac{\partial M}{\partial H}. \tag{10}$$

The susceptibilities of all nonferromagnetic materials, whether paramagnetic or diamagnetic, are so small as to be negligible for most practical purposes.

Thus far it has been assumed that a functional relation exists between the vector **P** or **M** and the applied field, and for this reason they may properly be called the *induced* polarizations. Under certain conditions, however, a magnetic field may be associated with a ferromagnetic body in the absence of any external excitation. The body is then said to be in a state of permanent magnetization. We shall maintain our initial assumption that the field both inside and outside the magnet is completely defined by the vectors **B** and **H**. But now the difference of these two vectors at an interior point is a *fixed* vector $\mathbf{M}_0$, which may be called the intensity of magnetization and which bears no functional relationship to **H**. On the contrary the magnetization $\mathbf{M}_0$ must be interpreted as the source of the field. If an external field is superposed on the field of a permanent magnet, the intensity of magnetization will be augmented by the induced polarization **M**. At any interior point we have, therefore,

$$\mathbf{B} = \mu_0(\mathbf{H} + \mathbf{M} + \mathbf{M}_0). \tag{11}$$

Of this induced polarization we can only say for the present that it is a function of the resultant **H** prevailing at the same point. The relation of the resultant field within the body to the intensity of an applied field generated by external sources depends not only on the magnetization $\mathbf{M}_0$ but also upon the shape of the body. There will be occasion to examine this matter more carefully in Chap. IV.

1.7. Conducting Media.—To Maxwell's equations there must now be added a third and last empirical relation between the current density and the field. We shall assume that at any point within a liquid or solid the current density is a function of the field **E**.

$$\mathbf{J} = \mathbf{J}(\mathbf{E}). \tag{12}$$

The distribution of current in an ionized, gaseous medium may depend also on the intensity of the magnetic field, but since electromagnetic phenomena in gaseous discharges are in general governed by a multitude of factors other than those taken into account in the present theory, we shall exclude such cases from further consideration.[1]

Throughout a remarkably wide range of conditions, in both solids and weakly ionized solutions, the relation (12) proves to be linear.

$$\mathbf{J} = \sigma \mathbf{E}. \tag{13}$$

The factor σ is called the *conductivity* of the medium. The distinction between good and poor conductors, or insulators, is relative and arbitrary. All substances exhibit conductivity to some degree but the range of observed values of σ is tremendous. The conductivity of copper, for example, is some 10^7 times as great as that of such a "good" conductor as sea water, and 10^{19} times that of ordinary glass. In Appendix III will be found an abbreviated table of the conductivities of representative materials.

Equation (13) is simply Ohm's law. Let us imagine, for example, a stationary distribution of current throughout the volume of any conducting medium. In virtue of the divergenceless character of the flow this distribution may be represented by closed streamlines. If a and b are two points on a particular streamline and $d\mathbf{s}$ is an element of its length, we have

$$\int_a^b \mathbf{E} \cdot d\mathbf{s} = \int_a^b \frac{\mathbf{J}}{\sigma} \cdot d\mathbf{s}. \tag{14}$$

A bundle of adjacent streamlines constitutes a current filament or tube. Since the flow is solenoidal, the current I through every cross section of the filament is the same. Let S be the cross-sectional area of the filament on a plane drawn normal to the direction of flow. S need not be infinitesimal, but is assumed to be so small that over its area the current density is uniform. Then $S\mathbf{J} \cdot d\mathbf{s} = I\, d\mathbf{s}$, and

$$\int_a^b \mathbf{E} \cdot d\mathbf{s} = I \int_a^b \frac{1}{\sigma S}\, ds. \tag{15}$$

The factor,

$$R = \int_a^b \frac{1}{\sigma S}\, ds, \tag{16}$$

[1] It is true that to a very slight degree the current distribution in a liquid or solid conductor may be modified by an impressed magnetic field, but the magnitude of this so-called Hall effect is so small that it may be ignored without incurring an appreciable error.

is equal to the resistance of the filament between the points a and b. The resistance of a linear section of homogeneous conductor of uniform cross section S and length l is

$$R = \frac{1}{\sigma}\frac{l}{S}, \tag{17}$$

a formula which is strictly valid only in the case of stationary currents

Within a region of nonvanishing conductivity there can be no permanent distribution of free charge. This fundamentally important theorem can be easily demonstrated when the medium is homogeneous and such that the relations between **D** and **E** and **J** and **E** are linear. By the equation of continuity,

$$\nabla \cdot \mathbf{J} + \frac{\partial \rho}{\partial t} = \nabla \cdot \sigma \mathbf{E} + \frac{\partial \rho}{\partial t} = 0. \tag{18}$$

On the other hand in a homogeneous medium

$$\nabla \cdot \mathbf{E} = \frac{1}{\epsilon}\rho, \tag{19}$$

which combined with (18) leads to

$$\frac{\partial \rho}{\partial t} + \frac{\sigma}{\epsilon}\rho = 0. \tag{20}$$

The density of charge at any instant is, therefore,

$$\rho = \rho_0 e^{-\frac{\sigma}{\epsilon}t}, \tag{21}$$

the constant of integration ρ_0 being equal to the density at the time $t = 0$. The initial charge distribution throughout the conductor decays exponentially with the time at every point and in a manner wholly independent of the applied field. If the charge density is initially zero, it remains zero at all times thereafter.

The time

$$\tau = \frac{\epsilon}{\sigma} \tag{22}$$

required for the charge at any point to decay to $1/e$ of its original value is called the *relaxation time.* In all but the poorest conductors τ is exceedingly small. Thus in sea water the relaxation time is about 2×10^{-10} sec.; even in such a poor conductor as distilled water it is not greater than 10^{-6} sec. In the best insulators, such as fused quartz, it may nevertheless assume values exceeding 10^{6} sec., an instance of the extraordinary range in the possible values of the parameter σ.

Let us suppose that at $t = 0$ a charge is concentrated within a small spherical region located somewhere in a conducting body. At every other point of the conductor the charge density is zero. The charge within the sphere now begins to fade away exponentially, but according to (21) no charge can reappear anywhere *within* the conductor. What becomes of it? Since the charge is conserved, the decay of charge within the spherical surface must be accompanied by an outward flow, or current. No charge can accumulate at any other interior point; hence the flow must be divergenceless. It will be arrested, however, on the outer surface of the conductor and it is here that we shall rediscover the charge that has been lost from the central sphere. This surface charge makes its appearance at the exact instant that the interior charge begins to decay, for the total charge is constant.

UNITS AND DIMENSIONS

1.8. The M.K.S. or Giorgi System.—An electromagnetic field thus far is no more than a complex of vectors subject to a postulated system of differential equations. To proceed further we must establish the physical dimensions of these vectors and agree on the units in which they are to be measured.

In the customary sense, an "absolute" system of units is one in which every quantity may be measured or expressed in terms of the three fundamental quantities mass, length, and time. Now in electromagnetic theory there is an essential arbitrariness in the matter of dimensions which is introduced with the factors ϵ_0 and μ_0 connecting **D** and **E**, **H** and **B** respectively in free space. No experiment has yet been imagined by means of which dimensions may be attributed to either ϵ_0 or μ_0 as an independent physical entity. On the other hand, it is a direct consequence of the field equations that the quantity

$$c = \frac{1}{\sqrt{\epsilon_0 \mu_0}} \tag{1}$$

shall have the dimensions of a velocity, and every arbitrary choice of ϵ_0 and μ_0 is subject to this restriction. The magnitude of this velocity cannot be calculated a priori, but by suitable experiment it may be measured. The value obtained by the method of Rosa and Dorsey of the Bureau of Standards and corrected by Curtis[1] in 1929 is

$$c = \frac{1}{\sqrt{\epsilon_0 \mu_0}} = 2.99790 \times 10^8 \qquad \text{meters/sec.,} \tag{2}$$

[1] Rosa and Dorsey, A New Determination of the Ratio of the Electrostatic Unit of Electricity, *Bur. Standards, Bull.* 3, p. 433, 1907. Curtis, *Bur. Standards J. Research*, **3**, 63, 1929.

or for all practical purposes

(3) $$c = 3 \times 10^8 \qquad \text{meters/sec.}$$

Throughout the early history of electromagnetic theory the absolute *electromagnetic system* of units was employed for all scientific investigations. In this system the centimeter was adopted as the unit of length, the gram as the unit of mass, the second as the unit of time, and as a fourth unit the factor μ_0 was placed arbitrarily equal to unity and considered dimensionless. The dimensions of ϵ_0 were then uniquely determined by (1) and it could be shown that the units and dimensions of every other quantity entering into the theory might be expressed in terms of centimeters, grams, seconds, and μ_0. Unfortunately, this absolute system failed to meet the needs of practice. The units of resistance and of electromotive force were, for example, far too small. To remedy this defect a *practical system* was adopted. Each unit of the practical system had the dimensions of the corresponding electromagnetic unit and differed from it in magnitude by a power of ten which, in the case of voltage and resistance at least, was wholly arbitrary. The practical units have the great advantage of convenient size and they are now universally employed for technical measurements and computations. Since they have been defined as arbitrary multiples of absolute units, they do not, however, constitute an absolute system. Now the quantities mass, length, and time are fundamental solely because the physicist has found it expedient to raise them to that rank. That there are other fundamental quantities is obvious from the fact that all electromagnetic quantities cannot be expressed in terms of these three alone. The restriction of the term "absolute" to systems based on mass, length, and time is, therefore, wholly unwarranted; one should ask only that such a system be self-consistent and that every quantity be defined in terms of a minimum number of basic, independent units. The antipathy of physicists in the past to the practical system of electrical units has been based not on any firm belief in the sanctity of mass, length, and time, but rather on the lack of self-consistency within that system.

Fortunately a most satisfactory solution has been found for this difficulty. In 1901 Giorgi,[1] pursuing an idea originally due to Maxwell, called attention to the fact that the practical system could be converted into an absolute system by an appropriate choice of fundamental units. It is indeed only necessary to choose for the unit of length the inter-

[1] GIORGI: Unità Razionali di Elettromagnetismo, *Atti dell' A.E.I.*, 1901. An historical review of the development of the practical system, including a report of the action taken at the 1935 meeting of the International Electrotechnical Commission and an extensive bibliography is given by Kennelly, *J. Inst. Elec. Engrs.*, **78**, 235–245, 1936. See also GLAZEBROOK, The M.K.S. System of Electrical Units, *J. Inst. Elec. Engrs.*, **78**, pp. 245–247.

national *meter*, for the unit of mass the *kilogram*, for the unit of time the *second*, and as a fourth unit any electrical quantity belonging to the practical system such as the coulomb, the ampere, or the ohm. From the field equations it is then possible to deduce the units and dimensions of every electromagnetic quantity in terms of these four fundamental units. Moreover the derived quantities will be related to each other exactly as in the practical system and may, therefore, be expressed in practical units. In particular it is found that the parameter μ_0 must have the value $4\pi \times 10^{-7}$, whence from (1) the value of ϵ_0 may be calculated. Inversely one might equally well *assume* this value of μ_0 as a fourth basic unit and then deduce the practical series from the field equations.

At a plenary session in June, 1935, the International Electrotechnical Commission adopted unanimously the m.k.s. system of Giorgi. Certain questions, however, still remain to be settled. No official agreement has as yet been reached as to the fourth fundamental unit. Giorgi himself recommended that the ohm, a material standard defined as the resistance of a specified column of mercury under specified conditions of pressure and temperature, be introduced as a basic quantity. If $\mu_0 = 4\pi \times 10^{-7}$ be chosen as the fourth unit and assumed dimensionless, all derived quantities may be expressed in terms of mass, length, and time alone, the dimensions of each being identical with those of the corresponding quantity in the absolute electromagnetic system and differing from them only in the size of the units. This assumption leads, however, to fractional exponents in the dimensions of many quantities, a direct consequence of our arbitrariness in clinging to mass, length, and time as the sole fundamental entities. In the absolute electromagnetic system, for example, the dimensions of charge are $\text{grams}^{\frac{1}{2}} \cdot \text{centimeters}^{\frac{1}{2}}$, an irrationality which can hardly be physically significant. These fractional exponents are entirely eliminated if we choose as a fourth unit the coulomb; for this reason, charge has been advocated at various times as a fundamental quantity quite apart from the question of its magnitude.[1] In the present volume we shall adhere exclusively to the meter-kilogram-second-coulomb system. A subsequent choice by the I.E.C. of some other electrical quantity as basic will in nowise affect the size of our units or the form of the equations.[2]

[1] See the discussion by WALLOT: *Elektrotechnische Zeitschrift*, Nos. 44–46, 1922. Also SOMMERFELD: "Ueber die Electromagnetischen Einheiten," pp. 157–165, Zeeman Verhandelingen, Martinus Nijhoff, The Hague, 1935; *Physik. Z.* **36**, 814–820, 1935.

[2] No ruling has been made as yet on the question of rationalization and opinion seems equally divided in favor and against. If one bases the theory on Maxwell's equations, it seems definitely advantageous to drop the factors 4π which in unrationalized systems stand before the charge and current densities. A rationalized system will be employed in this book.

To demonstrate that the proposed units do constitute a self-consistent system let us proceed as follows. The unit of current in the m.k.s. system is to be the absolute ampere and the unit of resistance is to be the absolute ohm. These quantities are to be such that the work expended per second by a current of 1 amp. passing through a resistance of 1 ohm is 1 joule (absolute). If R is the resistance of a section of conductor carrying a constant current of I amp., the work dissipated in heat in t sec. is

$$W = I^2Rt \qquad \text{joules.} \tag{4}$$

By means of a calorimeter the heat generated may be measured and thus one determines the relation of the unit of electrical energy to the unit quantity of heat. It is desired that the joule defined by (4) be identical with the joule defined as a unit of mechanical work, so that in the electrical as well as in the mechanical case

$$1 \text{ joule} \doteq 0.2389 \quad \text{gram-calorie (mean).} \tag{5}$$

Now we shall *define* the ampere on the basis of the equation of continuity (6), page 4, as the current which transports across any surface 1 coulomb in 1 sec. Then the ohm is a *derived* unit whose magnitude and dimensions are determined by (4):

$$1 \text{ ohm} = 1 \frac{\text{watt}}{\text{ampere}^2} = 1 \frac{\text{kilogram} \cdot \text{meter}^2}{\text{coulomb}^2 \cdot \text{second}}, \tag{6}$$

since 1 watt is equal to 1 joule/sec. The resistivity of a medium is defined as the resistance measured between two parallel faces of a unit cube. The reciprocal of this quantity is the conductivity. The dimensions of σ follow from Eq. (17), page 15.

$$1 \text{ unit of conductivity} = \frac{1}{\text{ohm} \cdot \text{meter}} = 1 \frac{\text{coulomb}^2 \cdot \text{second}}{\text{kilogram} \cdot \text{meter}^3}. \tag{7}$$

In the United States the reciprocal ohm is usually called the mho, although the name *siemens* has been adopted officially by the I.E.C. The unit of conductivity is therefore 1 siemens/meter.

The volt will be defined simply as 1 watt/amp., or

$$1 \text{ volt} = 1 \frac{\text{watt}}{\text{ampere}} = 1 \frac{\text{kilogram} \cdot \text{meter}^2}{\text{coulomb} \cdot \text{second}^2}. \tag{8}$$

Since the unit of current density is 1 amp./meter^2, we deduce from the relation $\mathbf{J} = \sigma \mathbf{E}$ that

$$1 \text{ unit of } \mathbf{E} = 1 \frac{\text{watt}}{\text{ampere} \cdot \text{meter}} = 1 \frac{\text{volt}}{\text{meter}} = 1 \frac{\text{kilogram} \cdot \text{meter}}{\text{coulomb} \cdot \text{second}^2}. \tag{9}$$

The power expended per unit volume by a current of density **J** is therefore $\mathbf{E} \cdot \mathbf{J}$ watts/meter³. It will be noted furthermore that the product of charge and electric field intensity **E** has the dimensions of force. Let a charge of 1 coulomb be placed in an electric field whose intensity is 1 volt/meter.

$$(10) \qquad 1 \text{ coulomb} \times 1 \frac{\text{volt}}{\text{meter}} = 1 \frac{\text{joule}}{\text{meter}} = 1 \frac{\text{kilogram} \cdot \text{meter}}{\text{second}^2}.$$

The unit of force in the m.k.s. system is called the *newton*, and is equivalent to 1 joule/meter, or 10^5 dynes.

The flux of the vector **B** shall be measured in *webers*,

$$(11) \qquad \Phi = \int_S \mathbf{B} \cdot \mathbf{n} \, da \qquad \text{webers},$$

and the intensity of the field **B**, or flux density, may therefore be expressed in webers per square meter. According to (25), page 8,

$$(12) \qquad \int_C \mathbf{E} \cdot d\mathbf{s} = -\frac{d\Phi}{dt} \qquad \frac{\text{webers}}{\text{second}}.$$

The line integral $\int_a^b \mathbf{E} \cdot d\mathbf{s}$ is measured in volts and is usually called the electromotive force (abbreviated e.m.f.) between the points a and b, although its value in a nonstationary field depends on the path of integration. The induced e.m.f. around any closed contour C is, therefore, equal to the rate of decrease of flux threading that contour, so that between the units there exists the relation

$$(13) \qquad 1 \text{ volt} = 1 \frac{\text{weber}}{\text{second}},$$

or

$$(14) \qquad 1 \text{ weber} = 1 \frac{\text{joule}}{\text{ampere}} = 1 \frac{\text{kilogram} \cdot \text{meter}^2}{\text{coulomb} \cdot \text{second}}.$$

It is important to note that the product of current and magnetic flux is an energy. Note also that the product of **B** and a velocity is measured in volts per meter, and is therefore a quantity of the same kind as **E**.

$$(15) \qquad 1 \text{ unit of } \mathbf{B} = 1 \frac{\text{weber}}{\text{meter}^2} = 1 \frac{\text{kilogram}}{\text{coulomb} \cdot \text{second}}.$$

$$(16) \quad 1 \text{ unit of } |\mathbf{B}|\,|\mathbf{v}| = 1 \frac{\text{weber}}{\text{meter}^2} \times 1 \frac{\text{meter}}{\text{second}} = 1 \frac{\text{volt}}{\text{meter}} = 1 \text{ unit of } |\mathbf{E}|.$$

The units which have been deduced thus far constitute an absolute system in the sense that each has been expressed in terms of the four basic quantities, mass, length, time, and charge. That this system is identical with the practical series may be verified by the substitutions

$$(17)\qquad 1 \text{ kilogram} = 10^3 \text{ grams},\qquad 1 \text{ meter} = 10^2 \text{ centimeters},\qquad 1 \text{ coulomb} = \tfrac{1}{10} \text{ abcoulomb}.$$

The numerical factors which now appear in each relation are observed to be those that relate the practical units to the absolute electromagnetic units. For example, from (6),

$$(18)\qquad 1 \text{ ohm} = 1\,\frac{\text{kilogram}\cdot\text{meter}^2}{\text{coulomb}^2\cdot\text{second}} = \frac{10^3 \text{ grams}\cdot 10^4 \text{ centimeters}^2}{10^{-2}\text{ abcoulomb}^2\cdot\text{seconds}} = 10^9 \quad \text{abohms};$$

and again from (8),

$$(19)\qquad 1 \text{ volt} = 1\,\frac{\text{kilogram}\cdot\text{meter}^2}{\text{coulomb}\cdot\text{second}^2} = \frac{10^3 \text{ grams}\cdot 10^4 \text{ centimeters}^2}{10^{-1}\text{ abcoulomb}\cdot\text{second}^2} = 10^8 \quad \text{abvolts}.$$

The series must be completed by a determination of the units and dimensions of the vectors **D** and **H**. Since $\mathbf{D} = \epsilon\mathbf{E}$, $\mathbf{H} = \frac{1}{\mu}\mathbf{B}$, it is necessary and sufficient that ϵ_0 and μ_0 be determined such as to satisfy Eq. (2) and such that the proper ratio of practical to absolute units be maintained. We shall represent mass, length, time, and charge by the letters M, L, T, and Q, respectively, and employ the customary symbol $[A]$ as meaning "the dimensions of A." Then from Eq. (31), page 9,

$$(20)\qquad \int_S \mathbf{D}\cdot\mathbf{n}\,da = q \qquad \text{coulombs}$$

and, hence,

$$(21)\qquad [\mathbf{D}] = \frac{\text{coulombs}}{\text{meter}^2} = \frac{Q}{L^2},$$

$$(22)\qquad [\epsilon_0] = \left[\frac{D}{\kappa_e E}\right] = \frac{\text{coulombs}}{\text{volt}\cdot\text{meter}} = \frac{Q^2T^2}{ML^3}.$$

The *farad*, a derived unit of capacity, is defined as the capacity of a conducting body whose potential will be raised 1 volt by a charge of 1 coulomb. It is equal, in other words, to 1 coulomb/volt. The parameter ϵ_0 in the m.k.s. system has dimensions, and may be measured in *farads per meter*.

By analogy with the electrical case, the line integral $\int_a^b \mathbf{H}\cdot d\mathbf{s}$ taken along a specified path is commonly called the magnetomotive force (abbreviated m.m.f.). In a stationary magnetic field

$$\int_C \mathbf{H} \cdot d\mathbf{s} = I \qquad \text{amperes,} \tag{23}$$

where I is the current determined by the flow of charge through any surface spanning the closed contour C. If the field is variable, I must include the displacement current as in (28), page 9. According to (23) a magnetomotive force has the dimensions of current. In practice, however, the current is frequently carried by the turns of a coil or winding which is linked by the contour C. If there are n such turns carrying a current I, the total current threading C is nI ampere-turns and it is customary to express magnetomotive force in these terms, although dimensionally n is a numeric.

$$[\text{m.m.f.}] = \text{ampere-turns,} \tag{24}$$

whence

$$[\mathbf{H}] = \frac{\text{ampere-turns}}{\text{meter}} = \frac{Q}{LT}. \tag{25}$$

It will be observed that the dimensions of **D** and those of **H** divided by a velocity are identical. For the parameter μ_0 we find

$$[\mu_0] = \left[\frac{B}{\kappa_m H}\right] = \frac{\text{volt} \cdot \text{second}}{\text{ampere} \cdot \text{meter}} = \frac{ML}{Q^2}. \tag{26}$$

As in the case of ϵ_0 it is convenient to express μ_0 in terms of a derived unit, in this case the *henry*, defined as 1 volt-second/amp. (The henry is that inductance in which an induced e.m.f. of 1 volt is generated when the inducing current is varying at the rate of 1 amp./sec.) The parameter μ_0 may, therefore, be measured in *henrys per meter*.

From (22) and (26) it follows now that

$$\left[\frac{1}{\mu_0 \epsilon_0}\right] = \frac{L^2}{T^2}, \tag{27}$$

and hence that our system is indeed dimensionally consistent with Eq. (2). Since it is known that in the rationalized, absolute c.g.s. electromagnetic system μ_0 is equal in magnitude to 4π, Eq. (26) fixes also its magnitude in the m.k.s. system.

$$\mu_0 = 4\pi \frac{\text{gram} \cdot \text{centimeters}}{\text{abcoulombs}^2} = 4\pi \frac{10^{-3}\ \text{kilogram} \cdot 10^{-2}\ \text{meter}}{10^2\ \text{coulombs}^2}, \tag{28}$$

or

$$\mu_0 = 4\pi \times 10^{-7} \frac{\text{kilogram} \cdot \text{meters}}{\text{coulombs}^2} = 1.257 \times 10^{-6} \frac{\text{henry}}{\text{meter}}. \tag{29}$$

The appropriate value of ϵ_0 is then determined from

$$(2)\qquad c = \frac{1}{\sqrt{\epsilon_0\mu_0}} = 2.998 \times 10^8 \frac{\text{meters}}{\text{second}}$$

to be

$$(30)\qquad \epsilon_0 = 8.854 \times 10^{-12} \frac{\text{coulomb}^2 \cdot \text{seconds}^2}{\text{kilogram} \cdot \text{meter}^3} = 8.854 \times 10^{-12} \frac{\text{farad}}{\text{meter}}.$$

It is frequently convenient to know the reciprocal values of these factors,

$$(31)\qquad \frac{1}{\mu_0} = 0.7958 \times 10^6 \frac{\text{meters}}{\text{henry}}, \qquad \frac{1}{\epsilon_0} = 0.1129 \times 10^{12} \frac{\text{meters}}{\text{farad}},$$

and the quantities

$$(32)\qquad \sqrt{\frac{\mu_0}{\epsilon_0}} = 376.6 \text{ ohms}, \qquad \sqrt{\frac{\epsilon_0}{\mu_0}} = 2.655 \times 10^{-3} \text{ mho},$$

recur constantly throughout the investigation of wave propagation.

In Appendix I there will be found a summary of the units and dimensions of electromagnetic quantities in terms of mass, length, time, and charge.

THE ELECTROMAGNETIC POTENTIALS

1.9. Vector and Scalar Potentials.—The analysis of an electromagnetic field is often facilitated by the use of auxiliary functions known as potentials. At every ordinary point of space, the field vectors satisfy the system

$$\text{(I)}\quad \nabla \times \mathbf{E} + \frac{\partial \mathbf{B}}{\partial t} = 0, \qquad \text{(III)}\quad \nabla \cdot \mathbf{B} = 0,$$

$$\text{(II)}\quad \nabla \times \mathbf{H} - \frac{\partial \mathbf{D}}{\partial t} = \mathbf{J}, \qquad \text{(IV)}\quad \nabla \cdot \mathbf{D} = \rho.$$

According to (III) the field of the vector **B** is always solenoidal. Consequently **B** can be represented as the curl of another vector $\mathbf{A}_0$.

$$(1)\qquad \mathbf{B} = \nabla \times \mathbf{A}_0.$$

However $\mathbf{A}_0$ is not uniquely defined by (1); for **B** is equal also to the curl of some vector **A**,

$$(2)\qquad \mathbf{B} = \nabla \times \mathbf{A},$$

where

$$(3)\qquad \mathbf{A} = \mathbf{A}_0 - \nabla\psi,$$

and ψ is any arbitrary scalar function of position.

If now **B** is replaced in (I) by either (1) or (2), we obtain, respectively,

(4) $$\nabla \times \left(\mathbf{E} + \frac{\partial \mathbf{A}_0}{\partial t}\right) = 0, \qquad \nabla \times \left(\mathbf{E} + \frac{\partial \mathbf{A}}{\partial t}\right) = 0.$$

Thus the fields of the vectors $\mathbf{E} + \frac{\partial \mathbf{A}_0}{\partial t}$ and $\mathbf{E} + \frac{\partial \mathbf{A}}{\partial t}$ are irrotational and equal to the gradients of two scalar functions ϕ_0 and ϕ.

(5) $$\mathbf{E} = -\nabla\phi_0 - \frac{\partial \mathbf{A}_0}{\partial t},$$

(6) $$\mathbf{E} = -\nabla\phi - \frac{\partial \mathbf{A}}{\partial t}.$$

The functions ϕ and ϕ_0 are obviously related by

(7) $$\phi = \phi_0 + \frac{\partial \psi}{\partial t}.$$

The functions **A** are *vector potentials* of the field, and the ϕ are *scalar potentials.* $\mathbf{A}_0$ and ϕ_0 designate one specific pair of potentials from which the field can be derived through (1) and (5). An infinite number of potentials leading to the same field can then be constructed from (3) and (7).

Let us suppose that the medium is homogeneous and isotropic, and that ϵ and μ are independent of field intensity.

(8) $$\mathbf{D} = \epsilon\mathbf{E}, \qquad \mathbf{B} = \mu\mathbf{H}.$$

In terms of the potentials

(9) $$\mathbf{D} = -\epsilon\left(\nabla\phi + \frac{\partial \mathbf{A}}{\partial t}\right), \qquad \mathbf{H} = \frac{1}{\mu}\nabla \times \mathbf{A},$$

which upon substitution into (II) and (IV) give

(10) $$\nabla \times \nabla \times \mathbf{A} + \mu\epsilon\nabla\frac{\partial \phi}{\partial t} + \mu\epsilon\frac{\partial^2 \mathbf{A}}{\partial t^2} = \mu\mathbf{J},$$

(11) $$\nabla^2\phi + \nabla \cdot \frac{\partial \mathbf{A}}{\partial t} = -\frac{1}{\epsilon}\rho.$$

All particular solutions of (10) and (11) lead to the same electromagnetic field when subjected to identical boundary conditions. They differ among themselves by the arbitrary function ψ. Let us impose now upon **A** and ϕ the supplementary condition

(12) $$\nabla \cdot \mathbf{A} + \mu\epsilon\frac{\partial \phi}{\partial t} = 0.$$

To do this it is only necessary that ψ shall satisfy

(13) $$\nabla^2\psi - \mu\epsilon\frac{\partial^2 \psi}{\partial t^2} = \nabla \cdot \mathbf{A}_0 + \mu\epsilon\frac{\partial \phi_0}{\partial t},$$

where ϕ_0 and $\mathbf{A}_0$ are particular solutions of (10) and (11). The potentials ϕ and $\mathbf{A}$ are now uniquely defined and are solutions of the equations

$$\text{(14)} \qquad \nabla \times \nabla \times \mathbf{A} - \nabla\nabla \cdot \mathbf{A} + \mu\epsilon \frac{\partial^2 \mathbf{A}}{\partial t^2} = \mu \mathbf{J},$$

$$\text{(15)} \qquad \nabla^2 \phi - \mu\epsilon \frac{\partial^2 \phi}{\partial t^2} = -\frac{1}{\epsilon} \rho.$$

Equation (14) reduces to the same form as (15) when use is made of the vector identity

$$\text{(16)} \qquad \nabla \times \nabla \times \mathbf{A} = \nabla\nabla \cdot \mathbf{A} - \nabla \cdot \nabla \mathbf{A}.$$

The last term of (16) can be interpreted as the Laplacian operating on the *rectangular* components of $\mathbf{A}$. In this case

$$\text{(17)} \qquad \nabla^2 \mathbf{A} - \mu\epsilon \frac{\partial^2 \mathbf{A}}{\partial t^2} = -\mu \mathbf{J}.$$

The expansion of the operator $\nabla \cdot \nabla \mathbf{A}$ in curvilinear systems will be discussed in Sec. 1.16, page 50.

The relations (2) and (6) for the vectors $\mathbf{B}$ and $\mathbf{E}$ are by no means general. To them may be added any particular solution of the *homogeneous* equations

$$\text{(I}a\text{)} \quad \nabla \times \mathbf{E} + \frac{\partial \mathbf{B}}{\partial t} = 0, \qquad \text{(III}a\text{)} \quad \nabla \cdot \mathbf{B} = 0,$$

$$\text{(II}a\text{)} \quad \nabla \times \mathbf{H} - \frac{\partial \mathbf{D}}{\partial t} = 0, \qquad \text{(IV}a\text{)} \quad \nabla \cdot \mathbf{D} = 0.$$

From the symmetry of this system it is at once evident that it can be satisfied identically by

$$\text{(18)} \qquad \mathbf{D} = -\nabla \times \mathbf{A}^*, \qquad \mathbf{H} = -\nabla\phi^* - \frac{\partial \mathbf{A}^*}{\partial t},$$

from which we construct

$$\text{(19)} \qquad \mathbf{E} = -\frac{1}{\epsilon} \nabla \times \mathbf{A}^*, \qquad \mathbf{B} = -\mu \left(\nabla\phi^* + \frac{\partial \mathbf{A}^*}{\partial t}\right).$$

The new potentials are subject only to the conditions

$$\text{(20)} \qquad \begin{aligned} \nabla^2 \mathbf{A}^* - \mu\epsilon \frac{\partial^2 \mathbf{A}^*}{\partial t^2} &= 0, \\ \nabla^2 \phi^* - \mu\epsilon \frac{\partial^2 \phi^*}{\partial t^2} &= 0, \\ \nabla \cdot \mathbf{A}^* + \mu\epsilon \frac{\partial \phi^*}{\partial t} &= 0. \end{aligned}$$

A general solution of the inhomogeneous system (I) to (IV) is, therefore,

$$\mathbf{B} = \nabla \times \mathbf{A} - \mu \frac{\partial \mathbf{A}^*}{\partial t} - \mu \nabla \phi^*, \tag{21}$$

$$\mathbf{E} = -\nabla \phi - \frac{\partial \mathbf{A}}{\partial t} - \frac{1}{\epsilon} \nabla \times \mathbf{A}^*, \tag{22}$$

provided μ and ϵ are constant.

The functions ϕ^* and $\mathbf{A}^*$ are potentials of a source distribution which is entirely external to the region considered. Usually ϕ^* and $\mathbf{A}^*$ are put equal to zero and the potentials of all charges, both distant and local, are represented by ϕ and $\mathbf{A}$.

At any point where the charge and current densities are zero a *possible* field is $\phi_0 = 0$, $\mathbf{A}_0 = 0$. The function ψ is now any solution of the homogeneous equation

$$\nabla^2 \psi - \mu\epsilon \frac{\partial^2 \psi}{\partial t^2} = 0. \tag{23}$$

Since at the same point the scalar potential ϕ satisfies the same equation, ψ may be chosen such that ϕ vanishes. *In this case the field can be expressed in terms of a vector potential alone.*

$$\mathbf{B} = \nabla \times \mathbf{A}, \qquad \mathbf{E} = -\frac{\partial \mathbf{A}}{\partial t}, \tag{24}$$

$$\nabla^2 \mathbf{A} - \mu\epsilon \frac{\partial^2 \mathbf{A}}{\partial t^2} = 0, \qquad \nabla \cdot \mathbf{A} = 0. \tag{25}$$

Concerning the units and dimensions of these new quantities, we note first that $\mathbf{E}$ is measured in volts/meter and that the scalar potential ϕ is therefore to be measured in volts. If q is a charge measured in coulombs, it follows that the product $q\phi$ represents an energy expressed in joules. From the relation $\mathbf{B} = \nabla \times \mathbf{A}$ it is clear that the vector potential $\mathbf{A}$ may be expressed in webers/meter, but equally well in either volt-seconds/meter or in joules/ampere. The product of current and vector potential is therefore an energy. The dimensions of $\mathbf{A}^*$ are found to be coulombs/meter, while ϕ^* will be measured in ampere-turns.

1.10. Homogeneous Conducting Media.—In view of the extreme brevity of the relaxation time it may be assumed that the density of free charge is always zero in the interior of a conductor. The field equations for a homogeneous, isotropic medium then reduce to

$$\text{(I}b\text{)}\quad \nabla \times \mathbf{E} + \frac{\partial \mathbf{B}}{\partial t} = 0, \qquad \text{(III}b\text{)}\quad \nabla \cdot \mathbf{B} = 0,$$

$$\text{(II}b\text{)}\quad \nabla \times \mathbf{H} - \frac{\partial \mathbf{D}}{\partial t} - \sigma \mathbf{E} = 0, \qquad \text{(IV}b\text{)}\quad \nabla \cdot \mathbf{D} = 0.$$

We are now free to express either **B** or **D** in terms of a vector potential. In the first alternative we have

$$\mathbf{B} = \nabla \times \mathbf{A}, \qquad \mathbf{E} = -\nabla\phi - \frac{\partial \mathbf{A}}{\partial t}. \tag{26}$$

If the vector and scalar potentials are subjected to the relation

$$\nabla \cdot \mathbf{A} + \mu\epsilon \frac{\partial \phi}{\partial t} + \mu\sigma\phi = 0, \tag{27}$$

a possible electromagnetic field may be constructed from any pair of solutions of the equations

$$\nabla^2 \mathbf{A} - \mu\epsilon \frac{\partial^2 \mathbf{A}}{\partial t^2} - \mu\sigma \frac{\partial \mathbf{A}}{\partial t} = 0, \tag{28}$$

$$\nabla^2 \phi - \mu\epsilon \frac{\partial^2 \phi}{\partial t^2} - \mu\sigma \frac{\partial \phi}{\partial t} = 0. \tag{29}$$

As in the preceding paragraph one will note that the field vectors are invariant to changes in the potentials satisfying the relations

$$\phi = \phi_0 + \frac{\partial \psi}{\partial t}, \qquad \mathbf{A} = \mathbf{A}_0 - \nabla\psi, \tag{30}$$

where ϕ_0, $\mathbf{A}_0$ are the potentials of a possible field and ψ is an arbitrary scalar function. In order that **A** and ϕ satisfy (27) it is only necessary that ψ be subjected to the additional condition,

$$\nabla^2 \psi - \mu\epsilon \frac{\partial^2 \psi}{\partial t^2} - \mu\sigma \frac{\partial \psi}{\partial t} = \nabla \cdot \mathbf{A}_0 + \mu\epsilon \frac{\partial \phi_0}{\partial t} + \mu\sigma\phi_0. \tag{31}$$

To a particular solution of (31) one is free to add any solution of the homogeneous equation

$$\nabla^2 \psi - \mu\epsilon \frac{\partial^2 \psi}{\partial t^2} - \mu\sigma \frac{\partial \psi}{\partial t} = 0. \tag{32}$$

Frequently it is convenient to choose ψ such that the scalar potential vanishes. The field within the conductor is then determined by a single vector **A**.

$$\mathbf{B} = \nabla \times \mathbf{A}, \qquad \mathbf{E} = -\frac{\partial \mathbf{A}}{\partial t}, \tag{33}$$

$$\nabla^2 \mathbf{A} - \mu\epsilon \frac{\partial^2 \mathbf{A}}{\partial t^2} - \mu\sigma \frac{\partial \mathbf{A}}{\partial t} = 0, \qquad \nabla \cdot \mathbf{A} = 0. \tag{34}$$

The field may also be defined in terms of potentials ϕ^* and $\mathbf{A}^*$ by

$$\mathbf{D} = -\nabla \times \mathbf{A}^*, \qquad \mathbf{H} = -\nabla\phi^* - \frac{\partial \mathbf{A}^*}{\partial t} - \frac{\sigma}{\epsilon}\mathbf{A}^*. \tag{35}$$

If ϕ^* and $\mathbf{A}^*$ are to satisfy (28) and (29), it is necessary that they be related by

$$\nabla \cdot \mathbf{A}^* + \mu\epsilon \frac{\partial \phi^*}{\partial t} = 0. \tag{36}$$

The field defined by (35) is invariant to all transformations of the potentials of the type

$$\phi^* = \phi_0^* + \frac{\partial \psi^*}{\partial t} + \frac{\sigma}{\epsilon}\psi^*, \qquad \mathbf{A}^* = \mathbf{A}_0^* - \nabla\psi^*, \tag{37}$$

where as above ϕ_0^* and $\mathbf{A}_0^*$ are the potentials of any possible electromagnetic field. To ensure the relation (36) it is only necessary that ψ^* be chosen such as to satisfy

$$\nabla^2\psi^* - \mu\epsilon \frac{\partial^2 \psi^*}{\partial t^2} - \mu\sigma \frac{\partial \psi^*}{\partial t} = \nabla \cdot \mathbf{A}_0^* + \mu\epsilon \frac{\partial \phi_0^*}{\partial t}. \tag{38}$$

Finally, by a proper choice of ψ^* the scalar potential ϕ^* may be made to vanish.

$$\mathbf{D} = -\nabla \times \mathbf{A}^*, \qquad \mathbf{H} = -\frac{\partial \mathbf{A}^*}{\partial t} - \frac{\sigma}{\epsilon}\mathbf{A}^*, \tag{39}$$

$$\nabla^2\mathbf{A}^* - \mu\epsilon \frac{\partial^2 \mathbf{A}^*}{\partial t^2} - \mu\sigma \frac{\partial \mathbf{A}^*}{\partial t} = 0, \qquad \nabla \cdot \mathbf{A}^* = 0. \tag{40}$$

1.11. The Hertz Vectors, or Polarization Potentials.—We have seen that the integration of Maxwell's equations may be reduced to the determination of a vector and a scalar potential, which in homogeneous media satisfy one and the same differential equation. It was shown by Hertz[1] that it is possible under ordinary conditions to define an electromagnetic field in terms of a single vector function.

Let us confine ourselves for the present to regions of an isotropic, homogeneous medium within which there are neither conduction currents nor free charges. The field equations then reduce to the homogeneous system (I*a*)–(IV*a*). We assume, for reasons which will become apparent, that the vector potential $\mathbf{A}$ is proportional to the time derivative of a vector $\mathbf{\Pi}$.

$$\mathbf{A} = \mu\epsilon \frac{\partial \mathbf{\Pi}}{\partial t}. \tag{41}$$

Consequently,

$$\mathbf{B} = \mu\epsilon\nabla \times \frac{\partial \mathbf{\Pi}}{\partial t}, \qquad \mathbf{E} = -\nabla\phi - \mu\epsilon \frac{\partial^2 \mathbf{\Pi}}{\partial t^2}, \tag{42}$$

[1] HERTZ, *Ann. Physik*, **36**, 1, 1888. The general solution is due to Righi: *Bologna Mem.*, (5) **9**, 1, 1901, and *Il Nuovo Cimento*, (5) **2**, 2, 1901.

and, when in turn this expression for **E** is introduced into (II*a*), it is found that

$$\frac{\partial}{\partial t}\left(\nabla \times \nabla \times \mathbf{\Pi} + \nabla \phi + \mu\epsilon \frac{\partial^2 \mathbf{\Pi}}{\partial t^2}\right) = 0. \tag{43}$$

We recall that at points where there is no charge, the scalar function ϕ is wholly arbitrary so long as it satisfies an equation such as (23). In the present instance it will be chosen such that

$$\phi = -\nabla \cdot \mathbf{\Pi}. \tag{44}$$

Then upon integrating (43) with respect to the time, we obtain

$$\nabla \times \nabla \times \mathbf{\Pi} - \nabla\nabla \cdot \mathbf{\Pi} + \mu\epsilon \frac{\partial^2 \mathbf{\Pi}}{\partial t^2} = \text{constant}. \tag{45}$$

The particular value of the constant does not affect the determination of the field and we are therefore free to place it equal to zero. Equation (IV*a*) is also satisfied, for the divergence of the curl of any vector vanishes identically. Then we may state that *every solution of the vector equation*

$$\nabla \times \nabla \times \mathbf{\Pi} - \nabla\nabla \cdot \mathbf{\Pi} + \mu\epsilon \frac{\partial^2 \mathbf{\Pi}}{\partial t^2} = 0 \tag{46}$$

determines an electromagnetic field through

$$\mathbf{B} = \mu\epsilon \nabla \times \frac{\partial \mathbf{\Pi}}{\partial t}, \qquad \mathbf{E} = \nabla\nabla \cdot \mathbf{\Pi} - \mu\epsilon \frac{\partial^2 \mathbf{\Pi}}{\partial t^2}. \tag{47}$$

The condition that ϕ shall satisfy (23) is fulfilled in virtue of (46). One may replace (46) by

$$\nabla^2 \mathbf{\Pi} - \mu\epsilon \frac{\partial^2 \mathbf{\Pi}}{\partial t^2} = 0, \tag{48}$$

provided ∇^2 is understood to operate on the *rectangular* components of $\mathbf{\Pi}$.

Since the vector **D** as well as **B** is solenoidal in a charge-free region, an alternative solution can be constructed of the form

$$\mathbf{A}^* = \mu\epsilon \frac{\partial \mathbf{\Pi}^*}{\partial t}, \qquad \phi^* = -\nabla \cdot \mathbf{\Pi}^*, \tag{49}$$

$$\mathbf{D} = -\mu\epsilon \nabla \times \frac{\partial \mathbf{\Pi}^*}{\partial t}, \qquad \mathbf{H} = \nabla\nabla \cdot \mathbf{\Pi}^* - \mu\epsilon \frac{\partial^2 \mathbf{\Pi}^*}{\partial t^2}, \tag{50}$$

where $\mathbf{\Pi}^*$ is any solution of (46) or (48).

From these results we conclude that the electromagnetic field within a region throughout which ϵ and μ are constant, ρ and **J** equal to zero, may be resolved into two partial fields, the one derived from the vector $\mathbf{\Pi}$

and the other from the vector Π^*. The origin of these fields lies exterior to the region. To determine the physical significance of the Hertz vectors it is now necessary to relate them to their sources; in other words, we must find the *inhomogeneous* equations from which (48) is derived.

Let us express the vector **D** in terms of **E** and the electric polarization **P**. According to (6), page 11, $\mathbf{D} = \epsilon_0\mathbf{E} + \mathbf{P}$. Then in place of (II*a*) and (IV*a*), we must now write

$$(51)\qquad \nabla \times \mathbf{H} - \epsilon_0 \frac{\partial \mathbf{E}}{\partial t} = \frac{\partial \mathbf{P}}{\partial t}, \qquad \nabla \cdot \mathbf{E} = -\frac{1}{\epsilon_0} \nabla \cdot \mathbf{P}.$$

It may be verified without difficulty that these two equations, as well as (I*a*) and (III*a*), are still identically satisfied by (47), provided only that ϵ be replaced by ϵ_0 and that Π be now any solution of

$$(52)\qquad \nabla^2 \Pi - \mu\epsilon_0 \frac{\partial^2 \Pi}{\partial t^2} = -\frac{1}{\epsilon_0} \mathbf{P}.$$

The source of the vector Π *and the electromagnetic field derived from it is a distribution of electric polarization* **P**. In due course we shall interpret the vector **P** as the electric dipole moment per unit volume of the medium. Since Π is associated with a distribution of electric dipoles, the partial field which it defines is sometimes said to be of *electric type*, and Π itself may be called the electric polarization potential.

In like manner it can be shown that the field associated with Π^* is set up by a distribution of magnetic polarization. According to (6), page 11, the vector **B** is related to **H** by $\mathbf{B} = \mu_0(\mathbf{H} + \mathbf{M})$, which when introduced into (I*a*) and (III*a*) gives

$$(53)\qquad \nabla \times \mathbf{E} + \mu_0 \frac{\partial \mathbf{H}}{\partial t} = -\mu_0 \frac{\partial \mathbf{M}}{\partial t}, \qquad \nabla \cdot \mathbf{H} = -\nabla \cdot \mathbf{M}.$$

Then these equations, as well as (II*a*) and (IV*a*), are satisfied identically by (50) if we replace there μ by μ_0 and prescribe that Π^* shall be a solution of

$$(54)\qquad \nabla^2 \Pi^* - \mu_0\epsilon \frac{\partial^2 \Pi^*}{\partial t^2} = -\mathbf{M}.$$

We shall show later that the polarization **M** may be interpreted as the density of a distribution of magnetic moment. The partial field derived from Π^* may be imagined to have its origin in magnetic dipoles and is said to be a field of *magnetic type*.

The electric polarization **P** may be induced in the dielectric by the field **E**, but it may also contain a part whose magnitude is controlled by wholly external factors. In the practical application of the theory one is interested usually only in this independent part $\mathbf{P}_0$, which will be shown to represent the electric moment of dipole oscillators activated by external power sources. The same is true for the magnetic polarization. To represent these conditions we shall write (6), page 11, in the modified form

$$\mathbf{D} = \epsilon\mathbf{E} + \mathbf{P}_0, \qquad \mathbf{H} = \frac{1}{\mu}\mathbf{B} - \mathbf{M}_0, \tag{55}$$

in which $\mathbf{P}_0$ and $\mathbf{M}_0$ are prescribed and independent of $\mathbf{E}$ and $\mathbf{H}$, *and where the induced polarizations of the medium have again been absorbed into the parameters ϵ and μ.* Then the electromagnetic field due to these distributions of $\mathbf{P}_0$ and $\mathbf{M}_0$ is determined by

$$\mathbf{E} = \nabla\nabla\cdot\mathbf{\Pi} - \mu\epsilon\frac{\partial^2\mathbf{\Pi}}{\partial t^2} - \mu\nabla\times\frac{\partial\mathbf{\Pi}^*}{\partial t}, \tag{56}$$

$$\mathbf{H} = \epsilon\nabla\times\frac{\partial\mathbf{\Pi}}{\partial t} + \nabla\nabla\cdot\mathbf{\Pi}^* - \mu\epsilon\frac{\partial^2\mathbf{\Pi}^*}{\partial t^2}, \tag{57}$$

when $\mathbf{\Pi}$ and $\mathbf{\Pi}^*$ are solutions of

$$\nabla^2\mathbf{\Pi} - \mu\epsilon\frac{\partial^2\mathbf{\Pi}}{\partial t^2} = -\frac{1}{\epsilon}\mathbf{P}_0, \qquad \nabla^2\mathbf{\Pi}^* - \mu\epsilon\frac{\partial^2\mathbf{\Pi}^*}{\partial t^2} = -\mathbf{M}_0. \tag{58}$$

In virtue of the second of Eqs. (58) and of the identity (16) we may also write (57) as

$$\mathbf{H} = \epsilon\nabla\times\frac{\partial\mathbf{\Pi}}{\partial t} + \nabla\times\nabla\times\mathbf{\Pi}^* - \mathbf{M}_0. \tag{59}$$

Since $\mathbf{B} = \nabla\times\mathbf{A}$, it is evident from this last relation that the vector potential $\mathbf{A}$ may be derived from the Hertzian vectors by putting

$$\mathbf{A} = \mu\epsilon\frac{\partial\mathbf{\Pi}}{\partial t} + \mu\nabla\times\mathbf{\Pi}^* - \nabla\psi, \tag{60}$$

where ψ is an arbitrary scalar function. The associated potential ϕ is

$$\phi = -\nabla\cdot\mathbf{\Pi} + \frac{\partial\psi}{\partial t}, \tag{61}$$

with ψ subject only to the condition that it satisfy

$$\nabla^2\psi - \mu\epsilon\frac{\partial^2\psi}{\partial t^2} = 0. \tag{62}$$

The extension of these equations to a homogeneous conducting medium follows without difficulty. The reader will verify by direct substitution that the system (I*b*)–(IV*b*), in a medium which is *free of* *fixed polarization* $\mathbf{P}_0$ *and* $\mathbf{M}_0$, is satisfied by

$$\mathbf{E} = \nabla \times \nabla \times \mathbf{\Pi} - \mu\nabla \times \frac{\partial \mathbf{\Pi}^*}{\partial t}, \tag{63}$$

$$\mathbf{H} = \nabla \times \left(\epsilon \frac{\partial \mathbf{\Pi}}{\partial t} + \sigma\mathbf{\Pi}\right) + \nabla \times \nabla \times \mathbf{\Pi}^*, \tag{64}$$

$$\begin{aligned} \nabla \times \nabla \times \mathbf{\Pi} - \nabla\nabla \cdot \mathbf{\Pi} + \mu\epsilon \frac{\partial^2 \mathbf{\Pi}}{\partial t^2} + \mu\sigma \frac{\partial \mathbf{\Pi}}{\partial t} = 0, \\ \nabla \times \nabla \times \mathbf{\Pi}^* - \nabla\nabla \cdot \mathbf{\Pi}^* + \mu\epsilon \frac{\partial^2 \mathbf{\Pi}^*}{\partial t^2} + \mu\sigma \frac{\partial \mathbf{\Pi}^*}{\partial t} = 0. \end{aligned} \tag{65}$$

1.12. Complex Field Vectors and Potentials.—It has been shown by Silberstein, Bateman, and others that the equations satisfied by the fields and potentials may be reduced to a particularly compact form by the construction of a complex vector whose real and imaginary parts are formed from the vectors defining the magnetic and electric fields.[1] The procedure has no apparent physical significance but frequently facilitates analysis.

Consider again a homogeneous, isotropic medium in which $\mathbf{D} = \epsilon\mathbf{E}$, $\mathbf{B} = \mu\mathbf{H}$. If now we define $\mathbf{Q}$ as a complex field vector by

$$\mathbf{Q} = \mathbf{B} + i\sqrt{\epsilon\mu}\,\mathbf{E}, \tag{66}$$

the Maxwell equations (I)–(IV) reduce to

$$\nabla \times \mathbf{Q} + i\sqrt{\epsilon\mu}\,\frac{\partial \mathbf{Q}}{\partial t} = \mu\mathbf{J}, \qquad \nabla \cdot \mathbf{Q} = i\sqrt{\frac{\mu}{\epsilon}}\,\rho. \tag{67}$$

The vector operation $\nabla \times \mathbf{Q}$ may be eliminated from (67) by the simple expedient of taking the curl of both members. By the identity (16) we obtain

$$\nabla\nabla \cdot \mathbf{Q} - \nabla^2\mathbf{Q} + i\sqrt{\epsilon\mu}\,\nabla \times \frac{\partial \mathbf{Q}}{\partial t} = \mu\nabla \times \mathbf{J}, \tag{68}$$

which, on replacing the curl and divergence of $\mathbf{Q}$ by their values from (67), reduces to

$$\nabla^2\mathbf{Q} - \epsilon\mu \frac{\partial^2 \mathbf{Q}}{\partial t^2} = -\mu\left(\nabla \times \mathbf{J} - i\sqrt{\epsilon\mu}\,\frac{\partial \mathbf{J}}{\partial t} - i\,\frac{1}{\sqrt{\epsilon\mu}}\,\nabla\rho\right). \tag{69}$$

When this last equation is resolved into its real and imaginary com-

[1] SILBERSTEIN, *Ann. phys.*, **22**, **24**, 1907. Also *Phil. Mag.* (6) **23**, 790, 1912. BATEMAN, "Electrical and Optical Wave Motion," Chap. I, Cambridge University Press.

ponents, one obtains the equations satisfied individually by the vectors **E** and **H**.

$$\nabla^2\mathbf{H} - \epsilon\mu \frac{\partial^2\mathbf{H}}{\partial t^2} = -\nabla \times \mathbf{J}, \tag{70}$$

$$\nabla^2\mathbf{E} - \epsilon\mu \frac{\partial^2\mathbf{E}}{\partial t^2} = \mu \frac{\partial\mathbf{J}}{\partial t} + \frac{1}{\epsilon}\nabla\rho. \tag{71}$$

Next, let us define **Q** in terms of complex vector and scalar potentials **L** and Φ by the equation

$$\mathbf{Q} = \nabla \times \mathbf{L} - i\sqrt{\epsilon\mu}\frac{\partial\mathbf{L}}{\partial t} - i\sqrt{\epsilon\mu}\nabla\Phi, \tag{72}$$

subject to the condition

$$\nabla \cdot \mathbf{L} + \epsilon\mu \frac{\partial\Phi}{\partial t} = 0. \tag{73}$$

It will be verified without difficulty that (72) is an integral of (67) provided the complex potentials satisfy the equations

$$\nabla^2\mathbf{L} - \epsilon\mu \frac{\partial^2\mathbf{L}}{\partial t^2} = -\mu\mathbf{J}, \tag{74}$$

$$\nabla^2\Phi - \epsilon\mu \frac{\partial^2\Phi}{\partial t^2} = -\frac{1}{\epsilon}\rho. \tag{75}$$

If the real and imaginary parts of these potentials are written in the form

$$\mathbf{L} = \mathbf{A} - i\sqrt{\frac{\mu}{\epsilon}}\mathbf{A}^*, \qquad \Phi = \phi - i\sqrt{\frac{\mu}{\epsilon}}\phi^*, \tag{76}$$

and substituted into (72), one finds again after separation of reals and imaginaries the general expressions for the field vectors deduced in Eqs. (21) and (22).

If the free currents and charges are everywhere zero in the region under consideration, Eq. (67) reduces to

$$\nabla \times \mathbf{Q} + i\sqrt{\epsilon\mu}\frac{\partial\mathbf{Q}}{\partial t} = 0, \qquad \nabla \cdot \mathbf{Q} = 0. \tag{77}$$

The electromagnetic field may now be expressed in terms of a single complex Hertzian vector $\mathbf{\Gamma}$.

$$\mathbf{Q} = \mu\epsilon\nabla \times \frac{\partial\mathbf{\Gamma}}{\partial t} + i\sqrt{\mu\epsilon}\nabla \times \nabla \times \mathbf{\Gamma}, \tag{78}$$

where $\mathbf{\Gamma}$ is any solution of

$$\nabla^2\mathbf{\Gamma} - \epsilon\mu \frac{\partial^2\mathbf{\Gamma}}{\partial t^2} = 0. \tag{79}$$

If, finally, $\mathbf{\Gamma}$ is defined as

$$\mathbf{\Gamma} = \mathbf{\Pi} - i\sqrt{\frac{\mu}{\epsilon}}\,\mathbf{\Pi}^* \tag{80}$$

and substituted into (78), one finds again after separation into real and imaginary parts exactly the expressions (47) and (50) for the electric and magnetic field vectors.

When the medium is conducting, the field equations are no longer symmetrical and the method fails. The difficulty may be overcome if the field varies harmonically. The time then enters explicitly as a factor such as $e^{\pm i\omega t}$. After differentiating with respect to time, the system (I*b*)–(IV*b*) may be made symmetrical by introducing a *complex inductive capacity* $\epsilon' = \epsilon \pm i\dfrac{\sigma}{\omega}$.

BOUNDARY CONDITIONS

1.13. Discontinuities in the Field Vectors.—The validity of the field equations has been postulated only for ordinary points of space; that is to say, for points in whose neighborhood the physical properties of the medium vary continuously. However, across any surface which bounds one body or medium from another there occur sharp changes in the parameters ϵ, μ, and σ. On a macroscopic scale these changes may usually be considered discontinuous and hence the field vectors themselves may be expected to exhibit corresponding discontinuities.

Let us imagine at the start that the surface S which bounds medium (1) from medium (2) has been replaced by a very thin transition layer within which the parameters ϵ, μ, σ vary rapidly but *continuously* from their values near S in (1) to their values near S in (2). Within this layer, as within the media (1) and (2), the field vectors and their first derivatives are continuous, bounded functions of position and time. Through the layer we now draw a small right cylinder, as indicated in Fig. 2*a*. The elements of the cylinder are normal to S and its ends lie in the surfaces of the layer so that they are separated by just the layer thickness Δl. Fixing our attention first on the field of the vector $\mathbf{B}$, we have

$$\oint \mathbf{B}\cdot\mathbf{n}\,da = 0, \tag{1}$$

when integrated over the walls and ends of the cylinder. If the base, whose area is Δa, is made sufficiently small, it may be assumed that $\mathbf{B}$ has a constant value over each end. Neglecting differentials of higher order we may approximate (1) by

$$(\mathbf{B}\cdot\mathbf{n}_1 + \mathbf{B}\cdot\mathbf{n}_2)\Delta a + \text{contributions of the walls} = 0. \tag{2}$$

The contribution of the walls to the surface integral is directly proportional to Δl. Now let the transition layer shrink into the surface S. In the limit, as $\Delta l \to 0$, the ends of the cylinder lie just on either side of S and the contribution from the walls becomes vanishingly small. The value of $\mathbf{B}$ at a point on S in medium (1) will be denoted by $\mathbf{B}_1$, while

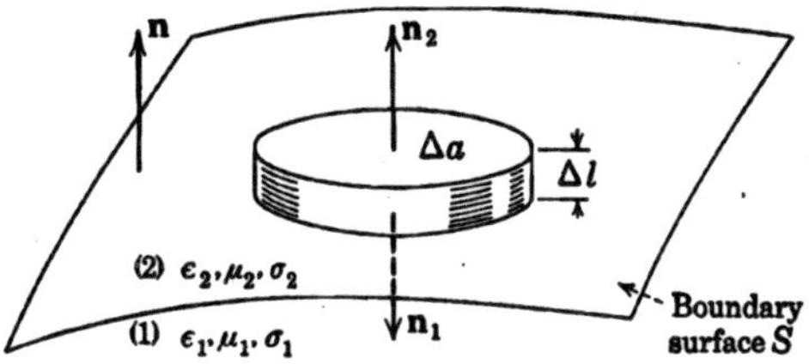

FIG. 2a.—For the normal boundary condition.

the corresponding value of $\mathbf{B}$ just across the surface in (2) will be denoted by $\mathbf{B}_2$. We shall also indicate the positive normal to S by a unit vector $\mathbf{n}$ drawn from (1) into (2). According to this convention medium (1) lies on the negative side of S, medium (2) on the positive side, and $\mathbf{n}_1 = -\mathbf{n}$. Then as $\Delta l \to 0$, $\Delta a \to 0$,

$$(\mathbf{B}_2 - \mathbf{B}_1) \cdot \mathbf{n} = 0; \tag{3}$$

the transition of the normal component of $\mathbf{B}$ *across any surface of discontinuity in the medium is continuous.* Equation (3) is a direct consequence of the condition $\nabla \cdot \mathbf{B} = 0$, and is sometimes called the surface divergence.

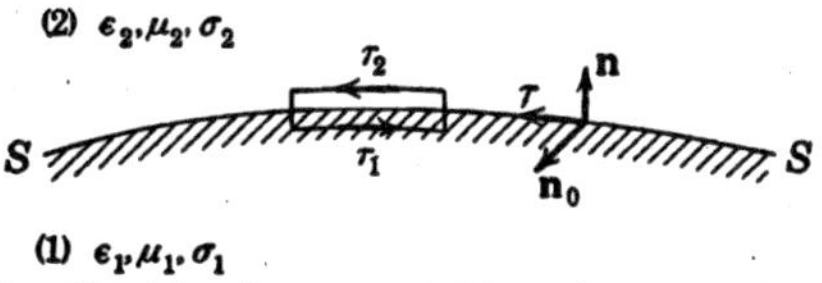

FIG. 2b.—For the tangential boundary condition.

The vector $\mathbf{D}$ may be treated in the same manner, but in this case the surface integral of the normal component over a closed surface is equal to the total charge contained within it.

$$\oint \mathbf{D} \cdot \mathbf{n}\, da = q. \tag{4}$$

The charge is distributed throughout the transition layer with a density ρ. As the ends of the cylinder shrink together, the total charge q remains constant, for it cannot be destroyed, and

$$q = \rho\, \Delta l\, \Delta a. \tag{5}$$

In the limit as $\Delta l \to 0$, the volume density ρ becomes infinite. It is then convenient to replace the product $\rho\, \Delta l$ by a *surface density* ω, defined as the charge per unit area. The transition of the normal component of the vector **D** across any surface S is now given by

$$(6) \qquad (\mathbf{D}_2 - \mathbf{D}_1) \cdot \mathbf{n} = \omega.$$

The presence of a layer of charge on S results in an abrupt change in the normal component of **D**, *the amount of the discontinuity being equal to the surface density measured in coulombs per square meter.*

Turning now to the behavior of the tangential components we replace the cylinder of Fig. 2*a* by a rectangular path drawn as in Fig. 2*b*. The sides of the rectangle of length Δs lie in either face of the transition layer and the ends which penetrate the layer are equal in length to its thickness Δl. This rectangle constitutes a contour C_0 about which

$$(7) \qquad \int_{C_0} \mathbf{E} \cdot d\mathbf{s} + \int_{S_0} \frac{\partial \mathbf{B}}{\partial t} \cdot \mathbf{n}_0 \, da = 0,$$

where S_0 is the area of the rectangle and $\mathbf{n}_0$ its positive normal. The direction of this positive normal is determined, as in Fig. 1, page 8, by the direction of circulation about C_0. Let $\boldsymbol{\tau}_1$ and $\boldsymbol{\tau}_2$ be unit vectors in the direction of circulation along the lower and upper sides of the rectangle as shown. Neglecting differentials of higher order, one may approximate (7) by

$$(8) \qquad (\mathbf{E} \cdot \boldsymbol{\tau}_1 + \mathbf{E} \cdot \boldsymbol{\tau}_2)\,\Delta s + \text{contributions from ends} = -\frac{\partial \mathbf{B}}{\partial t} \cdot \mathbf{n}_0 \,\Delta s\,\Delta l.$$

As the layer contracts to the surface S, the contributions from the segments at the ends, which are proportional to Δl, become vanishingly small. If $\mathbf{n}$ is again the positive normal to S drawn from (1) into (2), we may define the unit tangent vector $\boldsymbol{\tau}$ by

$$(9) \qquad \boldsymbol{\tau} = \mathbf{n}_0 \times \mathbf{n}.$$

Since

$$(10) \qquad \mathbf{n}_0 \times \mathbf{n} \cdot \mathbf{E} = \mathbf{n}_0 \cdot \mathbf{n} \times \mathbf{E},$$

we have in the limit as $\Delta l \to 0$, $\Delta s \to 0$,

$$(11) \qquad \mathbf{n}_0 \cdot \left[\mathbf{n} \times (\mathbf{E}_2 - \mathbf{E}_1) + \lim_{\Delta l \to 0} \left(\frac{\partial \mathbf{B}}{\partial t}\,\Delta l\right)\right] = 0.$$

The orientation of the rectangle – and hence also of $\mathbf{n}_0$ – is entirely arbitrary, from which it follows that the bracket in (11) must equal zero, or

$$(12) \qquad \mathbf{n} \times (\mathbf{E}_2 - \mathbf{E}_1) = -\lim_{\Delta l \to 0} \frac{\partial \mathbf{B}}{\partial t}\,\Delta l.$$

The field vectors and their derivatives have been assumed to be bounded; consequently the right-hand side of (12) vanishes with Δl.

$$\mathbf{n} \times (\mathbf{E}_2 - \mathbf{E}_1) = 0. \tag{13}$$

The transition of the tangential components of the vector **E** *through a surface of discontinuity is continuous.*

The behavior of **H** at the boundary may be deduced immediately from (12) and the field equation

$$\int_{C_0} \mathbf{H} \cdot d\mathbf{s} - \int_{S_0} \frac{\partial \mathbf{D}}{\partial t} \cdot \mathbf{n}_0 \, da = \int_{S_0} \mathbf{J} \cdot \mathbf{n}_0 \, da. \tag{14}$$

We have

$$\mathbf{n} \times (\mathbf{H}_2 - \mathbf{H}_1) = \lim_{\Delta l \to 0} \left(\frac{\partial \mathbf{D}}{\partial t} + \mathbf{J} \right) \Delta l. \tag{15}$$

The first term on the right of (15) vanishes as $\Delta l \to 0$ because **D** and its derivatives are bounded. If the current density **J** is finite, the second term vanishes as well. It may happen, however, that the current $I = \mathbf{J} \cdot \mathbf{n}_0 \, \Delta s \, \Delta l$ through the rectangle is squeezed into an infinitesimal layer on the surface S as the sides are brought together. It is convenient to represent this surface current by a surface density **K** defined as the limit of the product $\mathbf{J} \, \Delta \, l$ as $\Delta l \to 0$ and $\mathbf{J} \to \infty$. Then

$$\mathbf{n} \times (\mathbf{H}_2 - \mathbf{H}_1) = \mathbf{K}. \tag{16}$$

When the conductivities of the contiguous media are finite, there can be no surface current, for **E** is bounded and hence the product $\sigma \mathbf{E} \, \Delta l$ vanishes with Δl. In this case, which is the usual one,

$$\mathbf{n} \times (\mathbf{H}_2 - \mathbf{H}_1) = 0, \quad \text{(finite conductivity)}. \tag{17}$$

Not infrequently, however, it is necessary to assume the conductivity of a body to be infinite in order to simplify the analysis of its field. One must then apply (16) as a boundary condition rather than (17).

Summarizing, we are now able to supplement the field equations by four relations which determine the transition of an electromagnetic field from one medium to another separated by a surface of discontinuity.

$$\begin{aligned} \mathbf{n} \cdot (\mathbf{B}_2 - \mathbf{B}_1) &= 0, & \mathbf{n} \times (\mathbf{H}_2 - \mathbf{H}_1) &= \mathbf{K}, \\ \mathbf{n} \times (\mathbf{E}_2 - \mathbf{E}_1) &= 0, & \mathbf{n} \cdot (\mathbf{D}_2 - \mathbf{D}_1) &= \omega. \end{aligned} \tag{18}$$

From them follow immediately the conditions for the transition of the normal components of **E** and **H**.

$$\mathbf{n} \cdot \left(\mathbf{H}_2 - \frac{\mu_1}{\mu_2} \mathbf{H}_1 \right) = 0, \qquad \mathbf{n} \cdot \left(\mathbf{E}_2 - \frac{\epsilon_1}{\epsilon_2} \mathbf{E}_1 \right) = \frac{\omega}{\epsilon_2}. \tag{19}$$

Likewise the tangential components of **D** and **B** must satisfy

$$(20)\qquad \mathbf{n}\times\left(\mathbf{D}_2-\frac{\epsilon_2}{\epsilon_1}\mathbf{D}_1\right)=0,\qquad \mathbf{n}\times\left(\mathbf{B}_2-\frac{\mu_2}{\mu_1}\mathbf{B}_1\right)=\mu_2\mathbf{K}.$$

COORDINATE SYSTEMS

1.14. Unitary and Reciprocal Vectors.—It is one of the principal advantages of vector calculus that the equations defining properties common to all electromagnetic fields may be formulated without reference to any particular system of coordinates. To determine the peculiarities that distinguish a given field from all other possible fields, it becomes necessary, unfortunately, to resolve each vector equation into an equivalent scalar system in appropriate coordinates.

In a given region let

$$(1)\qquad u^1=f_1(x,y,z),\qquad u^2=f_2(x,y,z),\qquad u^3=f_3(x,y,z),$$

be three independent, continuous, single-valued functions of the rectangular coordinates $x.\ y,\ z$. These equations may be solved with respect to $x,\ y,\ z$, and give

$$(2)\qquad x=\varphi_1(u^1,u^2,u^3),\qquad y=\varphi_2(u^1,u^2,u^3),\qquad z=\varphi_3(u^1,u^2,u^3),$$

three functions which are also independent and continuous, and which are single-valued within certain limits. In general the functions φ_i as well as the functions f_i are continuously differentiable, but at certain singular points this property may fail and due care must be exercised in the application of general formulas.

With each point $P(x, y, z)$ in the region there is associated by means of (1) a triplet of values u^1, u^2, u^3; inversely (within limits depending on the boundaries of the region) there corresponds to each triplet u^1, u^2, u^3 a definite point. The functions u^1, u^2, u^3 are called *general* or *curvilinear coordinates*. Through each point P there pass three surfaces

$$(3)\qquad u^1=\text{constant},\qquad u^2=\text{constant},\qquad u^3=\text{constant},$$

called the coordinate surfaces. On each coordinate surface one coordinate is constant and two are variable. A surface will be designated by the coordinate which is constant. Two surfaces intersect in a curve, called a coordinate curve, along which two coordinates are constant and one is variable. A coordinate curve will be designated by the variable coordinate.

Let **r** denote the vector from an arbitrary origin to a variable point $P(x, y, z)$. The point, and consequently also its position vector **r**, may be considered functions of the curvilinear coordinates u^1, u^2, u^3.

$$(4)\qquad \mathbf{r}=\mathbf{r}(u^1,u^2,u^3).$$

A differential change in $\mathbf{r}$ due to small displacements along the coordinate curves is expressed by

$$(5)\qquad d\mathbf{r} = \frac{\partial \mathbf{r}}{\partial u^1}\,du^1 + \frac{\partial \mathbf{r}}{\partial u^2}\,du^2 + \frac{\partial \mathbf{r}}{\partial u^3}\,du^3.$$

Now if one moves unit distance along the u^1-curve, the change in $\mathbf{r}$ is directed tangentially to this curve and is equal to $\partial\mathbf{r}/\partial u^1$. The vectors

$$(6)\qquad \mathbf{a}_1 = \frac{\partial \mathbf{r}}{\partial u^1},\qquad \mathbf{a}_2 = \frac{\partial \mathbf{r}}{\partial u^2},\qquad \mathbf{a}_3 = \frac{\partial \mathbf{r}}{\partial u^3},$$

are known as the *unitary vectors* associated with the point P. They constitute a base system of reference for all other vectors associated with that particular point.

$$(7)\qquad d\mathbf{r} = \mathbf{a}_1\,du^1 + \mathbf{a}_2\,du^2 + \mathbf{a}_3\,du^3.$$

It must be carefully noted that *the unitary vectors are not necessarily of unit length*, and their dimensions will depend on the nature of the general coordinates.

The three base vectors $\mathbf{a}_1, \mathbf{a}_2, \mathbf{a}_3$ define a parallelepiped whose volume is

$$(8)\qquad V = \mathbf{a}_1\cdot(\mathbf{a}_2\times\mathbf{a}_3) = \mathbf{a}_2\cdot(\mathbf{a}_3\times\mathbf{a}_1) = \mathbf{a}_3\cdot(\mathbf{a}_1\times\mathbf{a}_2).$$

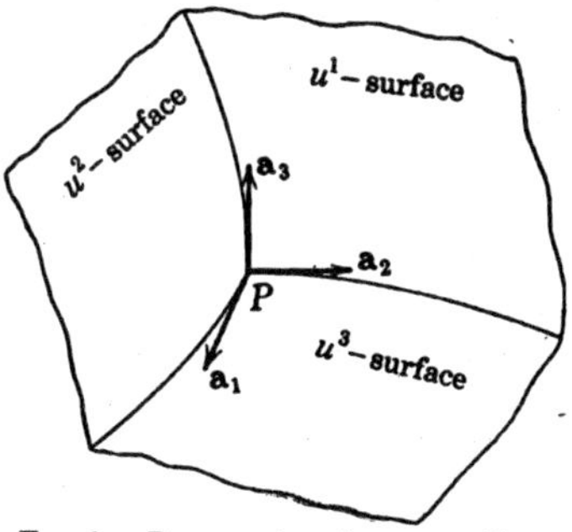

FIG. 3.—Base vectors for a curvilinear coordinate system.

The three vectors of a new triplet defined by

$$(9)\qquad \mathbf{a}^1 = \frac{1}{V}(\mathbf{a}_2\times\mathbf{a}_3),\qquad \mathbf{a}^2 = \frac{1}{V}(\mathbf{a}_3\times\mathbf{a}_1),\qquad \mathbf{a}^3 = \frac{1}{V}(\mathbf{a}_1\times\mathbf{a}_2),$$

are respectively perpendicular to the planes determined by the pairs $(\mathbf{a}_2, \mathbf{a}_3)$, $(\mathbf{a}_3, \mathbf{a}_1)$, $(\mathbf{a}_1, \mathbf{a}_2)$. Upon forming all possible scalar products of the form $\mathbf{a}^i\cdot\mathbf{a}_j$, it is easy to see that they satisfy the condition

$$(10)\qquad \mathbf{a}^i\cdot\mathbf{a}_j = \delta_{ij},$$

where δ_{ij} is a commonly used symbol denoting unity when $i = j$, and zero when $i \neq j$. The unitary vectors can be expressed in terms of the system $\mathbf{a}^1, \mathbf{a}^2, \mathbf{a}^3$ by relations identical in form.

$$(11)\qquad \mathbf{a}_1 = \frac{1}{V}(\mathbf{a}^2\times\mathbf{a}^3),\qquad \mathbf{a}_2 = \frac{1}{V}(\mathbf{a}^3\times\mathbf{a}^1),\qquad \mathbf{a}_3 = \frac{1}{V}(\mathbf{a}^1\times\mathbf{a}^2).$$

Any two sets of noncoplanar vectors related by the Eqs. (8) to (11) are said to constitute *reciprocal systems*. The triplets $\mathbf{a}^1, \mathbf{a}^2, \mathbf{a}^3$ are called *reciprocal unitary vectors* and they may serve as a base system quite as well as the unitary vectors themselves.

If the reciprocal unitary vectors are employed as a base system, the differential $d\mathbf{r}$ will be written

$$d\mathbf{r} = \mathbf{a}^1\,du_1 + \mathbf{a}^2\,du_2 + \mathbf{a}^3\,du_3. \tag{12}$$

The differentials du_1, du_2, du_3 are evidently components of $d\mathbf{r}$ in the directions defined by the new base vectors. The quantities u_1, u_2, u_3 are functions of the coordinates u^1, u^2, u^3, but the differentials du_1, du_2, du_3 are not necessarily perfect. On the contrary they are related to the differentials of the coordinates by a set of linear equations which in general are nonintegrable. Thus equating (7) and (12), we have

$$d\mathbf{r} = \sum_{i=1}^{3} \mathbf{a}_i\,du^i = \sum_{j=1}^{3} \mathbf{a}^j\,du_j. \tag{13}$$

Upon scalar multiplication of (13) by $\mathbf{a}^i$ and by $\mathbf{a}_j$ in turn, we find, thanks to (10):

$$du_j = \sum_{i=1}^{3} \mathbf{a}_j \cdot \mathbf{a}_i\,du^i, \qquad du^i = \sum_{j=1}^{3} \mathbf{a}^i \cdot \mathbf{a}^j\,du_j. \tag{14}$$

It is customary to represent the scalar products of the unitary vectors and those of the reciprocal unitary vectors by the symbols

$$g_{ij} = \mathbf{a}_i \cdot \mathbf{a}_j = g_{ji}, \tag{15}$$

$$g^{ij} = \mathbf{a}^i \cdot \mathbf{a}^j = g^{ji}. \tag{16}$$

The components of $d\mathbf{r}$ in the unitary and in the reciprocal base systems are then related by

$$du_j = \sum_{i=1}^{3} g_{ji}\,du^i, \qquad du^i = \sum_{j=1}^{3} g^{ij}\,du_j. \tag{17}$$

A fixed vector $\mathbf{F}$ at the point P may be resolved into components either with respect to the base system $\mathbf{a}_1$, $\mathbf{a}_2$, $\mathbf{a}_3$, or with respect to the reciprocal system $\mathbf{a}^1$, $\mathbf{a}^2$, $\mathbf{a}^3$.

$$\mathbf{F} = \sum_{i=1}^{3} f^i \mathbf{a}_i = \sum_{j=1}^{3} f_j \mathbf{a}^j. \tag{18}$$

The components of $\mathbf{F}$ in the unitary system are evidently related to those in its reciprocal system by

$$f_j = \sum_{i=1}^{3} g_{ji} f^i, \qquad f^i = \sum_{j=1}^{3} g^{ij} f_j, \tag{19}$$

and in virtue of the orthogonality of the base vectors $\mathbf{a}_j$ with respect to the reciprocal set $\mathbf{a}^i$ as expressed by (10), we may also write

$$f^i = \mathbf{F} \cdot \mathbf{a}^i, \qquad f_j = \mathbf{F} \cdot \mathbf{a}_j. \tag{20}$$

It follows from this that (18) is equivalent to

$$\mathbf{F} = \sum_{i=1}^{3} (\mathbf{F} \cdot \mathbf{a}^i)\mathbf{a}_i = \sum_{j=1}^{3} (\mathbf{F} \cdot \mathbf{a}_j)\mathbf{a}^j. \tag{21}$$

The quantities f^i are said to be the *contravariant* components of the vector $\mathbf{F}$, while the components f_j are called *covariant*. A small letter has been used to designate these components to avoid confusion with the components F_1, F_2, F_3 of $\mathbf{F}$ with respect to a base system coinciding with the $\mathbf{a}_i$ but of *unit* length. It has been noted previously that the length and dimensions of the unitary vectors depend on the nature of the curvilinear coordinates. An appropriate set of *unit* vectors which, like the unitary set $\mathbf{a}_i$, are tangent to the u^i-curves, is defined by

$$\mathbf{i}_1 = \frac{\mathbf{a}_1}{\sqrt{\mathbf{a}_1 \cdot \mathbf{a}_1}} = \frac{1}{\sqrt{g_{11}}}\mathbf{a}_1, \qquad \mathbf{i}_2 = \frac{1}{\sqrt{g_{22}}}\mathbf{a}_2, \qquad \mathbf{i}_3 = \frac{1}{\sqrt{g_{33}}}\mathbf{a}_3, \tag{22}$$

and, hence,

$$\mathbf{F} = F_1\mathbf{i}_1 + F_2\mathbf{i}_2 + F_3\mathbf{i}_3, \tag{23}$$

with

$$F_i = \sqrt{g_{ii}}\, f^i. \tag{24}$$

The F_i are of the same dimensions as the vector $\mathbf{F}$ itself.

The vector $d\mathbf{r}$ represents an infinitesimal displacement from the point $P(u^1, u^2, u^3)$ to a neighboring point whose coordinates are $u^1 + du^1$, $u^2 + du^2$, $u^3 + du^3$. The magnitude of this displacement, which constitutes a line element, we shall denote by ds. Then

$$ds^2 = d\mathbf{r} \cdot d\mathbf{r} = \sum_{i=1}^{3}\sum_{j=1}^{3} \mathbf{a}_i \cdot \mathbf{a}_j\, du^i\, du^j = \sum_{i=1}^{3}\sum_{j=1}^{3} \mathbf{a}^i \cdot \mathbf{a}^j\, du_i\, du_j; \tag{25}$$

or, in the notation of (15) and (16),

$$ds^2 = \sum_{i,j=1}^{3} g_{ij}\, du^i\, du^j = \sum_{i,j=1}^{3} g^{ij}\, du_i\, du_j. \tag{26}$$

The g_{ij} and g^{ij} appear here as coefficients of two differential quadratic forms expressing the length of a line element in the space of the general coordinates u^i or of its reciprocal set u_i. They are commonly called the *metrical coefficients*.

It is now a relatively simple matter to obtain expressions for elements of arc, surface, and volume in a system of curvilinear coordinates. Let $d\mathbf{s}_1$ be an infinitesimal displacement at $P(u^1, u^2, u^3)$ along the u^1-curve.

$$(27) \qquad d\mathbf{s}_1 = \mathbf{a}_1\, du^1, \qquad ds_1 = |d\mathbf{s}_1| = \sqrt{g_{11}}\, du^1.$$

Similarly, for elements of length along the u^2- and u^3-curves, we have

$$(28) \qquad ds_2 = \sqrt{g_{22}}\, du^2, \qquad ds_3 = \sqrt{g_{33}}\, du^3.$$

Consider next an infinitesimal parallelogram in the u^1-surface bounded by intersecting u^2- and u^3-curves as indicated in Fig. 4. The area of such an element is equal in magnitude to

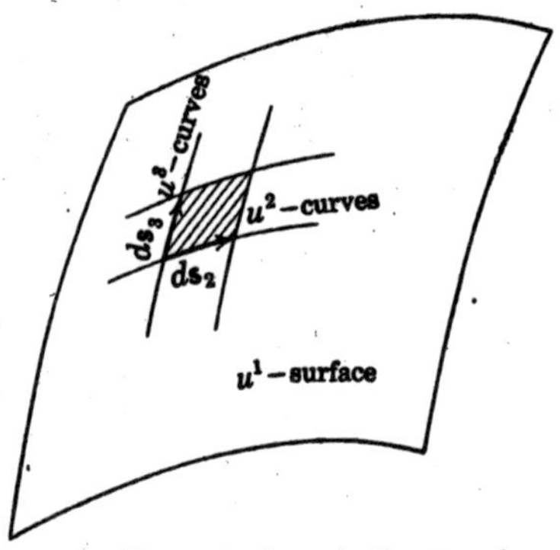

FIG. 4.—Element of area in the u^1-surface.

$$(29) \qquad \begin{aligned} da_1 &= |d\mathbf{s}_2 \times d\mathbf{s}_3| \\ &= |\mathbf{a}_2 \times \mathbf{a}_3|\, du^2\, du^3 \\ &= \sqrt{(\mathbf{a}_2 \times \mathbf{a}_3) \cdot (\mathbf{a}_2 \times \mathbf{a}_3)}\, du^2\, du^3. \end{aligned}$$

By a well-known vector identity

$$(30) \qquad \begin{aligned} &(\mathbf{a} \times \mathbf{b}) \cdot (\mathbf{c} \times \mathbf{d}) \\ &\quad = (\mathbf{a} \cdot \mathbf{c})(\mathbf{b} \cdot \mathbf{d}) - (\mathbf{a} \cdot \mathbf{d})(\mathbf{b} \cdot \mathbf{c}), \end{aligned}$$

where $\mathbf{a}, \mathbf{b}, \mathbf{c}, \mathbf{d}$ are any four vectors, and hence

$$(31) \qquad \begin{aligned} (\mathbf{a}_2 \times \mathbf{a}_3) \cdot (\mathbf{a}_2 \times \mathbf{a}_3) &= (\mathbf{a}_2 \cdot \mathbf{a}_2)(\mathbf{a}_3 \cdot \mathbf{a}_3) - (\mathbf{a}_2 \cdot \mathbf{a}_3)(\mathbf{a}_3 \cdot \mathbf{a}_2) \\ &= g_{22}g_{33} - g_{23}^2. \end{aligned}$$

For the area of an element in the u^1-surface we have, therefore,

$$(32) \qquad da_1 = \sqrt{g_{22}g_{33} - g_{23}^2}\, du^2\, du^3,$$

and similarly for elements in the u^2- and u^3-surfaces,

$$(33) \qquad \begin{aligned} da_2 &= \sqrt{g_{33}g_{11} - g_{31}^2}\, du^3\, du^1, \\ da_3 &= \sqrt{g_{11}g_{22} - g_{12}^2}\, du^1\, du^2. \end{aligned}$$

Finally, a volume element bounded by coordinate surfaces is written as

$$(34) \qquad dv = d\mathbf{s}_1 \cdot d\mathbf{s}_2 \times d\mathbf{s}_3 = \mathbf{a}_1 \cdot \mathbf{a}_2 \times \mathbf{a}_3\, du^1\, du^2\, du^3.$$

If now in (21) we let $\mathbf{F} = \mathbf{a}_2 \times \mathbf{a}_3$, we have

$$(35) \qquad \mathbf{a}_2 \times \mathbf{a}_3 = (\mathbf{a}^1 \cdot \mathbf{a}_2 \times \mathbf{a}_3)\mathbf{a}_1 + (\mathbf{a}^2 \cdot \mathbf{a}_2 \times \mathbf{a}_3)\mathbf{a}_2 + (\mathbf{a}^3 \cdot \mathbf{a}_2 \times \mathbf{a}_3)\mathbf{a}_3,$$

or, on replacing the $\mathbf{a}^i$ by their values from (8) and (9),

$$(36)\quad \mathbf{a}_1 \cdot \mathbf{a}_2 \times \mathbf{a}_3 = \frac{\mathbf{a}_1}{\mathbf{a}_1 \cdot \mathbf{a}_2 \times \mathbf{a}_3} \cdot [(\mathbf{a}_2 \times \mathbf{a}_3 \cdot \mathbf{a}_2 \times \mathbf{a}_3)\mathbf{a}_1 + (\mathbf{a}_3 \times \mathbf{a}_1 \cdot \mathbf{a}_2 \times \mathbf{a}_3)\mathbf{a}_2 + (\mathbf{a}_1 \times \mathbf{a}_2 \cdot \mathbf{a}_2 \times \mathbf{a}_3)\mathbf{a}_3].$$

The quantities within parentheses can be expanded by (30) and the terms arranged in the form

$$(37)\quad (\mathbf{a}_1 \cdot \mathbf{a}_2 \times \mathbf{a}_3)^2 = \mathbf{a}_1 \cdot \mathbf{a}_1[(\mathbf{a}_2 \cdot \mathbf{a}_2)(\mathbf{a}_3 \cdot \mathbf{a}_3) - (\mathbf{a}_2 \cdot \mathbf{a}_3)(\mathbf{a}_3 \cdot \mathbf{a}_2)] + \mathbf{a}_1 \cdot \mathbf{a}_2[(\mathbf{a}_2 \cdot \mathbf{a}_3)(\mathbf{a}_3 \cdot \mathbf{a}_1) - (\mathbf{a}_2 \cdot \mathbf{a}_1)(\mathbf{a}_3 \cdot \mathbf{a}_3)] + \mathbf{a}_1 \cdot \mathbf{a}_3[(\mathbf{a}_2 \cdot \mathbf{a}_1)(\mathbf{a}_3 \cdot \mathbf{a}_2) - (\mathbf{a}_2 \cdot \mathbf{a}_2)(\mathbf{a}_3 \cdot \mathbf{a}_1)].$$

If finally the scalar products in (37) are replaced by their g_{ij}, we obtain as an expression for a volume element

$$(38)\qquad dv = \sqrt{g}\, du^1\, du^2\, du^3,$$

in which

$$(39)\qquad g = \begin{vmatrix} g_{11} & g_{12} & g_{13} \\ g_{21} & g_{22} & g_{23} \\ g_{31} & g_{32} & g_{33} \end{vmatrix}.$$

A corresponding set of expressions for the elements of arc, area, and volume in the reciprocal base system may be obtained by replacing the g_{ij} by the g^{ij}, but they will not be needed in what follows.

Clearly the coefficients g_{ij} are sufficient to characterize completely the geometrical properties of space with respect to any curvilinear system of coordinates; it is therefore essential that we know how these coefficients may be determined. To unify our notation we shall represent the rectangular coordinates x, y, z of a point P by the letters x^1, x^2, x^3 respectively. Then

$$(40)\qquad ds^2 = (dx^1)^2 + (dx^2)^2 + (dx^3)^2.$$

In this most elementary of all systems the metrical coefficients are

$$(41)\qquad g_{ij} = \delta_{ij}, \qquad (\delta_{ii} = 1, \qquad \delta_{ij} = 0 \text{ when } i \neq j).$$

From the orthogonality of the coordinate planes and the definition (9), it is evident that the unitary and the reciprocal unitary vectors are identical, are of unit length, and are the base vectors customarily represented by the letters **i, j, k.**

Suppose now that the rectangular coordinates are related functionally to a set of general coordinates as in (2) by the equations

$$(42)\quad x^1 = x^1(u^1, u^2, u^3), \qquad x^2 = x^2(u^1, u^2, u^3), \qquad x^3 = x^3(u^1, u^2, u^3).$$

The differentials of the rectangular coordinates are linear functions of the differentials of the general coordinates, as we see upon differentiating Eqs. (42).

$$
\begin{aligned}
dx^1 &= \frac{\partial x^1}{\partial u^1} du^1 + \frac{\partial x^1}{\partial u^2} du^2 + \frac{\partial x^1}{\partial u^3} du^3, \\
dx^2 &= \frac{\partial x^2}{\partial u^1} du^1 + \frac{\partial x^2}{\partial u^2} du^2 + \frac{\partial x^2}{\partial u^3} du^3, \\
dx^3 &= \frac{\partial x^3}{\partial u^1} du^1 + \frac{\partial x^3}{\partial u^2} du^2 + \frac{\partial x^3}{\partial u^3} du^3.
\end{aligned} \tag{43}
$$

According to (26) and (40)

$$ds^2 = \sum_{i=1}^{3} \sum_{j=1}^{3} g_{ij}\, du^i\, du^j = \sum_{k=1}^{3} (dx^k)^2, \tag{44}$$

whence on squaring the differentials in (43) and equating coefficients of like terms we obtain

$$g_{ij} = \frac{\partial x^1}{\partial u^i}\frac{\partial x^1}{\partial u^j} + \frac{\partial x^2}{\partial u^i}\frac{\partial x^2}{\partial u^j} + \frac{\partial x^3}{\partial u^i}\frac{\partial x^3}{\partial u^j}. \tag{45}$$

1.15. The Differential Operators.—The gradient of a scalar function $\phi(u^1, u^2, u^3)$ is a fixed vector defined in direction and magnitude as the maximum rate of change of ϕ with respect to the cordinates. The variation in ϕ incurred during a displacement $d\mathbf{r}$ is, therefore,

$$d\phi = \nabla\phi \cdot d\mathbf{r} = \sum_{i=1}^{3} \frac{\partial \phi}{\partial u^i} du^i. \tag{46}$$

Now the du^i are the contravariant components of the displacement vector $d\mathbf{r}$, and hence by (20),

$$du^i = \mathbf{a}^i \cdot d\mathbf{r}. \tag{47}$$

This value for du^i introduced into (46) leads to

$$\left(\nabla\phi - \sum_{i=1}^{3} \mathbf{a}^i \frac{\partial \phi}{\partial u^i}\right) \cdot d\mathbf{r} = 0 \tag{48}$$

and, since the displacement $d\mathbf{r}$ is arbitrary, we find for the gradient of a scalar function in any system of curvilinear coordinates:

$$\nabla\phi = \sum_{i=1}^{3} \mathbf{a}^i \frac{\partial \phi}{\partial u^i}. \tag{49}$$

In this expression the reciprocal unitary vectors constitute the base system, but these may be replaced by the unitary vectors through the transformation

$$\mathbf{a}^i = \sum_{j=1}^{3} g^{ij}\mathbf{a}_j. \tag{50}$$

The divergence of a vector function $\mathbf{F}(u^1, u^2, u^3)$ at the point P may be deduced most easily from its definition in Eq. (9), page 4, as the limit of a surface integral of the normal component of $\mathbf{F}$ over a closed surface, per unit of enclosed volume. Consider those two ends of the volume element illustrated in Fig. 5 which lie in u^2-surfaces. The left end is located at u^2, the right at $u^2 + du^2$. The area of the face at u^2 is $(\mathbf{a}_1 \times \mathbf{a}_3)du^1\,du^3$, the order of the vectors being such that the normal is directed outward, *i.e.*, to the left. The net contribution of these two ends to the outward flux is, therefore,

$$[\mathbf{F}\cdot(\mathbf{a}_3 \times \mathbf{a}_1)\,du^1\,du^3]_{u^2+du^2} + [\mathbf{F}\cdot(\mathbf{a}_1 \times \mathbf{a}_3)\,du^1\,du^3]_{u^2}, \tag{51}$$

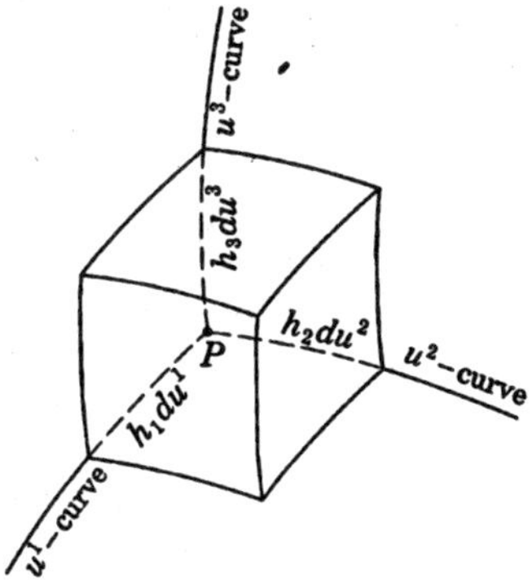

FIG. 5.—Element of volume in a curvilinear coordinate system.

the subscripts to the brackets indicating that the enclosed quantities are to be evaluated at $u^2 + du^2$ and u^2 respectively. For sufficiently small values of du^2, (51) may be approximated by the linear term of a Taylor expansion,

$$\frac{\partial}{\partial u^2}(\mathbf{F}\cdot\mathbf{a}_3 \times \mathbf{a}_1\,du^1\,du^2\,du^3), \tag{52}$$

$\mathbf{a}_1 \times \mathbf{a}_3$ having been replaced by $-\mathbf{a}_3 \times \mathbf{a}_1$. Now by (21), (20), and (37) we have

$$\mathbf{F}\cdot\mathbf{a}_3 \times \mathbf{a}_1 = \mathbf{F}\cdot\mathbf{a}^2(\mathbf{a}_2\cdot\mathbf{a}_3 \times \mathbf{a}_1) = f^2\sqrt{g}; \tag{53}$$

hence the contribution of the two ends to the surface integral is

$$\frac{\partial}{\partial u^2}(f^2\sqrt{g})\,du^1\,du^2\,du^3. \tag{54}$$

Analogous contributions result from the two remaining pairs of faces. These are to be measured per unit volume; hence we divide by $dv = \sqrt{g}\,du^1\,du^2\,du^3$ and pass to the limit $du^1 \to 0$, $du^2 \to 0$, $du^3 \to 0$, ensuring thereby the vanishing of all but the linear terms in the Taylor expansion. The divergence of a vector $\mathbf{F}$ referred to a system of curvilinear coordinates is, therefore,

$$\nabla\cdot\mathbf{F} = \frac{1}{\sqrt{g}}\sum_{i=1}^{3}\frac{\partial}{\partial u^i}(f^i\sqrt{g}). \tag{55}$$

The curl of the vector **F** is found in the same manner by calculating the line integral of **F** around an infinitesimal closed path. According to Eq. (21), page 7, the component of the curl in a direction defined by a unit normal **n** is

$$(\nabla \times \mathbf{F}) \cdot \mathbf{n} = \lim_{C \to 0} \frac{1}{S} \int_C \mathbf{F} \cdot d\mathbf{s}. \tag{56}$$

Let us take the line integral of **F** about the contour of a rectangular element of area located in the u^1-surface, as indicated in Fig. 6. The sides of the rectangle are $\mathbf{a}_2\, du^2$ and $\mathbf{a}_3\, du^3$. The direction of circulation is such that the positive normal is in the sense of the positive u^1-curve. The contribution from the sides parallel to u^3-curves is

$$(\mathbf{F} \cdot \mathbf{a}_3\, du^3)_{u^2+du^2} - (\mathbf{F} \cdot \mathbf{a}_3\, du^3)_{u^2};$$

from the bottom and top parallel to u^2-curves, we obtain

$$-(\mathbf{F} \cdot \mathbf{a}_2\, du^2)_{u^3+du^3} + (\mathbf{F} \cdot \mathbf{a}_2\, du^2)_{u^3}.$$

Approximating these differences by the linear terms of a Taylor expansion, we obtain for the line integral

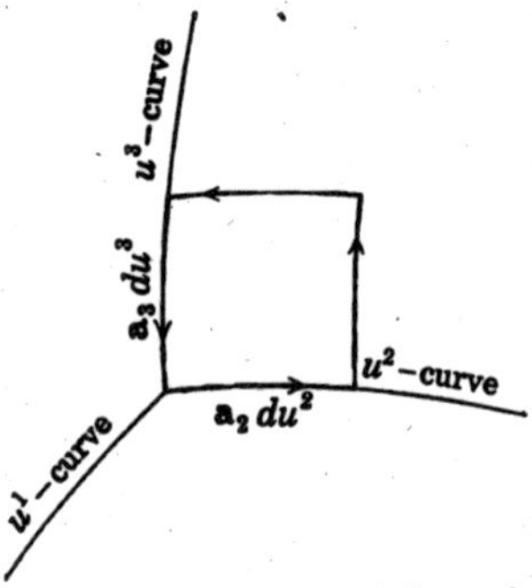

FIG. 6.—Calculation of the curl in curvilinear coordinates.

$$\left[\frac{\partial}{\partial u^2}(\mathbf{F} \cdot \mathbf{a}_3) - \frac{\partial}{\partial u^3}(\mathbf{F} \cdot \mathbf{a}_2)\right] du^2\, du^3. \tag{57}$$

This quantity must now be divided by the area of the rectangle, or $\sqrt{(\mathbf{a}_2 \times \mathbf{a}_3) \cdot (\mathbf{a}_2 \times \mathbf{a}_3)}\, du^2\, du^3$. As for the unit normal **n**, we note that the reciprocal vector $\mathbf{a}^1$, not the unitary vector $\mathbf{a}_1$, is always normal to the u^1-surface. Its magnitude must be unity; hence

$$\mathbf{n} = \frac{\mathbf{a}^1}{\sqrt{\mathbf{a}^1 \cdot \mathbf{a}^1}}. \tag{58}$$

These values introduced into (56) now lead to

$$(\nabla \times \mathbf{F}) \cdot \frac{\mathbf{a}^1}{\sqrt{\mathbf{a}^1 \cdot \mathbf{a}^1}} = \frac{1}{\sqrt{(\mathbf{a}_2 \times \mathbf{a}_3) \cdot (\mathbf{a}_2 \times \mathbf{a}_3)}} \left[\frac{\partial}{\partial u^2}(\mathbf{F} \cdot \mathbf{a}_3) - \frac{\partial}{\partial u^3}(\mathbf{F} \cdot \mathbf{a}_2)\right]. \tag{59}$$

By (9) and (37)

$$\mathbf{a}_2 \times \mathbf{a}_3 = [\mathbf{a}_1 \cdot (\mathbf{a}_2 \times \mathbf{a}_3)]\mathbf{a}^1 = \sqrt{g}\, a^1; \tag{60}$$

hence (59) reduces to

$$(\nabla \times \mathbf{F}) \cdot \mathbf{a}^1 = \frac{1}{\sqrt{g}} \left[\frac{\partial}{\partial u^2} (\mathbf{F} \cdot \mathbf{a}_3) - \frac{\partial}{\partial u^3} (\mathbf{F} \cdot \mathbf{a}_2) \right]. \tag{61}$$

The two remaining components of $\nabla \times \mathbf{F}$ are obtained from (61) by permutation of indices. Then by (21)

$$\nabla \times \mathbf{F} = \sum_{i=1}^{3} (\nabla \times \mathbf{F} \cdot \mathbf{a}^i)\mathbf{a}_i. \tag{62}$$

Remembering that $\mathbf{F} \cdot \mathbf{a}_i$ is the covariant component f_i, we have for the curl of a vector with respect to a set of general coordinates

$$\nabla \times \mathbf{F} = \frac{1}{\sqrt{g}} \left[\left(\frac{\partial f_3}{\partial u^2} - \frac{\partial f_2}{\partial u^3} \right) \mathbf{a}_1 + \left(\frac{\partial f_1}{\partial u^3} - \frac{\partial f_3}{\partial u^1} \right) \mathbf{a}_2 + \left(\frac{\partial f_2}{\partial u^1} - \frac{\partial f_1}{\partial u^2} \right) \mathbf{a}_3 \right]. \tag{63}$$

Finally, we consider the operation $\nabla^2\phi$, by which we must understand $\nabla \cdot \nabla\phi$. We need only let $\mathbf{F} = \nabla\phi$ in (55). The contravariant components of the gradient are

$$f^i = \mathbf{F} \cdot \mathbf{a}^i = \sum_{j=1}^{3} \mathbf{a}^i \cdot \mathbf{a}^j \frac{\partial \phi}{\partial u^j} = \sum_{j=1}^{3} g^{ij} \frac{\partial \phi}{\partial u^j}. \tag{64}$$

Then

$$\nabla \cdot \nabla\phi = \nabla^2\phi = \frac{1}{\sqrt{g}} \sum_{i=1}^{3} \sum_{j=1}^{3} \frac{\partial}{\partial u^i} \left(g^{ij} \sqrt{g} \frac{\partial \phi}{\partial u^j} \right). \tag{65}$$

1.16. Orthogonal Systems.—Thus far no restriction has been imposed on the base vectors other than that they shall be noncoplanar. Now it happens that in almost all cases only the orthogonal systems can be usefully applied, and these allow a considerable simplification of the formulas derived above. Oblique systems might well be of the greatest practical importance; but they lead, unfortunately, to partial differential equations which cannot be mastered by present-day analysis.

The unitary vectors $\mathbf{a}_1, \mathbf{a}_2, \mathbf{a}_3$ of an orthogonal system are by definition mutually perpendicular, whence it follows that $\mathbf{a}^i$ is parallel to $\mathbf{a}_i$ and is its reciprocal in magnitude.

$$\mathbf{a}^i = \frac{1}{\mathbf{a}_i \cdot \mathbf{a}_i} \mathbf{a}_i = \frac{1}{g_{ii}} \mathbf{a}_i. \tag{66}$$

Furthermore

$$\mathbf{a}_1 \cdot \mathbf{a}_2 = \mathbf{a}_2 \cdot \mathbf{a}_3 = \mathbf{a}_3 \cdot \mathbf{a}_1 = 0; \tag{67}$$

hence $g_{ij} = 0$, when $i \neq j$. It is customary in this orthogonal case to introduce the abbreviations

$$h_1 = \sqrt{g_{11}}, \qquad h_2 = \sqrt{g_{22}}, \qquad h_3 = \sqrt{g_{33}}, \tag{68}$$

$$g^{ii} = \frac{1}{g_{ii}} = \frac{1}{h_i^2}. \tag{69}$$

The h_i may be calculated from the formula

$$h_i^2 = \left(\frac{\partial x^1}{\partial u^i}\right)^2 + \left(\frac{\partial x^2}{\partial u^i}\right)^2 + \left(\frac{\partial x^3}{\partial u^i}\right)^2, \tag{70}$$

although their value is usually obvious from the geometry of the system. The elementary cell bounded by coordinate surfaces is now a rectangular box whose edges are

$$ds_1 = h_1\, du^1, \qquad ds_2 = h_2\, du^2, \qquad ds_3 = h_3\, du^3, \tag{71}$$

and whose volume is

$$dv = h_1 h_2 h_3\, du^1\, du^2\, du^3. \tag{72}$$

All off-diagonal terms of the determinant for g vanish and hence

$$\sqrt{g} = h_1 h_2 h_3. \tag{73}$$

The distinction between the contravariant and covariant components of a vector with respect to a unitary or reciprocal unitary base system is essential to an understanding of the invariant properties of the differential operators and of scalar and vector products. However, in a fixed reference system this distinction may usually be ignored. It is then convenient to express the vector **F** in terms of its components, or projections, F_1, F_2, F_3 on an orthogonal base system of *unit* vectors $\mathbf{i}_1, \mathbf{i}_2, \mathbf{i}_3$. By (22) and (66)

$$\mathbf{a}_i = h_i \mathbf{i}_i, \qquad \mathbf{a}^i = \frac{1}{h_i} \mathbf{i}_i. \tag{74}$$

In terms of the components F_i the contravariant and covariant components are

$$f^i = \frac{1}{h_i} F_i, \qquad f_i = h_i F_i. \tag{75}$$

Also

$$F = F_1 \mathbf{i}_1 + F_2 \mathbf{i}_2 + F_3 \mathbf{i}_3, \tag{76}$$

$$\mathbf{i}_j \cdot \mathbf{i}_k = \delta_{jk}. \tag{77}$$

The gradient, divergence, curl, and Laplacian in an orthogonal system of curvilinear coordinates can now be written down directly from the results of the previous section.

From (49) we have for the gradient

$$\nabla\phi = \sum_{j=1}^{3} \frac{1}{h_j} \frac{\partial \phi}{\partial u^j} \mathbf{i}_j. \tag{78}$$

According to (55) the divergence of a vector $\mathbf{F}$ is

$$\nabla \cdot \mathbf{F} = \frac{1}{h_1 h_2 h_3} \left[\frac{\partial}{\partial u^1} (h_2 h_3 F_1) + \frac{\partial}{\partial u^2} (h_3 h_1 F_2) + \frac{\partial}{\partial u^3} (h_1 h_2 F_3) \right]. \tag{79}$$

For the curl of $\mathbf{F}$ we have by (63)

$$\begin{aligned} \nabla \times \mathbf{F} = \frac{1}{h_2 h_3} \left[\frac{\partial}{\partial u^2} (h_3 F_3) - \frac{\partial}{\partial u^3} (h_2 F_2) \right] \mathbf{i}_1 + \frac{1}{h_3 h_1} \left[\frac{\partial}{\partial u^3} (h_1 F_1) \right. \\ \left. - \frac{\partial}{\partial u^1} (h_3 F_3) \right] \mathbf{i}_2 + \frac{1}{h_1 h_2} \left[\frac{\partial}{\partial u^1} (h_2 F_2) - \frac{\partial}{\partial u^2} (h_1 F_1) \right] \mathbf{i}_3. \end{aligned} \tag{80}$$

It may be remarked that (80) is the expansion of the determinant

$$\nabla \times \mathbf{F} = \frac{1}{h_1 h_2 h_3} \begin{vmatrix} h_1 \mathbf{i}_1 & h_2 \mathbf{i}_2 & h_3 \mathbf{i}_3 \\ \dfrac{\partial}{\partial u^1} & \dfrac{\partial}{\partial u^2} & \dfrac{\partial}{\partial u^3} \\ h_1 F_1 & h_2 F_2 & h_3 F_3 \end{vmatrix}. \tag{81}$$

Finally, the Laplacian of an *invariant scalar* ϕ is

$$\nabla^2 \phi = \frac{1}{h_1 h_2 h_3} \left[\frac{\partial}{\partial u^1} \left(\frac{h_2 h_3}{h_1} \frac{\partial \phi}{\partial u^1} \right) + \frac{\partial}{\partial u^2} \left(\frac{h_3 h_1}{h_2} \frac{\partial \phi}{\partial u^2} \right) + \frac{\partial}{\partial u^3} \left(\frac{h_1 h_2}{h_3} \frac{\partial \phi}{\partial u^3} \right) \right]. \tag{82}$$

By an invariant scalar is meant a quantity such as temperature or energy which is invariant to a rotation of the coordinate system. The components, or measure numbers, F_i of a vector $\mathbf{F}$ are scalars, but they transform with a transformation of the base vectors. Now in the analysis of the field we encounter frequently the operation

$$\nabla \times \nabla \times \mathbf{F} = \nabla\nabla \cdot \mathbf{F} - \nabla \cdot \nabla\mathbf{F}. \tag{83}$$

No meaning has been attributed as yet to $\nabla \cdot \nabla\mathbf{F}$. In a rectangular, Cartesian system of coordinates x^1, x^2, x^3, it is clear that this operation is equivalent to

$$\nabla \cdot \nabla\mathbf{F} = \nabla^2\mathbf{F} = \sum_{j=1}^{3} \left(\frac{\partial^2 F_j}{\partial (x^1)^2} + \frac{\partial^2 F_j}{\partial (x^2)^2} + \frac{\partial^2 F_j}{\partial (x^3)^2} \right) \mathbf{i}_j, \tag{84}$$

i.e., the Laplacian acting on the rectangular components of $\mathbf{F}$. In generalized coordinates $\nabla \times \nabla \times \mathbf{F}$ is represented by the determinant

$$\nabla \times \nabla \times \mathbf{F} = \begin{vmatrix} \frac{1}{h_2 h_3}\mathbf{i}_1 & \frac{1}{h_3 h_1}\mathbf{i}_2 & \frac{1}{h_1 h_2}\mathbf{i}_3 \\ \frac{\partial}{\partial u^1} & \frac{\partial}{\partial u^2} & \frac{\partial}{\partial u^3} \\ \frac{h_1}{h_2 h_3}\left[\frac{\partial}{\partial u^2}(h_3 F_3) - \frac{\partial}{\partial u^3}(h_2 F_2)\right] & \frac{h_2}{h_3 h_1}\left[\frac{\partial}{\partial u^3}(h_1 F_1) - \frac{\partial}{\partial u^1}(h_3 F_3)\right] & \frac{h_3}{h_1 h_2}\left[\frac{\partial}{\partial u^1}(h_2 F_2) - \frac{\partial}{\partial u^2}(h_1 F_1)\right] \end{vmatrix}. \tag{85}$$

The vector $\nabla \cdot \nabla \mathbf{F}$ may now be obtained by subtraction of (85) from the expansion of $\nabla\nabla \cdot \mathbf{F}$, and the result differs from that which follows a direct application of the Laplacian operator to the curvilinear components of $\mathbf{F}$.

1.17. The Field Equations in General Orthogonal Coordinates.—In any orthogonal system of curvilinear coordinates characterized by the coefficients h_1, h_2, h_3, the Maxwell equations can be resolved into a set of eight partial differential equations relating the scalar components of the field vectors.

$$\text{(I)}\quad \begin{aligned} &\frac{1}{h_2 h_3}\left[\frac{\partial}{\partial u^2}(h_3 E_3) - \frac{\partial}{\partial u^3}(h_2 E_2)\right] + \frac{\partial B_1}{\partial t} = 0. \\ &\frac{1}{h_3 h_1}\left[\frac{\partial}{\partial u^3}(h_1 E_1) - \frac{\partial}{\partial u^1}(h_3 E_3)\right] + \frac{\partial B_2}{\partial t} = 0. \\ &\frac{1}{h_1 h_2}\left[\frac{\partial}{\partial u^1}(h_2 E_2) - \frac{\partial}{\partial u^2}(h_1 E_1)\right] + \frac{\partial B_3}{\partial t} = 0. \end{aligned}$$

$$\text{(II)}\quad \begin{aligned} &\frac{1}{h_2 h_3}\left[\frac{\partial}{\partial u^2}(h_3 H_3) - \frac{\partial}{\partial u^3}(h_2 H_2)\right] - \frac{\partial D_1}{\partial t} = J_1. \\ &\frac{1}{h_3 h_1}\left[\frac{\partial}{\partial u^3}(h_1 H_1) - \frac{\partial}{\partial u^1}(h_3 H_3)\right] - \frac{\partial D_2}{\partial t} = J_2. \\ &\frac{1}{h_1 h_2}\left[\frac{\partial}{\partial u^1}(h_2 H_2) - \frac{\partial}{\partial u^2}(h_1 H_1)\right] - \frac{\partial D_3}{\partial t} = J_3. \end{aligned}$$

$$\text{(III)}\quad \frac{\partial}{\partial u^1}(h_2 h_3 B_1) + \frac{\partial}{\partial u^2}(h_3 h_1 B_2) + \frac{\partial}{\partial u^3}(h_1 h_2 B_3) = 0.$$

$$\text{(IV)}\quad \frac{\partial}{\partial u^1}(h_2 h_3 D_1) + \frac{\partial}{\partial u^2}(h_3 h_1 D_2) + \frac{\partial}{\partial u^3}(h_1 h_2 D_3) = h_1 h_2 h_3 \rho.$$

It is not feasible to solve this system simultaneously in such a manner as to separate the components of the field vectors and to obtain equations satisfied by each individually. In any given problem one must make the most of whatever advantages and peculiarities the various coordinate systems have to offer.

1.18. Properties of Some Elementary Systems.—An orthogonal coordinate system has been shown to be completely characterized by the three metrical coefficients, h_1, h_2, h_3. These parameters will now be determined for certain elementary systems and in a few cases the differential operators set down for convenient reference.

1. *Cylindrical Coordinates.*—Let P' be the projection of a point $P(x, y, z)$ on the z-plane and r, θ be the polar coordinates of P' in this plane (Fig. 7). The variables

$$u^1 = r, \qquad u^2 = \theta, \qquad u^3 = z, \tag{86}$$

are called circular cylindrical coordinates. They are related to the rectangular coordinates by the equations

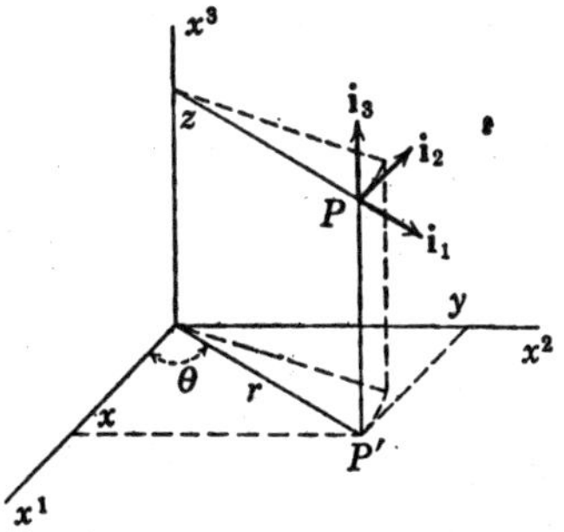

FIG. 7.—Coordinates of the circular cylinder.

$$x = r\cos\theta, \qquad y = r\sin\theta, \qquad z = z. \tag{87}$$

The coordinate surfaces are coaxial cylinders of circular cross section intersected orthogonally by the planes $\theta =$ constant and $z =$ constant. The infinitesimal line element is

$$ds^2 = dr^2 + r^2\,d\theta^2 + dz^2, \tag{88}$$

whence it is apparent that the metrical coefficients are

$$h_1 = 1, \qquad h_2 = r, \qquad h_3 = 1. \tag{89}$$

If ψ is any scalar and $\mathbf{F}$ a vector function we find:

$$\begin{aligned}
\nabla\psi &= \frac{\partial\psi}{\partial r}\mathbf{i}_1 + \frac{1}{r}\frac{\partial\psi}{\partial\theta}\mathbf{i}_2 + \frac{\partial\psi}{\partial z}\mathbf{i}_3,\\
\nabla\cdot\mathbf{F} &= \frac{1}{r}\frac{\partial}{\partial r}(rF_1) + \frac{1}{r}\frac{\partial F_2}{\partial\theta} + \frac{\partial F_3}{\partial z},\\
\nabla\times\mathbf{F} &= \left(\frac{1}{r}\frac{\partial F_3}{\partial\theta} - \frac{\partial F_2}{\partial z}\right)\mathbf{i}_1 + \left(\frac{\partial F_1}{\partial z} - \frac{\partial F_3}{\partial r}\right)\mathbf{i}_2 + \left[\frac{1}{r}\frac{\partial}{\partial r}(rF_2) - \frac{1}{r}\frac{\partial F_1}{\partial\theta}\right]\mathbf{i}_3,\\
\nabla^2\psi &= \frac{1}{r}\frac{\partial}{\partial r}\left(r\frac{\partial\psi}{\partial r}\right) + \frac{1}{r^2}\frac{\partial^2\psi}{\partial\theta^2} + \frac{\partial^2\psi}{\partial z^2}.
\end{aligned}\tag{90}$$

2. *Spherical Coordinates.*—The variables

(91) $$u^1 = r, \qquad u^2 = \theta, \qquad u^3 = \phi,$$

related to the rectangular coordinates by the transformation

(92) $$x = r \sin\theta \cos\phi, \qquad y = r \sin\theta \sin\phi, \qquad z = r \cos\theta,$$

are called the spherical coordinates of the point P. The coordinate surfaces, $r =$ constant, are concentric spheres intersected by meridian planes, $\phi =$ constant, and a family of cones, $\theta =$ constant. The unit vectors $\mathbf{i}_1$, $\mathbf{i}_2$, $\mathbf{i}_3$ are drawn in the direction of *increasing* r, θ, and ϕ such as to constitute a right-hand base system, as indicated in Fig. 8. The line element is

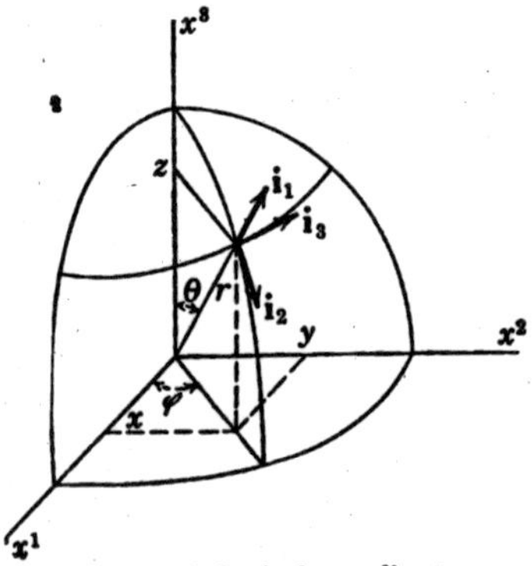

FIG. 8.—Spherical coordinates.

(93) $$ds^2 = dr^2 + r^2\, d\theta^2 + r^2 \sin^2\theta\, d\phi^2,$$

whence for the metrical coefficients we obtain

(94) $$h_1 = 1, \qquad h_2 = r, \qquad h_3 = r \sin\theta.$$

These values lead to

(95)
$$\nabla\psi = \frac{\partial\psi}{\partial r}\mathbf{i}_1 + \frac{1}{r}\frac{\partial\psi}{\partial\theta}\mathbf{i}_2 + \frac{1}{r\sin\theta}\frac{\partial\psi}{\partial\phi}\mathbf{i}_3,$$
$$\nabla\cdot\mathbf{F} = \frac{1}{r^2}\frac{\partial}{\partial r}(r^2F_1) + \frac{1}{r\sin\theta}\frac{\partial}{\partial\theta}(\sin\theta F_2) + \frac{1}{r\sin\theta}\frac{\partial F_3}{\partial\phi},$$
$$\nabla\times\mathbf{F} = \frac{1}{r\sin\theta}\left[\frac{\partial}{\partial\theta}(\sin\theta F_3) - \frac{\partial F_2}{\partial\phi}\right]\mathbf{i}_1 + \frac{1}{r}\left[\frac{1}{\sin\theta}\frac{\partial F_1}{\partial\phi} - \frac{\partial}{\partial r}(rF_3)\right]\mathbf{i}_2 + \frac{1}{r}\left[\frac{\partial}{\partial r}(rF_2) - \frac{\partial F_1}{\partial\theta}\right]\mathbf{i}_3,$$
$$\nabla^2\psi = \frac{1}{r^2}\frac{\partial}{\partial r}\left(r^2\frac{\partial\psi}{\partial r}\right) + \frac{1}{r^2\sin\theta}\frac{\partial}{\partial\theta}\left(\sin\theta\frac{\partial\psi}{\partial\theta}\right) + \frac{1}{r^2\sin^2\theta}\frac{\partial^2\psi}{\partial\phi^2}.$$

3. *Elliptic Coordinates.*—Let two fixed points P_1 and P_2 be located at $x = c$ and $x = -c$ on the x-axis and let r_1 and r_2 be the distances of a variable point P in the z-plane from P_1 and P_2. Then the variables

(96) $$u^1 = \xi, \qquad u^2 = \eta, \qquad u^3 = z,$$

defined by equations

(97) $$\xi = \frac{r_1 + r_2}{2c}, \qquad \eta = \frac{r_1 - r_2}{2c},$$

c = focus

are called elliptic coordinates. From these relations it is evident that

(98) $$\xi \geqq 1, \qquad -1 \leqq \eta \leqq 1.$$

The coordinate surface, $\xi =$ constant, is a cylinder of elliptic cross section, whose foci are P_1 and P_2. The semimajor and semiminor axes of an ellipse ξ are given by

(99) $$a = c\xi, \qquad b = c\sqrt{\xi^2 - 1},$$

and the eccentricity is

(100) $$e = \frac{c}{a} = \frac{1}{\xi}.$$

The surfaces, $\eta =$ constant, represent a family of confocal hyperbolic

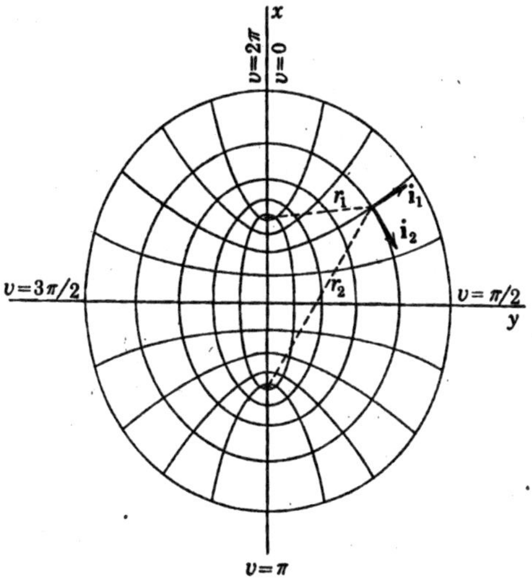

FIG. 9.—Coordinates of the elliptic cylinder. Ambiguity of sign is avoided by placing $\xi = \cosh u$, $\eta = \cos v$.

cylinders of two sheets as illustrated in Fig. 9. The equations of these two confocal systems are

(101) $$\frac{x^2}{\xi^2} + \frac{y^2}{\xi^2 - 1} = c^2, \qquad \frac{x^2}{\eta^2} - \frac{y^2}{1 - \eta^2} = c^2,$$

from which we deduce the transformation

(102) $$x = c\xi\eta, \qquad y = c\sqrt{(\xi^2 - 1)(1 - \eta^2)}, \qquad z = z.$$

The variable η corresponds to the cosine of an angle measured from the x-axis and the unit vectors $\mathbf{i}_1$, $\mathbf{i}_2$ of a right-hand base system are therefore drawn as indicated in Fig. 9, with $\mathbf{i}_3$ normal to the page and directed from the reader.

The metrical coefficients are calculated from (102) and (70), giving

$$h_1 = c\sqrt{\frac{\xi^2 - \eta^2}{\xi^2 - 1}}, \qquad h_2 = c\sqrt{\frac{\xi^2 - \eta^2}{1 - \eta^2}}, \qquad h_3 = 1. \tag{103}$$

4. *Parabolic Coordinates.*—If r, θ are polar coordinates of a variable point in the z-plane, one may define two mutually orthogonal families of parabolas by the equations

$$\xi = \sqrt{2r}\sin\frac{\theta}{2}, \qquad \eta = \sqrt{2r}\cos\frac{\theta}{2}. \tag{104}$$

The surfaces, $\xi =$ constant and $\eta =$ constant, are intersecting parabolic

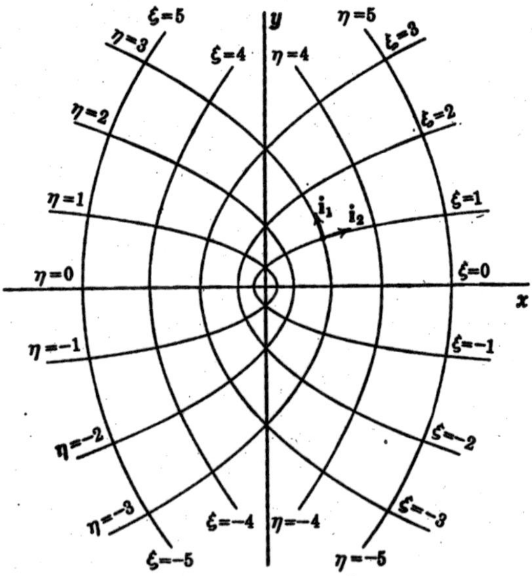

FIG. 10.—Parabolic coordinates.

cylinders whose elements are parallel to the z-axis as shown in Fig. 10. The parameters

$$u^1 = \xi, \qquad u^2 = \eta, \qquad u^3 = -z \tag{105}$$

are called parabolic coordinates. Upon replacing r and θ in (104) by rectangular coordinates we find

$$\xi^2 = \sqrt{x^2 + y^2} - x, \qquad \eta^2 = \sqrt{x^2 + y^2} + x, \tag{106}$$

whence for the transformation from rectangular to parabolic coordinates we have

$$x = \tfrac{1}{2}(\eta^2 - \xi^2), \qquad y = \xi\eta, \qquad z = -z. \tag{107}$$

The unit vectors $\mathbf{i}_1$ and $\mathbf{i}_2$ are directed as shown in Fig. 10, with $\mathbf{i}_3$ normal to the page and away from the reader. The calculation of the metrical coefficients from (107) and (70) leads to

$$h_1 = h_2 = \sqrt{\xi^2 + \eta^2}, \qquad h_3 = 1. \tag{108}$$

5. *Bipolar Coordinates.*—Let P_1 and P_2 be two fixed points in any z-plane with the coordinates $(a, 0)$, $(-a, 0)$ respectively. If ξ is a parameter, the equation

$$(x - a \coth \xi)^2 + y^2 = a^2 \operatorname{csch}^2 \xi, \tag{109}$$

describes two families of circles whose centers lie on the x-axis. These two families are symmetrical with respect to the y-axis as shown in

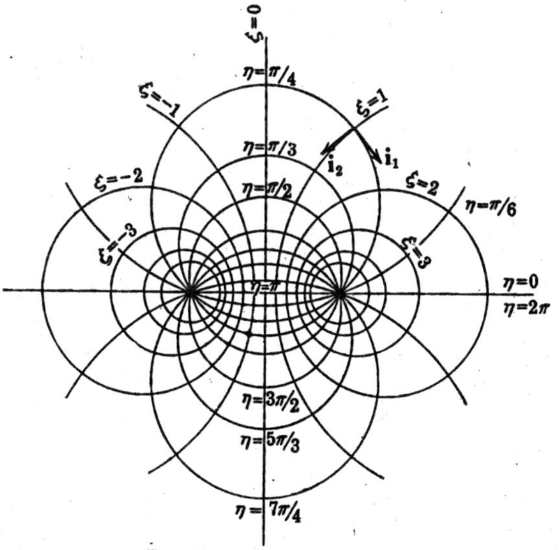

FIG. 11.—Bipolar coordinates.

Fig. 11. The point P_1 at $(a, 0)$ corresponds to $\xi = +\infty$, whereas its image P_2 at $(-a, 0)$ is approached when $\xi = -\infty$. The locus of (109), when $\xi = 0$, coincides with the y-axis. The orthogonal set is likewise a family of circles whose centers all lie on the y-axis and all of which pass through the fixed points P_1 and P_2. They are defined by the equation,

$$x^2 + (y - a \cot \eta)^2 = a^2 \csc^2 \eta, \tag{110}$$

wherein the parameter η is confined to the range $0 \leqq \eta \leqq 2\pi$. In order that the coordinates of a point P in a given quadrant shall be single-valued, each circle of this family is separated into two segments by the points P_1 and P_2. A value less than π is assigned to the arc above the x-axis, while the lower arc is denoted by a value of η equal to π plus the value of η assigned to the upper segment of the same circle.

The variables

$$u^1 = \xi, \qquad u^2 = \eta, \qquad u^3 = z, \tag{111}$$

are called bipolar coordinates. From (109) and (110) the transformation to rectangular coordinates is found to be

$$x = \frac{a \sinh \xi}{\cosh \xi - \cos \eta}, \qquad y = \frac{a \sin \eta}{\cosh \xi - \cos \eta}, \qquad z = z. \tag{112}$$

The unit vectors $\mathbf{i}_1$ and $\mathbf{i}_2$ are in the direction of increasing ξ and η as indicated in Fig. 11, while $\mathbf{i}_3$ is directed away from the reader along the z-axis. The calculation of the metrical coefficients yields

$$h_1 = h_2 = \frac{a}{\cosh \xi - \cos \eta}, \qquad h_3 = 1. \tag{113}$$

6. *Spheroidal Coordinates.*—The coordinates of the elliptic cylinder were generated by translating a system of confocal ellipses along the z-axis. The spheroidal coordinates are obtained by rotation of the ellipses about an axis of symmetry. Two cases are to be distinguished, according to whether the rotation takes place about the major or about the minor axis. In Fig. 9 the major axes are oriented along the x-axis of a rectangular system. If the figure is rotated about this axis, a set of confocal prolate spheroids is generated whose orthogonal surfaces are hyperboloids of two sheets. If ϕ measures the angle of rotation from the y-axis in the x-plane and r the perpendicular distance of a point from the x-axis, so that

$$y = r \cos \phi, \qquad z = r \sin \phi, \tag{114}$$

then the variables

$$u^1 = \xi, \qquad u^2 = \eta, \qquad u^3 = \phi, \tag{115}$$

defined by (97) and (114) are called prolate spheroidal coordinates. In place of (101) we have for the equations of the two confocal systems

$$\frac{x^2}{\xi^2} + \frac{r^2}{\xi^2 - 1} = c^2, \qquad \frac{x^2}{\eta^2} - \frac{r^2}{1 - \eta^2} = c^2, \tag{116}$$

from which we deduce

$$(117)\quad x = c\xi\eta, \qquad y = c\sqrt{(\xi^2 - 1)(1 - \eta^2)}\cos\phi, \qquad z = c\sqrt{(\xi^2 - 1)(1 - \eta^2)}\sin\phi,$$

$$(118)\qquad \xi \geqq 1, \qquad -1 \leqq \eta \leqq 1, \qquad 0 \leqq \phi \leqq 2\pi.$$

A calculation of the metrical coefficients gives

$$(119)\quad h_1 = c\sqrt{\frac{\xi^2 - \eta^2}{\xi^2 - 1}}, \quad h_2 = c\sqrt{\frac{\xi^2 - \eta^2}{1 - \eta^2}}, \quad h_3 = c\sqrt{(\xi^2 - 1)(1 - \eta^2)}.$$

When the ellipses of Fig. 9 are rotated about the y-axis, the spheroids are oblate and the focal points P_1, P_2 describe a circle in the plane $y = 0$. Let r, ϕ, y, be cylindrical coordinates about the y-axis,

$$(120)\qquad z = r\cos\phi, \qquad x = r\sin\phi.$$

If by P_1 and P_2 we now understand the points where the focal ring of radius c intercepts the plane ϕ = constant, the variables ξ and η are still defined by (97); but for the equations of the coordinate surfaces we have

$$(121)\qquad \frac{r^2}{\xi^2} + \frac{y^2}{\xi^2 - 1} = c^2, \qquad \frac{r^2}{\eta^2} - \frac{y^2}{1 - \eta^2} = c^2,$$

from which we deduce the transformation from oblate spheroidal coordinates

$$(122)\qquad u^1 = \xi, \qquad u^2 = \eta, \qquad u^3 = \phi,$$

to rectangular coordinates

$$(123)\quad x = c\xi\eta\sin\phi, \qquad y = c\sqrt{(\xi^2 - 1)(1 - \eta^2)}, \qquad z = c\xi\eta\cos\phi.$$

The surfaces, ξ = constant, are oblate spheroids, whereas the orthogonal family, η = constant, are hyperboloids of one sheet. The metrical coefficients are

$$(124)\qquad h_1 = c\sqrt{\frac{\xi^2 - \eta^2}{\xi^2 - 1}}, \qquad h_2 = c\sqrt{\frac{\xi^2 - \eta^2}{1 - \eta^2}}, \qquad h_3 = c\xi\eta.$$

The practical utility of spheroidal coordinates may be surmised from the fact that as the eccentricity approaches unity the prolate spheroids become rod-shaped, whereas the oblate spheroids degenerate into flat, elliptic disks. In the limit, as the focal distance $2c$ and the eccentricity approach zero, the spheroidal coordinates go over into spherical coordinates, with $\xi \to r$, $\eta \to \cos\theta$.

7. *Paraboloidal Coordinates.*—Another set of rotational coordinates may be obtained by rotating the parabolas of Fig. 10 about their axis

of symmetry. The variables

$$u^1 = \xi, \qquad u^2 = \eta, \qquad u^3 = \phi, \tag{125}$$

defined by

$$x = \xi\eta \cos \phi, \qquad y = \xi\eta \sin \phi, \qquad z = \tfrac{1}{2}(\xi^2 - \eta^2), \tag{126}$$

are called paraboloidal coordinates. The surfaces, $\xi =$ constant, $\eta =$ constant, are paraboloids of revolution about an axis of symmetry which in this case has been taken coincident with the z-axis. The plane, $y = 0$, is cut by these surfaces along the curves

$$x^2 = 2\xi^2\left(\frac{\xi^2}{2} - z\right), \qquad x^2 = 2\eta^2\left(\frac{\eta^2}{2} + z\right), \tag{127}$$

which are evidently parabolas whose foci are located at the origin and whose parameters are ξ^2 and η^2. The metrical coefficients are

$$h_1 = h_2 = \sqrt{\xi^2 + \eta^2}, \qquad h_3 = \xi\eta. \tag{128}$$

8. *Ellipsoidal Coordinates.*—The equation

$$\frac{x^2}{a^2} + \frac{y^2}{b^2} + \frac{z^2}{c^2} = 1, \qquad (a > b > c), \tag{129}$$

is that of an ellipsoid whose semiprincipal axes are of length a, b, c. Then

$$\begin{aligned}
&\frac{x^2}{a^2 + \xi} + \frac{y^2}{b^2 + \xi} + \frac{z^2}{c^2 + \xi} = 1, && (\xi > -c^2),\\
&\frac{x^2}{a^2 + \eta} + \frac{y^2}{b^2 + \eta} + \frac{z^2}{c^2 + \eta} = 1, && (-c^2 > \eta > -b^2),\\
&\frac{x^2}{a^2 + \zeta} + \frac{y^2}{b^2 + \zeta} + \frac{z^2}{c^2 + \zeta} = 1, && (-b^2 > \zeta > -a^2),
\end{aligned} \tag{130}$$

are the equations respectively of an ellipsoid, a hyperboloid of one sheet, and a hyperboloid of two sheets, all confocal with the ellipsoid (129). Through each point of space there will pass just one surface of each kind, and to each point there will correspond a unique set of values for ξ, η, ζ. The variables

$$u^1 = \xi, \qquad u^2 = \eta, \qquad u^3 = \zeta, \tag{131}$$

are called ellipsoidal coordinates. The surface, $\xi =$ constant, is a hyperboloid of one sheet and $\eta =$ constant, a hyperboloid of two sheets. The transformation to rectangular coordinates is obtained by solving

(130) simultaneously for x, y, z. This gives

$$
\begin{aligned}
x &= \pm\left[\frac{(\xi + a^2)(\eta + a^2)(\zeta + a^2)}{(b^2 - a^2)(c^2 - a^2)}\right]^{\frac{1}{2}}, \\
y &= \pm\left[\frac{(\xi + b^2)(\eta + b^2)(\zeta + b^2)}{(c^2 - b^2)(a^2 - b^2)}\right]^{\frac{1}{2}}, \\
z &= \pm\left[\frac{(\xi + c^2)(\eta + c^2)(\zeta + c^2)}{(a^2 - c^2)(b^2 - c^2)}\right]^{\frac{1}{2}}.
\end{aligned} \tag{132}
$$

The mutual orthogonality of the three families of surfaces may be verified by calculating the coefficients g_{ij} from (132) by means of (45). They are zero when $i \neq j$; for the diagonal terms we find

$$
\begin{aligned}
h_1 &= \frac{1}{2}\left[\frac{(\xi - \eta)(\xi - \zeta)}{(\xi + a^2)(\xi + b^2)(\xi + c^2)}\right]^{\frac{1}{2}}, \\
h_2 &= \frac{1}{2}\left[\frac{(\eta - \zeta)(\eta - \xi)}{(\eta + a^2)(\eta + b^2)(\eta + c^2)}\right]^{\frac{1}{2}}, \\
h_3 &= \frac{1}{2}\left[\frac{(\zeta - \xi)(\zeta - \eta)}{(\zeta + a^2)(\zeta + b^2)(\zeta + c^2)}\right]^{\frac{1}{2}},
\end{aligned} \tag{133}
$$

It is convenient to introduce the abbreviation

$$
R_s = \sqrt{(s + a^2)(s + b^2)(s + c^2)}, \qquad (s = \xi, \eta, \zeta). \tag{134}
$$

For the Laplacian of a scalar ψ we then have

$$
\begin{aligned}
\nabla^2\psi = \frac{4}{(\xi - \eta)(\xi - \zeta)(\eta - \zeta)}\bigg[(\eta - \zeta)R_\xi \frac{\partial}{\partial \xi}\left(R_\xi \frac{\partial \psi}{\partial \xi}\right) \\
+ (\zeta - \xi)R_\eta \frac{\partial}{\partial \eta}\left(R_\eta \frac{\partial \psi}{\partial \eta}\right) + (\xi - \eta)R_\zeta \frac{\partial}{\partial \zeta}\left(R_\zeta \frac{\partial \psi}{\partial \zeta}\right)\bigg].
\end{aligned} \tag{135}
$$

THE FIELD TENSORS

1.19. Orthogonal Transformations and Their Invariants.—In the theory of relativity one undertakes the formulation of the laws of physics, and in particular the equations of the electromagnetic field, such that they are invariant to transformations of the system of reference. Although in the present volume we shall have no occasion to examine the foundations of the relativity theory, it will nevertheless prove occasionally advantageous to employ the symmetrical, four-dimensional notation introduced by Minkowski and Sommerfeld and to deduce the Lorentz transformation with respect to which the field equations are invariant. To discover quantities which are invariant to a transformation from one system of general curvilinear coordinates to another, it is essential that one distinguish between the covariant and contravariant components of vectors and between unitary and reciprocal unitary base systems. For our present purposes it will be sufficient, however, to confine the discussion to systems of rectangular, Cartesian coordinates in which, as we have seen, covariant and contravariant components are identical.[1]

Let $\mathbf{i}_1$, $\mathbf{i}_2$, $\mathbf{i}_3$ be three orthogonal, unit base vectors defining a rectangular coordinate system X whose origin is located at the fixed point O, and let $\mathbf{r}$ be the position vector of any point P with respect to O.

$$\mathbf{r} = x_1\mathbf{i}_1 + x_2\mathbf{i}_2 + x_3\mathbf{i}_3, \tag{1}$$

and since

$$\mathbf{i}_j \cdot \mathbf{i}_k = \delta_{jk}, \tag{2}$$

the coordinates of P in the system X are

$$x_k = \mathbf{r} \cdot \mathbf{i}_k. \tag{3}$$

Suppose now that $\mathbf{i}'_1$, $\mathbf{i}'_2$, $\mathbf{i}'_3$ are the base vectors of a second rectangular system X' whose origin coincides with O and which, therefore, differs from X only by a rotation of the coordinate axes. Since

$$\mathbf{r} = x'_1\mathbf{i}'_1 + x'_2\mathbf{i}'_2 + x'_3\mathbf{i}'_3, \tag{4}$$

the coordinates of P with respect to X' are

$$x'_j = \mathbf{r} \cdot \mathbf{i}'_j = x_1\mathbf{i}_1 \cdot \mathbf{i}'_j + x_2\mathbf{i}_2 \cdot \mathbf{i}'_j + x_3\mathbf{i}_3 \cdot \mathbf{i}'_j; \tag{5}$$

each coordinate of P in X' is a linear function of its coordinates in X, whereby the coefficients

$$a_{jk} = \mathbf{i}'_j \cdot \mathbf{i}_k \tag{6}$$

of the linear form are clearly the direction cosines of the coordinate axes of X' with respect to the axes of X. A rotation of a rectangular coordinate system effects a change in the coordinates of a point which may be represented by the linear transformation

$$x'_j = \sum_{k=1}^{3} a_{jk}x_k, \qquad (j = 1, 2, 3). \tag{7}$$

The coefficients a_{jk} are subject to certain conditions which are a consequence of the fact that the distance from O to P, that is to say, the

[1] This section is based essentially on the following papers: MINKOWSKI, *Ann. Physik*, **47**, 927, 1915; SOMMERFELD, *Ann. Physik*, **32**, 749, 1910 and **33**, 649, 1910; MIE, *Ann. Physik*, **37**, 511, 1912; PAULI, Relativitätstheorie, in the Encyklopädie der mathematischen Wissenschaften, Vol. V, part 2, p. 539, 1920.

magnitude of **r**, is independent of the orientation of the coordinate system.

$$\sum_{j=1}^{3} (x_j)^2 = \sum_{j=1}^{3} (x'_j)^2. \tag{8}$$

$$\sum_{j=1}^{3} (x'_j)^2 = \sum_{j=1}^{3} \left(\sum_{i=1}^{3} a_{ji}x_i\right)\left(\sum_{k=1}^{3} a_{jk}x_k\right) = \sum_{i=1}^{3}\sum_{k=1}^{3} x_ix_k \left(\sum_{j=1}^{3} a_{ji}a_{jk}\right), \tag{9}$$

whence it follows that

$$\sum_{j=1}^{3} a_{ji}a_{jk} = \delta_{ik} = \begin{matrix} 1, & \text{when } i = k, \\ 0, & \text{when } i \neq k. \end{matrix} \tag{10}$$

Equation (10) expresses in fact the relations which must prevail among the cosines of the angles between coordinate axes in order that they be rectangular and which are, therefore, known as *conditions of orthogonality*. The system (7), when subject to (10), is likewise called an *orthogonal transformation*. As a direct consequence of (10), it may be shown that the square of the determinant $|a_{jk}|$ is equal to unity and hence $|a_{jk}| = \pm 1$. Any set of coefficients a_{jk} which satisfy (10) define an orthogonal transformation in the sense that the relation (8) is preserved. Geometrically the transformation (7) represents a rotation only when the determinant $|a_{jk}| = +1$. The orthogonal transformation whose determinant is equal to -1 corresponds to an inversion followed by a rotation.

Since the determinant of an orthogonal transformation does not vanish, the x_k may be expressed as linear functions of the x'_j. These relations are obtained most simply by writing as in (5):

$$x_k = \mathbf{r} \cdot \mathbf{i}_k = x'_1\mathbf{i}'_1 \cdot \mathbf{i}_k + x'_2\mathbf{i}'_2 \cdot \mathbf{i}_k + x'_3\mathbf{i}'_3 \cdot \mathbf{i}_k, \tag{11}$$

or

$$x_k = \sum_{j=1}^{3} a_{jk}x'_j, \qquad (k = 1, 2, 3), \tag{12}$$

whence it follows from (8) that

$$\sum_{j=1}^{3} a_{ij}a_{kj} = \delta_{ik}. \tag{13}$$

Let **A** be any fixed vector in space, so that

$$\mathbf{A} = \sum_{k=1}^{3} A_k\mathbf{i}_k = \sum_{k=1}^{3} A'_k\mathbf{i}'_k. \tag{14}$$

The component A'_j of this vector with respect to the system X' is given by

$$A'_j = \mathbf{A} \cdot \mathbf{i}'_j = \sum_{k=1}^{3} A_k\mathbf{i}_k \cdot \mathbf{i}'_j = \sum_{k=1}^{3} a_{jk}A_k; \tag{15}$$

thus the rectangular components of a fixed vector upon rotation of the coordinate system transform like the coordinates of a point. Now while every vector has in general three scalar components, it does not follow that any three scalar quantities constitute the components of a vector. *In order that three scalars A_1, A_2, A_3 may be interpreted as the components of a vector, it is necessary that they transform like the coordinates of a point.*

Among scalar quantities one must distinguish the *variant* from the *invariant*. Quantities such as temperature, pressure, work, and the like are independent of the orientation of the coordinate system and are, therefore, called invariant scalars. On the other hand the coordinates of a point, and the measure numbers, or components, of a vector have only magnitude, but they transform with the coordinate system itself. We know that the product $\mathbf{A} \cdot \mathbf{B}$ of two vectors $\mathbf{A}$ and $\mathbf{B}$ is a scalar, but a scalar of what kind? In virtue of (12) and (13) we have

$$\text{(16)} \qquad \mathbf{A} \cdot \mathbf{B} = \sum_{k=1}^{3} A_k B_k = \sum_{k=1}^{3} \left(\sum_{j=1}^{3} a_{jk} A'_j \right) \left(\sum_{i=1}^{3} a_{ik} B'_i \right) = \sum_{j=1}^{3} A'_j B'_j;$$

the scalar product of two vectors is invariant to an orthogonal transformation of the coordinate system.

Let ϕ be an invariant scalar and consider the triplet of quantities

$$\text{(17)} \qquad B_i = \frac{\partial \phi}{\partial x_i}, \qquad (i = 1, 2, 3).$$

Now by (12),

$$\text{(18)} \qquad \frac{\partial x_k}{\partial x'_i} = a_{ik},$$

and hence

$$\text{(19)} \qquad B'_i = \frac{\partial \phi}{\partial x'_i} = \sum_{k=1}^{3} \frac{\partial \phi}{\partial x_k} \frac{\partial x_k}{\partial x'_i} = \sum_{k=1}^{3} a_{ik} B_k;$$

the B_i transform like the components of a vector and therefore the gradient of ϕ,

$$\text{(20)} \qquad \nabla \phi = \sum_{k=1}^{3} \frac{\partial \phi}{\partial x_k} \mathbf{i}_k,$$

calculated at a point P, is a fixed vector associated with that point.

Let A_i be a rectangular component of a vector $\mathbf{A}$, and

$$\text{(21)} \qquad B_i = \frac{\partial A_i}{\partial x_i}.$$

Then by (18) and (15),

$$\text{(22)} \qquad B'_i = \frac{\partial A'_i}{\partial x'_i} = \sum_{k=1}^{3} a_{ik} \frac{\partial A'_i}{\partial x_k} = \sum_{k=1}^{3} \sum_{j=1}^{3} a_{ik} a_{ij} \frac{\partial A_j}{\partial x_k},$$

whence, from (10), it follows that

$$\sum_{i=1}^{3} B_i' = \sum_{k=1}^{3}\sum_{j=1}^{3}\left(\sum_{i=1}^{3} a_{ik}a_{ij}\right)\frac{\partial A_j}{\partial x_k} = \sum_{j=1}^{3} B_j; \tag{23}$$

the divergence of a vector is invariant to an orthogonal transformation of the coordinate system.

Lastly, since the gradient of an invariant scalar is a vector and since the divergence of a vector is invariant, it follows that *the Laplacian*

$$\nabla^2\phi = \nabla \cdot \nabla\phi = \sum_{i=1}^{3} \frac{\partial^2\phi}{\partial x_i^2} \tag{24}$$

is invariant to an orthogonal transformation.

The transformation properties of vectors may be extended to manifolds of more than three dimensions. Let x_1, x_2, x_3, x_4 be the rectangular coordinates of a point P with respect to a reference system X in a four-dimensional continuum. The location of P with respect to a fixed origin O is determined by the vector

$$\mathbf{r} = \sum_{j=1}^{4} x_j\mathbf{i}_j, \qquad \mathbf{i}_j \cdot \mathbf{i}_k = \delta_{jk}. \tag{25}$$

The linear transformation

$$x_j' = \sum_{k=1}^{4} a_{jk}x_k, \qquad (j = 1, 2, 3, 4), \tag{26}$$

will be called orthogonal if the coefficients satisfy the conditions

$$\sum_{j=1}^{4} a_{ji}a_{jk} = \delta_{ik}. \tag{27}$$

The characteristic property of an orthogonal transformation is that it leaves the sum of the squares of the coordinates invariant:

$$\sum_{j=1}^{4} (x_j)^2 = \sum_{j=1}^{4} (x_j')^2. \tag{28}$$

The square of the determinant formed from the a_{jk} is readily shown to be positive and equal to unity, and hence the determinant itself may equal ± 1. However, if (26) is to include the identical transformation

$$x_j' = x_j, \qquad (j = 1, 2, 3, 4), \tag{29}$$

it is obvious that the determinant must be positive. Henceforth we shall confine ourselves to the subgroup of orthogonal transformations characterized by (27) and the condition

$$|a_{jk}| = +1. \tag{30}$$

The transformation then corresponds geometrically to a rotation of the coordinate axes.

A *four-vector* is now defined as any set of four variant scalars A_i $(i = 1, 2, 3, 4)$ which transform with a rotation of the coordinate system like the coordinates of a point.

$$A'_j = \sum_{k=1}^{4} a_{jk} A_k, \qquad A_k = \sum_{j=1}^{4} a_{jk} A'_j. \tag{31}$$

It is then easy to show, as above, that the scalar product of two four-vectors and the four-dimensional divergence of a four-vector are invariant to a rotation of the coordinate system.

$$\mathbf{A} \cdot \mathbf{B} = \sum_{k=1}^{4} A_k B_k = \sum_{j=1}^{4} A'_j B'_j, \tag{32}$$

$$\square \cdot \mathbf{A} = \sum_{i=1}^{4} \frac{\partial A_i}{\partial x_i} = \sum_{j=1}^{4} \frac{\partial A'_j}{\partial x'_j}. \tag{33}$$

Furthermore the derivatives of a scalar,

$$B_i = \frac{\partial \phi}{\partial x_i} \qquad (i = 1, 2, 3, 4) \tag{34}$$

transform like the components of a four-vector and hence the four-dimensional Laplacian of an invariant scalar,

$$\square^2 \phi = \sum_{i=1}^{4} \frac{\partial^2 \phi}{\partial x_i^2} = \sum_{j=1}^{4} \frac{\partial^2 \phi}{\partial x_j'^2}, \tag{35}$$

is also invariant to an orthogonal transformation.

1.20. Elements of Tensor Analysis.—Although most physical quantities may be classified either as scalars, having only magnitude, or as vectors, characterized by magnitude and direction, there are certain entities which cannot be properly represented by either of these terms. The displacement of the center of gravity of a metal rod, for example, may be defined by a vector; but the rod may also be stretched along the axis by application of a tension at the two ends without displacing the center at all. The quantity employed to represent this stretching must thus indicate a *double* direction. The inadequacy of the vector concept becomes all the more apparent when one attempts the description of a volume deformation, taking into account the lateral contraction of the rod. In the present section we shall deal only with the simpler aspects of tensor calculus, which is the appropriate tool for the treatment of such problems.

In a three-dimensional continuum let each rectangular component of a vector **B** be a linear function of the components of a vector **A**.

$$\begin{aligned} B_1 &= T_{11}A_1 + T_{12}A_2 + T_{13}A_3, \\ B_2 &= T_{21}A_1 + T_{22}A_2 + T_{23}A_3, \\ B_3 &= T_{31}A_1 + T_{32}A_2 + T_{33}A_3. \end{aligned} \tag{36}$$

In order that this association of the components of **B** with the components of **A** in the system X be preserved as the coordinates are rotated, it is necessary that the coefficients T_{jk} transform in a specific manner. The T_{jk} are therefore *variant scalars*. A tensor—or more properly, a tensor of rank two—will now be defined as a linear transformation of the components of a vector **A** into the components of a vector **B** which is invariant to rotations of the coordinate system. The nine coefficients T_{jk} of the linear transformation are called the tensor components.

To determine the manner in which a tensor component must transform we write first (36) in the abbreviated form

$$B_j = \sum_{k=1}^{3} T_{jk}A_k, \qquad (j = 1, 2, 3). \tag{37}$$

If (37) is to be invariant to the transformation defined by

$$x'_j = \sum_{k=1}^{3} a_{jk}x_k, \qquad \sum_{j=1}^{3} a_{ji}a_{jk} = \delta_{ik}, \tag{38}$$

then the T_{jk} must transform to T'_{il} such that

$$B'_i = \sum_{l=1}^{3} T'_{il}A'_l, \qquad (i = 1, 2, 3). \tag{39}$$

Multiply (37) by a_{ij} and sum over the index j.

$$\sum_{j=1}^{3} a_{ij}B_j = \sum_{j=1}^{3}\sum_{k=1}^{3} a_{ij}T_{jk}A_k. \tag{40}$$

But

$$B'_i = \sum_{j=1}^{3} a_{ij}B_j, \qquad A_k = \sum_{l=1}^{3} a_{lk}A'_l, \tag{41}$$

and, hence,

$$B'_i = \sum_{l=1}^{3}\left(\sum_{j=1}^{3}\sum_{k=1}^{3} a_{ij}a_{lk}T_{jk}\right)A'_l = \sum_{l=1}^{3} T'_{il}A'_l. \tag{42}$$

The components of a tensor of rank two transform according to the law

$$T'_{il} = \sum_{j=1}^{3} \sum_{k=1}^{3} a_{ij} a_{lk} T_{jk}, \qquad (i, l = 1, 2, 3); \tag{43}$$

inversely, *any set of nine quantities which transform according to* (43) *constitutes a tensor.*

By an analogous procedure one can show that the reciprocal transformation is

$$T_{jk} = \sum_{i=1}^{3} \sum_{l=1}^{3} a_{ij} a_{lk} T'_{il}. \tag{44}$$

If the order of the indices in all the components of a tensor may be changed with no resulting change in the tensor itself, so that $T_{jk} = T_{kj}$, the tensor is said to be completely symmetric. A tensor is completely antisymmetric if an interchange of the indices in each component results in a change in sign of the tensor. The diagonal terms T_{jj} of an antisymmetric tensor evidently vanish, while for the off-diagonal terms, $T_{jk} = -T_{kj}$. It is clear from (43) that if $T_{jk} = T_{kj}$, then also $T'_{il} = T'_{li}$. Likewise if $T_{jk} = -T_{kj}$, it follows that $T'_{il} = -T'_{li}$. The symmetric or antisymmetric character of a tensor is invariant to a rotation of the coordinate system.

The sum or difference of two tensors is constructed from the sums or differences of their corresponding components. If $^2\mathbf{R}$ is the sum of the tensors $^2\mathbf{S}$ and $^2\mathbf{T}$,[1] its components are by definition

$$R_{jk} = S_{jk} + T_{jk}, \qquad (j, k = 1, 2, 3). \tag{45}$$

In virtue of the linear character of (43) the quantities R_{jk} transform like the S_{jk} and T_{jk} and, therefore, constitute the components of a tensor $^2\mathbf{R}$. From this rule it follows that any asymmetric tensor may be represented as the sum of a symmetric and an antisymmetric tensor. Assuming $^2\mathbf{R}$ to be the given asymmetric tensor, we construct a symmetric tensor $^2\mathbf{S}$ from the components

$$S_{jk} = \tfrac{1}{2}(R_{jk} + R_{kj}) = S_{kj} \tag{46}$$

and an antisymmetric tensor $^2\mathbf{T}$ from the components

$$T_{jk} = \tfrac{1}{2}(R_{jk} - R_{kj}) = -T_{kj}. \tag{47}$$

Then by (45) the sum of $^2\mathbf{S}$ and $^2\mathbf{T}$ so constructed is equal to $^2\mathbf{R}$.

In a three-dimensional manifold an antisymmetric tensor reduces to three independent components and in this sense resembles a vector. The

[1] Tensors of second rank will be indicated by a superscript as shown.

tensor (36), for example, reduces in this case to

$$
\begin{aligned}
B_1 &= 0 - T_{21}A_2 + T_{13}A_3, \\
B_2 &= T_{21}A_1 + 0 - T_{32}A_3, \\
B_3 &= -T_{13}A_1 + T_{32}A_2 + 0.
\end{aligned} \tag{48}
$$

These, however, are the components of a vector,

$$\mathbf{B} = \mathbf{T} \times \mathbf{A}, \tag{49}$$

wherein the vector **T** has the components

$$T_1 = T_{32}, \qquad T_2 = T_{13}, \qquad T_3 = T_{21}. \tag{50}$$

Now it will be recalled that in vector analysis it is customary to distinguish *polar vectors*, such as are employed to represent translations and mechanical forces, from *axial vectors* with which there are associated directions of rotation. Geometrically, a polar vector is represented by a displacement or line, whereas an axial vector corresponds to an area. A typical axial vector is that which results from the vector or cross product of two polar vectors, and we must conclude from the above that an axial vector is in fact an antisymmetric tensor and its components should properly be denoted by two indices rather than one. Thus for the components of $\mathbf{T} = \mathbf{A} \times \mathbf{B}$ we write

$$T_{jk} = A_jB_k - A_kB_j = -T_{kj}, \qquad (j, k = 1, 2, 3). \tag{51}$$

If the coordinate system is rotated, the components of **A** and **B** are transformed according to

$$A_j = \sum_{l=1}^{3} a_{lj}A'_l, \qquad B_k = \sum_{i=1}^{3} a_{ik}B'_i. \tag{52}$$

Upon introducing these values into (51) we find

$$A_jB_k - A_kB_j = \sum_{l=1}^{3}\sum_{i=1}^{3} a_{lj}a_{ik}(A'_lB'_i - A'_iB'_l), \tag{53}$$

a relation which is identical with (44) and which demonstrates that the components of a "vector product" of two vectors transform like the components of a tensor. The essential differences in the properties of polar vectors and the properties of those axial vectors by means of which one represents angular velocities, moments, and the like, are now clear: axial vectors are vectors only in their manner of composition, not in their law of transformation. It is important to add that an antisymmetric tensor can be represented by an axial or pseudo-vector only in a three-dimensional space, and then only in rectangular components.

Since the cross product of two vectors is in fact an antisymmetric tensor, one should anticipate that the same is true of the curl of a vector. That the quantity $\partial A_j/\partial x_k$, where A_j is a component of a vector **A**, is the component of a tensor is at once evident from Eq. (22).

$$\frac{\partial A_i'}{\partial x_l'} = \sum_{j=1}^{3}\sum_{k=1}^{3} a_{ij}a_{lk}\frac{\partial A_j}{\partial x_k}. \tag{54}$$

The components of $\nabla \times \mathbf{A}$,

$$T_{jk} = \frac{\partial A_k}{\partial x_j} - \frac{\partial A_j}{\partial x_k} = -T_{kj}, \qquad (j, k = 1, 2, 3), \tag{55}$$

therefore, transform like the components of an antisymmetric tensor.

The divergence of a tensor is defined as the operation

$$(\text{div }{}^2\mathbf{T})_j = \sum_{k=1}^{3}\frac{\partial T_{jk}}{\partial x_k} = B_j, \qquad (j = 1, 2, 3). \tag{56}$$

The quantities B_j are easily shown to transform like the components of a vector.

$$\frac{\partial T_{il}'}{\partial x_l'} = \sum_{k=1}^{3} a_{lk}\frac{\partial T_{il}'}{\partial x_k} = \sum_{k=1}^{3}\sum_{j=1}^{3} a_{lk}a_{ij}a_{lk}\frac{\partial T_{jk}}{\partial x_k}, \tag{57}$$

or, on summing over l and applying the conditions of orthogonality,

$$B_i' = \sum_{l=1}^{3}\frac{\partial T_{il}'}{\partial x_l'} = \sum_{j=1}^{3} a_{ij}\left(\sum_{k=1}^{3}\frac{\partial T_{jk}}{\partial x_k}\right) = \sum_{j=1}^{3} a_{ij}B_j. \tag{58}$$

The divergence of a tensor of second rank is a vector, or tensor of first rank. The divergence of a vector is an invariant scalar, or tensor of zero rank. These are examples of a process known in tensor analysis as contraction.

As in the case of vectors, the tensor concept may be extended to manifolds of four dimensions. Any set of 16 quantities which transform according to the law,

$$T_{il}' = \sum_{j=1}^{4}\sum_{k=1}^{4} a_{ij}a_{lk}T_{jk}, \qquad (i, l = 1, 2, 3, 4), \tag{59}$$

or its reciprocal,

$$T_{jk} = \sum_{i=1}^{4}\sum_{l=1}^{4} a_{ij}a_{lk}T_{il}', \qquad (j, k = 1, 2, 3, 4), \tag{60}$$

will be called a tensor of second rank in a four-dimensional manifold. As in the three-dimensional case, the tensor is said to be completely symmetric if $T_{jk} = T_{kj}$, and completely antisymmetric if $T_{jk} = -T_{kj}$, with $T_{jj} = 0$. In virtue of its definition it is evident that an antisymmetric four-tensor contains only six independent components. Upon expanding (59) and replacing T_{kj} by $-T_{jk}$, then re-collecting terms, we obtain as the transformation formula of an antisymmetric tensor the relation

$$(61) \qquad T'_{il} = \sum_{j=1}^{4}\sum_{k=1}^{4}(a_{ij}a_{lk} - a_{ik}a_{lj})T_{jk} = \sum_{j=1}^{4}\sum_{k=1}^{4}\begin{vmatrix} a_{ij} & a_{ik} \\ a_{lj} & a_{lk} \end{vmatrix} T_{jk}, \quad (k > j).$$

Any six quantities that transform according to this rule constitute an antisymmetric four-tensor or, as it is frequently called, a *six-vector*.

In three-space the vector product is represented geometrically by the area of a parallelogram whose sides are defined by two vectors drawn from a common origin. The components of this product are then the projections of the area on the three coordinate planes. By analogy, the vector product in four-space is *defined* as the "area" of a parallelogram formed by two four-vectors, A and B, drawn from a common origin. The components of this extended product are now the projections of the parallelogram on the *six* coordinate planes, whose areas are

$$(62) \qquad T_{jk} = A_jB_k - A_kB_j = -T_{kj}, \qquad (j, k = 1, 2, 3, 4).$$

The vector product of two four-vectors is therefore an antisymmetric four-tensor, or six-vector.

If again $\mathbf{A}$ is a four-vector, the quantities

$$(63) \qquad T_{jk} = \frac{\partial A_k}{\partial x_j} - \frac{\partial A_j}{\partial x_k} = -T_{kj}, \qquad (j, k = 1, 2, 3, 4),$$

can be shown as in (54) to transform like the components of an antisymmetric tensor. The T_{jk} may be interpreted as the components of the curl of a four-vector.

As in the three-dimensional case the divergence of a four-tensor is defined by

$$(64) \qquad (\text{div}\ {}^2\mathbf{T})_j = \sum_{k=1}^{4}\frac{\partial T_{jk}}{\partial x_k}, \qquad (j = 1, 2, 3, 4),$$

a set of quantities which are evidently the components of a four-vector.

1.21. The Space-time Symmetry of the Field Equations.—A remarkable symmetry of form is apparent in the equations of the electromagnetic field when one introduces as independent variables the four lengths

$$(65) \qquad x_1 = x, \qquad x_2 = y, \qquad x_3 = z, \qquad x_4 = ict,$$

where c is the velocity of light in free space. When expanded in rectangular coordinates, the equations

$$\text{(II)}\quad \nabla \times \mathbf{H} - \frac{\partial \mathbf{D}}{\partial t} = \mathbf{J}, \qquad \text{(IV)}\quad \nabla \cdot \mathbf{D} = \rho,$$

are represented by the system

$$\begin{aligned} 0 + \frac{\partial H_3}{\partial x_2} - \frac{\partial H_2}{\partial x_3} - ic\frac{\partial D_1}{\partial x_4} &= J_1, \\ -\frac{\partial H_3}{\partial x_1} + 0 + \frac{\partial H_1}{\partial x_3} - ic\frac{\partial D_2}{\partial x_4} &= J_2, \\ \frac{\partial H_2}{\partial x_1} - \frac{\partial H_1}{\partial x_2} + 0 - ic\frac{\partial D_3}{\partial x_4} &= J_3, \\ ic\frac{\partial D_1}{\partial x_1} + ic\frac{\partial D_2}{\partial x_2} + ic\frac{\partial D_3}{\partial x_3} + 0 &= ic\rho. \end{aligned} \tag{66}$$

We shall treat the right-hand members of this system as the components of a "four-current" density,

$$J_1 = J_x, \qquad J_2 = J_y, \qquad J_3 = J_z, \qquad J_4 = ic\rho, \tag{67}$$

and introduce in the left-hand members a set of dependent variables defined by

$$\begin{array}{llll} G_{11} = 0 & G_{12} = H_3 & G_{13} = -H_2 & G_{14} = -icD_1 \\ G_{21} = -H_3 & G_{22} = 0 & G_{23} = H_1 & G_{24} = -icD_2 \\ G_{31} = H_2 & G_{32} = -H_1 & G_{33} = 0 & G_{34} = -icD_3 \\ G_{41} = icD_1 & G_{42} = icD_2 & G_{43} = icD_3 & G_{44} = 0. \end{array} \tag{68}$$

Then in the reference system X, Eqs. (II) and (IV) reduce to

$$\sum_{k=1}^{4} \frac{\partial G_{jk}}{\partial x_k} = J_j, \qquad (j = 1, 2, 3, 4). \tag{69}$$

Only six of the G_{jk} are independent, and the resemblance of this set of quantities to the components of an antisymmetric four-tensor is obvious. Since the divergence of a four-tensor is a four-vector, it follows from (69) that *if* the G_{jk} constitute a tensor, then the J_k are the components of a four-vector; inversely, if we can show that $\mathbf{J}$ is indeed a four-vector, we may then infer the tensor character of ${}^2\mathbf{G}$. However, we have as yet offered no evidence to justify such an assumption. In the preceding sections it was shown that the vector or tensor properties of sets of scalar quantities are determined by the manner in which they transform on passing from one reference system to another. Evidently an orthogonal transformation of the coordinates x_k corresponds to a simultaneous change in both the space coordinates x, y, z *and* the time t, and only recourse to experiment will tell us how the field intensities may be expected to transform under such circumstances. In Sec. 1.22 we shall set forth briefly the experimental facts which lead one to conclude that the J_k *are* components of a four-vector, and the G_{jk} the components of a field tensor; in the interim we regard (69) and the deductions that follow below merely as concise and symmetrical expressions of the field equations in a fixed system of coordinates.

The two homogeneous equations

$$\text{(I)}\quad \nabla \times \mathbf{E} + \frac{\partial \mathbf{B}}{\partial t} = 0, \qquad \text{(III)}\quad \nabla \cdot \mathbf{B} = 0,$$

are represented by the system

$$\begin{aligned} 0 + \frac{\partial E_3}{\partial x_2} - \frac{\partial E_2}{\partial x_3} + ic\,\frac{\partial B_1}{\partial x_4} &= 0, \\ -\frac{\partial E_3}{\partial x_1} + 0 + \frac{\partial E_1}{\partial x_3} + ic\,\frac{\partial B_2}{\partial x_4} &= 0, \\ \frac{\partial E_2}{\partial x_1} - \frac{\partial E_1}{\partial x_2} + 0 + ic\,\frac{\partial B_3}{\partial x_4} &= 0, \\ -\frac{\partial B_1}{\partial x_1} - \frac{\partial B_2}{\partial x_2} - \frac{\partial B_3}{\partial x_3} + 0 &= 0. \end{aligned} \tag{70}$$

After division of the first three of these equations by ic, an antisymmetrical array of components is defined as follows:

$$\begin{array}{llll} F_{11} = 0 & F_{12} = B_3 & F_{13} = -B_2 & F_{14} = -\dfrac{i}{c}E_1 \\ F_{21} = -B_3 & F_{22} = 0 & F_{23} = B_1 & F_{24} = -\dfrac{i}{c}E_2 \\ F_{31} = B_2 & F_{32} = -B_1 & F_{33} = 0 & F_{34} = -\dfrac{i}{c}E_3 \\ F_{41} = \dfrac{i}{c}E_1 & F_{42} = \dfrac{i}{c}E_2 & F_{43} = \dfrac{i}{c}E_3 & F_{44} = 0. \end{array} \tag{71}$$

Then all the equations of (70) are contained in the system

$$\frac{\partial F_{ij}}{\partial x_k} + \frac{\partial F_{ki}}{\partial x_j} + \frac{\partial F_{jk}}{\partial x_i} = 0, \tag{72}$$

where i, j, k are any three of the four numbers 1, 2, 3, 4.

The arrays (68) and (71) are congruent in the sense that in each the real components pertain to the magnetic field, while the imaginary components are associated with the electric field. To indicate this partition it is convenient to represent the sets of components by the symbols

$$^2\mathbf{F} = \left(\mathbf{B}, -\frac{i}{c}\mathbf{E}\right), \qquad ^2\mathbf{G} = (\mathbf{H}, -ic\mathbf{D}). \tag{73}$$

Now the field equations may be defined equally well in terms of the "dual" systems:

(74) $$ {}^2\mathbf{F}^* = \left(-\frac{i}{c}\mathbf{E}, \mathbf{B}\right), \qquad {}^2\mathbf{G}^* = (-ic\mathbf{D}, \mathbf{H}), $$

or

(75) $$ \begin{array}{llll} F^*_{11} = 0 & F^*_{12} = -\dfrac{i}{c}E_3 & F^*_{13} = \dfrac{i}{c}E_2 & F^*_{14} = B_1 \\ F^*_{21} = \dfrac{i}{c}E_3 & F^*_{22} = 0 & F^*_{23} = -\dfrac{i}{c}E_1 & F^*_{24} = B_2 \\ F^*_{31} = -\dfrac{i}{c}E_2 & F^*_{32} = \dfrac{i}{c}E_1 & F^*_{33} = 0 & F^*_{34} = B_3 \\ F^*_{41} = -B_1 & F^*_{42} = -B_2 & F^*_{43} = -B_3 & F^*_{44} = 0, \end{array} $$

and a corresponding system for the components of ${}^2\mathbf{G}^*$. Upon introducing these values into (I) and (III), and (II) and (IV), respectively, we obtain

(76) $$ \sum_{k=1}^{4} \frac{\partial F^*_{jk}}{\partial x_k} = 0, \qquad (j = 1, 2, 3, 4), $$

(77) $$ \frac{\partial G^*_{ij}}{\partial x_k} + \frac{\partial G^*_{ki}}{\partial x_j} + \frac{\partial G^*_{jk}}{\partial x_i} = J_l, \quad (i, j, k, l = 1, 2, 3, 4). $$

It has been pointed out by various writers that this last representation is artificial, in that (74) implies that **E** is an axial vector in three-space and **B** a polar vector, whereas the contrary is known to be true. The representation

(72) $$ \frac{\partial F_{ij}}{\partial x_k} + \frac{\partial F_{ki}}{\partial x_j} + \frac{\partial F_{jk}}{\partial x_i} = 0, \qquad (i, j, k = 1, 2, 3, 4), $$

(69) $$ \sum_{k=1}^{4} \frac{\partial G_{jk}}{\partial x_k} = J_j, \qquad (j = 1, 2, 3, 4), $$

must in this sense be considered the "natural" form of the field equations. To these we add the equation of continuity,

(V) $$ \nabla \cdot \mathbf{J} + \frac{\partial \rho}{\partial t} = 0, $$

which in four-dimensional notation becomes

(78) $$ \sum_{k=1}^{4} \frac{\partial J_k}{\partial x_k} = 0. $$

If the components F_{jk} are defined in terms of the components of a "four-potential" $\mathbf{\Phi}$ by

(79) $$ F_{jk} = \frac{\partial \Phi_k}{\partial x_j} - \frac{\partial \Phi_j}{\partial x_k}, \qquad (j, k = 1, 2, 3, 4), $$

one may readily verify that Eq. (72) is satisfied identically. Now in three-space the vectors **E** and **B** are derived from a vector and scalar potential.

$$E = -\nabla\phi - \frac{\partial \mathbf{A}}{\partial t}, \qquad \mathbf{B} = \nabla \times \mathbf{A}; \tag{80}$$

or, in component form,

$$-\frac{i}{c}E_j = \frac{\partial}{\partial x_j}\left(\frac{i}{c}\phi\right) - \frac{\partial A_j}{\partial x_4}, \qquad B_i = \frac{\partial A_k}{\partial x_j} - \frac{\partial A_j}{\partial x_k}, \tag{81}$$

where the indices i, j, k are to be taken in cyclical order. Clearly all these equations are comprised in the system (79) if we define the components of the four-potential by

$$\Phi_1 = A_x, \qquad \Phi_2 = A_y, \qquad \Phi_3 = A_z, \qquad \Phi_4 = \frac{i}{c}\phi. \tag{82}$$

As in three dimensions, the four-potential is useful only if we can determine from it the field ${}^2\mathbf{G}(\mathbf{H}, -ic\mathbf{D})$ as well as the field ${}^2\mathbf{F}(\mathbf{B}, -\frac{i}{c}\mathbf{E})$. Some supplementary condition must, therefore, be imposed upon $\mathbf{\Phi}$ in order that it satisfy (69) as well as (72); thus it is necessary that the components G_{jk} be related functionally to the F_{jk}. We shall confine the discussion here to the usual case of a *homogeneous, isotropic medium* and assume the relations to be *linear*. To preserve symmetry of notation it will be convenient to write the proportionality factor which characterizes the medium as γ_{jk}, so that

$$G_{jk} = \gamma_{jk}F_{jk}; \tag{83}$$

but it is clear from (68) and (71) that[1]

$$\gamma_{jk} = \frac{1}{\mu} \text{ when } j, k = 1, 2, 3, \qquad \gamma_{jk} = \epsilon c^2 \text{ when } j \text{ or } k = 4. \tag{84}$$

These coefficients are in fact components of a symmetrical tensor, and with a view to subsequent needs the diagonal terms are given the values

$$\gamma_{jj} = \frac{1}{\mu}, \; (j = 1, 2, 3), \qquad \gamma_{44} = \mu\epsilon^2c^4. \tag{85}$$

Equation (69) is now to be replaced by

$$\sum_{k=1}^{4} \gamma_{jk}\frac{\partial F_{jk}}{\partial x_k} = J_j, \qquad (j = 1, 2, 3, 4). \tag{86}$$

[1] A medium which is anisotropic in either its electrical properties or its magnetic properties may be represented as in (83) provided the coordinate system is chosen to coincide with the principal axes. This also is the case if its principal axes of electrical anisotropy coincide with those of magnetic anisotropy.

Upon introducing (79) we find that (86) is satisfied, provided $\mathbf{\Phi}$ is a solution of

$$\sum_{k=1}^{4} \frac{\partial^2}{\partial x_k^2}(\gamma_{jk}\Phi_j) = -J_j, \qquad (j = 1, 2, 3, 4), \tag{87}$$

subject to the condition

$$\sum_{k=1}^{4} \frac{\partial}{\partial x_k}(\gamma_{jk}\Phi_k) = 0, \qquad (j = 1, 2, 3, 4). \tag{88}$$

This last relation is evidently equivalent to

$$\nabla \cdot \mathbf{A} + \mu\epsilon \frac{\partial \phi}{\partial t} = 0, \tag{89}$$

and (87) comprises the two equations

$$\nabla^2 A_j - \mu\epsilon \frac{\partial^2 A_j}{\partial t^2} = -\mu J_j, \qquad (j = 1, 2, 3),$$

$$\nabla^2 \phi - \mu\epsilon \frac{\partial^2 \phi}{\partial t^2} = -\frac{1}{\epsilon}\rho. \tag{90}$$

In free space $\mu_0\epsilon_0 = c^{-2}$, $\gamma_{jk} = \mu_0^{-1}$, for all values of the indices. Equations (87) and (88) then reduce to the simple form:

$$\sum_{k=1}^{4} \frac{\partial^2 \Phi_j}{\partial x_k^2} = -\mu_0 J_j, \qquad \sum_{k=1}^{4} \frac{\partial \Phi_k}{\partial x_k} = 0, \qquad (j = 1, 2, 3, 4). \tag{91}$$

1.22. The Lorentz Transformation.—The physical significance of these results is of vastly greater importance than their purely formal elegance. A series of experiments, the most decisive being the celebrated investigation of Michelson and Morley,[1] have led to the establishment of two fundamental postulates as highly probable, if not absolutely certain. According to the first of these, called the *relativity postulate*, it is impossible to detect by means of physical measurements made within a reference system X a uniform translation relative to a second system X'. That the earth is moving in an orbit about the sun we know from observations on distant stars; but if the earth were enveloped in clouds, no measurement on its surface would disclose a uniform translational motion in space. The course of natural phenomena must therefore be unaffected by a nonaccelerated motion of the coordinate systems to which they are referred, and all reference systems moving linearly and uniformly relative to each other are equivalent. For our present needs we shall state the relativity postulate as follows: *When properly formulated, the laws of*

[1] MICHELSON and MORLEY, *Am. J. Sci.*, **3**, 34, 1887.

physics are invariant to a transformation from one reference system to another moving with a linear, uniform relative velocity. A direct consequence of this postulate is that the components of all vectors or tensors entering into an equation must transform in the same way, or *covariantly.* The existence of such a principle restricted to uniform translations was established for classical mechanics by Newton, but we are indebted to Einstein for its extension to electrodynamics.

The second postulate of Einstein is more remarkable: *The velocity of propagation of an electromagnetic disturbance in free space is a universal constant c which is independent of the reference system.* This proposition is evidently quite contrary to our experience with mechanical or acoustical waves in a material medium, where the velocity is known to depend on the relative motion of medium and observer. Many attempts have been made to interpret the experimental evidence without recourse to this radical assumption, the most noteworthy being the electrodynamic theory of Ritz.[1] The results of all these labors indicate that although a constant velocity of light is *not* necessary to account for the negative results of the Michelson-Morley experiment, this postulate alone is consistent with that experiment *and* other optical phenomena.[2]

Let us suppose, then, that a source of light is fixed at the origin O of a system of coordinates $X(x, y, z)$. At the instant $t = 0$, a spherical wave is emitted from this source. An observer located at the point x, y, z in X will first note the passage of the wave at the instant ct, and the equation of a point on the wave front is therefore

$$x^2 + y^2 + z^2 - c^2t^2 = 0. \tag{92}$$

The observer, however, is free to measure position and time with respect to a second reference frame $X'(x', y', z')$ which is moving along a straight line with a uniform velocity relative to O. For simplicity we shall assume the origin O' to coincide with O at the instant $t = 0$. According to the second postulate the light wave is propagated in X' with the same velocity as in X, and the equation of the wave front in X' is

$$x'^2 + y'^2 + z'^2 - c^2t'^2 = 0. \tag{93}$$

By t' we must understand the time as measured by an observer in X' with instruments located in that system. Here, then, is the key to the transformation that connects the coordinates x, y, z, t of an observation or event in X with the coordinates x', y', z', t' of the same event in X': it must be linear and must leave the quadratic form (92) invariant. The linearity follows from the requirement that a uniform, linear motion of a particle in X should also be linear in X'.

[1] RITZ, *Ann. chim. phys.*, **13**, 145, 1908.

[2] An account of these investigations will be found in Pauli's article, *loc. cit.*, p. 549.

Let

(94) $$x_1 = x, \qquad x_2 = y, \qquad x_3 = z, \qquad x_4 = ict,$$

be the components of a vector **R** in a four-dimensional manifold $X(x_1, x_2, x_3, x_4)$.

(95) $$R^2 = x_1^2 + x_2^2 + x_3^2 + x_4^2.$$

The postulate on the constancy of the velocity c will be satisfied by the group of transformations which leaves this length invariant. But in Sec. 1.19 it was shown that (95) is invariant to the group of rotations in four-space and we conclude, therefore, that the transformations which take one from the coordinates of an event in X to the coordinates of that event in X' are of the form

(26) $$x_j' = \sum_{k=1}^{4} a_{jk} x_k, \qquad (j = 1, 2, 3, 4),$$

where

(27) $$\sum_{j=1}^{4} a_{ji} a_{jk} = \delta_{ik}, \qquad (i, k = 1, 2, 3, 4),$$

the determinant $|a_{jk}|$ being equal to unity.

We have now to find these coefficients. The calculation will be simplified if we assume that the rotation involves only the axes x_3 and x_4, and the resultant lack of generality is inconsequential. We take, therefore, $x_1' = x_1$, $x_2' = x_2$, and write down the coefficient matrix as follows:

(96)

	x_1	x_2	x_3	x_4
x_1'	1	0	0	0
x_2'	0	1	0	0
x_3'	0	0	a_{33}	a_{34}
x_4'	0	0	a_{43}	a_{44}

The conditions of orthogonality reduce to

(97) $$a_{33}^2 + a_{43}^2 = 1, \qquad a_{34}^2 + a_{44}^2 = 1, \qquad a_{33}a_{34} + a_{43}a_{44} = 0.$$

If we put $a_{33} = \alpha$, $a_{34} = i\alpha\beta$, we find from (97) that $a_{44} = \pm\alpha$, $a_{43} = \mp i\alpha\beta$, $\alpha\sqrt{1 - \beta^2} = \pm 1$. Only the upper sign is consistent with the requirement that the determinant of the coefficients be positive unity, and this in turn is the necessary condition that the group shall contain the identical transformation. In terms of the single parameter β the coefficients are

(98) $$a_{33} = a_{44} = \frac{1}{\sqrt{1 - \beta^2}}, \qquad a_{34} = -a_{43} = \frac{i\beta}{\sqrt{1 - \beta^2}};$$

for the transformation itself, we have

$$
(99) \qquad x_1' = x_1, \qquad x_2' = x_2,
$$
$$
x_3' = \frac{1}{\sqrt{1-\beta^2}}(x_3 + i\beta x_4), \qquad x_4' = \frac{1}{\sqrt{1-\beta^2}}(x_4 - i\beta x_3).
$$

Reverting to the original space-time manifold this is equivalent to

$$
(100) \qquad \begin{aligned} x' &= x, & y' &= y, \\ z' &= \frac{1}{\sqrt{1-\beta^2}}(z - \beta ct), & t' &= \frac{1}{\sqrt{1-\beta^2}}\left(t - \frac{\beta}{c} z\right). \end{aligned}
$$

The parameter β may be determined by considering x', y', z' to be the coordinates of a fixed point in X'. The coordinates of this point with respect to X are x, y, z. Since $dz' = 0$, it follows that

$$
(101) \qquad \frac{dz}{dt} = v = \beta c, \qquad \beta = \frac{v}{c},
$$

and hence the rotation defined in (96) and (97) is equivalent to a translation of the system X' along the z-axis with the constant velocity v relative to the unprimed system X.

The transformation

$$
(102) \qquad \begin{aligned} x' &= x, & y' &= y, \\ z' &= \frac{1}{\sqrt{1-\frac{v^2}{c^2}}}(z - vt), & t' &= \frac{1}{\sqrt{1-\frac{v^2}{c^2}}}\left(t - \frac{v}{c^2} z\right), \end{aligned}
$$

obtained from (100) by substitution of the value for β, or its inversion,

$$
(103) \qquad z = \frac{1}{\sqrt{1-\frac{v^2}{c^2}}}(z' + vt'), \qquad t = \frac{1}{\sqrt{1-\frac{v^2}{c^2}}}\left(t' + \frac{v}{c^2} z'\right),
$$

has been named for Lorentz, who was the first to show that Maxwell's equations are invariant with respect to the change of variables defined by (102), but *not* invariant under the "Galilean transformation:"

$$
(104) \qquad z' = z - vt, \qquad t' = t.
$$

All known electromagnetic phenomena may be properly accounted for if the position *and time* coordinates of an event in a moving system X' be related to the coordinates of that event in an arbitrarily fixed system X by a Lorentz transformation. The Galilean transformation of classical mechanics represents the limit approached by (102) when $v \ll c$, and may be interpreted as the relativity principle appropriate to a world in which electromagnetic forces are propagated with infinite velocity.

1.23. Transformation of the Field Vectors to Moving Systems.—We shall not dwell upon the manifold consequences of the Lorentz transformation; the Fitzgerald-Lorentz contraction, the modified concept of simultaneity, the variation in apparent mass, the upper limit c which is imposed upon the velocity of matter, belong properly to the theory of relativity. The application of the principles of relativity to the equations of the electromagnetic field is essential, however, to an understanding of the four-dimensional formulation of Sec. 1.21.

The Lorentz transformation has been deduced from the postulate on the constancy of the velocity of light and has been shown to be equivalent to a rotation in a space $x_1, x_2, x_3, x_4 = ict$. Now according to the relativity postulate, the laws of physics, when properly stated, must have the same form in all systems moving with a relative, uniform motion; otherwise, it would obviously be possible to detect such a motion. In Secs. 1.19 and 1.20 it was shown that the curl, divergence, and Laplacian of vectors and tensors in a four-dimensional manifold are invariant to a rotation of the coordinate system. Therefore, to ensure the invariance of the field equations under a Lorentz transformation it is only necessary to assume that the four-current **J** and the four-potential **Φ** do indeed transform like vectors, and that the quantities $^2\mathbf{F}$, $^2\mathbf{G}$ transform like tensors. In other words, we base the vector and tensor character of these four-dimensional quantities directly on the two postulates.

The four-current **J** satisfies the equation

$$\square \cdot \mathbf{J} = \sum_{k=1}^{4} \frac{\partial J_k}{\partial x_k} = 0. \tag{78}$$

Under a rotation of the coordinate system the components transform as

$$J'_j = \sum_{k=1}^{4} a_{jk} J_k, \tag{105}$$

or, upon introducing the values for a_{jk} from (98),

$$J'_x = J_x, \qquad J'_y = J_y,$$
$$J'_z = \frac{1}{\sqrt{1-\beta^2}}(J_z - v\rho), \qquad \rho' = \frac{1}{\sqrt{1-\beta^2}}\left(\rho - \frac{v}{c^2}J_z\right), \tag{106}$$

with its inverse transformation

$$J_z = \frac{1}{\sqrt{1-\beta^2}}(J'_z + v\rho'), \qquad \rho = \frac{1}{\sqrt{1-\beta^2}}\left(\rho' + \frac{v}{c^2}J'_z\right). \tag{107}$$

We shall assume henceforth that the reference system X' is fixed within a material body which moves with the constant velocity v relative to the system X. This latter may usually be assumed at rest with respect to the earth. If the velocity v is very much less than the speed of light, Eq. (107) is approximately equal to

$$J_z = J'_z + v\rho', \qquad \rho = \rho'. \tag{108}$$

An observer on the moving body measures a charge density ρ' and a current density J'_z; his colleague at rest in X finds the current J'_z augmented by the *convection current* $v\rho'$.

In like manner the relations between the electric and magnetic vectors defining a given field in a fixed and in a moving system are obtained directly from the rule (61) for the transformation of the components of an antisymmetric tensor. Upon substitution of the appropriate values for the coefficients a_{jk}, one obtains for the components of ${}^2\mathbf{F}$:

$$\begin{aligned}
F'_{12} &= a_{11}a_{22}F_{12} = F_{12},\\
F'_{13} &= a_{11}a_{33}F_{13} + a_{11}a_{34}F_{14} = \frac{1}{\sqrt{1-\beta^2}}(F_{13} + i\beta F_{14}),\\
F'_{14} &= a_{11}a_{43}F_{13} + a_{11}a_{44}F_{14} = \frac{1}{\sqrt{1-\beta^2}}(F_{14} - i\beta F_{13}),\\
F'_{23} &= a_{22}a_{33}F_{23} + a_{22}a_{34}F_{24} = \frac{1}{\sqrt{1-\beta^2}}(F_{23} + i\beta F_{24}),\\
F'_{24} &= a_{22}a_{43}F_{23} + a_{22}a_{44}F_{24} = \frac{1}{\sqrt{1-\beta^2}}(F_{24} - i\beta F_{23}),\\
F'_{34} &= (a_{33}a_{44} - a_{34}a_{43})F_{34} = F_{34},
\end{aligned} \tag{109}$$

and, hence,

$$\begin{aligned}
B'_x &= \frac{1}{\sqrt{1-\beta^2}}\left(B_x + \frac{v}{c^2}E_y\right), & E'_x &= \frac{1}{\sqrt{1-\beta^2}}(E_x - vB_y),\\
B'_y &= \frac{1}{\sqrt{1-\beta^2}}\left(B_y - \frac{v}{c^2}E_x\right), & E'_y &= \frac{1}{\sqrt{1-\beta^2}}(E_y + vB_x),\\
B'_z &= B_z, & E'_z &= E_z.
\end{aligned} \tag{110}$$

The restriction to translations along the z-axis may be discarded by writing $\mathbf{v}$ as a vector representing the translational velocity of X' (the moving body) in any direction with respect to a fixed system X. Since in (110) the orientation of the z-axis was arbitrary, we have in general

$$\begin{aligned}
B'_\parallel &= B_\parallel, & E'_\parallel &= E_\parallel,\\
B'_\perp &= \frac{1}{\sqrt{1-\beta^2}}\left(\mathbf{B} - \frac{1}{c^2}\mathbf{v}\times\mathbf{E}\right)_\perp, & &\\
& & E'_\perp &= \frac{1}{\sqrt{1-\beta^2}}(\mathbf{E} + \mathbf{v}\times\mathbf{B})_\perp,
\end{aligned} \tag{111}$$

where $\|$ denotes components parallel, $\perp$ components perpendicular to the axis of translation. Dropping terms in c^{-2}, as is justifiable whenever the body is moving with a velocity $v \ll 3 \times 10^8$ meters/second, we obtain the approximate formulas

$$(112) \qquad \begin{aligned} B'_{\|} &= B_{\|}, & E'_{\|} &= E_{\|}, \\ B'_{\perp} &= B_{\perp}, & E'_{\perp} &= E_{\perp} + (\mathbf{v} \times \mathbf{B})_{\perp}. \end{aligned}$$

The implication of these results is striking indeed: the electric and magnetic fields **E** and **B** have no independent existence as separate entities. The fundamental complex is the field tensor ${}^2\mathbf{F} = \left(\mathbf{B}, -\frac{i}{c}\mathbf{E}\right)$; the resolution into electric and magnetic components is wholly relative to the motion of the observer. When at rest with respect to permanent magnets or stationary currents, one measures a purely magnetic field **B**. An observer within a moving body or system X', on the other hand, notes approximately the same magnetic field, but in addition an electrostatic field of intensity $\mathbf{E}' = \mathbf{v} \times \mathbf{B}$. Or, inversely, the moving body may carry a fixed charge. To an observer on the body, moving with the charge, the field is purely electrostatic, whereas his colleague aground finds a magnetic field in company with the electric, identifying quite rightly the moving charge with a current.

From the tensor ${}^2\mathbf{G} = (\mathbf{H}, -ic\mathbf{D})$ are calculated in like fashion the transformations of the vectors **H** and **D** from a fixed to a moving system.

$$(113) \qquad \begin{aligned} H'_{\|} &= H_{\|}, & D'_{\|} &= D_{\|}, \\ H'_{\perp} &= \frac{1}{\sqrt{1-\beta^2}} (\mathbf{H} - \mathbf{v} \times \mathbf{D})_{\perp}, & D'_{\perp} &= \frac{1}{\sqrt{1-\beta^2}} \left(\mathbf{D} + \frac{1}{c^2} \mathbf{v} \times \mathbf{H}\right)_{\perp}. \end{aligned}$$

The invariance of Maxwell's equations to uniform translations amounts to this: if the vector functions **E, B, H, D** define an electromagnetic field in a system X, the equations

$$(114) \qquad \begin{aligned} \nabla' \times \mathbf{E}' + \frac{\partial \mathbf{B}'}{\partial t'} &= 0, & \nabla' \cdot \mathbf{B}' &= 0, \\ \nabla' \times \mathbf{H}' - \frac{\partial \mathbf{D}'}{\partial t'} &= \mathbf{J}', & \nabla' \cdot \mathbf{D}' &= \rho', \end{aligned}$$

are satisfied in a system X' which moves with the constant velocity **v** relative to X, the operator ∇' implying that differentiation is to be effected with respect to the variables x', y', z'. An observer at rest in X' interprets the vectors $\mathbf{E}'$, $\mathbf{B}'$, $\mathbf{H}'$, $\mathbf{D}'$ as the intensities of an electromagnetic field satisfying Maxwell's equations. Clearly the ratios of **D** to **E** and **H** to **B** are not preserved in both systems. The macroscopic parameters

ϵ, μ, σ are also subject to transformation, which may be ascribed to an actual change in the structure of matter in motion. In practice one is interested usually in *the mechanical and electromotive forces, measured in the fixed system X, which act on moving matter*, rather than in the transformed field intensities **E**′, **B**′, **H**′, **D**′. The determination of these forces and of the differential equations which they satisfy within the framework of the relativity theory was accomplished by Minkowski in the course of his investigation on the electrodynamics of moving bodies.

The vector character of the four-potential is demonstrated by Eq. (79) which expresses the field tensor $^2\mathbf{F}$ as the curl of $\mathbf{\Phi}$. Under a Lorentz transformation

$$\Phi'_j = \sum_{k=1}^{4} a_{jk}\Phi_k \qquad (j = 1, 2, 3, 4), \tag{115}$$

or, in terms of vector and scalar potentials,

$$\begin{aligned} A'_x &= A_x, & A'_y &= A_y \\ A'_z &= \frac{1}{\sqrt{1-\beta^2}}\left(A_z - \frac{v}{c^2}\phi\right), & \phi' &= \frac{1}{\sqrt{1-\beta^2}}(\phi - vA_z). \end{aligned} \tag{116}$$

As in the case of the field vectors, the resolution into vector and scalar potentials in three-space is determined by the relative motion of the observer.

In conclusion it may be remarked that a rotation of the coördinate system leaves invariant the scalar product of any two vectors. It was in fact from the required invariance of the quantity

$$\mathbf{R}\cdot\mathbf{R} = R^2 = x^2 + y^2 + z^2 - c^2t^2, \tag{117}$$

that we deduced the Lorentz transformation. Since the current density **J** and the potential $\mathbf{\Phi}$ have been shown to be four-vectors, it follows that the quantities

$$\begin{aligned} J^2 &= J_x^2 + J_y^2 + J_z^2 - c^2\rho^2, \\ \Phi^2 &= A_x^2 + A_y^2 + A_z^2 - \frac{\phi^2}{c^2}, \\ \mathbf{\Phi}\cdot\mathbf{J} &= A_xJ_x + A_yJ_y + A_zJ_z - \phi\rho, \end{aligned} \tag{118}$$

are true scalar invariants in a space-time continuum. There are, moreover, certain other scalar invariants of fundamental importance to the general theory of the electromagnetic field. From the transformation formula Eq. (59) the reader will verify that if $^2\mathbf{S}$ and $^2\mathbf{T}$ are two tensors of second rank, the sums

$$\sum_{j=1}^{4}\sum_{k=1}^{4} S_{jk}T_{jk} = \text{invariant}, \qquad \sum_{j=1}^{4}\sum_{k=1}^{4} S_{jk}T_{kj} = \text{invariant}, \tag{119}$$

are invariant to a rotation of the coordinate axes. These quantities may be interpreted as scalar products of the two tensors. Let us form first the scalar product of ${}^2\mathbf{F}$ with itself. According to (71) and (119) we find

$$\sum_{j=1}^{4}\sum_{k=1}^{4} F_{jk}^2 = 2\left(B^2 - \frac{1}{c^2}E^2\right) = \text{invariant.} \tag{120}$$

Next we construct the scalar product of ${}^2\mathbf{F}$ with its dual ${}^2\mathbf{F}^*$ defined in (75).

$$\sum_{j=1}^{4}\sum_{k=1}^{4} F_{jk}F_{jk}^* = -4\frac{i}{c}\mathbf{B}\cdot\mathbf{E} = \text{invariant.} \tag{121}$$

From the tensor ${}^2\mathbf{G}$ and its dual ${}^2\mathbf{G}^*$ may be constructed the invariants

$$\sum_{j=1}^{4}\sum_{k=1}^{4} G_{jk}^2 = 2(H^2 - c^2D^2) = \text{invariant,} \tag{122}$$

$$\sum_{j=1}^{4}\sum_{k=1}^{4} G_{jk}G_{jk}^* = -4ic\mathbf{H}\cdot\mathbf{D} = \text{invariant.} \tag{123}$$

Proceeding in the same fashion, we obtain

$$\sum_{j=1}^{4}\sum_{k=1}^{4} F_{jk}G_{jk} = \sum_{j=1}^{4}\sum_{k=1}^{4} F_{jk}^*G_{jk}^* = 2(\mathbf{B}\cdot\mathbf{H} - \mathbf{E}\cdot\mathbf{D}) = \text{invariant,} \tag{124}$$

$$\sum_{j=1}^{4}\sum_{k=1}^{4} F_{jk}G_{jk}^* = \sum_{j=1}^{4}\sum_{k=1}^{4} F_{jk}^*G_{jk} = -2i\left(c\mathbf{B}\cdot\mathbf{D} + \frac{1}{c}\mathbf{E}\cdot\mathbf{H}\right) = \text{invariant.} \tag{125}$$

The invariance of these quantities in configuration space is trivial; they are set apart from other scalar products by the fact that they preserve the same value in every system moving with a uniform relative velocity.

CHAPTER II

STRESS AND ENERGY

To translate the mathematical structure developed in the preceding pages into experiments which can be conducted in the laboratory, we must calculate the mechanical forces exerted in the field upon elements of charge and current or upon bodies of neutral matter. In the present chapter it will be shown how by an appropriate definition of the vectors **E** and **B** these forces may be deduced directly from the Maxwell equations. In the course of this investigation we shall have to take account of the elastic properties of material media. A brief digression on the analysis of elastic stress and strain will provide an adequate basis for the treatment of the body and surface forces exerted by electric or magnetic fields.

STRESS AND STRAIN IN ELASTIC MEDIA

2.1. The Elastic Stress Tensor.—Let us suppose that a given solid or fluid body of matter is in static equilibrium under the action of a specified system of applied forces. Within this body we isolate a finite volume V by means of a closed surface S, as indicated in Fig. 12.

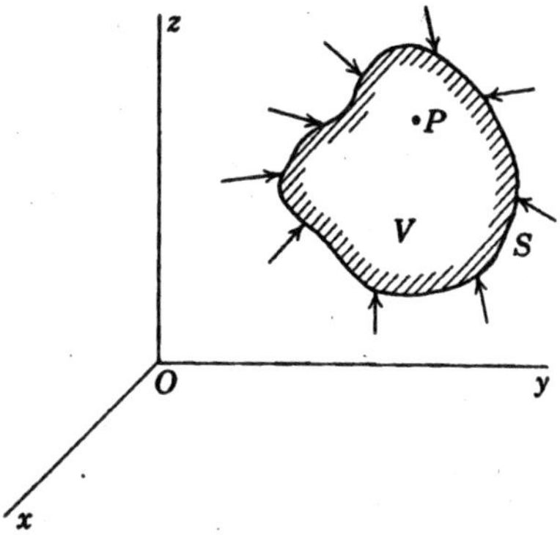

FIG. 12.—A region V bounded by a surface S in an elastic medium under stress.

Since equilibrium has been assumed for the body and all its parts, the resultant force **F** exerted on the matter within S must be zero. Contributing to this resultant are *volume* or *body forces*, such as gravity, and *surface forces* exerted by elements of matter just outside the enclosed region on contiguous elements within. Throughout V, therefore, we suppose force to be distributed with a density **f** per unit volume, while the force exerted by matter outside S on a unit area of S will be represented by the vector **t**. The components of **t** are evidently normal pressures or tensions and tangential shears. The condition of translational equilibrium is expressed by the equation

$$\int_V \mathbf{f}\,dv + \int_S \mathbf{t}\,da = 0. \tag{1}$$

To ensure rotational equilibrium it is necessary also that the resultant torque be zero, or

$$\int_V \mathbf{r} \times \mathbf{f}\, dv + \int_S \mathbf{r} \times \mathbf{t}\, da = 0, \tag{2}$$

where $\mathbf{r}$ is the radius vector from an arbitrary origin O to an element of volume or surface.

The transition from the integral relations (1) and (2) to an equivalent differential or local system expressing these same conditions in the immediate neighborhood of an arbitrary point $P(x, y, z)$ may be accomplished by a method employed in Chap. I. The surface S is shrunk about P and the components of $\mathbf{t}$ are expanded in Taylor series. The integral can then be evaluated over the surface and on passing to the limit the conditions of equilibrium are obtained in terms of the derivatives of $\mathbf{t}$ at P. This labor may be avoided by applying theorems already at our disposal. Let $\mathbf{n}$ be the unit, outward normal to an element of S. There are then three vectors $\mathbf{X}$, $\mathbf{Y}$, $\mathbf{Z}$ which satisfy the equations

$$t_x = \mathbf{X} \cdot \mathbf{n}, \qquad t_y = \mathbf{Y} \cdot \mathbf{n}, \qquad t_z = \mathbf{Z} \cdot \mathbf{n}. \tag{3}$$

The quantity X_n is clearly the x-component of force acting outward on a unit element of area whose orientation is fixed by the normal $\mathbf{n}$. The expansion of the scalar products in the form

$$\begin{aligned} t_x &= X_x n_x + X_y n_y + X_z n_z, \\ t_y &= Y_x n_x + Y_y n_y + Y_z n_z, \\ t_z &= Z_x n_x + Z_y n_y + Z_z n_z \end{aligned} \tag{4}$$

may also be interpreted as a linear transformation of the components of $\mathbf{n}$ into the components of $\mathbf{t}$, the components n_x, n_y, n_z being the direction cosines of $\mathbf{n}$ with respect to the coordinate axes. The equilibrium of the x-components of forces acting on matter within S is now expressed by

$$\int_V f_x\, dv + \int_S \mathbf{X} \cdot \mathbf{n}\, da = 0, \tag{5}$$

which in virtue of the divergence theorem and the arbitrariness of V is equivalent to the condition that at all points within S

$$f_x + \nabla \cdot \mathbf{X} = 0. \tag{6}$$

For the y- and z-components we have, likewise,

$$f_y + \nabla \cdot \mathbf{Y} = 0, \qquad f_z + \nabla \cdot \mathbf{Z} = 0. \tag{7}$$

The rotational equilibrium expressed by (2) imposes further conditions upon the nine components of the three vectors $\mathbf{X}$, $\mathbf{Y}$, $\mathbf{Z}$. The x-component of this equation is, for example,

$$\int_V (yf_z - zf_y)\, dv + \int_S (yt_z - zt_y)\, da = 0. \tag{8}$$

Introduction of the values of t_z and t_y defined in (3) leads to

$$\int_V (yf_z - zf_y)\,dv + \int_S (y\mathbf{Z} - z\mathbf{Y}) \cdot \mathbf{n}\,da = 0, \tag{9}$$

which, again thanks to the divergence theorem and the arbitrariness of V, is equivalent to

$$yf_z - zf_y + \nabla \cdot (y\mathbf{Z} - z\mathbf{Y}) = 0. \tag{10}$$

But

$$\nabla \cdot (y\mathbf{Z}) = y\nabla \cdot \mathbf{Z} + \mathbf{Z} \cdot \nabla y = y\nabla \cdot \mathbf{Z} + Z_y. \tag{11}$$

Equation (10) reduces to

$$y(f_z + \nabla \cdot \mathbf{Z}) - z(f_y + \nabla \cdot \mathbf{Y}) + Z_y - Y_z = 0; \tag{12}$$

or, on taking account of (7),

$$Z_y = Y_z. \tag{13}$$

In like manner there may be derived from the y- and z-components of (2) the symmetry relations

$$X_y = Y_x, \qquad X_z = Z_x. \tag{14}$$

The nine components $X_x, X_y, \ldots Z_z$, representing forces exerted on unit elements of area, are called *stresses*. The diagonal terms X_x, Y_y, Z_z act in a direction normal to the surface element and are, therefore, pressures or tensions. The remaining six components are shearing stresses acting in the plane of the element. These nine quantities constitute the components of a symmetrical tensor, as is evident from (4) and the fact that $\mathbf{t}$ and $\mathbf{n}$ are true vectors (*cf.* Sec. 1.20). For the sake of a condensed notation we shall henceforth represent the components of the stress tensor by T_{jk}, where

$$X_x = T_{11}, \qquad X_y = T_{12}, \qquad \ldots \qquad Y_x = T_{21}, \qquad \ldots \qquad Z_z = T_{33}. \tag{15}$$

In order that a fluid or solid medium under stress shall be in static equilibrium it is necessary that at every point

$$\mathbf{f} + \operatorname{div} {}^2\mathbf{T} = 0, \qquad T_{jk} = T_{kj}. \tag{16}$$

Imagine an infinitesimal plane element of area containing a point (x, y, z) in a stressed medium. The stresses acting across this surface element will, in general, be both normal and tangential, but there are three distinct orientations with respect to which the stress is purely normal. Now if the resultant force $\mathbf{t}$ acting on unit area of a plane element is in the direction of the positive normal $\mathbf{n}$, one may put

$$\mathbf{t} = \lambda\mathbf{n}, \tag{17}$$

where λ is an unknown scalar function of the coordinates x, y, z, of the element. When this condition is imposed upon Eqs. (4) one obtains the homogeneous system

$$\lambda n_j = \sum_{k=1}^{3} T_{jk} n_k, \qquad (j = 1, 2, 3). \tag{18}$$

In order that these homogeneous equations be self-consistent, it is necessary that their determinant shall vanish; whence it follows that the scalar function λ can be determined from

$$\begin{vmatrix} T_{11} - \lambda & T_{12} & T_{13} \\ T_{21} & T_{22} - \lambda & T_{23} \\ T_{31} & T_{32} & T_{33} - \lambda \end{vmatrix} = 0, \tag{19}$$

provided, of course, that the stress components T_{jk} with respect to some arbitrary coordinate system are known. The secular equation (19) has three roots, $\lambda_a, \lambda_b, \lambda_c$, and these fix, through Eq. (18), three orientations of the surface element which we shall designate respectively as $\mathbf{n}^{(a)}, \mathbf{n}^{(b)}, \mathbf{n}^{(c)}$. If the roots are distinct, the preferred directions defined by the unit vectors $\mathbf{n}^{(a)}, \mathbf{n}^{(b)}, \mathbf{n}^{(c)}$—called the principal axes of stress—are mutually perpendicular. Let us consider, for example, $\mathbf{n}^{(a)}$ and $\mathbf{n}^{(b)}$. According to (18),

$$\lambda_a n_j^{(a)} = \sum_{k=1}^{3} T_{jk} n_k^{(a)}, \qquad \lambda_b n_j^{(b)} = \sum_{k=1}^{3} T_{jk} n_k^{(b)}. \tag{20}$$

Multiply the first of these equations by $n_j^{(b)}$, the second by $n_j^{(a)}$, subtract the second from the first and sum over j.

$$(\lambda_a - \lambda_b) \sum_{j=1}^{3} n_j^{(a)} n_j^{(b)} = \sum_{j=1}^{3} \sum_{k=1}^{3} T_{jk} (n_k^{(a)} n_j^{(b)} - n_k^{(b)} n_j^{(a)}). \tag{21}$$

The right-hand sum vanishes, leaving

$$(\lambda_a - \lambda_b) \mathbf{n}^{(a)} \cdot \mathbf{n}^{(b)} = 0. \tag{22}$$

If $\lambda_a \neq \lambda_b$, the vector $\mathbf{n}^{(a)}$ must be orthogonal to $\mathbf{n}^{(b)}$ as stated.

The physical significance of the principal axes may be made clear in another manner. At a point P in a stressed medium the symmetrical tensor ${}^2\mathbf{T}$ associates with a unit vector $\mathbf{n}$ the resultant force $\mathbf{t}$ acting on a unit surface element normal to $\mathbf{n}$. Let the origin of a rectangular coordinate system be located at P. The scale of length in this new system is arbitrary and we may suppose that the components of $\mathbf{n}$, drawn from P, are ξ, η, ζ. Now the scalar product of the vectors $\mathbf{t}$ and $\mathbf{n}$, which we shall call Φ, is a quadratic in ξ, η, ζ.

$$\Phi(\xi, \eta, \zeta) = \mathbf{t} \cdot \mathbf{n} = T_{11}\xi^2 + T_{22}\eta^2 + T_{33}\zeta^2 + 2T_{12}\xi\eta + 2T_{23}\eta\zeta + 2T_{31}\zeta\xi. \tag{23}$$

The surface Φ = constant is called the *stress quadric*. By a rotation of the coordinate axes (23) can be reduced to the square terms alone, and it is clear that the principal axes $\mathbf{n}^{(a)}$, $\mathbf{n}^{(b)}$, $\mathbf{n}^{(c)}$ must coincide with the principal axes of the quadric. One will observe, furthermore, that

$$\mathbf{t} = \tfrac{1}{2}\nabla\Phi. \tag{24}$$

This property enables one to find the stress across any surface element at P by graphical construction. We shall suppose the stress to be such that the quadric is an ellipsoid and draw about P the surface $\Phi = 1$. The radius vector from P to any point P' on this surface is $\mathbf{n}$. According to (24) the resultant force acting on an element at P whose normal is $\mathbf{n}$ is in the direction of $\nabla\Phi$, or normal to $\Phi = 1$ at P'. The magnitude of $\mathbf{t}$ according to (23) is equal to the reciprocal of the projection NP' indicated

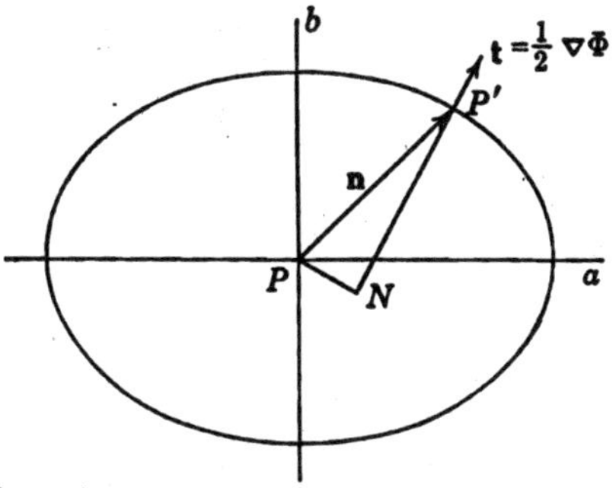

FIG. 13.—Graphical determination of stress on an element of area at the point P from the stress quadric.

in Fig. 13. Along the principal axes, a, b, c, the normal to the surface $\Phi = 1$ coincides with the vector $\mathbf{n}$ and the stress on the element at P is purely normal. If these principal stresses are known, the quadric can be constructed and the stress on an arbitrarily oriented element determined graphically.

2.2. Analysis of Strain.—If surface and volume forces are applied to a perfectly rigid body, the resultant motion may be described in terms of a translation of the body as a whole and a rotation about its center of mass. If, however, the body is deformable, its parts suffer also relative displacements which increase until equilibrium is reestablished by internal forces evoked in the deformation. The changes in the relative positions of the parts of a body under stress are called *strains*.

In an unstressed, continuous medium a point P is located with respect to an arbitrary, fixed origin by the radius vector $\mathbf{r}$. Under the influence of applied forces the medium undergoes a deformation, in the course of which the matter located initially at P moves to a point P' at $\mathbf{r}' = \mathbf{r} + \mathbf{s}$. The displacement $\mathbf{s}$ corresponding to a given deformation depends on the coordinates of the initial point P; we assume $\mathbf{s}$ to be a continuous function of $\mathbf{r}$. Consider now any neighboring point P_1 located by the radius vector $\mathbf{r}_1$. The initial position of P_1 with respect to P is fixed by the vector $\delta\mathbf{r}$, where

$$\delta\mathbf{r} = \mathbf{r}_1 - \mathbf{r}. \tag{25}$$

The displacement of P_1 during the deformation is $\mathbf{s}_1$, a function of $\mathbf{r}_1$. The *relative* displacement of two neighboring points P and P_1 occasioned by the deformation is therefore

$$\delta\mathbf{s} = \mathbf{s}_1 - \mathbf{s} = \mathbf{s}(\mathbf{r} + \delta\mathbf{r}) - \mathbf{s}(\mathbf{r}). \tag{26}$$

Since $\mathbf{s}(\mathbf{r} + \delta\mathbf{r})$ is continuous, it may be expanded in a Taylor series; for points in a region sufficiently small about P we need retain only the linear term

$$\delta\mathbf{s} = (\delta\mathbf{r} \cdot \nabla)\mathbf{s}; \tag{27}$$

or, in component form,

$$\delta s_j = \sum_{k=1}^{3} \frac{\partial s_j}{\partial x_k}\,\delta x_k, \qquad (j = 1, 2, 3). \tag{28}$$

(The indices 1, 2, 3 again replace the subscripts x, y, z for convenience in summing.)

The nine derivatives $\partial s_j/\partial x_k$ are the components of a tensor (*cf.* Eq. (54), Sec. 1.20) which is in general asymmetrical. However, it can be resolved after the manner of Eqs. (46) and (47), page 66, into a symmetric part,

$$a_{jk} = \frac{1}{2}\left(\frac{\partial s_j}{\partial x_k} + \frac{\partial s_k}{\partial x_j}\right), \qquad (j, k = 1, 2, 3), \tag{29}$$

and an antisymmetric part,

$$b_{jk} = \frac{1}{2}\left(\frac{\partial s_j}{\partial x_k} - \frac{\partial s_k}{\partial x_j}\right), \qquad (j, k = 1, 2, 3). \tag{30}$$

The components of this antisymmetric tensor are evidently identical with the components of an axial vector $\mathbf{b}$, defined by

$$\mathbf{b} = \tfrac{1}{2}\nabla \times \mathbf{s}, \tag{31}$$

$$b_1 = b_{32}, \qquad b_2 = b_{13}, \qquad b_3 = b_{21}. \tag{32}$$

The relative displacement $\delta\mathbf{s}$ can likewise be split up into a part $\delta\mathbf{s}'$ associated with the symmetric tensor a_{jk}, and a part $\delta\mathbf{s}''$ associated with the antisymmetric tensor b_{jk}. Then, by (28) and (30),

$$\delta\mathbf{s}'' = \mathbf{b} \times \delta\mathbf{r} = \tfrac{1}{2}(\nabla \times \mathbf{s}) \times \delta\mathbf{r}. \tag{33}$$

Physically this implies that the matter contained within an infinitesimal volume about P is subjected to a rotation as a rigid solid; the rotation takes place about an instantaneous axis through P in the direction of $\nabla \times \mathbf{s}$, and in circular measure is equal to $\frac{1}{2}|\nabla \times \mathbf{s}|$. This local rotation is brought about by stresses whose curl does not vanish.

Whereas the relative positions of points in the immediate vicinity of P are preserved by the local rotation (33), the symmetric tensor (29) defines a local deformation—an actual stretching and twisting. We ask first whether there are any vectors $\delta\mathbf{r}$ drawn from P whose direction is unchanged after the deformation; that is to say, are there any points P_1, Fig. 14, which in the course of the deformation will move along the line defined by $\delta\mathbf{r}$? The necessary condition is that

$$(34) \qquad \delta\mathbf{s}' = \lambda\,\delta\mathbf{r},$$

where λ is an unknown scalar function of the coordinates of P; or

$$(35) \qquad \lambda\,\delta x_j = \sum_{k=1}^{3} a_{jk}\delta x_k, \quad (j = 1, 2, 3).$$

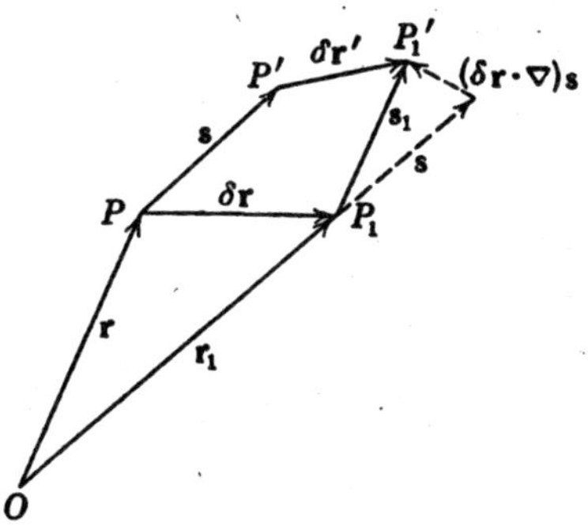

FIG. 14.—Vectors characterizing the deformation of a continuous medium.

The parameter λ is found from the condition that the determinant of this homogeneous system shall vanish and, as in the analogous case of the stress tensor discussed in the preceding paragraph, it is easy to show that the three roots fix three principal axes which in general are mutually orthogonal. Along these principal axes of strain, and along them only, the deformation consists of a pure stretching. The nature of the deformation along any other axes can be visualized with the aid of a strain quadric such as (23). Let the point P_1, Fig. 14, have the coordinates ξ_1, ξ_2, ξ_3 with respect to P and construct the surface

$$(36) \qquad \Psi = \delta\mathbf{r} \cdot \delta\mathbf{s}' = \sum_{j=1}^{3}\sum_{k=1}^{3} a_{jk}\xi_j\xi_k = \text{constant}.$$

If all lines issuing from P are extended, or if all are contracted, the quadric is an ellipsoid; if some lines are extended and others contracted, the surface is an hyperboloid. The relative displacement of P_1 with respect to P is given by

$$(37) \qquad \delta\mathbf{s}' = \tfrac{1}{2}\nabla\Psi.$$

The radius vector from P to P_1 on the surface $\Psi = 1$ is $\delta\mathbf{r}$. In Fig. 15 it has been assumed that this surface is an ellipsoid. The direction of the relative displacement $\delta\mathbf{s}'$ due to the deformation is that of the normal erected at P_1 and its magnitude is equal to the reciprocal of the projection

NP_1 of the radius vector $\delta\mathbf{r}$ on the normal. Obviously the vectors $\delta\mathbf{s}'$ and $\delta\mathbf{r}$ are parallel only along the principal axes of the ellipsoid. These axes represent also the directions of maximum and minimum contraction or extension.

There is an interesting physical interpretation to be given to the tensor components a_{jk}. If before a deformation the relative position of two points P and P_1, Fig. 14, is fixed by $\delta\mathbf{r}$, their relative position after the deformation is $\delta\mathbf{r}' = \delta\mathbf{r} + \delta\mathbf{s}'$, the prime over $\delta\mathbf{s}'$ indicating that local rotations are excluded. Suppose now that $\delta\mathbf{r}$ is directed along

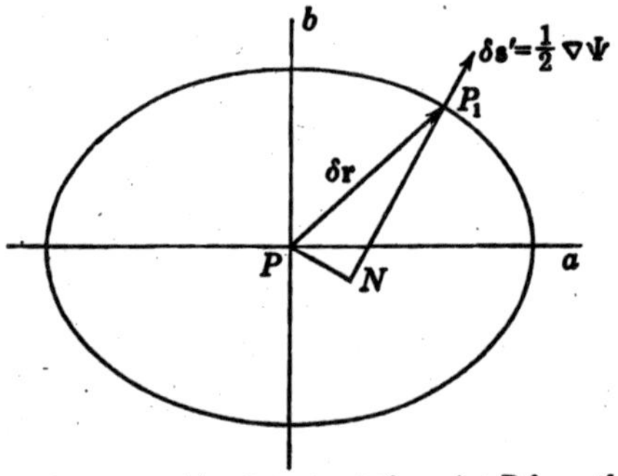

FIG. 15.—Graphical determination of strain at the point P from the strain quadric.

the x_1-axis, so that $\delta x_2 = \delta x_3 = 0$. Then $\delta\mathbf{r}'$ is a vector whose components are

$$(38) \qquad \delta x_1' = (1 + a_{11})\,\delta x_1, \qquad \delta x_2' = a_{21}\,\delta x_1, \qquad \delta x_3' = a_{31}\,\delta x_1.$$

The absolute value of the distance between P and P_1 after deformation is

$$(39) \qquad |\delta\mathbf{r}'| = \sqrt{(1 + a_{11})^2 + a_{21}^2 + a_{31}^2}\,\delta x_1,$$

and the *relative* change in length is

$$(40) \qquad \frac{|\delta r'| - \delta x_1}{\delta x_1} = a_{11} = \frac{\partial s_1}{\partial x_1},$$

obtained by expanding (39) and discarding all terms of higher order than the first. Similar expressions can be found for deformations of line elements lying initially along the x_2- and x_3-axes, whence we conclude that *the diagonal components* $a_{jj} = \partial s_j/\partial x_j$ *represent extensions of linear elements which in the unstrained state are parallel to the coordinate axes.*

Again let $\delta\mathbf{a}$ be a linear element which in the unstrained state is parallel to the x_1-axis and $\delta\mathbf{b}$ be a linear element initially parallel to the x_2-axis. A local deformation transforms the vector $\delta\mathbf{a} = \mathbf{i}_1\,\delta x_1$ into an element

$$(41) \qquad \delta\mathbf{a}' = \mathbf{i}_1(1 + a_{11})\,\delta x_1 + \mathbf{i}_2 a_{21}\,\delta x_1 + \mathbf{i}_3 a_{31}\,\delta x_1,$$

while at the same time $\delta \mathbf{b} = \mathbf{i}_2\, \delta x_2$ is transformed into

$$\delta \mathbf{b}' = \mathbf{i}_1 a_{12}\, \delta x_2 + \mathbf{i}_2(1 + a_{22})\, \delta x_2 + \mathbf{i}_3 a_{32}\, \delta x_2. \tag{42}$$

In the initial state $\delta \mathbf{a}$ and $\delta \mathbf{b}$ are mutually orthogonal, but after the deformation, they form an angle which differs but little from $\pi/2$ and which we shall denote by $\frac{\pi}{2} - \theta_{12}$.

$$\delta \mathbf{a}' \cdot \delta \mathbf{b}' = |\delta \mathbf{a}'|\, |\delta \mathbf{b}'| \cos\left(\frac{\pi}{2} - \theta_{12}\right). \tag{43}$$

But

$$\cos\left(\frac{\pi}{2} - \theta_{12}\right) = \sin \theta_{12} \simeq \theta_{12}, \tag{44}$$

and, hence,

$$\theta_{12} = a_{12}(1 + a_{11}) + a_{21}(1 + a_{22}) + a_{31}a_{32}; \tag{45}$$

or, on neglecting terms of higher order than the first,

$$\theta_{12} = a_{12} + a_{21} = \frac{\partial s_1}{\partial x_2} + \frac{\partial s_2}{\partial x_1}. \tag{46}$$

Thus, in general, *the coefficient a_{jk}, with $j \neq k$, measures one-half the cosine of the angle made by two linear elements after a deformation, which in the unstrained state were directed along a pair of orthogonal coordinate axes.*

Conforming to this interpretation, it is customary to write the components of the deformation tensor in a slightly altered manner. The coefficients defined by

$$\begin{aligned}
e_{11} &= \frac{\partial s_1}{\partial x_1} & e_{12} &= \frac{\partial s_1}{\partial x_2} + \frac{\partial s_2}{\partial x_1} = e_{21} \\
e_{22} &= \frac{\partial s_2}{\partial x_2} & e_{23} &= \frac{\partial s_2}{\partial x_3} + \frac{\partial s_3}{\partial x_2} = e_{32} \\
e_{33} &= \frac{\partial s_3}{\partial x_3} & e_{31} &= \frac{\partial s_3}{\partial x_1} + \frac{\partial s_1}{\partial x_3} = e_{13}
\end{aligned} \tag{47}$$

are called the *components of strain*, and the components of the relative displacement of any two points P and P_1 due to a local deformation are therefore

$$\begin{aligned}
\delta s_1' &= e_{11}\, \delta x_1 + \tfrac{1}{2}e_{12}\, \delta x_2 + \tfrac{1}{2}e_{13}\, \delta x_3, \\
\delta s_2' &= \tfrac{1}{2}e_{21}\, \delta x_1 + e_{22}\, \delta x_2 + \tfrac{1}{2}e_{23}\, \delta x_3, \\
\delta s_3' &= \tfrac{1}{2}e_{31}\, \delta x_1 + \tfrac{1}{2}e_{32}\, \delta x_2 + e_{33}\, \delta x_3.
\end{aligned} \tag{48}$$

If the origin of the arbitrary coordinate system x_1, x_2, x_3 is located at the point P and the axes then rotated into coincidence with the principal axes of the strain ellipsoid associated with that point, all components of strain are reduced to zero with the exception of those lying on the main diagonal. We shall measure distances along the principal axes of strain by the coordinates u_1, u_2, u_3. Then the components of any local deformation with respect to P are

$$\delta s'_{u_1} = e_1\,\delta u_1, \qquad \delta s'_{u_2} = e_2\,\delta u_2, \qquad \delta s'_{u_3} = e_3\,\delta u_3, \tag{49}$$

where

$$e_1 = \frac{\partial s_{u_1}}{\partial u_1}, \qquad e_2 = \frac{\partial s_{u_2}}{\partial u_2}, \qquad e_3 = \frac{\partial s_{u_3}}{\partial u_3}, \tag{50}$$

and where δu_1, δu_2, δu_3 are the components of $\delta\mathbf{r}$ with respect to the principal axes. The coefficients e_1, e_2, e_3 are called the principal strains, or principal extensions.

Associated with the deformation of an infinitesimal region there is a change in the element of volume. If $\delta\mathbf{a}$, $\delta\mathbf{b}$, $\delta\mathbf{c}$ are three vectors defining a parallelepiped, its volume is given by

$$\delta V = (\delta\mathbf{a} \times \delta\mathbf{b}) \cdot \delta\mathbf{c} = \begin{vmatrix} \delta a_1 & \delta a_2 & \delta a_3 \\ \delta b_1 & \delta b_2 & \delta b_3 \\ \delta c_1 & \delta c_2 & \delta c_3 \end{vmatrix}. \tag{51}$$

Without loss of generality we may choose for this initial element a rectangular block whose sides are parallel to the coordinate axes.

$$\delta\mathbf{a} = \mathbf{i}_1\,\delta x_1, \qquad \delta\mathbf{b} = \mathbf{i}_2\,\delta x_2, \qquad \delta\mathbf{c} = \mathbf{i}_3\,\delta x_3. \tag{52}$$

The deformation transforms these vectors as in (41) and (42), whence for the volume after strain we find

$$\delta V' = (\delta\mathbf{a}' \times \delta\mathbf{b}') \cdot \delta\mathbf{c}' = \begin{vmatrix} 1 + a_{11} & a_{21} & a_{31} \\ a_{12} & 1 + a_{22} & a_{32} \\ a_{13} & a_{23} & 1 + a_{33} \end{vmatrix} \delta x_1\,\delta x_2\,\delta x_3. \tag{53}$$

Expansion of this determinant and discard of all terms of higher order than the first lead to

$$\frac{\delta V' - \delta V}{\delta V} = a_{11} + a_{22} + a_{33} = \nabla \cdot \mathbf{s} \tag{54}$$

for the change of volume per unit volume—a quantity called the *cubical dilatation.*

In summary, the analysis has shown that the most general displacement of particles in the neighborhood of a point may be resolved into a translation and a local rotation, upon which there is superposed a deformation characterized by the strain components e_{jk} and accompanied by a change in volume. The rotational component of the total strain can be induced only by rotational stresses—force functions, that is to say, whose curl is not everywhere zero.

2.3. Elastic Energy and the Relations of Stress to Strain.—Under the influence of external forces the particles composing an elastic fluid or solid suffer relative displacements. These strains, in turn, evoke internal forces which, in the case of static equilibrium, eventually compensate the applied stresses. The work done by the applied volume and surface forces in displacing every point of the medium an amount $\delta\mathbf{s}$ is

$$\delta W = \int \mathbf{f} \cdot \delta\mathbf{s}\, dv + \int \mathbf{t} \cdot \delta\mathbf{s}\, da. \tag{55}$$

According to Eq. (3)

$$\mathbf{t} \cdot \delta\mathbf{s} = \delta s_x \mathbf{X} \cdot \mathbf{n} + \delta s_y \mathbf{Y} \cdot \mathbf{n} + \delta s_z \mathbf{Z} \cdot \mathbf{n}, \tag{56}$$

and, hence,

$$\int \mathbf{t} \cdot \delta\mathbf{s}\, da = \int \nabla \cdot (\mathbf{X}\, \delta s_x + \mathbf{Y}\, \delta s_y + \mathbf{Z}\, \delta s_z)\, dv. \tag{57}$$

Now

$$\nabla \cdot (\mathbf{X}\, \delta s_x) = \delta s_x \nabla \cdot \mathbf{X} + \mathbf{X} \cdot \nabla\, \delta s_x, \tag{58}$$

and by (6) $f_x = -\nabla \cdot \mathbf{X}$. Thus (55) reduces to

$$\delta W = \int (\mathbf{X} \cdot \nabla\, \delta s_x + \mathbf{Y} \cdot \nabla\, \delta s_y + \mathbf{Z} \cdot \nabla\, \delta s_z)\, dv. \tag{59}$$

The components of the integrand may, however, be written

$$\mathbf{X} \cdot \nabla\, \delta s_x = T_{11}\delta\left(\frac{\partial s_x}{\partial x}\right) + T_{12}\delta\left(\frac{\partial s_x}{\partial y}\right) + T_{13}\delta\left(\frac{\delta s_x}{\partial z}\right), \text{ etc.,} \tag{60}$$

which in the notation of Eqs. (47) leads to

$$\delta W = \int (T_{11}\, \delta e_{11} + T_{22}\, \delta e_{22} + T_{33}\, \delta e_{33} + T_{12}\, \delta e_{12} + T_{23}\, \delta e_{23} + T_{31}\, \delta e_{31})\, dv \tag{61}$$

for the work done by the applied stresses against the elastic restoring forces in the course of an infinitesimal change in the state of strain. If it be assumed that this change takes place so slowly that the variation in kinetic energy may be neglected, and that no heat is added or lost in the process, then δW must be equal to the increase in the potential energy U stored up in the elastically deformed medium. The elastic energy stored in unit volume will be denoted by u, whence

$$\delta W = \delta U = \int \delta u\, dv. \tag{62}$$

The energy density u at any point must depend on the local state of strain and it may, therefore, be assumed that

$$u = u(e_{11}, e_{12}, \cdots e_{33}), \tag{63}$$

or

$$\delta U = \int \left(\frac{\partial u}{\partial e_{11}} \delta e_{11} + \frac{\partial u}{\partial e_{22}} \delta e_{22} + \frac{\partial u}{\partial e_{33}} \delta e_{33} + \frac{\partial u}{\partial e_{12}} \delta e_{12} + \frac{\partial u}{\partial e_{23}} \delta e_{23} + \frac{\partial u}{\partial e_{31}} \delta e_{31} \right) dv. \tag{64}$$

The strain components e_{jk} are arbitrary; hence it follows that the components of applied stress (which in static equilibrium are equal and opposite to the induced elastic stresses) can be derived from a scalar potential function,

$$T_{jk} = \frac{\partial u}{\partial e_{jk}}. \tag{65}$$

The existence of an elastic potential, which has been demonstrated here for the case of quasi-stationary, adiabatic deformations, may be shown for isothermal changes as well.

When the deformations are not excessive, the components of elastic stress may be expressed as linear functions of the components of strain.

$$\begin{aligned} T_{11} &= c_{11}e_{11} + c_{12}e_{22} + c_{13}e_{33} + c_{14}e_{12} + c_{15}e_{23} + c_{16}e_{31}, \\ &\cdots\cdots\cdots\cdots\cdots\cdots\cdots\cdots \\ T_{12} &= c_{41}e_{11} + c_{42}e_{22} + c_{43}e_{33} + c_{44}e_{12} + c_{45}e_{23} + c_{46}e_{31}, \\ &\cdots\cdots\cdots\cdots\cdots\cdots\cdots\cdots \end{aligned} \tag{66}$$

The coefficients c_{mn} are called the *elastic constants* of the medium. The elastic potential u is in this case a homogeneous quadratic function of the strain components. The existence of a scalar function $u(e_{jk})$ satisfying (65) and (66) imposes on the elastic constants the conditions

$$c_{mn} = c_{nm} \qquad (n, m = 1, 2, \cdots, 6), \tag{67}$$

whereby the number of parameters necessary to specify the relation of stress to strain in an anisotropic medium reduces from 36 to 21. Any further reduction is accomplished by taking advantage of possible symmetries in the structure of the medium. If, in particular, the substance is elastically isotropic, the quadratic form u must be invariant to orthogonal transformations of the coordinate system, and the number of independent constants then reduces to two. The elastic potential assumes the form

$$u = \tfrac{1}{2}\lambda_1(e_{11} + e_{22} + e_{33})^2 + \lambda_2(e_{11}^2 + e_{22}^2 + e_{33}^2 + \tfrac{1}{2}e_{12}^2 + \tfrac{1}{2}e_{23}^2 + \tfrac{1}{2}e_{31}^2), \tag{68}$$

whence. by (65)

$$\begin{aligned} T_{11} &= \lambda_1(e_{11} + e_{22} + e_{33}) + 2\lambda_2 e_{11}, & T_{12} &= \lambda_2 e_{12}, \\ T_{22} &= \lambda_1(e_{11} + e_{22} + e_{33}) + 2\lambda_2 e_{22}, & T_{23} &= \lambda_2 e_{23}, \\ T_{33} &= \lambda_1(e_{11} + e_{22} + e_{33}) + 2\lambda_2 e_{33}, & T_{31} &= \lambda_2 e_{31}. \end{aligned} \tag{69}$$

Now the conditions of static equilibrium are expressed by

$$\text{(16)} \qquad \mathbf{f} + \nabla \cdot {}^2\mathbf{T} = 0.$$

Upon introducing the components (69) into the expanded divergence and expressing the strain components e_{jk} in terms of the deformation $\mathbf{s}$, we find

$$\text{(70)} \qquad \mathbf{f} + (\lambda_1 + \lambda_2)\nabla\nabla \cdot \mathbf{s} + \lambda_2 \nabla^2 \mathbf{s} = 0$$

as the equation determining the deformation of an isotropic body subjected to a volume force $\mathbf{f}$. The constants of integration are evaluated in terms of data specifying either the displacement of points on the bounding surface, or the distribution of stress over that surface.

The parameters λ_1 and λ_2 are positive quantities. λ_2 is called the *rigidity*, or *shear modulus*, for it measures the strains induced by tangential, or shearing, stresses. No simple physical meaning can be attached to λ_1; however, it may be defined in terms of the better known constants E and σ, where *Young's modulus* E is the ratio of a simple longitudinal tension to the elongation per unit length which it produces, and where the *Poisson ratio* σ is the ratio of lateral contraction to longitudinal extension of a bar under tension. Then

$$\text{(71)} \qquad \lambda_1 = \frac{E\sigma}{(1 + \sigma)(1 - 2\sigma)}, \qquad \lambda_2 = \frac{E}{2(1 + \sigma)}.$$

An ideal fluid supports no shearing stress, and hence $\lambda_2 = 0$. The stress acting on any element of a closed surface within the fluid is, therefore, a normal pressure,

$$\text{(72)} \qquad T_{11} = T_{22} = T_{33} = -p, \qquad T_{12} = T_{23} = T_{31} = 0,$$

the negative sign indicating that the stress is directed inward. The components of strain e_{11}, e_{22}, e_{33} are all equal and

$$\text{(73)} \qquad p = -\lambda_1(e_{11} + e_{22} + e_{33}) = -\lambda_1 \nabla \cdot \mathbf{s},$$

while the equation of equilibrium reduces to

$$\text{(74)} \qquad \mathbf{f} - \nabla p = 0.$$

Let V_0 be the initial volume of an element of fluid, V_1 its volume at a pressure p_1, and V_2 its volume at a pressure p_2. From the definition (54) of the cubical dilatation $\nabla \cdot \mathbf{s}$ we have

$$\text{(75)} \qquad p_1 = -\lambda_1 \frac{V_1 - V_0}{V_0}, \qquad p_2 = \lambda_1 \frac{V_2 - V_0}{V_0},$$

and for sufficiently small changes,

$$\text{(76)} \qquad p_2 - p_1 = -\lambda_1 \frac{V_2 - V_1}{V_0}.$$

For an infinitesimal change in pressure

$$dp = -\lambda_1 \frac{dV}{V} = \lambda_1 \frac{d\tau}{\tau}, \tag{77}$$

where τ is the density of the fluid. The reciprocal of λ_1 is in this case called the compressibility.

ELECTROMAGNETIC FORCES ON CHARGES AND CURRENTS

2.4. Definition of the Vectors E and B.—An electromagnetic field is defined according to our initial hypotheses by four vectors **E, B, D** and **H** satisfying Maxwell's equations. The physical nature of these vectors will be expressed in terms of experiments by means of which they may be measured. Now it is easy to show that $\rho\mathbf{E}$ and $\mathbf{J} \times \mathbf{B}$ are quantities whose dimensions are those of force per unit volume. For by Sec. 1.8

$$[\rho\mathbf{E}] = \frac{\text{coulombs}}{\text{meter}^3} \times \frac{\text{volts}}{\text{meter}} = \frac{\text{kilograms}}{\text{second}^2 \cdot \text{meter}^2} \tag{1}$$

$$[\mathbf{J} \times \mathbf{B}] = \frac{\text{amperes}}{\text{meter}^2} \times \frac{\text{webers}}{\text{meter}^2} = \frac{\text{kilograms}}{\text{second}^2 \cdot \text{meter}^2}. \tag{2}$$

We are, therefore, free to *define* the vectors **E** and **B** as forces exerted by the field on unit elements respectively of charge and current. More precisely, we shall suppose that charge is distributed throughout a volume V with a macroscopically continuous density ρ. Then **E** is defined such that the net mechanical force acting on the charge is

$$\mathbf{F}_e = \int_V \rho\mathbf{E}\, dv, \tag{3}$$

distributed with a volume density

$$\mathbf{f}_e = \rho\mathbf{E}. \tag{4}$$

If V is sufficiently small, the field within this region at any instant may be assumed homogeneous, so that in the limit

$$\mathbf{F}_e = \mathbf{E} \int_V \rho\, dv = q\mathbf{E}, \qquad (V \to 0). \tag{5}$$

In like manner the association of the magnetic vector **B** with the force exerted on a unit volume element of current through the equation

$$\mathbf{f}_m = \mathbf{J} \times \mathbf{B}, \tag{6}$$

leads to

$$\mathbf{F}_m = \int_V \mathbf{J} \times \mathbf{B}\, dv \tag{7}$$

as the net force exerted on a volume distribution of current. If the current is confined to a linear conductor of sufficiently small cross section,

B may be assumed homogeneous and the current lines parallel to an element of length $d\mathbf{s}$. The force acting on a linear current element is then

$$dF_m = I\,d\mathbf{s} \times \mathbf{B}. \tag{8}$$

Whereas the force exerted by an electric field on an element of charge is directed along the vector **E**, the force exerted on an element of current in a magnetic field is normal to the plane defined by the element $d\mathbf{s}$ and the vector **B**.

This arbitrariness in our definition of **E** and **B** is inevitable. The mutual forces exerted by charges upon charges or currents upon currents can be measured, but the field vectors themselves are not independent entities accessible to direct observation. The definitions of **E** and **B** based on Eqs. (3) and (7) have been shown by purely dimensional considerations to be compatible with Maxwell's equations. In the following we shall have to show that the properties of a field of the vectors **E** and **B** defined in this manner and satisfying these equations are in complete accord with experiment. From the forces exerted by charges and currents may be determined the work necessary to establish a field; from energy relations in turn it will be possible to deduce the forces exerted on ponderable elements of neutral matter.

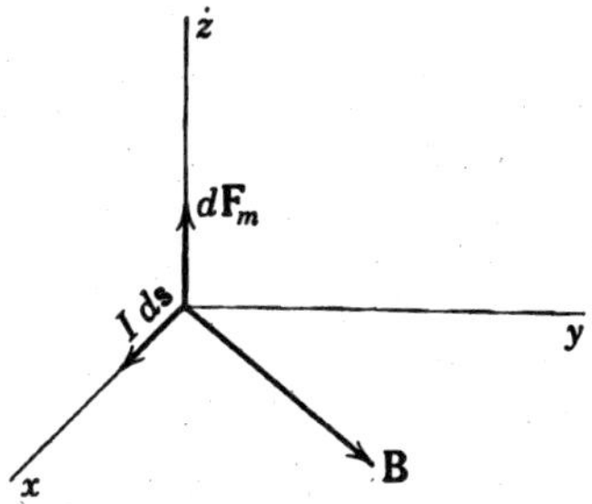

FIG. 16.—Direction of force exerted on a current element $I\,ds$ in a magnetic field **B**.

2.5. The Electromagnetic Stress Tensor in Free Space.—Let us suppose that a certain bounded region of space contains charge and current distributions but is free of all neutral dielectric or magnetic materials. The field is produced in part by the charges and currents within the region, in part by sources which are exterior to it. At every interior point

$$\text{(I)}\quad \nabla \times \mathbf{E} + \frac{\partial \mathbf{B}}{\partial t} = 0, \qquad \text{(III)}\quad \nabla \cdot \mathbf{B} = 0,$$

$$\text{(II)}\quad \nabla \times \mathbf{B} - \mu_0\epsilon_0 \frac{\partial \mathbf{E}}{\partial t} = \mu_0 \mathbf{J}, \qquad \text{(IV)}\quad \nabla \cdot \mathbf{E} = \frac{1}{\epsilon_0}\rho.$$

Let (I) be multiplied vectorially by $\epsilon_0\mathbf{E}$, (II) by the vector **B**. Upon adding and transposing terms we find

$$\epsilon_0(\nabla \times \mathbf{E}) \times \mathbf{E} + \frac{1}{\mu_0}(\nabla \times \mathbf{B}) \times \mathbf{B} = \mathbf{J} \times \mathbf{B} + \epsilon_0 \frac{\partial}{\partial t}(\mathbf{E} \times \mathbf{B}). \tag{9}$$

In a rectangular system of coordinates the first term of (9) may be represented by the determinant

$$(10)\qquad (\nabla \times \mathbf{E}) \times \mathbf{E} = \begin{vmatrix} \mathbf{i} & \mathbf{j} & \mathbf{k} \\ \frac{\partial E_z}{\partial y} - \frac{\partial E_y}{\partial z} & \frac{\partial E_x}{\partial z} - \frac{\partial E_z}{\partial x} & \frac{\partial E_y}{\partial x} - \frac{\partial E_x}{\partial y} \\ E_x & E_y & E_z \end{vmatrix}$$

The x-component of this vector is

$$(11)\qquad [(\nabla \times \mathbf{E}) \times \mathbf{E}] \cdot \mathbf{i} = E_x \frac{\partial E_x}{\partial x} + E_y \frac{\partial E_x}{\partial y} + E_z \frac{\partial E_x}{\partial z} - E_x \frac{\partial E_x}{\partial x} - E_y \frac{\partial E_y}{\partial x} - E_z \frac{\partial E_z}{\partial x}$$
$$= \frac{\partial}{\partial x}\left(E_x^2 - \frac{1}{2} E^2\right) + \frac{\partial}{\partial y}(E_y E_x) + \frac{\partial}{\partial z}(E_z E_x) - E_x \nabla \cdot \mathbf{E}.$$

Now the quantities

$$(12)\qquad S_{11}^{(e)} = \epsilon_0 (E_x^2 - \tfrac{1}{2} E^2), \qquad S_{12}^{(e)} = \epsilon_0 E_x E_y, \qquad S_{13}^{(e)} = \epsilon_0 E_x E_z,$$

transform like the components of a tensor [Eq. (43), page 66] and the first three terms on the right of (11) constitute therefore the x-component of the divergence of a tensor ${}^2\mathbf{S}^{(e)}$. The remaining components are calculated from the y- and z-components of $\epsilon_0(\nabla \times \mathbf{E}) \times \mathbf{E}$, such that we are led to the identity

$$(13)\qquad \epsilon_0 (\nabla \times \mathbf{E}) \times \mathbf{E} = \operatorname{div} {}^2\mathbf{S}^{(e)} - \epsilon_0 \mathbf{E}\, \nabla \cdot \mathbf{E},$$

the components of ${}^2\mathbf{S}^{(e)}$ being tabulated below.

TABLE I.—COMPONENTS $S_{jk}^{(e)}$ OF THE TENSOR ${}^2\mathbf{S}^{(e)}$ IN FREE SPACE

j \ k	1	2	3
1	$\epsilon_0 E_x^2 - \frac{\epsilon_0}{2} E^2$	$\epsilon_0 E_x E_y$	$\epsilon_0 E_x E_z$
2	$\epsilon_0 E_y E_x$	$\epsilon_0 E_y^2 - \frac{\epsilon_0}{2} E^2$	$\epsilon_0 E_y E_z$
3	$\epsilon_0 E_z E_x$	$\epsilon_0 E_z E_y$	$\epsilon_0 E_z^2 - \frac{\epsilon_0}{2} E^2$

The transformation of $(\nabla \times \mathbf{B}) \times \mathbf{B}$ is effected similarly, giving

$$(14)\qquad \frac{1}{\mu_0} (\nabla \times \mathbf{B}) \times \mathbf{B} = \operatorname{div} {}^2\mathbf{S}^{(m)} - \frac{1}{\mu_0} \mathbf{B}\, \nabla \cdot \mathbf{B},$$

where the components of ${}^2\mathbf{S}^{(m)}$ are as represented in Table II.

TABLE II.—COMPONENTS $S_{jk}^{(m)}$ OF THE TENSOR ${}^2\mathbf{S}^{(m)}$ IN FREE SPACE

$j \backslash k$	1	2	3
1	$\frac{1}{\mu_0} B_x^2 - \frac{1}{2\mu_0} B^2$	$\frac{1}{\mu_0} B_x B_y$	$\frac{1}{\mu_0} B_x B_z$
2	$\frac{1}{\mu_0} B_y B_x$	$\frac{1}{\mu_0} B_y^2 - \frac{1}{2\mu_0} B^2$	$\frac{1}{\mu_0} B_y B_z$
3	$\frac{1}{\mu_0} B_z B_x$	$\frac{1}{\mu_0} B_z B_y$	$\frac{1}{\mu_0} B_z^2 - \frac{1}{2\mu_0} B^2$

Upon replacing the first two terms of (9) by (13) and (14) and taking account of (III) and (IV), we obtain an identity of the form

$$\text{div}\ {}^2\mathbf{S} = \mathbf{E}\rho + \mathbf{J} \times \mathbf{B} + \epsilon_0 \frac{\partial}{\partial t} (\mathbf{E} \times \mathbf{B}), \tag{15}$$

wherein the components of the tensor ${}^2\mathbf{S}$ are

$$S_{jk} = S_{jk}^{(e)} + S_{jk}^{(m)}, \tag{16}$$

and where

$$\text{div}\ {}^2\mathbf{S} = \sum_{j=1}^{3} \sum_{k=1}^{3} \mathbf{i}_j \frac{\partial S_{jk}}{\partial x_k}. \tag{17}$$

Equation (15) is a relation through which the forces exerted on elements of charge and current at any point in otherwise empty space may be expressed in terms of the vectors **E** and **B** alone.

Let us integrate this identity over a volume V. Now the integral of the divergence of a tensor throughout V is equal to the integral of a vector over the surface bounding V. To demonstrate this tensor analogue of the vector divergence theorem, let **n** be the outward unit normal at a point on the bounding surface and consider S_{11} to be the x-component of a vector whose y- and z-components are zero. The component of this vector in the direction of the normal **n** is $S_{11}n_x$, whence by the divergence theorem

$$\int S_{11} n_x \, da = \int \frac{\partial S_{11}}{\partial x} \, dv. \tag{18}$$

Likewise it is apparent that

$$\begin{aligned} \int S_{12} n_y \, da &= \int \frac{\partial S_{12}}{\partial y} \, dv, \\ \int S_{13} n_z \, da &= \int \frac{\partial S_{13}}{\partial z} \, dv. \end{aligned} \tag{19}$$

Upon adding these three identities, we obtain

$$\int (S_{11}n_x + S_{12}n_y + S_{13}n_z)\, da = \int \left(\frac{\partial S_{11}}{\partial x} + \frac{\partial S_{12}}{\partial y} + \frac{\partial S_{13}}{\partial z}\right) dv. \tag{20}$$

The integrand on the right is the x-component of div ${}^2\mathbf{S}$, and consequently

$$t_x = S_{11}n_x + S_{12}n_y + S_{13}n_z \tag{21}$$

is the x-component of a vector $\mathbf{t}$ which according to (20) is to be integrated over the surface bounding V. Proceeding similarly for the other components, we have the general theorem:

$$\int \mathbf{t}\, da = \int \operatorname{div} {}^2\mathbf{S}\, dv, \tag{22}$$

where the components of $\mathbf{t}$ are

$$t_j = \sum_{k=1}^{3} S_{jk}n_k, \qquad (j = 1, 2, 3), \tag{23}$$

or, in abbreviated notation,

$$\mathbf{t} = {}^2\mathbf{S} \cdot \mathbf{n}. \tag{24}$$

Applied to Eq. (15), the divergence theorem (22) leads to

$$\int {}^2\mathbf{S} \cdot \mathbf{n}\, da = \mathbf{F}_e + \mathbf{F}_m + \epsilon_0 \frac{\partial}{\partial t} \int \mathbf{E} \times \mathbf{B}\, dv \tag{25}$$

with $\mathbf{F}_e$ and $\mathbf{F}_m$ representing, as in Eqs. (3) and (7), the resultant forces acting respectively on the charge and the current contained within V.

Consider first stationary distributions of charge and current. The fields are then independent of time and the third term on the right of (25) is zero. Equation (25) now states that the force exerted on stationary charges or currents can be expressed as the integral of a vector over any regular surface enclosing these charges and currents. It does *not* state that the volume forces $\mathbf{F}_e$ and $\mathbf{F}_m$ are maintained in equilibrium by the force ${}^2\mathbf{S} \cdot \mathbf{n}$ distributed over the surface. The equilibrium must be established with mechanical forces of some other type, and in fact it will be shown shortly that a charge distribution cannot possibly be maintained in static equilibrium under the action of electrical forces alone. To be more specific, let us imagine a stationary charge distribution to be divided into two parts by an arbitrary closed surface Σ. The force exerted by the external charges on those within Σ must in some manner be transmitted across this surface. The net force on the interior charges may, according to (25), be correctly calculated on the assumption that the force transmitted across an element of area da is ${}^2\mathbf{S}^{(e)} \cdot \mathbf{n}\, da$, and the components $S_{jk}^{(e)}$ are hence effective stresses in the electrostatic field.

Associated with every point in the field there is a stress quadric from which may be determined the normal and tangential components of transmitted stress on an element whose normal is **n**. To find the principal axes of this quadric with respect to the direction of the field we shall adopt the procedure described in Sec. 2.1. The secular determinant (19) of that section assumes in the present instance the form

$$\begin{vmatrix} \epsilon_0(E_x^2 - \frac{1}{2}E^2) - \lambda & \epsilon_0 E_x E_y & \epsilon_0 E_x E_z \\ \epsilon_0 E_y E_x & \epsilon_0(E_y^2 - \frac{1}{2}E^2) - \lambda & \epsilon_0 E_y E_z \\ \epsilon_0 E_z E_x & \epsilon_0 E_z E_y & \epsilon_0(E_z^2 - \frac{1}{2}E^2) - \lambda \end{vmatrix} = 0. \tag{26}$$

When expanded and reduced by taking account of the relation

$$E_x^2 + E_y^2 + E_z^2 = E^2,$$

Eq. (26) proves equivalent to

$$8\lambda^3 + 4E^2\lambda^2 - 2E^4\lambda - E^6 = 0. \tag{27}$$

The roots of (26) are, therefore,

$$\lambda_a = \frac{\epsilon_0}{2} E^2, \qquad \lambda_b = \lambda_c = -\frac{\epsilon_0}{2} E^2, \tag{28}$$

from which it is apparent that the stress quadric has an axis of symmetry.

Let $\mathbf{n}^{(a)}$ be a unit vector fixing the direction of the principal axis associated with λ_a. According to (18) page 86, the components of $\mathbf{n}^{(a)}$ with respect to an arbitrary reference system must satisfy

$$\begin{aligned} (E_x^2 - E^2)n_x^{(a)} + E_x E_y n_y^{(a)} + E_x E_z n_z^{(a)} &= 0, \\ E_y E_x n_x^{(a)} + (E_y^2 - E^2)n_y^{(a)} + E_y E_z n_z^{(a)} &= 0, \\ E_z E_x n_x^{(a)} + E_z E_y n_y^{(a)} + (E_z^2 - E^2)n_z^{(a)} &= 0. \end{aligned} \tag{29}$$

From the theory of homogeneous equations it is known that the ratios of the unknowns $n_x^{(a)}$, $n_y^{(a)}$, $n_z^{(a)}$ are as the ratios of the minors of the determinant of the system, whence one may easily verify from (29) that

$$n_x^{(a)} : n_y^{(a)} : n_z^{(a)} = E_x : E_y : E_z. \tag{30}$$

The major axis of the electric stress quadric at any point in the field is directed along the vector **E** *at that point.* The stress transmitted across an element of surface whose normal is oriented in this direction is a simple tension,

$$\mathbf{t}^{(a)} = \frac{\epsilon_0}{2} E^2 \mathbf{n}^{(a)}. \tag{31}$$

The stress across any element of surface containing the vector **E**—*i.e.*, an element whose normal is at right angles to the lines of force—is also normal but negative, and corresponds therefore to a compression

$$\mathbf{t}^{(b)} = -\frac{\epsilon_0}{2} E^2 \mathbf{n}^{(b)}, \qquad \mathbf{t}^{(c)} = -\frac{\epsilon_0}{2} E^2 \mathbf{n}^{(c)}. \tag{32}$$

Suppose, finally, that the normal of a surface element in the field is oriented in an arbitrary direction specified by **n**. Let the z-axis of a coordinate system located at the point in question be drawn parallel to **E** and choose the x-axis perpendicular to the plane through **E** and **n**. The angle made by **n** with the direction of **E** will be called θ. Then

$$E_x = E_y = 0, \qquad |\mathbf{E}| = E_z,$$
$$n_x = 0, \qquad n_y = \sin\theta, \qquad n_z = \cos\theta,$$

whence according to Eq. (23) the stress components are

$$(33) \qquad t_x = 0, \qquad t_y = -\frac{\epsilon_0}{2} E^2 \sin\theta, \qquad t_z = \frac{\epsilon_0}{2} E^2 \cos\theta.$$

The absolute value of the stress transmitted across any surface element, whatever its orientation, is therefore

$$(34) \qquad |\mathbf{t}| = \frac{\epsilon_0}{2} E^2.$$

Furthermore **t** lies in the plane of **E** and **n** in a direction such that **E** bisects the angle between **n** and **t** as illustrated in Fig. 17.

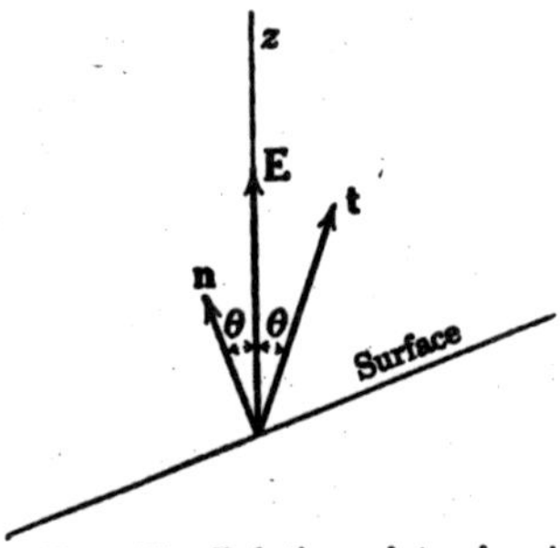

FIG. 17.—Relation of tension **t** transmitted across an element of surface in an electrostatic field to the field intensity **E**.

In the light of this representation it is easy to comprehend the efforts of Faraday and Maxwell to reduce the problem of electric and magnetic fields of force to that of an elastic continuum. To both, the concept of a force propagated from one point in space to another without the intervention of a supporting medium appeared wholly untenable, and in the absence of anything more tangible an all-pervading "ether" was eventually postulated to fill that role. There was then attributed to the stress components of the field, even in space free of matter, a physical reality, and a valiant attempt was made to associate with the ether properties analogous to the strains of elastic media. These efforts did not bear fruit. Subsequent research has shown that electromagnetic phenomena may be formulated without employment of any fixed reference system, and that there are no grounds for the assumption that force can be propagated only by actual contact of contiguous elements of matter or ether. On the modern view, the representation of an electrostatic field in terms of the stress components $S_{jk}^{(e)}$ has no essential physical reality. All that can be said—and all that it is necessary to say—is that the mutual forces between elements of charge can be correctly calculated on the assumption that there exists throughout the field a fictitious state of stress as described.

Since the tensors ${}^2\mathbf{S}^{(m)}$ and ${}^2\mathbf{S}^{(e)}$ are of identical form, the discussion of the foregoing paragraphs applies in its entirety to the field of a stationary distribution of current. The force exerted by external sources on the current within an arbitrary closed surface is obtained by integrating over this surface a stress whose absolute value is

$$|\mathbf{t}| = \frac{1}{2\mu_0} B^2, \tag{35}$$

and whose direction with respect to the orientation of a surface element and the direction of the field $\mathbf{B}$ is as in Fig. 17.

If in Eq. (23) the components S_{jk} are introduced from Tables I and II, it is apparent that the stress transmitted across an element of area whose positive normal is $\mathbf{n}$ may be written vectorially as

$$\mathbf{t}^{(e)} = {}^2\mathbf{S}^{(e)} \cdot \mathbf{n} = \epsilon_0(\mathbf{E} \cdot \mathbf{n})\mathbf{E} - \frac{\epsilon_0}{2} E^2\mathbf{n}, \tag{36}$$

$$\mathbf{t}^{(m)} = {}^2\mathbf{S}^{(m)} \cdot \mathbf{n} = \frac{1}{\mu_0} (\mathbf{B} \cdot \mathbf{n})\mathbf{B} - \frac{1}{2\mu_0} B^2\mathbf{n}. \tag{37}$$

Over any volume bounded by a regular, closed surface Σ we have therefore for *stationary* distributions of charge and current:

$$\int_\Sigma \left[\epsilon_0(\mathbf{E} \cdot \mathbf{n})\mathbf{E} - \frac{\epsilon_0}{2} E^2\mathbf{n}\right] da = \int_V \rho\mathbf{E}\, dv, \tag{38}$$

$$\int_\Sigma \left[\frac{1}{\mu_0} (\mathbf{B} \cdot \mathbf{n})\mathbf{B} - \frac{1}{2\mu_0} B^2\mathbf{n}\right] da = \int_V \mathbf{J} \times \mathbf{B}\, dv. \tag{39}$$

2.6. Electromagnetic Momentum.—In a stationary field the net force transmitted across a closed surface Σ bounding a region containing neither charge nor current is zero. If, however, the field is variable, it is clear from Eq. (25) that this is not the case. How, then, are we to interpret the apparent action of a force on volume elements of empty space? Since the dimensions of $\epsilon_0\mathbf{E} = \mathbf{D}$ are QL^{-2}, and those of $\mathbf{B} = \mu_0\mathbf{H}$ are $MQ^{-1}T^{-1}$, it is evident that the quantity

$$\mathbf{g} = \epsilon_0\mathbf{E} \times \mathbf{B} = \frac{1}{c^2} \mathbf{E} \times \mathbf{H} \tag{40}$$

is dimensionally a *momentum per unit volume*. The dimension Q drops out of the product $\mathbf{D} \times \mathbf{B}$ and this conclusion, therefore, is not simply a consequence of the particular system of units employed. The identity

$$\int_\Sigma {}^2\mathbf{S} \cdot \mathbf{n}\, da = \frac{\partial}{\partial t} \int_V \mathbf{g}\, dv \tag{41}$$

can be interpreted on the hypothesis that there is associated with an electromagnetic field a momentum distributed with a density $\mathbf{g}$. The total momentum of the field contained within V is

$$\mathbf{G} = \int_V \mathbf{g}\, dv \qquad \text{kg.-meters/sec.,} \tag{42}$$

and (41) now states that the force transmitted across Σ is accounted for by an increase in the momentum of the field within Σ. The vector ${}^2\mathbf{S} \cdot \mathbf{n}$ measures the *inward* flow of momentum per unit area through Σ, while the quantity $-S_{jk}$ may be interpreted as *the momentum which in unit time crosses in the j-direction a unit element of surface whose normal is oriented along the k-axis.*

A direct consequence of this hypothesis is the conclusion that Newton's third law and the principle of conservation of momentum are strictly valid only when the momentum of an electromagnetic field is taken into account along with that of the matter which produces it. Let us suppose that within the closed surface Σ there are charges distributed with a density ρ, and that the motion of these charges may be specified by a current density $\mathbf{J}$. The force exerted on the charged matter within Σ is then

$$\mathbf{F}_e + \mathbf{F}_m = \int (\rho \mathbf{E} + \mathbf{J} \times \mathbf{B})\, dv = \frac{d}{dt} \mathbf{G}_{\text{mech}}, \tag{43}$$

where $\mathbf{G}_{\text{mech}}$ is the total linear momentum of the moving, ponderable charges. The conservation of momentum theorem for a system composed of charges and field within a bounded region is, therefore, expressed according to Eq. (25) by

$$\frac{d}{dt}(\mathbf{G}_{\text{mech}} + \mathbf{G}_{\text{electromag}}) = \int_\Sigma {}^2\mathbf{S} \cdot \mathbf{n}\, da. \tag{44}$$

If the surface Σ is extended to enclose the entire field, the right-hand side of (44) must vanish, and in this case

$$\mathbf{G}_{\text{mech}} + \mathbf{G}_{\text{electromag}} = \text{constant}. \tag{45}$$

There appears to be associated with an electromagnetic field an *inertia property* similar to that of ponderable matter.

ELECTROSTATIC ENERGY

2.7. Electrostatic Energy as a Function of Charge Density.—A finite charge concentrated in a region so small as to be of negligible extent relative to other macroscopic dimensions will be referred to as a *point charge*. Now the force exerted on such a point charge q in the field of a stationary charge distribution is $q\mathbf{E}$, and the work done in a displacement of q from a point $\mathbf{r} = \mathbf{r}_1$ to a second point $\mathbf{r} = \mathbf{r}_2$ is

$$W = q \int_{\mathbf{r}_1}^{\mathbf{r}_2} \mathbf{E} \cdot d\mathbf{s}. \tag{1}$$

Since the curl of $\mathbf{E}$ vanishes at every point in an electrostatic field, the vector $\mathbf{E}$ is equal to the negative gradient of a scalar potential ϕ, and we have

$$\mathbf{E} \cdot d\mathbf{s} = -\nabla\phi \cdot d\mathbf{s} = -d\phi, \tag{2}$$

where $d\phi$ is the change in potential along an element $d\mathbf{s}$ of the path of integration. From this it follows clearly that the work done in a displacement of q from $\mathbf{r}_1$ to $\mathbf{r}_2$ is independent of the choice of path and is a function solely of the initial and terminal values of the potential.

$$W = -q \int_{\mathbf{r}_1}^{\mathbf{r}_2} d\phi = q[\phi(\mathbf{r}_1) - \phi(\mathbf{r}_2)]. \tag{3}$$

In particular, the work done in the course of a displacement about a closed path is zero. A field of force is said to be *conservative* if the work done in a displacement of a system of particles from one configuration to another depends only on the initial and final configurations and is independent of the sequence of infinitesimal changes by which the finite displacement is effected. The conservative nature of the electrostatic field is established by (3), or more precisely by the condition $\nabla \times \mathbf{E} = 0$. Displacements and variations in a static field must be understood to occur so slowly as to be equivalent to a sequence of stationary states. Such changes are said in thermodynamics to be *reversible.*

In Chap. III it will be shown that if all the sources of an electrostatic field are located at finite distances from some arbitrary origin the potential and field intensities become vanishingly small at points which are sufficiently remote. The work done by a charge q as it recedes from an initial point $\mathbf{r} = \mathbf{r}_1$ to $\mathbf{r}_2 = \infty$ is, therefore,

$$W = q\phi(\mathbf{r}). \tag{4}$$

Obviously the scalar potential itself may be interpreted as the work done *against* the forces of the field in bringing a unit charge from infinity to the point $\mathbf{r}$, or as the work returned *by* the system as a unit element of charge recedes to infinity.

$$\phi(x, y, z) = -\int_{\infty}^{(x,y,z)} \mathbf{E} \cdot d\mathbf{s}. \tag{5}$$

We shall use the term *energy* of an electrostatic system somewhat loosely to mean the work done on the system in carrying its elements of charge from infinity to the specified distribution by a sequence of reversible steps. It will be assumed that the temperature of all dielectric or magnetic matter in the field is held constant.[1]

The energy of a point charge q_2 in the field of a single point source q_1 is

$$U = q_2\phi_{21}, \tag{6}$$

where ϕ_{21} is the potential at q_2 due to q_1. Now the work done in bringing q_2 from infinity to a terminal point in the field of q_1 would be returned were q_1 allowed next to recede to infinity, and therefore

$$U = q_1\phi_{12}. \tag{7}$$

The mutual energy of the two charges may consequently be expressed by the symmetrical relation

$$U = \tfrac{1}{2}(q_1\phi_{12} + q_2\phi_{21}). \tag{8}$$

If first q_2 and then q_3 be introduced into the field of q_1, the energy is

$$U = q_2\phi_{21} + q_3(\phi_{31} + \phi_{32}), \tag{9}$$

which in virtue of the reciprocal relations between pairs is equivalent to

$$U = \tfrac{1}{2}(\phi_{12} + \phi_{13})q_1 + \tfrac{1}{2}(\phi_{21} + \phi_{23})q_2 + \tfrac{1}{2}(\phi_{31} + \phi_{32})q_3. \tag{10}$$

By induction it follows that the energy of a closed system of n point charges is

$$U = \frac{1}{2}\sum_{\substack{i=1 \\ i\neq j}}^{n}\sum_{j=1}^{n} \phi_{ij}q_i = \frac{1}{2}\sum_{i=1}^{n} \phi_i q_i, \tag{11}$$

where ϕ_i is the potential at q_i due to the remaining $n - 1$ charges of the system.

Note that (11) is valid only if the system is complete, or closed. If on the contrary the n charges are situated in an external field of potential ϕ_0, a term appears which does not involve the factor $\frac{1}{2}$. In this case

$$U = \sum_{i=1}^{n} \phi_0 q_i + \frac{1}{2}\sum_{i=1}^{n} \phi_i q_i. \tag{12}$$

Let us consider now a region of space V within which there are to be found fixed conductors and dielectric matter. Within the dielectrics charge is distributed with a volume density ρ, while over the surfaces of these dielectrics and on the conductors there may be thin layers of charge of surface density ω. At a point (x, y, z) within V the potential of the distribution plus that of possible sources situated outside V is ϕ. The work required to increase the charge at (x, y, z) by an infinitesimal

[1] The energy in question is in fact the *free energy* in a thermodynamic sense.

amount δq is $\phi\ \delta q$, and the increase in energy resulting from an increase in the volume density of charge by an amount $\delta\rho$, or the surface density by an amount $\delta\omega$, at every point in V is

$$\delta U_1 = \int \phi\ \delta\rho\ dv + \int \phi\ \delta\omega\ da. \tag{13}$$

The second integral is to be extended over all surfaces within V bearing charge. On the other hand, the addition of an element of charge δq at (x, y, z) increases the potential at all points *both inside and outside* V by an amount $\delta\phi$, with a resultant increase of the energy of the charges already existing within V of

$$\delta U_2 = \int \rho\ \delta\phi\ dv + \int \omega\ \delta\phi\ da. \tag{14}$$

The work δU_1 done on the system in building up the charge density is equal to the potential energy δU_2 stored in the field provided the system is closed; provided, that is, that the region V of integration is extended to include *all* charges contributing to the field. In that case

$$\delta U = \delta U_1 = \delta U_2,$$

$$\delta U = \frac{1}{2}\int (\phi\ \delta\rho + \rho\ \delta\phi)\ dv + \frac{1}{2}\int (\phi\ \delta\omega + \omega\ \delta\phi)\ da, \tag{15}$$

which upon integration leads to

$$U = \frac{1}{2}\int \phi\rho\ dv + \frac{1}{2}\int \phi\omega\ da \tag{16}$$

as the electrostatic energy of a charge system referred to the zero state $\rho = \omega = 0$.[1]

2.8. Electrostatic Energy as a Function of Field Intensity.—Let us imagine for a moment that conductors have been eliminated from the field and that all surface discontinuities at the boundaries of dielectrics are replaced by thin but continuous transition layers. Charge is distributed throughout the dielectric with a density ρ, but we shall assume that this distribution is confined to a region of finite extent: the potential and intensities of the field vanish at infinity. Regions free of matter are, of course, to be considered as dielectrics of unit inductive capacity. The work required to increase the charge density at every point in the field by an amount $\delta\rho$ is

$$\delta U = \int \phi\ \delta\rho\ dv, \tag{17}$$

where ϕ is the potential due to the initial distribution ρ. Now the increment in charge density is related to a variation of the vector **D**

[1] The convergence of these integrals will be demonstrated in Chap. III.

by the equation

$$\delta\rho = \delta(\nabla \cdot \mathbf{D}) = \nabla \cdot (\delta\mathbf{D}). \tag{18}$$

Furthermore,

$$\phi\nabla \cdot (\delta\mathbf{D}) = \nabla \cdot (\phi\, \delta\mathbf{D}) - \delta\mathbf{D} \cdot \nabla\phi = \nabla \cdot (\phi\, \delta\mathbf{D}) + \mathbf{E} \cdot \delta\mathbf{D}, \tag{19}$$

and hence (17) is equivalent to

$$\delta U = \int \mathbf{E} \cdot \delta\mathbf{D}\, dv + \int \nabla \cdot (\phi\, \delta\mathbf{D})\, dv, \tag{20}$$

or, upon application of the divergence theorem, to

$$\delta U = \int_V \mathbf{E} \cdot \delta\mathbf{D}\, dv + \int_S \phi\, \delta\mathbf{D} \cdot \mathbf{n}\, da, \tag{21}$$

where S is a closed surface bounding a volume V. This region V need not include all the charges that contribute to the field. If, however, we allow the surface S to expand into a sphere of infinite radius about some arbitrary origin, the contribution of the surface integral vanishes; for ϕ will be shown later to diminish as $1/r$ at sufficiently large distances from the origin, and $\mathbf{D}$ as $1/r^2$. The surface S increases with increasing radius as r^2, and the surface integral therefore vanishes as $1/r$. The increment of energy stored in the electrostatic field can be calculated from the integral

$$\delta U = \int \mathbf{E} \cdot \delta\mathbf{D}\, dv \tag{22}$$

extended over all space.

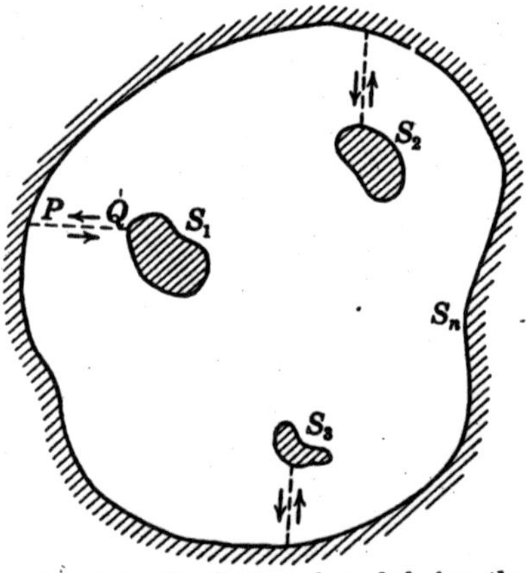

FIG. 18.—Conductors bounded by the surfaces S_1, S_2, . . . , embedded in a dielectric medium.

In practice the charge is rarely distributed throughout the volume of a dielectric but is spread in a thin layer of density ω over the surfaces of conductors. These conductors may be considered as electrodes of condensers, and the increase in the energy of the field is the result, for example, of work done by the electromotive forces of batteries in the process of building up the charge on the electrodes. Let us suppose that there are n conductors, whose surfaces we shall denote by S_i, embedded in a dielectric of infinite extent. One of these surfaces, let us say S_n, will be assumed to enclose all the others as illustrated in Fig. 18. To ensure the continuity of the potential and field necessary for the application of the divergence theorem, surfaces of abrupt change in the properties of the dielectric may again be replaced by thin transition layers without in any way affecting the generality of the results. The work required to increase the charge density on the surfaces S_i by an amount $\delta\omega$ is

$$\delta U = \sum_{i=1}^{n} \int_{S_i} \phi \, \delta\omega \, da_i. \tag{23}$$

The sum of these n surface integrals can be represented as a single integral by the artifice of drawing finlike surfaces from the interior conductors to S_n as indicated in Fig. 18. The integration starts at a point P on S_n, extends on one side of the fin to Q, then over S_1 and then returns to P over the opposite face of the fin. The surface PQ carries no charge and contributes nothing to the integral. The interior of S_n is thus reduced to a simply connected region[1] bounded by a single surface $S = \sum_{i=1}^{n} S_i$.

$$\delta U = \int_S \phi \, \delta\omega \, da. \tag{24}$$

In Sec. 1.13 it was shown that at any surface of discontinuity the normal and tangential components of the vectors $\mathbf{D}$ and $\mathbf{E}$ satisfy the conditions

$$\mathbf{n} \cdot (\mathbf{D}_2 - \mathbf{D}_1) = \omega, \qquad \mathbf{n} \times (\mathbf{E}_2 - \mathbf{E}_1) = 0. \tag{25}$$

The positive normals to the surfaces S_i are directed into the dielectric, and since $\mathbf{n}$ in Sec. 1.13 was drawn from medium (1) into medium (2), the index (1) now denotes the interior of the conductors whereas (2) refers to the dielectric. In Chap. III we shall have occasion to consider in some detail the electrostatic properties of conductors, but for the present we need only accept the elementary fact that the field at any interior point of a conductor is zero. Were it otherwise, a movement of free charge would occur, contrary to the assumption of a stationary state. The interior and surface of a conductor are, therefore, a region of constant potential. The charge density at interior points is zero [Eq. (21), page 15] and whatever charge the conductor carries is distributed on the surface in such a way as to bring about the vanishing of the interior field. Since $\mathbf{E}_1 = \mathbf{D}_1 = 0$,

$$\mathbf{n} \cdot \mathbf{D}_2 = \omega, \qquad \mathbf{n} \times \mathbf{E}_2 = 0; \tag{26}$$

at a point just outside the conductor the tangential component of $\mathbf{E}$ is zero and the normal component of $\mathbf{D}$ is equal to the surface density of charge.

Upon introducing the first of these relations into (24) and dropping the index, we obtain

$$\delta U = \int_S \phi \, \delta\mathbf{D} \cdot \mathbf{n} \, da. \tag{27}$$

[1] See Sec. 4.2, p. 227.

To express this result as an integral extended throughout the volume of dielectric bounded by the surfaces S_i we again apply the divergence theorem, but must note that **n** in (27) is directed *into* the dielectric; the sign must therefore be reversed.

$$(28) \quad \delta U = -\int \nabla \cdot (\phi\, \delta \mathbf{D})\, dv = \int \mathbf{E} \cdot \delta \mathbf{D}\, dv - \int \phi \nabla \cdot (\delta \mathbf{D})\, dv.$$

If the density of charge throughout the dielectric is zero or constant, $\nabla \cdot (\delta \mathbf{D}) = 0$, and we find as in (22) that the work expended by the batteries in building up the electrode charges by an amount $\delta\omega$ is

$$(29) \qquad \delta U = \int \mathbf{E} \cdot \delta \mathbf{D}\, dv,$$

an integral extended over the entire space occupied by the field, and that this work is stored in the field as electrostatic energy.

Since the energy δU appears to be stored in the field, as the potential energy of an extended spring is in some manner stored within it, it is not unreasonable to suppose that the electrostatic energy is distributed throughout the field with a *density*

$$(30) \qquad \delta u = \mathbf{E} \cdot \delta \mathbf{D}. \qquad \text{joules/meter}^3$$

It is difficult either to justify or disprove such a hypothesis. The transformation from a surface integral to a volume integral is obviously not unique, for there might be added to $\phi\, \delta\mathbf{D}$ in (27) any vector whose normal component integrated over S is zero. Furthermore, it may be questioned whether the term "energy density" has any physical significance. Energy is a function of the configuration of a system as a whole. The objection has been stated in a rather quaint way by Mason and Weaver,[1] who suggest that it is no more sensible to inquire about the location of energy than to declare that the beauty of a painting is distributed over the canvas in a specified manner. However ingenious, such an analogy seems not entirely well founded. The energy of an inhomogeneously stressed elastic medium is certainly concentrated principally in regions of greatest strain and in this case the elastic energy per unit volume has a very definite physical sense. Granting that the analogy of the electrostatic to the elastic field is not a close one, and that we can be *certain* only of the correctness of the expression (29) integrated over the entire field, it is nevertheless plausible to assume that the energy is localized in the more intense regions of the field in the manner prescribed by (30).

To find the total energy stored in a field, the increment δU must be integrated from the initial state $\mathbf{D} = 0$ to the final value $\mathbf{D}$.

[1] MASON and WEAVER, "The Electromagnetic Field," p. 266, University of Chicago Press, 1929.

$$U = \int_V \int_0^D \mathbf{E} \cdot \delta\mathbf{D}\, dv. \tag{31}$$

In case the medium is *isotropic* and *linear*, such that ϵ in the relation $\mathbf{D} = \epsilon\mathbf{E}$ is a function possibly of position but not of $\mathbf{E}$, we have

$$\mathbf{E} \cdot \delta\mathbf{D} = \frac{\epsilon}{2}\,\delta\mathbf{E}^2,$$

and hence

$$U = \frac{1}{2}\int_V \epsilon\mathbf{E}^2\, dv. \tag{32}$$

2.9. A Theorem on Vector Fields.—Let $\mathbf{P}$ and $\mathbf{Q}$ be two vector functions of position which throughout all space satisfy the conditions

$$\nabla \times \mathbf{P} = 0, \qquad \nabla \cdot \mathbf{Q} = 0, \tag{33}$$

and which are continuous and have continuous derivatives everywhere except on a closed, regular surface S_1. The transition of the tangential components of $\mathbf{P}$ and of the normal components of $\mathbf{Q}$ across the surface S_1 is assumed continuous, but arbitrary discontinuities are permissible in the normal components of $\mathbf{P}$ and the tangential components of $\mathbf{Q}$. The prescribed conditions over S_1 are, therefore,

$$\mathbf{n} \times (\mathbf{P}_+ - \mathbf{P}_-) = 0, \qquad \mathbf{n} \cdot (\mathbf{Q}_+ - \mathbf{Q}_-) = 0. \tag{34}$$

The unit normal $\mathbf{n}$ is drawn outward from S_1 and vectors in the immediate neighborhood of this outer face are denoted by the subscript $+$, whereas those located just inside the surface are denoted by the subscript $-$. Finally, it is assumed that the sources of the fields $\mathbf{P}$ and $\mathbf{Q}$ are located at finite distances from an arbitrary origin and that $\mathbf{P}$ and $\mathbf{Q}$ vanish at infinity such that

$$\lim_{r\to\infty} r\mathbf{P} = 0, \qquad \lim_{r\to\infty} r\mathbf{Q} = 0. \tag{35}$$

Then it can be shown that *the integral over all space of the scalar product of an irrotational vector* $\mathbf{P}$ *and a solenoidal vector* $\mathbf{Q}$ *is zero, provided* $\mathbf{P}$ *and* $\mathbf{Q}$ *and their derivatives are continuous everywhere except on a finite number of closed surfaces across which the discontinuities are as specified in* (34). For since $\nabla \times \mathbf{P} = 0$, we may write $\mathbf{P} = -\nabla\phi$ and, hence,

$$\mathbf{P} \cdot \mathbf{Q} = -\nabla \cdot (\phi\mathbf{Q}) + \phi\nabla \cdot \mathbf{Q}. \tag{36}$$

We shall denote the volume enclosed by the surface S_1 by V_1, and that exterior to it by V_2, so that the complete field of $\mathbf{P}$ and $\mathbf{Q}$ is $V_1 + V_2$. The last term of (36) vanishes and therefore

$$\int_{V_1+V_2} \mathbf{P} \cdot \mathbf{Q}\, dv = -\int_{V_1} \nabla \cdot (\phi\mathbf{Q})\, dv - \int_{V_2} \nabla \cdot (\phi\mathbf{Q})\, dv. \tag{37}$$

To apply the divergence theorem to this expression we observe that V_1

is bounded by S_1, and that V_2 is bounded on the interior by S_1 and on the exterior by a surface S which recedes to infinity. In view of the behavior of **P** and **Q** at infinity as specified by (35), the integral of $\phi\mathbf{Q}\cdot\mathbf{n}$ over S is zero. The positive normal to a closed surface is conventionally directed outward from the enclosed volume; the two integrals of $\phi\mathbf{Q}\cdot\mathbf{n}$ over S_1 are therefore of opposite sign but of equal magnitude in virtue of (34). Equation (37) transforms to

$$(38)\qquad \int_{V_1+V_2} \mathbf{P}\cdot\mathbf{Q}\,dv = -\int_{S_1} \phi\mathbf{Q}_+\cdot\mathbf{n}\,da + \int_{S_1} \phi\mathbf{Q}_-\cdot\mathbf{n}\,da = 0,$$

as stated. The extension of this theorem to a finite number of closed surfaces S_i of discontinuity is elementary.

2.10. Energy of a Dielectric Body in an Electrostatic Field.—The useful theorem of the preceding section may be applied to the solution of the following problem. Let us suppose that an electrostatic field $\mathbf{E}_1$ has been established in a dielectric medium. To simplify matters we shall assume that this medium is *isotropic* and *linear*, and hence characterized by an inductive capacity ϵ_1 which is either constant or at the most a scalar function of position. A nonconducting body whose inductive capacity is ϵ_2, is now introduced into the field $\mathbf{E}_1$, while the sources of $\mathbf{E}_1$ are maintained strictly constant. We wish to know the energy of the foreign dielectric body due to its position in the field.

The initial energy U_1, representing the total work done in establishing the initial field, is obtained by evaluating

$$(39)\qquad U_1 = \frac{1}{2}\int \mathbf{E}_1\cdot\mathbf{D}_1\,dv$$

over all space. After the introduction of the body the modified field at any point is **E**, and the difference $\mathbf{E}_2 = \mathbf{E} - \mathbf{E}_1$ is thus the field resulting from the polarization of the body. The volume occupied by the foreign body we denote by V_1, that of the medium exterior to it by V_2. The energy of the field in this new state is

$$(40)\qquad U_2 = \frac{1}{2}\int_{V_1+V_2} \mathbf{E}\cdot\mathbf{D}\,dv,$$

and the change

$$(41)\qquad U = U_2 - U_1 = \frac{1}{2}\int_{V_1+V_2} (\mathbf{E}\cdot\mathbf{D} - \mathbf{E}_1\cdot\mathbf{D}_1)\,dv$$

must be the energy of the body in the external field $\mathbf{E}_1$, and consequently equal to the work done in introducing it. Equation (41) is equivalent to

$$(42)\qquad U = \frac{1}{2}\int_{V_1+V_2} \mathbf{E}\cdot(\mathbf{D} - \mathbf{D}_1)\,dv + \frac{1}{2}\int_{V_1+V_2} (\mathbf{E} - \mathbf{E}_1)\cdot\mathbf{D}_1\,dv.$$

The curl of **E** is zero everywhere, and the divergence of $\mathbf{D} - \mathbf{D}_1$ is zero since the initial source distribution is fixed. Across the surface which bounds the medium ϵ_2 from the medium ϵ_1 the conditions

$$\mathbf{n} \times (\mathbf{E}_+ - \mathbf{E}_-) = 0, \qquad \mathbf{n} \cdot [(\mathbf{D} - \mathbf{D}_1)_+ - (\mathbf{D} - \mathbf{D}_1)_-] = 0, \tag{43}$$

are satisfied provided the surface carries no charge. Then by the theorem of Sec. 2.9 the first integral of (42) is zero, leaving

$$U = \frac{1}{2}\int_{V_1} (\mathbf{E} - \mathbf{E}_1) \cdot \mathbf{D}_1\, dv + \frac{1}{2}\int_{V_2} (\mathbf{E} - \mathbf{E}_1) \cdot \mathbf{D}_1\, dv. \tag{44}$$

Since $\mathbf{D}_1 = \epsilon_1 \mathbf{E}_1$, the second integral of (44) is equivalent to

$$\frac{1}{2}\int_{V_2} (\mathbf{E} - \mathbf{E}_1) \cdot \mathbf{D}_1\, dv = \frac{1}{2}\int_{V_2} (\mathbf{D} - \mathbf{D}_1) \cdot \mathbf{E}_1\, dv. \tag{45}$$

The conditions of Sec. 2.9 are satisfied by $\nabla \times \mathbf{E}_1 = 0$, $\nabla \cdot (\mathbf{D} - \mathbf{D}_1) = 0$, so that

$$\int_{V_1+V_2} \mathbf{E}_1 \cdot (\mathbf{D} - \mathbf{D}_1)\, dv = \int_{V_1} \mathbf{E}_1 \cdot (\mathbf{D} - \mathbf{D}_1)\, dv + \int_{V_2} \mathbf{E}_1 \cdot (\mathbf{D} - \mathbf{D}_1)\, dv = 0, \tag{46}$$

and hence

$$\frac{1}{2}\int_{V_2} \mathbf{E}_1 \cdot (\mathbf{D} - \mathbf{D}_1)\, dv = -\frac{1}{2}\int_{V_1} \mathbf{E}_1 \cdot (\mathbf{D} - \mathbf{D}_1)\, dv. \tag{47}$$

Upon introducing (47) into (44), we obtain an expression for the energy of a dielectric body embedded in a dielectric medium in terms of an integral extended, not over all space, but *over its own volume alone.*

$$U = \frac{1}{2}\int_{V_1} (\mathbf{E}_2 \cdot \mathbf{D}_1 - \mathbf{E}_1 \cdot \mathbf{D}_2)\, dv, \tag{48}$$

or, since $\mathbf{D}_1 = \epsilon_1 \mathbf{E}_1$, $\mathbf{D} = \epsilon_2 \mathbf{E}$ within V_1,

$$U = \frac{1}{2}\int_{V_1} (\mathbf{E} \cdot \mathbf{D}_1 - \mathbf{E}_1 \cdot \mathbf{D})\, dv = \frac{1}{2}\int_{V_1} (\epsilon_1 - \epsilon_2)\mathbf{E} \cdot \mathbf{E}_1\, dv. \tag{49}$$

In case the external medium is free space, the inductive capacity ϵ_1 reduces to ϵ_0. Then, since the resultant field within the body is related to its polarization according to Sec. 1.6, by

$$\mathbf{D} = \epsilon_0 \mathbf{E} + \mathbf{P} = \epsilon_2 \mathbf{E}, \qquad \mathbf{P} = (\epsilon_2 - \epsilon_0)\mathbf{E}, \tag{50}$$

U may be written

$$U = -\frac{1}{2}\int_{V_1} \mathbf{P} \cdot \mathbf{E}_1\, dv. \tag{51}$$

The potential energy of a dielectric body in free space and in a fixed external field $\mathbf{E}_1$ *is equal to* $-\frac{1}{2}\mathbf{P}\cdot\mathbf{E}_1$ *per unit volume.*

It is the variation of this potential energy U expressed by (49) or (51) with respect to a small displacement that will lead us to the mechanical forces exerted on polarized dielectrics. It will be observed that the energy is negative if $\epsilon_2 > \epsilon_1$, and decreases in this case with decreasing ϵ_1, increasing ϵ_2, or increasing field intensity $\mathbf{E}_1$. We may anticipate, therefore, that if $\epsilon_2 > \epsilon_1$, the body will be impelled toward regions of intense field or diminishing inductive capacity ϵ_1. On the other hand if $\epsilon_1 > \epsilon_2$, as might be the case of a solid immersed in a liquid of high inductive capacity, the forces exerted on the body will tend to expel it from the field.

A direct consequence of (49) is the theorem that *any increase in the inductive capacity of a dielectric results in a decrease in the total energy of the field.* Let us suppose again that $\mathbf{E}_1$ is the field of a *fixed* set of charges in a dielectric medium whose inductive capacity is $\epsilon = \epsilon(x, y, z)$. If at every point ϵ is increased by an infinitesimal amount $\delta\epsilon$, the consequent variation in the electrostatic energy will according to (49) be equal to

$$\delta U = -\frac{1}{2}\int \delta\epsilon\, E^2\, dv, \tag{52}$$

for the product $\delta\epsilon\mathbf{E}\cdot\mathbf{E}_1$ will then differ from $\delta\epsilon\, E^2$ by an infinitesimal of second order.

The energy of an electrostatic field is now completely determined by the distribution of charge ρ and ω, and by the inductive capacity $\epsilon(x, y, z)$. Equation (52) expresses the variation in energy resulting from a slight change in the properties of the dielectric, in the course of which the charges are held constant.

$$U = -\frac{1}{2}\int\int_0^\epsilon E^2\, d\epsilon\, dv, \qquad \text{(constant charge)}. \tag{53}$$

The variation

$$\delta U = \int \phi\, \delta\rho\, dv + \int \phi\, \delta\omega\, da = \int \mathbf{E}\cdot\delta\mathbf{D}\, dv \tag{54}$$

expresses, on the other hand, the increment of energy resulting from a small change in the density of charge, in the course of which the properties of the dielectric are held constant.

2.11. Thomson's Theorem.—*Charges placed on a system of fixed conductors embedded in a dielectric will distribute themselves on the surfaces of these conductors such that the energy of the resultant electrostatic field is a minimum.*

The proof of this and the following theorems will be confined to the case of a linear, isotropic dielectric. Let us suppose that there are n conductors bounded by the surfaces S_i, $(i = 1, 2, \ldots, n)$, each bearing a charge q_i. Discontinuities in the properties of the dielectric may be replaced by thin transition layers without affecting the generality of the theorem and we shall, therefore, assume the inductive capacity ϵ of the medium to be a continuous but otherwise arbitrary function of position. It may also be assumed that there is a density of charge throughout the volume of the dielectric, although such a condition does not often occur in practice.

At every point within the dielectric the field of the charges in equilibrium must satisfy the conditions:

$$\nabla \cdot \mathbf{D} = \rho, \qquad \nabla \times \mathbf{E} = 0, \qquad \mathbf{E} = -\nabla\phi; \tag{55}$$

over the surface of each conductor S_i

$$\phi_i = \text{constant}, \qquad \int_{S_i} \mathbf{D} \cdot \mathbf{n}\, da = q_i; \tag{56}$$

at infinity the potential vanishes as $1/r$.[1]

Suppose that ϕ', $\mathbf{E}'$, $\mathbf{D}'$ is any other possible electrostatic field; it satisfies the conditions (55), but not necessarily (56), and is known to differ somewhere, if not everywhere, from ϕ, $\mathbf{E}$, $\mathbf{D}$. Since the volume distribution ρ and the *total* charge on the conductors is fixed we have

$$\nabla \cdot (\mathbf{D}' - \mathbf{D}) = 0, \qquad \int_{S_i} (\mathbf{D}' - \mathbf{D}) \cdot n\, da = 0. \tag{57}$$

If U and U' are the electrostatic energies of the two fields, their difference is

$$U' - U = \frac{1}{2}\int \mathbf{E}' \cdot \mathbf{D}'\, dv - \frac{1}{2}\int \mathbf{E} \cdot \mathbf{D}\, dv, \tag{58}$$

or, since $\mathbf{D}' = \epsilon\mathbf{E}'$,

$$U' - U = \frac{1}{2}\int (\mathbf{E}' - \mathbf{E}) \cdot (\mathbf{D}' - \mathbf{D})\, dv + \int \mathbf{E} \cdot (\mathbf{D}' - \mathbf{D})\, dv. \tag{59}$$

The second term on the right vanishes, for we see that on putting $\mathbf{E} = \mathbf{P}$, $\mathbf{D}' - \mathbf{D} = \mathbf{Q}$, the conditions of the theorem demonstrated in Sec. 2.9 are satisfied. There remains

$$U' - U = \frac{1}{2}\int \epsilon(E' - E)^2\, dv, \tag{60}$$

which is an essentially positive quantity. The theorem is proved, for if $\mathbf{E}'$ differs in any region of space from $\mathbf{E}$ the resultant energy U' will be greater than U. The condition of electrostatic equilibrium is character-

[1] See Sec. 3.5, p. 167.

ized by a minimum value of electrostatic energy and one therefore concludes that for the determination of equilibrium *the energy U plays the same role in electrostatics as the potential energy in mechanics.*

2.12. Earnshaw's Theorem.—*A charged body placed in an electrostatic field cannot be maintained in stable equilibrium under the influence of the electric forces alone.*

We shall suppose that the initial field is generated by a set of charges q_i, $(i = 1, \ldots, n)$, distributed on n fixed conductors whose bounding surfaces are denoted by S_i. These conductors are embedded in a dielectric whose inductive capacity ϵ may be a continuous function of position but in which there is no volume distribution of charge. We note first that neither the potential nor any of its partial derivatives can assume a maximum or a minimum value at a point within the dielectric; for if ϕ is to be an absolute maximum it is necessary that the three partial derivatives $\partial^2\phi/\partial x^2$, $\partial^2\phi/\partial y^2$, $\partial^2\phi/\partial z^2$ shall all be negative at the point in question, but this condition is incompatible with Laplace's equation

$$\text{(61)} \qquad \frac{\partial^2\phi}{\partial x^2} + \frac{\partial^2\phi}{\partial y^2} + \frac{\partial^2\phi}{\partial z^2} = 0.$$

Likewise the existence of an absolute minimum requires that these three derivatives shall be positive, which again is inconsistent with (61). The same argument applies also to the derivatives of the potential.

Let us suppose now that a charge q_0 is placed on a conducting surface S_0. The distributions on all the conductors are momentarily assumed to be fixed and S_0 is introduced into the field of the other n charges. If ω_0 is the surface density of charge on S_0, the energy of this conductor is

$$\text{(62)} \qquad U_0 = \frac{1}{2}\int_{S_0} \phi\omega_0 \, da,$$

where ϕ is the potential of the initial field. Let x, y, z be the coordinates of any point which is fixed with respect to S_0 and ξ, η, ζ, those of any point on the surface. The potential on S_0 due to the other n charges may be represented by a Taylor series in terms of its value and the values of its derivatives at x, y, z,

$$\text{(63)} \qquad \phi(\xi, \eta, \zeta) = \phi(x, y, z) + \frac{\partial\phi}{\partial x}(\xi - x) + \frac{\partial\phi}{\partial y}(\eta - y) + \frac{\partial\phi}{\partial z}(\zeta - z) + \cdots$$

hence the energy, too, may be referred to the potential and its derivatives at this point. But since ϕ cannot be a minimum at x, y, z, it is always possible to displace the conductor S_0 in such a way that the energy U is decreased. If after this displacement the charges, which thus far have been assumed to be "frozen" on the surfaces S_0, S_i, are released these surfaces will again become equipotentials and by Thomson's theorem the energy of the field will be still further diminished. A minimum value of the energy function $U_0(x, y, z)$ does not exist in the electrostatic field and consequently the conductor S_0 is never in static equilibrium.

2.13. Theorem on the Energy of Uncharged Conductors.—*The introduction of an uncharged conductor into the field of a fixed set of charges diminishes the total energy of the field.*

The conditions are the same as in the previous theorem but the surface S_0 now carries no charge. Let **E, D** be the vectors characterizing the field before, **E′, D′** the field after the introduction of the conductor S_0. The change in energy is

$$U - U' = \frac{1}{2}\int \mathbf{E}\cdot\mathbf{D}\,dv - \frac{1}{2}\int \mathbf{E}'\cdot\mathbf{D}'\,dv. \tag{64}$$

The volume integrals are to be extended through all space, but since the field vanishes in the interior of the surfaces S_0, S_i, the contribution from these regions is zero. Let V be the volume of the field exterior to the n conductors S_i *before* the introduction of S_0, and V_0 the volume bounded by S_0. Then

$$V_1 = V - V_0 \tag{65}$$

represents the volume of the region occupied by the field *after* the introduction of S_0, and Eq. (64) may be written in the form

$$\begin{aligned} U - U' &= \frac{1}{2}\int_V \mathbf{E}\cdot\mathbf{D}\,dv - \frac{1}{2}\int_{V_1} \mathbf{E}'\cdot\mathbf{D}'\,dv \\ &= \frac{1}{2}\int_{V_0} \mathbf{E}\cdot\mathbf{D}\,dv + \frac{1}{2}\int_{V_1} (\mathbf{E}\cdot\mathbf{D} - \mathbf{E}'\cdot\mathbf{D}')\,dv \\ &= \frac{1}{2}\int_{V_0} \mathbf{E}\cdot\mathbf{D}\,dv + \frac{1}{2}\int_{V_1} (\mathbf{E} - \mathbf{E}')\cdot(\mathbf{D} - \mathbf{D}')\,dv \\ &\qquad + \int_{V_1} \mathbf{E}'\cdot(\mathbf{D} - \mathbf{D}')\,dv. \end{aligned} \tag{66}$$

The last integral on the right can be shown to vanish. For

$$\mathbf{E}'\cdot(\mathbf{D} - \mathbf{D}') = -\nabla\phi'\cdot(\mathbf{D} - \mathbf{D}') = -\nabla\cdot[\phi'(\mathbf{D} - \mathbf{D}')] + \phi'\,\nabla\cdot(\mathbf{D} - \mathbf{D}'),$$

and $\nabla\cdot(\mathbf{D} - \mathbf{D}') = 0$. By the divergence theorem,

$$\begin{aligned} \int_{V_1} \mathbf{E}'\cdot(\mathbf{D} - \mathbf{D}')\,dv &= \int_{V_1} \nabla\cdot[\phi'(\mathbf{D}' - \mathbf{D})]\,dv \\ &= \sum_{i=1}^{n} \phi_i' \int_{S_i} (\mathbf{D}' - \mathbf{D})\cdot\mathbf{n}\,da, \end{aligned} \tag{67}$$

where the ϕ_i' are the potentials of the conducting surfaces S_i. The total charge on each surface is constant, so that

$$\int_{S_i} \mathbf{D}' \cdot \mathbf{n}\, da = \int_{S_i} \mathbf{D} \cdot \mathbf{n}\, da = q_i, \tag{68}$$

and hence

$$\int_{S_i} (\mathbf{D}' - \mathbf{D}) \cdot \mathbf{n}\, da = 0. \tag{69}$$

The difference between the initial and final energies is, therefore,

$$U - U' = \frac{1}{2} \int_{V_0} \epsilon E^2\, dv + \frac{1}{2} \int_{V_1} \epsilon (E - E')^2\, dv, \tag{70}$$

an essentially positive quantity.

MAGNETOSTATIC ENERGY

2.14. Magnetic Energy of Stationary Currents.—Let us consider a stationary distribution of current confined to a finite region of space. This current may be supported by conducting matter, or result from the convection of charges in free space. The equation of continuity reduces to $\nabla \cdot \mathbf{J} = 0$, in virtue of which we may imagine the distribution to be resolved into current lines closing upon themselves. A current tube or filament may be constructed from the current lines passing through an infinitesimal element of area. The tube is bounded by those lines which pass through points on the contour of the element. At every point on the surface of a tube the flow is tangential; no current leaves the tube and consequently the net charge transported across every cross section of the tube in a given time is the same.

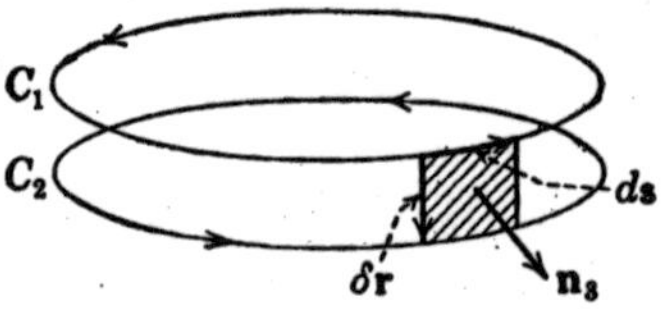

FIG. 19.—Illustrating displacement of a current filament.

We shall calculate first the potential energy of a single, isolated current filament in the field $\mathbf{B}$ of fixed external sources. The current carried by the filament is I and it follows an initial contour designated in Fig. 19 as C_1. The force exerted by the field on a linear element of the filament is

$$\mathbf{f} = I\, d\mathbf{s} \times \mathbf{B}. \tag{1}$$

Suppose now that the filament C_1 is translated and deformed in such a manner that every element $d\mathbf{s}$ is displaced by the infinitesimal amount $\delta\mathbf{r}$ into the contour C_2. The displacement $\delta\mathbf{r}$ is assumed to be a continuous function of position about C_1 but is otherwise arbitrary. The work done by the force (1) in the displacement of the element $d\mathbf{s}$ is

$$\mathbf{f} \cdot \delta\mathbf{r} = I(d\mathbf{s} \times \mathbf{B}) \cdot \delta\mathbf{r} = I\mathbf{B} \cdot (\delta\mathbf{r} \times d\mathbf{s}). \tag{2}$$

Let S_1 be any regular surface spanning the contour C_1. A second surface S_2 is drawn to span C_2, but in such a way as to pass through C_1 and then coincide with S_1. The surface

$$S_3 = S_2 - S_1 \tag{3}$$

is, therefore, a band or ribbon of width $\delta\mathbf{r}$ bounded by the curves C_1 and C_2. If da_3 is an element of S_3 we have clearly

$$\delta\mathbf{r} \times d\mathbf{s} = \mathbf{n}_3\, da_3, \tag{4}$$

where $\mathbf{n}_3$ is the positive unit normal to S_3. The positive faces of S_1 and S_2 are determined by the usual convention that an observer, circulating about the contours in the direction of the current, shall have the positive face at his left. The magnetic fluxes threading C_1 and C_2 are

$$\Phi_1 = \int_{S_1} \mathbf{B} \cdot \mathbf{n}_1\, da_1, \qquad \Phi_2 = \int_{S_2} \mathbf{B} \cdot \mathbf{n}_2\, da_2; \tag{5}$$

hence the net change in flux resulting from the displacement is

$$\delta\Phi = \Phi_2 - \Phi_1 = \int_{S_3} \mathbf{B} \cdot \mathbf{n}_3\, da_3. \tag{6}$$

On the other hand the total work done by the mechanical forces is obtained by integrating (2) around the closed contour C_1, whence in virtue of (4) we find

$$\delta W = \oint_{C_1} \mathbf{f} \cdot \delta\mathbf{r} = I\, \delta\Phi. \tag{7}$$

If now it be assumed that in the course of the virtual displacement $\delta\mathbf{r}$ the external sources and the current I *are maintained strictly constant*, then the work done by the mechanical forces is compensated by a decrease in a potential energy function U.

$$\delta U = -\delta W = -I\, \delta\Phi, \tag{8}$$

or

$$U = -I\Phi = -I \int_S \mathbf{B} \cdot \mathbf{n}\, da, \tag{9}$$

where S represents any surface spanning the current filament. Inversely, the mechanical forces and torques on a current filament in a magnetostatic field can be determined from a variation of the function U while holding constant the current I and the strength of the sources.

If $\delta\mathbf{r}$ is a real rather than a virtual displacement, work must be expended to keep the current constant. The change $\delta\Phi$ in the flux induces an electromotive force

$$V = \int_C \mathbf{E} \cdot d\mathbf{s} = -\frac{\delta\Phi}{\delta t}, \tag{10}$$

where δt is the time required to effect the displacement. This induced e.m.f. must be counterbalanced by an equal and opposite applied e.m.f. V'. The work expended by the applied voltage on the circuit in the time δt is

$$V'I\,\delta t = +I\,\delta\Phi. \tag{11}$$

The work done by the transverse mechanical forces in a small displacement of a linear circuit is exactly compensated by the energy expended by longitudinal electromotive forces necessary to maintain the current constant. The total work done on the circuit is zero.

Let us suppose now that the source of the field **B** is a second current filament. Thus we imagine for the moment that the current distribution consists of just two isolated, closed filaments I_1 and I_2. The fields generated by these currents are respectively $\mathbf{B}_1$ and $\mathbf{B}_2$. The potential energy of circuit I_2 in the external field $\mathbf{B}_1$ is

$$U_{21} = -I_2 \int_{S_2} \mathbf{B}_1 \cdot \mathbf{n}_2\, da_2, \tag{12}$$

where S_2 is any surface spanning the filament I_2. Likewise the potential energy of circuit I_1 in the field of I_2 is

$$U_{12} = -I_1 \int_{S_1} \mathbf{B}_2 \cdot \mathbf{n}_1\, da_1. \tag{13}$$

U_{12} and U_{21} are scalar functions of position and circuit configuration whose derivatives give the forces and torques exerted by one filament on the other. From the equality of action and reaction it follows that

$$U_{12} = U_{21} = U, \tag{14}$$

and hence the mutual potential energy may be expressed as

$$U = -\tfrac{1}{2}I_1\Phi_1 - \tfrac{1}{2}I_2\Phi_2, \tag{15}$$

where Φ_1 is the magnetic flux threading the filament I_1.

The mutual energy U, which reduces to zero when the separation of the filaments becomes infinite, is *not* equal to the total work that must be done in approaching them from infinity to some finite mutual configuration, and therefore does not represent the total energy of the system. For let us suppose that I_2 is allowed a small displacement under the forces exerted by I_1, with a resultant decrease $-\delta U$ in the potential energy. To maintain the current I_2 constant during the displacement an equal amount of work δW must be done on the circuit I_2 to compensate the effect of the induced e.m.f. Thus far the net change in energy is zero. But now we must note that the displacement of I_2 results in a change in the flux threading I_1 and gives rise, therefore, to an induced e.m.f. V opposing the current I_1. If I_1 is to be maintained constant, a voltage V

must be impressed on the circuit I_1 compensating V_1 and doing work at the rate I_1V_1'. The induced e.m.f., however, depends only on the *relative* motion of I_1 and I_2. The work done by the electromotive forces induced in circuit (1) by the displacement of circuit (2) must equal that which would be done were (2) fixed and an equal and opposite displacement imparted to (1). The work done on (1) is therefore δW. In sum, when the mechanical forces acting on the two filaments are allowed to do work, there is a decrease in the mutual potential energy of amount $-\delta U$, but this is offset by the work $2\delta W$ which must be done *on* the system to maintain I_1 and I_2 constant. If the total magnetic energy of the system be denoted by T, the variation in T associated with a relative displacement *at constant current* is

$$\delta T = 2\delta W - \delta U = \delta W = -\delta U, \tag{16}$$

or, by (15),

$$T = \tfrac{1}{2}I_1\Phi_1 + \tfrac{1}{2}I_2\Phi_2. \tag{17}$$

In general, if the current system is composed of n distinct current filaments, the magnetic energy is

$$T = \frac{1}{2}\sum_{i=1}^{n} I_i\Phi_i, \tag{18}$$

where Φ_i is the flux threading circuit i due to the other $n - 1$ circuits.

Equation (18) is an expression for the work necessary to bring n closed current filaments from an initial position at infinity to some specified finite configuration. The cross section of any filament is very small but does not vanish, and a filament in the sense that we have used it here is not, therefore, a line singularity. Each filament or tube may be subdivided into a bundle of thinner filaments carrying fractions of the initial current. In the limit of vanishing cross section, the current carried by the filament is infinitesimally small, as is also the energy necessary to establish it. Consequently the total energy of a current distribution is simply the mutual energy of the infinitesimal filaments into which the distribution can be resolved.

Now in practice a current distribution in a conducting medium cannot be established by the mathematically simple expedient of collecting together current filaments from infinity, and although (18) proves to be correct for the magnetic energy of n circuits in a medium for which the relation of **H** to **B** is linear, this is not necessarily true in general. In place of a continuous distribution let us consider for the moment n linear circuits embedded in any magnetic material. The resistance of the ith circuit is R_i and at any instant the current it carries is I_i. To each circuit there is now applied an external e.m.f. V_i' generated by chemical or mechanical means. These applied voltages give rise to variations in the currents and corresponding variations in the magnetic flux threading each circuit. If V_i is the voltage induced by a variation in Φ_i, the relation of current to total e.m.f. in the circuit at any instant is

$$V_i' + V_i = R_iI_i; \tag{19}$$

or, since $V_i = -\frac{d\Phi_i}{dt}$,

$$V_i' = R_iI_i + \frac{d\Phi_i}{dt}. \tag{20}$$

The power expended by the impressed voltage V_i' is $V_i'I_i$, and consequently the work done on the ith circuit in the interval δt is

$$\delta W_i = R_iI_i^2\,\delta t + I_i\,\delta\Phi_i. \tag{21}$$

Of this work the amount $R_iI_i^2\,\delta t$ is dissipated as heat, while the quantity $I_i\,\delta\Phi_i$ is stored as magnetic energy.[1] A variation in the magnetic energy of the n filaments is therefore related to increments in the fluxes by

$$\delta T = \sum_{i=1}^{n} I_i\,\delta\Phi_i, \tag{22}$$

and the total energy expended on the system, apart from that dissipated in ohmic heat loss within the conductors, as the currents are slowly increased from zero to their final values, is

$$T = \sum_{i=1}^{n} \int_{\Phi_{i0}}^{\Phi_i} I_i\,\delta\Phi_i. \tag{23}$$

Φ_{i0} is the flux threading the ith circuit at the initial instant when all currents are zero. This initial flux will be zero if there is no remnant magnetization of the medium about the circuits. Equation (23) can be interpreted as available energy stored in the magnetic field only when the relation of I_i to Φ_i is single-valued (no hysteresis), and (23) reduces to (18) only when the relation of $\mathbf{H}$ to $\mathbf{B}$ and consequently of I_i to Φ_i is linear, such that $I_i\,\delta\Phi_i = \Phi_i\,\delta I_i$.

It is a simple matter to extend (23) from a finite number of current filaments to a continuous distribution of current. Within this distribution let us choose arbitrarily a surface element $d\sigma$. The current lines passing through points of $d\sigma$ constitute a current tube, which, in virtue of the stationary character of the distribution, closes upon itself. If $\mathbf{J}$ is the current density, the scalar product $\mathbf{J}\cdot\mathbf{n}\,d\sigma = dI$ is constant

[1] A portion of the energy $I_i\,\delta\Phi_i$ may, however, be made unavailable because of hysteresis effects. See Sec. 2.16.

over every cross section of the tube. Let $\mathbf{B}'$ be the magnetic field at any point within the tube produced by all those current filaments which lie outside. The increment in the energy of the current dI due to an infinitesimal increase in the field of the *external* filaments is

$$\delta(dT) = dI\,\delta\Phi' = dI \int \delta\mathbf{B}' \cdot \mathbf{n}\,da, \tag{24}$$

the integral to be extended over any surface bounded by the contour of the filament dI. If $\mathbf{A}'$ is the vector potential of the field $\mathbf{B}'$, then $\mathbf{B}' = \nabla \times \mathbf{A}'$ and (24) may be transformed by Stokes' theorem to a line integral following the closed contour of the filament.

$$\delta(dT) = \mathbf{J} \cdot \mathbf{n}\,d\sigma \oint \delta\mathbf{A}' \cdot d\mathbf{s}. \tag{25}$$

Now the total vector potential $\mathbf{A}$ along any central line of the tube dI is equal to $\mathbf{A}'$ plus the contribution of the current $dI = \mathbf{J} \cdot \mathbf{n}\,d\sigma$ itself; but as the cross-sectional area $d\sigma \to 0$, the latter contribution vanishes, so that in the limit $\mathbf{A}'$ may be replaced by $\mathbf{A}$. Since furthermore the current density vector $\mathbf{J}$, the unit vector $\mathbf{n}$ normal to $d\sigma$, and the element of length $d\mathbf{s}$ along the tube are parallel to one another within a tube of infinitesimal cross section, one may write in place of (25):

$$\delta(dT) = \oint \mathbf{J} \cdot \delta\mathbf{A}\,d\sigma\,ds. \tag{26}$$

The product $d\sigma\,ds = dv$ represents the volume of an infinitesimal length of tube. The total increment in the energy of the distribution is to be found by summing the contributions of all the tubes into which it has been resolved, and this summation is clearly equivalent to an integration of the product $\mathbf{J} \cdot \delta\mathbf{A}$ over the entire volume occupied by current.

$$\delta T = \int \mathbf{J} \cdot \delta\mathbf{A}\,dv. \tag{27}$$

The work required to set up a continuous current distribution by means of applied electromotive forces is, therefore, in general

$$T = \int \int_{\mathbf{A}_0}^{\mathbf{A}} \mathbf{J} \cdot \delta\mathbf{A}\,dv. \tag{28}$$

In case the relation between the current and the vector potential which it produces is linear, this reduces to

$$T = \frac{1}{2} \int \mathbf{J} \cdot \mathbf{A}\,dv. \tag{29}$$

2.15. Magnetic Energy as a Function of Field Intensity.—We shall suppose that discontinuities in the magnetic properties of matter in the field can be represented by layers of rapid but continuous transition. The field may be due in part to currents, in part to permanent magnets or residually magnetized matter, but all sources are located within a finite radius of some arbitrary origin. It will be demonstrated in Chap. IV that under these circumstances the vectors **A** and **B** vanish at infinity as r^{-1} and r^{-2}. The current density **J** at any point is related to the vector **H** at the same point by

$$\mathbf{J} = \nabla \times \mathbf{H}. \tag{30}$$

Furthermore, by a well-known identity,

$$\begin{aligned} \mathbf{J} \cdot \delta\mathbf{A} = \delta\mathbf{A} \cdot \nabla \times \mathbf{H} &= \nabla \cdot (\mathbf{H} \times \delta\mathbf{A}) + \mathbf{H} \cdot \nabla \times \delta\mathbf{A} \\ &= \nabla \cdot (\mathbf{H} \times \delta\mathbf{A}) + \mathbf{H} \cdot \delta\mathbf{B}. \end{aligned} \tag{31}$$

Upon introducing (31) into (28) and applying the divergence theorem, we obtain

$$T = \int_V \int_{\mathbf{B}_0}^{\mathbf{B}} \mathbf{H} \cdot d\mathbf{B}\, dv + \int_S \int_{\mathbf{A}_0}^{\mathbf{A}} (\mathbf{H} \times d\mathbf{A}) \cdot \mathbf{n}\, da, \tag{32}$$

where V is any volume bounded by a surface S enclosing all the sources of the field. If the surface S is allowed to recede toward infinity, the second integral of (32) vanishes, for the integrand diminishes as r^{-3} whereas S grows only as r^2. Therefore the work done by impressed e.m.fs. (such as might be derived from batteries or generators) in building up a magnetic field from the initial value $\mathbf{B}_0$ to the final value **B** can be represented by the integral

$$T = \int dv \int_{\mathbf{B}_0}^{\mathbf{B}} \mathbf{H} \cdot d\mathbf{B} \tag{33}$$

extended over all space. It is to be emphasized that (33) is the energy associated with the establishment of a current distribution in the presence of magnetic materials, and does not include the internal energy of permanent magnets or the mutual energy of systems of permanent magnets.

The magnetic properties of all materials exclusive of the ferromagnetic group differ but slightly from those of free space; the relation between **B** and **H** is linear within wide limits of field intensity, the factor μ in the relation $\mathbf{B} = \mu\mathbf{H}$ is very nearly equal to μ_0, and there is no appreciable remnant magnetism. Under these circumstances the work done in building up the field to a value **B** is returned as the field is again decreased to zero. Equation (33) integrates to

$$T = \frac{1}{2} \int \mu H^2\, dv, \tag{34}$$

which we interpret as the energy stored in the magnetic field. As in the corresponding electrostatic case we may suppose this energy to be distributed throughout the field with a density $\frac{1}{2}\mu H^2$ joules/meter.[3]

2.16. Ferromagnetic Materials.—The relation of **H** to **B** for ferromagnetic substances is in general nonlinear and multivalued. In an initial, unmagnetized state the vectors **B** and **H** are zero. If the field is now built up slowly by means of impressed electromotive forces applied to conducting circuits, the function $\mathbf{B} = \mathbf{B}(\mathbf{H})$ at any point in the ferro-

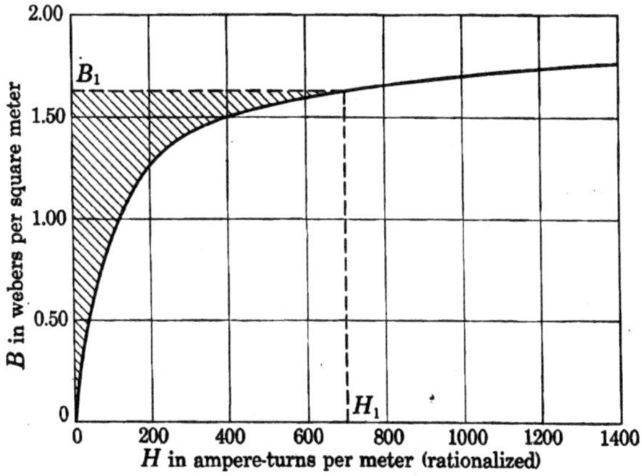

FIG. 20.—Typical magnetization curve for annealed sheet steel.

magnetic material follows a curve of the form indicated by Fig. 20. According to (33) the work done in magnetizing a unit volume of the substance is represented by the shaded area in the figure. Were the function $\mathbf{B}(\mathbf{H})$ single-valued, a decrease in the field from $\mathbf{H}_1$, $\mathbf{B}_1$ to zero would follow the same curve and the entire energy (33) would be available for useful work. Actually, the return usually follows a path such as that indicated in Fig. 21. Starting at $\mathbf{H}_1$, the field is decreased until $\mathbf{H} = 0$. The associated value of $\mathbf{B} = \mathbf{B}_2$, however, is still positive. To reduce **B** to zero, negative values must be imparted to **H**, meaning physically that **H** must be increased in the opposite direction. At $\mathbf{H} = -\mathbf{H}_3$ the vector **B** is zero, and as **H** continues to increase in negative value a point is eventually reached where simultaneously $\mathbf{H} = -\mathbf{H}_1$, $\mathbf{B} = -\mathbf{B}_1$. The return to the positive values $\mathbf{H}_1$, $\mathbf{B}_1$ follows the symmetrical path through $\mathbf{B} = -\mathbf{B}_2$, $\mathbf{H} = 0$ and $\mathbf{B} = 0$, $\mathbf{H} = +\mathbf{H}_3$. At all points along the segment $\mathbf{B}_1\mathbf{B}_2$ the value of **B** is greater than its initial value on the dotted curve for the same value of **H**; the change of **B** lags behind that of **H** and the substance is said to exhibit *hysteresis*.

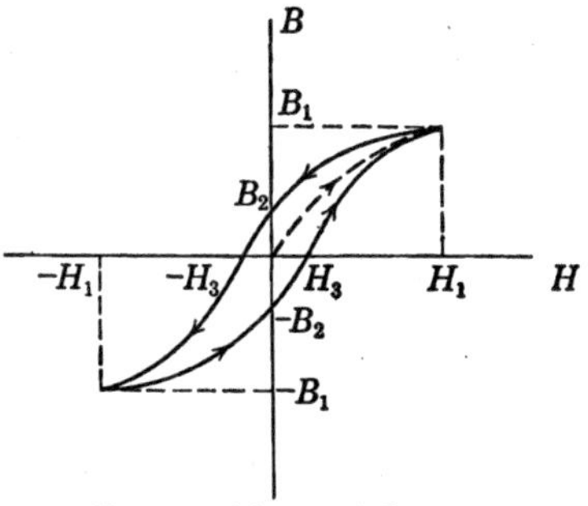

FIG. 21.—Hysteresis loop.

Let w be the work done *per unit volume of magnetic material* in changing the field from the value $\mathbf{B}_1$ to $\mathbf{B}$.

$$w = \int_{\mathbf{B}_1}^{\mathbf{B}} \mathbf{H} \cdot d\mathbf{B} = \mathbf{H} \cdot \mathbf{B} \Big|_{\mathbf{B}_1}^{\mathbf{B}} - \int_{\mathbf{H}_1}^{\mathbf{H}} \mathbf{B} \cdot d\mathbf{H}. \tag{35}$$

If the variation of the field is carried through a complete cycle following the hysteresis loop from $\mathbf{B}_1$ through $\mathbf{B}_2$, $-\mathbf{B}_1$, $-\mathbf{B}_2$ and returning to $\mathbf{B}_1$, the net work done per unit volume is

$$w = -\oint \mathbf{B} \cdot d\mathbf{H}, \tag{36}$$

a quantity evidently represented by the enclosed area of the hysteresis loop illustrated in Fig. 21. The net work done per cycle throughout the entire field is

$$Q = -\int dv \oint \mathbf{B} \cdot d\mathbf{H}. \tag{37}$$

Q is the hysteresis loss, an irretrievable fraction of the field energy dissipated in heat.

2.17. Energy of a Magnetic Body in a Magnetostatic Field.—Let us suppose that a magnetic field $\mathbf{B}_1$ from *fixed* sources has been established in a magnetic medium. We shall assume that the relation of $\mathbf{B}_1$ to $\mathbf{H}_1$ is linear and that the medium is isotropic. Then $\mathbf{B}_1 = \mu_1 \mathbf{H}_1$, where μ_1 is at most a scalar function of position which reduces to a constant in case the medium is homogeneous. The energy of the field is

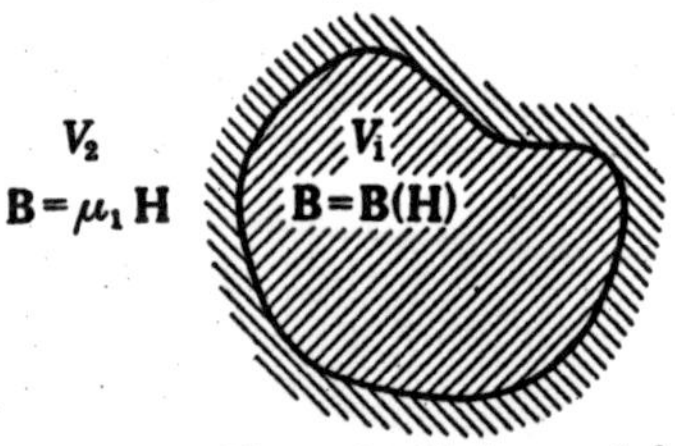

FIG. 22.—The region V_1 is occupied by a magnetic body embedded in the homogeneous, isotropic medium V_2.

$$T_1 = \frac{1}{2} \int \mathbf{H}_1 \cdot \mathbf{B}_1 \, dv, \tag{38}$$

extended over all space. Now reduce the intensity of the sources to zero and introduce into a suitable cavity formed in the first medium a body which we shall assume to be unmagnetized but whose magnetic properties are otherwise arbitrary. As in Sec. 2.10 the volume occupied by the embedded body will be denoted by V_1, Fig. 22, and the entire region exterior to it by V_2. If the external sources are currents, the work which must be done in order to restore them to their initial intensity is

$$T_2 = \int_{V_1+V_2} dv \int_0^{\mathbf{B}} \mathbf{H} \cdot d\mathbf{B} = \frac{1}{2} \int_{V_2} \mathbf{H} \cdot \mathbf{B} \, dv + \int_{V_1} dv \int_0^{\mathbf{B}} \mathbf{H} \cdot d\mathbf{B}. \tag{39}$$

The ultimate field $\mathbf{B}$ differs at every point from the initial field $\mathbf{B}_1$ by an amount $\mathbf{B}_2 = \mathbf{B} - \mathbf{B}_1$ arising from the change in polarization of the matter contained within V_1. The increase in the work necessary to build up the current strength to the initial value is

$$(40)\quad T = T_2 - T_1 = \frac{1}{2}\int_{V_2} (\mathbf{H}\cdot\mathbf{B} - \mathbf{H}_1\cdot\mathbf{B}_1)\,dv + \int_{V_1} dv\left(\int_0^{\mathbf{B}} \mathbf{H}\cdot d\mathbf{B} - \frac{1}{2}\mathbf{H}_1\cdot\mathbf{B}_1\right).$$

It has been assumed that throughout the region V_2 exterior to the body the relations between **B** and **H** are linear:

$$(41)\quad \mathbf{B}_1 = \mu_1\mathbf{H}_1, \qquad \mathbf{B} = \mu_1\mathbf{H} \qquad \text{in } V_2.$$

The first integral in (40) is therefore equivalent to

$$(42)\quad \frac{1}{2}\int_{V_2} (\mathbf{H}\cdot\mathbf{B} - \mathbf{H}_1\cdot\mathbf{B}_1)\,dv = \frac{1}{2}\int_{V_2} (\mathbf{H} - \mathbf{H}_1)\cdot(\mathbf{B} + \mathbf{B}_1)\,dv.$$

By hypothesis the initial and final current distributions in the source are identical, so that in both V_1 and V_2 the conditions

$$(43)\quad \nabla\cdot(\mathbf{B} + \mathbf{B}_1) = 0, \qquad \nabla\times(\mathbf{H} - \mathbf{H}_1) = 0,$$

are satisfied, while over the surface bounding V_1 we have according to (18), page 37,

$$(44)\quad \begin{aligned} \mathbf{n}\cdot[(\mathbf{B} + \mathbf{B}_1)_+ - (\mathbf{B} + \mathbf{B}_1)_-] &= 0, \\ \mathbf{n}\times[(\mathbf{H} - \mathbf{H}_1)_+ - (\mathbf{H} - \mathbf{H}_1)_-] &= 0. \end{aligned}$$

The theorem of Sec. 2.9 may be applied, giving

$$(45)\quad \frac{1}{2}\int_{V_1+V_2} (\mathbf{H} - \mathbf{H}_1)\cdot(\mathbf{B} + \mathbf{B}_1)\,dv = \frac{1}{2}\int_{V_1} (\mathbf{H} - \mathbf{H}_1)\cdot(\mathbf{B} + \mathbf{B}_1)\,dv + \frac{1}{2}\int_{V_2} (\mathbf{H} - \mathbf{H}_1)\cdot(\mathbf{B} + \mathbf{B}_1)\,dv = 0,$$

or

$$(46)\quad \frac{1}{2}\int_{V_2} (\mathbf{H} - \mathbf{H}_1)\cdot(\mathbf{B} + \mathbf{B}_1)\,dv = -\frac{1}{2}\int_{V_1} (\mathbf{H} - \mathbf{H}_1)\cdot(\mathbf{B} + \mathbf{B}_1)\,dv.$$

The additional work required to build up the currents in the presence of the body can thus be expressed in terms of integrals extended over the volume occupied by the body. Upon introducing (46) into (40), we obtain

$$(47)\quad T = \frac{1}{2}\int_{V_1}\left(\mathbf{H}_1\cdot\mathbf{B} - \mathbf{H}\cdot\mathbf{B}_1 - \mathbf{H}\cdot\mathbf{B} + 2\int_0^{\mathbf{B}} \mathbf{H}\cdot d\mathbf{B}\right)dv.$$

If hysteresis effects are negligible (**H** is a single-valued function of **B**), the quantity T may be interpreted as the energy of the body in the magnetic field of any system of constant sources; and the variation of T resulting from a virtual displacement of the body determines the mechanical forces exerted upon it. If furthermore the magnetic properties of the material within V_1 can be characterized by a permeability μ_2, so that $\mathbf{B} = \mu_2\mathbf{H}$, then (47) reduces to

$$(48)\quad T = \frac{1}{2}\int_{V_1} (\mathbf{H}_1 \cdot \mathbf{B} - \mathbf{H} \cdot \mathbf{B}_1)\, dv = \frac{1}{2}\int_{V_1} (\mu_2 - \mu_1)\mathbf{H} \cdot \mathbf{H}_1\, dv.$$

The body was assumed to be initially free of residual magnetism; consequently the vectors **B** and **H** within V_1 are related to the induced magnetic polarization **M** by Sec. 1.6,

$$(49)\qquad \mathbf{B} = \mu_0(\mathbf{H} + \mathbf{M}),$$

and, when $\mathbf{B} = \mu_2\mathbf{H}$, this leads to

$$(50)\qquad \mathbf{M} = \left(\frac{\mu_2}{\mu_0} - 1\right)\mathbf{H} = \left(\frac{1}{\mu_0} - \frac{1}{\mu_2}\right)\mathbf{B}.$$

If, therefore, the cavity V_1 was initially free of magnetic matter we may put $\mu_1 = \mu_0$ and write (48) in the form

$$(51)\qquad T = \frac{1}{2}\int_{V_1} \mathbf{M} \cdot \mathbf{B}_1\, dv,$$

an expression which corresponds to (51), page 113, for the electrical case in all but algebraic sign. This distinction, however, is fundamental. Whereas in the electrostatic case the work done by the external forces in a virtual displacement is accompanied by a decrease of the potential energy U, we shall learn soon that the mechanical forces exerted on a magnetized body are to be determined from the *increase* in T, so that T in this sense behaves as a kinetic rather than a potential energy.

A useful analogue of Eq. (52), page 114, may be deduced for the magnetostatic field from (48). Suppose again that $\mathbf{B}_1$ is the field of *fixed* sources—either currents or permanent magnets—and that at every point in space the permeability is specified by $\mu = \mu(x, y, z)$, a continuous function of position. If now the permeability is varied by an infinitesimal amount $\delta\mu$, the consequent change in the magnetic energy is

$$(52)\qquad \delta T = \frac{1}{2}\int \delta\mu\, H^2\, dv;$$

therefore, the magnetic energy of a distribution of matter in a fixed field is

$$(53)\qquad T = \frac{1}{2}\int dv \int_{\mu_0}^{\mu} H^2\, d\mu, \qquad (\mu = \mu_0\kappa_m)$$

provided the matter is initially unmagnetized and μ is independent of $\mathbf{H}$. For paramagnetic material $\mu > \mu_0$, $\kappa_m > 1$, but in the case of diamagnetic matter $\mu < \mu_0$ and the magnetic energy of the initial field is diminished by the introduction of diamagnetic bodies.

2.18. Potential Energy of a Permanent Magnet.—A rigorous analysis of the energy of material bodies in a magnetic field becomes very much more difficult when these bodies are permanently magnetized. In that case there is a residual field, which we shall denote by $\mathbf{B}_0$, $\mathbf{H}_0$, associated with the magnet even in the absence of external sources. The relation of $\mathbf{B}_0$ to $\mathbf{H}_0$ may still be expressed in the form

$$\mathbf{B}_0 = \mu_0(\mathbf{H}_0 + \mathbf{M}_0), \tag{54}$$

but $\mathbf{M}_0$, the intensity of magnetization, is now quite independent of $\mathbf{H}_0$, being determined solely by the previous history of the specimen. If an external field is applied, there will be induced an additional component of magnetization which, as in the past, we denote by $\mathbf{M}$. This induced polarization $\mathbf{M}$ depends primarily on the *resultant* field $\mathbf{H}$ within the magnet (vanishing as $\mathbf{H}$ reduces to $\mathbf{H}_0$), but also on the state of the magnet and its permanent or residual magnetization $\mathbf{M}_0$.

$$\mathbf{B} = \mu_0[\mathbf{H} + \mathbf{M}(\mathbf{H}, \mathbf{M}_0) + \mathbf{M}_0]. \tag{55}$$

The resultant field $\mathbf{H}$ at an interior point is determined furthermore by the shape of the magnet as well as by the intensity and distribution of the external sources.

We shall content ourselves with the derivation of a simple and frequently used expression for the potential energy of a system of permanent magnets. The proof rests on assumptions which are approximately fulfilled in practice; namely, that the magnetization $\mathbf{M}_0$ is absolutely rigid and that the induced magnetization $\mathbf{M}$ in one magnet arising from the external field of another is of negligible intensity with respect to $\mathbf{M}_0$. The field at all points, both inside and outside a magnetized body occupying a volume V_1, is exactly that which would be produced by a stationary volume distribution of current throughout V_1 of density

$$\mathbf{J} = \nabla \times \mathbf{M}_0, \tag{56}$$

together with a current distribution on the surface S bounding V_1 of density[1]

$$\mathbf{K} = \mathbf{M}_0 \times \mathbf{n}, \tag{57}$$

where $\mathbf{n}$ is the unit outward normal to S. As a direct consequence of the analysis of Sec. 2.14 and in particular of Eq. (9), it follows that the potential energy of the magnet V_1 in the field of other permanent magnets or of

[1] See Secs. 1.6 and 4.10.

constant currents is

$$U = -\int_{V_1} \mathbf{J} \cdot \mathbf{A}\, dv - \int_S \mathbf{K} \cdot \mathbf{A}\, da, \tag{58}$$

or, in virtue of (56) and (57),

$$U = -\int_{V_1} (\nabla \times \mathbf{M}_0) \cdot \mathbf{A}\, dv - \int_S (\mathbf{M}_0 \times \mathbf{n}) \cdot \mathbf{A}\, da. \tag{59}$$

Now

$$\mathbf{A} \cdot \nabla \times \mathbf{M}_0 = \nabla \cdot (\mathbf{M}_0 \times \mathbf{A}) + \mathbf{M} \cdot \nabla \times \mathbf{A}, \tag{60}$$

$$(\mathbf{M} \times \mathbf{A}) \cdot \mathbf{n} = (\mathbf{n} \times \mathbf{M}_0) \cdot \mathbf{A} = -(\mathbf{M}_0 \times \mathbf{n}) \cdot \mathbf{A}, \tag{61}$$

so that upon applying the divergence theorem and putting $\mathbf{B} = \nabla \times \mathbf{A}$, (59) reduces to

$$U = -\int_{V_1} \mathbf{M}_0 \cdot \mathbf{B}\, dv. \tag{62}$$

In (62) one may consider $d\mathbf{m} = \mathbf{M}_0\, dv$ to be the magnetic moment of an element dv of the magnet. Its potential energy in the resultant field $\mathbf{B}$ is $dU = -\mathbf{B} \cdot d\mathbf{m}$. The resultant field $\mathbf{B}$ is composed of the initial field $\mathbf{B}_1$ of all external sources, plus the field $\mathbf{B}_2$ due to all other elements of the same magnet V_1. Therefore the work necessary to construct the magnet by collecting together permanently magnetized elements in the absence of an external field should be

$$U_2 = -\int_{V_1} \mathbf{M}_0 \cdot \mathbf{B}_2\, dv. \tag{63}$$

These elements must be held together by forces of a nonmagnetic character. The work necessary to introduce the magnet as a rigid whole from infinity to a point within the *external* field $\mathbf{B}_1$ is then

$$U_1 = -\int_{V_1} \mathbf{M}_0 \cdot \mathbf{B}_1\, dv, \tag{64}$$

and the force exerted on a unit volume of the magnet by the external sources is

$$\mathbf{f} = +\nabla(\mathbf{M}_0 \cdot \mathbf{B}_1). \tag{65}$$

The difference of (64) and (51) is accounted for when we note that (64) is only the potential energy of the magnetized body in the external field, while (51) includes the work involved in building up the magnetization from zero to $\mathbf{M}$, and is based on the assumption that there exists a linear relation between $\mathbf{B}_1$ and $\mathbf{H}_1$, and consequently between $\mathbf{B}_1$ and $\mathbf{M}$.

ENERGY FLOW

2.19. Poynting's Theorem.—In the preceding sections of this chapter it has been shown how the work done in bringing about small variations in the intensity or distribution of charge and current sources may be expressed in terms of integrals of the field vectors extended over all space. The form of these integrals suggests, but does not prove, the hypothesis that electric and magnetic energies are distributed throughout the field with volume densities respectively

$$(1)\qquad u = \int_0^{\mathbf{D}} \mathbf{E} \cdot d\mathbf{D}, \qquad w = \int_0^{\mathbf{B}} \mathbf{H} \cdot d\mathbf{B}.$$

The derivation of these results was based on the assumption of *reversible* changes; the building up of the field was assumed to take place so slowly that it might be represented by a succession of stationary states. It is essential that we determine now whether or not such expressions for the energy density remain valid when the fields are varying at an arbitrary rate. It is apparent, furthermore, that if our hypothesis of an energy distribution throughout the field is at all tenable, a change of field intensity and energy density must be associated with a *flow* of energy from or toward the source.

A relation between the rate of change of the energy stored in the field and the energy flow can be deduced as a general integral of the field equations.

$$\text{(I)}\quad \nabla \times \mathbf{E} + \frac{\partial \mathbf{B}}{\partial t} = 0, \qquad \text{(III)}\quad \nabla \cdot \mathbf{B} = 0,$$

$$\text{(II)}\quad \nabla \times \mathbf{H} - \frac{\partial \mathbf{D}}{\partial t} = \mathbf{J}, \qquad \text{(IV)}\quad \nabla \cdot \mathbf{D} = \rho.$$

We note that $\mathbf{E} \cdot \mathbf{J}$ has the dimensions of power expended per unit volume (watts per cubic meter) and this suggests scalar multiplication of (II) by $\mathbf{E}$.

$$(2)\qquad \mathbf{E} \cdot \nabla \times \mathbf{H} - \mathbf{E} \cdot \frac{\partial \mathbf{D}}{\partial t} = \mathbf{E} \cdot \mathbf{J}.$$

In order that each term in (I) may have the dimensions of energy per unit volume per unit time it must be multiplied by $\mathbf{H}$.

$$(3)\qquad \mathbf{H} \cdot \nabla \times \mathbf{E} + \mathbf{H} \cdot \frac{\partial \mathbf{B}}{\partial t} = 0.$$

Upon subtracting (2) from (3) and applying the identity

$$(4)\qquad \nabla \cdot (\mathbf{E} \times \mathbf{H}) = \mathbf{H} \cdot \nabla \times \mathbf{E} - \mathbf{E} \cdot \nabla \times \mathbf{H},$$

we obtain

(5) $$\nabla \cdot (\mathbf{E} \times \mathbf{H}) + \mathbf{E} \cdot \mathbf{J} = -\mathbf{E} \cdot \frac{\partial \mathbf{D}}{\partial t} - \mathbf{H} \cdot \frac{\partial \mathbf{B}}{\partial t}.$$

Finally, let us integrate (5) over a volume V bounded by a surface S.

(6) $$\int_S (\mathbf{E} \times \mathbf{H}) \cdot \mathbf{n}\, da + \int_V \mathbf{E} \cdot \mathbf{J}\, dv = -\int_V \left(\mathbf{E} \cdot \frac{\partial \mathbf{D}}{\partial t} + \mathbf{H} \cdot \frac{\partial \mathbf{B}}{\partial t}\right) dv.$$

This result was first derived by Poynting in 1884, and again in the same year by Heaviside. Its customary interpretation is as follows. We assume that the formal expressions for densities of energy stored in the electromagnetic field are the same as in the stationary regime. Then the right-hand side of (6) represents the rate of decrease of electric and magnetic energy stored within the volume. The loss of available stored energy must be accounted for by the terms on the left-hand side of (6). Let σ be the conductivity of the medium and $\mathbf{E}'$ the intensity of *impressed* electromotive forces such as arise in a region of chemical activity—the interior of a battery, for example. Then

(7) $$\mathbf{J} = \sigma(\mathbf{E} + \mathbf{E}'), \qquad \mathbf{E} = \frac{\mathbf{J}}{\sigma} - \mathbf{E}',$$

and hence

(8) $$\int_V \mathbf{E} \cdot \mathbf{J}\, dv = \int_V \frac{1}{\sigma} J^2\, dv - \int_V \mathbf{E}' \cdot \mathbf{J}\, dv.$$

The first term on the right of (8) represents the power dissipated in Joule heat—an irreversible transformation. The second term expresses the power expended by the flow of charge against the impressed forces, the negative sign indicating that these impressed forces are doing work *on* the system, offsetting in part the Joule loss and tending to increase the energy stored in the field. If, finally, all material bodies in the field are absolutely rigid, thereby excluding possible transformations of electromagnetic energy into elastic energy of a stressed medium, the balance can be maintained only by a flow of electromagnetic energy across the surface bounding V. This, according to Poynting, is the significance of the surface integral in (6). The diminution of electromagnetic energy stored in V is partly accounted for by the Joule heat loss, partly compensated by energy introduced through impressed forces; the remainder flows outward across the bounding surface S, representing a loss measured in joules per second, or watts, by the integral

(9) $$\int_S \mathbf{S} \cdot \mathbf{n}\, da = \int_S (\mathbf{E} \times \mathbf{H}) \cdot \mathbf{n}\, da.$$

The *Poynting vector* $\mathbf{S}$ defined by

(10) $$\mathbf{S} = \mathbf{E} \times \mathbf{H} \qquad \text{watts/meter}^2,$$

may be interpreted as the *intensity* of energy flow at a point in the field; *i.e.*, the energy per second crossing a unit area whose normal is oriented in the direction of the vector $\mathbf{E} \times \mathbf{H}$.

It has been tacitly assumed that the medium is free of hysteresis effects. In case the relation between **B** and **H** is multivalued, energy to the amount Q [Eq. (37), page 126] is dissipated in the medium in the course of every complete cycle of the hysteresis loop. If the field is harmonic in time with a frequency ν, there will be ν cycles/sec. and consequently the hysteresis will also participate in the diminution of magnetic energy at the rate of νQ joules/sec.

In the absence of ferromagnetic materials the relations between the field intensities are usually linear, and if the media are also isotropic Poynting's theorem in its differential form reduces to

$$\nabla \cdot (\mathbf{E} \times \mathbf{H}) + \frac{1}{\sigma} J^2 + \frac{\partial}{\partial t}\left(\frac{\epsilon}{2} E^2 + \frac{\mu}{2} H^2\right) = \mathbf{E}' \cdot \mathbf{J}. \tag{11}$$

As a general integral of the field equations, the validity of Poynting's theorem is unimpeachable. Its physical interpretation, however, is open to some criticism. The remark has already been made that from a volume integral representing the *total* energy of a field no rigorous conclusion can be drawn with regard to its distribution. The energy of the electrostatic field was first expressed as the sum of two volume integrals. Of these one was transformed by the divergence theorem into a surface integral which was made to vanish by allowing the surface to recede to the farther limits of the field. Inversely, the divergence of any vector function vanishing properly at infinity may be added to the conventional expression $u = \frac{1}{2}\mathbf{E} \cdot \mathbf{D}$ for the density of electrostatic energy without affecting its total value. A similar indefiniteness appears in the magnetostatic case.

A question may also be raised as to the propriety of assuming that $\mathbf{E} \cdot \frac{\partial \mathbf{D}}{\partial t}$ and $\mathbf{H} \cdot \frac{\partial \mathbf{B}}{\partial t}$ represent the rate of change of energy density for rapid as well as quasi-static changes. Such an assumption seems plausible, but we must note in passing that a transformation of the energy function expressed in terms of the field vectors to an expression in terms of the densities of charge and current leads now to difficulties with surface integrals. Let

$$\frac{\partial W}{\partial t} = \int \left(\mathbf{E} \cdot \frac{\partial \mathbf{D}}{\partial t} + \mathbf{H} \cdot \frac{\partial \mathbf{B}}{\partial t}\right) dv \tag{12}$$

be the rate at which work is done on the system by external forces. Upon introducing the potentials

$$\text{(13)} \qquad \mathbf{E} = -\nabla\phi - \frac{\partial \mathbf{A}}{\partial t}, \qquad \mathbf{B} = \nabla \times \mathbf{A},$$

and applying the identities

$$\text{(14)} \qquad \begin{aligned} \nabla\phi \cdot \frac{\partial \mathbf{D}}{\partial t} &= \nabla \cdot \left(\phi \frac{\partial \mathbf{D}}{\partial t}\right) - \phi \frac{\partial}{\partial t} \nabla \cdot \mathbf{D}, \\ \mathbf{H} \cdot \nabla \times \frac{\partial \mathbf{A}}{\partial t} &= \nabla \cdot \left(\frac{\partial \mathbf{A}}{\partial t} \times \mathbf{H}\right) - \frac{\partial \mathbf{A}}{\partial t} \cdot \nabla \times \mathbf{H}, \end{aligned}$$

we obtain

$$\text{(15)} \qquad \frac{\partial W}{\partial t} = \int \left(\phi \frac{\partial \rho}{\partial t} + \mathbf{J} \cdot \frac{\partial \mathbf{A}}{\partial t}\right) dv - \int_S \left(\phi \frac{\partial \mathbf{D}}{\partial t} + \mathbf{H} \times \frac{\partial \mathbf{A}}{\partial t}\right) \cdot \mathbf{n}\, da,$$

where S is a surface enclosing the entire electromagnetic field. Now in the stationary or quasi-stationary state the potentials can be shown to vanish at infinity as r^{-1} and the fields as r^{-2}; the integral over an infinite surface then vanishes. It will be shown in due course, however, that the fields of variable sources vanish only as r^{-1} and in this case the last term of (15) cannot be discarded by the simple expedient of extending the integral over an infinitely remote surface. On the other hand we know that fields and potentials are propagated with a *finite* velocity. If, therefore, the field was first established at some finite instant of the past, a surface may be imagined whose elements are so distant from the source that the field has not yet arrived. The intensity over S is then strictly zero and under these circumstances

$$\text{(16)} \qquad \frac{\partial W}{\partial t} = \int \left(\phi \frac{\partial \rho}{\partial t} + \mathbf{J} \cdot \frac{\partial \mathbf{A}}{\partial t}\right) dv.$$

Finally, it must be granted that even though the total flow of energy through a closed surface may be represented correctly by (9), one cannot conclude definitely that the intensity of energy flow at a point is $\mathbf{S} = \mathbf{E} \times \mathbf{H}$; for there might be added to this quantity any vector integrating to zero over a closed surface without affecting the total flow.

The classical interpretation of Poynting's theorem appears to rest to a considerable degree on hypothesis. Various alternative forms of the theorem have been offered from time to time,[1] but none of these has the advantage of greater plausibility or greater simplicity to recommend it, and it is significant that thus far no other interpretation has contributed anything of value to the theory. The hypothesis of an energy density

[1] MACDONALD, "Electric Waves," Cambridge University Press, 1902. LIVENS, "The Theory of Electricity," Cambridge University Press, pp. 238 *ff.*, 1926. MASON and WEAVER, "The Electromagnetic Field," University of Chicago Press, pp. 264 *ff.*, 1929.

in the electromagnetic field and a flow of intensity $\mathbf{S} = \mathbf{E} \times \mathbf{H}$ has, on the other hand, proved extraordinarily fruitful. A theory is not an absolute truth but a self-consistent analytical formulation of the relations governing a group of natural phenomena. By this standard there is every reason to retain the Poynting-Heaviside viewpoint until a clash with new experimental evidence shall call for its revision.

2.20. The Complex Poynting Vector.—If we now let $h = u + w$ represent the density of electromagnetic energy at any instant and $Q = \frac{1}{\sigma} J^2 - \mathbf{E}' \cdot \mathbf{J}$ the power expended per unit volume in thermochemical activity, the Poynting theorem for a field free from hysteresis effects may be written

$$\nabla \cdot \mathbf{S} + \frac{\partial h}{\partial t} + Q = 0. \tag{17}$$

In a stationary field h is independent of the time, so that (17) reduces to

$$\nabla \cdot \mathbf{S} + Q = 0. \tag{18}$$

Q may be positive or negative as the work done by the impressed electromotive forces $\mathbf{E}'$ is less or greater than the energy dissipated in heat. Accordingly the energy flows from or toward a volume element depending on its action as an energy source or sink.

The sources and their fields in most practical applications of electromagnetic theory are periodic functions of the time. The mean value of the energy density h is constant and $\partial \bar{h}/\partial t = \overline{\partial h/\partial t} = 0$, the bar indicating a mean value obtained by averaging over a period. In the case of periodic fields, therefore,

$$\overline{\nabla \cdot \mathbf{S}} + \bar{Q} = 0, \qquad \text{or} \qquad \int_S \overline{\mathbf{S}} \cdot \mathbf{n}\, da + \int_V \bar{Q}\, dv = 0. \tag{19}$$

When there are no sources within V, the energy dissipated in heat throughout V is equal to the mean value of the inward flow across the surface S.

The advantages of complex quantities for the treatment of periodic states are too well known to be in need of detailed exposition. The reader may be reminded, however, that certain precautions must be observed when dealing with products and squares. Throughout the remainder of this book we shall usually represent a harmonic variation in time as a complex function of the coordinates multiplied by the factor $e^{-i\omega t}$. Thus if A is the quantity in question, we write

$$A = A_0 e^{-i\omega t} = (\alpha + i\beta) e^{-i\omega t} = (\alpha + i\beta)(\cos \omega t - i \sin \omega t), \tag{20}$$

where α and β are real functions of the coordinates x, y, z. The conjugate of A is obtained by replacing $i = \sqrt{-1}$ by $-i$, and is indicated by the sign $\tilde{A}$.

$$\tilde{A} = (\alpha - i\beta)e^{i\omega t}. \tag{21}$$

Although it is convenient to employ complex quantities in the course of analytical operations, physical entities must finally be represented by real functions. If A satisfies a linear equation with real coefficients, then both its real and its imaginary parts are also solutions and either may be chosen at the conclusion of the calculation to represent the physical state. However, in the case of squares and products, we must first take the real parts of the factors and then multiply, for the product of the real parts of two complex quantities is not equal to the real part of their product. We shall indicate the real part of A by $Re(A)$ and the imaginary part by $Im(A)$.

$$\begin{aligned} Re(A) &= \alpha \cos \omega t + \beta \sin \omega t = \sqrt{\alpha^2 + \beta^2} \sin (\omega t + \phi), \\ Im(A) &= \beta \cos \omega t - \alpha \sin \omega t = \sqrt{\alpha^2 + \beta^2} \cos (\omega t + \phi), \\ \phi &= \tan^{-1} \frac{\alpha}{\beta}. \end{aligned} \tag{22}$$

The square of the amplitude—or magnitude—of A is obtained by multiplication with its conjugate.

$$A\tilde{A} = \alpha^2 + \beta^2. \tag{23}$$

The real part of A is also given by

$$Re(A) = \frac{A + \tilde{A}}{2}. \tag{24}$$

The product of the real parts of two complex quantities A_1 and A_2 is, therefore, equal to

$$\begin{aligned} Re(A_1) \cdot Re(A_2) &= \tfrac{1}{4}(A_1 + \tilde{A}_1)(A_2 + \tilde{A}_2) \\ &= \tfrac{1}{4}(A_1A_2 + \tilde{A}_1\tilde{A}_2 + A_1\tilde{A}_2 + A_2\tilde{A}_1). \end{aligned} \tag{25}$$

The time average of a periodic function A is defined by

$$\bar{A} = \frac{1}{\tau} \int_0^\tau A\, dt, \tag{26}$$

where τ is the period. If A is a simple harmonic function of the time, its mean value is of course zero. The mean value of such functions as $\cos \omega t \sin \omega t$ also vanishes, and we have consequently from (25) the result

$$\overline{Re(A_1) \cdot Re(A_2)} = \tfrac{1}{4}(A_1\tilde{A}_2 + A_2\tilde{A}_1) = \tfrac{1}{2}(\alpha_1\alpha_2 + \beta_1\beta_2), \tag{27}$$

or

$$\overline{Re(A_1) \cdot Re(A_2)} = \tfrac{1}{2}Re(A_1\tilde{A}_2). \tag{28}$$

According to the preceding formulas the mean intensity of the energy flow in a harmonic electromagnetic field is

$$\bar{\mathbf{S}} = \overline{Re(\mathbf{E}) \times Re(\mathbf{H})} = \tfrac{1}{2}Re(\mathbf{E} \times \tilde{\mathbf{H}}), \tag{29}$$

the real part of the complex vector $\frac{1}{2}\mathbf{E} \times \tilde{\mathbf{H}}$. The properties of this so-called *complex Poynting vector* are interesting. We shall denote it by

$$\mathbf{S}^* = \tfrac{1}{2}\mathbf{E} \times \tilde{\mathbf{H}}. \tag{30}$$

Let us suppose that the medium is defined by the constants ϵ, μ, σ and that the field contains the time only in the factor $e^{-i\omega t}$. Then the Maxwell equations in regions free of impressed electromotive forces $\mathbf{E}'$ are

$$\nabla \times \mathbf{E} = i\omega\mu\mathbf{H}, \qquad \nabla \times \mathbf{H} = (\sigma - i\omega\epsilon)\mathbf{E}. \tag{31}$$

The conjugate of the second equation is

$$\nabla \times \tilde{\mathbf{H}} = (\sigma + i\omega\epsilon)\tilde{\mathbf{E}}, \tag{32}$$

which when united with the first as in Sec. 2.19 leads to

$$\nabla \cdot \mathbf{S}^* = -\tfrac{1}{2}\sigma\mathbf{E} \cdot \tilde{\mathbf{E}} + i\omega\left(\frac{\mu}{2}\mathbf{H} \cdot \tilde{\mathbf{H}} - \frac{\epsilon}{2}\mathbf{E} \cdot \tilde{\mathbf{E}}\right), \tag{33}$$

or in virtue of (28) and (23),

$$\nabla \cdot \mathbf{S}^* = -\bar{Q} + i2\omega(\bar{w} - \bar{u}). \tag{34}$$

The divergence of the real part of $\mathbf{S}^*$ determines the energy dissipated in heat per unit volume per second, whereas the divergence of the imaginary part is equal to 2ω times the difference of the mean values of magnetic and electric densities. Throughout any region of the field bounded by a surface S we have

$$Re \int_S \mathbf{S}^* \cdot \mathbf{n}\, da = \text{total energy dissipated}, \tag{35}$$

and

$$Im \int_S \mathbf{S}^* \cdot \mathbf{n}\, da = 2\omega \times \text{difference of the mean values of magnetic and electric energies.} \tag{36}$$

FORCES ON A DIELECTRIC IN AN ELECTROSTATIC FIELD

2.21. Body Forces in Fluids.—Let us consider an electrostatic field arising from charges located on the surfaces of conductors embedded in an isotropic dielectric. To simplify the analysis we shall assume for the moment that throughout the entire field the dielectric medium has no discontinuities other than those occurring at the surfaces of the conductors. It will be assumed furthermore that $\mathbf{D} = \epsilon\mathbf{E}$, where ϵ is a continuous function of position. The total electrostatic energy of the field is therefore

$$U = \frac{1}{2} \int \epsilon E^2 \, dv \tag{1}$$

extended over all space.

The position of any point in the dielectric with respect to a fixed reference origin is specified by the vector $\mathbf{r}$. Every point of the dielectric is now subjected to an arbitrary, infinitesimal displacement $\mathbf{s}(\mathbf{r})$. It is supposed, however, that the conductors are held rigid and the displacement of dielectric particles in the neighborhood of the conducting surfaces is necessarily tangential. The displacement or strain results in a variation in the parameter ϵ with a corresponding change in the electrostatic energy equal to

$$\delta U = -\frac{1}{2} \int \delta\epsilon E^2 \, dv. \tag{2}$$

At the same time a slight readjustment of the charge distribution will occur over the surfaces of the conductors. Initially the surface charges were in static equilibrium and the initial energy (1), according to Thomson's theorem, Sec. 2.11, is a minimum with respect to infinitesimal variations in charge distribution. The variation in energy associated with the redistribution is, therefore, an infinitesimal of second order which may be neglected with respect to (2).

Consider the dielectric material contained initially in a volume dv_1. The displacement is accompanied by a deformation, so that after the strain this element of matter may occupy a volume Eq. (54), page 92.

$$dv_2 = (1 + \nabla \cdot \mathbf{s}) \, dv_1. \tag{3}$$

The mass of the element is conserved and hence if the density of matter is denoted by τ we have

$$\tau_1 \, dv_1 = \tau_2(1 + \nabla \cdot \mathbf{s}) \, dv_1, \tag{4}$$

or, for an infinitesimal change,

$$\delta\tau = \tau_2 - \tau_1 = -\tau \nabla \cdot \mathbf{s}. \tag{5}$$

The inductive capacity ϵ is a function of position in the dielectric and also of the density τ. The element which after the displacement finds itself at the fixed point $\mathbf{r}$ was located before the displacement at the point $\mathbf{r} - \mathbf{s}$, and the contribution to $\delta\epsilon$ arising from the inhomogeneity of the dielectric is therefore $-\mathbf{s} \cdot \nabla\epsilon$. If it be assumed that ϵ *depends only on* $\mathbf{r}$ *and* τ, the total variation is

$$\delta\epsilon = -\mathbf{s} \cdot \nabla\epsilon + \frac{\partial\epsilon}{\partial\tau} \delta\tau = -\mathbf{s} \cdot \nabla\epsilon - \tau \frac{\partial\epsilon}{\partial\tau} \nabla \cdot \mathbf{s}. \tag{6}$$

The change in electrostatic energy associated with the deformation is

$$\delta U = \frac{1}{2} \int \left(E^2 \nabla \epsilon \cdot \mathbf{s} + E^2 \tau \frac{\partial \epsilon}{\partial \tau} \nabla \cdot \mathbf{s} \right) dv. \tag{7}$$

Now

$$E^2 \tau \frac{\partial \epsilon}{\partial \tau} \nabla \cdot \mathbf{s} = \nabla \cdot \left(E^2 \tau \frac{\partial \epsilon}{\partial \tau} \mathbf{s} \right) - \mathbf{s} \cdot \nabla \left(E^2 \tau \frac{\partial \epsilon}{\partial \tau} \right), \tag{8}$$

and hence

$$\delta U = \int \left[\frac{1}{2} E^2 \nabla \epsilon - \frac{1}{2} \nabla \left(E^2 \tau \frac{\partial \epsilon}{\partial \tau} \right) \right] \cdot \mathbf{s}\, dv + \int \frac{1}{2} \nabla \cdot \left(E^2 \tau \frac{\partial \epsilon}{\partial \tau} \mathbf{s} \right) dv. \tag{9}$$

These volume integrals are to be extended over the entire field. The surfaces of the conductors, within which the field is zero, shall be denoted by $S_1, S_2, \ldots, S_n$. A surface S_0 is drawn to enclose within it all conductors and all parts of the dielectric where the field is of appreciable intensity. Within the volume bounded externally by S_0 and internally by $S_1, S_2, \ldots, S_n$ the function $E^2 \tau \frac{\partial \epsilon}{\partial \tau} \mathbf{s}$ is continuous. Thus the divergence theorem may be applied to the second integral of (9), giving

$$\int \frac{1}{2} \nabla \cdot \left(E^2 \tau \frac{\partial \epsilon}{\partial \tau} \mathbf{s} \right) dv = \sum_{j=0}^{n} \int_{S_j} E^2 \tau \frac{\partial \epsilon}{\partial \tau} \mathbf{s} \cdot \mathbf{n}\, da. \tag{10}$$

Over S_0 the intensity of $\mathbf{E}$ is zero, while over the rigid conductors $S_1, \ldots, S_n$ the normal component of $\mathbf{s}$ has been assumed zero. The integral (10) therefore vanishes.

If one neglects the effect of gravitational action, it may be assumed that the only body force $\mathbf{f}$ is that exerted by the field on elements of dielectric. The work done by this force per unit volume during the displacement is $\mathbf{f} \cdot \mathbf{s}$, and hence according to the principle of conservation of energy,

$$\int \mathbf{f} \cdot \mathbf{s}\, dv = -\int \left[\frac{1}{2} E^2 \nabla \epsilon - \frac{1}{2} \nabla \left(E^2 \tau \frac{\partial \epsilon}{\partial \tau} \right) \right] \cdot \mathbf{s}\, dv. \tag{11}$$

The displacement $\mathbf{s}$ is arbitrary and we find

$$\mathbf{f} = -\frac{1}{2} E^2 \nabla \epsilon + \frac{1}{2} \nabla \left(E^2 \tau \frac{\partial \epsilon}{\partial \tau} \right). \tag{12}$$

In case there are also charges distributed throughout the dielectric, a term $\rho \mathbf{E}$ must be added.

The last term of (12) is associated with the deformation of the dielectric. The assumption that ϵ can be expressed as a function of position and density alone is admissible for liquids and gases but is not necessarily valid in the case of solids. To a very good approximation the relation of dielectric constant κ, ($\epsilon = \kappa\epsilon_0$) to density in gases, liquids, and even some solids is expressed by the Clausius-Mossotti law,

$$\frac{\kappa - 1}{\kappa + 2} = C\tau, \qquad \text{or} \qquad \kappa - 1 = \frac{3C\tau}{1 - C\tau}. \tag{13}$$

where C is a constant determined by the nature of the dielectric. Through a simple calculation this leads to

$$\tau \frac{\partial \epsilon}{\partial \tau} = \frac{\epsilon_0}{3}(\kappa - 1)(\kappa + 2). \tag{14}$$

The force exerted by the field on a unit volume of a liquid or gas is, therefore,

$$\mathbf{f} = -\frac{\epsilon_0}{2} E^2 \nabla\kappa + \frac{\epsilon_0}{6} \nabla [E^2(\kappa - 1)(\kappa + 2)]. \tag{15}$$

2.22. Body Forces in Solids.—The volume force (12) exerted on a dielectric by an electrostatic field has been derived on the assumption that the variation in the inductive capacity ϵ during an infinitesimal displacement can be accounted for by the inhomogeneity of the dielectric and the change in density associated with the deformation. It is clear, however, that deformations in solids may occur without any accompanying change of volume, and that a rigorous theory must therefore express the variation of ϵ in terms of the components of strain. Since the dielectric is presumed to be in static equilibrium, the mechanical forces exerted by the field will be balanced by elastic forces induced during the deformation. Our problem is to find these forces and to determine the resultant deformation of the dielectric due to the applied field. We shall confine the investigation to media whose *electric and elastic properties in the unstrained state are isotropic.* Since the variations in ϵ are dependent on the components of strain, it is hardly to be expected that a solid will remain electrically isotropic after the strain is applied. Our first task is to set up an expression for the electrostatic energy of an anisotropic dielectric.

In Sec. 2.8 it was shown that the electrostatic energy density within a dielectric is

$$u = \int_0^{\mathbf{D}} \mathbf{E} \cdot d\mathbf{D} \tag{16}$$

without regard to the relation of $\mathbf{D}$ to $\mathbf{E}$. We shall now assume that in an anisotropic medium the components of $\mathbf{D}$ are linear functions of the components of $\mathbf{E}$.

$$D_j = \sum_{j=1}^{3} \epsilon_{jk} E_k, \qquad (j = 1, 2, 3). \tag{17}$$

Since the quantities E_k, D_j transform like the components of vectors it is clear that the ϵ_{jk} are the components of a tensor of second rank. By a rotation of the coordinate system this tensor may be referred to principal axes defined by the unit vectors **a**, **b**, **c** such that

$$D_a = \epsilon_a E_a, \qquad D_b = \epsilon_b E_b, \qquad D_c = \epsilon_c E_c. \tag{18}$$

Then

$$\mathbf{E} \cdot d\mathbf{D} = \epsilon_a E_a\, dE_a + \epsilon_b E_b\, dE_b + \epsilon_c E_c\, dE_c = \mathbf{D} \cdot d\mathbf{E}. \tag{19}$$

The energy density in an anisotropic, linear medium is, therefore,

$$u = \frac{1}{2}\mathbf{E} \cdot \mathbf{D} = \frac{1}{2}\sum_j \sum_k \epsilon_{jk} E_j E_k, \qquad (\epsilon_{jk} = \epsilon_{kj}). \tag{20}$$

We have next to calculate the change in the energy resulting from a variation of the parameters ϵ_{jk} while holding the charges fixed. A review of the proof of Sec. 2.10 shows that it may be adapted directly to our present needs. To avoid confusion of subscripts we shall replace the index 1 in Sec. 2.10 by a prime to denote initial conditions. Thus the dielectric in the field was characterized initially by the coefficients ϵ'_{jk}. Within a region V_1 these values are changed to ϵ_{jk}. The total change in the energy of the field, according to Eq. (49), page 113, is

$$U = \frac{1}{2}\int_{V_1} (\mathbf{E} \cdot \mathbf{D}' - \mathbf{E}' \cdot \mathbf{D})\, dv, \tag{21}$$

or in virtue of (17)

$$\begin{aligned} U = -\frac{1}{2}\int_{V_1} [(\epsilon_{11} - \epsilon'_{11})E_1E'_1 + (\epsilon_{22} - \epsilon'_{22})E_2E'_2 + (\epsilon_{33} - \epsilon'_{33})E_3E'_3 \\ + (\epsilon_{12} - \epsilon'_{12})(E_2E'_1 + E_1E'_2) + (\epsilon_{23} - \epsilon'_{23})(E_3E'_2 + E_2E'_3) \\ + (\epsilon_{31} - \epsilon'_{31})(E_1E'_3 + E_3E'_1)]\, dv. \end{aligned} \tag{22}$$

In case the variations in ϵ_{jk} are infinitesimal, this becomes

$$\begin{aligned} \delta U = -\frac{1}{2}\int_{V_1} (\delta\epsilon_{11}E_1^2 + \delta\epsilon_{22}E_2^2 + \delta\epsilon_{33}E_3^2 + 2\delta\epsilon_{12}E_1E_2 \\ + 2\delta\epsilon_{23}E_2E_3 + 2\delta\epsilon_{31}E_3E_1)\, dv. \end{aligned} \tag{23}$$

Finally, the parameters ϵ_{jk} must be related to the components of strain. For sufficiently small deformations we may put

$$\delta\epsilon_{jk} = \sum_{l=1}^{3}\sum_{m=1}^{3} a_{lm}^{jk}\, \delta e_{lm}, \qquad a_{lm}^{jk} = \frac{\partial \epsilon_{jk}}{\partial e_{lm}}. \tag{24}$$

The 81 coefficients a_{lm}^{jk} are the components of a tensor of rank four. In virtue of the relations $\epsilon_{jk} = \epsilon_{kj}$, $e_{lm} = e_{ml}$, we have

$$a_{lm}^{jk} = a_{lm}^{kj} = a_{ml}^{jk}. \tag{25}$$

Then of the $81 - 9 = 72$ nondiagonal terms, only 36 are independent and the total number of independent coefficients is reduced to $36 + 9 = 45$. To reduce the system still further use must be made of symmetry conditions and we now introduce the limitation to a medium which is initially isotropic,[1] though not necessarily homogeneous.

The variation in the density of the electrostatic energy due to a small deformation is

$$\delta u = -\frac{1}{2}\sum_{j=1}^{3}\sum_{k=1}^{3}\sum_{l=1}^{3}\sum_{m=1}^{3} a_{lm}^{jk}E_jE_k\,\delta e_{lm}. \tag{26}$$

The coefficients a_{lm}^{jk} have fixed values characteristic of the dielectric at each point, but varying from point to point in case the medium is inhomogeneous. Now since the dielectric is assumed to be initially isotropic, Eq. (26) must be invariant to a reversal in the direction of any coordinate axis and to the interchange of any two coordinate axes. Thus a reversal of the axis x_j reverses the signs of E_j and e_{jk}, $k \neq j$, but leaves all other factors unchanged. As a consequence, certain coefficients must vanish if δu is to be unaffected by the reversal or exchange of axes. In fact it is evident that all but three classes of coefficients are zero, namely:

$$a_{jj}^{jj} = a_1, \qquad a_{kk}^{jj} = a_2, \qquad a_{jk}^{jk} = a_3, \qquad (j \neq k). \tag{27}$$

Equation (26) is thereby reduced to

$$\begin{aligned}\delta u = -\tfrac{1}{2}\{&a_1(E_1^2\,\delta e_{11} + E_2^2\,\delta e_{22} + E_3^2\,\delta e_{33})\\ &+ a_2[(E_2^2 + E_3^2)\,\delta e_{11} + (E_3^2 + E_1^2)\,\delta e_{22} + (E_1^2 + E_2^2)\,\delta e_{33}]\\ &+ 4a_3(E_1E_2\,\delta e_{12} + E_2E_3\,\delta e_{23} + E_3E_1\,\delta e_{31})\}.\end{aligned} \tag{28}$$

Now δu must also be invariant to a rotation of the coordinate axes and this implies a further relation between the parameters a_1, a_2, a_3. The condition is easily found by rewriting (28) in the form

$$\begin{aligned}\delta u = -\frac{1}{2}\Bigg[&(a_1 - a_2 - 4a_3)\sum_{j=1}^{3} E_j^2\,\delta e_{jj} + a_2E^2\sum_{j=1}^{3}\delta e_{jj}\\ &+ 4a_3(E_1^2\,\delta e_{11} + E_2^2\,\delta e_{22} + E_3^2\,\delta e_{33} + E_1E_2\,\delta e_{12}\\ &+ E_2E_3\,\delta e_{23} + E_3E_1\,\delta e_{31})\Bigg],\end{aligned} \tag{29}$$

in which $E^2 = E_1^2 + E_2^2 + E_3^2$. The sum $\sum_{j=1}^{3} e_{jj}$ is, according to (54), page 92, the cubical dilatation, a quantity not dependent on the coordinate system. The middle term of (29) is, therefore, invariant to a rotation of the reference system. The same is true also of the last term; for if the strains e_{jk} are replaced by the coefficients $2a_{jk}$ of Sec. 2.2, this last

[1] The anisotropic case is discussed by POCKELS, Encyklopädie der mathematischen Wissenschaften, Vol. V, Part II, Teubner, 1906.

term is an invariant quadric similar in form to the strain quadric of Eq. (36), page 89. Only the first term of (29) is variant with the coordinates and it is necessary that[1]

$$a_1 - a_2 - 4a_3 = 0, \qquad a_3 = \tfrac{1}{4}(a_1 - a_2). \tag{30}$$

The variation in the density of electrostatic energy due to the local deformation, or pure strain, of an isotropic dielectric is finally

$$\begin{aligned}\delta u_s = -\tfrac{1}{2}[(a_1E_1^2 + a_2E_2^2 + a_2E_3^2)\,\delta e_{11} + (a_2E_1^2 + a_1E_2^2 + a_2E_3^2)\,\delta e_{22} \\ + (a_2E_1^2 + a_2E_2^2 + a_1E_3^2)\,\delta e_{33} \\ + (a_1 - a_2)(E_1E_2\,\delta e_{12} + E_2E_3\,\delta e_{23} + E_3E_1\,\delta e_{31})].\end{aligned} \tag{31}$$

The subscript s has been added to emphasize the fact that this is only that portion of the total variation which arises from a pure strain. The most general deformation of an elastic medium, it will be remembered, is composed of a translation, a local rotation defined by (33), page 88, and the local strain defined by (29), page 88, of which we have taken account in (31). In an anisotropic dielectric the local rotation gives rise to variations in the tensor components ϵ_{jk} and consequently a system of torques must act on each volume element.[2] In the initially isotropic medium considered here these rotational variations are absent. The variation in the density of electrostatic energy associated with an infinitesimal translation $\delta\mathbf{s}$ of an inhomogeneous dielectric was calculated in Sec. 2.21.

$$\delta u_t = \tfrac{1}{2}E^2\nabla\epsilon \cdot \delta\mathbf{s}, \tag{32}$$

where ϵ is the inductive capacity in the unstrained, isotropic state.

A deformation of the dielectric occasions a variation in the *elastic energy* as well as in the electrostatic energy. According to (68), page 94,

$$\begin{aligned}\delta u_e = (\lambda_1\nabla\cdot\mathbf{s} + 2\lambda_2e_{11})\,\delta e_{11} + (\lambda_1\nabla\cdot\mathbf{s} + 2\lambda_2e_{22})\,\delta e_{22} \\ + (\lambda_1\nabla\cdot\mathbf{s} + 2\lambda_2e_{33})\,\delta e_{33} + \lambda_2(e_{12}\,\delta e_{12} + e_{23}\,\delta e_{23} + e_{31}\,\delta e_{31}).\end{aligned} \tag{33}$$

Now the work done by the mechanical forces acting on the dielectric within the volume V_1 and over the surface S_1 bounding V_1 during an infinitesimal displacement $\delta\mathbf{s}$ must equal the decrease in the *total* available energy, elastic plus electrostatic. We write, therefore, for the energy balance:

[1] The procedure followed here in reducing the constants of electrostriction to two is identical with that which leads to the reduction of the elastic constants c_{jk}, Sec. 2.3, to the two parameters λ_1 and λ_2. See, for example, LOVE, "Treatise on the Mathematical Theory of Elasticity," Chap. VI, 4th ed., Cambridge Press, 1927. The determination of these constants for various classes of anisotropic crystals was made by Voigt, "Kompendium der theoretischen Physik," Vol. I, pp. 143–144, Leipzig, 1895. His results are cited by Love, *loc. cit.*

[2] POCKELS, *loc. cit.*, p. 353.

$$\int_{V_1} \mathbf{f} \cdot \delta\mathbf{s}\, dv + \int_{S_1} \mathbf{t} \cdot \delta\mathbf{s}\, da = -\delta \int_{V_1} (u_e + u_s + u_t)\, dv. \tag{34}$$

It will be convenient to resolve the total body force per unit volume **f** arbitrarily into the two components **f**′ and **f**″, where

$$\mathbf{f}' = -\tfrac{1}{2}E^2 \nabla\epsilon \tag{35}$$

is the contribution resulting from the inhomogeneity of the dielectric and where **f**″ is the force associated with a pure strain. The latter must satisfy the relation

$$\int_{V_1} \mathbf{f}'' \cdot \delta\mathbf{s}\, dv + \int_{S_1} \mathbf{t}'' \cdot \delta\mathbf{s}\, da = -\delta \int_{V_1} (u_e + u_s)\, dv. \tag{36}$$

The resolution into translational and strain components of body and surface force is such that

$$\int_{V_1} \mathbf{f}'\, dv + \int_{S_1} \mathbf{t}'\, da = 0, \qquad \int_{V_1} \mathbf{f}''\, dv + \int_{S_1} \mathbf{t}''\, da = 0. \tag{37}$$

Then by Sec. 2.3, the left-hand side of (36) may be transformed to

$$\delta W'' = \int_{V_1} (T''_{11}\, \delta e_{11} + T''_{22}\, \delta e_{22} + T''_{33}\, \delta e_{33} + T''_{12}\, \delta e_{12} + T''_{23}\, \delta e_{23} + T''_{31}\, \delta e_{31})\, dv. \tag{38}$$

The variations in the strain components are arbitrary and hence on equating coefficients of corresponding terms from (31) and (33) we obtain

$$\begin{aligned}
T''_{11} &= -(\lambda_1 \nabla \cdot \mathbf{s} + 2\lambda_2 e_{11}) + \tfrac{1}{2}(a_1 E_1^2 + a_2 E_2^2 + a_2 E_3^2),\\
T''_{22} &= -(\lambda_1 \nabla \cdot \mathbf{s} + 2\lambda_2 e_{22}) + \tfrac{1}{2}(a_2 E_1^2 + a_1 E_2^2 + a_2 E_3^2),\\
T''_{33} &= -(\lambda_1 \nabla \cdot \mathbf{s} + 2\lambda_2 e_{33}) + \tfrac{1}{2}(a_2 E_1^2 + a_2 E_2^2 + a_1 E_3^2),\\
T''_{12} &= -\lambda_2 e_{12} + \frac{a_1 - a_2}{2} E_1 E_2,\\
T''_{23} &= -\lambda_2 e_{23} + \frac{a_1 - a_2}{2} E_2 E_3,\\
T''_{31} &= -\lambda_2 e_{31} + \frac{a_1 - a_2}{2} E_3 E_1.
\end{aligned} \tag{39}$$

These are the components of a stress tensor whose negative divergence gives us the resultant body force associated with a strain. If the body force is solely of electrical origin, as will be the case when gravitational action is neglected, the divergence of the elastic stresses will vanish and we obtain for the x-component of force:

$$f_1'' = -\frac{1}{2}\frac{\partial}{\partial x_1}(a_1 E_1^2 + a_2 E_2^2 + a_2 E_3^2) - \frac{1}{2}\frac{\partial}{\partial x_2}[(a_1 - a_2)E_1 E_2] - \frac{1}{2}\frac{\partial}{\partial x_3}[(a_1 - a_2)E_1 E_3], \tag{40}$$

with analogous expressions for f_2'' and f_3''. To these are added the components of $\mathbf{f}'$ from (35) to obtain the total body force exerted by an electrostatic field in an isotropic dielectric.

The parameters a_1 and a_2 must in general be determined for a given substance by measurement. Physically the parameter a_1 expresses the increment of ϵ corresponding to an elongation parallel to the lines of field intensity, while a_2 determines this increment for strains at right angles to these lines.

If the dielectric within V_1 is *homogeneous and contains no charge,* the gradient of ϵ is zero and the vector $\mathbf{E}$ satisfies $\nabla \cdot \mathbf{E} = 0$, $\nabla \times E = 0$. In this case $\mathbf{f}' = 0$ and the resultant force $\mathbf{f}'' = \mathbf{f}$ reduces to

$$\mathbf{f} = -\tfrac{1}{4}(a_1 + a_2)\nabla E^2. \tag{41}$$

In a liquid or gas the shearing strains e_{jk} are all zero and hence by Eqs. (24) and (27) $a_3 = a_{jk}^{jk} = 0$, $a_1 = a_2 = a$. Furthermore there can be no preferred directions in the dielectric properties of a fluid and consequently $\epsilon_{jk} = 0$ when $j \neq k$, $\epsilon_{11} = \epsilon_{22} = \epsilon_{33} = \epsilon$. In place of (24) we may write:

$$\delta\epsilon = \frac{\partial\epsilon}{\partial e_{11}}\,\delta e_{11} + \frac{\partial\epsilon}{\partial e_{22}}\,\delta e_{22} + \frac{\partial\epsilon}{\partial e_{33}}\,\delta e_{33}. \tag{42}$$

Denoting the cubical dilatation by $\Delta = e_{11} + e_{22} + e_{33}$, we have

$$a = \frac{\partial\epsilon}{\partial e_{jj}} = \frac{\partial\epsilon}{\partial\Delta}, \qquad \delta\epsilon = a\,\delta\Delta. \tag{43}$$

Now by (5)

$$\frac{\partial\epsilon}{\partial\Delta} = \frac{\partial\epsilon}{\partial\tau}\frac{\partial\tau}{\partial\Delta} = -\tau\frac{\partial\epsilon}{\partial\tau}, \tag{44}$$

and consequently the total force per unit volume exerted by an electrostatic field on a fluid dielectric is

$$\mathbf{f} = -\frac{1}{2}E^2\nabla\epsilon + \frac{1}{2}\nabla\left(E^2\tau\frac{\partial\epsilon}{\partial\tau}\right), \tag{45}$$

as was deduced directly in Sec. 2.21.

The derivation of the volume forces from an energy principle as in the foregoing paragraphs seems to have been proposed first by Korteweg[1] and developed by Helmholtz[2] and others. A complete account of the theory with references to the older literature is given by Pockels.[3] The energy method has been criticized by Larmor[4] and Livens[5] who propose

[1] KORTEWEG, *Wied. Ann.*, **9**, 1880.

[2] HELMHOLTZ, *Wied. Ann.*, **13**, 385, 1881.

[3] POCKELS, *Arch. Math. Phys.*, (2) **12**, 57–95, 1893, and in the Encyklopädie der mathematischen Wissenschaften, Vol. V, Part II, pp. 350–392, 1906.

[4] LARMOR, *Phil. Trans.*, **A. 190**, 280, 1897.

[5] LIVENS, *Phil. Mag.*, **32**, 162, 1916, and in his text "The Theory of Electricity," p. 93, Cambridge University Press, 1926.

alternative expressions for the forces. These criticisms, however, do not appear to be well founded. Objections to the particular form of the energy integral employed by Helmholtz are satisfied by a more careful procedure. Livens has undertaken a generalization of the energy method to media in which the relation between **D** and **E** is nonlinear in order to show that it leads to absurd results; in so doing he has omitted the essential term associated with the deformation. The alternative expressions for the volume force proposed by Livens may on the other hand be derived very simply, in both the electric and magnetic cases, by assuming that polarized matter is equivalent to a region occupied by a charge of density $\rho' = -\nabla \cdot \mathbf{P}$ and current of density $\mathbf{J}' = \frac{\partial \mathbf{P}}{\partial t} + \nabla \times \mathbf{M}$ (*cf.* Sec. 1.6). The forces may then be calculated exactly as in Sec. 2.5. Such a procedure would be justifiable were the medium absolutely rigid. According to the Livens theory one calculates the force exerted by the field on the polarized matter, and then introduces these forces into the equations of elasticity to determine the deformation. The deformation, however, affects the polarizations and the two parts of the problem cannot, in general, be handled separately in any such fashion. Under certain conditions, particularly in fluids, the two theories lead to identical results, but in most circumstances they differ. There appears to be little reason to doubt that the energy method of Korteweg and Helmholtz is fundamentally sound.

2.23. The Stress Tensor.—We shall suppose again that S_1 is a closed surface drawn within an isotropic dielectric under electrical stress. The properties of the dielectric are assumed to be continuous across this surface and at all interior points; it is not, in other words, a surface of discontinuity bounding a whole dielectric body. The total force exerted on the matter and charge within S_1 is

$$\mathbf{F} = \int_{V_1} \left(\rho \mathbf{E} - \frac{1}{2} E^2 \nabla \epsilon + \mathbf{f}'' \right) dv. \tag{46}$$

We wish to show now that the force **F** can be represented by a surface integral over S_1. To this end it is only necessary to express the integrand as the divergence of a tensor. The term $\mathbf{f}''$ appears already in this form in (40) and therefore needs no further transformation. In the first term we replace ρ by $\nabla \cdot \mathbf{D}$ and then make use of the identity

$$E^2 \nabla \epsilon = \nabla (\epsilon E^2) - 2(\mathbf{D} \cdot \nabla)\mathbf{E}, \tag{47}$$

which holds only when $\nabla \times \mathbf{E} = 0$. It follows without difficulty that the components of the vector $\rho \mathbf{E} - \frac{1}{2} E^2 \nabla \epsilon$ are

$$\rho E_j - \frac{1}{2} E^2 \frac{\partial \epsilon}{\partial x_j} = \sum_{k=1}^{3} \frac{\partial}{\partial x_k} (E_j D_k) - \frac{1}{2} \frac{\partial}{\partial x_j} (\mathbf{E} \cdot \mathbf{D}), \tag{48}$$

and the integrand of (46) can thus be expressed as the divergence of a tensor ${}^2\mathbf{S}$,

$$\rho E - \tfrac{1}{2}E^2\nabla\epsilon + \mathbf{f}'' = \nabla \cdot {}^2\mathbf{S}, \tag{49}$$

whose components S_{jk} are tabulated below.

$$\begin{aligned}
S_{11} &= \frac{\epsilon}{2}(E_1^2 - E_2^2 - E_3^2) - \frac{1}{2}(a_1E_1^2 + a_2E_2^2 + a_2E_3^2),\\
S_{12} &= \epsilon E_1E_2 + \frac{a_2 - a_1}{2}E_1E_2 = S_{21},\\
S_{13} &= \epsilon E_1E_3 + \frac{a_2 - a_1}{2}E_1E_3 = S_{31},\\
S_{22} &= \frac{\epsilon}{2}(E_2^2 - E_3^2 - E_1^2) - \frac{1}{2}(a_2E_1^2 + a_1E_2^2 + a_2E_3^2),\\
&\cdots\cdots\cdots\cdots\cdots\cdots\cdots\cdots
\end{aligned} \tag{50}$$

Upon applying the divergence theorem expressed in Eqs. (22) and (23), page 100, it appears that the force $\mathbf{F}$ exerted on the charge and dielectric matter within S_1 is equivalent to the integral over S_1 of a surface force of density

$$\mathbf{t} = \left(\epsilon + \frac{a_2 - a_1}{2}\right)\mathbf{E}(\mathbf{E}\cdot\mathbf{n}) - \frac{\epsilon + a_2}{2}E^2\mathbf{n}, \tag{51}$$

where $\mathbf{n}$ is as usual the outward unit normal to an element of S_1. To maintain equilibrium an equal and opposite force per unit area must be exerted by the material outside S_1 on the particles composing S_1.

The tensor components (50) and the surface force $\mathbf{t}$ differ from the corresponding expressions developed for free space in Sec. 2.5 in that ϵ_0 is replaced by ϵ and the deformation is accounted for by the constants a_1 and a_2, thus giving rise to additional terms which may well be important. At any point on the surface S_1 the vector $\mathbf{t}$ lies in the plane of $\mathbf{E}$ and $\mathbf{n}$, but is no longer oriented such that $\mathbf{E}$ bisects the angle between $\mathbf{n}$ and $\mathbf{t}$ as was the case illustrated by Fig. 17, page 102.

2.24. Surfaces of Discontinuity.—Throughout the previous analysis it has been assumed that abrupt discontinuities in the properties of the medium, such as must occur across surfaces bounding dielectric bodies, may be replaced by layers of rapid but continuous transition. As the transition layer becomes vanishingly thin, a discontinuity is generated that gives rise to infinite values for the gradient of ϵ. But the volume within the layer also vanishes and we must seek the limit of the force $\mathbf{f}$ per unit volume to learn what total force is exerted on the resultant surface of discontinuity.

In Fig. (23) the crosshatched area represents a section of a transition layer between dielectric media (1) and (2). The force exerted by the field on the matter and charge within this layer can be found by integrating (51) over the two faces. The two essential parameters that characterize an isotropic dielectric will be denoted as

$$\alpha = \epsilon + \frac{a_2 - a_1}{2}, \qquad \beta = \frac{\epsilon + a_2}{2}. \tag{52}$$

The force acting on unit area in the direction from medium (1) towards (2) as the thickness of the layer approaches zero is then

$$\mathbf{t} = [\alpha\mathbf{E}(\mathbf{E}\cdot\mathbf{n})]_2 + [\alpha\mathbf{E}(\mathbf{E}\cdot\mathbf{n})]_1 - [\beta E^2\mathbf{n}]_2 - [\beta E^2\mathbf{n}]_1, \tag{53}$$

the subscripts indicating that the values are to be taken on either side of S. It should be noted that $\mathbf{n}_1 = -\mathbf{n}_2$.

Very few reliable measurements of the constants a_1 and a_2 for solids have been reported in the literature. The traction $\mathbf{t}$ may, however, be computed in the case of an interface separating two fluids to which the Clausius-Mossotti law can be applied. According to Eqs. (43) and (14),

$$a_1 = a_2 = -\tau\frac{\partial\epsilon}{\partial\tau} = -\frac{\epsilon_0}{3}(\kappa - 1)(\kappa + 2). \tag{54}$$

Then for a fluid,

$$\alpha = \epsilon, \qquad \beta = -\frac{\epsilon_0}{6}(\kappa^2 - 2\kappa - 2). \tag{55}$$

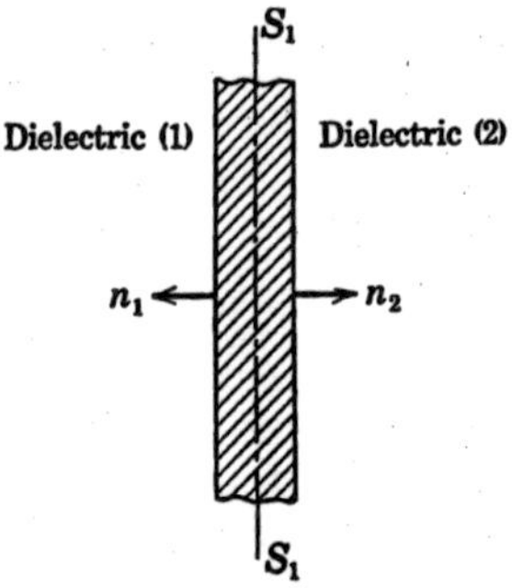

FIG. 23.—Section of a transition layer separating two dielectric media.

At the interface the tangential components of $\mathbf{E}$ are continuous, $E_{t1} = E_{t2}$, and in the absence of surface charge we have also $\epsilon_1 E_{n1} = \epsilon_2 E_{n2}$. The field $\mathbf{E}$ in (53) may, therefore, be expressed in terms of its intensity in either medium. The values of α and β from (55) introduced into (53) lead to

$$\mathbf{t} = \frac{\epsilon_0}{6}\left\{\frac{1}{\kappa_2}(\kappa_1^2\kappa_2^2 + 4\kappa_1^2\kappa_2 - 6\kappa_1\kappa_2^2 - 3\kappa_1^2 + 2\kappa_1 + 2)E_{n1}^2 + [\kappa_2(\kappa_2 - 2) - \kappa_1(\kappa_1 - 2)]E_{t1}^2\right\}\mathbf{n}_2. \tag{56}$$

The traction exerted by the field on the interface of two fluids is *normal* to the surface, directed from medium (1) toward (2), a necessary consequence, of course, of the fact that a fluid supports no shearing stress.

In case medium (2) is air, κ_2 may be placed approximately equal to unity. Equation (56) then reduces to

$$\mathbf{t} = \frac{\epsilon_0(\kappa_1 - 1)^2}{6}(2E_{n1}^2 - E_{t1}^2)\mathbf{n}_2. \tag{57}$$

The traction is a maximum when **E** is normal to the surface, and is in the direction of diminishing inductive capacity (*cf.* page 114). However, if **E** lies in the surface, $E_{n1} = 0$, and the traction becomes negative, implying a pressure exerted by the field on dielectric κ_1.

Another case of interest is that in which medium (1) is a conductor. Then $\mathbf{E}_1 = 0$, $E_{t2} = 0$ and (53) reduces to

$$\mathbf{t} = (\alpha - \beta)E_n^2\mathbf{n} = \frac{\epsilon - a_1}{2} E_n^2, \tag{58}$$

all quantities having their values just outside the metal surface in the dielectric (1). In case this dielectric is a fluid,

$$\mathbf{t} = \frac{1}{2}\left(\epsilon + \tau\frac{\partial \epsilon}{\partial \tau}\right) E_n^2\mathbf{n} = \frac{1}{2}\frac{\partial}{\partial \tau}(\epsilon\tau)E_n^2\mathbf{n}. \tag{59}$$

These expressions are valid whether or not the conductors are charged, for E_n is related to the surface charge by $\epsilon E_n = \omega$. *But Eqs.* (58) *and* (59) *do not represent the resultant force per unit area exerted on the surface,* for they will be compensated to a certain degree by stresses or pressures generated by the field within the dielectric. These we shall now investigate. Only in case medium (1) is free space or a gas of negligible inductive capacity can this internal pressure be ignored. We have then $\tau \dfrac{\partial \epsilon}{\partial \tau} = 0$ and (59) becomes

$$\mathbf{t} = \tfrac{1}{2}\epsilon_0 E_n^2\mathbf{n} = \tfrac{1}{2}\omega E_n\mathbf{n}, \tag{60}$$

where ω is the charge per unit area.

2.25. Electrostriction.—The elastic deformation of a dielectric under the forces exerted by an electrostatic field is called *electrostriction*. The displacement **s** at any interior point of an isotropic substance must satisfy Eq. (70), page 95:

$$\mathbf{f} + (\lambda_1 + \lambda_2)\nabla\nabla \cdot \mathbf{s} + \lambda_2\nabla^2\mathbf{s} = 0, \tag{61}$$

where now the body force $\mathbf{f} = \mathbf{f}' + \mathbf{f}''$ is given in general by (35) and (40). To these there may be added, when the occasion demands, a force $\mathbf{E}\rho$ to account for a volume charge, and a gravitational force τg. Solutions of (61) must be found in an appropriate system of coordinates. These solutions are subject to boundary conditions over a surface bounding the dielectric. On this surface either the applied stresses or the displacement must be specified. Thus if the surface is free, stresses will be exerted upon it by the field and from these may be determined the boundary values of **s**.

In solids the effect is small and easily masked by deformations arising from extraneous causes. The attractive forces between metal electrodes in contact with the dielectric may bring about deformations within the volume of the dielectric which have nothing to do with electrostriction. Experimental data on the subject are not abundant and frequently contradictory. The theory of electrostriction in a cylindrical condenser, however, has been developed in considerable detail. The method described above was applied by Adams,[1] and his results were later reconciled by Kemble[2] with the earlier work of Sacerdote.[3] An extensive list of references is given by Cady in the International Critical Tables.[4]

The electrostriction of fluids is more amenable to calculation, for the shearing modulus λ_2 is then zero and λ_1 is equal to the reciprocal of the compressibility. According to (73), page 95, the pressure within the fluid is related to the volume dilatation by $p = -\lambda_1 \nabla \cdot \mathbf{s}$. This expression, together with the body force $\mathbf{f}$ from Eq. (45), introduced into (61), leads to the relation

$$(62)\qquad -\frac{1}{2} E^2 \nabla \epsilon + \frac{1}{2} \nabla \left(E^2 \tau \frac{\partial \epsilon}{\partial \tau} \right) - \nabla p = 0$$

between pressure and field intensity at any point in the fluid. To integrate this equation the functional dependence of both pressure and inductive capacity on the density τ must be specified. In a chemically homogeneous liquid or gas, one may assume that $p = p(\tau)$, $\epsilon = \epsilon(\tau)$, and that the dependence on position is implicit in the independent variable τ. Then

$$(63)\qquad \begin{aligned} \nabla \left(E^2 \frac{\partial \epsilon}{\partial \tau} \tau \right) &= \tau \nabla \left(E^2 \frac{d\epsilon}{d\tau} \right) + E^2 \frac{d\epsilon}{d\tau} \nabla \tau, \\ \nabla \epsilon &= \frac{d\epsilon}{d\tau} \nabla \tau, \end{aligned}$$

and hence (62) reduces to

$$(64)\qquad \frac{1}{2} \nabla \left(E^2 \frac{d\epsilon}{d\tau} \right) = \frac{1}{\tau} \nabla p.$$

A scalar function $P(p)$ of pressure is now defined such that

$$(65)\qquad P(p) = \int_{p_0}^{p} \frac{1}{\tau(p)}\, dp, \qquad \nabla P = \frac{1}{\tau} \nabla p,$$

where p_0 is the pressure at a point in the fluid where $\mathbf{E} = 0$. If $d\mathbf{s}$ is an element of length along any path connecting two points whose pressures are respectively p and p_0, then $\nabla P \cdot d\mathbf{s} = dP$, and (64) is satisfied by

$$(66)\qquad \int_{p_0}^{p} \frac{1}{\tau}\, dp = \frac{1}{2} E^2 \frac{d\epsilon}{d\tau}.$$

[1] ADAMS, *Phil. Mag.*, (6) **22**, 889, 1911.
[2] KEMBLE, *Phys. Rev.*, (2) **7**, 614, 1916.
[3] SACERDOTE, *Jour. physique*, (3) **8**, 457, 1899; **10**, 196, 1901.
[4] International Critical Tables, Vol. VI, p. 207, 1929.

In liquids $dp = \lambda_1 \frac{d\tau}{\tau}$, Eq. (77), page 96, or

$$\tau = \tau_0 e^{\frac{p-p_0}{\lambda_1}}. \tag{67}$$

The parameter λ_1, the reciprocal of the compressibility, is a very large quantity and hence $\tau \simeq \tau_0$. For *liquids*, therefore,

$$p - p_0 = \frac{1}{2} E^2 \tau \frac{d\epsilon}{d\tau}, \tag{68}$$

or, upon applying the Clausius-Mossotti law,

$$p - p_0 = \frac{\epsilon_0}{6} E^2(\kappa - 1)(\kappa + 2). \tag{69}$$

In gases (67) is replaced by the relation $p = \tau \frac{RT}{M}$, where R is the gas constant, T the absolute temperature, and M the molecular weight.[1]

2.26. Force on a Body Immersed in a Fluid.—The results of our analysis may be summed up profitably by a consideration of the following

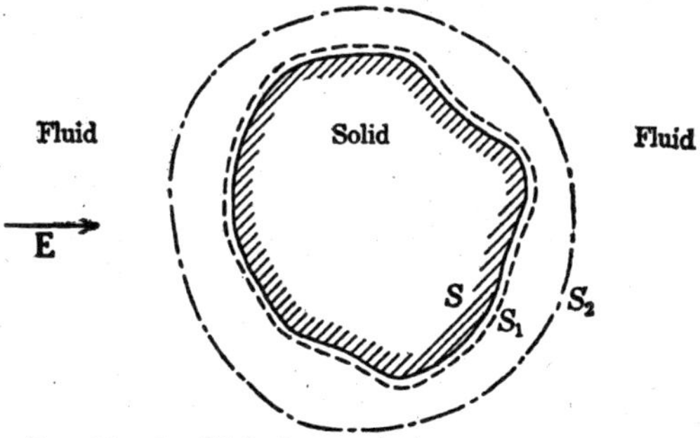

FIG. 24.—A solid body immersed in a fluid dielectric.

problem. In Fig. 24, a dielectric or conducting body is shown immersed in a fluid. An electrostatic field is applied and we desire an expression for the *resultant force on the entire body*. Let us draw a surface S_1 enclosing the body and located in the fluid just outside the boundary S. On every volume element of the solid there is a force of density

$$\mathbf{f} = \rho \mathbf{E} - \tfrac{1}{2} E^2 \nabla \epsilon + \mathbf{f}'', \tag{70}$$

where $\mathbf{f}''$ is given by Eq. (40), and on each element of the surface of discontinuity there is a traction given by (53). Rather than evaluate the integrals of $\mathbf{f}$ and $\mathbf{t}$ over the volume and bounding surface S, we need only

[1] ABRAHAM, BECKER, and DOUGALL, "The Classical Theory of Electricity and Magnetism," p. 98, Blackie, 1932.

calculate the integral of $\mathbf{t}$ from Eq. (51) over the surface S_1. Since S_1 lies in the fluid, $a_1 = a_2 = -\tau \frac{\partial \epsilon}{\partial \tau}$ and (51) reduces to

$$\mathbf{t} = \epsilon \mathbf{E}(\mathbf{E} \cdot \mathbf{n}) - \frac{\epsilon}{2} E^2 \mathbf{n} + \frac{\tau}{2} \frac{\partial \epsilon}{\partial \tau} E^2 \mathbf{n}, \tag{71}$$

where $\mathbf{n}$ is the unit normal directed from the solid into the fluid and ϵ and τ apply to the fluid. Now the integral of this traction over S_1 gives the force exerted *directly* by the field on the volume and surface of the body. But the field, as we have just seen, also generates a pressure (68) in the fluid which at every point on S_1 acts normally *inward*, *i.e.*, toward the solid. The resultant force per unit area transmitted across S_1 is, therefore, equal to (71) diminished by (68), and the net force exerted by the field on the solid—the force which must be compensated by exterior supports—is

$$\mathbf{F} = \int_{S_1} \left[\epsilon \mathbf{E}(\mathbf{E} \cdot \mathbf{n}) - \frac{\epsilon}{2} E^2 \mathbf{n} \right] da. \tag{72}$$

Thus to obtain the resultant force on an *entire* body we need not know the constants a_1 and a_2 within either the solid or the fluid, for the forces with which they are associated are compensated *locally* by elastic stresses. Furthermore, since the fluid is in equilibrium, it is not essential that S_1 lie in the immediate neighborhood of the body surface S. The net force on the liquid contained between S_1 and an arbitrary enclosing surface S_2 is zero and consequently

$$\int_{S_1} \left[\epsilon \mathbf{E}(\mathbf{E} \cdot \mathbf{n}) - \frac{\epsilon}{2} E^2 \mathbf{n} \right] da = \int_{S_2} \left[\epsilon \mathbf{E}(\mathbf{E} \cdot \mathbf{n}) - \frac{\epsilon}{2} E^2 \mathbf{n} \right] da, \tag{73}$$

if we adhere to the convention that $\mathbf{n}$ is directed *outward* from the enclosed region. The surface S_2 may, therefore, be chosen in any manner that will facilitate integration, provided only that no foreign bodies are intersected or enclosed.

In case the solid is a conductor, the field $\mathbf{E}$ is normal at the boundary. The resultant force on an isolated conductor is then

$$\mathbf{F} = \frac{1}{2} \int_{S_1} \epsilon E^2 \mathbf{n} \, da = \frac{1}{2} \int_{S_1} \omega \mathbf{E} \, da, \tag{74}$$

where ω is the charge density on S. This last is true, of course, *only* for a surface S_1 just outside the conductor.

It should be evident now that the force on one solid embedded in another can be calculated only when the nature of the contact over their common surface is specified; the problem is complicated by the fact that tangential shearing stresses as well as normal pressures must be compensated by elastic stresses across the boundary.

FORCES IN THE MAGNETOSTATIC FIELD

2.27. Nonferromagnetic Materials.—The analysis of Sec. 2.22 may be applied directly to the magnetostatic field in all cases where the relation between **B** and **H** is of the form

$$B_j = \sum_{k=1}^{3} \mu_{jk} H_k \qquad (j = 1, 2, 3), \tag{1}$$

with the components μ_{jk} of the permeability tensor functions possibly of position but independent of field intensity. The change in magnetic energy of an isotropic body resulting from a variation in μ is given by Eq. (52), Sec. 2.17.

$$\delta T = \frac{1}{2} \int \delta\mu \, H^2 \, dv, \tag{2}$$

and the generalization of this expression to an anisotropic body occupying the volume V_1 is

$$\delta T = \frac{1}{2} \int_{V_1} (\delta\mu_{11} H_1^2 + \delta\mu_{22} H_2^2 + \delta\mu_{33} H_3^2 + 2\delta\mu_{12} H_1 H_2 + 2\delta\mu_{23} H_2 H_3 + 2\delta\mu_{31} H_3 H_1) \, dv. \tag{3}$$

This variation may be expressed in terms of the components of strain, assuming a linear relation between the components of the permeability and strain tensors.

$$\delta\mu_{jk} = \sum_{l=1}^{3} \sum_{m=1}^{3} b_{lm}^{jk} \, \delta e_{lm}, \qquad b_{lm}^{jk} = \frac{\partial \mu_{jk}}{\partial e_{lm}}. \tag{4}$$

If we assume further that *in the unstrained state the medium is isotropic*, the coefficients b_{lm}^{jk} reduce to two,

$$b_1 = \frac{\partial \mu_{jj}}{\partial e_{jj}}, \qquad b_2 = \frac{\partial \mu_{jj}}{\partial e_{kk}}, \qquad \frac{1}{4}(b_1 - b_2) = \frac{\partial \mu_{jk}}{\partial e_{jk}}, \tag{5}$$

so that the variation in magnetic energy due to a pure strain is

$$\delta T = \frac{1}{2} \int_{V_1} [(b_1 H_1^2 + b_2 H_2^2 + b_2 H_3^2) \, \delta e_{11} + (b_2 H_1^2 + b_1 H_2^2 + b_2 H_3^2) \, \delta e_{22} + (b_2 H_1^2 + b_2 H_2^2 + b_1 H_3^2) \, \delta e_{33} + (b_1 - b_2)(H_1 H_2 \, \delta e_{12} + H_2 H_3 \, \delta e_{23} + H_3 H_1 \, \delta e_{31})] \, dv. \tag{6}$$

Equation (6) differs formally from (31), page 143, only in algebraic sign. Now the electrostatic energy U represents the work done in building up the field against the mutual forces between elements of charge. The magnetic energy T, on the other hand, represents work done against the mutual forces exerted by elements of current, *plus* the work done against induced electromotive forces. The work required to bring about a small displacement or deformation of a body in a magnetic field is done partly against the mechanical forces acting on the body, partly on the current sources to maintain them constant.[1] When the work done against the induced electromotive forces is subtracted from the total magnetic energy, there remains the potential energy of the mutual mechanical forces, and this we found in Sec. 2.14 to be equal and opposite to T.

$$\delta U = -\delta T. \tag{7}$$

The forces exerted by the field on the body within V_1 are to be calculated from (6) *with sign reversed.*

In complete analogy with the electrostatic case we now find that the body force exerted by a magnetic field on a medium which in the unstrained state is isotropic, whose permeability is independent of field intensity, which is free of residual magnetism, but which may carry a current of density $\mathbf{J}$, is

$$f_i = \mu(\mathbf{J} \times \mathbf{H})_i - \frac{1}{2} H^2 \frac{\partial \mu}{\partial x_i} - \frac{1}{2} \sum_{k=1}^{3} \frac{\partial}{\partial x_k} [(b_1 - b_2)H_i H_k] - \frac{1}{2} \frac{\partial}{\partial x_i} (b_2 H^2). \tag{8}$$

If the medium is homogeneous and carries no current, (8) reduces to

$$\mathbf{f} = -\tfrac{1}{4}(b_1 + b_2)\nabla H^2. \tag{9}$$

In a gas or liquid $b_1 = b_2 = -\tau \frac{\partial \mu}{\partial \tau}$, and (8) can be written in the form

$$\mathbf{f} = \mu \mathbf{J} \times \mathbf{H} - \frac{1}{2} H^2 \nabla \mu + \frac{1}{2} \nabla \left(H^2 \tau \frac{\partial \mu}{\partial \tau} \right). \tag{10}$$

As in the electrostatic case, the body force whose components are given by (8) can be expressed as the divergence of a tensor ${}^2\mathbf{S}$,

$$\mathbf{f} = \nabla \cdot {}^2\mathbf{S}, \tag{11}$$

whose components are

$$S_{ij} = \left(\mu + \frac{b_2 - b_1}{2} \right) H_j^2 - \frac{\mu + b_2}{2} H^2,$$

[1] This remark does not apply if the source is a permanent magnet; in that case the energy of the field is not given by $T = \frac{1}{2}\int B \cdot H \, dv$.

(12)
$$S_{jk} = \left(\mu + \frac{b_2 - b_1}{2}\right) H_j H_k, \qquad (j \neq k).$$

The force exerted by the magnetic field on the matter within a closed surface S_1 is, therefore, the same as would result from the application over S_1 of a force of surface density

(13)
$$\mathbf{t} = \left(\mu + \frac{b_2 - b_1}{2}\right) \mathbf{H}\,(\mathbf{H} \cdot \mathbf{n}) - \frac{\mu + b_2}{2} H^2\mathbf{n}.$$

Likewise the force per unit area exerted on a surface of discontinuity can be obtained by calculating (13) on either side of the surface. The result is the magnetic equivalent of (53). A case of particular importance is that of a magnetic body immersed in a fluid whose permeability to a sufficient approximation is independent of density and equal to μ_0, the value in free space. Let the body be represented by the region (1) in Fig. 23 and the fluid by the region (2). The force per unit area on the bounding surface is then found to be

(14)
$$\mathbf{t} = \frac{\mu_1 - \mu_0}{2}\left[H_1^2 + \left(\frac{\mu_1 - \mu_0}{\mu_0}\right) H_{n1}^2\right] \mathbf{n}_2 + \frac{b_2}{2} H_1^2\mathbf{n}_2 + \frac{b_1 - b_2}{2} H_{n1}\mathbf{H}_1,$$

where the constants b_1, b_2 apply to the magnetic body and $\mathbf{H}_1$ is measured just inside its surface.

Not very much is known about the parameters b_1 and b_2, although there are indications that they may be very large. They must, of course, be determined if the elastic deformation of a body is to be calculated. In most practical problems, fortunately, one is interested only in the net force acting on the body as a whole. The forces arising from deformations are compensated locally by induced elastic stresses and consequently terms involving b_1 and b_2 will drop out. As in the electric case, the resultant force exerted by a magnetostatic field on a nonferromagnetic body immersed in any fluid is obtained by evaluating the integral

(15)
$$\mathbf{F} = \int_{S_1} \left[\mu\mathbf{H}(\mathbf{H} \cdot \mathbf{n}) - \frac{\mu}{2} H^2\mathbf{n}\right] da$$

over any surface enclosing the body.

2.28. Ferromagnetic Materials.—The preceding formulas apply to ferromagnetic substances in sufficiently weak fields. If, however, the permeability μ depends markedly on the intensity of the field, the energy of a magnetic body can no longer be represented in the form of Eq. (2) or (3) and we must use in their place the integral derived in Sec. 2.17.

(16)
$$T = \frac{1}{2}\int_{V_1}\left(\mathbf{H}_1 \cdot \mathbf{B} - \mathbf{H} \cdot \mathbf{B}_1 - \mathbf{H} \cdot \mathbf{B} + 2\int_0^{\mathbf{B}} \mathbf{H} \cdot d\mathbf{B}\right) dv.$$

An analysis of the volume and surface forces for such a case has been made by Pockels,[1] who also treats briefly the problem of magnetostriction. The phenomena of magnetostriction are governed, however, by several important factors other than the simple elastic deformation considered here. A specimen of iron, for example, which in the large appears isotropic, exhibits under the microscope a fine-grained structure. The properties of the individual grains or microcrystals are strongly anisotropic. Of the same order of magnitude as these grains but not necessarily identified with them are also groups or domains of atoms, each domain acting as a permanent magnet. In the unmagnetized state the orientation of the magnetic domains is random and the net magnetic moment is zero. A weak applied field disturbs these little magnets only slightly from their initial positions of equilibrium. A small resultant moment is induced and under these circumstances the behavior of the iron may be wholly analogous to a polarizable dielectric in an electric field. As the intensity of the field is increased, however, the domains begin to flip over suddenly to new positions of equilibrium in line with the applied field, with a consequent change in the elastic properties of the specimen. A dilatation in weak fields may be followed by a contraction as the field becomes more intense, quite contrary to what would be predicted by the magnetostriction theory when applied to strictly isotropic solids. We are confronted here with a problem in which the macroscopic behavior of matter cannot be treated apart from its microscopic structure.

FORCES IN THE ELECTROMAGNETIC FIELD

2.29. Force on a Body Immersed in a Fluid.—The expressions

$$(1) \qquad \delta U = -\frac{1}{2}\int E^2\,\delta\epsilon\,dv, \qquad \delta T = \frac{1}{2}\int H^2\,\delta\mu\,dv,$$

for the energy of a body in a stationary electric or magnetic field have been based in Secs. 2.10 and 2.17 on the irrotational character of the vectors $\mathbf{E}$ and $\mathbf{H} - \mathbf{H}_1$ as well as upon their proper behavior at infinity. In a variable field these conditions are not satisfied and therefore (1) cannot be applied to the determination of the force on a body or an element of a body without a thorough revision of the proof. At best the analysis will contain some element of hypothesis, for our assumption that the quantities $\mathbf{E}\cdot\frac{\partial \mathbf{D}}{\partial t}$ and $\mathbf{H}\cdot\frac{\partial \mathbf{B}}{\partial t}$ represent the energy densities in an electromagnetic field is, after all, only a plausible interpretation of Poynting's theorem.

[1] POCKELS, *loc. cit.*, p. 369.

In Sec. 2.5 it was shown that the total force transmitted by an electromagnetic field across any closed surface Σ in *free space* is expressed by the integral

$$(2)\qquad \mathbf{F} = \int_\Sigma \left[\epsilon_0(\mathbf{E}\cdot\mathbf{n})\mathbf{E} + \frac{1}{\mu_0}(\mathbf{B}\cdot\mathbf{n})\mathbf{B} - \frac{1}{2}\left(\epsilon_0 E^2 + \frac{1}{\mu_0}B^2\right)\mathbf{n}\right] da.$$

Subsequently we were able to demonstrate for stationary fields that if the surface Σ is drawn entirely within a fluid *which supports no shearing stress*, the net force is given by (2) if only we replace ϵ_0 and μ_0 by their appropriate values in the fluid.

$$(3)\qquad \mathbf{F} = \int_\Sigma \left[\epsilon(\mathbf{E}\cdot\mathbf{n})\mathbf{E} + \mu(\mathbf{H}\cdot\mathbf{n})\mathbf{H} - \frac{1}{2}(\epsilon E^2 + \mu H^2)\mathbf{n}\right] da.$$

This is the resultant force on the charge, current, and matter *within* Σ. The matter within Σ need not be fluid and there may be sharp surfaces of discontinuity in its physical properties; the relations between **E** and **D**, and **B** and **H**, however, are assumed to be linear.

Now since (2) is valid in the dynamic as well as in the stationary regimes, there is reason to suppose that (3) may be applied also to variable fields. It is not difficult to marshal support for such a hypothesis, particularly from the theory of relativity. However, the right-hand side of (3) must now be interpreted in the sense of Sec. 2.6 not as the force exerted by the field on the matter within Σ but as the inward flow of momentum per unit time through Σ. We denote the total mechanical momentum of the matter within Σ, including the ponderable charges, by $\mathbf{G}_{\text{mech}}$, and the electromagnetic momentum of the field by $\mathbf{G}_e$. Then

$$(4)\qquad \frac{d}{dt}(\mathbf{G}_{\text{mech}} + \mathbf{G}_e) = \int_\Sigma \left[\epsilon(\mathbf{E}\cdot\mathbf{n})\mathbf{E} + \mu(\mathbf{H}\cdot\mathbf{n})\mathbf{H} - \frac{1}{2}(\epsilon E^2 + \mu H^2)\mathbf{n}\right] da.$$

The surface integral (4) may be transformed into a volume integral. For in the first place:

$$(5)\qquad \frac{1}{2}\int_\Sigma \epsilon E^2\mathbf{n}\,da = \frac{1}{2}\int_V \nabla(\epsilon E^2)\,dv.$$

Then

$$(6)\qquad \tfrac{1}{2}\nabla(\epsilon E^2) = \tfrac{1}{2}E^2\nabla\epsilon + (\mathbf{D}\cdot\nabla)\mathbf{E} + \mathbf{D}\times\nabla\times\mathbf{E}.$$

$$(7)\qquad (\mathbf{D}\cdot\nabla)\mathbf{E} = \frac{\partial}{\partial x}(D_x\mathbf{E}) + \frac{\partial}{\partial y}(D_y\mathbf{E}) + \frac{\partial}{\partial z}(D_z\mathbf{E}) - \mathbf{E}\nabla\cdot\mathbf{D}.$$

$$(8)\qquad \int_V \left[\frac{\partial}{\partial x}(D_xE_x) + \frac{\partial}{\partial y}(D_yE_x) + \frac{\partial}{\partial z}(D_zE_x)\right] dv = \int_\Sigma (\mathbf{D}\cdot\mathbf{n})E_x\,da,$$

and this obviously cancels the x-component of the vector $\epsilon(\mathbf{E}\cdot\mathbf{n})\mathbf{E}$ in

(3). Proceeding similarly with the magnetic terms and then making use of the relations $\nabla \times \mathbf{E} = -\partial\mathbf{B}/\partial t$, $\nabla \times \mathbf{H} = \dfrac{\partial \mathbf{D}}{\partial t} + \mathbf{J}$, $\nabla \cdot \mathbf{D} = \rho$, we are led to

$$(9) \quad \frac{d}{dt}(\mathbf{G}_{\text{mech}} + \mathbf{G}_e) = \int_V \left[\rho\mathbf{E} + \mathbf{J} \times \mathbf{B} - \frac{1}{2}E^2\nabla\epsilon - \frac{1}{2}H^2\nabla\mu + \frac{\partial}{\partial t}(\mathbf{D} \times \mathbf{B})\right] dv.$$

The increase in the mechanical momentum is the result of the forces exerted by the field on charges and neutral matter.

$$(10) \qquad \mathbf{F} = \frac{d}{dt}\mathbf{G}_{\text{mech}}.$$

If, therefore, the right-hand side of (9) can be split into two parts identified with $\mathbf{G}_{\text{mech}}$ and $\mathbf{G}_e$, the force can be determined. Just how this resolution is to be made is by no means obvious and various hypotheses have been suggested.[1] According to Poynting's theorem the flow of energy, even within ponderable matter, is determined by the vector

$$(11) \qquad \mathbf{S} = \mathbf{E} \times \mathbf{H} \qquad \text{joules/sec.-meter}^2.$$

Abraham and von Laue take for the density of electromagnetic momentum

$$(12) \qquad \mathbf{g}_e = \mu_0\epsilon_0\mathbf{S} = \frac{1}{c^2}\mathbf{S} \qquad \text{kg./sec.-meter}^2.$$

Then according to this hypothesis, the resultant force on the charges, currents, and polarized matter within Σ is

$$(13) \quad \mathbf{F} = \int_V \left(\rho\mathbf{E} + \mathbf{J} \times \mathbf{B} - \frac{1}{2}E^2\nabla\epsilon - \frac{1}{2}H^2\nabla\mu + \frac{\kappa_m\kappa_e - 1}{c^2}\frac{\partial \mathbf{S}}{\partial t}\right) dv,$$

or

$$(14) \quad \mathbf{F} = \int_\Sigma \left[\epsilon(\mathbf{E} \cdot \mathbf{n})\mathbf{E} + \mu(\mathbf{H} \cdot \mathbf{n})\mathbf{H} - \frac{1}{2}(\epsilon E^2 + \mu H^2)\mathbf{n}\right] da - \frac{1}{c^2}\frac{d}{dt}\int_V \mathbf{E} \times \mathbf{H}\, dv.$$

Practically, the exact form of the electromagnetic momentum term is of no great importance, for the factor $1/c^2$ makes it far too small to be easily detected.

[1] PAULI, Encyklopädie der mathematischen Wissenschaften, Vol. V, Part 2, pp. 662–667, 1906.

On the grounds of (13) it is sometimes stated that the force exerted by an electromagnetic field on a unit volume of isotropic matter is

$$\mathbf{f} = \rho\mathbf{E} + \mathbf{J} \times \mathbf{B} - \frac{1}{2} E^2 \nabla\epsilon - \frac{1}{2} H^2 \nabla\mu + \frac{\kappa_m\kappa_e - 1}{c^2} \frac{\partial \mathbf{S}}{\partial t}. \tag{15}$$

Such a conclusion is manifestly incorrect, for (15) does not include the forces associated with the deformation. As previously noted, these strictive forces are compensated locally by elastic stresses and do not enter into the integrals (13) and (14) for the resultant force necessary to maintain the body as a whole in equilibrium.

CHAPTER III

THE ELECTROSTATIC FIELD

From fundamental equations and general theorems we turn our attention in the following chapters to the structure and properties of specific fields. The simplest of these are the fields associated with stationary distributions of charge. Of all branches of our subject, however, the properties of electrostatic fields have received by far the most adequate and abundant treatment. In the present chapter we shall touch only upon the more outstanding of these properties and of the methods which have been developed for their analysis.

GENERAL PROPERTIES OF AN ELECTROSTATIC FIELD

3.1. Equations of Field and Potential.—The equations satisfied by the field of a stationary charge distribution follow directly from Maxwell's equations when all derivatives with respect to time are placed equal to zero. We have, then, at all regular points of an electrostatic field:

$$\text{(I)}\quad \nabla \times \mathbf{E} = 0, \qquad \text{(II)}\quad \nabla \cdot \mathbf{D} = \rho.$$

According to (I) the line integral of the field intensity **E** around any closed path is zero and the field is conservative.

The conservative nature of the field is a necessary and sufficient condition for the existence of a scalar potential whose gradient is **E**.

$$\mathbf{E} = -\nabla\phi. \tag{1}$$

The algebraic sign is arbitrary but has been chosen negative to conform with the convention which directs the vector **E** outward from a positive charge. Equation (1) does not define the potential uniquely, for there might be added to ϕ any constant ϕ_0 without invalidating the condition

$$\nabla \times \nabla(\phi + \phi_0) \equiv 0. \tag{2}$$

In Chap. II it was shown that the scalar potential of an electrostatic field might be interpreted as the work required to bring a unit positive charge from infinity to a point (x, y, z) within the field:

$$\phi(x, y, z) = -\int_{\infty}^{(x, y, z)} \mathbf{E} \cdot d\mathbf{r}. \tag{3}$$

We shall show below that the field of a system of charges confined to a finite region of space vanishes at infinity. The condition that ϕ shall vanish at infinity, therefore, fixes the otherwise arbitrary constant ϕ_0.

Those surfaces on which ϕ is constant are called equipotential surfaces, or simply *equipotentials*. At every point on an equipotential the field intensity $\mathbf{E}$ is normal to the surface. For let

$$\phi(x, y, z) = \text{constant} \tag{4}$$

be an equipotential, and take the first differential.

$$d\phi = \frac{\partial \phi}{\partial x}\,dx + \frac{\partial \phi}{\partial y}\,dy + \frac{\partial \phi}{\partial z}\,dz = 0. \tag{5}$$

The differentials dx, dy, dz, are the components of a vector displacement $d\mathbf{r}$ along which we wish to determine the change in ϕ, and since $d\phi = 0$, this vector must lie in the surface, $\phi = \text{constant}$. The partials $\partial\phi/\partial x$, $\partial\phi/\partial y$, $\partial\phi/\partial z$, on the other hand, are rates of change along the x-, y-, z-axes respectively and as such have been shown to be the components of another vector, namely the gradient

$$\nabla\phi = \mathbf{i}\,\frac{\partial \phi}{\partial x} + \mathbf{j}\,\frac{\partial \phi}{\partial y} + \mathbf{k}\,\frac{\partial \phi}{\partial z} = -\mathbf{E}. \tag{6}$$

$d\phi$ is manifestly the scalar product of these two and, since this product vanishes, the vectors must be orthogonal. An exception occurs at those points in which the three partial derivatives vanish simultaneously. The field intensity is zero and the points are said to be points of equilibrium.

The orthogonal trajectories of the equipotential surfaces constitute a family of lines which at every point of the field are tangent to the vector $\mathbf{E}$. They are the *lines of force*. It is frequently convenient to represent graphically the field of a given system of charges by sketching the projection of these lines on some plane through the field. Let $d\mathbf{s}$ represent a small displacement along a line of force, where

$$d\mathbf{s} = \mathbf{i}\,dx' + \mathbf{j}\,dy' + \mathbf{k}\,dz', \tag{7}$$

the primes being introduced to avoid confusion with a variable point (x, y, z) on an equipotential. Then, since by definition the lines of force are everywhere tangent to the field-intensity vector, the rectangular components of $d\mathbf{s}$ and $\mathbf{E}(x', y', z')$ must be proportional.

$$\mathbf{E}_x = \lambda\,dx', \qquad E_y = \lambda\,dy', \qquad E_z = \lambda\,dz'. \tag{8}$$

The differential equations of the lines of force are, therefore,

$$\frac{dx'}{E_x(x', y', z')} = \frac{dy'}{E_y(x', y', z')} = \frac{dz'}{E_z(x', y', z')}. \tag{9}$$

The relations between the components of $\mathbf{D}$ and those of $\mathbf{E}$ are almost invariably linear. If the medium is also isotropic, one may put

$$\mathbf{D} = \epsilon\mathbf{E} = -\epsilon\nabla\phi, \tag{10}$$

whence, by (II), ϕ must satisfy

$$\nabla \cdot (\epsilon \nabla \phi) = \epsilon \nabla^2 \phi + \nabla \epsilon \cdot \nabla \phi = -\rho. \tag{11}$$

In case the medium is homogeneous, ϕ must be a solution of *Poisson's equation,*

$$\nabla^2 \phi = -\frac{1}{\epsilon} \rho. \tag{12}$$

At points of the field which are free of charge (12) reduces to *Laplace's equation,*

$$\nabla^2 \phi = 0. \tag{13}$$

The fundamental problem of electrostatics is to determine a scalar function $\phi(x, y, z)$ that satisfies at every point in space the Poisson equation, and on prescribed surfaces fulfills the necessary boundary conditions. A much simpler inverse problem is occasionally encountered: Given the potential as an empirical function of the coordinates representing experimental data, to find a system of charges that would produce such a potential. The density of the necessary *continuous* distribution is immediately determined by carrying out the differentiation indicated by Eq. (12). There will in general, however, be supplementary *point* charges, whose presence and nature are not disclosed by Poisson's equation. At such points the potential becomes infinite and, inversely, one may expect to find point charges or systems of point charges located at the singularities of the potential function. The nature of these systems, or multipoles as they are called, is the subject of a later section, but a simple example may serve to illustrate the situation.

Let the assumed potential be

$$\phi = \frac{1}{4\pi\epsilon} \frac{e^{-ar}}{r}, \tag{14}$$

where r is the radial distance from the origin to the point of observation and a is a constant. By virtue of the spherical symmetry of this function, Poissons's equation, when written in spherical coordinates, reduces to

$$\frac{1}{r^2} \frac{\partial}{\partial r} \left(r^2 \frac{\partial \phi}{\partial r} \right) = -\frac{1}{\epsilon} \rho; \tag{15}$$

on differentiation one finds for the density of the required continuous charge distribution

$$\rho = -\frac{a^2}{4\pi} \frac{e^{-ar}}{r}. \tag{16}$$

The charge contained within a sphere of radius r is obtained by integrating ρ over the volume, or

$$\int^r \rho \, dv = e^{-ar}(ar + 1) - 1. \tag{17}$$

If the radius is made infinite, the total charge of the continuous distribution is seen to be

$$\int_0^\infty \rho\, dv = -1. \tag{18}$$

But this is *not* the total charge required to establish the potential (14). For at $r = 0$ the potential becomes infinite and we must look for a point charge located at the singularity. To verify this we need only apply (II) in its integral form. If q is a point charge at the origin,

$$\int_S \mathbf{D} \cdot \mathbf{n}\, da = q + \int_V \rho\, dv, \tag{19}$$

the surface integral extending over a sphere S bounding the volume V.

$$D_r = -\epsilon \frac{\partial \phi}{\partial r} = \frac{1}{4\pi}\left(\frac{a}{r} + \frac{1}{r^2}\right) e^{-ar}, \tag{20}$$

$$\int D_r\, da = (ar + 1)e^{-ar}, \tag{21}$$

and, hence,

$$q = +1. \tag{22}$$

The potential defined by (14) arises, therefore, from a positive unit charge located at the origin and an equal negative charge distributed about it with a density ρ, the system as a whole being neutral.

3.2. Boundary Conditions.—The transition of the field vectors across a surface of discontinuity in the medium was investigated in Sec. 1.13 and the results of that section apply directly to the electrostatic case. The two media may be supposed to meet at a surface S and the unit normal $\mathbf{n}$ is drawn from medium (1) into medium (2), so that (1) lies on the negative side of S, medium (2) on the positive side. Then

$$\mathbf{n} \times (\mathbf{E}_2 - \mathbf{E}_1) = 0, \qquad \mathbf{n} \cdot (\mathbf{D}_2 - \mathbf{D}_1) = \omega, \tag{23}$$

where ω is the density of any surface charge distributed over S.

It will be convenient to introduce the unit vector $\mathbf{t}$ tangent to the surface S. The derivatives $\partial\phi/\partial\mathbf{n}$ and $\partial\phi/\partial\mathbf{t}$ represent respectively the rates of change of ϕ in the normal and in a tangential direction. Then the boundary conditions (23) can be expressed in terms of the potential by

$$\left(\frac{\partial \phi}{\partial \mathbf{t}}\right)_2 - \left(\frac{\partial \phi}{\partial \mathbf{t}}\right)_1 = 0, \qquad \epsilon_2\left(\frac{\partial \phi}{\partial \mathbf{n}}\right)_2 - \epsilon_1\left(\frac{\partial \phi}{\partial \mathbf{n}}\right)_1 = -\omega. \tag{24}$$

From the conservative nature of the field it follows also that the potential itself must be continuous across S, for the work required to carry a small charge from infinity to either of two adjacent points located on opposite sides of S must be the same Hence

$$\phi_1 = \phi_2. \tag{25}$$

The two conditions

$$\phi_1 = \phi_2, \qquad \epsilon_2 \left(\frac{\partial \phi}{\partial \mathbf{n}}\right)_2 - \epsilon_1 \left(\frac{\partial \phi}{\partial \mathbf{n}}\right)_1 = -\omega, \tag{26}$$

are independent.

Conductors play an especially prominent role in electrostatics. For the purposes of a purely macroscopic theory it is sufficient to consider a conductor as a closed domain within which charge moves freely. If the conductor is a metal or electrolyte, the flow of charge is directly proportional to the intensity $\mathbf{E}$ of the electric field: $\mathbf{J} = \sigma \mathbf{E}$. Charge is free to move on the surface of a conductor but can leave it only under the influence of very intense external fields or at high temperatures (thermal emission). If the conductor is in electrostatic equilibrium all flow of charge has ceased, whence it is evident that *at every interior point of a conductor in an electrostatic field the resultant field intensity* $\mathbf{E}$ *is zero, and at every point on its surface the tangential component of* $\mathbf{E}$ *is zero. Furthermore the electrostatic potential* ϕ *within a conductor is constant and the surface of every conductor is an equipotential.* Let us suppose that an uncharged conductor is introduced into a fixed external field $\mathbf{E}_0$. In the first instant there occurs a transient current. According to Sec. 1.7, no charge can accumulate at an interior point, but a redistribution will occur over the surface such that the surface density at any point is ω, subject to the condition

$$\int_S \omega \, da = 0. \tag{27}$$

This surface distribution gives rise to an induced or secondary field of intensity $\mathbf{E}_1$. Equilibrium is attained when the distribution is such that at every interior point

$$\mathbf{E}_0 + \mathbf{E}_1 = 0. \tag{28}$$

Likewise, if a charge q is placed on an isolated conductor, the charge will distribute itself over the surface with a density ω subject to (28) and the condition

$$\int_S \omega \, da = q. \tag{29}$$

We shall denote the interior of a conductor in electrostatic equilibrium by the index (1) and the exterior dielectric by (2). Then at the surface S

$$\mathbf{E}_1 = \mathbf{D}_1 = 0, \qquad \mathbf{n} \times \mathbf{E}_2 = 0, \qquad \mathbf{n} \cdot \mathbf{D}_2 = \omega, \tag{30}$$

or in terms of the potential,

$$\phi = \text{constant}, \qquad \epsilon_2 \frac{\partial \phi}{\partial \mathbf{n}} = -\omega. \tag{31}$$

If a solution of Laplace's equation can be found which is constant over the given conductors, the surface density of charge may be determined by calculating the normal derivative of the potential.

CALCULATION OF THE FIELD FROM THE CHARGE DISTRIBUTION

3.3. Green's Theorem.—Let V be a closed region of space bounded by a regular surface S, and let ϕ and ψ be two scalar functions of position which together with their first and second derivatives[1] are continuous throughout V and on the surface S. Then the divergence theorem applied to the vector $\psi\nabla\phi$ gives

$$\int_V \nabla \cdot (\psi\nabla\phi)\, dv = \int_S (\psi\nabla\phi) \cdot \mathbf{n}\, da. \tag{1}$$

Upon expanding the divergence to

$$\nabla \cdot (\psi\nabla\phi) = \nabla\psi \cdot \nabla\phi + \psi\nabla \cdot \nabla\phi = \nabla\psi \cdot \nabla\phi + \psi\nabla^2\phi, \tag{2}$$

and noting that

$$\nabla\phi \cdot \mathbf{n} = \frac{\partial\phi}{\partial n}, \tag{3}$$

where $\partial\phi/\partial n$ is the derivative in the direction of the positive normal, we obtain what is known as *Green's first identity:*

$$\int_V \nabla\psi \cdot \nabla\phi\, dv + \int_V \psi\nabla^2\phi\, dv = \int_S \psi \frac{\partial\phi}{\partial n}\, da. \tag{4}$$

If in particular we place $\psi = \phi$ and let ϕ be a solution of Laplace's equation, Eq. (4) reduces to

$$\int_V (\nabla\phi)^2\, dv = \int_S \phi \frac{\partial\phi}{\partial n}\, da. \tag{5}$$

Next let us interchange the roles of the functions ϕ and ψ; *i.e.*, apply the divergence theorem to the vector $\phi\nabla\psi$.

$$\int_V \nabla\phi \cdot \nabla\psi\, dv + \int_V \phi\nabla^2\psi\, dv = \int_S \phi \frac{\partial\psi}{\partial n}\, da. \tag{6}$$

Upon subtracting (6) from (4) a relation between a volume integral and a surface integral is obtained of the form

$$\int_V (\psi\nabla^2\phi - \phi\nabla^2\psi)\, dv = \int_S \left(\psi \frac{\partial\phi}{\partial n} - \phi \frac{\partial\psi}{\partial n}\right) da, \tag{7}$$

known as *Green's second identity* or also frequently as *Green's theorem.*

[1] This condition is more stringent than is necessary. The second derivative of one function ψ need not be continuous.

3.4. Integration of Poisson's Equation.—By means of Green's theorem the potential at a fixed point (x', y', z') within the volume V can be expressed in terms of a volume integral plus a surface integral over S. Let us suppose that charge is distributed with a volume density $\rho(x, y, z)$. We shall assume that $\rho(x, y, z)$ is bounded but is an otherwise arbitrary function of position. An arbitrary, regular surface S is now drawn enclosing a volume V, Fig. 25. It is not necessary that S enclose all the charge, or even any of it. Let O be an arbitrary origin and $x = x'$, $y = y'$, $z = z'$, a fixed point of observation within V. The potential at this point due to the entire charge distribution is $\phi(x', y', z')$. For

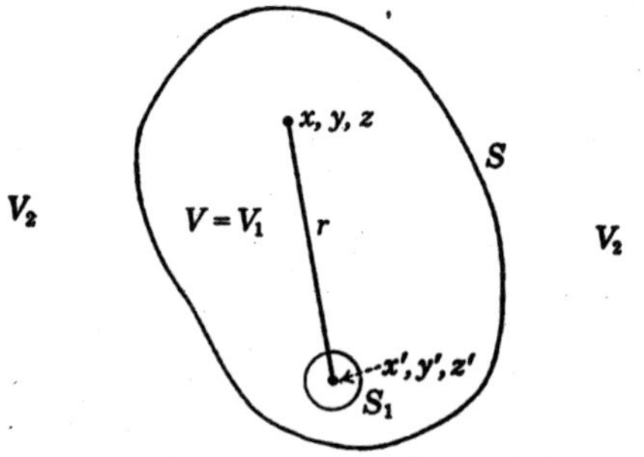

FIG. 25.—Application of Green's theorem to a region V bounded externally by the surface S and internally by the sphere S_1.

the function ψ we shall choose a spherically symmetrical solution of Laplace's equation,

$$\psi(x, y, z; x', y', z') = \frac{1}{r}, \tag{8}$$

where r is the distance from a variable point (x, y, z) within V to the fixed point (x', y', z').

$$r = \sqrt{(x' - x)^2 + (y' - y)^2 + (z' - z)^2}. \tag{9}$$

This function ψ, however, fails to satisfy the necessary conditions of continuity at $r = 0$. To exclude the singularity, a small sphere of radius r_1 is circumscribed about (x', y', z') as a center. The volume V is then bounded externally by S and internally by the sphere S_1. Within V both ϕ and ψ now satisfy the requirements of Green's theorem and furthermore $\nabla^2\psi = 0$. Thus (7) reduces to

$$\int_V \frac{\nabla^2\phi}{r}\, dv = \int_{S+S_1} \left[\frac{1}{r}\frac{\partial\phi}{\partial n} - \phi\frac{\partial}{\partial n}\left(\frac{1}{r}\right)\right] da, \tag{10}$$

the surface integral to be extended over both S and S_1. Over the sphere S_1 the positive normal is directed radially toward the center (x', y', z'), since this is *out* of the volume V. Over S_1, therefore,

$$\frac{\partial\phi}{\partial n} = -\frac{\partial\phi}{\partial r}, \qquad \left[\frac{\partial}{\partial n}\left(\frac{1}{r}\right)\right]_{r=r_1} = \frac{1}{r_1^2}. \tag{11}$$

Since r_1 is constant, the contribution of the sphere to the right-hand side of (10) is

$$-\frac{1}{r_1}\int_{S_1}\frac{\partial\phi}{\partial r}\,da - \frac{1}{r_1^2}\int_{S_1}\phi\,da.$$

If $\bar{\phi}$ and $\overline{\partial\phi}/\partial r$ denote mean values of ϕ and $\partial\phi/\partial r$ on S_1, this contribution is

$$-\frac{1}{r_1}4\pi r_1^2\,\overline{\frac{\partial\phi}{\partial r}} - \frac{1}{r_1^2}4\pi r_1^2\bar{\phi},$$

which in the limit $r_1 \to 0$ reduces to $-4\pi\phi(x', y', z')$. Upon introducing this value into (10) the potential at any interior point (x', y', z') is

$$\phi(x', y', z') = -\frac{1}{4\pi}\int_V \frac{\nabla^2\phi}{r}\,dv + \frac{1}{4\pi}\int_S\left[\frac{1}{r}\frac{\partial\phi}{\partial n} - \phi\frac{\partial}{\partial n}\left(\frac{1}{r}\right)\right]da, \tag{12}$$

or in terms of the charge density when the medium is homogeneous, $\nabla^2\phi = -\frac{1}{\epsilon}\rho$,

$$\phi(x', y', z') = \frac{1}{4\pi\epsilon}\int_V \frac{\rho}{r}\,dv + \frac{1}{4\pi}\int_S\left[\frac{1}{r}\frac{\partial\phi}{\partial n} - \phi\frac{\partial}{\partial n}\left(\frac{1}{r}\right)\right]da. \tag{13}$$

In case the region V bounded by S contains no charge, (13) reduces to

$$\phi(x', y', z') = \frac{1}{4\pi}\int_S\left[\frac{1}{r}\frac{\partial\phi}{\partial n} - \phi\frac{\partial}{\partial n}\left(\frac{1}{r}\right)\right]da. \tag{14}$$

It is apparent that the surface integrals in (13) and (14) represent the contribution to the potential at (x', y', z') of all charges which are exterior to S. If the values of ϕ and its normal derivative over S are known, the potential at any interior point can be determined by integration. Equation (14) may be interpreted therefore as a solution of Laplace's equation within V satisfying specified conditions over the boundary. The integral

$$\phi(x', y', z') = \frac{1}{4\pi\epsilon}\int_V \frac{\rho}{r}\,dv \tag{15}$$

is a particular solution of Poisson's equation valid at (x', y', z'); the general solution is obtained by adding the integral (14) of the homogeneous equation $\nabla^2\phi = 0$. If there are no charges exterior to S, the surface integral must vanish.

3.5. Behavior at Infinity.—Let us suppose that every element of a charge distribution is located within a finite distance of some arbitrary origin O. We imagine this distribution to be confined within the interior of a surface S_1 which also contains the origin O. The distances r, r_1 and R are indicated in Fig. 26.

$$r = (R^2 + r_1^2 - 2r_1R\cos\theta)^{\frac{1}{2}}. \tag{16}$$

The potential at any point $P(x', y', z')$ outside S_1 is then

$$\phi(x', y', z') = \frac{1}{4\pi\epsilon}\int_{V_1}\frac{\rho(x, y, z)\,dv}{(R^2 + r_1^2 - 2r_1R\cos\theta)^{\frac{1}{2}}}. \tag{17}$$

As P recedes to infinity, the terms r_1^2 and $2r_1R\cos\theta$ become negligible

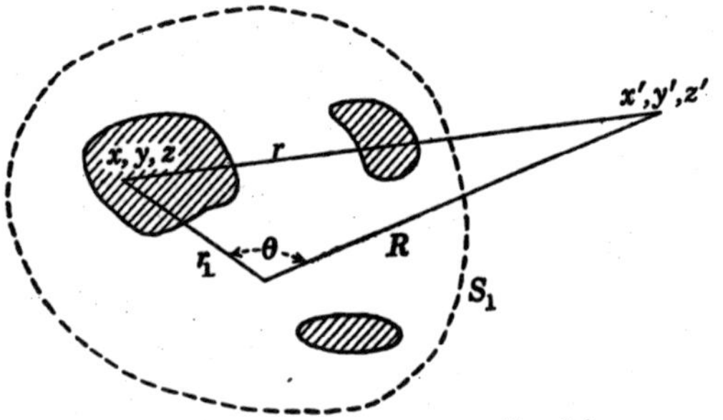

FIG. 26.—Figure to accompany Sec. 3.5.

with respect to R^2, and in the limit as $R \rightarrow \infty$ we find

$$\lim_{R\rightarrow\infty}\phi = \frac{1}{4\pi\epsilon R}\int_{V_1}\rho\,dv = \frac{1}{4\pi\epsilon}\frac{q}{R}, \tag{18}$$

where q is the total charge of the system. A potential function is said to be *regular at infinity* if $R\phi$ is bounded as $R \rightarrow \infty$. The field intensity $\mathbf{E} = -\nabla\phi$ at great distances is directed radially from O and the function $R^2|\mathbf{E}|$ is bounded.

All real charge systems are contained within domains of finite extent and their fields are, therefore, regular at infinity. Frequently, however, the analysis of a problem is simplified by assuming the external field to be parallel. Such a field does not vanish at infinity and can only arise from sources located at an infinite distance from the origin.

It is important to note that a closed surface S divides all space into two volumes, an interior V_1 and an exterior V_2, and that if the functions ϕ and ψ are regular at infinity Green's theorem applies to the external region V_2 as well as to V_1. For certainly the theorem applies to a closed region bounded internally by S and externally by another surface S_2. If now S_2 recedes towards infinity, the quantities $\psi\frac{\partial\phi}{\partial n}$ and $\phi\frac{\partial\psi}{\partial n}$ vanish as $1/r^3$ and the integral over S_2 approaches zero. Consequently

$$\int_{V_2} (\psi \nabla^2 \phi - \phi \nabla^2 \psi)\, dv = \int_S \left(\psi \frac{\partial \phi}{\partial n} - \phi \frac{\partial \psi}{\partial n}\right) da, \tag{19}$$

but the positive normal at points on S is now directed *out* of V_2 and therefore *into* V_1. Let us suppose, for example, that a charge system is confined entirely to the region V_1 within S. The potential at any point (x', y', z') in V_2 outside S may be calculated from (17), or equally well from a knowledge of ϕ and $\partial\phi/\partial n$ on S, applying (14) as indicated in Fig. 27.

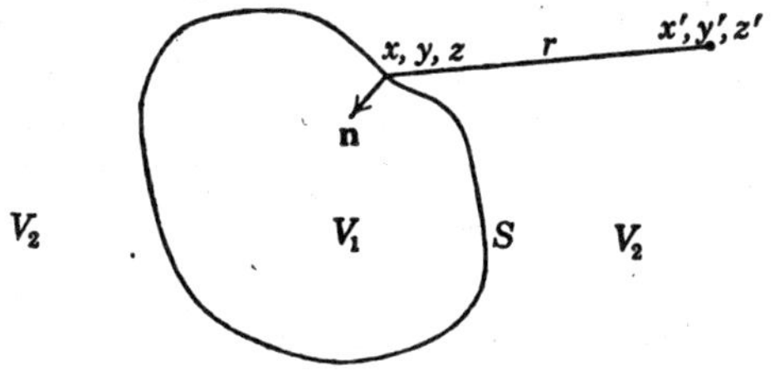

FIG. 27.—Application of Green's theorem to the exterior V_2 of a closed surface S.

3.6. Coulomb Field.—According to (18) the potential of a charge q in a homogeneous medium at distances very great relative to the dimensions of the charge itself approaches the value

$$\phi(x', y', z') = \frac{1}{4\pi\epsilon} \frac{q}{r}. \tag{20}$$

If q is located at (x, y, z), the distance from q to the point of observation (x', y', z') is

$$r = \sqrt{(x' - x)^2 + (y' - y)^2 + (z' - z)^2}. \tag{21}$$

Let now $\mathbf{r}^0$ represent a unit vector directed from the source; *i.e.*, from (x, y, z) towards (x', y', z'). Then

$$\nabla' \left(\frac{1}{r}\right) = -\nabla \left(\frac{1}{r}\right) = -\frac{1}{r^2} \mathbf{r}^0, \tag{22}$$

where the prime above the gradient operator denotes differentiation with respect to the variables (x', y', z') at the point of observation. The field intensity at this point is

$$\mathbf{E}(x', y', z') = -\nabla' \phi = \frac{1}{4\pi\epsilon} \frac{q}{r^2} \mathbf{r}^0. \tag{23}$$

The field of a point charge is inversely proportional to the square of the distance and is directed radially outward when q is positive. This is the law established experimentally by Coulomb and Cavendish which is usually taken as a point of departure for the theory of electrostatics. The term "point charge" is employed here in the sense of a charge whose dimensions are negligible with respect to r. Mathematically one may imagine the dimensions of q to grow vanishingly small while the density ρ is increased such that q is maintained constant. In this fashion a point singularity is generated and (23) is then valid at all points except $r = 0$. There is no reason to believe that such singularities exist in nature, but it is convenient to interpret a field at sufficient distances as that which might be generated by systems of mathematical point charges. We shall have occasion to develop this concept in the subsequent section.

The potential at (x', y', z') is obtained by integrating the contributions of charge elements $dq = \rho\, dv$ over all space.

$$d\phi(x', y', z') = \frac{1}{4\pi\epsilon} \frac{\rho(x, y, z)}{r}\, dv, \tag{24}$$

$$d\mathbf{E} = -\frac{1}{4\pi\epsilon} \rho(x, y, z) \nabla' \left(\frac{1}{r}\right) dv = \frac{1}{4\pi\epsilon} \rho(x, y, z) \nabla \left(\frac{1}{r}\right) dv. \tag{25}$$

The field intensity due to a complete charge distribution in a homogeneous, isotropic medium is, therefore,

$$\mathbf{E}(x', y', z') = \frac{1}{4\pi\epsilon} \int \rho(x, y, z) \nabla \left(\frac{1}{r}\right) dv. \tag{26}$$

3.7. Convergence of Integrals.—The proof of convergence of the integrals for potential and field intensity is implicit in the method by which they have been derived, but an alternative treatment will be described which may serve as a model for proofs of this kind. Since the element of integration at (x, y, z) can coincide with (x', y', z'), with the result that the integrand becomes infinite at this point, it is not obvious that the integral has a meaning. It will be shown that, although an improper integral of this type cannot be defined in the ordinary manner as the limit of a sum, it can by suitable definition be made to converge absolutely to a finite value.

Let the point $Q(x', y', z')$ at which the field is to be determined be surrounded by a closed surface S of arbitrary shape, thus dividing the total volume V occupied by charge into two parts: a portion V_1 representing the volume within S and a portion V_2 external to S. Throughout V_2 the integral of Eq. (26) is bounded and consequently the charge *outside* S contributes a finite amount to the resultant field intensity at Q, a contribution $\mathbf{E}_2$,

$$\mathbf{E}_2(x', y', z') = \frac{1}{4\pi\epsilon} \int_{V_2} \rho(x, y, z) \nabla \left(\frac{1}{r}\right) dv. \tag{27}$$

The contribution of the charge *within* S to the field intensity at Q may be called $\mathbf{E}_1$. If now the field intensity $\mathbf{E} = \mathbf{E}_1 + \mathbf{E}_2$ at Q is to have a definite significance, it is necessary that $\mathbf{E}_1$ vanish in the limit as S shrinks about Q. A moment's reflection will indicate that this is indeed the case for, although the denominator of the integrand vanishes as the square of the distance, the element of charge in the numerator is proportional to the volume V_1 which vanishes as the cube of a linear dimension. It has been tacitly assumed that the charge density ρ is finite at every point in the volume V. There exists, therefore, a number m such that $|\rho| < m$, and $|\rho/r| < m/r$, at every point in V. It follows, furthermore, for this upper bound m that

$$\left|\rho\nabla\left(\frac{1}{r}\right)\right| < \frac{m}{r^2}, \tag{28}$$

and

$$\left|\int_{V_1} \rho\nabla\left(\frac{1}{r}\right) dv\right| < m \int_{V_1} \frac{dv}{r^2}. \tag{29}$$

The surface S bounding the volume V_1 is of arbitrary form, and so, to avoid the awkwardness of evaluating the integral over this region, a sphere of radius a concentric with Q is circumscribed about V_1. The volume element of integration is positive; consequently the integral extended throughout V_1 must be less than, or at the most equal to, the integral extended throughout the volume enclosed by the circumscribed sphere.

$$\int_{V_1} \frac{dv}{r^2} \leqq \frac{4}{3}\pi a, \tag{30}$$

which evidently vanishes with a. The contribution of the charge within S to the field at Q becomes vanishingly small as S shrinks about Q and hence (26) converges for interior as well as exterior points of a charge distribution.

The potential integral

$$\phi(x', y', z') = \frac{1}{4\pi\epsilon} \int \frac{\rho(x, y, z)}{r}\, dv \tag{31}$$

is also an improper integral when the point of observation is taken within the charge but, since the denominator of the integrand vanishes only as the first power of r, the proof above holds here a fortiori. If ρ is a bounded, integrable function of position, (31) is a continuous function of the coordinates x', y', z', has continuous first derivatives, and satisfies the condition $\mathbf{E} = -\nabla'\phi$ everywhere. It can be shown, furthermore, that *if ρ and all its derivatives of order less than n are continuous, the potential ϕ has continuous derivatives of all orders less than $n + 1$.*[1]

EXPANSION OF THE POTENTIAL IN SPHERICAL HARMONICS

3.8. Axial Distributions of Charge.—We shall suppose first that an element of charge q is located at the point $z = \zeta$ on the z-axis of a rectangular coordinate system whose origin is at O. We wish to express the potential of q at any other point P in terms of the coordinates of P with respect to the origin O. The rectangular coordinates of P are x, y, z, but since the field is symmetric about the z-axis it will be sufficient to locate P in terms of the two polar coordinates r and θ, Fig. 28. The distance from q to P is r_2 and the potential at P is, therefore,

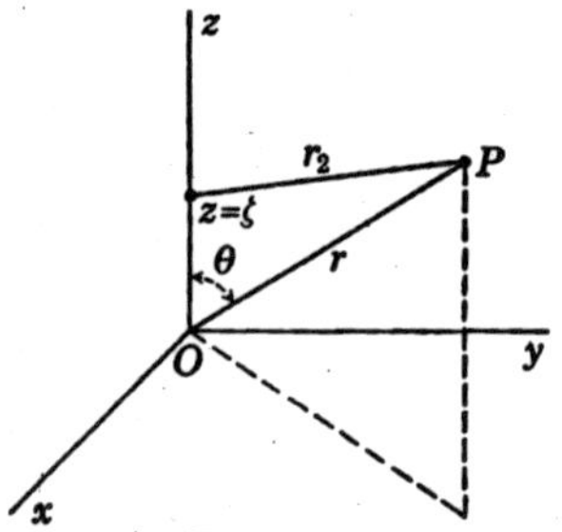

FIG. 28.—Figure to accompany Sec. 3.8.

$$\phi(r, \theta) = \frac{q}{4\pi\epsilon}\frac{1}{r_2}, \tag{1}$$

the medium being assumed homogeneous and isotropic.

$$r_2 = (r^2 + \zeta^2 - 2r\zeta \cos \theta)^{\frac{1}{2}}. \tag{2}$$

There are now two cases to be considered. The first and perhaps less common is that in which P lies within a sphere drawn from O as a center through ζ. Then $r < \zeta$ and we shall write

$$\frac{1}{r_2} = \frac{1}{\zeta}\left[1 + \left(\frac{r}{\zeta}\right)^2 - 2\frac{r}{\zeta}\cos \theta\right]^{-\frac{1}{2}}. \tag{3}$$

The bracket may be expanded by the binomial theorem if

$$\left|\left(\frac{r}{\zeta}\right)^2 - 2\frac{r}{\zeta}\cos \theta\right| < 1.$$

If furthermore $\left|\frac{r}{\zeta}\right|^2 + \left|2\frac{r}{\zeta}\cos \theta\right| < 1$, the resultant series converges absolutely and consequently the various powers may be multiplied out and the terms rearranged at will. If the terms of the series are now ordered in ascending powers of r/ζ, we find

$$\frac{1}{r_2} = \frac{1}{\zeta}\left[1 + \frac{r}{\zeta}\cos \theta + \left(\frac{r}{\zeta}\right)^2\left(\frac{3}{2}\cos^2 \theta - \frac{1}{2}\right) + \cdots\right], \tag{4}$$

[1] See for example KELLOGG, "Foundations of Potential Theory," Chap. VI, Springer, 1929, or PHILLIPS, "Vector Analysis," pp. 122 *ff.*, Wiley, 1933.

which shall be written in the abbreviated form

(5) $$\frac{1}{r_2} = \frac{1}{\zeta}\sum_{n=0}^{\infty} P_n(\cos\theta)\left(\frac{r}{\zeta}\right)^n, \qquad (r < \zeta).$$

The coefficients of r/ζ are polynomials in $\cos\theta$ and are known as the *Legendre polynomials.*

(6) $$\begin{aligned} P_0(\cos\theta) &= 1, \\ P_1(\cos\theta) &= \cos\theta, \\ P_2(\cos\theta) &= \tfrac{1}{2}(3\cos^2\theta - 1) = \tfrac{1}{4}(3\cos 2\theta + 1), \\ P_3(\cos\theta) &= \tfrac{1}{2}(5\cos^3\theta - 3\cos\theta) = \tfrac{1}{8}(5\cos 3\theta + 3\cos\theta). \\ &\cdots\cdots\cdots\cdots\cdots\cdots\cdots\cdots \end{aligned}$$

The absolute value of the coefficients P_n is never greater than unity; hence the expansion converges absolutely provided $r < |\zeta|$.

In the second case P lies outside the sphere of radius ζ, such that $r > \zeta$. The corresponding expansion is obtained by interchanging r and ζ in (3) and (5).

(7) $$\frac{1}{r_2} = \frac{1}{r}\left[1 + \left(\frac{\zeta}{r}\right)^2 - 2\frac{\zeta}{r}\cos\theta\right]^{-\frac{1}{2}},$$

(8) $$\frac{1}{r_2} = \frac{1}{r}\sum_{n=0}^{\infty} P_n(\cos\theta)\left(\frac{\zeta}{r}\right)^n, \qquad (r > \zeta).$$

This last result may be obtained in a slightly different manner. Consider the inverse distance from the point $z = \zeta$ to P as a function of ζ and expand in a Taylor series about the origin, $\zeta = 0$.

(9) $$f(\zeta) = \frac{1}{r_2} = (r^2 + \zeta^2 - 2r\zeta\cos\theta)^{-\frac{1}{2}},$$

(10) $$f(\zeta) = f(0) + \zeta\left[\frac{\partial f(\zeta)}{\partial\zeta}\right]_{\zeta=0} + \frac{\zeta^2}{2!}\left[\frac{\partial^2 f(\zeta)}{\partial\zeta^2}\right]_{\zeta=0} + \cdots.$$

Now in rectangular coordinates, r_2 is

(11) $$r_2 = [x^2 + y^2 + (z - \zeta)^2]^{\frac{1}{2}},$$

and

(12) $$\frac{\partial}{\partial\zeta}\left(\frac{1}{r_2}\right) = -\frac{\partial}{\partial z}\left(\frac{1}{r_2}\right).$$

Hence,

(13) $$\left[\frac{\partial^n f(\zeta)}{\partial\zeta^n}\right]_{\zeta=0} = (-1)^n\left[\frac{\partial^n f(\zeta)}{\partial z^n}\right]_{\zeta=0} = (-1)^n\frac{\partial^n f(0)}{\partial z^n},$$

and, since $f(0) = 1/r$, we have

(14) $$f(\zeta) = \frac{1}{r_2} = \frac{1}{r} - \zeta\frac{\partial}{\partial z}\left(\frac{1}{r}\right) + \cdots \zeta^n\frac{(-1)^n}{n!}\frac{\partial^n}{\partial z^n}\left(\frac{1}{r}\right) + \cdots.$$

The potential at a point P outside the sphere through q can be written in either of the forms

$$\phi = \frac{q}{4\pi\epsilon} \sum_{n=0}^{\infty} \zeta^n \frac{P_n(\cos\theta)}{r^{n+1}} = \frac{q}{4\pi\epsilon} \sum_{n=0}^{\infty} \zeta^n \frac{(-1)^n}{n!} \frac{\partial^n}{\partial z^n}\left(\frac{1}{r}\right), \tag{15}$$

from which it is apparent that

$$\frac{P_n(\cos\theta)}{r^{n+1}} = \frac{(-1)^n}{n!} \frac{\partial^n}{\partial z^n}\left(\frac{1}{r}\right). \tag{16}$$

Finally, let us suppose that charge is distributed continuously along a length l of the z-axis with a density $\rho = \rho(\zeta)$. The potential at a sufficiently great distance from the origin is

$$\phi(r, \theta) = \sum_{n=0}^{\infty} \phi_n = \frac{1}{4\pi\epsilon} \sum_{n=0}^{\infty} \int_0^l \rho(\zeta)\zeta^n \, d\zeta \, \frac{P_n(\cos\theta)}{r^{n+1}}, \qquad (r > l). \tag{17}$$

The leading term of this expansion,

$$\phi_0 = \frac{1}{4\pi\epsilon} \frac{1}{r} \int_0^l \rho(\zeta) \, d\zeta = \frac{q}{4\pi\epsilon} \frac{1}{r}, \tag{18}$$

where q is now the *total* charge on the line, is evidently *the Coulomb potential of a point charge q located at the origin.* However, the density may conceivably assume negative as well as positive values such that the net charge

$$q = \int_0^l \rho(\zeta) \, d\zeta \tag{19}$$

is zero. The dominant term approached by the potential when $r \gg l$ is then

$$\phi_1 = \frac{1}{4\pi\epsilon} \frac{P_1(\cos\theta)}{r^2} \int_0^l \rho(\zeta)\zeta \, d\zeta = \frac{p}{4\pi\epsilon} \frac{\cos\theta}{r^2}. \tag{20}$$

The quantity

$$p = \int_0^l \rho(\zeta)\zeta \, d\zeta \tag{21}$$

is called the *dipole moment* of the distribution. In general, we shall write

$$\phi_n = \frac{1}{4\pi\epsilon} \, p^{(n)} \frac{P_n(\cos\theta)}{r^{n+1}}, \tag{22}$$

and define

$$p^{(n)} = \int_0^l \rho(\zeta)\zeta^n \, d\zeta \tag{23}$$

as an axial multipole of nth order.

3.9. The Dipole.—In order that the potential of a linear charge distribution may be represented by a dipole it is necessary that the net charge be zero—the system as a whole is neutral—and that the distance to the point of observation be very great relative to the length of the line. We have seen that the potential ϕ_0 is that which would be generated by a mathematical point charge located at the origin. ϕ_0 has a singularity at $r = 0$, for a true point charge implies an infinite density. We now ask whether a configuration of point charges can be constructed which will give rise to the dipole potential ϕ_1.

Let us place a point charge $+q$ at a point $z = l$ on the z-axis and an equal negative charge $-q$ at the origin. According to Eq. (8) the potential at a distant point is

$$\phi = \frac{q}{4\pi\epsilon}\left(\frac{1}{r_2} - \frac{1}{r}\right) = \frac{ql}{4\pi\epsilon}\frac{\cos\theta}{r^2} + \text{higher order terms.} \tag{24}$$

The product $p = ql$ is evidently the dipole moment of the configuration Suppose now that $l \to 0$, but at the same time q is increased in magnitude in such a manner that the product p remains constant. Then in the limit a double-point singularity is generated whose potential is

$$\phi = \frac{p}{4\pi\epsilon}\frac{\cos\theta}{r^2} = -\frac{p}{4\pi\epsilon}\frac{\partial}{\partial z}\left(\frac{1}{r}\right) \tag{25}$$

everywhere but at the origin. A direction has been associated with a point. The dipole moment is in fact a *vector* $\mathbf{p}$ directed, in this case, along the z-axis. The unit vector directed along r from the dipole towards the point of observation is again $\mathbf{r}^0$, and the potential is, therefore,[1]

$$\phi(x, y, z) = \frac{1}{4\pi\epsilon}\frac{\mathbf{p}\cdot\mathbf{r}^0}{r^2} = -\frac{1}{4\pi\epsilon}\mathbf{p}\cdot\nabla\left(\frac{1}{r}\right). \tag{26}$$

The field of a dipole is cylindrically symmetrical about the axis; hence in any meridian plane the radial and transverse components of field intensity are

$$\begin{aligned} E_r &= -\frac{\partial\phi}{\partial r} = \frac{1}{2\pi\epsilon}\frac{p\cos\theta}{r^3}, \\ E_\theta &= -\frac{1}{r}\frac{\partial\phi}{\partial\theta} = \frac{1}{4\pi\epsilon}\frac{p\sin\theta}{r^3}. \end{aligned} \tag{27}$$

[1] In (26) $r = \sqrt{x^2 + y^2 + z^2}$ and $\nabla(1/r)$ implies differentiation at the point of observation. If the dipole were located at (x, y, z) and the potential measured at (x', y', z'), we should have $\phi(x', y', z') = -\frac{1}{4\pi\epsilon}\mathbf{p}\cdot\nabla'\left(\frac{1}{r}\right) = +\frac{1}{4\pi\epsilon}\mathbf{p}\cdot\nabla\left(\frac{1}{r}\right)$. See Eq. (22), p. 169.

The potential energy of a dipole in an external field is most easily determined from the potential energies of its two point charges. Let a charge $+q$ be located at a point a, and a charge $-q$ at b displaced from the first by an amount $\mathbf{l}$, in an *external* field whose potential is $\phi(x, y, z)$. The potential energy of the system is then

$$U = q\phi(a) - q\phi(b), \tag{28}$$

or as $b \to a$,

$$U = q\,d\phi = q\mathbf{l} \cdot \nabla\phi = -\mathbf{p} \cdot \mathbf{E} = -pE \cos \theta, \tag{29}$$

where θ is the angle made by the dipole with the external field $\mathbf{E}$.

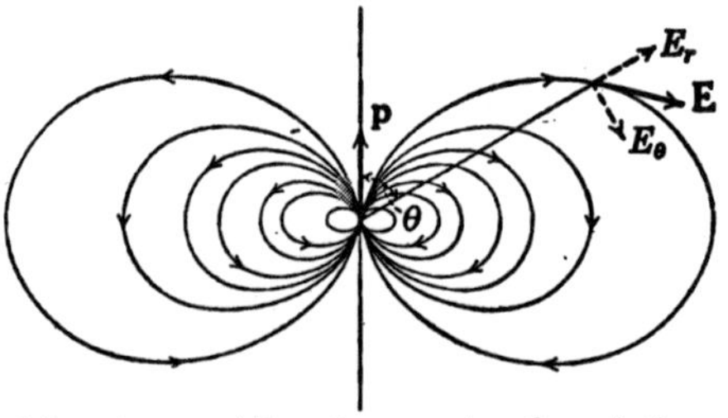

FIG. 29.—Lines of force in a meridian plane passing through the axis of a dipole p.

The force exerted on the dipole by the external field is equal to the negative gradient of U when the orientation is fixed.

$$\mathbf{F} = \nabla(\mathbf{p} \cdot \mathbf{E})_{\theta\ \text{constant}}. \tag{30}$$

On the other hand a change in orientation at a fixed point of the field also leads to a variation in potential energy. The torque exerted on a dipole by an external field is, therefore,

$$T = -\frac{\partial U}{\partial \theta} = -pE \sin \theta, \tag{31}$$

or vectorially,

$$\mathbf{T} = \mathbf{p} \times \mathbf{E}. \tag{32}$$

3.10. Axial Multipoles.—Let us refer again to Eq. (17) for the potential of a linear distribution of charge. This expansion is valid at all points outside a sphere whose diameter is the charged line. Now the first term ϕ_0 of the series is just the potential that would be produced by a point charge q located at the center of the sphere. The second term ϕ_1 represents the potential of a dipole $\mathbf{p}$ located at the same origin. We shall show that the remaining terms ϕ_n may likewise be interpreted as the potentials of higher order charge singularities clustered at the center.

The dipole, whose moment we now denote as $p^{(1)}$, was constructed by placing a negative charge $-q$ at the origin and a positive charge $+q$ at a point $z = l_0$ along the axis. The two are then allowed to coalesce such that $p^{(1)} = ql_0$ remains constant. The potential of the resultant singularity is

$$\phi_1 = -\frac{1}{4\pi\epsilon} p^{(1)} \frac{\partial}{\partial z}\left(\frac{1}{r}\right). \tag{33}$$

The singularity of next higher order is constructed by locating a dipole of negative moment $-p^{(1)}$ at the origin and displacing from it another equal positive moment by the small amount $\mathbf{l}_1$. For the present we confine ourselves to the special case wherein the axes of both dipoles as well as the displacement $\mathbf{l}_1$ are directed along the z-axis. The potential of the resultant configuration is

$$\phi_2 = \phi_1 + l_1\left(\frac{\partial \phi_1}{\partial \zeta}\right)_{\zeta=0} + \cdots - \phi_1, \tag{34}$$

or by virtue of (12)

$$\phi_2 = \frac{1}{4\pi\epsilon} p^{(1)} l_1 \frac{\partial^2}{\partial z^2}\left(\frac{1}{r}\right) + \cdots. \tag{35}$$

An axial *quadrupole moment* is defined as the product

$$p^{(2)} = 2(p^{(1)}l_1) = 2(ql_0l_1). \tag{36}$$

The mathematical quadrupole is generated by letting $l_0 \to 0$, $l_1 \to 0$, $q \to \infty$ such that the product (36) remains finite. The potential of this configuration is then strictly

$$\phi_2 = \frac{1}{4\pi\epsilon} \frac{p^{(2)}}{2!} \frac{\partial^2}{\partial z^2}\left(\frac{1}{r}\right) \tag{37}$$

at all points excluding $r = 0$.

By induction one constructs charge singularities, or multipoles, of yet higher order. In each case a multipole of order $n - 1$ and negative moment $-p^{(n-1)}$ is located at the origin and an equal positive multipole $p^{(n-1)}$ displaced from it by $\mathbf{l}_n$. In the general case to be dealt with below, the displacements are arbitrary in direction; if, as in the present special case, all are along the same straight line, the multipole is said to be axial. The potential is given approximately by the first term of a Taylor series, which again by (12) may be written

$$\phi_n = \frac{(-1)^n}{4\pi\epsilon} \frac{p^{(n-1)}l_{n-1}}{(n-1)!} \frac{\partial^n}{\partial z^n}\left(\frac{1}{r}\right) + \cdots. \tag{38}$$

The n-pole moment is defined as the limit of the product

$$p^{(n)} = np^{(n-1)}l_{n-1} \tag{39}$$

when $l_n \to 0$ and $p^{(n-1)} \to \infty$ in such a manner that $p^{(n-1)}l_n$ remains finite. The potential of a point charge of nth order is then

$$\phi_n = \frac{p^{(n)}}{4\pi\epsilon}\frac{(-1)^n}{n!}\frac{\partial^n}{\partial z^n}\left(\frac{1}{r}\right), \qquad (n = 0, 1, \cdots). \tag{40}$$

The potential of an arbitrary distribution of charge along a line is identical, outside a sphere whose diameter coincides with the line, with the potential of a system of multipoles located at the origin. The moments of these multipoles can be determined from the charge distribution by (23).

3.11. Arbitrary Distributions of Charge.—We shall consider next a charge which is distributed in a completely arbitrary manner in any finite region of a homogeneous, isotropic dielectric, or upon the surfaces of conductors embedded in such a dielectric. The region containing charge lies within a sphere of finite radius R drawn about the origin O of a coordinate system. We wish to express the potential of the distribution at a point P *outside the sphere* in terms of the coordinates of P with respect to O.

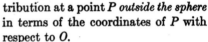

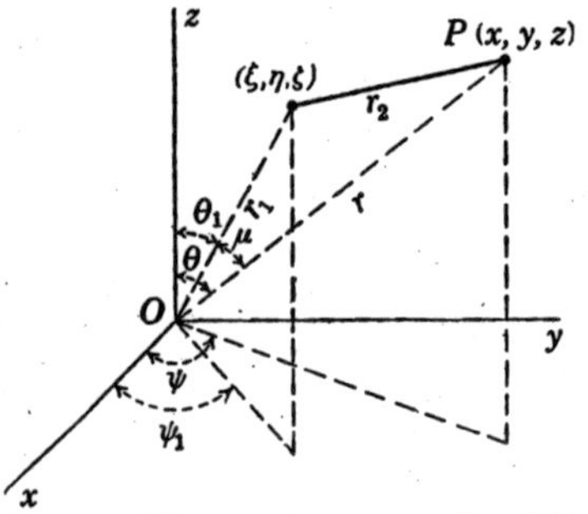

FIG. 30.—Figure to accompany Sec. 3.11.

The rectangular coordinates of P are x, y, z and its polar coordinates r, θ, ψ, where

$$x = r\sin\theta\cos\psi, \qquad y = r\sin\theta\sin\psi, \qquad z = r\cos\theta. \tag{41}$$

We calculate first the potential at P of an element of charge $dq = \rho\, dv$ located at the variable point ξ, η, ζ. If r_2, Fig. 30, is the distance from dq to P, the contribution of this element is

$$d\phi = \frac{1}{4\pi\epsilon}\frac{dq}{r_2} = \frac{1}{4\pi\epsilon}\frac{dq}{\sqrt{(x-\xi)^2 + (y-\eta)^2 + (z-\zeta)^2}}. \tag{42}$$

The denominator of this function can be expanded about the origin in powers of ξ, η, ζ exactly as in Eq. (14), and one obtains

$$\begin{aligned} d\phi = \frac{1}{4\pi\epsilon}dq\Bigg\{\frac{1}{r} &- \left[\xi\frac{\partial}{\partial x}\left(\frac{1}{r}\right) + \eta\frac{\partial}{\partial y}\left(\frac{1}{r}\right) + \zeta\frac{\partial}{\partial z}\left(\frac{1}{r}\right)\right] \\ &+ \frac{1}{2}\left[\xi^2\frac{\partial^2}{\partial x^2}\left(\frac{1}{r}\right) + \eta^2\frac{\partial^2}{\partial y^2}\left(\frac{1}{r}\right) + \zeta^2\frac{\partial^2}{\partial z^2}\left(\frac{1}{r}\right) + 2\xi\eta\frac{\partial^2}{\partial x\,\partial y}\left(\frac{1}{r}\right)\right. \\ &\left. + 2\eta\zeta\frac{\partial^2}{\partial y\,\partial z}\left(\frac{1}{r}\right) + 2\zeta\xi\frac{\partial^2}{\partial z\,\partial x}\left(\frac{1}{r}\right)\right] - \cdots\Bigg\}. \end{aligned} \tag{43}$$

The expansion converges provided $r_1 = \sqrt{\xi^2 + \eta^2 + \zeta^2} < R$, $r > R$.

Next we integrate (43) over the entire charge distribution; *i.e.*, with respect to the coordinates ξ, η, ζ, noting that $r = \sqrt{x^2 + y^2 + z^2}$ is independent of this integration. Once again we shall write ϕ as the sum of partial potentials ϕ_n, $\phi = \sum_{n=0}^{\infty} \phi_n$. Then the first term of the expansion is

$$\phi_0 = \frac{1}{4\pi\epsilon} \frac{p^{(0)}}{r}, \tag{44}$$

where

$$p^{(0)} = q = \int \rho(\xi, \eta, \zeta)\, dv \tag{45}$$

is the total charge. *At sufficiently large distances relative to R the potential of an arbitrary distribution of charge may be represented approximately by the Coulomb potential of a point charge located at the origin.*

The second term, which is dominant when the net charge is zero, is

$$\phi_1 = -\frac{1}{4\pi\epsilon} \mathbf{p}^{(1)} \cdot \nabla\left(\frac{1}{r}\right), \tag{46}$$

where $\mathbf{p}^{(1)}$ is a vector dipole moment whose rectangular components are

$$p_x^{(1)} = \int \rho\xi\, dv, \qquad p_y^{(1)} = \int \rho\eta\, dv, \qquad p_z^{(1)} = \int \rho\zeta\, dv. \tag{47}$$

The potential of any distribution whose net charge is zero may be approximated at large distances by the potential of a dipole located at the origin whose components are determined by (47).

In like manner the partial potential ϕ_2 arises from a *quadrupole* whose components are

$$\begin{aligned} p_{xx}^{(2)} &= \int \rho\xi^2\, dv, & p_{yy}^{(2)} &= \int \rho\eta^2\, dv, & p_{zz}^{(2)} &= \int \rho\zeta^2\, dv, \\ p_{xy}^{(2)} &= \int \rho\xi\eta\, dv, & p_{yz}^{(2)} &= \int \rho\eta\zeta\, dv, & p_{zx}^{(2)} &= \int \rho\zeta\xi\, dv. \end{aligned} \tag{48}$$

Whereas the dipole moment is a vector, the quantities defined in (48) constitute the *components of a tensor of second rank.* The multipole moments of higher order determined from subsequent terms of the expansion (43) are likewise tensors of higher rank.

3.12. General Theory of Multipoles.—Let $\mathbf{l}_i$ be a vector drawn from the origin O to the point $Q(\xi, \eta, \zeta)$ whose direction cosines with the coordinate axes x, y, z are respectively α_i, β_i, γ_i. Let $\frac{p^{(i)}}{4\pi\epsilon} f_i(x, y, z)$ be the potential at the point $P(x, y, z)$ due to a charge singularity $-p^{(i)}$ of any order located at the origin. If now a charge singularity of equal magnitude but positive sign be placed at the outer extremity of the vector $\mathbf{l}_i$, the potential at P due to the pair of multiple points will be

$$\phi_{i+1} = \frac{p^{(i)}}{\epsilon_0} [f_i(x - \alpha_i l_i, y - \beta_i l_i, z - \gamma_i l_i) - f_i(x, y, z)], \tag{49}$$

for it is to be noted that the change in potential at (x, y, z) arising from a displacement of the source from the origin to a point Q whose coordinates are $\xi = \alpha_i l_i$, $\eta = \beta_i l_i$, $\zeta = \gamma_i l_i$, is identical with that resulting from an equal but opposite displacement of P to a point $(x - \xi, y - \eta, z - \zeta)$ during which the source remains fixed at the origin. Expanding by Taylor's theorem, we have

$$\phi_{i+1}(x, y, z) = -\frac{p^{(i)} l_i}{\epsilon_0}\left(\alpha_i \frac{\partial f_i}{\partial x} + \beta_i \frac{\partial f_i}{\partial y} + \gamma_i \frac{\partial f_i}{\partial z}\right) + \text{terms of higher order in } l_i. \tag{50}$$

The expression in parentheses is evidently the derivative of f_i with respect to the direction specified by the vector $\mathbf{l}_i$.

$$\frac{\partial f_i}{\partial l_i} = \alpha_i \frac{\partial f_i}{\partial x} + \beta_i \frac{\partial f_i}{\partial y} + \gamma_i \frac{\partial f_i}{\partial z}. \tag{51}$$

The multipole moment of order $i + 1$ is again defined in terms of the moment of order i by the limit of

$$p^{(i+1)} = (i + 1) p^{(i)} l_i, \qquad (i = 0, 1, \cdots), \tag{52}$$

as $l_i \to 0$ and $p^{(i)} \to \infty$ in such a manner that their product remains finite. The potential at any point (x, y, z) due to a multipole of moment $p^{(i+1)}$ located at the origin is then

$$\phi_{i+1} = -\frac{1}{4\pi\epsilon} \frac{p^{(i+1)}}{(i + 1)} \frac{\partial f_i}{\partial l_i}, \qquad (i = 0, 1, \cdots). \tag{53}$$

A single point charge may be considered as a singularity of zero order and "moment" $p^{(0)} = q$.

$$\phi_0 = \frac{1}{4\pi\epsilon} p^{(0)} \left(\frac{1}{r}\right), \qquad f_0 = \left(\frac{1}{r}\right). \tag{54}$$

The potential of a dipole oriented in any direction specified by the vector $\mathbf{l}_0$ is consequently

$$\phi_1 = -\frac{1}{4\pi\epsilon} p^{(1)} \frac{\partial}{\partial l_0}\left(\frac{1}{r}\right). \tag{55}$$

The quadrupole is next generated by displacing the dipole parallel to its axis in a direction specified by the vector $\mathbf{l}_1$. For its potential we find from (53) and (55)

$$\phi_2 = \frac{1}{4\pi\epsilon} \frac{p^{(2)}}{2!} \frac{\partial^2}{\partial l_0\, \partial l_1}\left(\frac{1}{r}\right). \tag{56}$$

Thus by induction the potential of the general multipole of nth order is

$$\phi_n = \frac{(-1)^n}{4\pi\epsilon} \frac{p^{(n)}}{n!} \frac{\partial^n}{\partial l_0\, \partial l_1 \cdots \partial l_n}\left(\frac{1}{r}\right). \tag{57}$$

The construction of a quadrupole and an octupole is visualized in Figs. 31*a*, 31*b*.

The result of applying the operator

$$\frac{\partial}{\partial l_i} = \alpha_i \frac{\partial}{\partial x} + \beta_i \frac{\partial}{\partial y} + \gamma_i \frac{\partial}{\partial z} \tag{58}$$

n times to the function $\frac{1}{r} = (x^2 + y^2 + z^2)^{-\frac{1}{2}}$ must be of the form

$$\phi_n = \frac{1}{4\pi\epsilon} p^{(n)} \frac{Y_n}{r^{n+1}}, \tag{59}$$

where Y_n is a function of the cosines of the angles μ_i between $\mathbf{r}$ and the n axes $\mathbf{l}_i$, and of the direction cosines α_i, β_i, γ_i of the angles made by the

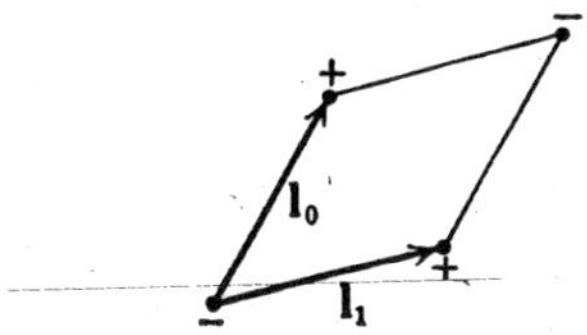

FIG. 31*a*.—Quadrupole.

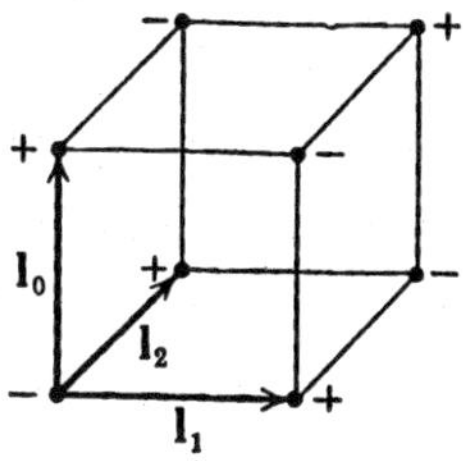

FIG. 31*b*.—Octupole.

vectors $\mathbf{l}_i$ with the coordinate axes. But the angles μ_i are related to the spherical coordinates θ, ψ of P by the equation

$$\cos \mu_i = (\alpha_i \cos \psi + \beta_i \sin \psi) \sin \theta + \gamma_i \cos \theta, \tag{60}$$

which may be demonstrated by calculating the distance $QP = r_2$, Fig. 30, both in terms of $r_1 = l_i$, r, and the angle μ_i, and in terms of its projections on the coordinate axes. Thus Y_n is a function of the angles θ and ψ, and hence of position on the surface of a unit sphere. For the dipole term we have

$$\begin{aligned} \phi_1 &= -\frac{1}{4\pi\epsilon} p^{(1)} \left[\alpha_0 \frac{\partial}{\partial x}\left(\frac{1}{r}\right) + \beta_0 \frac{\partial}{\partial y}\left(\frac{1}{r}\right) + \gamma_0 \frac{\partial}{\partial z}\left(\frac{1}{r}\right)\right] \\ &= \frac{1}{4\pi\epsilon} \frac{p^{(1)}}{r^3} (\alpha_0 x + \beta_0 y + \gamma_0 z). \end{aligned} \tag{61}$$

Upon replacing rectangular by spherical coordinates, Eqs. (41), this becomes

$$\phi_1 = \frac{1}{4\pi\epsilon} \frac{p^{(1)}}{r^2} [(\alpha_0 \cos \psi + \beta_0 \sin \psi) \sin \theta + \gamma_0 \cos \theta], \tag{62}$$

or

$$\phi_1 = \frac{1}{4\pi\epsilon} \frac{p^{(1)}}{r^2} [(\alpha_0 \cos \psi + \beta_0 \sin \psi) P_1^1(\cos \theta) + \gamma_0 P_1^0(\cos \theta)], \tag{63}$$

where the functions $P_n^m(\cos\theta)$ are the *associated Legendre functions* defined by

$$(64) \qquad P_n^m(\cos\theta) = \sin^m\theta \frac{d^m}{d(\cos\theta)^m} P_n(\cos\theta).$$

Between the four constants $p^{(1)}$, α_0, β_0, γ_0, there exists one relation, namely $\alpha_0^2 + \beta_0^2 + \gamma_0^2 = 1$, and there remain, consequently, three arbitrary parameters which determine the magnitude and orientation of the dipole.

Proceeding to the quadrupole term, we obtain

$$(65) \quad \phi_2 = \frac{1}{4\pi\epsilon}\frac{p^{(2)}}{r^3}\left\{\left(\gamma_0\gamma_1 - \frac{\alpha_0\alpha_1 + \beta_0\beta_1}{2}\right)\frac{1}{2}(3\cos^2\theta - 1)\right.$$
$$+ \frac{1}{2}\left[(\beta_0\gamma_1 + \beta_1\gamma_0)\sin\psi + \frac{1}{2}(\gamma_0\alpha_1 + \gamma_1\alpha_0)\cos\psi\right]\frac{3}{2}\sin 2\theta$$
$$\left. + \left[\frac{1}{4}(\alpha_0\beta_1 + \alpha_1\beta_0)\sin 2\psi + \frac{1}{4}(\alpha_0\alpha_1 + \beta_0\beta_1)\cos 2\psi\right] 3\sin^2\theta\right\},$$

or, with reference to (64),

$$(66) \quad \phi_2 = \frac{1}{4\pi\epsilon}\frac{p^{(2)}}{r^3}[a_{20}P_2^0(\cos\theta) + (a_{21}\cos\psi + b_{21}\sin\psi)P_2^1(\cos\theta)$$
$$+ (a_{22}\cos 2\psi + b_{22}\sin 2\psi)P_2^2(\cos\theta)].$$

Between the six direction cosines there are two relations of the form $\alpha_i^2 + \beta_i^2 + \gamma_i^2 = 1$; consequently there are, together with the arbitrary moment $p^{(2)}$, five completely arbitrary constants which are sufficient to determine the magnitude and orientation of the quadrupole.

A solution of Laplace's equation is called a *harmonic function*. Since the coordinates r, θ, ψ are arbitrary, each term ϕ_n in the expansion of ϕ must itself satisfy Laplace's equation and is, therefore, harmonic. Generalizing, we conclude that the harmonic function ϕ_n representing the potential of a multipole of nth order is a homogeneous polynomial of nth degree in x, y, z. Upon transformation to spherical coordinates r, θ, ψ, the function ϕ_n can be expressed in the form (59), where the spherical surface harmonic $Y_n(\theta, \psi)$ is now explicitly

$$(67) \quad Y_n(\theta, \psi) = \sum_{m=0}^{n} Y_{nm}(\theta, \psi) = \sum_{m=0}^{n}(a_{nm}\cos m\psi + b_{nm}\sin m\psi)P_n^m(\cos\theta).$$

The harmonic of nth order contains $2n + 1$ arbitrary constants sufficient to determine the moment and orientation of the axes of the corresponding multipole. If charge be distributed in arbitrary manner within a sphere of finite radius, the potential at all points outside the sphere must be harmonic and it must furthermore be regular at infinity. These conditions are satisfied by

$$\phi(r, \theta, \psi) = \frac{1}{4\pi\epsilon} \sum_{n=0}^{\infty} p^{(n)} \frac{Y_n(\theta, \psi)}{r^{n+1}}. \tag{68}$$

If the charge distribution is known, the components of the multipole moments are determined as in Sec. 3.11; if the distribution is unknown but an external field is specified, these constants are obtained from the boundary conditions as we shall shortly see.

DIELECTRIC POLARIZATION

3.13. Interpretation of the Vectors P and Π.—Although the two vectors **E** and **D** are sufficient to characterize completely an electrostatic field in any medium, it is convenient to introduce a third vector **P** defined in Sec. 1.6 as the difference of these two fields,

$$\mathbf{P} = \mathbf{D} - \epsilon_0 \mathbf{E}. \tag{1}$$

The vector **P** vanishes in free space and therefore is definitely associated with the constitution of the dielectric. The divergence equation then becomes

$$\nabla \cdot \mathbf{E} = \frac{1}{\epsilon_0} (\rho - \nabla \cdot \mathbf{P}), \tag{2}$$

from which one concludes that the effect of a dielectric on the field may be accounted for by an equivalent charge distribution whose volume density is

$$\rho' = -\nabla \cdot \mathbf{P}. \tag{3}$$

At every interior point of the dielectric the potential satisfies a modified Poisson equation,

$$\nabla^2 \phi = -\frac{1}{\epsilon_0} (\rho + \rho'). \tag{4}$$

The validity of (4) is no longer contingent upon the isotropy and homogeneity of the medium. If, however, there are surfaces of discontinuity, such as the boundaries of a dielectric, the vector **D** is subject to the condition

$$\mathbf{n} \cdot (\mathbf{D}_2 - \mathbf{D}_1) = \omega, \tag{5}$$

$$\mathbf{n} \cdot (\mathbf{E}_2 - \mathbf{E}_1) = \frac{1}{\epsilon_0} \omega - \frac{1}{\epsilon_0} \mathbf{n} \cdot (\mathbf{P}_2 - \mathbf{P}_1) = \frac{1}{\epsilon_0} (\omega + \omega'). \tag{6}$$

The primary sources of an electrostatic field are the "real" charges whose volume and surface densities are respectively ρ and ω. The effect of *rigid* dielectric bodies on the potential may be completely accounted for by distributions of induced, or "bound," charges of volume density ρ'

and a density

$$\omega' = -\mathbf{n} \cdot (\mathbf{P}_2 - \mathbf{P}_1) \tag{7}$$

over each surface of discontinuity, the unit vector $\mathbf{n}$ being drawn from medium (1) into (2). If (1) is dielectric and (2) free space, the induced surface charge is simply $\omega' = \mathbf{n} \cdot \mathbf{P}_1$, where $\mathbf{n}$ is the outward normal. In the case of a transition from a dielectric (2) to a conductor (1) bearing a free surface charge ω, there results a net surface charge $\omega - \mathbf{n} \cdot \mathbf{P}_2$, since $\mathbf{P}_1$ vanishes within the conductor.

The potential at any fixed point either within a dielectric or exterior to it can be expressed by

$$\phi(x', y', z') = \frac{1}{4\pi\epsilon_0} \int \frac{\rho - \nabla \cdot \mathbf{P}}{r} \, dv + \frac{1}{4\pi\epsilon_0} \int \frac{\omega + \mathbf{n} \cdot (\mathbf{P}_1 - \mathbf{P}_2)}{r} \, da. \tag{8}$$

The surface integrals are to be extended over all surfaces of discontinuity. It need scarcely be remarked that this analysis in most cases is purely formal, for in order to know $\mathbf{P}$ it is usually necessary to calculate first the potential.

The physical significance of the vector $\mathbf{P}$ becomes apparent if we transform (8) with the help of the identity

$$\frac{\nabla \cdot \mathbf{P}}{r} = \nabla \cdot \left(\frac{\mathbf{P}}{r}\right) - \mathbf{P} \cdot \nabla \left(\frac{1}{r}\right). \tag{9}$$

The dielectric may be resolved into partial volumes V_i bounded by surfaces S_i of discontinuity. For each partial volume

$$\int_{V_i} \nabla \cdot \left(\frac{\mathbf{P}}{r}\right) dv = \int_{S_i} \frac{\mathbf{P}_i \cdot \mathbf{n}_i}{r} \, da. \tag{10}$$

On summing these integrals over the partial volumes and remembering that $\mathbf{n}_i$ is directed out of V_i and hence *in* to a contiguous V_k, one finds that (8) reduces to

$$\phi(x', y', z') = \frac{1}{4\pi\epsilon_0} \int \frac{\rho}{r} \, dv + \frac{1}{4\pi\epsilon_0} \int \frac{\omega}{r} \, da + \frac{1}{4\pi\epsilon_0} \int \mathbf{P} \cdot \nabla \left(\frac{1}{r}\right) dv. \tag{11}$$

The third integral is now clearly the potential produced by a continuous distribution of dipole moment, and $\mathbf{P}$ is, therefore, to be interpreted as the *dipole moment per unit volume,* or *polarization* of the dielectric. According to (11) a dielectric body in the primary field of external sources gives rise to an induced secondary field such as is generated by a distribution of dipoles of moment $\mathbf{P}$ per unit volume.

In a purely macroscopic theory this is really all that can be said about the polarization vector. Macroscopically a charged conductor behaves as though the charge were distributed with a continuous density, though in fact we know it to be constituted of discrete electronic charges. Likewise, neutral matter under the influence of an external field acts as though dipoles were distributed as a continuous function of position. Now we know that neutral atoms and molecules are in fact constituted of equal numbers of positive and negative elementary charges. A microscopic electrodynamics must show how the electrical moments of each atom—or at least their most probable values—can be calculated. Then by a suitable process of averaging over large-scale volume elements the polarization per unit volume will be determined from the atomic moments and the transition effected from microscopic to macroscopic quantities.

A word at this point is in order concerning the physical significance of the Hertz vector $\mathbf{\Pi}$ introduced in a more general connection in Sec. 1.11. We shall suppose that the density of free charge ρ within the dielectric is zero. If the potential ϕ is expressed as the divergence of a vector function $\mathbf{\Pi}$,

$$\phi = -\nabla \cdot \mathbf{\Pi}, \tag{12}$$

then Poisson's equation assumes the form $\nabla \cdot \left(\nabla^2 \mathbf{\Pi} + \frac{1}{\epsilon_0} \mathbf{P}\right) = 0$, and this condition will certainly be satisfied if

$$\nabla^2 \mathbf{\Pi} = -\frac{1}{\epsilon_0} \mathbf{P}. \tag{13}$$

The vector $\mathbf{\Pi}$ is determined from the polarization by evaluating the integral

$$\mathbf{\Pi}\,(x', y', z') = \frac{1}{4\pi\epsilon_0} \int \frac{\mathbf{P}(x, y, z)}{r}\, dv, \tag{14}$$

whence $\mathbf{\Pi}$ is sometimes referred to as the *polarization potential.*

DISCONTINUITIES OF INTEGRALS OCCURRING IN POTENTIAL THEORY

3.14. Volume Distributions of Charge and Dipole Moment.—According to the foregoing analysis, a conductor is simply a region of zero field bounded by a surface bearing a layer of charge of density ω, while a rigid dielectric may be represented either by an equivalent volume distribution of density $\rho' = -\nabla \cdot \mathbf{P}$ bounded by a surface layer ω', or as a region occupied by a continuous distribution of dipoles whose moment per unit volume is $\mathbf{P}$. The boundary conditions, which heretofore have been deduced from the field equations, must evidently follow directly from the analytic properties of the integrals expressing the potential and its derivatives in terms of volume and surface densities of charge and dipole moment.

In Sec. 3.7 it was established that the improper integrals

(1) $$\phi(x', y', z') = \frac{1}{4\pi\epsilon_0}\int \frac{\rho(x, y, z)}{r}\, dv,$$

(2) $$\mathbf{E}(x', y', z') = \frac{1}{4\pi\epsilon_0}\int \rho(x, y, z)\nabla\left(\frac{1}{r}\right) dv,$$

are convergent and continuous functions of x', y', z' provided only that the charge density is bounded and piecewise continuous; i.e., that the volume occupied by the charge can be resolved into a finite number of partial volumes within each of which ρ is finite and continuous. Thus on either side of a surface bounding a discontinuous change in ρ, ϕ and $\mathbf{E}$ will have the same values, although the derivatives of $\mathbf{E}$ will in general be dis-

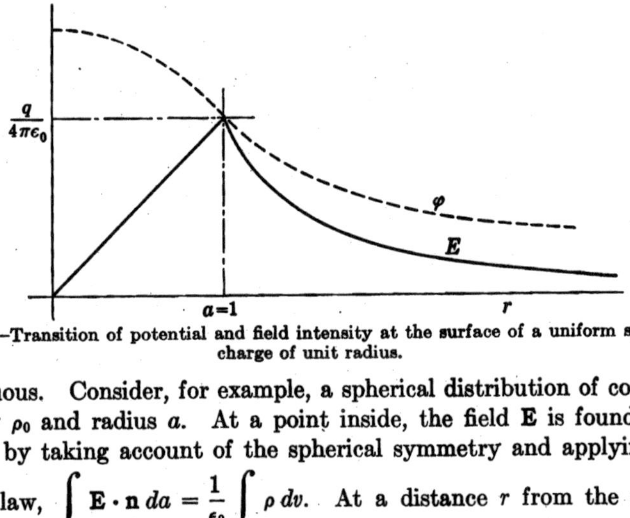

FIG. 32.—Transition of potential and field intensity at the surface of a uniform spherical charge of unit radius.

continuous. Consider, for example, a spherical distribution of constant density ρ_0 and radius a. At a point inside, the field $\mathbf{E}$ is found most simply by taking account of the spherical symmetry and applying the Gauss law, $\int \mathbf{E}\cdot\mathbf{n}\, da = \frac{1}{\epsilon_0}\int \rho\, dv$. At a distance r from the origin, $r < a$, we have

(3) $$E = \frac{1}{3\epsilon_0}\rho_0 r = \frac{1}{4\pi\epsilon_0} q \frac{r}{a^3}, \qquad (r < a),$$

where q is the total charge contained within the sphere. Likewise at any external point

(4) $$E = \frac{1}{4\pi\epsilon_0}\frac{q}{r^2}, \qquad (r > a).$$

The potential is calculated from the integral $\phi = -\int_\infty^r E\, dr$, giving

(5) $$\phi = -\frac{q}{4\pi\epsilon_0}\int_\infty^r \frac{1}{r^2}\, dr = \frac{1}{4\pi\epsilon_0}\frac{q}{r}, \qquad (r > a),$$
$$\phi = -\frac{q}{4\pi\epsilon_0}\int_\infty^a \frac{1}{r^2}\, dr - \frac{q}{4\pi\epsilon_0}\int_a^r r\, dr = \frac{q}{8\pi\epsilon_0}\left(\frac{3}{a} - \frac{r^2}{a^3}\right), \quad (r < a).$$

In Fig. 32, ϕ and its first derivative E are plotted as functions of r. Note that although E passes continuously through the surface $r = a$, its slope—the second derivative of ϕ—suffers an abrupt change because of the failure in continuity of the density ρ.

Consider next a volume distribution of dipole moment of density $\mathbf{P}$. The potential at any exterior point is

$$\phi(x', y', z') = \frac{1}{4\pi\epsilon_0} \int \mathbf{P} \cdot \nabla \left(\frac{1}{r}\right) dv. \tag{6}$$

This can be resolved into three integrals of the type

$$\phi_1 = \frac{1}{4\pi\epsilon_0} \int P_x \frac{\partial}{\partial x} \left(\frac{1}{r}\right) dv = -\frac{1}{4\pi\epsilon_0} \frac{\partial}{\partial x'} \int \frac{P_x}{r} dv. \tag{7}$$

From what has just been proved regarding (1) it follows that (6) *converges within the dipole distribution as well as at points exterior to it, and is a continuous function of position provided only* $\mathbf{P}(x, y, z)$ *is bounded and piecewise continuous.* Since a dielectric body is equivalent to a region of dipole moment, we have here proof of the continuity of the potential across a dielectric surface without recourse to an energy principle.

Across a surface of discontinuity in $\mathbf{P}$, the normal derivative of (6), and consequently the normal component of $\mathbf{E}$, is discontinuous. The magnitude of this discontinuity can be determined most readily by reverting from the volume distribution of moment to the equivalent volume and surface-charge distribution defined in Sec. 3.13. The vector $\mathbf{E}$ passes continuously into the volume charge, but we shall now show that its normal component suffers an abrupt change in passing through a layer of surface charge.

3.15. Single-layer Charge Distributions.—Let charge be distributed over a surface S with a density ω which we shall assume to be a bounded, piecewise continuous function of position on S. The potential at any point not on S is

$$\phi(x', y', z') = \frac{1}{4\pi\epsilon_0} \int_S \frac{\omega(x, y, z)}{r} da, \tag{8}$$

where r as usual is drawn from the charge element $\omega(x, y, z)\, da$ to the point of observation. If now (x', y', z') lies *on* the surface, the integral (8) is improper and its finiteness and continuity must be examined. About the point (x', y', z') on S let us circumscribe a circle of radius a. If the radius is sufficiently small, the circular disk thus defined may be assumed plane. Now the potential at (x', y', z') due to surface charges outside the disk is a bounded and continuous function of position in the vicinity of (x', y', z'). Call this portion of the potential ϕ_2. There remains a contribution ϕ_1 due to the charge on the disk itself. The surface density ω is bounded, and we proceed as in Sec. 3.7. There exists a number m such that at every point on the disk $|\omega| < m$, $|\omega/r| < m/r$.

$$|\phi_1| = \frac{1}{4\pi\epsilon_0}\left|\int_{S_1} \frac{\omega}{r}\, da\right| < \frac{m}{2\epsilon_0}\int_0^a \frac{1}{r}\, r dr = \frac{ma}{2\epsilon_0}, \tag{9}$$

where S_1 indicates that the surface integral is to be extended over the disk. The resultant potential at an arbitrary point (x', y', z') on the surface is, hence,

$$\phi = \phi_1 + \phi_2. \tag{10}$$

Both ϕ_1 and ϕ_2 have been shown to be bounded, and ϕ_2 is also continuous. But ϕ_1 vanishes with the area of the disk and consequently, since ϕ differs as little as desired from a continuous function, it is itself continuous. The specialization of the disk to circular form is no restriction on the generality of the proof, since the circle may be considered to be circumscribed about a disk of arbitrary shape. *The potential due to a surface distribution of charge is a bounded, continuous function of position at all points, both on and off the surface.* The function defined by (8), therefore, passes continuously through the surface.

The integral expression of the field intensity,

$$\mathbf{E}(x', y', z') = \frac{1}{4\pi\epsilon_0}\int_S \omega(x, y, z)\nabla\left(\frac{1}{r}\right) da, \tag{11}$$

is continuous and has continuous derivatives of all orders at points not on the surface, but suffers an abrupt change as the point (x', y', z') passes through S. The nature of the discontinuity may be determined directly from (11),[1] but we shall content ourselves here with the simple method employed in Sec. 1.13 based on the divergence and rotational properties of $\mathbf{E}$. The transition of the vector $\mathbf{E}$ through the layer S is subject to a discontinuity defined by

$$\mathbf{E}_+ - \mathbf{E}_- = \frac{1}{\epsilon_0}\,\omega\mathbf{n}, \tag{12}$$

where $\mathbf{n}$ is the unit normal drawn outward from the positive face of the surface. If now by ω we understand the true plus the bound charge, $\omega - \mathbf{n}\cdot(\mathbf{P}_+ - \mathbf{P}_-)$, then (12) is equivalent to

$$(\epsilon_0\mathbf{E}_+ + \mathbf{P}_+)\cdot\mathbf{n} - (\epsilon_0\mathbf{E}_- + \mathbf{P}_-)\cdot\mathbf{n} = (\mathbf{D}_+ - \mathbf{D}_-)\cdot\mathbf{n} = \omega. \tag{13}$$

This specifies at the same time the transition of $\mathbf{E}$ at a surface of discontinuity in a dipole distribution.

3.16. Double-layer Distributions.—It frequently happens that the potential of a charge distribution is identical with that which might be

[1] *Cf.* KELLOGG, *loc. cit.*; PHILLIPS, *loc. cit.*, Chap VI.

produced by a layer of dipoles distributed over a surface. We imagine such a surface distribution to be generated by spreading positive charge of density $+\omega$ over the *positive* side of a regular surface S, and an identical distribution of opposite sign on its negative side. The result is a double layer of charge separated by the infinitesimal distance l. The dipole moment per unit area, or surface density $\boldsymbol{\tau}$, is a vector directed along the positive normal to S and is defined as the limit

$$\boldsymbol{\tau} = \mathbf{n} \lim (\omega l) \text{ as } l \to 0, \qquad \omega \to \infty. \tag{14}$$

The dipole moment corresponding to an element of area da on the double layer is $\tau\, da$, and its contribution to the potential at a fixed point (x', y', z') not on the surface is

$$d\phi = \frac{1}{4\pi\epsilon_0} \frac{\tau \cos\theta}{r^2}\, da = \frac{1}{4\pi\epsilon_0} \boldsymbol{\tau} \cdot \nabla \left(\frac{1}{r}\right) da. \tag{15}$$

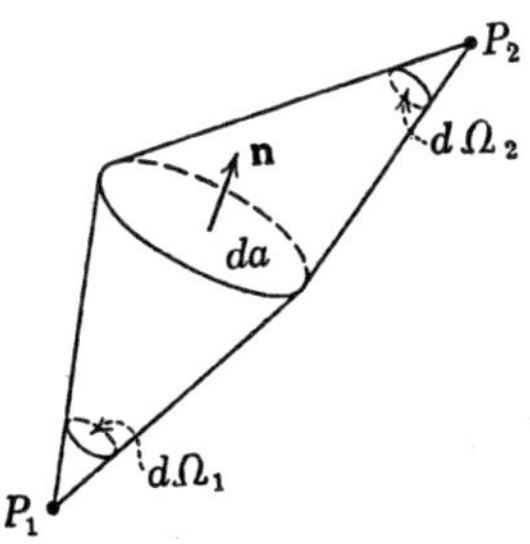

FIG. 33.—The element of area da subtends a positive solid angle at P_1 and a negative angle at P_2.

Now $\dfrac{\cos\theta}{r^2}\, da$ measures exactly the solid angle $d\Omega$ at the point of observation (x', y', z') subtended by the element of area da. This angle is positive if the radius vector $\mathbf{r}$ drawn from (x', y', z') to the element da makes an acute angle with the positive normal $\mathbf{n}$. Thus in Fig. 33 the element da subtends a positive angle at P_1 and a negative one at P_2. The potential due to the entire distribution may be written

$$\phi(x', y', z') = \frac{1}{4\pi\epsilon_0} \int_S \boldsymbol{\tau}(x, y, z) \cdot \nabla \left(\frac{1}{r}\right) da = -\frac{1}{4\pi\epsilon_0} \int_\Omega \tau\, d\Omega, \tag{16}$$

where Ω is the solid angle subtended at (x', y', z') by the surface S. That the second integral must be preceded by a negative sign is evident, if one notes that $d\Omega$ is positive when the layer is viewed from the lower or negative side, where the potential is certainly negative.

The potential ϕ has distinct values on opposite sides of a double layer. Suppose first that the surface S is closed and that the distribution is of constant density, so that τ may be taken outside the integral sign. The positive layer lies on the outer side of S so that $\boldsymbol{\tau}$ has the same direction as the positive normal. There is a well-known theorem on solid angles which applies to this case: "If S forms the complete boundary of a three-dimensional region, the total solid angle subtended by S at P is zero, if P lies outside the region, and 4π if P lies inside."[1] The potential at any interior point of S is, therefore, $\phi_- = -\dfrac{1}{\epsilon_0}\tau$; at any exterior point

[1] On the analytic properties of solid angles see Phillips, *loc. cit.*, p. 112.

$\phi_+ = 0$. The difference in potential on either side of the double layer is

$$\phi_+ - \phi_- = \frac{1}{\epsilon_0}\tau. \tag{17}$$

Next we observe that (17) represents correctly the discontinuity in ϕ as one traverses the double layer also when S is an open surface. For S may be closed by adding an arbitrary surface S'. At every point within or without this closed surface the resultant potential can be resolved into two parts, a fraction ϕ due to the distribution on S, and a fraction ϕ', due to that on S'. Inside the closed surface the resultant potential is again $-\frac{1}{\epsilon_0}\tau$; outside, it vanishes. Thus, as we traverse the surface S, $\phi + \phi'$ changes discontinuously by an amount $\frac{1}{\epsilon_0}\tau$; but ϕ', due to the layer S', is certainly continuous across S and hence the entire discontinuity is in ϕ, as specified by (17).

It remains to show that (17) is also valid when τ is a function of position on S. About an arbitrary point P on the surface S draw a circle of radius a. Let the radius a be so small that, over the area enclosed by the circle, τ may be assumed to have a constant value τ_0. The potential ϕ in the neighborhood of P may again be resolved into two parts, a fraction ϕ' due to the infinitesimal circular disk and a fraction ϕ'' due to that portion of the dipole layer lying outside the circle. ϕ'' is continuous at P. ϕ' on the other hand suffers a discontinuous jump of $\frac{1}{\epsilon_0}\tau_0$ on crossing the circular disk. The resultant potential $\phi = \phi' + \phi''$ must, therefore, exhibit a discontinuity of the same amount and, in the limit as $a \to 0$, we find that (17) holds for variable distributions if for τ we take the value at the point of transition through S.

These results might be interpreted to mean that the work done in moving a unit positive charge around a closed path which passes once through a surface bearing a uniform dipole distribution of density τ_0 is $\pm\frac{1}{\epsilon_0}\tau_0$, depending on the direction of circuitation. The potential in presence of such a double layer is a multivalued function; for to its value at (x', y', z') one may add $\frac{1}{\epsilon_0}m\tau$, where m is any positive or negative integer. To obtain a different value of ϕ one need only return to the initial point after traversing the surface S. All this appears to be in contradiction to the conservative nature of the electrostatic field. As a matter of fact, the mathematical double layer constitutes a singular surface which has no true counterpart in nature. We shall show below by means of Green's theorem that the potential within a closed domain bounded by a surface S due to external charges is identical with that which might be produced by a certain dipole distribution over S. This equivalent double layer, however, does not lead to correct values of ϕ *outside* S.

The application of the integral $\oint \mathbf{E} \cdot \mathbf{n}\, da = \frac{1}{\epsilon_0} q$ to a small right cylinder, the ends of which lie on either side of the double layer, after the manner of Sec. 1.13, indicates at once that there is no discontinuity in the normal component of the field vector $\mathbf{E}$ across the surface since the total charge within the cylinder is now zero.

$$(\mathbf{E}_+ - \mathbf{E}_-) \cdot \mathbf{n} = 0. \tag{18}$$

The line integral of $\mathbf{E}$ around a closed path, however, is no longer necessarily zero; consequently we anticipate a possible discontinuity in the tangential components. Let the contour start at 1, Fig. 34, at which point the dipole density is τ. The difference in potential between points 1 and 2 is, according to (17), $\phi_1 - \phi_2 = \frac{1}{\epsilon_0} \tau$. (Note that the intensity $\mathbf{E}$ is given by the *derivative* of the potential, not the *difference* in potential across the discontinuity. The normal derivative is continuous.)

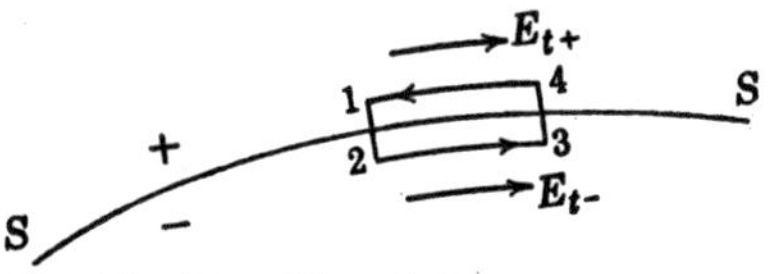

FIG. 34.—Transition of the tangential components of **E** across a double layer.

If $\Delta \mathbf{l}$ is the length of the path tangential to the surface, the density at 3 is $\tau + \nabla\tau \cdot \Delta \mathbf{l}$, and the potential difference between 4 and 3 is

$$\phi_4 - \phi_3 = \frac{1}{\epsilon_0} (\tau + \nabla\tau \cdot \Delta \mathbf{l}).$$

Now the quantity $(\phi_2 - \phi_1) + (\phi_3 - \phi_2) + (\phi_4 - \phi_3) + (\phi_1 - \phi_4)$ is identically zero; hence,

$$-\frac{1}{\epsilon_0} \tau + (\phi_3 - \phi_2) + \frac{1}{\epsilon_0} (\tau + \nabla\tau \cdot \Delta \mathbf{l}) + (\phi_1 - \phi_4) = 0. \tag{19}$$

Abbreviating $\phi_4 - \phi_1 = \Delta\phi_+$, $\phi_3 - \phi_2 = \Delta\phi_-$, (19) reduces to

$$\frac{\Delta\phi_+}{\Delta l} - \frac{\Delta\phi_-}{\Delta l} = \frac{1}{\epsilon_0} \nabla\tau \cdot \mathbf{t}, \tag{20}$$

where $\mathbf{t}$ is a unit vector tangent to S. The limit of $-\Delta\phi/\Delta l$ as $l \to 0$ is the component of $\mathbf{E}$ in the direction of $\mathbf{t}$, so that in virtue of the continuity of the normal component of $\mathbf{E}$ the transition of the field vector is specified by

$$\mathbf{E}_+ - \mathbf{E}_- = -\frac{1}{\epsilon_0} \nabla\tau. \tag{21}$$

Clearly $\nabla\tau$ signifies here the gradient of τ in the surface and is, therefore, a vector tangent to S. The proof is subject to the assumption that τ and its first derivatives are continuous over S.

3.17. Interpretation of Green's Theorem.—In Sec. 3.4 it was shown that the potential at any interior point of a region V bounded by a closed, regular surface S could be expressed in the form

$$\phi(x', y', z') = \frac{1}{4\pi\epsilon}\int_V \frac{\rho}{r}\,dv + \frac{1}{4\pi}\int_S \frac{1}{r}\frac{\partial\phi}{\partial n}\,da - \frac{1}{4\pi}\int_S \phi\frac{\partial}{\partial n}\left(\frac{1}{r}\right)da. \tag{22}$$

From the analysis of the preceding paragraph we are led to interpret this result in the following way. The volume integral represents, of course, the contribution of the charge within S, and the surface integrals account for all charges exterior to it. However, the first of these surface integrals is also equivalent to the potential of a single layer of charge distributed over S with a density

$$\omega = \epsilon\frac{\partial\phi}{\partial n}, \tag{23}$$

and the second can evidently be interpreted as the potential of a double layer on S whose density is

$$\tau = \epsilon\phi. \tag{24}$$

The charges outside S may be replaced by an equivalent single and double layer, the densities of which are specified by (23) *and* (24), *without modifying in any way the potential at an interior point.* The potential outside S produced by these surface distributions corresponds in no way, however, to that arising from the true distribution. On the contrary, it is most important to note that these equivalent surface layers, which give rise to the proper value of the potential at all interior points of S, *are just those required to reduce both potential and field* **E** *to zero at every point outside.* We observe first that the single layer gives rise to a discontinuity in the normal derivative of ϕ equal to

$$\left(\frac{\partial\phi}{\partial n}\right)_- - \left(\frac{\partial\phi}{\partial n}\right)_+ = \frac{1}{\epsilon}\,\omega. \tag{25}$$

The normal derivative in Eq. (23) must be calculated on the inner or negative side of S, since this alone belongs to V. Replacing $(\partial\phi/\partial n)_-$ by its value (23), it follows that $(\partial\phi/\partial n)_+ = 0$. Furthermore, the discontinuity in potential due to the double layer is specified by (17), which together with (24) shows that $\phi_+ = 0$. That ϕ and its derivatives vanish *everywhere* outside S is readily shown by applying (22) to the volume V_2 exterior to S. Since there are now no charges in V_2, $\nabla^2\phi = 0$. At infinity the potential is regular, and consequently the potential within

V_2 is determined solely by the values of ϕ_+ and $(\partial\phi/\partial n)_+$ on S. But these have just been shown to vanish. The function $\phi(x', y', z')$ of (22) is continuous and has continuous derivatives. The field intensity **E** outside S is, therefore, also zero.

From the foregoing discussion it is clear that one may always close off any portion of an electrostatic field by a surface, reducing the field and potential outside to zero, and taking account of the effect of external charges on the field within by proper single- and double-layer distributions on the bounding surface. It is instructive to consider these results from the standpoint of the field intensity **E**. The single layer introduces the proper discontinuity in the normal component E_n, but does not affect the transition of the tangential component. The double layer, on the other hand, in no wise affects the transition of the normal component, but may be adjusted to introduce the proper discontinuity in E_t, or according to (21),

$$(26) \qquad E_{t-} = \frac{1}{\epsilon}\frac{d\tau}{dl}, \qquad E_{t+} = 0,$$

where l is any direction tangent to S. If in particular the surface S is an equipotential, $\mathbf{n} \times \mathbf{E} = 0$, and no dipole distribution is necessary; the field inside S due to external charges can then be accounted for by a single layer of density $\omega = \epsilon \dfrac{\partial\phi}{\partial n}$ on the equipotential.

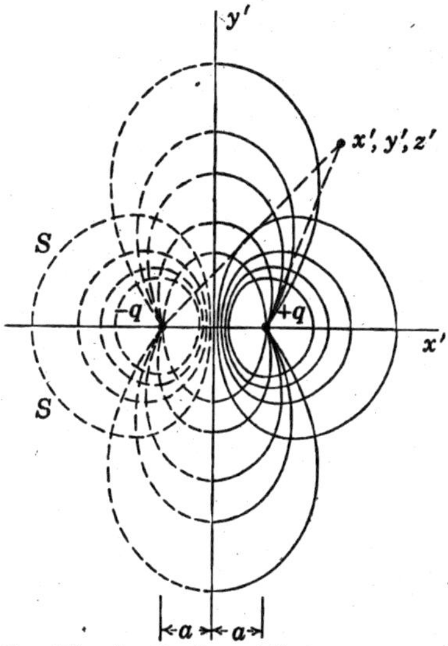

FIG. 35.—Application of Green's theorem to the theory of images.

3.18. Images.—An important application of these principles is to be found in the theory of images. The equipotential surfaces of a pair of equal point charges, one positive and the other negative in a homogeneous dielectric of inductive capacity ϵ, form a family of spheres whose centers lie along the line joining the charges. Let the surface S be represented by any equipotential about $-q$, Fig. 35. This surface thus divides all space into two distinct regions, and in view of the regularity of the potential at infinity, Green's theorem applies to both, S being the bounding surface in either case; *i.e.*, either region may be taken as the "interior" of S. If, therefore, the charge $-q$ be removed, the field in the region occupied by $+q$ is unmodified if a charge of density ω is spread over S as specified by (23). Inversely, if the charge $+q$ be located as shown with respect to a conducting sphere, a surface charge ω will be induced upon it such that it becomes an equipotential. The contribution of this induced charge to the field outside the conductor S is now determined most simply by replacing the surface distribution by the equivalent point charge $-q$. The charge $-q$ is said to be the image of $+q$ with respect to the given sphere. In case the equipotential surface is the median plane $x' = 0$, the calculation is extremely simple. The potential at any point (x', y', z') to the right of $x' = 0$ is

$$(27) \quad \phi(x', y', z') = \frac{q}{4\pi\epsilon}\left[\frac{1}{\sqrt{(x'-a)^2 + y'^2 + z'^2}} - \frac{1}{\sqrt{(x'+a)^2 + y'^2 + z'^2}}\right].$$

The normal derivative at $x' = 0$ must be taken in the direction of the negative x'-axis, since the region on the right of $x' = 0$ is to be considered the interior of the equipotential.

$$(28) \qquad \left(-\frac{\partial\phi}{\partial x'}\right)_{x'=0} = -\frac{1}{2\pi\epsilon}\frac{aq}{(a^2 + y'^2 + z'^2)^{\frac{3}{2}}};$$

hence by (23) the charge density is

$$(29) \qquad \omega = -\frac{1}{2\pi}\frac{aq}{r^3},$$

where $r^2 = a^2 + y'^2 + z'^2$.[1]

BOUNDARY-VALUE PROBLEMS

3.19. Formulation of Electrostatic Problems.—The analysis of the preceding sections enables one to calculate the potential at any point in an electrostatic field when the distribution of charge and polarization is completely specified. In practice, however, the problem is rarely so elementary. Ordinarily only certain external sources, or an applied field, are given from which the polarization of dielectrics and the surface charge distribution on conductors must be determined such as to satisfy the boundary conditions over surfaces of discontinuity.

Among electrostatic problems of this type are to be recognized two classes: the *homogeneous* boundary-value problem and the *inhomogeneous* problem. To illustrate the first, consider an isolated conductor embedded in a dielectric. A charge is placed on the conductor and we wish to know its distribution over the surface and the potential of the conductor with respect to earth or to infinity. At all points outside the conductor the

[1] On the method of images, see JEANS, "Mathematical Theory of Electricity and Magnetism," 5th ed., Chap. VIII, Cambridge University Press; or MASON and WEAVER, "The Electromagnetic Field," pp. 109*ff*., University of Chicago Press, 1929.

potential must satisfy Laplace's equation. At infinity it must vanish (regularity), and over the surface of the conductor it must assume a constant value. We shall show that these conditions are sufficient to determine ϕ uniquely. The density of surface charge can then be determined from the normal derivative of ϕ, subject to the condition that $\int \omega \, da$ over the surface of the conductor must equal the total charge.

An inhomogeneous problem is represented by the case of a dielectric or conducting body introduced into the *fixed* field of external sources. A charge is then induced on the surface of a conductor which will distribute itself in such a manner that the *resultant* potential is constant over its surface. The integral $\int \omega \, da$ is now zero. Likewise there will be induced in the dielectrics a polarization whose field is superposed on the primary field to give a resultant field satisfying the boundary conditions.

A schedule can be drawn up of the conditions which must be satisfied in every boundary-value problem. To simplify matters we shall assume henceforth that the dielectrics are isotropic and homogeneous except across a finite number of surfaces of discontinuity.

(1) $\nabla^2\phi = 0$ *at all points not on a boundary surface or within external sources;*

(2) ϕ *is continuous everywhere, including boundaries of dielectrics or of conductors, but excluding surfaces bearing a double layer;*

(3) ϕ *is finite everywhere, except at external point charges introduced as primary sources;*

(4) $\epsilon_2 \left(\frac{\partial \phi}{\partial n}\right)_2 - \epsilon_1 \left(\frac{\partial \phi}{\partial n}\right)_1 = 0$ *across a surface bounding two dielectrics;*

(5) $\epsilon \frac{\partial \phi}{\partial n} = -\omega$ *at the interface of a conductor and dielectric;*

(6) *On the surface of a conductor either*

   (a) ϕ *is a known constant* ϕ_i, *or*

   (b) ϕ *is an unknown constant and*

$$\int_S \epsilon \frac{\partial \phi}{\partial n} \, da = -q_i;$$

(7) ϕ *is regular at infinity provided all sources are within a finite distance of the origin.*

In (4) it is assumed that the interface of the dielectrics bears no charge, as is almost invariably the case. The normal is directed from (1) to (2), and in (5) from conductor into dielectric.

An electrostatic problem consists in finding among all possible solutions of Laplace's equation the particular one that will satisfy the conditions of the above schedule over the surfaces of specified conductors and dielectrics.

3.20. Uniqueness cf Solution.—Let ϕ be a function which is harmonic (satisfies Laplace's equation) and which has continuous first and second derivatives throughout a region V and over its bounding surface S. According to Green's first identity, (5), page 165, ϕ satisfies

$$\int_V (\nabla\phi)^2\,dv = \int_S \phi\,\frac{\partial\phi}{\partial n}\,da. \tag{1}$$

Suppose now that $\phi = 0$ on the surface S. Then $\int_V (\nabla\phi)^2\,dv = 0$. The integrand is an essentially positive quantity and consequently $\nabla\phi$ must vanish throughout V. This is possible only if ϕ is constant. Over the boundary $\phi = 0$ and, since by hypothesis ϕ is continuous throughout V, it follows that $\phi = 0$ over the entire region.

Now let ϕ_1 and ϕ_2 be two functions that are harmonic throughout the closed region V and let

$$\phi = \phi_1 - \phi_2. \tag{2}$$

Then if ϕ_1 and ϕ_2 are equal over the boundary S, their difference vanishes identically throughout V. *A function which is harmonic and which possesses continuous first- and second-order derivatives in a closed, regular region V is uniquely determined by its values on the boundary S.*

Consider next a system of conductors embedded in a homogeneous dielectric whose potentials are specified. We wish to show that the potential at every point in space is thereby uniquely determined. The reasoning of the preceding paragraph is applied to a volume V which is bounded interiorly by the surfaces of the conductors, and on the exterior by a sphere of very large radius R. Let us suppose that there are two solutions, ϕ_1 and ϕ_2, that satisfy the prescribed boundary conditions. Then on the surfaces of the conductors, $\phi = \phi_1 - \phi_2 = 0$. Since ϕ_1 and ϕ_2 are assumed to be solutions of the problem, they must both satisfy the conditions of Sec. 3.19; hence their difference ϕ is harmonic, has the value zero over the conductors, and is regular at infinity. The surface integral on the right-hand side of (1) must now be extended over both the interior and exterior boundaries. Over the interior boundary $\phi = 0$; hence the integral vanishes. On the outer sphere $\partial\phi/\partial n = \partial\phi/\partial R$. If R approaches infinity, ϕ vanishes as $1/R$ and $\partial\phi/\partial R$ as $1/R^2$. The integrand $\phi\,\frac{\partial\phi}{\partial n}$ thus vanishes as $1/R^3$, whereas the area of the sphere becomes infinite as R^2. The surface integral over the exterior boundary is, therefore, zero in the limit $R \to \infty$. It follows too that the volume integral in (1) must vanish when extended over the entire space exterior to the conductors, and we conclude as before that if the two functions ϕ_1

and ϕ_2 are identical on the boundaries they are identical everywhere: there is only one potential function that assumes the specified constant values over a given set of conductors.

The left-hand side of (1) may also be made to vanish by specifying that $\partial\phi/\partial n$ shall be zero over the enclosing boundary S. Then throughout V we have again $\nabla\phi = 0$, whence it follows that ϕ is constant everywhere, although not necessarily zero since the condition $\partial\phi/\partial n = 0$ does not imply the vanishing of ϕ on S. One concludes, as above, that if the normal derivatives $\partial\phi_1/\partial n$ and $\partial\phi_2/\partial n$ of two solutions are identical on the boundaries, the solutions themselves can differ only by a constant. In other words, *the potential is uniquely determined except for an additive constant by the values of the normal derivative on the boundaries.* But the normal derivative of the potential is in turn proportional to the surface charge density and, consequently, there is only one solution corresponding to a given set of charges on the conductors.

In case there are dielectric bodies present in the field the requirement that the first derivatives of ϕ, as well as ϕ itself, shall be continuous is no longer satisfied and (1) cannot be applied directly. However, the region outside the conductors may be resolved into partial volumes V_i bounded by the surfaces S_i within which the dielectric is homogeneous. To each of these regions in turn, (1) is then applied. The potential is continuous across any surface S_i and the derivatives on one side of S_i are fixed in terms of the derivatives on the other. It is easy to see that also in this more general case *the electrostatic problem is completely determined by the values either of the potentials or of the charges specified on the conductors of the system.*

3.21. Solution of Laplace's Equation.—It should be apparent that the fundamental task in solving an electrostatic problem is the determination of a solution of Laplace's equation in a form that will enable one to satisfy the boundary conditions by adjusting arbitrary constants. There are a certain number of special methods, such as the method of images, which can sometimes be applied for this purpose. Apart from the theory of integral equations, the only procedure that is both practical and general in character is the method known as "separation of the variables." Let us suppose that the surface S bounding a conductor or dielectric body satisfies the equation

$$f_1(x, y, z) = C. \tag{3}$$

We now introduce a set of orthogonal, curvilinear coordinates u^1, u^2, u^3, as in Sec. 1.14, such that one coordinate surface, say $u^1 = C$, coincides with the prescribed boundary (3). If then a harmonic function $\phi(u^1, u^2, u^3)$ can be found in this coordinate system, it is evident that the normal derivative at any point on the boundary is proportional to the derivative of ϕ with respect to u^1, and that the derivatives with respect to u^2 and u^3 are tangential.

According to (82), page 49, Laplace's equation in curvilinear coordinates is

$$\sum_{i=1}^{3} \frac{\partial}{\partial u^i}\left(\frac{\sqrt{g}}{h_i^2}\frac{\partial \phi}{\partial u^i}\right) = 0, \qquad \sqrt{g} = h_1h_2h_3. \tag{4}$$

Let us suppose that the scale factors h_i satisfy the condition

$$\frac{\sqrt{g}}{h_i^2} = M_i f_1(u^1)f_2(u^2)f_3(u^3). \tag{5}$$

Each of the product functions $f_i(u^i)$ depends on the variable u^i alone, but the factor M_i does not contain u^i. It may, however, depend on the other two. We next assume that ϕ likewise may be expressed as a product of three functions of one variable each.

$$\phi = F_1(u^1)F_2(u^2)F_3(u^3). \tag{6}$$

Then (4) can be written in the form

$$\sum_{i=1}^{3} \frac{M_i}{f_iF_i}\frac{\partial}{\partial u^i}\left(f_i\frac{\partial F_i}{\partial u^i}\right) = 0. \tag{7}$$

If finally *the* M_i *are rational functions*, Eq. (7) can be resolved into three ordinary differential equations. The method is best described by example.

1. *Cylindrical Coordinates.*—From (1), page 51, we have

$$\sqrt{g} = r, \qquad \frac{\sqrt{g}}{h_1^2} = r, \qquad \frac{\sqrt{g}}{h_2^2} = \frac{1}{r}, \qquad \frac{\sqrt{g}}{h_3^2} = r, \tag{8}$$

whence by inspection

$$f_1 = r, \qquad f_2 = f_3 = 1, \qquad M_1 = M_3 = 1, \qquad M_2 = \frac{1}{r^2}. \tag{9}$$

Equation (7) becomes

$$\frac{1}{F_1r}\frac{\partial}{\partial r}\left(r\frac{\partial F_1}{\partial r}\right) + \frac{1}{r^2F_2}\frac{\partial^2 F_2}{\partial \theta^2} + \frac{1}{F_3}\frac{\partial^2 F_3}{\partial z^2} = 0. \tag{10}$$

The first two terms of (10) do not contain z; the last term is independent of r and ϕ. A change in z cannot affect the first two terms and therefore the last term must be constant if (10) is to be satisfied identically for any range of z.

$$\frac{1}{F_3}\frac{d^2F_3}{dz^2} = -C_1. \tag{11}$$

The arbitrary constant C_1, called the *separation constant*, has been chosen negative purely as a matter of convenience and the partial derivative has been changed to a total derivative since F_3 is a function of z alone.

Upon replacing the third term of (10) by $-C_1$ and multiplying by r^2 one obtains

$$\frac{r}{F_1}\frac{\partial}{\partial r}\left(r\frac{\partial F_1}{\partial r}\right) + \frac{1}{F_2}\frac{\partial^2 F_2}{\partial \theta^2} = r^2 C_1. \tag{12}$$

It is again apparent that the second term of (12) is constant, leading to two ordinary equations,

$$\frac{d^2F_2}{d\theta^2} + C_2 F_2 = 0, \tag{13}$$

$$\frac{1}{r}\frac{d}{dr}\left(r\frac{dF_1}{dr}\right) - \left(C_1 + \frac{C_2}{r^2}\right)F_1 = 0. \tag{14}$$

The equations for F_2 and F_3 are satisfied by exponential functions of real or imaginary argument depending on the algebraic sign attributed to the separation constants; F_1 is a *Bessel function*. In the not uncommon case of a potential ϕ which is independent of z, we find F_3 constant, $C_1 = 0$, and in place of (14).

$$r\frac{d}{dr}\left(r\frac{dF_1}{dr}\right) = C_2. \tag{15}$$

2. *Spherical Coordinates*.—From (2), page 52, we have

$$\sqrt{g} = r^2 \sin\theta, \qquad \frac{\sqrt{g}}{h_1^2} = r^2\sin\theta, \qquad \frac{\sqrt{g}}{h_2^2} = \sin\theta, \qquad \frac{\sqrt{g}}{h_3^2} = \frac{1}{\sin\theta}, \tag{16}$$

whence

$$\begin{aligned} f_1 &= r^2, & f_2 &= \sin\theta, & f_3 &= 1, \\ M_1 &= 1, & M_2 &= \frac{1}{r^2}, & M_3 &= \frac{1}{r^2\sin^2\theta}. \end{aligned} \tag{17}$$

Equation (7) reduces to

$$\frac{1}{F_1 r^2}\frac{\partial}{\partial r}\left(r^2\frac{\partial F_1}{\partial r}\right) + \frac{1}{r^2\sin\theta\, F_2}\frac{\partial}{\partial\theta}\left(\sin\theta\frac{\partial F_2}{\partial\theta}\right) + \frac{1}{r^2\sin^2\theta\, F_3}\frac{\partial^2 F_3}{\partial\psi^2} = 0, \tag{18}$$

where ψ has in this case been employed in place of ϕ to represent the azimuthal angle. Separation of the variables leads to the three ordinary equations

$$\frac{1}{F_1}\frac{d}{dr}\left(r^2\frac{dF_1}{dr}\right) = C_1, \tag{19a}$$

$$\frac{\sin\theta}{F_2}\frac{d}{d\theta}\left(\sin\theta\frac{dF_2}{d\theta}\right) = C_2 - C_1\sin^2\theta, \tag{19b}$$

$$\frac{d^2F_3}{d\psi^2} + C_2 F_3 = 0. \tag{19c}$$

Of these three, only the *Legendre equation* (19b) is of any complexity.

3. *Elliptic Coordinates.*—According to (3), page 52, we have

$$(20)\quad \frac{\sqrt{g}}{h_1^2} = \sqrt{\frac{\xi^2 - 1}{1 - \eta^2}}, \qquad \frac{\sqrt{g}}{h_2^2} = \sqrt{\frac{1 - \eta^2}{\xi^2 - 1}}, \qquad \frac{\sqrt{g}}{h_3^2} = \frac{c^2(\xi^2 - \eta^2)}{\sqrt{(\xi^2 - 1)(1 - \eta^2)}},$$

which by inspection lead to

$$(21)\quad \begin{aligned} f_1 &= \sqrt{\xi^2 - 1}, & f_2 &= \sqrt{1 - \eta^2}, & f_3 &= 1, \\ M_1 &= \frac{1}{1 - \eta^2}, & M_2 &= \frac{1}{\xi^2 - 1}, & M_3 &= \frac{c^2(\xi^2 - \eta^2)}{(\xi^2 - 1)(1 - \eta^2)}. \end{aligned}$$

The Laplace equation for the elliptic cylinder is

$$(22)\quad \frac{\sqrt{\xi^2 - 1}}{F_1} \frac{\partial}{\partial \xi}\left(\sqrt{\xi^2 - 1}\,\frac{\partial F_1}{\partial \xi}\right) + \frac{\sqrt{1 - \eta^2}}{F_2} \frac{\partial}{\partial \eta}\left(\sqrt{1 - \eta^2}\,\frac{\partial F_2}{\partial \eta}\right) + \frac{c^2(\xi^2 - \eta^2)}{F_3} \frac{\partial^2 F_3}{\partial z^2} = 0,$$

which upon separation gives

$$(23a)\quad \frac{1}{F_3} \frac{d^2 F_3}{dz^2} = -C_1,$$

$$(23b)\quad \frac{\sqrt{\xi^2 - 1}}{F_1} \frac{d}{d\xi}\left(\sqrt{\xi^2 - 1}\,\frac{dF_1}{d\xi}\right) - c^2\xi^2 C_1 = C_2,$$

$$(23c)\quad \frac{\sqrt{1 - \eta^2}}{F_2} \frac{d}{d\eta}\left(\sqrt{1 - \eta^2}\,\frac{dF_2}{d\eta}\right) + c^2\eta^2 C_1 = -C_2.$$

Both F_1 and F_2 are *Mathieu functions*, but simplify notably when $C_1 = 0$ as is the case when ϕ is uniform along the length of the cylinder.

4. *Spheroidal Coordinates.*—According to (6), page 56, we have

$$(24)\quad \frac{\sqrt{g}}{h_1^2} = c(\xi^2 - 1), \qquad \frac{\sqrt{g}}{h_2^2} = c(1 - \eta^2), \qquad \frac{\sqrt{g}}{h_3^2} = \frac{c(\xi^2 - \eta^2)}{(\xi^2 - 1)(1 - \eta^2)},$$

$$(25)\quad \begin{aligned} f_1 &= \xi^2 - 1, & f_2 &= 1 - \eta^2, & f_3 &= 1, \\ M_1 &= \frac{c}{1 - \eta^2}, & M_2 &= \frac{c}{\xi^2 - 1}, & M_3 &= \frac{c(\xi^2 - \eta^2)}{[(\xi^2 - 1)(1 - \eta^2)]^2}. \end{aligned}$$

Laplace's equation reduces to

$$(26)\quad \frac{1}{F_1} \frac{\partial}{\partial \xi}\left[(\xi^2 - 1)\frac{\partial F_1}{\partial \xi}\right] + \frac{1}{F_2} \frac{\partial}{\partial \eta}\left[(1 - \eta^2)\frac{\partial F_2}{\partial \eta}\right] + \frac{\xi^2 - \eta^2}{(\xi^2 - 1)(1 - \eta^2)F_3} \frac{\partial^2 F_3}{\partial \psi^2} = 0,$$

which separates into the three ordinary equations

$$(27a)\quad \frac{1}{F_3} \frac{d^2 F_3}{d\psi^2} = -C_1,$$

$$(27b)\quad \frac{1}{F_1} \frac{d}{d\xi}\left[(\xi^2 - 1)\frac{dF_1}{d\xi}\right] - \frac{C_1}{\xi^2 - 1} = C_2,$$

$$(27c)\quad \frac{1}{F_2} \frac{d}{d\eta}\left[(1 - \eta^2)\frac{dF_2}{d\eta}\right] - \frac{C_1}{1 - \eta^2} = -C_2.$$

Equations (27*b*) and (27*c*), which obviously are identical, are satisfied by the associated Legendre functions.

The criteria stated on page 198 for the separability of Laplace's equation are not the most general known, but they include all coordinate systems of common use. A more general set of separable coordinate surfaces, of which the toroidal surfaces formed by rotating a set of bipolar coordinates is an example, has been investigated by Bôcher.[1]

It would lead us too far afield to discuss the properties of the various functions satisfying the equations enumerated above and the methods of determining the arbitrary constants. We shall have constant occasion throughout the remainder of this volume to employ Legendre and Bessel functions. The reader is referred to the classic treatises of Hobson[2] and Watson[3], and to Whittaker and Watson.[4] Detailed application of boundary-value methods to electrostatics will be found in Jeans[5] and in Smythe.[6]

PROBLEM OF THE SPHERE

3.22. Conducting Sphere in Field of a Point Charge.—Consider a conducting sphere of radius r_1 whose center is located at the origin of the coordinate system. The sphere is embedded in a homogeneous, isotropic dielectric of inductive capacity ϵ_2. At $z = \zeta > r_1$ on the z-axis, there is located a point charge q, Fig. 36. We wish to find the potential and the distribution of charge on the sphere.

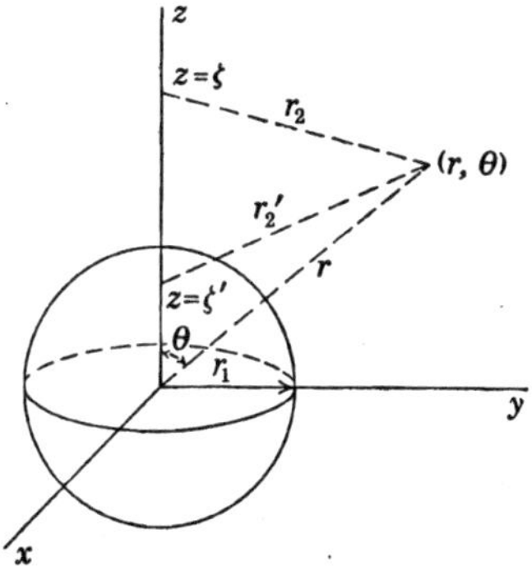

FIG. 36.—Sphere in the field of a point charge located at $z = \zeta$.

Let ϕ_0 be the potential of the source q and ϕ_1 the potential of the induced charge distribution on the sphere. The resultant potential at any point outside the sphere is $\phi = \phi_0 + \phi_1$. The induced potential ϕ_1 must be *single-valued*, a condition satisfied only when the separation constant C_2, Eq. (19*c*), page 199, is the square of an integer: $C_2 = m^2$, $m = 0, \pm 1, \pm 2, \ldots$. Likewise the only solutions of (19*b*) which are finite and single-valued over the sphere are the associated Legendre function

[1] "Ueber die Reihenentwickelungen der Potentialtheorie," Dissertation, 1894.

[2] HOBSON, "Spherical and Ellipsoidal Harmonics," Cambridge University Press, 1931.

[3] WATSON, "Treatise on the Theory of Bessel Functions," Cambridge University Press, 1922.

[4] WHITTAKER and WATSON, "Modern Analysis," Cambridge University Press, 1922.

[5] JEANS, *loc. cit.*

[6] SMYTHE, "Static and Dynamic Electricity," McGraw-Hill, 1939.

$P_n^m(\cos\theta)$, imposing on C_1 the value $C_1 = n(n+1)$, $n = 0, 1, 2, \ldots$. Under these circumstances (19a) is satisfied by either r^n or r^{-n-1}. The condition that the potential be single-valued is fulfilled by the function

$$(1) \qquad \phi_1 = \sum_{n=0}^{\infty}\sum_{m=0}^{n}\left(a_{nm}r^n + \frac{b_{nm}}{r^{n+1}}\right)P_n^m(\cos\theta)e^{im\psi},$$

where a_{nm}, b_{nm} are arbitrary constants. But ϕ_1 must also be regular at infinity, which necessitates our placing $a_{nm} = 0$. Furthermore, the primary potential ϕ_0 is symmetric about the z-axis; consequently $m = 0$ in this case. The potential of the induced distribution is, therefore, represented by the series

$$(2) \qquad \phi_1 = \sum_{n=0}^{\infty}\frac{b_n}{r^{n+1}}P_n(\cos\theta).$$

The expansion of the primary potential ϕ_0 in spherical coordinates was carried out in Sec. 3.8. When $r < \zeta$,

$$(3) \qquad \phi_0 = \frac{1}{4\pi\epsilon_2}\frac{q}{r_2} = \frac{1}{4\pi\epsilon}\frac{q}{\zeta}\sum_{n=0}^{\infty}\left(\frac{r}{\zeta}\right)^n P_n(\cos\theta),$$

and the resultant potential on the surface $r = r_1$ is

$$(4) \qquad \phi(r_1, \theta) = \sum_{n=0}^{\infty}\left[\frac{1}{4\pi\epsilon_2}\frac{q}{\zeta}\left(\frac{r_1}{\zeta}\right)^n + \frac{b_n}{r_1^{n+1}}\right]P_n(\cos\theta) = \phi_s.$$

Now ϕ_s is a constant, and since (4) must hold for all values of θ, it follows that the coefficients of $P_n(\cos\theta)$ must vanish for all values of n greater than zero. The coefficients b_n are thus determined from the set of relations

$$(5) \qquad b_0 = r_1\phi_s - \frac{q}{4\pi\epsilon_2}\frac{r_1}{\zeta}, \qquad b_n = -\frac{q}{4\pi\epsilon_2}\frac{qr_1^{2n+1}}{\zeta^{n+1}}, \qquad (n > 0).$$

At any point outside the sphere

$$(6) \qquad \phi = \frac{1}{4\pi\epsilon_2}\frac{q}{r_2} + \frac{r_1\phi_s}{r} - \frac{q}{4\pi\epsilon_2}\sum_{n=0}^{\infty}\frac{r_1^{2n+1}}{\zeta^{n+1}}\frac{P_n(\cos\theta)}{r^{n+1}}.$$

To determine the charge density, we compute the normal derivative on the surface.

$$(7) \qquad \left(\frac{\partial\phi}{\partial r}\right)_{r=r_1} = \frac{q}{4\pi\epsilon_2}\sum_{n=0}^{\infty}(2n+1)\frac{r_1^{n-1}}{\zeta^{n+1}}P_n(\cos\theta) - \frac{\phi_s}{r_1},$$

and the induced charge density is

$$(8)\qquad \omega = -\epsilon_2\left(\frac{\partial\phi}{\partial r}\right)_{r=r_1} = -\frac{q}{4\pi}\sum_{n=0}^{\infty}(2n+1)\frac{r_1^{n-1}}{\zeta^{n+1}}P_n(\cos\theta) + \frac{\epsilon_2\phi_s}{r_1}.$$

The total charge on the sphere is

$$(9)\qquad q_1 = \int_0^{\pi}\int_0^{2\pi}\omega r_1^2 \sin\theta\, d\theta\, d\psi.$$

Now a well-known property of the Legendre functions is their *orthogonality*.

$$(10)\qquad \int_0^{\pi} P_n(\cos\theta)P_m(\cos\theta)\sin\theta\, d\theta = 0, \qquad n \neq m.$$

We may take $m = 0$, $P_0(\cos\theta) = 1$, and learn that $P_n(\cos\theta)$ vanishes when integrated from 0 to π if $n > 0$.

$$(11)\qquad q_1 = -q\frac{r_1}{\zeta} + 4\pi\epsilon_2 r_1\phi_s.$$

The potential of the sphere is, therefore,

$$(12)\qquad \phi_s = \frac{1}{4\pi\epsilon_2}\frac{q_1}{r_1} + \frac{1}{4\pi\epsilon_2}\frac{q}{\zeta},$$

q_1 representing an excess charge that has been placed on the isolated sphere. If the sphere is grounded, ϕ_s may be put equal to zero.

It is interesting to observe that the potential ϕ_1 of the *induced* distribution at any point outside the sphere is that which would be produced by a charge $4\pi\epsilon_2 b_0 = q_1$, a dipole moment $4\pi\epsilon_2 b_1 = -q\dfrac{r_1^3}{\zeta^2}$, etc., all located at the origin and oriented along the z-axis (Sec. 3.8). There is, however, another simple interpretation. The point $z = \zeta'$, Fig. 36, where $\zeta\zeta' = r_1^2$, is said to be the inverse of $z = \zeta$ with respect to the sphere. The reciprocal distance from this inverse point to the point of observation is, by (8), page 173,

$$(13)\qquad \frac{1}{r_2'} = \frac{1}{r}\sum_{n=0}^{\infty}\left(\frac{\zeta'}{r}\right)^n P_n(\cos\theta) = \sum_{n=0}^{\infty}\frac{r_1^{2n}}{\zeta^n}\frac{P_n(\cos\theta)}{r^{n+1}}.$$

Thus the resultant potential (6) may be written

$$(14)\qquad 4\pi\epsilon_2\phi = \frac{q}{r_2} - \frac{qr_1}{\zeta}\frac{1}{r_2'} + \frac{q_1}{r} + \frac{qr_1}{\zeta}\frac{1}{r}.$$

Outside the sphere the potential is that of a charge q at $z = \zeta$, an *image charge* $q' = -q\dfrac{r_1}{\zeta}$ located at the inverse point $z = \zeta'$, a charge q_1 (which is zero when the sphere is uncharged) located at the origin, and a charge $q\frac{r_1}{\zeta}$ at the origin which raises the potential of the floating sphere to the proper value in the external field.

3.23. Dielectric Sphere in Field of a Point Charge.—At any point outside the sphere, whose conductivity is zero and whose inductive capacity is ϵ_1, the potential is

$$\phi^+ = \phi_0 + \phi_1^+ = \frac{1}{4\pi\epsilon_2}\frac{q}{r_2} + \sum_{n=0}^{\infty} b_n \frac{P_n(\cos\theta)}{r^{n+1}}. \tag{15}$$

The notation ϕ^+ will be employed to denote a potential or field at points outside, or on the positive side, of a closed surface. The expansion of ϕ_1 in inverse powers of r does not hold within the sphere, for the potential must be finite everywhere. We resort, therefore, to an alternative solution of Laplace's equation obtained from (1) by putting the coefficients b_{nm} equal to zero. At any interior point the resultant potential is

$$\phi^- = \sum_{n=0}^{\infty} a_n r^n P_n(\cos\theta), \qquad (r < r_1). \tag{16}$$

ϕ^- includes the contribution of the charge q as well as that of the induced polarization, for the singularity occasioned by this point charge lies outside the region to which (16) is confined. In the neighborhood of the surface, $r < |\zeta|$, so that ϕ_0 can be expanded as in (3). Just outside the sphere,

$$\phi^+ = \sum_{n=0}^{\infty}\left[\frac{1}{4\pi\epsilon_2}\frac{q}{\zeta}\left(\frac{r}{\zeta}\right)^n + \frac{b_n}{r^{n+1}}\right]P_n(\cos\theta). \tag{17}$$

Across the surface

$$\phi^+ = \phi^-, \qquad \epsilon_2\frac{\partial\phi^+}{\partial r} = \epsilon_1\frac{\partial\phi^-}{\partial r}, \qquad (r = r_1). \tag{18}$$

A calculation of the coefficients from these boundary conditions leads to

$$\begin{aligned} a_n &= \frac{q}{4\pi\zeta^{n+1}}\frac{2n+1}{n\epsilon_1 + (n+1)\epsilon_2}, \\ b_n &= \frac{q}{4\pi}\frac{r_1^{2n+1}}{\zeta^{n+1}}\frac{\epsilon_2 - \epsilon_1}{\epsilon_2}\frac{n}{n\epsilon_1 + (n+1)\epsilon_2}. \end{aligned} \tag{19}$$

The potential at any point outside the sphere is

$$\phi^+ = \frac{1}{4\pi\epsilon_2}\frac{q}{r_2} + \frac{q}{4\pi}\frac{\epsilon_2 - \epsilon_1}{\epsilon_2}\sum_{n=0}^{\infty}\frac{n}{n\epsilon_1 + (n+1)\epsilon_2}\frac{r_1^{2n+1}}{\zeta^{n+1}}\frac{P_n(\cos\theta)}{r^{n+1}}, \tag{20}$$

while at an interior point

$$\phi^- = \frac{q}{4\pi}\sum_{n=0}^{\infty}\frac{2n+1}{n\epsilon_1 + (n+1)\epsilon_2}\frac{r^n}{\zeta^{n+1}}P_n(\cos\theta). \tag{21}$$

It is important to observe that a region of infinite inductive capacity behaves like an uncharged conductor. As ϵ_1 becomes very large, it will be noted that the first term in the series (20) corresponding to $n = 0$ vanishes, so that in the limit one obtains (6) for the case $q_1 = 0$.

3.24. Sphere in a Parallel Field.—As the point source q recedes from the origin, the field in the proximity of the sphere becomes homogeneous and parallel. We shall consider the case of a sphere embedded in a dielectric of inductive capacity ϵ_2 under the influence of a uniform, parallel, external field $\mathbf{E}_0$ directed along the positive z-axis. The primary potential is then

$$\phi_0 = -E_0 z = -E_0 r \cos\theta = -E_0 r P_1(\cos\theta). \tag{22}$$

Note that ϕ_0 is no longer regular at infinity, for the source itself is infinitely remote. The potential outside the sphere due to either induced surface charge or polarization is again

$$\phi_1^+ = \sum_{n=0}^{\infty} b_n \frac{P_n(\cos\theta)}{r^{n+1}}. \tag{23}$$

If the sphere is conducting, the resultant potential on its surface and throughout its interior is a constant ϕ_s.

$$\phi_s = -E_0 r_1 P_1(\cos\theta) + \sum_{n=0}^{\infty} b_n \frac{P_n(\cos\theta)}{r_1^{n+1}}. \tag{24}$$

ϕ_s is independent of θ; whence it follows that

$$b_0 = r_1\phi_s, \qquad b_1 = r_1^3 E_0, \qquad b_n = 0 \qquad \text{when } n > 1. \tag{25}$$

$$\phi^+ = -E_0 r \cos\theta + E_0 r_1^3 \frac{\cos\theta}{r^2} + \phi_s \frac{r_1}{r}. \tag{26}$$

The charge density and the total charge are respectively

$$\omega = 3\epsilon_2 E_0 \cos\theta + \frac{\epsilon_2\phi_s}{r_1}, \qquad q_1 = 4\pi r_1 \epsilon_2 \phi_s. \tag{27}$$

The potential of the induced surface charge is, therefore, that of a dipole of moment $p = 4\pi\epsilon_2 E_0 r_1^3$; a moment, in other words, which is proportional to the volume of the sphere. To this is added the potential of q_1 in case the sphere is charged.

If the sphere is a dielectric of inductive capacity ϵ_1, the potential at an interior point is of the form (16). To satisfy the boundary conditions (18), the coefficients must now be

$$\begin{aligned} &a_0 = b_0 = 0, \qquad a_1 = \frac{-3\epsilon_2}{\epsilon_1 + 2\epsilon_2} E_0, \qquad b_1 = \frac{\epsilon_1 - \epsilon_2}{\epsilon_1 + 2\epsilon_2} r_1^3 E_0, \\ &a_n = b_n = 0, \qquad \text{when} \quad n > 1. \end{aligned} \tag{28}$$

The resultant potential is then

$$\begin{aligned} \phi^+ &= -E_0 r \cos\theta + \frac{\epsilon_1 - \epsilon_2}{\epsilon_1 + 2\epsilon_2} r_1^3 E_0 \frac{\cos\theta}{r^2}, \\ \phi^- &= -\frac{3\epsilon_2}{\epsilon_1 + 2\epsilon_2} E_0 r \cos\theta. \end{aligned} \tag{29}$$

Within the sphere the field is parallel and uniform.

$$E^- = -\frac{\partial\phi}{\partial z} = \frac{3\epsilon_2}{\epsilon_1 + 2\epsilon_2} E_0. \tag{30}$$

The dielectric constant κ_1 of the sphere may be either larger or smaller than κ_2. Thus the field within a spherical cavity excised from a homogeneous dielectric κ_2 is

$$E^- = \frac{3\kappa_2}{1 + 2\kappa_2} E_0 > E_0. \tag{31}$$

Next we note that the induced field outside is that of a dipole oriented along the z-axis whose moment is

$$p = 4\pi\epsilon_2 \frac{\epsilon_1 - \epsilon_2}{\epsilon_1 + 2\epsilon_2} r_1^3 E_0. \tag{32}$$

Apparently even a spherical cavity behaves like a dipole. This effect may be readily accounted for by recalling that the walls of the cavity bear a bound charge of density $\omega' = -\mathbf{n} \cdot \mathbf{P}_2$, where $\mathbf{P}_2$ is the polarization of the external medium.

In the case of a dielectric sphere in air, $\epsilon_2 = \epsilon_0$. The polarization of the sphere is then

$$\mathbf{P}_1 = \epsilon_0(\kappa_1 - 1)\mathbf{E}^- = 3\frac{\kappa_1 - 1}{\kappa_1 + 2}\epsilon_0 \mathbf{E}_0, \tag{33}$$

and its dipole moment

$$\mathbf{p} = \frac{4}{3}\pi r_1^3 \mathbf{P}_1 = 4\pi r_1^3 \frac{\kappa_1 - 1}{\kappa_1 + 2}\epsilon_0 \mathbf{E}_0. \tag{34}$$

The energy of this polarized sphere in the external field is

$$U_1 = -\frac{1}{2}\int_V \mathbf{P}_1 \cdot \mathbf{E}_0 \, dv = -2\pi r_1^3 \frac{\kappa_1 - 1}{\kappa_1 + 2}\epsilon_0 E_0^2 = -\frac{1}{2}\mathbf{p} \cdot \mathbf{E}_0. \tag{35}$$

The dielectric polarization modifies the field within the sphere. It will be convenient to express this modification directly in terms of $\mathbf{P}_1$. A *depolarizing factor* L is defined by

$$\mathbf{E}^- = \mathbf{E}_0 - L\mathbf{P}_1. \tag{36}$$

From (30) and the relation $\mathbf{P}_1 = \epsilon_0(\kappa_1 - 1)\mathbf{E}^-$, one calculates for a sphere in a parallel external field

$$L = \frac{1}{3\epsilon_0} \frac{\kappa_1 - \kappa_2}{\kappa_2(\kappa_1 - 1)}. \tag{37}$$

In the case of a sphere immersed in air $\kappa_2 = 1$, $L = \dfrac{1}{3\epsilon_0}$.

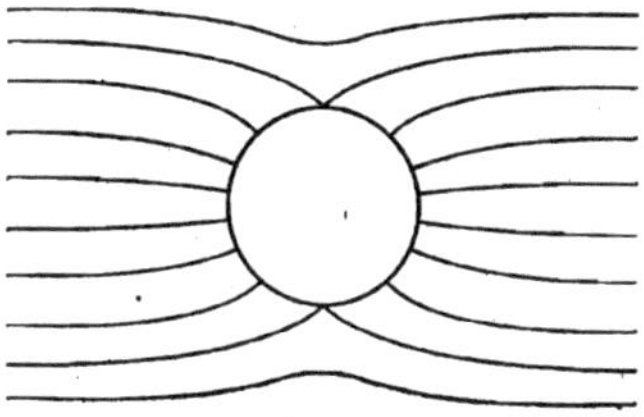

FIG. 37a.—Conducting sphere in a parallel field. The external medium is air.

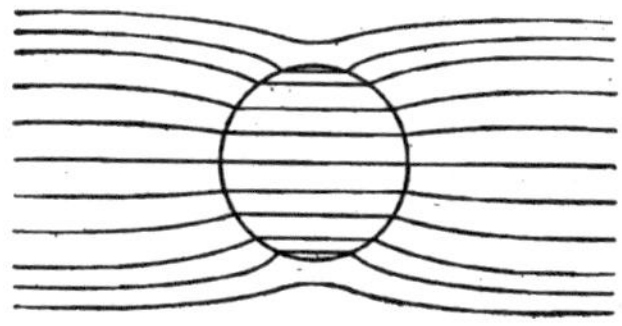

FIG. 37b.—Dielectric sphere in a parallel field. The external medium is air.

PROBLEM OF THE ELLIPSOID

3.25. Free Charge on a Conducting Ellipsoid.—In ellipsoidal coordinates Laplace's equation reduces by virtue of Eq. (135), page 59, to

$$(\eta - \zeta)R_\xi \frac{\partial}{\partial \xi}\left(R_\xi \frac{\partial \phi}{\partial \xi}\right) + (\zeta - \xi)R_\eta \frac{\partial}{\partial \eta}\left(R_\eta \frac{\partial \phi}{\partial \eta}\right) + (\xi - \eta)R_\zeta \frac{\partial}{\partial \zeta}\left(R_\zeta \frac{\partial \phi}{\partial \zeta}\right) = 0. \tag{1}$$

The properties of the *ellipsoidal harmonics* that satisfy this equation have been extensively studied, but we shall construct here only certain elementary solutions that will prove sufficient for the problems in view.

Consider first a conducting ellipsoid embedded in a homogeneous dielectric ϵ_2. The semiprincipal axes of the ellipsoid are a, b, c. It carries a total charge q, and we assume initially that there is no external field. We wish to know the potential and the distribution of charge over the conducting surface.

To solve this problem a potential function must be found which satisfies (1), which is regular at infinity, and which is constant over the given ellipsoid. Now ξ is the parameter of a family of ellipsoids all confocal with the standard surface $\xi = 0$ whose axes have the specified values a, b, c. The variables η and ζ are the parameters of confocal hyperboloids and as such serve to measure position on any ellipsoid $\xi =$ constant. On the surface $\xi = 0$, therefore, ϕ must be independent of η and ζ. If we can find a function depending only on ξ which satisfies (1) and behaves properly at infinity, it can be adjusted to represent the potential correctly at any point outside the ellipsoid $\xi = 0$.

Let us assume, then, that $\phi = \phi(\xi)$. Laplace's equation reduces to

$$(2)\qquad \frac{\partial}{\partial\xi}\left(R_\xi \frac{\partial\phi}{\partial\xi}\right) = 0, \qquad R_\xi = \sqrt{(\xi + a^2)(\xi + b^2)(\xi + c^2)},$$

which on integration leads to

$$(3)\qquad \phi(\xi) = C_1 \int_\xi^\infty \frac{d\xi}{R_\xi},$$

where C_1 is an arbitrary constant. The choice of the upper limit is such as to ensure the proper behavior at infinity. When ξ becomes very large, R_ξ approaches $\xi^{\frac{3}{2}}$ and

$$(4)\qquad \phi \simeq \frac{2C_1}{\sqrt{\xi}}, \qquad (\xi \to \infty).$$

On the other hand, the equation of an ellipsoid can be written in the form

$$(5)\qquad \frac{x^2}{1 + \frac{a^2}{\xi}} + \frac{y^2}{1 + \frac{b^2}{\xi}} + \frac{z^2}{1 + \frac{c^2}{\xi}} = \xi.$$

If $r^2 = x^2 + y^2 + z^2$ is the distance from the origin to any point on the ellipsoid ξ, it is apparent that as ξ becomes very large $\xi \to r^2$ and hence at great distances from the origin

$$(6)\qquad \phi \simeq \frac{2C_1}{r}.$$

The solution (3) is, therefore, regular at infinity. Moreover (6) enables us to determine at once the value of C_1; for it has been shown that, whatever the distribution, the dominant term of the expansion at remote points is the potential of a point charge at the origin equal to the total charge of the distribution—in this case q. Hence $C_1 = \frac{q}{8\pi\epsilon_2}$, and the potential at any point is

$$(7)\qquad \phi(\xi) = \frac{q}{8\pi\epsilon_2} \int_\xi^\infty \frac{d\xi}{R_\xi}.$$

The equipotential surfaces are the ellipsoids ξ = constant. Equation (7) is an elliptic integral and its values have been tabulated.[1]

To obtain the normal derivative we must remember that distance along a curvilinear coordinate u^1 is measured not by du^1 but by $h_1\,du^1$ (Sec. 1.16). In ellipsoidal coordinates

$$(8)\qquad h_1 = \frac{1}{2}\frac{\sqrt{(\xi - \eta)(\xi - \zeta)}}{R_\xi},$$

$$(9)\qquad \frac{\partial\phi}{\partial n} = \frac{1}{h_1}\frac{\partial\phi}{\partial\xi} = -\frac{q}{4\pi\epsilon_2}\frac{1}{\sqrt{(\xi - \eta)(\xi - \zeta)}}.$$

[1] See for example Jahnke-Emde, "Tables of Functions," 2d ed., Teubner, 1933.

The density of charge over the surface $\xi = 0$ is

$$\omega = -\epsilon_2 \left(\frac{\partial \phi}{\partial n}\right)_{\xi=0} = \frac{q}{4\pi} \frac{1}{\sqrt{\eta\zeta}}. \tag{10}$$

If now in the three equations (132), page 59, defining x, y, z in terms of ξ, η, ζ, we put $\xi = 0$, it may be easily verified that

$$\frac{x^2}{a^4} + \frac{y^2}{b^4} + \frac{z^2}{c^4} = \frac{\eta\zeta}{a^2b^2c^2}. \qquad (\xi = 0). \tag{11}$$

Consequently, the charge density in rectangular coordinates is

$$\omega = \frac{q}{4\pi abc} \frac{1}{\sqrt{\dfrac{x^2}{a^4} + \dfrac{y^2}{b^4} + \dfrac{z^2}{c^4}}}. \tag{12}$$

Several special cases are of interest. If two of the axes are equal, the body is a spheroid and (7) can be integrated in terms of elementary functions. Thus, if $a = b > c$, the spheroid is oblate and

$$\phi = \frac{q}{8\pi\epsilon_2} \int_\xi^\infty \frac{d\xi}{(\xi + a^2)\sqrt{\xi + c^2}} = \frac{q}{4\pi\sqrt{a^2 - c^2}} \tan^{-1} \sqrt{\frac{a^2 - c^2}{\xi + c^2}}. \tag{13}$$

When $c = 0$, the spheroid degenerates into a circular disk. On the other hand, if $a > b = c$ the spheroid is prolate and we find for the potential

$$\phi = \frac{1}{8\pi\epsilon_2} \frac{q}{\sqrt{a^2 - b^2}} \ln \frac{\sqrt{\xi + a^2} + \sqrt{a^2 - b^2}}{\sqrt{\xi + a^2} - \sqrt{a^2 - b^2}}. \tag{14}$$

The eccentricity of a prolate spheroid is $e = \sqrt{1 - \left(\dfrac{b}{a}\right)^2}$. As $e \to 1$, the spheroid degenerates into a long thin rod.

3.26. Conducting Ellipsoid in a Parallel Field.—We assume first that a uniform, parallel field $\mathbf{E}_0$ is directed along the x-axis, and consequently along the major axis of the ellipsoid. The potential of the applied field is

$$\phi_0 = -E_0 x = -E_0 \left[\frac{(\xi + a^2)(\eta + a^2)(\zeta + a^2)}{(b^2 - a^2)(c^2 - a^2)}\right]^{\frac{1}{2}}, \tag{15}$$

the value of x in ellipsoidal coordinates being substituted from Eqs. (132), page 59. This primary potential is clearly a solution of Laplace's equation in the form of a product of three functions,

$$\phi_0 = C_1 F_1(\xi) F_2(\eta) F_3(\zeta), \qquad C_1 = -\frac{E_0}{\sqrt{(b^2 - a^2)(c^2 - a^2)}}. \tag{16}$$

It is not, however, regular at infinity.

Now if the boundary conditions are to be satisfied, the potential ϕ_1 of the induced distribution must vary functionally over every surface of the family ξ = constant in exactly the same manner as ϕ_0. It differs from ϕ_0 in its regularity at infinity. We presume, therefore, that ϕ_1 is a function of the form

$$\phi_1 = C_2 G_1(\xi) F_2(\eta) F_3(\zeta), \tag{17}$$

where

$$F_2(\eta) = \sqrt{\eta + a^2}, \qquad F_3(\zeta) = \sqrt{\zeta + a^2}. \tag{18}$$

To find the equation satisfied by $G_1(\xi)$ we need only substitute (17) and (18) into (1), obtaining as a result

$$R_\xi \frac{d}{d\xi}\left(R_\xi \frac{dG_1}{d\xi}\right) - \left(\frac{b^2 + c^2}{4} + \frac{\xi}{2}\right) G_1 = 0. \tag{19}$$

Equation (19) is an ordinary equation of the second order and as such possesses two independent solutions. One of these we know already to be $F_1 = \sqrt{\xi + a^2}$. There is a theorem[1] which states that if one solution of a second-order linear equation is known, an independent solution can be determined from it by integration. If y_1 is a solution of

$$\frac{d^2y}{dx^2} + p(x)\frac{dy}{dx} + q(x)y = 0, \tag{20}$$

then an independent solution y_2 is given by

$$y_2 = y_1 \int \frac{e^{-\int p\,dx}}{y_1^2}\,dx. \tag{21}$$

In the present instance

$$p(\xi) = \frac{1}{R_\xi}\frac{dR_\xi}{d\xi} = \frac{d}{d\xi}\ln R_\xi; \tag{22}$$

hence,

$$G_1(\xi) = F_1 \int \frac{d\xi}{F_1^2 R_\xi}. \tag{23}$$

The limits of integration are arbitrary, but $G_1(\xi)$ is easily shown to vanish properly at infinity if the upper limit is made infinite. The potential of the induced charge is, therefore,

$$\phi_1 = \phi_0 \frac{C_2}{C_1} \int_\xi^\infty \frac{d\xi}{(\xi + a^2)R_\xi}. \tag{24}$$

[1] See for example Ince, "Ordinary Differential Equations," p. 122, Longmans, 1927.

The constant C_2 is determined finally from the condition that on the ellipsoid $\xi = 0$ the potential is a constant ϕ_s.

$$\phi_s = \phi_0 \left[1 + \frac{C_2}{C_1} \int_0^\infty \frac{d\xi}{(\xi + a^2)R_\xi} \right]. \tag{25}$$

At any external point the potential is

$$\phi = \phi_0 + \frac{\phi_s - \phi_0}{\displaystyle\int_0^\infty \frac{d\xi}{(\xi + a^2)R_\xi}} \int_\xi^\infty \frac{d\xi}{(\xi + a^2)R_\xi}. \tag{26}$$

As in the analogous problem of the sphere, the constant ϕ_s can be calculated in terms of the total charge on the ellipsoid. The integrals occurring in (26) are *elliptic* and of the *second kind*.[1]

In case the field is parallel to either of the two minor axes it is only necessary to replace the parameter a^2 above by b^2 or c^2. Thus the potential about a conducting ellipsoid oriented arbitrarily with respect to the axis of a uniform, parallel field $\mathbf{E}_0$ can be found by resolving $\mathbf{E}_0$ into three components parallel to the principal axes of the ellipsoid and then superposing the resulting three solutions of the type (26).

3.27. Dielectric Ellipsoid in Parallel Field.—It is now a simple matter to calculate the perturbation of a uniform, parallel field due to a dielectric ellipsoid. We shall assume that the inductive capacity of the ellipsoid is ϵ_1, and that it is embedded in a homogeneous medium whose inductive capacity is again ϵ_2 The applied $\mathbf{E}_0$ is directed arbitrarily with respect to the reference system and has the components E_{0x}, E_{0y}, E_{0z} along the axes of the ellipsoid.

Consider first the component field E_{0x}. Outside the ellipsoid the resultant potential must exhibit the same general functional behavior as in the preceding example, and will differ from it only in the value of the constant C_2. In this region, therefore,

$$\phi^+ = \phi_0 + \phi_1^+ = F_1(\xi)F_2(\eta)F_3(\zeta) \left[C_1 + C_2 \int_\xi^\infty \frac{ds}{(s + a^2)R_s} \right]. \tag{27}$$

The variable s has replaced ξ under the integral to avoid confusion with the lower limit.

The interior of the ellipsoid corresponds to the range $-c^2 \leq \xi \leq 0$ if $a \geq b \geq c$. In this region ϕ^- must vary with η and ζ as determined by the function $F_2(\eta)F_3(\zeta)$ and, since (19) has only two independent solutions, the dependence on ξ must be represented by either $F_1(\xi)$ or $G_1(\xi)$ to satisfy Laplace's equation. But the function $G_1(\xi)$ is infinite at $\xi = -c^2$,

[1] See for example Whittaker and Watson, "Modern Analysis," 4th ed., pp. 512*ff*., Cambridge University Press, 1927.

whereas $F_1(\xi)$ is finite at all points within the surface $\xi = 0$. Thus the potential within this region must have the form

$$\phi^- = C_3 F_1(\xi) F_2(\eta) F_3(\zeta), \tag{28}$$

where C_3 is an undetermined constant.

The constants C_2 and C_3 are to be adjusted to satisfy the boundary conditions

$$\phi^+ = \phi^-, \qquad \epsilon_2 \left[\frac{1}{h_1}\frac{\partial \phi^+}{\partial \xi}\right]_{\xi=0} = \epsilon_1 \left[\frac{1}{h_1}\frac{\partial \phi^-}{\partial \xi}\right]_{\xi=0}. \tag{29}$$

The first of these leads to

$$C_3 = C_1 + C_2 \int_0^\infty \frac{ds}{(s+a^2)R_s}; \tag{30}$$

the second gives

$$C_2 = \frac{abc}{2}\frac{(\epsilon_2 - \epsilon_1)}{\epsilon_2} C_3. \tag{31}$$

Since $\phi_0 = -E_{0x}x$, one finds that the potential at any interior point of the ellipsoid is

$$\phi^- = -\frac{E_{0x}x}{1 + \frac{abc}{2\epsilon_2}(\epsilon_1 - \epsilon_2)A_1}, \qquad A_1 = \int_0^\infty \frac{ds}{(s+a^2)R_s}, \tag{32}$$

and the field intensity is

$$E_x^- = \frac{E_{0x}}{1 + \frac{abc}{2\epsilon_2}(\epsilon_1 - \epsilon_2)A_1}. \tag{33}$$

The components along the other two axes are found in an identical manner. The potential of the applied field is in total

$$\phi_0 = -E_{0x}x - E_{0y}y - E_{0z}z, \tag{34}$$

and the resultant potential at an interior point is

$$\phi^- = -\left[\frac{E_{0x}x}{1 + \frac{abc}{2\epsilon_2}(\epsilon_1 - \epsilon_2)A_1} + \frac{E_{0y}y}{1 + \frac{abc}{2\epsilon_2}(\epsilon_1 - \epsilon_2)A_2} + \frac{E_{0z}z}{1 + \frac{abc}{2\epsilon_2}(\epsilon_1 - \epsilon_2)A_3}\right] \tag{35}$$

in which the constants A_2 and A_3 are defined by

$$A_2 = \int_0^\infty \frac{ds}{(s+b^2)R_s}, \qquad A_3 = \int_0^\infty \frac{ds}{(s+c^2)R_s}. \tag{36}$$

The remaining components of the perturbed field are

$$E_y^- = \frac{E_{0y}}{1 + \frac{abc}{2\epsilon_2}(\epsilon_1 - \epsilon_2)A_2}, \qquad E_z^- = \frac{E_{0z}}{1 + \frac{abc}{2\epsilon_2}(\epsilon_1 - \epsilon_2)A_3}. \tag{37}$$

We are led to a conclusion of great practical importance: *If the applied field is initially uniform and parallel, the resultant field within the ellipsoid is also uniform and parallel, whatever the orientation of the axes.* The vector $\mathbf{E}^-$, however, is not in general parallel to $\mathbf{E}_0$, for the constants A_1, A_2, A_3 are equal only in degenerate cases.

Outside the ellipsoid the resultant potential associated with the primary component E_{0x} is given by (27),

$$\phi^+ = \phi_0 \frac{1 + \frac{abc}{2}\frac{\epsilon_1 - \epsilon_2}{\epsilon_2}\int^{\xi} \frac{ds}{(s + a^2)R_s}}{1 + \frac{abc}{2}\frac{\epsilon_1 - \epsilon_2}{\epsilon_2}\int_0^{\infty} \frac{ds}{(s + a^2)R_s}}, \tag{38}$$

with corresponding potentials for the fields E_{0y} and E_{0z}. If in (38) we let $\epsilon_1 \to \infty$, we find that ϕ^+ reduces to (26) for the case $\phi_s = 0$. *The potential outside a body of infinite inductive capacity is the same as that outside a grounded conductor of the same shape.*

The induced polarization $\mathbf{P}_1$ within the ellipsoid tends to *decrease* the applied field. The *depolarizing factors* L_1, L_2, L_3 are defined by the relations

$$E_x^- = E_{0x} - L_1 P_{1x}, \qquad E_y^- = E_{0y} - L_2 P_{1y}, \qquad E_z^- = E_{0z} - L_3 P_{1z}. \tag{39}$$

Putting $\mathbf{P}_1 = (\epsilon_1 - \epsilon_0)\mathbf{E}^-$ and introducing the components of $\mathbf{E}^-$ from (33) and (37) leads to

$$L_j = \frac{abc}{2\epsilon_2}\frac{\kappa_1 - \kappa_2}{\kappa_1 - 1} A_j, \qquad (j = 1, 2, 3). \tag{40}$$

Most commonly the medium outside the ellipsoid is free space, $\kappa_2 = 1$. In that case the depolarizing factors depend only on the form of the ellipsoid.

$$L_j = \frac{abc}{2\epsilon_0} A_j, \qquad (j = 1, 2, 3). \tag{41}$$

The polarization may then be determined from the simple formula

$$P_x = \frac{E_0}{L_1 + \frac{1}{\epsilon_1 - \epsilon_0}}, \tag{42}$$

with corresponding expressions for P_y and P_z.

3.28. Cavity Definitions of E and D.—The field inside an ellipsoidal cavity is given by (33) and (37) when $\epsilon_1 = \epsilon_0$. There are two cases of considerable interest: that of a disk-like cavity whose plane is normal to the direction of the applied field, and that of a needle-shaped cavity parallel to the field.

We shall consider first the applied field to be oriented along the z-axis. Then as c becomes very small the ellipsoid degenerates into the disk illustrated in Fig. 38*a*. There is no loss of generality in assuming the disk circular, $a = b$, and in this case the elliptic integral A_3 reduces to a more elementary function.

$$(43)\quad A_3 = \int_0^\infty \frac{ds}{(s+a^2)(s+c^2)^{\frac{3}{2}}} = \frac{2}{(a^2-c^2)^{\frac{3}{2}}}\left(\frac{\sqrt{a^2-c^2}}{c} - \tan^{-1}\frac{\sqrt{a^2-c^2}}{c}\right).$$

The limit of the product a^2cA_3 as $c \to 0$ is 2. The field inside the cavity is purely transverse, and since $\kappa_1 = 1$,

$$(44)\quad \mathbf{E}^- = \kappa_2\mathbf{E}_0 = \frac{1}{\epsilon_0}\mathbf{D}_0.$$

Apart from the proportionality constant ϵ_0, *the field intensity* $\mathbf{E}^-$ *within the disk-shaped cavity is equal to the vector* $\mathbf{D}_0$ *of the initial field in the dielectric.*

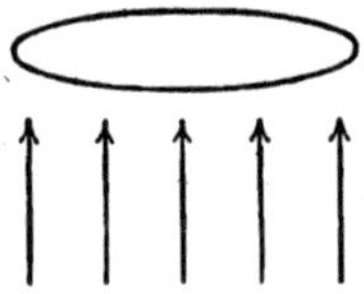

FIG. 38*a*.

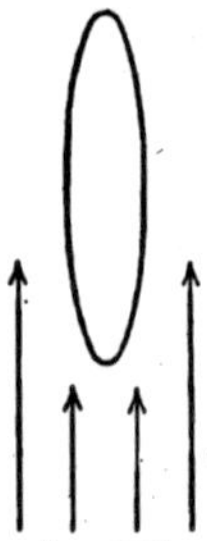

FIG. 38*b*.

FIGS. 38*a* and *b*.—Illustrating the cavity definitions of the vectors **E** and **D**. The arrows indicate the direction of the applied field.

In Fig. 38*b* the field is directed along the major axis of the prolate spheroid $a > b = c$.

$$(45)\quad A_1 = \int_0^\infty \frac{ds}{(s+b^2)(s+a^2)^{\frac{3}{2}}} = -\frac{1}{a^3e^3}\left(2e + \ln\frac{1+e}{1-e}\right),$$

where $e = \sqrt{1 - \frac{b^2}{a^2}}$ is the eccentricity. As $e \to 1$, the spheroid degenerates into a long, needle-shaped cavity and the product ab^2A_1 approaches zero. At any point inside this cavity $\mathbf{E}^- = \mathbf{E}_0$: *the field intensity* $\mathbf{E}^-$ *is exactly the same as that prevailing initially in the dielectric.*

These *cavity definitions* of the vectors **E** and **D** in a ponderable medium were introduced by Lord Kelvin. Obviously a direct measurement of **E** or **D** in terms of the force or torque exerted on a small test body is not feasible within a solid dielectric, and if a cavity is excised the field within it will depend on the shape of the hole. Our calculations have shown, however, that the electric field intensity within the dielectric is exactly that which might be measured within the needle-shaped opening of Fig. 38*b*, whereas the field measured within the flat slit of Fig. 38*a* differs from **D** within the dielectric only by a constant factor.

3.29. Torque Exerted on an Ellipsoid.—The intensity of an electrostatic field may be measured by introducing a small test body of known shape and inductive capacity suspended by a torsion fiber and observing the torque. Inversely, if the intensity of the field is known, such an experiment may be employed to determine the inductive capacity or susceptibility of a sample of dielectric matter. In general the field and polarization throughout the interior of the probe are nonuniform and an accurate computation is difficult or impossible. On the other hand, the advantages of an ellipsoidal test body for these purposes are obvious. The polarization of the entire probe is constant and the applied torque depends essentially only on its volume and its inductive capacity.

According to Eq. (49), page 113, the energy of a dielectric body in an external field is

$$U = \frac{\epsilon_2 - \epsilon_1}{2} \int_V \mathbf{E}^- \cdot \mathbf{E}_0 \, dv, \tag{46}$$

where as above $\mathbf{E}^-$ denotes the resultant field inside the body and $\mathbf{E}_0$ the initial field. If the body is ellipsoidal and $\mathbf{E}_0$ is homogeneous, we have by (33) and (37):

$$\mathbf{E}^- \cdot \mathbf{E}_0 = \frac{E_{0x}^2}{1 + \frac{abc}{2}\left(\frac{\kappa_1 - \kappa_2}{\kappa_2}\right) A_1} + \frac{E_{0y}^2}{1 + \frac{abc}{2}\left(\frac{\kappa_1 - \kappa_2}{\kappa_2}\right) A_2} + \frac{E_{0z}^2}{1 + \frac{abc}{2}\left(\frac{\kappa_1 - \kappa_2}{\kappa_2}\right) A_3} \tag{47}$$

and, since the volume of an ellipsoid is $\frac{4}{3}\pi abc$,

$$U = \tfrac{2}{3}\pi abc(\epsilon_2 - \epsilon_1)\mathbf{E}^- \cdot \mathbf{E}_0. \tag{48}$$

This energy depends not only upon the intensity of the initial field but also upon the orientation of the principal axes with respect to the field. Let the vector $\delta\omega$ represent a virtual angular displacement of the ellipsoid about its center and $\mathbf{T}$ the resultant torque exerted by the field. Both $\mathbf{T}$ and $\delta\omega$ are axial vectors (page 67), and the components $\delta\omega_x$, $\delta\omega_y$, $\delta\omega_z$ are the angles of rotation about the axes x, y, z, respectively. The work done in the course of such a rotation is

$$\delta W = \mathbf{T} \cdot \delta\omega = T_x \, \delta\omega_x + T_y \, \delta\omega_y + T_z \, \delta\omega_z. \tag{49}$$

This work must be compensated by a decrease in the potential energy U. In virtue of the homogeneous, quadratic character of (47), we may write

$$\delta U = \tfrac{2}{3}\pi abc(\epsilon_2 - \epsilon_1)\delta(\mathbf{E}^- \cdot \mathbf{E}_0) = \tfrac{4}{3}\pi abc(\epsilon_2 - \epsilon_1)\mathbf{E}^- \cdot \delta\mathbf{E}_0. \tag{50}$$

The reference axes have been chosen to coincide with the principal axes of the ellipsoid. Relative to this system fixed in the body the variation of $\mathbf{E}_0$ corresponding to a virtual rotation $\delta\omega$ is

$$\delta\mathbf{E}_0 = \mathbf{E}_0 \times \delta\omega, \tag{51}$$

whence for the energy balance we obtain

$$\begin{aligned}\delta U &= \tfrac{4}{3}\pi abc(\epsilon_2 - \epsilon_1)\mathbf{E}^- \cdot (\mathbf{E}_0 \times \delta\omega) \\ &= \tfrac{4}{3}\pi abc(\epsilon_2 - \epsilon_1)(\mathbf{E}^- \times \mathbf{E}_0) \cdot \delta\omega = -\delta W,\end{aligned} \tag{52}$$

and from the arbitrariness of the components of rotation it follows that the torque exerted by the field is

$$\mathbf{T} = \tfrac{4}{3}\pi abc(\epsilon_1 - \epsilon_2)\mathbf{E}^- \times \mathbf{E}_0. \tag{53}$$

The components of this torque are

$$\begin{aligned}
T_x &= \tfrac{2}{3}\pi(abc)^2 \frac{(\epsilon_1 - \epsilon_2)^2}{\epsilon_2} E_{0y}E_{0z} \left[\frac{A_3 - A_2}{\left(1 + \frac{abc}{2}\frac{\kappa_1 - \kappa_2}{\kappa_2}A_2\right)\left(1 + \frac{abc}{2}\frac{\kappa_1 - \kappa_2}{\kappa_2}A_3\right)}\right], \\
T_y &= \tfrac{2}{3}\pi(abc)^2 \frac{(\epsilon_1 - \epsilon_2)^2}{\epsilon_2} E_{0z}E_{0x} \left[\frac{A_1 - A_3}{\left(1 + \frac{abc}{2}\frac{\kappa_1 - \kappa_2}{\kappa_2}A_3\right)\left(1 + \frac{abc}{2}\frac{\kappa_1 - \kappa_2}{\kappa_2}A_1\right)}\right], \\
T_z &= \tfrac{2}{3}\pi(abc)^2 \frac{(\epsilon_1 - \epsilon_2)^2}{\epsilon_2} E_{0x}E_{0y} \left[\frac{A_2 - A_1}{\left(1 + \frac{abc}{2}\frac{\kappa_1 - \kappa_2}{\kappa_2}A_1\right)\left(1 + \frac{abc}{2}\frac{\kappa_1 - \kappa_2}{\kappa_2}A_2\right)}\right].
\end{aligned} \tag{54}$$

To investigate the stability of the ellipsoid we must determine the relative magnitudes of the constants A_1, A_2, A_3. In the first place, it is clear from their definition, (32) and (36), that all three are positive, whatever the values of a, b, c. It is easy to show, furthermore, that the order of their relative magnitudes is the inverse of the order of the three parameters. That is, if $a > b > c$, $A_1 < A_2 < A_3$. Next, one finds that the sum of the three integrals can be reduced to a simple integral when $u = R_s^2$ is introduced as a new variable, giving

$$A_1 + A_2 + A_3 = \frac{2}{abc}; \tag{55}$$

whence from the essentially positive character of the constants it follows that

$$0 \leq A_j \leq \frac{2}{abc}, \qquad (j = 1, 2, 3), \tag{56}$$

$$1 + \frac{abc}{2}\frac{\kappa_1 - \kappa_2}{\kappa_2} A_j > 0, \qquad (j = 1, 2, 3). \tag{57}$$

The denominators of (54) are, therefore, positive both when $\epsilon_1 > \epsilon_2$ and when $\epsilon_1 < \epsilon_2$, and *the direction of the components of torque is independent of the relative magnitudes of* ϵ_1 *and* ϵ_2. If the applied field is parallel to any of the three principal axes, all components of torque are zero, so that these constitute three directions of equilibrium.

The stability of equilibrium depends on the direction of the torque and this, we see, depends solely on the sign of $A_j - A_k$, which in turn is determined by the relative magnitudes of the axes a, b, c. Thus $A_3 - A_2$ is positive if $b > c$, negative when $c > b$. The components of torque are such as to rotate the longest axis into the direction of the field by the shortest route. *An ellipsoid whose major axis is oriented along the applied field is in stable equilibrium; the equilibrium positions of the minor axes are unstable.*

Problems

1. The coordinates ξ, η, ζ are obtained from the rectangular coordinates x, y, z by the transformation

$$\xi + i\eta = f(x + iy), \qquad \zeta = z,$$

where f is any analytic function of the complex variable $x + iy$. Demonstrate the following properties of this transformation:

a. The differential line element is $ds^2 = h^2(d\xi^2 + d\eta^2) + d\zeta^2$, where $h = 1/|f'|$, f' denoting differentiation with respect to the variable $x + iy$;
b. The system ξ, η, ζ is orthogonal;
c. The transformation is conformal, so that every infinitesimal figure in the xy-plane is mapped as a geometrically similar figure on the $\xi\eta$-plane.

Show that in these coordinates the Laplacian of a scalar function ϕ assumes the form

$$\nabla^2\phi = \frac{1}{h^2}\left(\frac{\partial^2\phi}{\partial\xi^2} + \frac{\partial^2\phi}{\partial\eta^2}\right) + \frac{\partial^2\phi}{\partial z^2}$$

and find expressions for the divergence and curl of a vector.

2. With reference to Problem 1, discuss the coordinates defined by the following transformations:

$$\xi + i\eta = \ln\,(x + iy), \tag{1}$$
$$x + iy = a\cosh\,(\xi + i\eta), \tag{2}$$
$$x + iy = (\xi + i\eta)^2, \tag{3}$$
$$\xi + i\eta = (x + iy)^2, \tag{4}$$

(5)
$$\xi + i\eta = \ln \frac{x + iy + a}{x + iy - a},$$

(6)
$$x + iy = ia \cot \frac{(\xi + i\eta)}{2}.$$

3. A set of *ring* or *toroidal coordinates* λ, μ, ψ are defined by the relations

$$x = r \cos \psi, \qquad y = r \sin \psi, \qquad z = \frac{\sin \mu}{\cosh \lambda + \cos \mu},$$

where

$$r = \frac{\sinh \lambda}{\cosh \lambda + \cos \mu}.$$

Show that the surfaces, ψ = constant, are meridian planes through the z-axis, that the surfaces, λ = constant, are the toruses whose meridian sections are the circles

$$x^2 + z^2 - 2x \coth \lambda + 1 = 0,$$

and that the surfaces, μ = constant, are spheres whose meridian sections are the circles

$$x^2 + z^2 + 2z \tan \mu - 1 = 0.$$

Show that the system is orthogonal apart from certain exceptional points, and find these points. Find the expression for the differential element of length and show that

$$\nabla^2 \phi = \frac{\sinh^2 \lambda}{r^3} \left[\frac{\partial}{\partial \lambda} \left(r \frac{\partial \phi}{\partial \lambda} \right) + \frac{\partial}{\partial \mu} \left(r \frac{\partial \phi}{\partial \mu} \right) + \frac{r}{\sinh^2 \lambda} \frac{\partial^2 \phi}{\partial \psi^2} \right].$$

Is Laplace's equation separable in these coordinates?

4. Let $F(x, y, z) = \lambda$ represent a family of surfaces such that F has continuous partial derivatives of first and second orders. Show that a necessary and sufficient condition that these surfaces may be equipotentials is

$$\frac{\nabla^2 F}{(\nabla F)^2} = f(\lambda),$$

where $f(\lambda)$ is a function of λ only. Show that if this condition is fulfilled the potential is

$$\phi = c_1 \int e^{-\int f(\lambda)\, d\lambda}\, d\lambda + c_2,$$

where c_1 and c_2 are constants.

5. Show that $\psi = F(z + ix \cos u + iy \sin u)$ is a solution of Laplace's equation

$$\frac{\partial^2 \psi}{\partial x^2} + \frac{\partial^2 \psi}{\partial y^2} + \frac{\partial^2 \psi}{\partial z^2} = 0$$

in three dimensions for all values of the parameter u and for any analytic function F. Show further that any linear combination of $2n + 1$ independent particular solutions can be expressed by the integral

$$\psi = \int_{-\pi}^{\pi} (z + ix \cos u + iy \sin u)^n f_n(u)\, du,$$

where $f_n(u)$ is a rational function of e^{iu}, and finally that every solution of Laplace's equation which is analytic within some spherical domain can be expressed as an integral of the form

$$\psi = \int_{-\pi}^{\pi} f(z + ix \cos u + iy \sin u, u)\, du.$$

(See Whittaker and Watson, "Modern Analysis," Chap. XVIII.)

6. Charge is distributed along an infinite straight line with a constant density of q coulombs/meter. Show that the field intensity at any point, whose distance from the line is r, is

$$E_r = \frac{1}{2\pi\epsilon}\frac{q}{r}$$

and that this field is the negative gradient of a potential function

$$\phi(x, y) = \frac{1}{2\pi\epsilon} q \ln \frac{r_0}{r},$$

where r_0 is an arbitrary constant representing the radius of a cylinder on which $\phi = 0$.

From these results show that if charge is distributed in a two-dimensional space with a density $\omega(x, y)$ the potential at any point in the xy-plane is

$$\phi(x', y') = \frac{1}{2\pi\epsilon}\int \omega \ln \frac{r_0}{r}\, da,$$

where $r = \sqrt{(x' - x)^2 + (y' - y)^2}$ and that $\phi(x, y)$ satisfies

$$\frac{\partial^2\phi}{\partial x^2} + \frac{\partial^2\phi}{\partial y^2} = -\frac{1}{\epsilon}\,\omega(x, y).$$

Show furthermore that this potential function is not in general regular at infinity, but that if the total charge is zero, so that $\int \omega\, da = 0$, then $r\phi$ is bounded as $r \to \infty$.

7. Charge is distributed over an area of finite extent in a two-dimensional space with a density $\omega(x, y)$. Express the potential at an external point x', y' as a power series in r, the distance from an arbitrary origin to the fixed point x', y'. Show that the successive terms are the potentials of a series of two-dimensional multipoles.

8. Let C be any closed contour in the xy-plane bounding an area S and let ϕ be a two-dimensional scalar potential function. By Green's theorem show that

$$2\pi u(x', y') = \int_S \ln r\, \nabla^2\phi\, da - \int_C \left(\ln r \frac{\partial\phi}{\partial n} - \phi \frac{\partial}{\partial n} \ln r\right) ds,$$

where $u = \phi(x', y')$ if x', y' is an interior point, and $u = 0$ if x', y' is exterior. In this formula

$$\nabla^2\phi = \frac{\partial^2\phi}{\partial x^2} + \frac{\partial^2\phi}{\partial y^2}, \qquad r = \sqrt{(x' - x)^2 + (y' - y)^2}$$

and n is the normal drawn outward from the contour.

9. A circle of radius a is drawn in a two-dimensional space. Position on the circle is specified by an angle θ'. The potential on the circle is a given function $\phi(a, \theta')$.

Show that at any point outside the circle whose polar coordinates are r, θ the potential is

$$\phi(r, \theta) = \frac{1}{2\pi}\int_0^{2\pi} \phi(a, \theta')\frac{r^2 - a^2}{r^2 + a^2 - 2ar\cos(\theta - \theta')}\,d\theta.$$

This integral is due to Poisson.

10. An infinite line charge of density q per unit length runs parallel to a conducting cylinder. Show that the induced field is that of a properly located image. Calculate the charge distribution on the cylinder.

11. The radii of two infinitely long conducting cylinders of circular cross section are respectively a and b and the distance between centers is c, with $c > a + b$. The external medium is a fluid whose inductive capacity is ϵ. The potential difference between the cylinders is V volts. Obtain expressions for the charge density and the mechanical force exerted on one cylinder by the other per unit length. Solve by introducing bipolar coordinates into Laplace's equation and again by application of the method of images.

12. An infinite dielectric cylinder is placed in a parallel, uniform field $\mathbf{E}_0$ which is normal to the axis. Calculate the induced dipole moment per unit length and the depolarizing factor L (p. 206).

13. Two point charges are immersed in an infinite, homogeneous, dielectric fluid. Show that the coulomb force exerted by one charge on the other can be found by evaluating the integral

$$\mathbf{F} = \int \left[\epsilon\mathbf{E}(\mathbf{E}\cdot\mathbf{n}) - \frac{\epsilon}{2}E^2\mathbf{n}\right] da,$$

[Eq. (72), p. 152] over any infinite plane intersecting the line which joins the two charges.

14. Two fluid dielectrics whose inductive capacities are ϵ_1 and ϵ_2 meet at an infinite plane surface. Two charges q_1 and q_2 are located a distance a from the plane on opposite sides and the line joining them is normal to the plane. Calculate the forces acting on q_1 and q_2 and account for the fact that they are unequal.

15. A conducting sphere of radius r_1, carrying a charge Q, is in the field of a point charge q of the same sign. Assume $q \ll Q$. Plot the force exerted by the sphere on q as a function of distance from the center and calculate the point at which the direction reverses.

16. A charge q is located within a spherical, conducting shell of radius b at a point whose distance from the center is a, with $a < b$. Calculate the potential of the sphere, the charge density on its internal surface, and the force exerted on q. Assume the sphere to be insulated and uncharged.

17. Find the distribution of charge giving rise to a two-dimensional field whose potential at any point in the xy-plane is

$$\phi = \frac{1}{2\pi\epsilon_0}\left(\tan^{-1}\frac{1+x}{y} + \tan^{-1}\frac{1-x}{y}\right).$$

18. Show that the electrostatic potential ϕ is uniquely determined by the values of either ϕ or $\partial\phi/\partial n$ on the surfaces of conductors embedded in an isotropic, inhomogeneous dielectric.

19. A set of n conducting bodies is embedded in a dielectric medium which is isotropic but not necessarily homogeneous. If the charges $q_1, q_2, \ldots, q_n$ are placed on these conductors the corresponding potentials on the conductors are $\phi_1, \phi_2, \ldots,$

ϕ_n. Likewise the charges $q_1', q_2', \ldots, q_n'$ give rise to the potentials $\phi_1', \phi_2', \ldots, \phi_n'$. Show that

$$\sum_{i=1}^{n} q_i\phi_i' = \sum_{i=1}^{n} q_i'\phi_i.$$

20. Find the charge in coulombs on a metal sphere of radius a meters necessary to raise the potential one volt with respect to infinity.

21. An infinite plane surface divides two half spaces, the one occupied by a dielectric of inductive capacity ϵ, the other by air. A charge q is located on the air side a distance a from the surface. Obtain expressions for the potential at points in the air and in the dielectric and show that the potential in either space can be represented in terms of images.

What is the density ω' of induced, or bound, charge on the surface of the dielectric?

Calculate the force acting on q and the work done in withdrawing the charge to infinity.

22. The radii of two metal spheres are respectively a and b and the separation of their centers is c, where $c > a + b$. A charge q_a is placed on the first, thus raising its potential to a specified value ϕ_a. The potential ϕ_b of the second sphere is maintained at zero by grounding. Obtain expressions for the charges q_a and q_b and their distribution over the spheres by building up a series of images at the inverse points of the spheres (p. 203).

The method is due to Lord Kelvin. See the discussion by F. Noether in Riemann-Weber's "Differential und Integralgleichungen der Physik," Vol. II, p. 281, 1927.

23. Two infinite, parallel, conducting planes coincide with the surfaces $x = 0$, $x = a$. The planes are grounded. A charge q is located on the x-axis at the point $x = c$, where $0 < c < a$. Show that the density of induced charge on the plane $x = 0$ is

$$\omega(x, y) = \frac{q}{4\pi} \sum_{n=-\infty}^{\infty} \left\{ \frac{2na - c}{[(2na - c)^2 + r^2]^{\frac{3}{2}}} - \frac{2na + c}{[(2na + c)^2 + r^2]^{\frac{3}{2}}} \right\},$$

where r is the distance from the origin to the point x, y on the plane. What is the density on the plane $x = a$? Show that the total charge on the two planes is $-q$. (Kellogg.)

24. Prove that the surface charge density at any point on a charged ellipsoidal conductor is proportional to the perpendicular distance from the center of the ellipsoid to the plane tangent to the ellipsoid at the point.

25. Equation (14), p. 209, gives the potential of a charged prolate spheroid. Show that as the length $\to \infty$ and the eccentricity $\to 1$, this function approaches the logarithmic potential characteristic of a two-dimensional space. Apply the theorem of Problem 24 to find the charge per unit length.

26. Charge is distributed with constant density within a thin ellipsoidal shell bounded by the two similar concentric ellipsoidal surfaces whose principal axes are a, b, c and $a\left(1 + \frac{\delta}{2}\right)$, $b\left(1 + \frac{\delta}{2}\right)$, $c\left(1 + \frac{\delta}{2}\right)$. Show that the field is zero at all internal points and obtain an expression for the potential at external points.

27. The major axis of a conducting, prolate spheroid is 20 cm. in length and its eccentricity is e. The spheroid is suspended in air and carries a total charge of q coulombs. The maximum field intensity at the surface shall at no point exceed 3×10^6 volts/meter. Plot the maximum value of charge that can be placed on the spheroid subject to this condition as a function of eccentricity.

28. The effect of lightning conductors has been studied by Larmor (*Proc. Roy. Soc.*, A, **90**, 312, 1914) through the following problem. A hemispheroidal rod projects above a flat conducting plane. The potential ϕ_0 of the charged cloud above is uniform in the neighborhood of the rod. Find a potential function ϕ which is constant over the spheroidal surface and over the plane. Plot the ratio of potential gradient at the tip of the rod to intensity of the applied field against eccentricity of the rod. Assume the length of the rod (semimajor axis) to be fixed and equal to one meter. Discuss the lateral influence of the rod on the field as a function of eccentricity.

29. A very thin, circular, metal disk of radius a carries a charge q. Show that the density of charge per unit area is

$$\omega = \frac{q}{2\pi a} \frac{1}{\sqrt{a^2 - r^2}}$$

where r is the radial distance from the center on the disk.

Calculate the field intensity at the surface of the disk and plot volts per meter per coulomb against r/a. Assume the disk to be suspended in air.

30. A spheroidal, dielectric body is suspended in a uniform electrostatic field. The length of the major axis of the spheroid is twice that of its minor axis, and the specific inductive capacity is 4. The external medium is air. For what orientation with respect to the axis of the applied field is the torque a maximum? Calculate the value of this maximum torque per unit volume of material, per volt of applied field intensity.

31. A stationary current distribution is established in a medium which is isotropic but not necessarily homogeneous. Show that the medium will in general acquire a volume distribution of charge whose density is

$$\rho = -\frac{1}{\sigma}(\sigma\nabla\epsilon - \epsilon\nabla\sigma)\cdot\nabla\phi,$$

where σ and ϵ are respectively the conductivity and inductive capacity of the medium.

32. Two homogeneous, isotropic media characterized by the constants ϵ_1, σ_1, and ϵ_2, σ_2, meet at a surface S. A stationary current crosses S from one medium to the other. If the angles made by a line of flow with the normal at the point of transition are ϕ_1 and ϕ_2, show that each line is refracted according to the law

$$\sigma_2 \cot \phi_2 = \sigma_1 \cot \phi_1.$$

Show also that a charge appears on S whose surface density is

$$\omega = \left(\frac{\epsilon_2}{\sigma_2} - \frac{\epsilon_1}{\sigma_1}\right)\mathbf{J}\cdot\mathbf{n}.$$

33. In many practical current-distribution problems it can be assumed that a system of perfectly conducting electrodes is embedded in a poorly conducting medium which is either homogeneous or in which discontinuities occur only across specified surfaces. The electrodes are maintained at fixed potentials and either these potentials or the electrode currents are specified. Show that the current distribution is uniquely determined by a potential function ϕ satisfying the following conditions:

(*a*) $\nabla^2\phi = 0$ at all ordinary points of the medium;
(*b*) $\phi = \phi_i$, a constant on the ith electrode;
(*c*) At a surface of discontinuity in the medium the transition is governed by

$$\phi_1 = \phi_2,\quad \sigma_2\left(\frac{\partial\phi}{\partial n}\right)_2 = \sigma_1\left(\frac{\partial\phi}{\partial n}\right)_1;$$

(d) Regularity at infinity;

(e) Condition (b) may be replaced by $I_i = -\oint_{S_i} \sigma \frac{\partial \phi}{\partial n}\, da$, where σ is the conductivity of the medium just outside the electrode whose bounding surface is S_i and where I_i is the total current leaving this electrode.

34. A system of electrodes is embedded in a conducting medium. The electrodes are maintained at the fixed potentials $\phi_1, \phi_2, \ldots, \phi_n$, and the currents leaving the electrodes are $I_1, I_2, \ldots, I_n$. Show that the total Joule heat generated in the medium is

$$Q = \sum_{i=1}^{n} \phi_i I_n.$$

35. Prove that for a system of stationary currents driven by electromotive forces the current will distribute itself in such a way that the heat generated will be less for the actual distribution than for any other which coincides with the actual distribution in the region occupied by the driving forces and which is solenoidal everywhere else. Assume that the medium is isotropic and that the relation between field intensity and current density is linear.

36. A circular disk electrode of radius a is in contact with an infinite half space of conductivity σ, such as the surface of the earth. The distance to all other electrodes is large relative to the radius a. Show that the current distribution is determined by the conditions:

(a) $\nabla^2\phi = 0$ within the conducting half space;
(b) $\partial\phi/\partial n = 0$ at the boundary excluding the disk;
(c) $\frac{\partial \phi}{\partial n} = -\frac{I}{2\pi\sigma a\sqrt{a^2 - r^2}}$ on the disk, where r is the radial distance from the center.

37. A stationary current is carried by a curved, conducting sheet of uniform thickness. The sheet is assumed to be so thin that the current distribution is essentially two-dimensional. If ξ and η are curvilinear coordinates on the surface, the geometry of the surface is defined by the element of arc

$$ds^2 = g_{11}\, d\xi^2 + 2g_{12}\, d\xi\, d\eta + g_{22}\, d\eta^2,$$

the properties of the g_{ik} having been defined in Sec. 1.14. The scalar potential determining the current distribution satisfies the Laplace equation

$$\frac{\partial}{\partial \xi} \frac{g_{22}\frac{\partial \phi}{\partial \xi} - g_{12}\frac{\partial \phi}{\partial \eta}}{\sqrt{g_{11}g_{22} - g_{12}^2}} + \frac{\partial}{\partial \eta} \frac{g_{11}\frac{\partial \phi}{\partial \eta} - g_{12}\frac{\partial \phi}{\partial \xi}}{\sqrt{g_{11}g_{22} - g_{12}^2}} = 0.$$

Define a conjugate function ψ by the relation

$$d\psi = -\frac{g_{11}\frac{\partial \phi}{\partial \eta} - g_{12}\frac{\partial \phi}{\partial \xi}}{\sqrt{g_{11}g_{22} - g_{12}^2}}\, d\xi + \frac{g_{22}\frac{\partial \phi}{\partial \xi} - g_{12}\frac{\partial \phi}{\partial \eta}}{\sqrt{g_{11}g_{22} - g_{12}^2}}\, d\eta$$

and show that ψ satisfies the same equation as ϕ. The functions ϕ and ψ are potential and streamline functions on the surface, and $\Phi = \phi + i\psi$ will be called a complex function on the surface. Now choose ξ and η so that $\zeta = \xi + i\eta$ is itself such a function. Show that in this case ϕ satisfies the simple equation

$$\frac{\partial^2\phi}{\partial\xi^2} + \frac{\partial^2\phi}{\partial\eta^2} = 0$$

and that ϕ and ψ are related by

$$\frac{\partial\phi}{\partial\xi} = \frac{\partial\psi}{\partial\eta}, \qquad \frac{\partial\phi}{\partial\eta} = -\frac{\partial\psi}{\partial\xi},$$

while the line element reduces to $ds^2 = g_{11}(d\xi^2 + d\eta^2)$. Thus the complex functions of the surface are analytic functions of each other and the transformation $\xi + i\eta = x + iy$ maps the surface conformally on to the complex z-plane. If the conformal mapping of a surface on to the plane is known, the current distribution problem reduces to the solution of Laplace's equation in rectangular coordinates. Inversely, if a current distribution over a curved surface can be determined experimentally, a conformal mapping of the surface on the plane is found. (Kirchhoff.)

38. Referring to Problem 37, show that

$$z = ae^{i\beta}\tan\frac{\theta}{2}$$

is a conformal, stereographic projection of a sphere of radius a on the complex plane, where $z = x + iy$ and β is the equatorial angle or longitude and θ the colatitude on the sphere. Show further that

$$\Phi = \phi + i\psi = f\left(2ae^{i\beta}\tan\frac{\theta}{2}\right) = f(z)$$

is a complex potential function on the sphere where f is any analytic function of z.

39. A steady current I enters a conducting spherical shell of radius a and surface conductivity σ at a point on the surface and leaves at a similar point diametrically opposite. Find the potential at any point and the equation of the streamlines.

40. A cylindrical condenser is formed of two concentric metal tubes. The space between the tubes is occupied by a solid dielectric of inductive capacity ϵ which, however, is not in direct contact with the tubes, but separated from them by thin layers of a nonconducting intermediary fluid of inductive capacity ϵ'. The solid dielectric extends so far beyond the ends of the metal tubes that the stray field is of negligible intensity. A constant potential difference is maintained between the metal electrodes. Obtain an expression for the change in length per unit length of dielectric. (Kemble, *Phys. Rev.*, **7**, 614, 1916.)

CHAPTER IV

THE MAGNETOSTATIC FIELD

The methods that have been developed for the analysis of electrostatic fields apply largely to the magnetostatic field as well. Every magnetostatic field can be represented by an electrostatic field of identical structure produced by dipole distributions and fictive double layers. The equivalence, however, is purely formal. There is no quantity in magnetostatics corresponding to free charge, and the surface singularity generated by a double layer of electric charge does not actually exist. A double layer leads in fact to a multivalued potential and consequently to a nonconservative field. Whatever the analytical advantages of the electrostatic analogy may be, it is well to remember that the physical structure of a field due to stationary distributions of current differs fundamentally from that of any configuration of electric charges.

GENERAL PROPERTIES OF A MAGNETOSTATIC FIELD

4.1. Field Equations and the Vector Potential.—The equations satisfied by the magnetic vectors of a stationary field are obtained by placing the time derivatives in Maxwell's equations equal to zero.

$$\text{(I)}\quad \nabla \times \mathbf{H} = \mathbf{J}, \qquad \text{(II)}\quad \nabla \cdot \mathbf{B} = 0.$$

To these must be added the equation of continuity which now reduces to

$$\text{(III)} \qquad \nabla \cdot \mathbf{J} = 0.$$

The current distribution in a stationary field is solenoidal; all current lines either close upon themselves, or start and terminate at infinity. From (II) it follows likewise that all lines of the vector **B**—lines of magnetic flux, as they are commonly called—close upon themselves. Suppose that the flux density of the field produced by a current filament I_1 is $\mathbf{B}_1$. All lines of this field link the circuit I_1. A fraction of the flux $\mathbf{B}_1$ may, however, thread also a second current filament I_2. The concept of "flux linkages" commonly employed in the practical analysis of electromagnetic problems is based on these solenoidal properties of current and flux in the stationary state. The extension to slowly varying fields (quasi-stationary state) is justifiable only so far as $\partial\rho/\partial t$ remains negligible with respect to $\nabla \cdot \mathbf{J}$.

Upon applying Stokes' theorem to (I), we obtain the equivalent integral equation,

$$\int_C \mathbf{H} \cdot d\mathbf{s} = \int_S \mathbf{J} \cdot \mathbf{n}\, da = I, \tag{Ia}$$

where S is any surface spanning the contour C, and I is the total current through this surface. The line integral of $\mathbf{H}$ around a closed path is equal to the current linked: *the magnetic field is nonconservative.*

Every solenoidal field admits a representation in terms of a vector potential. As in Sec. 1.9, Eq. (II) is identically satisfied by

$$\mathbf{B} = \nabla \times \mathbf{A}. \tag{1}$$

The vector $\mathbf{A}$ is now chosen to satisfy (I), and the relation $\mathbf{H} = H(\mathbf{B})$ between the magnetic vectors must therefore be specified. If the medium is nonferromagnetic, the relation is linear; if moreover it is homogeneous and isotropic, we put $\mathbf{B} = \mu\mathbf{H}$ and arrive at

$$\nabla \times \nabla \times \mathbf{A} = \mu\mathbf{J} \tag{2}$$

for the vector potential.

In Sec. 1.9, it was shown that there may be added to $\mathbf{A}$ the gradient of any scalar function ψ without affecting the definition (1). By a proper choice of ψ it is always possible to effect the vanishing of the divergence of $\mathbf{A}$. The vector potential is uniquely defined by (2) plus the condition

$$\nabla \cdot \mathbf{A} = 0. \tag{3}$$

In rectangular coordinates, (2) then reduces to

$$\nabla^2\mathbf{A} = -\mu\mathbf{J}, \tag{4}$$

where it is understood that the Laplacian operates on each *rectangular* component of $\mathbf{A}$.

4.2. The Scalar Potential.—The existence of a scalar potential function associated with the electrostatic field is a direct consequence of the irrotational character of the field vector $\mathbf{E}$. If double-layer distributions of charge are excluded, the curl of $\mathbf{E}$ is everywhere zero; hence one may express $\mathbf{E}$ as the negative gradient of a scalar ϕ. Furthermore, since the line integral of $\mathbf{E}$ vanishes about *every* closed path of the region, the potential is a single-valued function of the coordinates.

A region of space, enclosed by boundaries, is said to be connected when it is possible to pass from any one point of the region to any other by an infinite number of paths, each of which lies wholly within the region. Any two paths which can by continuous deformation be made to coincide without ever passing out of the region are said to be mutually reconcilable. Any closed curve or surface is said to be reducible if by continuous variation it can be contracted to a point without passing out of the region. Two reconcilable paths when combined form a reducible circuit. If finally the region is such that all paths joining any two points are reconcilable, or such that all circuits drawn within it are reducible, it is said to be simply connected. The electrostatic field of a volume distribution of charge constitutes a simply connected region of space within which every closed curve is reducible.

Contrast this geometrical property of the electrostatic field structure with the magnetic field of a stationary distribution of current. We shall call the region of space occupied by current V_1, while the region within which $\mathbf{J} = 0$ will be denoted as V_2. Figure 39 represents a plane section through the field, the shaded areas indicating, for example, the cross

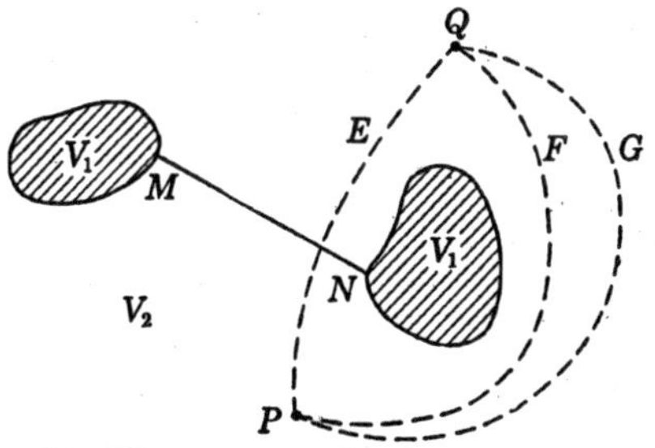

FIG. 39.—Illustrating a doubly connected space.

sections of conductors carrying current. In the region V_1 the curl of $\mathbf{H}$ does not vanish and consequently no scalar potential exists. Throughout V_2 we find

$$\nabla \times \mathbf{H} = 0, \tag{5}$$

and so within this region $\mathbf{H}$ may be expressed as the negative gradient of a scalar function $\phi^*(x, y, z)$.

$$\mathbf{H} = -\nabla\phi^*. \tag{6}$$

The line integral of $\mathbf{H}$ along any contour wholly within V_2 connecting two points P and Q is

$$\int_P^Q \mathbf{H} \cdot d\mathbf{s} = -\int_P^Q \nabla\phi^* \cdot d\mathbf{s} = \phi^*(P) - \phi^*(Q). \tag{7}$$

The paths by which one may reach Q from P are not, however, all mutually reconcilable. A closed curve composed of two segments such as PEQ and QFP is irreducible, since on contraction to a point it must necessarily penetrate the region V_1. Thus the region V_2 occupied by the magnetostatic field exterior to the current distribution is multiply connected. The scalar potential ϕ^* is a multivalued function of position, for to its value at any point P one may add a factor nI by performing n complete circuitations about the current I. However V_2 may be rendered simply connected and the potential single-valued by the introduction of a cut or barrier which prevents a circuit from being closed in such a manner as to link a current. There are, of course, an infinite number of ways in which such an imagined barrier might be placed but, if the current is constituted of a bundle of current filaments forming a single, closed current loop, it is most simple to think of it as a surface spanning the loop. The line MN in Fig. 39 represents the trace of such a surface. In the "cut" space the line integral of **H** about every closed path vanishes, for the barrier surface excludes current linkages, and consequently the potential function ϕ^* is single-valued. At points lying on either side and infinitely close to the barrier the values of ϕ^* differ by I. There is a discontinuity in the potential function thus defined equal to

$$\phi^*_+ - \phi^*_- = I, \tag{8}$$

where the subscripts refer to the positive and negative sides of the surface. *According to Sec.* 3.16 *the discontinuity in ϕ^* across the barrier is equivalent to that of a surface dipole layer of constant moment I per unit area.* Only when all currents are zero and the sources of the field are permanent magnets does the potential become a single-valued function of position.

4.3. Poisson's Analysis.—At every point where the current density is zero, and in particular at interior points of a permanent magnet, we have

$$\mathbf{B} = -\mu_0 \nabla \phi^* + \mu_0(\mathbf{M} + \mathbf{M}_0). \tag{9}$$

Since **B** is divergenceless, the scalar potential satisfies a Poisson equation

$$\nabla^2 \phi^* = -\rho^*, \qquad \rho^* = -\nabla \cdot (\mathbf{M} + \mathbf{M}_0). \tag{10}$$

The quantity ρ^* is the magnetic analogue of "bound charge" density, which in the older literature was frequently called the density of Poisson's "ideal magnetic matter."

Across any surface of discontinuity in the medium the magnetic vectors satisfy, according to Sec. 1.13, the conditions

$$\mathbf{n} \cdot (\mathbf{B}_2 - \mathbf{B}_1) = 0, \qquad \mathbf{n} \times (\mathbf{H}_2 - \mathbf{H}_1) = 0. \tag{11}$$

The equivalent conditions imposed on the scalar potential are, therefore,

$$\left(\frac{\partial \phi^*}{\partial n}\right)_2 - \left(\frac{\partial \phi^*}{\partial n}\right)_1 = -\omega^*, \qquad \left(\frac{\partial \phi^*}{\partial t}\right)_2 - \left(\frac{\partial \phi^*}{\partial t}\right)_1 = 0, \tag{12}$$

$$\omega^* = \mathbf{n} \cdot [(\mathbf{M} + \mathbf{M}_0)_1 - (\mathbf{M} + \mathbf{M}_0)_2]. \tag{13}$$

Within any closed region containing permanent magnets and polarizable matter, but throughout which the conduction current density **J** *is zero, the magnetostatic problem is mathematically equivalent to an electrostatic problem.*

The calculation of the potential ϕ^* from the densities ρ^* and ω^* has been described in the previous chapter.

$$\text{(14)} \quad \phi^*(x', y', z') = \frac{1}{4\pi}\int \rho^*(x, y, z)\,\frac{1}{r}\,dv + \frac{1}{4\pi}\int \omega^*(x, y, z)\,\frac{1}{r}\,da,$$

where as usual $r = \sqrt{(x' - x)^2 + (y' - y)^2 + (z' - z)^2}$. To evaluate the integrals one must know both the permanent magnetization $\mathbf{M}_0$ and the induced magnetization $\mathbf{M}$; but the induced magnetization is a function of the field intensity and consequently of ϕ^* itself. If, however, the permanently magnetized bodies are not too close to one another, the induced magnetization can usually be neglected with respect to $\mathbf{M}_0$. Some assumption is then made regarding the magnetization $\mathbf{M}_0$ and the fixed or primary field of the sources calculated by means of (14). This primary field induces a magnetization $\mathbf{M}$ in neighboring polarizable bodies—soft iron, for example. The calculation of the induced or secondary field is a boundary-value problem to be discussed later.

Let us calculate now the *magnetic moment* of a magnetized piece of matter. By analogy with the electrostatic case, Sec. 3.11, this moment is defined by the integral

$$\text{(15)} \qquad \mathbf{m} = \int_V \mathbf{r}_1\, \rho^*(\xi, \eta, \zeta)\, dv + \int_S \mathbf{r}_1\, \omega^*(\xi, \eta, \zeta)\, da,$$

where $\mathbf{r}_1 = \mathbf{i}\xi + \mathbf{j}\eta + \mathbf{k}\zeta$, Fig. 30, page 178, is the vector drawn from the origin to the element of "magnetic charge," and S the boundary of a body whose volume is V. Upon replacing ρ^* and ω^* by their values defined in (10) and (13), then transforming by means of the identity

$$\xi\nabla \cdot \mathbf{M} = \nabla \cdot (\xi\mathbf{M}) - \mathbf{M} \cdot \nabla\xi,$$

which applies to each component of $\mathbf{r}_1$, we find that (15) reduces to

$$\text{(16)} \qquad \mathbf{m} = \int_V (\mathbf{M} + \mathbf{M}_0)\, dv.$$

The magnetic moment of a body is equal to the volume integral of its magnetization.

Finally, if Eq. (14) is transformed by (9), page 184, after ρ^* and ω^* have been replaced by (10) and (13), one obtains for the potential

$$\text{(17)} \qquad \phi^*(x', y', z') = \frac{1}{4\pi}\int_V (\mathbf{M} + \mathbf{M}_0) \cdot \nabla\left(\frac{1}{r}\right) dv.$$

If the dimensions of the magnetized body are sufficiently small with respect to the distance to the observer at (x', y', z'), the variation of

$\nabla 1/r$ within V may be neglected and we are led to an expression for the scalar potential of a magnetic dipole,

$$\phi^*(x', y', z') = \frac{1}{4\pi}\mathbf{m}\cdot\nabla\left(\frac{1}{r}\right) = -\frac{1}{4\pi}\mathbf{m}\cdot\nabla'\left(\frac{1}{r}\right). \tag{18}$$

CALCULATION OF THE FIELD OF A CURRENT DISTRIBUTION

4.4. The Biot-Savart Law.—It has been shown that in a homogeneous, isotropic medium of constant permeability μ the vector potential satisfies the system of equations

$$\nabla^2 A_j = -\mu J_j, \qquad \nabla\cdot\mathbf{A} = 0, \qquad (j = 1, 2, 3). \tag{1}$$

The A_j are *rectangular components* of the vector **A**, but the restriction to rectangular coordinates does not necessarily apply to the scalar operator ∇^2. Considering the A_j and J_j as variant scalars, the theory developed in Sec. 3.4 for the integration of Poisson's equation may be applied directly to (1). If the current distribution can be circumscribed by a sphere of finite radius each component of the vector potential can be expressed as an integral,

$$A_j(x', y', z') = \frac{\mu}{4\pi}\int J_j(x, y, z)\,\frac{1}{r}\,dv, \qquad (j = 1, 2, 3), \tag{2}$$

extended over all space. These components are now recombined to give a vector.

$$\mathbf{A}(x', y', z') = \frac{\mu}{4\pi}\int \mathbf{J}(x, y, z)\,\frac{1}{r}\,dv, \tag{3}$$

where $r = \sqrt{(x' - x)^2 + (y' - y)^2 + (z' - z)^2}$. Moreover, it follows from the discussion of Sec. 3.7 that if **J** is a bounded, integrable function, then **A** is a continuous function of the coordinates (x', y', z') possessing continuous first derivatives at every point inside and outside the current distribution. If **J** and all its derivatives of order less than n are finite and continuous, the vector function **A** has continuous derivatives of all orders less than $n + 1$.

The vector potential defined by (3) satisfies the condition $\nabla\cdot\mathbf{A} = 0$ provided the current distribution is spatially bounded. The integral is regular and consequently may be differentiated under the sign of integration.

$$\nabla'\cdot\mathbf{A} = \frac{\mu}{4\pi}\int \mathbf{J}(x, y, z)\cdot\nabla'\left(\frac{1}{r}\right)dv = -\frac{\mu}{4\pi}\int \mathbf{J}(x, y, z)\cdot\nabla\left(\frac{1}{r}\right)dv. \tag{4}$$

$$\mathbf{J}\cdot\nabla\left(\frac{1}{r}\right) = \nabla\cdot\frac{\mathbf{J}}{r} - \frac{1}{r}\nabla\cdot\mathbf{J} = \nabla\cdot\frac{\mathbf{J}}{r} \tag{5}$$

in virtue of the condition $\nabla \cdot \mathbf{J} = 0$. Then

$$\nabla' \cdot \mathbf{A} = -\frac{\mu}{4\pi} \int \nabla \cdot \frac{\mathbf{J}}{r}\, dv = -\frac{\mu}{4\pi} \int_S \frac{\mathbf{J} \cdot \mathbf{n}}{r}\, da = 0, \tag{6}$$

for, since all current lines close upon themselves within a finite domain, the surface S may be chosen such that there is no normal component of flow.

Note that the contribution $d\mathbf{A}$ to the vector potential is directed parallel to the element of current $\mathbf{J}\, dv$. Equation (3) can be simplified in the case of currents in linear conductors; *i.e.*, conductors whose cross section da is so small with respect to the distance r that the density vector may be assumed uniform over the cross-sectional area and directed along the wire.

$$\mathbf{J}\, dv = \mathbf{J}\, da\, ds = I\, d\mathbf{s}, \tag{7}$$

where I is the total current carried by the conductor and ds is an element of conductor length. Since the steady current I must be constant throughout the circuit, Eq. (3) reduces to

$$\mathbf{A}(x', y', z') = \frac{\mu I}{4\pi} \int_C \frac{1}{r}\, d\mathbf{s}, \tag{8}$$

in which the integral is to be extended around the complete circuit C.

The field can be obtained directly from (3) by calculating the curl of $\mathbf{A}$ at the point (x', y', z'). In virtue of the convergence of the integral and the continuity of its first derivatives the differential operator may be introduced under the sign of integration.

$$\mathbf{H}(x', y', z') = \frac{1}{4\pi} \int \nabla' \times \left[\frac{\mathbf{J}(x, y, z)}{r}\right] dv. \tag{9}$$

Expansion of the integrand according to elementary rules yields

$$\nabla' \times \left[\frac{1}{r}\mathbf{J}(x, y, z)\right] = \frac{1}{r}\nabla' \times \mathbf{J}(x, y, z) + \nabla'\left(\frac{1}{r}\right) \times \mathbf{J}(x, y, z). \tag{10}$$

But the current density is a function of the running coordinates x, y, z, whereas the differentiation applies only to x', y', z'; consequently the first term on the right is zero, giving for the field vector

$$\begin{aligned} \mathbf{H}(x', y', z') &= \frac{1}{4\pi} \int \nabla'\left(\frac{1}{r}\right) \times \mathbf{J}(x, y, z)\, dv \\ &= \frac{1}{4\pi} \int \mathbf{J}(x, y, z) \times \nabla\left(\frac{1}{r}\right) dv. \end{aligned} \tag{11}$$

If $\mathbf{r}^0$ is a unit vector directed from the current element toward the point of observation, (11) is equivalent to

$$\mathbf{H}(x', y', z') = \frac{1}{4\pi}\int \frac{\mathbf{J} \times \mathbf{r}^0}{r^2}\, dv, \tag{12}$$

which for linear circuits assumes the form

$$\mathbf{H}(x', y', z') = \frac{I}{4\pi}\int_C \frac{\mathbf{s}^0 \times \mathbf{r}^0}{r^2}\, ds, \tag{13}$$

where $\mathbf{s}^0$ is a unit vector in the direction of the current element $I\,ds$.

The law expressed by (13) is frequently formulated with the statement that each linear current element $I\,ds$ contributes an amount

$$d\mathbf{H} = \frac{1}{4\pi}\frac{\mathbf{s}^0 \times \mathbf{r}^0}{r^2} I\, ds \tag{14}$$

to the total field. This is essentially the result deduced from the experiments of Biot and Savart in 1820. There is no very cogent reason to reject the formula (14), other than the fact that this resolution of (13) into differential elements is not unique; for there may be added to (14) any vector function integrating to zero around a closed circuit. All stationary currents are composed of closed filaments, so that it is difficult to imagine an experiment to determine the contribution of an isolated element. Obviously no error can result from an application of (14) to the analysis of a field when it is understood that these contributions are eventually to be summed around the circuit.

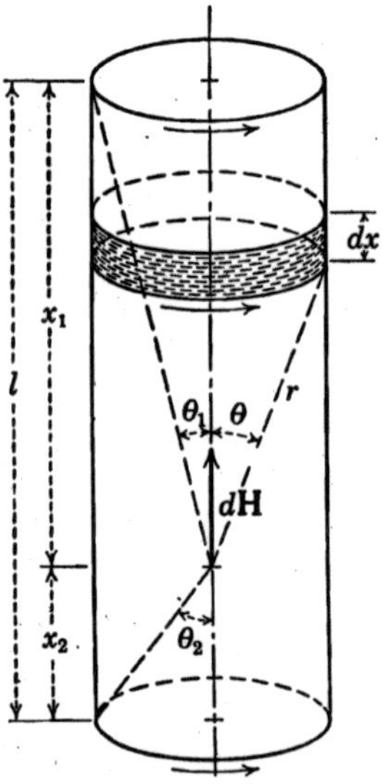

FIG. 40.—Solenoidal current distribution.

Whereas the vector potential due to a differential element of current is directed parallel to the element, the field vector $\mathbf{H}$ is oriented normal to a plane containing both the current element and the line joining it with the point of observation; *i.e.*, to the plane defined by the unit vectors $\mathbf{s}^0$ and $\mathbf{r}^0$. If θ is the angle which $\mathbf{r}^0$ makes with $\mathbf{s}^0$, the intensity of the field is

$$dH = \frac{I}{4\pi}\frac{\sin\theta}{r^2}\, ds. \tag{15}$$

The field along the axis of a solenoid, Fig. 40, may, for example, be calculated approximately by assuming an equivalent current sheet of negligible thickness. If there are n turns/meter carrying a current of I amp., the current density of the sheet is $K = In$. The contribution of a solenoidal element of radius a and length dx at a distance x along the axis from the point of observation is

$$dH = \frac{nI}{2} \frac{a^2}{(a^2 + x^2)^{\frac{3}{2}}} dx = \frac{nI}{2} d(\cos \theta). \tag{16}$$

These contributions are directed along the axis. At any point on the axis, the field is

$$H = \frac{nI}{2} \left(\frac{x_2}{\sqrt{x_2^2 + a^2}} - \frac{x_1}{\sqrt{x_1^2 + a^2}} \right) = \frac{nI}{2} (\cos \theta_2 + \cos \theta_1). \tag{17}$$

At the center this reduces to

$$\mathbf{H} = nI \frac{l}{\sqrt{l^2 + 4a^2}}. \tag{18}$$

Another elementary case of some interest is that of a straight wire of infinite length and circular cross section. The current I is assumed to be distributed uniformly over the cross section. The simplest solution is obtained by taking advantage of the cylindrical symmetry of the field to apply the integral law $\oint \mathbf{H} \cdot d\mathbf{s} = I$. If r is the radial distance from the axis and a the radius of the wire, it is obvious that

$$\begin{aligned} 2\pi r H &= \pi r^2 J = \frac{r^2}{a^2} I, \qquad H = \frac{I}{2\pi a^2} r, \qquad (r < a), \\ 2\pi r H &= I, \qquad H = \frac{I}{2\pi r}, \qquad (r > a). \end{aligned} \tag{19}$$

The field outside the wire, which is independent of its radius, can also be obtained from the Biot-Savart law. If, however, one attempts to calculate the vector potential by means of (8), it will be noted that the integral diverges owing to the fact that the current distribution is in this case not confined to a region of finite extent. The vector potential is actually

$$A_r = A_\theta = 0, \qquad A_z = \frac{\mu I}{2\pi} \ln \frac{1}{r}, \qquad (r > a), \tag{20}$$

as may be verified by calculating the curl in cylindrical coordinates (page 51), and this function becomes infinite as $r \to \infty$.

4.5. Expansion of the Vector Potential.—Following the example set in Sec. 3.11 for the scalar potential, we shall determine an expansion of the vector potential of a stationary current distribution in terms of the coordinates of a point relative to a fixed origin. To ensure regularity at infinity it will be assumed that the entire distribution can be circumscribed by a sphere of finite radius R drawn from the origin; we shall consider here only points of observation exterior to this sphere. The current distribution is completely arbitrary, but it must be assumed that the permeability μ is everywhere constant and uniform. Practically this excludes only ferromagnetic materials and admits conductors of any other nature and arbitrary shape embedded in dielectrics of any sort.

The notation will be essentially that adopted in Sec. 3.11 and illustrated in Fig. 30; for convenience in summing we shall let the coordinates of the fixed point P be x_1, x_2, x_3, and those of the current element ξ_1, ξ_2, ξ_3. The distance from the origin O to P is $r = \sqrt{x_1^2 + x_2^2 + x_3^2}$ and that from the element $\mathbf{J}(\xi_1, \xi_2, \xi_3)\,dv$ to P is

$$r_2 = \sqrt{\sum_{j=1}^{3} (x_j - \xi_j)^2}.$$

At P the vector potential is, therefore,

$$\mathbf{A}(x_j) = \frac{\mu}{4\pi} \int_V \frac{\mathbf{J}(\xi_j)}{r_2}\,dv, \tag{21}$$

extended over the volume V bounded by the sphere of radius R. If $\mathbf{r}_1$ is the vector whose components are ξ_j, the expansion of $1/r_2$ in a Taylor series about O leads to

$$\frac{1}{r_2} = \frac{1}{r} - \sum_{j=1}^{3} \xi_j \frac{\partial}{\partial x_j}\left(\frac{1}{r}\right) + \frac{1}{2}\sum_{j=1}^{3}\sum_{k=1}^{3} \xi_j \xi_k \frac{\partial^2}{\partial x_j\,\partial x_k}\left(\frac{1}{r}\right) - \cdots. \tag{22}$$

Consequently the vector potential can be represented by the expansion

$$\mathbf{A}(x_j) = \sum_{n=0}^{\infty} \mathbf{A}^{(n)} = \frac{\mu}{4\pi}\frac{1}{r}\int \mathbf{J}(\xi_j)\,dv - \frac{\mu}{4\pi}\int \left[\mathbf{r}_1 \cdot \nabla\left(\frac{1}{r}\right)\right] \mathbf{J}(\xi_j)\,dv$$
$$+ \frac{\mu}{4\pi}\int \left\{\mathbf{r}_1 \cdot \nabla\left[\mathbf{r}_1 \cdot \nabla\left(\frac{1}{r}\right)\right]\right\} \mathbf{J}(\xi)\,dv - \cdots. \tag{23}$$

Note that $\mathbf{J}$ and $\mathbf{r}_1$ are functions of the ξ_j with respect to which the integration is to be effected; the function $1/r$ depends only on the x_j, and the operator ∇ applies to these coordinates alone.

In virtue of its stationary character the current distribution may be resolved into filaments, all of which close upon themselves within V. Let us single out any one of these filaments. We suppose the infinitesimal cross section to be da_1 and the current carried by this linear circuit to be $I = |\mathbf{J}|\,da_1$. A point on the circuit is located by the vector $\mathbf{r}_1$ drawn from the origin, Fig. 41, and an element of length along the circuit in the direction of the current is $d\mathbf{r}_1$. Then for each filament or tube

(24) $$\mathbf{A}^{(0)} = \frac{\mu}{4\pi}\frac{I}{r}\int_C d\mathbf{r}_1 = 0,$$

where C is the closed filament contour. ($\mathbf{A}^{(0)}$ would not necessarily vanish were V to contain only a portion of the current distribution.)

The dominant term of the expansion is, therefore, the second, which we transform as follows.

(25) $$\left[\mathbf{r}_1 \cdot \nabla\left(\frac{1}{r}\right)\right] d\mathbf{r}_1 = \frac{1}{2}(\mathbf{r}_1 \times d\mathbf{r}_1) \times \nabla\left(\frac{1}{r}\right) + \frac{1}{2} d\left\{\left[\mathbf{r}_1 \cdot \nabla\left(\frac{1}{r}\right)\right]\mathbf{r}_1\right\}.$$

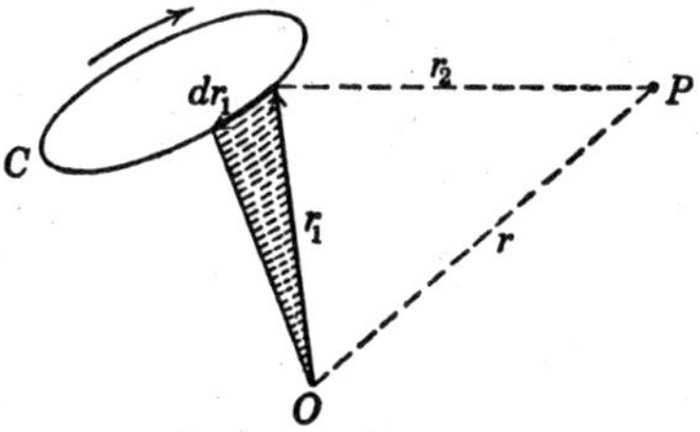

FIG. 41.—Closed current filament following the contour C.

Again the total differential vanishes when integrated about a closed filament, while $\frac{1}{2}\mathbf{r}_1 \times d\mathbf{r}_1 = \mathbf{n}\,da$ is the vector area of the infinitesimal triangle shown shaded in Fig. 41.

(26) $$\mathbf{A}^{(1)} = -\frac{\mu_0}{4\pi} I \int \mathbf{n}\,da \times \nabla\left(\frac{1}{r}\right).$$

The quantity

(27) $$\mathbf{m} = I\int \mathbf{n}\,da = \frac{I}{2}\int_C \mathbf{r}_1 \times d\mathbf{r}_1$$

is by definition the *magnetic dipole moment* of the circuit. Since the surface integral may be extended over any regular surface spanning the contour, the magnetic moment depends only on the current and the form of the contour.

The dipole potential of the total distribution is obtained by summation over all the current filaments. By definition the magnetic dipole moment of a volume distribution of current with respect to the origin O is

(28) $$\mathbf{m} = \frac{1}{2}\int_V \mathbf{r}_1 \times \mathbf{J}\,dv.$$

The *magnetization* $\mathbf{M}$ of a region carrying current can likewise be defined as the *magnetic moment per unit volume.*

(29) $$\mathbf{M} = \frac{d\mathbf{m}}{dv} = \frac{1}{2}\mathbf{r}_1 \times \mathbf{J}.$$

It is clear that magnetization with respect to a given origin will occur whenever the current density has a component transverse to the vector

$\mathbf{r}_1$; that is to say, whenever the charge has a motion of rotation about O as a center. If ρ is the charge per unit volume and $\mathbf{v}$ its effective velocity, so that $\mathbf{J} = \rho\mathbf{v}$, the magnetization is

$$\mathbf{M} = \tfrac{1}{2}\mathbf{r}_1 \times \rho\mathbf{v}, \tag{30}$$

which is identical with angular momentum per unit volume when ρ is interpreted as the density of mass.

The remaining terms of the expansion (23) are the vector potentials of magnetic multipoles of higher order. Two distinct definitions of magnetic moment have been presented: the one in terms of Poisson's equivalent magnetic charge, wholly analogous to the electrostatic case, Sec. 4.3; the other in terms of current distribution. The identity of these two viewpoints must now be demonstrated.

4.6. The Magnetic Dipole.—According to Sec. 4.3, a magnetized piece of matter has a dipole moment defined by (15), page 229, and at sufficient distances its scalar potential is given by (18) of the same section. Our first task is to determine the equivalent vector potential; *i.e.*, a vector function whose curl leads to the same field as the gradient of the scalar ϕ^*. We shall again let (x', y', z') be the fixed point of observation and locate the dipole at (x, y, z). Then

$$\mathbf{H}(x', y', z') = -\nabla'\phi^* = \frac{1}{4\pi}\nabla'\left[\mathbf{m}\cdot\nabla'\left(\frac{1}{r}\right)\right], \tag{31}$$

and, consequently,

$$\nabla' \times \mathbf{A} = \frac{\mu}{4\pi}\nabla'\left[\mathbf{m}\cdot\nabla'\left(\frac{1}{r}\right)\right]. \tag{32}$$

To expand the right-hand side of (32), note that $\mathbf{m}$ is a constant vector or at most a function of x, y, z, whereas ∇' applies only to the coordinates x', y', z'. Furthermore

$$\nabla' \times \nabla'\left(\frac{1}{r}\right) = 0, \qquad \nabla'\cdot\nabla'\left(\frac{1}{r}\right) = 0. \tag{33}$$

It may be verified now without difficulty that

$$\nabla'\left[\mathbf{m}\cdot\nabla'\left(\frac{1}{r}\right)\right] = -\nabla'\times\left[\mathbf{m}\times\nabla'\left(\frac{1}{r}\right)\right]; \tag{34}$$

hence,

$$\nabla' \times \mathbf{A} = -\frac{\mu}{4\pi}\nabla'\times\left[\mathbf{m}\times\nabla'\left(\frac{1}{r}\right)\right], \tag{35}$$

$$\mathbf{A} + \frac{\mu}{4\pi}\mathbf{m}\times\nabla'\left(\frac{1}{r}\right) = \nabla' f(x', y', z'), \tag{36}$$

where f is an undetermined scalar function. The divergence of the left side vanishes, and consequently f may be any solution of Laplace's equation. However, f contributes nothing to the field **H** and we are at liberty to place it equal to zero. The vector potential of a magnetic dipole is, therefore,

$$\mathbf{A}(x', y', z') = -\frac{\mu}{4\pi}\mathbf{m} \times \nabla'\left(\frac{1}{r}\right) = \frac{\mu}{4\pi}\mathbf{m} \times \nabla\left(\frac{1}{r}\right). \tag{37}$$

Equation (37) is identical with the vector potential attributed to a current dipole moment in the preceding paragraph. *At a sufficient distance the field of any source—magnetic matter or stationary current—reduces to that of a magnetic dipole.* This amounts to no more than a restatement of the fact that a magnetized region is fully equivalent to a current distribution of density $\mathbf{J}' = \nabla \times \mathbf{M}$, and gives mathematical support to Ampère's interpretation of magnetism in terms of infinitesimal circulating currents.

4.7. Magnetic Shells.—We have shown thus far that the potential of any closed linear current can be expanded into a series of multipole potentials. By allowing the circuit contour to shrink to infinitesimal dimensions about some point (x, y, z), the potential can be represented everywhere but at the singularity (x, y, z) by the dipole term alone. The magnetic moment of the resultant *elementary linear current* is

$$d\mathbf{m} = I\mathbf{n}\, da, \tag{38}$$

where **n** is the positive normal to the infinitesimal plane area da. The elementary moment depends on the current and the area enclosed by the circuit, not upon its particular form.

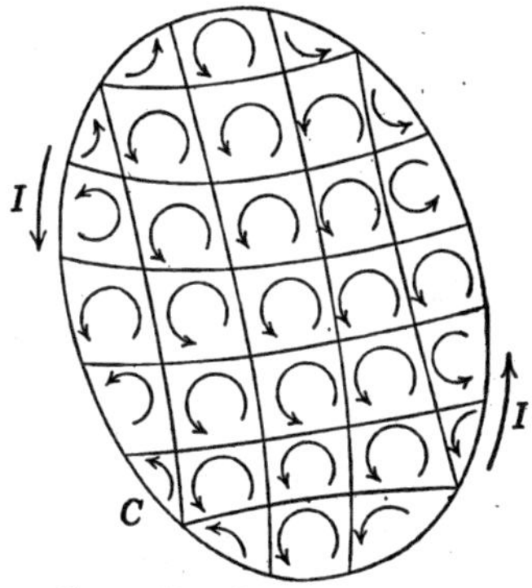

FIG. 42.—Resolution of the current I about the contour C into elementary currents.

Any linear current of arbitrary size and configuration can be resolved into a system of elementary currents. Let S, Fig. 42, be any regular surface spanning the circuit contour C and draw a network of intersecting lines terminating on C. Imagine now that about the contour of each elementary area there is a current I equal in magnitude to the current in C and circulating in the same sense. The magnetic field due to this network of currents is everywhere identical with that of the current I in the simple contour C; for the currents along the common boundaries of adjacent surface elements cancel one another except on the outer contour. As the fineness of the division is increased, the field of each mesh current approaches more closely that of a dipole whose axis is oriented in the direction of the positive normal. Clearly in the limit the field of the network, and consequently that of the current I about C, is identical with that of a dipole distribution, or *double layer*, over the surface S. The density, or moment per unit area, of the equivalent surface distribution is constant and equal to

$$\tau = \frac{d\mathbf{m}}{da} = I\mathbf{n}. \tag{39}$$

In effect every linear current acts then as a *magnetized shell*. The scalar potential of such a shell has been determined in Sec. 3.16 in connection with double layers of charge.

$$\phi^*(x', y', z') = -\frac{1}{4\pi}\int \tau\, d\Omega = -I\Omega, \tag{40}$$

where Ω is the solid angle from (x', y', z') subtended by S. Moreover the potential of a double layer has been shown to be multivalued. The discontinuity suffered by ϕ^* as one traverses S is by (17), page 190,

$$\phi_+^* - \phi_-^* = \tau = I; \tag{41}$$

consequently, the line integral of $\mathbf{H}$ about any closed path piercing S a single time is

$$\oint \mathbf{H} \cdot d\mathbf{s} = -\oint d\phi^* = I. \tag{42}$$

The surface S bearing the double layer or magnetic shell is completely arbitrary and evidently equivalent mathematically to the barrier or "cut" by means of which the field was reduced in Sec. 4.2 to a simply connected region.

A DIGRESSION ON UNITS AND DIMENSIONS

4.8. Fundamental Systems.—In Sec. 1.8 a dimensional analysis of electromagnetic quantities was based directly on Maxwell's equations and the adoption of the m.k.s. units was advocated on the grounds that they constitute a system which is both "absolute" and practical. Historically the development of the various systems followed another course. As long as electrostatic and magnetostatic phenomena were considered to be wholly independent of one another, it was natural that two independent, absolute systems should exist for the measurement of electric and magnetic quantities. Discovery of the law of induction by Faraday and proof much later that current is no more than charge in motion established a connection between the two groups of phenomena and imposed certain conditions on the choice of otherwise arbitrary constants.

The units and dimensions of electrostatic quantities have been commonly based on Coulomb's law: the force between two charges is proportional to the product of their magnitudes and inversely proportional to the square of the distance between them.

$$F = k_1 \frac{q_1 q_2}{r^2}. \tag{1}$$

Since the presence of matter does not affect the dimensions of charge, we assume henceforth that all forces are measured in free space. In "rationalized" units the proportionality constant k_1 is chosen equal to $1/4\pi\epsilon_0$; in "unrationalized" systems the factor 4π is dropped. The constant is arbitrary and not necessarily dimensionless, so that the dimensions of charge can be expressed only in terms of ϵ_0. From (1) we obtain

$$[q] = \epsilon_0^{\frac{1}{2}} M^{\frac{1}{2}} L^{\frac{3}{2}} T^{-1}; \tag{2}$$

from the equation of continuity it is evident that the dimensions of current are those of charge divided by time.

$$[I] = \epsilon_0^{\frac{1}{2}} M^{\frac{1}{2}} L^{\frac{3}{2}} T^{-2}. \tag{3}$$

From these it is a simple matter to deduce the dimensions of all other purely electrical quantities.

The properties of magnetic matter can be described more naturally in terms of magnetic moment $\mathbf{m}$ than by the introduction of fictitious "magnetic charges." The torque exerted on a magnet or current loop of moment $\mathbf{m}_2$ in the field $\mathbf{B}$ produced by a source of moment $\mathbf{m}_1$ is

$$\mathbf{T} = \mathbf{m}_2 \times \mathbf{B}, \tag{4}$$

and the vector $\mathbf{B}$ is expressed in terms of the source $\mathbf{m}_1$ by (31), page 236. From these two relations one may readily verify that the dimensions of magnetic moment are

$$[\mathbf{m}] = \mu_0^{-\frac{1}{2}} M^{\frac{1}{2}} L^{\frac{5}{2}} T^{-1}. \tag{5}$$

Equation (5) is a relation deduced directly from an expression for the mutual torques exerted by two magnetic dipoles, the proper magnetic counterpart of Coulomb's law. But this is not the only experimental law involving magnetic moment. It is a matter of observation that the moment $\mathbf{m}$ of a small loop or coil carrying a current I is proportional to the current and to the area of the loop. The constant of proportionality, which we shall call $1/\gamma$, is, like ϵ_0 and μ_0, arbitrary in size and dimensions so that $\mathbf{m}$ is subject also to the dimensional equation

$$[\mathbf{m}] = \gamma^{-1}[I]L^2 = \gamma^{-1}\epsilon_0^{\frac{1}{2}} M^{\frac{1}{2}} L^{\frac{7}{2}} T^{-2}. \tag{6}$$

The relation of magnetic moment to current and of current to electric charge thus imposes a condition upon the otherwise arbitrary constants

ϵ_0, μ_0 and γ. Whatever the choice of these three factors, their dimensions must be such that (5) and (6) are consistent. Upon equating (5) and (6), it follows that

$$\frac{\gamma}{\sqrt{\epsilon_0\mu_0}} = \frac{L}{T} = c. \tag{7}$$

The magnitude of the characteristic velocity c must be determined by experiment. In Chap. I it appeared as the velocity of propagation of fields and potentials in free space; now it occurs as a ratio of electric and magnetic units. To determine c without recourse to a direct measurement of velocity, a quantity is chosen that can be calculated in one system of units and measured in another. The capacity of a condenser, for example, can be calculated if its geometrical form is sufficiently simple, and the electrostatic capacity of any other condenser may be measured in terms of this primary standard by means of a capacity bridge. On the other hand a ballistic galvanometer measures the charge on the condenser in terms of the magnetic effects produced when the condenser is discharged through the galvanometer. The ratio of the two values of capacity can be shown to be

$$\frac{C_e}{C_m} = \frac{\gamma^2}{\epsilon_0\mu_0} = c^2, \tag{8}$$

where C_e is the calculated electrostatic value and C_m the measured value in magnetic units. The accepted value of c is approximately 3×10^8 meters/sec.

The various dimensional systems which at one time or another have been employed in the literature on electromagnetic theory are obtained by ascribing arbitrary values to any two of the three constants ϵ_0, μ_0 and γ_0; the third is then determined by (7). We shall review the most important of these systems briefly.

1. *The Electrostatic System.*—If $\epsilon_0 = 1$, $\gamma = 1$, all quantities, both electric and magnetic, are expressed in electrostatic units. For μ_0 one obtains from (7) the value

$$\mu_0 = \tfrac{1}{9} \times 10^{-20} \qquad (\text{sec./cm.})^2. \tag{9}$$

The c.g.s. units in this system are based on a definition of charge from Coulomb's law: two unit charges of like sign concentrated at points 1 cm. distant repel one another with a force of 1 dyne.

2. *The electromagnetic system* is obtained by putting

$$\mu_0 = 1, \qquad \gamma = 1, \qquad \epsilon_0 = \tfrac{1}{9} \times 10^{-20} \quad (\text{sec./cm.})^2. \tag{10}$$

The ratio of charge measured in the two systems is

$$\frac{q_e}{q_m} = c, \tag{11}$$

where q_e is the magnitude of a charge measured in electrostatic units, q_m the magnitude of the same charge measured in e.m.u.

3. *The m.k.s. system*, described in detail in Sec. 1.8, is essentially an electromagnetic system in which the units of mass and length have been adjusted such that the units of electric and magnetic quantities conform to practical standards. In the form adopted in the present volume, the coulomb is introduced as an independent unit, leading to

$$(12) \quad \mu_0 = 4\pi \times 10^{-7} \text{ henry/meter}, \qquad \gamma = 1, \qquad \epsilon_0 = 8.854 \times 10^{-12} \text{ farad/meter}.$$

An alternative form of the m.k.s. system puts

$$(13) \quad \mu_0 = 4\pi \times 10^{-7}, \qquad \gamma = 1, \qquad \epsilon_0 = 8.854 \times 10^{-12} \quad (\text{sec./meter})^2,$$

which is dimensionally identical with the electromagnetic system.

4. *The Gaussian system* has until recently been most commonly employed in scientific literature. It is a mixture of the electrostatic and electromagnetic systems obtained by putting

$$(14) \qquad \epsilon_0 = 1, \qquad \mu_0 = 1, \qquad \gamma = c.$$

The factors ϵ_0, μ_0 drop out of all equations and electrical quantities are expressed in electrostatic units, magnetic quantities in magnetic units. The advantages of all this are rather dubious. It is fully as difficult to recall the proper position of the factor c entering into electromagnetic equations written in Gaussian units as it is to retain the factors ϵ_0 and μ_0 from the outset. The m.k.s. system is quite as much an "absolute" or "scientific" system as the Gaussian, and there are few to whom the terms of the c.g.s. system occur more readily than volts, ohms, and amperes. All in all it would seem that the sooner the old favorite is discarded, the sooner will an end be made to a wholly unnecessary source of confusion.

Any equation appearing in this book can be transformed to Gaussian units by replacing ϵ_0 and μ_0 each by $1/c$; $\mathbf{J}$ and ρ by $\mathbf{J}/c$ and ρ/c respectively.

4.9. Coulomb's Law for Magnetic Matter.—In the older literature the magnetic system of units was commonly based on a law of force between magnetic "poles." The "charge" or pole strength q^* was defined such that the product of q^* and the length of a dipole was equal to its moment $\mathbf{m}$. The dimensions of q^* based on (5) are then

$$(15) \qquad [q^*] = \mu_0^{-\frac{1}{2}} M^{\frac{1}{2}} L^{\frac{3}{2}} T^{-1},$$

while for the law of force corresponding to Coulomb's law we obtain

$$(16) \qquad F = \frac{\mu_0}{4\pi} \frac{q_1^* q_2^*}{r^2}.$$

This expression differs from the one commonly assumed in that μ_0 *appears in the numerator rather than the denominator.*[1] At the root of this puzzling result is the fact that the force on an element of current, as deduced from Maxwell's equations, Sec. 2.4, is $\mathbf{J} \times \mathbf{B}$, *not* $\mathbf{J} \times \mathbf{H}$. Consequently the torque on a dipole is $\mathbf{m} \times \mathbf{B}$ and the force on a magnetic "charge" is $\mathbf{F} = q^*\mathbf{B}$.

To make (16) conform strictly to the electrostatic model it has been customary to write the factor μ in the denominator and to define the vector $\mathbf{H}$ by the equation $\mathbf{F} = q^*\mathbf{H}$. This procedure, however, leads to dimensional inconsistencies in the field equations unless μ_0 is assumed to be a numeric and there is little to be said in its favor. The parallel nature of the vectors $\mathbf{E}$ and $\mathbf{B}$ and the vectors $\mathbf{D}$ and $\mathbf{H}$ has been recognized by most recent writers.

MAGNETIC POLARIZATION

4.10. Equivalent Current Distributions.—The equivalence of a magnetized body and a distribution of current has been noted previously on various occasions. Let us see just how this distribution in a body of volume V bounded by a surface S is to be determined.

We shall suppose that the intensity of magnetization, or polarization, $\mathbf{M}$ includes also the permanent or residual magnetization $\mathbf{M}_0$ if any is present. At any point outside the body the vector potential is by (37), page 237, equal to

$$\mathbf{A}(x', y', z') = \frac{\mu_0}{4\pi}\int_V \mathbf{M}(x, y, z) \times \nabla\left(\frac{1}{r}\right) dv. \tag{1}$$

By virtue of the vector identity

$$\mathbf{M} \times \nabla\left(\frac{1}{r}\right) = \frac{\nabla \times \mathbf{M}}{r} - \nabla \times \left(\frac{\mathbf{M}}{r}\right), \tag{2}$$

and the fundamental formula for the transformation of the volume integral of the curl of a vector to the surface integral of its tangential component,

$$\int_V \nabla \times \left(\frac{\mathbf{M}}{r}\right) dv = \int \frac{\mathbf{n} \times \mathbf{M}}{r}\, da, \tag{3}$$

we see that the right-hand side of (1) may be resolved into two parts.

$$\mathbf{A}(x', y', z') = \frac{\mu_0}{4\pi}\int_V \frac{\nabla \times \mathbf{M}}{r}\, dv + \frac{\mu_0}{4\pi}\int_S \frac{\mathbf{M} \times \mathbf{n}}{r}\, da. \tag{4}$$

[1] SOMMERFELD, "Ueber die elektromagnetischen Einheiten," p. 161, Zeeman, Verhandelingen, Martinus Nijhoff, The Hague, 1935.

The vector potential due to a magnetized body is exactly the same as would be produced by volume and surface currents whose densities are

$$\mathbf{J}' = \nabla \times \mathbf{M}, \qquad \mathbf{K}' = \mathbf{M} \times \mathbf{n}. \tag{5}$$

The validity of this result at interior as well as exterior points is easily demonstrated. For the surface integral certainly converges at all points not on S, and the volume integral can be resolved into three scalar components, each of which is equivalent to the scalar potential of a charge distribution.

We recall that the magnetic polarization was introduced initially, Sec. 1.6, as the difference of the vectors $\mathbf{B}$ and $\mathbf{H}$.

$$\mathbf{M} = \frac{1}{\mu_0}\mathbf{B} - \mathbf{H}. \tag{6}$$

At a surface of discontinuity between two magnetic media the field vectors satisfy the boundary conditions of Sec. 1.13; namely,

$$\mathbf{n} \cdot (\mathbf{B}_2 - \mathbf{B}_1) = 0, \qquad \mathbf{n} \times (\mathbf{H}_2 - \mathbf{H}_1) = \mathbf{K}, \tag{7}$$

in which the surface current density is zero whenever the conductivity is finite. Then by (6) these conditions are equivalent to

$$\mathbf{n} \cdot (\mathbf{B}_2 - \mathbf{B}_1) = 0, \qquad \mathbf{n} \times (\mathbf{B}_2 - \mathbf{B}_1) = \mu_0(\mathbf{K} + \mathbf{K}'), \tag{8}$$

where

$$\mathbf{K}' = (\mathbf{M}_1 - \mathbf{M}_2) \times \mathbf{n}. \tag{9}$$

4.11. Field of Magnetized Rods and Spheres.—A steel rod or cylinder of circular cross section uniformly magnetized in the axial direction will serve as an example. The length of the cylinder is l, its radius a, and the magnetization is $\mathbf{M}_0$. Since $\mathbf{M}_0$ is assumed constant, the volume density of equivalent current $\mathbf{J}'$ is zero. On the cylindrical wall the current density is $\mathbf{K}' = \mathbf{M}_0 \times \mathbf{n}$, so that the magnetized rod is equivalent to a solenoidal coil of the same dimensions. The vector potential and field at any internal or external point can now be calculated from the series expansion derived in Sec. 4.5. The value of $\mathbf{B}$ along the axis has already been computed in Sec. 4.4. In particular, we find at the center

$$\mathbf{B} = \mu_0\mathbf{M}_0 \frac{l}{\sqrt{l^2 + 4a^2}}. \tag{10}$$

For an actual magnet the validity of our assumption of uniform magnetization is dubious, particularly in the vicinity of the edges. If, however, the length of the cylinder is either very large or very small with respect to the radius, the approximation is usually a good one. Thus, if $l \gg a$ so that the rod is needle-shaped, the field at the center reduces to

$$\mathbf{B} = \mu_0 \mathbf{M}_0. \tag{11}$$

In case $l \ll a$, it degenerates to a disk, with

$$\mathbf{B} = 0. \tag{12}$$

If the cylinder is soft iron instead of steel in an external field $\mathbf{H}_0$, the permanent polarization $\mathbf{M}_0$ must be replaced by the induced magnetization $\mathbf{M}$ and (10) is to be interpreted as the induced field $\mathbf{B}_1$. At the center of a long needle in a parallel field $\mathbf{H}_0$, the resultant fields are

$$\mathbf{B} = \mathbf{B}_0 + \mathbf{B}_1 = \mathbf{B}_0 + \mu_0 \mathbf{M}; \qquad \mathbf{H} = \frac{1}{\mu_0}\mathbf{B} - \mathbf{M} = \mathbf{H}_0; \tag{13}$$

the field $\mathbf{H}$ in the interior of the needle is the same as that which existed initially. In the case of the disk

$$\mathbf{B} = \mathbf{B}_0, \qquad \mathbf{H} = \mathbf{H}_0 - \mathbf{M}. \tag{14}$$

The assumption of a uniform magnetization of a soft-iron cylinder with squarely cut ends is obviously a direct violation of the boundary conditions on the vectors $\mathbf{B}$ and $\mathbf{H}$. On the other hand the induced magnetization of a sphere or ellipsoid in a parallel external field must be uniform, for the problem is wholly analogous to the electrostatic case considered in Secs. 3.24 and 3.27. The induced or secondary field $\mathbf{B}_1$ can now be attributed to a surface current circulating in zones concentric with the axis of magnetization, but the density will evidently vary with the latitude. If θ is the angle made by the outward normal with the direction of $\mathbf{M}$,

$$K' = |\mathbf{M} \times \mathbf{n}| = M \sin\theta. \tag{15}$$

By the Biot-Savart law the component of $\mathbf{B}_1^-$ in the direction of $\mathbf{M}$ is

$$dB_1^- = \frac{\mu_0 K}{2}\sin^2\theta\, d\theta = \frac{\mu_0 M}{2}\sin^3\theta\, d\theta. \tag{16}$$

When integrated this leads to

$$\mathbf{B}_1^- = \frac{\mu_0 \mathbf{M}}{2}\int_0^\pi \sin^3\theta\, d\theta = \frac{2}{3}\mu_0 \mathbf{M} \tag{17}$$

for the induced field at the center, while for the total field one obtains

$$\mathbf{B}^- = \mathbf{B}_0 + \mathbf{B}_1^- = \mu_0(\mathbf{H}_0 + \tfrac{2}{3}\mathbf{M}), \qquad \mathbf{H}^- = \mathbf{H}_0 - \tfrac{1}{3}\mathbf{M}. \tag{18}$$

An important corollary to this result is the value of the field at the center of a spherical cavity cut out of a *rigidly* and *uniformly* magnetized medium. The surface current distribution is the same but, since the normal is now directed into the sphere, the circulation is in the opposite direction. The induced field at the center is $-\frac{2}{3}\mu_0\mathbf{M}_0$, and the resultant field is

$$\text{(19)} \qquad \mathbf{B}^- = \mathbf{B}_0 - \tfrac{2}{3}\mu_0\mathbf{M}_0 = \mu_0(\mathbf{H}_0 + \tfrac{1}{3}\mathbf{M}_0), \qquad \mathbf{H}^- = \mathbf{H}_0 + \tfrac{1}{3}\mathbf{M}_0,$$

where $\mathbf{H}_0$ and $\mathbf{B}_0$ are the initial fields *within the medium.* Obviously these fields do not satisfy the customary boundary conditions on the normal and tangential components, and they are possible only when the cavity is excised *after* the rigid magnetization of the material.

DISCONTINUITIES OF THE VECTORS A AND B

4.12. Surface Distributions of Current.—It follows directly from the investigation of Sec. 3.14 that the vectors **A** and **B** are everywhere continuous functions of position provided the current is distributed in volume with a density which is bounded and piecewise continuous. When, however, the conductivity of a body is assumed infinite, or the magnetized state of the body is to be represented in terms of an equivalent current distribution, it becomes necessary to take account of surface currents across which the transition of the vector **B** is discontinuous. Let us imagine a current with a finite volume density **J** confined to a lamellar region of thickness t. Suppose now that the thickness is allowed to shrink, but that throughout the process the total current traversing its cross section is maintained at a constant value. As $t \to 0$, the *volume* density **J** necessarily approaches infinity. The *surface current density* is defined as the limit of a product:

$$\text{(1)} \qquad \mathbf{K} = \lim t\mathbf{J}, \qquad \text{as} \qquad t \to 0, \mathbf{J} \to \infty.$$

Because of the infinite value of **J**, this distribution constitutes a *surface singularity.*

We shall assume that **K** is a bounded, piecewise continuous function of position on a regular surface S. The vector potential at any point (x', y', z') not on S is

$$\text{(2)} \qquad \mathbf{A}(x', y', z') = \frac{\mu_0}{4\pi}\int_S \frac{\mathbf{K}(x, y, z)}{r}\, da,$$

in which r is, as usual, drawn from the element $\mathbf{K}(x, y, z)\, da$ to the fixed point (x', y', z'). Each rectangular component of **A** is an integral of the form representing the electrostatic potential of a surface distribution of charge. In Sec. 3.15 it was shown that such integrals are bounded, continuous functions of the coordinates (x', y', z') at all points in space, including those lying on S. *Both the normal and tangential components of the vector potential pass continuously through a single surface current layer.*

Turning our attention now to the vector **B**, we have according to Eq. (11), page 231, at a point not on S

$$\mathbf{B}(x', y', z') = \frac{\mu_0}{4\pi} \int_S \mathbf{K}(x, y, z) \times \nabla\left(\frac{1}{r}\right) da. \tag{3}$$

Upon resolving **K** into its rectangular components,

$$\mathbf{K} = \mathbf{i}K_x + \mathbf{j}K_y + \mathbf{k}K_z,$$

it appears that (3) is equivalent to

$$\mathbf{B}(x', y', z') = \frac{\mu_0}{4\pi}\left[\mathbf{i} \times \int_S K_x \nabla\left(\frac{1}{r}\right) da + \mathbf{j} \times \int_S K_y \nabla\left(\frac{1}{r}\right) da + \mathbf{k} \times \int_S K_z \nabla\left(\frac{1}{r}\right) da\right]. \tag{4}$$

Now an integral of the type

$$\mathbf{E}(x', y', z') = \frac{1}{4\pi} \int \omega(x, y, z) \nabla\left(\frac{1}{r}\right) da \tag{5}$$

may be interpreted, apart from a dimensional factor ϵ_0, as the field due to a charge of density ω spread over the surface S; and, as was shown in Sec. 3.15, it exhibits a discontinuity as the point (x', y', z') traverses the surface equal to

$$\mathbf{E}_+ - \mathbf{E}_- = \omega\mathbf{n}, \tag{6}$$

where ω is the local value of the surface density at the point of transition. Upon applying this result to (4) it follows immediately that

$$\mathbf{B}_+ - \mathbf{B}_- = \mu_0 \mathbf{K} \times \mathbf{n}. \tag{7}$$

The normal component of **B** *passes continuously through a surface layer of current.*

$$\mathbf{n} \cdot (\mathbf{B}_+ - \mathbf{B}_-) = 0. \tag{8}$$

The tangential components of **B** *experience a discontinuous jump defined by*

$$\mathbf{n} \times (\mathbf{B}_+ - \mathbf{B}_-) = \mu_0 \mathbf{n} \times (\mathbf{K} \times \mathbf{n}). \tag{9}$$

The right-hand side, by a well-known identity, transforms to

$$\mathbf{n} \times (\mathbf{K} \times \mathbf{n}) = (\mathbf{n} \cdot \mathbf{n})\mathbf{K} - (\mathbf{n} \cdot \mathbf{K})\mathbf{n}. \tag{10}$$

Since, however, **K** lies n the surface and is hence orthogonal to the unit normal **n**, (9) reduces to

$$\mathbf{n} \times (\mathbf{B}_+ - \mathbf{B}_-) = \mu_0 \mathbf{K}. \tag{11}$$

The significance of this analysis is made clearer by comparison with the boundary conditions deduced directly from Maxwell's equations in

Sec. 1.13. According to Sec. 4.10 a magnetized medium is equivalent to a volume distribution of current of density $\mathbf{J}' = \nabla \times \mathbf{M}$, but discontinuities in the magnetization must be accounted for by surface layers of current whose density is $\mathbf{K}' = \mathbf{n} \times (\mathbf{M}_+ - \mathbf{M}_-)$. The discontinuity in the field across such a surface is

$$\mathbf{n} \times (\mathbf{B}_+ - \mathbf{B}_-) = \mu_0[\mathbf{K} + \mathbf{n} \times (\mathbf{M}_+ - \mathbf{M}_-)], \tag{12}$$

or, since $\frac{1}{\mu_0}\mathbf{B} - \mathbf{M} = \mathbf{H}$,

$$\mathbf{n} \times (\mathbf{H}_+ - \mathbf{H}_-) = \mathbf{K}. \tag{13}$$

The continuity of the normal components of **B** across the boundary is ensured by (8).

4.13. Surface Distributions of Magnetic Moment.—The equivalence of a linear circuit and a surface bearing a magnetic dipole distribution was demonstrated in Sec. 4.7. We shall give some further attention to the discontinuities exhibited by the potential and field in the neighborhood of such singular surfaces without restriction to the case of normal magnetization. The vector M shall now represent the surface density of magnetization, the magnetic moment of the surface S per unit area. The direction of M with respect to the positive normal $\mathbf{n}$ is arbitrary.

At any point not on S the vector potential of the distribution is by (37), page 237, equal to

$$\mathbf{A}(x', y', z') = \frac{\mu_0}{4\pi}\int_S \mathsf{M}(x, y, z) \times \nabla\left(\frac{1}{r}\right) da. \tag{14}$$

We need but note the similarity of form which this expression bears to Eq. (3) to conclude that the vector potential experiences a discontinuity on passing through a surface bearing a distribution of magnetic moment, the amount of the discontinuity being

$$\mathbf{A}_+ - \mathbf{A}_- = \mu_0\mathsf{M} \times \mathbf{n}. \tag{15}$$

The normal component of **A** *is continuous,*

$$\mathbf{n} \cdot (\mathbf{A}_+ - \mathbf{A}_-) = 0, \tag{16}$$

but the tangential component is continuous only in case the magnetization is normal to the surface. Otherwise

$$\mathbf{n} \times (\mathbf{A}_+ - \mathbf{A}_-) = \mu_0\mathsf{M} - \mu_0(\mathbf{n} \cdot \mathsf{M})\mathbf{n}. \tag{17}$$

To investigate the discontinuity exhibited by the magnetic vector **B** as it passes through a surface layer of magnetic moment it will be advantageous to employ an expression of the form of Eq. (31), page 236, which we shall now write as

$$\mathbf{B}(x', y', z') = -\frac{\mu_0}{4\pi} \nabla' \int_S \mathsf{M}(x, y, z) \cdot \nabla \frac{1}{r}\, da. \tag{18}$$

The surface polarization M may be resolved into a normal and a tangential component by means of the identity

$$\mathsf{M} = (\mathbf{n} \cdot \mathsf{M})\mathbf{n} + (\mathbf{n} \times \mathsf{M}) \times \mathbf{n}, \tag{19}$$

so that in place of (18) one has

$$\begin{aligned} \mathbf{B}(x', y', z') = &-\frac{\mu_0}{4\pi} \nabla' \int_S (\mathbf{n} \cdot \mathsf{M})\mathbf{n} \cdot \nabla \frac{1}{r}\, da \\ &- \frac{\mu_0}{4\pi} \nabla' \int_S [(\mathbf{n} \times \mathsf{M}) \times \mathbf{n}] \cdot \nabla \frac{1}{r}\, da. \end{aligned} \tag{20}$$

On comparison with (16), page 189, it is clear that the first integral may be interpreted as the scalar potential due to a double layer of moment $\boldsymbol{\tau} = (\mathsf{M} \cdot \mathbf{n})\mathbf{n}$, while the negative gradient of this potential at the point (x', y', z') gives the field intensity. Now it was shown in Sec. 3.16 that the normal components of the field intensity approach the same limit on either side of such a distribution, but that the tangential components cross it discontinuously as expressed by Eq. (21), page 191, the amount of the jump in the present case being

$$B_{t+} - B_{t-} = -\mu_0 \frac{d}{dl} (\mathsf{M} \cdot \mathbf{n}). \tag{21}$$

Thus the component of **B** along any direction tangent to a surface bearing a distribution of magnetic moment experiences a discontinuity on traversing the surface which is proportional to the rate of change of the normal component of the polarization in that same direction.

Turning our attention next to the contribution of the second integral in (20), we reduce it to a more familiar form by certain transformations of the integrand. The identity

$$[(\mathbf{n} \times \mathsf{M}) \times \mathbf{n}] \cdot \nabla \frac{1}{r} = (\mathbf{n} \times \mathsf{M}) \cdot \left(\mathbf{n} \times \nabla \frac{1}{r}\right) \tag{22}$$

can easily be verified. Next let us resolve $(\mathbf{n} \times \mathsf{M})$ into scalar components, so that $\mathbf{n} \times \mathsf{M} = \mathbf{i}f_1 + \mathbf{j}f_2 + \mathbf{k}f_3$. We have then

$$\begin{aligned} [(\mathbf{n} \times \mathsf{M}) \times \mathbf{n}] \cdot \nabla \frac{1}{r} &= \mathbf{i} \cdot \left(\mathbf{n} \times f_1 \nabla \frac{1}{r}\right) + \mathbf{j} \cdot \left(\mathbf{n} \times f_2 \nabla \frac{1}{r}\right) \\ &+ \mathbf{k} \cdot \left(\mathbf{n} \times f_3 \nabla \frac{1}{r}\right) = \mathbf{i} \cdot \left(\mathbf{n} \times \nabla \frac{f_1}{r}\right) + \mathbf{j} \cdot \left(\mathbf{n} \times \nabla \frac{f_2}{r}\right) \\ &+ \mathbf{k} \cdot \left(\mathbf{n} \times \nabla \frac{f_3}{r}\right) - \mathbf{i} \cdot \left(\mathbf{n} \times \frac{\nabla f_1}{r}\right) - \mathbf{j} \cdot \left(\mathbf{n} \times \frac{\nabla f_2}{r}\right) \\ &- \mathbf{k} \cdot \left(\mathbf{n} \times \frac{\nabla f_3}{r}\right). \end{aligned} \tag{23}$$

Now if $f(x, y, z)$ be any function which is continuous and has continuous first and second derivatives over a regular surface S and on its contour C, then

$$\int_S \mathbf{n} \times \nabla f \, da = \int_C f \, d\mathbf{s}. \tag{24}$$

It follows, therefore, that if the tangential components of M satisfy the prescribed conditions over S, the three surface integrals whose integrands are of the type $\mathbf{i} \cdot \left(\mathbf{n} \times \nabla \frac{f_1}{r}\right)$ can be transformed to integrals along any closed contour on S, so that

$$\begin{aligned} \mathbf{i} \cdot \int_S \left(\mathbf{n} \times \nabla \frac{f_1}{r}\right) da + \mathbf{j} \cdot \int_S \left(\mathbf{n} \times \nabla \frac{f_2}{r}\right) da \\ + \mathbf{k} \cdot \int_S \left(\mathbf{n} \times \nabla \frac{f_3}{r}\right) da = \int_C \left(\mathbf{i}\frac{f_1}{r} + \mathbf{j}\frac{f_2}{r} + \mathbf{k}\frac{f_3}{r}\right) \cdot d\mathbf{s} \\ = \int_C \frac{(\mathbf{n} \times \mathsf{M})}{r} \cdot d\mathbf{s}. \end{aligned} \tag{25}$$

The remaining three terms of the integrand may be reduced by the relation

$$\begin{aligned} -\mathbf{i} \cdot \left(\mathbf{n} \times \frac{\nabla f_1}{r}\right) - \mathbf{j} \cdot \left(\mathbf{n} \times \frac{\nabla f_2}{r}\right) - \mathbf{k} \cdot \left(\mathbf{n} \times \frac{\nabla f_3}{r}\right) \\ = \frac{\mathbf{n}}{r} \cdot (\mathbf{i} \times \nabla f_1 + \mathbf{j} \times \nabla f_2 + \mathbf{k} \times \nabla f_3) = -\frac{\mathbf{n} \cdot \nabla \times (\mathbf{n} \times \mathsf{M})}{r}. \end{aligned} \tag{26}$$

The second term on the right-hand side of (20) has, therefore, been resolved into a contour integral plus a surface integral, or

$$\begin{aligned} -\frac{\mu_0}{4\pi} \nabla' \int_S [(\mathbf{n} \times \mathsf{M}) \times \mathbf{n}] \cdot \nabla \frac{1}{r} \, da = -\frac{\mu_0}{4\pi} \nabla' \int_C \frac{\mathbf{n} \times \mathsf{M}}{r} \cdot d\mathbf{s} \\ + \frac{\mu_0}{4\pi} \nabla' \int_S \frac{\mathbf{n} \cdot \nabla \times (\mathbf{n} \times \mathsf{M})}{r} \, da. \end{aligned} \tag{27}$$

Now the contour integral is continuous and continuously differentiable at points on S which are interior to the contour. On the other hand, the last term of (27) might be interpreted as the field intensity at a point (x', y', z') due to a surface charge distribution of density

$$-\mathbf{n} \cdot \nabla \times (\mathbf{n} \times \mathsf{M}).$$

Consequently, one must conclude that the tangential component of surface magnetization leads to a discontinuity in the normal components of $\mathbf{B}$ at the surface S of amount

$$B_{n+} - B_{n-} = -\mu_0 \mathbf{n} \cdot \nabla \times (\mathbf{n} \times \mathsf{M}). \tag{28}$$

Differentiation is restricted to directions tangent to the surface. Consequently the vector $\nabla(\mathsf{M}\cdot\mathbf{n})$ lies in the surface, while the vector $\nabla\times(\mathbf{n}\times\mathsf{M})$ is normal to it. *The discontinuities exhibited by the vector* **B** *in its transition through a surface distribution of magnetic moment can be expressed by the single formula:*

$$\mathbf{B}_+ - \mathbf{B}_- = -\mu_0[\nabla(\mathbf{n}\cdot\mathsf{M}) + \nabla\times(\mathbf{n}\times\mathsf{M})]. \tag{29}$$

INTEGRATION OF THE EQUATION $\nabla\times\nabla\times\mathbf{A} = \mu\mathbf{J}$

4.14. Vector Analogue of Green's Theorem.—The classical treatment of the vector potential is based on a resolution into rectangular components. On the assumption that $\nabla\cdot\mathbf{A} = 0$, each component can be shown to satisfy Poisson's equation and the methods developed for the analysis of the electrostatic potential are applicable.

The possibility of integrating the equation $\nabla\times\nabla\times\mathbf{A} = \mu\mathbf{J}$ directly by means of a set of vector identities wholly analogous to those of Green for scalar functions appears to have been overlooked. Let V be a closed region of space bounded by a regular surface S, and let **P** and **Q** be two vector functions of position which together with their first and second derivatives are continuous throughout V and on the surface S. Then, if the divergence theorem be applied to the vector $\mathbf{P}\times\nabla\times\mathbf{Q}$, we have

$$\int_V \nabla\cdot(\mathbf{P}\times\nabla\times\mathbf{Q})\,dv = \int_S (\mathbf{P}\times\nabla\times\mathbf{Q})\cdot\mathbf{n}\,da. \tag{1}$$

Upon expanding the integrand of the volume integral one obtains the vector analogue of Green's first identity, page 165,

$$\int_V (\nabla\times\mathbf{P}\cdot\nabla\times\mathbf{Q} - \mathbf{P}\cdot\nabla\times\nabla\times\mathbf{Q})\,dv = \int_S (\mathbf{P}\times\nabla\times\mathbf{Q})\cdot\mathbf{n}\,da. \tag{2}$$

The analogue of Green's second identity is obtained by an interchange of the roles of **P** and **Q** in (2) followed by subtraction from (2). As a result

$$\int_V (\mathbf{Q}\cdot\nabla\times\nabla\times\mathbf{P} - \mathbf{P}\cdot\nabla\times\nabla\times\mathbf{Q})\,dv = \int_S (\mathbf{P}\times\nabla\times\mathbf{Q} - \mathbf{Q}\times\nabla\times\mathbf{P})\cdot\mathbf{n}\,da. \tag{3}$$

4.15. Application to the Vector Potential.—We shall assume that the volume density of current $\mathbf{J}(x, y, z)$ is a bounded but otherwise arbitrary function of position. The regular surface S bounding a volume V need not necessarily contain within it the entire source distribution, or even any part of it. As in Sec. 3.4 we shall choose O as an arbitrary origin and $x = x'$, $y = y'$, $z = z'$ as a fixed point within V.

Now let $\mathbf{P}$ represent the vector potential $\mathbf{A}$ subject to the conditions

$$\nabla \times \nabla \times \mathbf{A} = \mu \mathbf{J}, \qquad \nabla \cdot \mathbf{A} = 0, \tag{4}$$

where μ is the permeability of a medium assumed homogeneous and isotropic. The choice of Q may be made in either of two ways. It will be recalled that in the scalar case the Green's function ψ satisfied Laplace's equation $\nabla^2 \psi = 0$ and could be interpreted as the potential at (x', y', z') due to a charge $4\pi\epsilon$ located at (x, y, z). Likewise the vectorial Green's function $\mathbf{Q}$ may be chosen to represent the vector potential at (x', y', z') produced by a current of density $4\pi/\mu$ located at (x, y, z) and directed arbitrarily along a line determined by the unit vector $\mathbf{a}$.

$$\mathbf{Q}(x, y, z; x', y', z') = \frac{\mathbf{a}}{r}, \qquad r = \sqrt{(x' - x)^2 + (y' - y)^2 + (z' - z)^2}. \tag{5}$$

However the divergence of this function is not zero and consequently (5) fails to satisfy the condition $\nabla \times \nabla \times \mathbf{Q} = 0$. On the other hand, the vector potential arising from a distribution of magnetic moment has been shown to satisfy $\nabla \times \nabla \times \mathbf{A} = 0$ everywhere and an appropriate Green's function for the present problem is, therefore,

$$\mathbf{Q} = \nabla\left(\frac{1}{r}\right) \times \mathbf{a} = \nabla \times \frac{\mathbf{a}}{r}. \tag{6}$$

Obviously (6) may be interpreted as the vector potential of a magnetic dipole of moment $\dfrac{4\pi}{\mu}\,\mathbf{a}$.

Either (5) or (6) may be applied to the integration of (4) but the necessary transformations turn out to be simpler in the case of (5) in spite of the divergence trouble. We have in fact

$$\nabla \times \mathbf{Q} = \nabla\left(\frac{1}{r}\right) \times \mathbf{a}, \qquad \nabla \times \nabla \times \mathbf{Q} = \nabla\left[\mathbf{a} \cdot \nabla\left(\frac{1}{r}\right)\right], \tag{7}$$

$$\mathbf{P} \cdot \nabla \times \nabla \times \mathbf{Q} = \mathbf{A} \cdot \nabla\left[\mathbf{a} \cdot \nabla\left(\frac{1}{r}\right)\right] = \nabla \cdot \left[\mathbf{a} \cdot \nabla\left(\frac{1}{r}\right) \mathbf{A}\right]. \tag{8}$$

In these transformations and those that follow it is to be kept in mind that $\mathbf{a}$ is a constant vector. The left-hand side of the identity (3) can now be written

$$\int_V \left\{\frac{\mathbf{a}}{r} \cdot \mathbf{J} - \nabla \cdot \left[\mathbf{a} \cdot \nabla\left(\frac{1}{r}\right) \mathbf{A}\right]\right\} dv \\ = \mathbf{a} \cdot \int_V \frac{\mathbf{J}}{r}\, dv - \mathbf{a} \cdot \int_S (\mathbf{A} \cdot \mathbf{n}) \nabla\left(\frac{1}{r}\right) da. \tag{9}$$

Proceeding to the transformation of the surface integrals, we have

$$(10)\quad (\mathbf{P} \times \nabla \times \mathbf{Q}) \cdot \mathbf{n} = \left\{\mathbf{A} \times \left[\nabla\left(\frac{1}{r}\right) \times \mathbf{a}\right]\right\} \cdot \mathbf{n} = \mathbf{a} \cdot \nabla\left(\frac{1}{r}\right) \times (\mathbf{A} \times \mathbf{n}),$$

$$(11)\quad (\mathbf{Q} \times \nabla \times \mathbf{P}) \cdot \mathbf{n} = \left(\frac{\mathbf{a}}{r} \times \nabla \times \mathbf{A}\right) \cdot \mathbf{n} = \mathbf{a} \cdot \frac{\mathbf{B} \times \mathbf{n}}{r},$$

in which $\mathbf{B}$ replaces $\nabla \times \mathbf{A}$. The identity (3) becomes

$$(12)\quad \mu \int_V \frac{\mathbf{J}}{r}\, dv = \int_S (\mathbf{A} \cdot \mathbf{n}) \nabla\left(\frac{1}{r}\right) da + \int_S \nabla\left(\frac{1}{r}\right) \times (\mathbf{A} \times \mathbf{n})\, da + \int_S \frac{\mathbf{n} \times \mathbf{B}}{r}\, da.$$

Now the validity of this relation has been established only for regions within which both $\mathbf{P}$ and $\mathbf{Q}$ are continuous and possess continuous first and second derivatives. $\mathbf{Q}$, however, has a singularity at $r = 0$ and consequently this point must be excluded. About the point (x', y', z') a small sphere of radius r_1 is circumscribed. The volume V is now bounded by the surface S_1 of the sphere and an outer enveloping surface S as indicated in Fig. 25, page 166. Since $\nabla(1/r) = \mathbf{r}^0/r^2$, the surface integrals over S_1 may be written

$$\frac{1}{r_1^2} \int_{S_1} \mathbf{r}^0 (\mathbf{A} \cdot \mathbf{n})\, da + \frac{1}{r_1^2} \int_{S_1} \mathbf{r}^0 \times (\mathbf{A} \times \mathbf{n})\, da + \frac{1}{r_1} \int_{S_1} \mathbf{n} \times \mathbf{B}\, da.$$

The integrand of the middle term is transformed to

$$(13)\quad \mathbf{r}^0 \times (\mathbf{A} \times \mathbf{n}) = (\mathbf{r}^0 \cdot \mathbf{n})\mathbf{A} - (\mathbf{A} \cdot \mathbf{n})\mathbf{r}^0 + \mathbf{A} \times (\mathbf{r}^0 \times \mathbf{n}),$$

and since on the sphere $\mathbf{r}^0 \cdot \mathbf{n} = 1$, $\mathbf{r}^0 \times \mathbf{n} = 0$, the surface integrals over S_1 reduce to

$$\frac{1}{r_1^2} \int_{S_1} \mathbf{A}\, da + \frac{1}{r_1} \int_{S_1} \mathbf{n} \times \mathbf{B}\, da.$$

If $\bar{\mathbf{A}}$ and $\overline{\mathbf{n} \times \mathbf{B}}$ denote the mean values of the vectors $\mathbf{A}$ and $\mathbf{n} \times \mathbf{B}$ over the surface of the sphere, these integrals have the value

$$\frac{\bar{\mathbf{A}}}{r_1^2}\, 4\pi r_1^2 + \frac{\overline{\mathbf{n} \times \mathbf{B}}}{r_1}\, 4\pi r_1^2,$$

which in the limit as $r_1 \to 0$ reduces to $4\pi \mathbf{A}(x', y', z')$. Upon introducing this result into (12) and transposing, we find the value of the vector potential at any fixed point expressed in terms of a volume integral and of surface integrals over an outer boundary which is again denoted by S.

$$\mathbf{A}(x', y', z') = \frac{\mu}{4\pi}\int_V \frac{\mathbf{J}(x, y, z)}{r}\,dv - \frac{1}{4\pi}\int_S \frac{\mathbf{n}\times\mathbf{B}}{r}\,da - \frac{1}{4\pi}\int_S (\mathbf{n}\times\mathbf{A})\times\nabla\left(\frac{1}{r}\right)da - \frac{1}{4\pi}\int_S (\mathbf{n}\cdot\mathbf{A})\nabla\left(\frac{1}{r}\right)da. \tag{14}$$

The proof that the divergence of (14) at the point (x', y', z') is zero is left to the reader. Clearly the surface integrals represent the contribution to the vector potential of all sources that are exterior to the surface S. At all points with n V, the vector $\mathbf{A}(x', y', z')$ defined by (14) is continuous and has continuous derivatives of all orders. Across the surface S, however, it is apparent from the form of the surface integrals that $\mathbf{A}$ and its derivatives will exhibit certain discontinuities. We shall show in fact that outside S the vector $\mathbf{A}$ is zero everywhere.

The first surface integral in (14) may be interpreted as the contribution to the vector potential of a surface current

$$\mathbf{K} = -\frac{1}{\mu}\mathbf{n}\times\mathbf{B}_-, \tag{15}$$

in which the subscript of $\mathbf{B}_-$ emphasizes that this value of $\mathbf{B}$ is taken just inside the surface S. Now in Sec. 4.12 it was shown that a surface layer of current does not affect the transition of the vector potential, but gives rise to a discontinuity in $\mathbf{B}$ of amount

$$\mathbf{n}\times(\mathbf{B}_+ - \mathbf{B}_-) = \mu\mathbf{K}. \tag{16}$$

Upon replacing $\mathbf{K}$ by its value from (15) it is clear that just outside S

$$\mathbf{n}\times\mathbf{B}_+ = 0 \tag{17}$$

In like manner the second surface integral is equivalent to the vector potential of a distribution of magnetic surface polarization of density

$$\mathsf{M} = \frac{1}{\mu}\mathbf{A}_-\times\mathbf{n}. \tag{18}$$

The normal component of $\mathbf{A}$ passes continuously through such a layer, but the tangential component is reduced discontinuously to zero, as we see on substituting (18) into (17), page 247

$$\mathbf{n}\times(\mathbf{A}_+ - \mathbf{A}_-) = -\mathbf{n}\times\mathbf{A}_- + [\mathbf{n}\cdot(\mathbf{n}\times\mathbf{A}_-)]\mathbf{n}, \tag{19}$$

whence

$$\mathbf{n}\times\mathbf{A}_+ = 0. \tag{20}$$

The mathematical significance of the last term of (14) is apparent, but it is difficult to imagine a physical distribution of current or magnetic moment of the type called for. This is the form of integral which we have associated with the field intensity of a surface charge of density $(\mathbf{A}_- \cdot \mathbf{n})$, and which leads to a discontinuity in the normal component specified by

$$\mathbf{n} \cdot (\mathbf{A}_+ - \mathbf{A}_-) = \mathbf{n} \cdot \mathbf{A}_-, \tag{21}$$

as a consequence of which we conclude that

$$\mathbf{n} \cdot \mathbf{A}_+ = 0. \tag{22}$$

Thus far we have demonstrated that on the positive side of the closed surface S the tangential and normal components of $\mathbf{A}$ and the tangential component of $\mathbf{B}$ are everywhere zero. It follows at once, however, that the normal component of $\mathbf{B}$ must also vanish over the positive side of S; for the normal component of the curl $\mathbf{A}$ involves only partial derivatives in directions tangential to the surface. Furthermore, we need but apply (14) itself to the region external to S to prove that $\mathbf{A}$, and consequently $\mathbf{B}$, must vanish everywhere. Current and magnetic polarization are absent outside V, since their effect is represented by the surface integrals. Then, since $\mathbf{n} \times \mathbf{B}_+$, $\mathbf{n} \times \mathbf{A}_+$ and $\mathbf{n} \cdot \mathbf{A}_+$ are all zero, it follows from (14) that $\mathbf{A}(x', y', z')$ must vanish at all points outside S.

When $\mathbf{Q} = \nabla(1/r) \times \mathbf{a}$ is chosen in place of (5) as a Green's function, it can be shown without great difficulty that

$$\mathbf{B}(x', y', z') = \frac{\mu}{4\pi} \int_V \mathbf{J} \times \nabla\left(\frac{1}{r}\right) dv - \frac{1}{4\pi} \int_S (\mathbf{n} \times \mathbf{B}) \times \nabla\left(\frac{1}{r}\right) da - \frac{1}{4\pi} \int_S (\mathbf{n} \cdot \mathbf{B}) \nabla\left(\frac{1}{r}\right) da. \tag{23}$$

This is the extension of the Biot-Savart law to a region of finite extent bounded by a surface S. The contribution of currents or magnetic matter outside S to the field within is accounted for by the two surface integrals.

BOUNDARY-VALUE PROBLEMS

4.16. Formulation of the Magnetostatic Problem.—A homogeneous, isotropic body is introduced into the constant field of a fixed and specified system of currents or permanent magnets. Our problem is to determine the resultant field both inside and outside the body. In case the current density at all points within the body is zero, the secondary field arising from the induced magnetization can be represented everywhere by a single-valued scalar potential ϕ_1^* and the methods developed for the treatment of electrostatic problems apply in full. A schedule for the solution may be drawn up as follows.

The scalar potential of the primary field is ϕ_0^*. In case the primary source is a current distribution ϕ_0^* is multivalued but this in no way affects the determination of the induced field ϕ_1^*. The resultant potential is $\phi^* = \phi_0^* + \phi_1^*$. The permeability of the body will be denoted by μ_1 and that of the homogeneous medium in which it is embedded by μ_2. Then a function ϕ_1^* must be constructed such that:

(1) $\nabla^2\phi_1^* = 0$, *at all points not on the boundary;*
(2) ϕ_1^* *is finite and continuous everywhere including the boundary;*
(3) *Across the boundary the normal derivatives of the resultant potential* ϕ^* *satisfy the condition*

$$\mu_2\left(\frac{\partial\phi^*}{\partial n}\right)_+ - \mu_1\left(\frac{\partial\phi^*}{\partial n}\right)_- = 0,$$

the subscripts + *and* − *implying that the derivative is calculated outside or inside the boundary surface respectively. The induced potential* ϕ_1^* *is, therefore, subject to the condition*

$$\mu_2\left(\frac{\partial\phi_1^*}{\partial n}\right)_+ - \mu_1\left(\frac{\partial\phi_1^*}{\partial n}\right)_- = (\mu_1 - \mu_2)\frac{\partial\phi_0^*}{\partial n} = f,$$

in which f *is a known function of position on the boundary satisfying the condition*

(4) $\int_S f\,da = 0$;
(5) *At infinity* ϕ_1^* *must vanish at least as* $1/r^2$, *so that* $r^2\phi_1^*$ *remains finite as* $r \to \infty$, *for there is no free magnetic charge and consequently* ϕ_1^* *must vanish as the potential of a dipole or multipole of higher order.*

In case the body carries a current, the interior field cannot be represented by a scalar potential and the boundary-value problem must be solved in terms of a vector potential. Such a case arises, for example, when an iron wire carrying a current is introduced into an external magnetic field. The distribution of the current in the stationary state is unaffected by the magnetic field. Its determination is in fact an electrostatic problem. The vector potential of the primary sources is $\mathbf{A}_0$ while the potential of the induced and permanent magnetization of the body and of the current which it may carry will be denoted by $\mathbf{A}_1$. This function $\mathbf{A}_1$ is subject to the following conditions:

(1) $\nabla \times \nabla \times \mathbf{A}_1 = \mu_1\mathbf{J}$, *at points inside the body where the current density is* $\mathbf{J}$;

(2) $\nabla \times \nabla \times \mathbf{A}_1 = 0$, *at all other points not on the boundary* S;
(3) $\nabla \cdot \mathbf{A}_1 = 0$, *at all points not on* S;
(4) $\mathbf{A}_1$ *is finite and continuous everywhere and passes continuously through the boundary surface (Secs. 4.10 and 4.12)*;
(5) *Across the boundary the normal derivatives of the potential* $\mathbf{A}_1$*—as well as those of the total potential* $\mathbf{A}$*—satisfy*

$$\left(\frac{\partial \mathbf{A}_1}{\partial n}\right)_+ - \left(\frac{\partial \mathbf{A}_1}{\partial n}\right)_- = \mu_0(\mathbf{M}_+ - \mathbf{M}_-) \times \mathbf{n},$$

in which $\mathbf{n}$ *is the outward normal and where* $\mathbf{M}_-$ *is the polarization of the body and* $\mathbf{M}_+$ *that of the medium just outside the boundary.*

Since $\mathbf{M}_+$ and $\mathbf{M}_-$ are determined at least in part by the field itself this relation is usually of no assistance in the determination of $\mathbf{A}_1$. In its place the customary boundary condition on the tangential components of the *total* field must be applied.

(6) $\mathbf{n} \times (\mathbf{H}_+ - \mathbf{H}_-) = \mathbf{n} \times \left(\mathbf{B}_+ - \frac{\mu_2}{\mu_1}\mathbf{B}_-\right) = 0$, *which imposes a relation between the derivatives of* $\mathbf{A}$ *in a specified coordinate system;*
(7) *As* $r \to \infty$ *the product* $r\mathbf{A}_1$ *remains finite.*

4.17. Uniqueness of Solution.—The proof that there is only one function ϕ_1^* satisfying the conditions scheduled above was presented in Sec. 3.20. A corresponding uniqueness theorem for the vector potential may be deduced from the identity (2) of page 250. Let us put $\mathbf{P} = \mathbf{Q} = \mathbf{A}$ and assume first that within V_1 bounded by S the current density is zero. Then $\nabla \times \nabla \times \mathbf{A} = 0$ and

$$\int_{V_1} (\nabla \times \mathbf{A})^2 \, dv = -\int_S \mathbf{A} \cdot (\mathbf{n} \times \nabla \times \mathbf{A}) \, da. \tag{1}$$

From the essentially positive character of the integrand on the left it follows that if $\mathbf{A}$ is zero over the surface S, then $\mathbf{B} = \nabla \times \mathbf{A}$ is zero everywhere within the volume V_1. Hence $\mathbf{A}$ is either constant or at most equal to the gradient of some scalar ψ. But since $\mathbf{A}$ is zero on S, the normal derivative $\partial\psi/\partial n$ is also zero over this surface and it was shown in Sec. 3.20 that this condition entails a constant value of ψ throughout V_1. Consequently, if $\mathbf{A}$ vanishes over a closed surface, it vanishes also at every point of the interior volume. It is also clear that the vector function $\mathbf{A}$ is uniquely determined in V_1 by its values on S. For if there existed two vectors $\mathbf{A}_1$ and $\mathbf{A}_2$ which assumed the specified values over the boundary, their difference must vanish not only over S but also throughout V_1.

The condition $\nabla \times \mathbf{A} = 0$, $\mathbf{A} = \nabla\psi$, throughout V_1 can be established also by the vanishing of $\nabla \times \mathbf{A}$, or of the tangential vector

$\mathbf{n} \times \nabla \times \mathbf{A}$ over S. In this case, however, it does not necessarily follow that $\mathbf{A}$ is everywhere zero. If the two functions $\nabla \times \mathbf{A}_1 = \mathbf{B}_1$ and $\nabla \times \mathbf{A}_2 = \mathbf{B}_2$ are identical on S, then $\mathbf{B}_1$ and $\mathbf{B}_2$ are identical at all interior points and $\mathbf{A}_1$ and $\mathbf{A}_2$ can differ at most by the gradient of a scalar function.

In case there are currents present within V_1 the vector potential $\mathbf{A}$ is resolved into a part $\mathbf{A}'$ due to these currents and a part $\mathbf{A}''$ due to external sources. The vector $\mathbf{A}'$ is uniquely determined by the current distribution, while the values of $\mathbf{A}''$ or its curl over S determine $\mathbf{A}''$ at all interior points.

The vector potential is regular at infinity and consequently the proof applies directly to the region V_2 exterior to S. *The vector* $\mathbf{B}$ *is uniquely determined within any domain by the values of its tangential component* $\mathbf{n} \times \mathbf{B}$ *over the boundary.*

PROBLEM OF THE ELLIPSOID

4.18. Field of a Uniformly Magnetized Ellipsoid.—An ellipsoid whose semiprincipal axes are a, b, c is uniformly and permanently magnetized. The direction of magnetization is arbitrary, but since the magnetization vector can be resolved into three components parallel to the principal axes we need consider only the case in which $\mathbf{M}_0$ is constant and parallel to the a-axis.

In view of the uniformity of magnetization $\rho^* = -\nabla \cdot \mathbf{M}_0 = 0$ at all points inside the ellipsoid. The potential ϕ^* of the magnet is due to a "surface charge" of density $\omega^* = \mathbf{n} \cdot \mathbf{M}_0$. The external medium is assumed in this case to be empty space. The problem is now fully equivalent to that of the polarized dielectric ellipsoid treated in Sec. 3.27. From Eqs. (27), (32), and (42) the potential ϕ_1 due to the polarization P_x may be found in terms of P_x and the parameters of the ellipsoid. Upon dropping the factor ϵ_0 and replacing P_x by M_{0x} one obtains

$$(1) \qquad \phi_-^* = \frac{abc}{2} A_1 M_{0x} x, \qquad A_1 = \int_0^\infty \frac{ds}{(s + a^2) R_s},$$

as the magnetic scalar potential at points inside the ellipsoid; at all external points

$$(2) \qquad \phi_+^* = \frac{abc}{2} M_{0x} x \int_\xi^\infty \frac{ds}{(s + a^2) R_s}.$$

The field inside the ellipsoid is

$$(3) \qquad H_x^- = -\frac{\partial \phi_-^*}{\partial x} = -\frac{abc}{2} A_1 M_{0x}.$$

Consequently if the magnetization is parallel to a principal axis, the field **H** inside the ellipsoid is constant, parallel to the same axis, *but opposed to the direction of magnetization.*

4.19. Magnetic Ellipsoid in a Parallel Field.—An ellipsoid of homogeneous, isotropic material is introduced into a fixed and uniform parallel field $\mathbf{H}_0$. The ellipsoid is free of residual magnetism and we shall suppose that for sufficiently weak fields its permeability μ_1 is constant, as is also the permeability μ_2 of the external medium.

The determination of the field can again be adopted directly from the electrostatic case, for upon reference to Sec. 4.16 it will be noted that the boundary conditions imposed upon ϕ^* are identical with those to be satisfied by ϕ when μ is substituted for ϵ. The resultant magnetic potential at an interior point of the ellipsoid is, therefore,

$$(4) \quad \phi_-^* = -\frac{H_{0x}x}{1+\frac{abc}{2\mu_2}(\mu_1-\mu_2)A_1} - \frac{H_{0y}y}{1+\frac{abc}{2\mu_2}(\mu_1-\mu_2)A_2} - \frac{H_{0z}z}{1+\frac{abc}{2\mu_2}(\mu_1-\mu_2)A_3};$$

outside the ellipsoid ϕ_+^* is obtained from (38), page 213, by replacing ϵ by μ and adding the contributions induced by H_{0y} and H_{0z}.

The torque exerted by the field on the ellipsoid may likewise be determined directly from (54), page 216, after a similar substitution. The importance of these results is fundamental. There are experimental methods of measuring small torques with great precision. The magnetic field near the center of a long solenoid is very nearly parallel and uniform, and its intensity in terms of the current can be calculated. Thus the permeability or susceptibility of the test sample can be determined with great accuracy.

CYLINDER IN A PARALLEL FIELD

4.20. Calculation of the Field.—As an example of the application of the vector potential, let us consider the following simple problem. A cylinder or wire of circular cross section and permeability μ_1 is embedded in a medium of permeability μ_2. The wire is infinitely long, has a radius a, and carries a current I. The external field $\mathbf{B}_0$ is directed transverse to the axis of the wire and is everywhere parallel and uniform. We wish to calculate first the resultant field at points inside and outside the cylinder.

The x-axis of the reference system will be chosen parallel to the vector $\mathbf{B}_0$, while the z-axis is made to coincide with the axis of the cylinder and the direction of the current, Fig. 43. A vector potential is to be found which satisfies the conditions

$$(1) \qquad \nabla \times \nabla \times \mathbf{A} = 0, \qquad (r > a),$$
$$(2) \qquad \nabla \times \nabla \times \mathbf{A} = \mu_1 \mathbf{J}, \qquad (r < a).$$

The current density has only a z-component.

$$(3) \qquad J_x = J_y = 0, \qquad J_z = \frac{I}{\pi a^2} \qquad (r < a).$$

The vector potential may be resolved into two parts: the potential $\mathbf{A}_0$ of the applied field, and the secondary potential $\mathbf{A}_1$ due in part to the

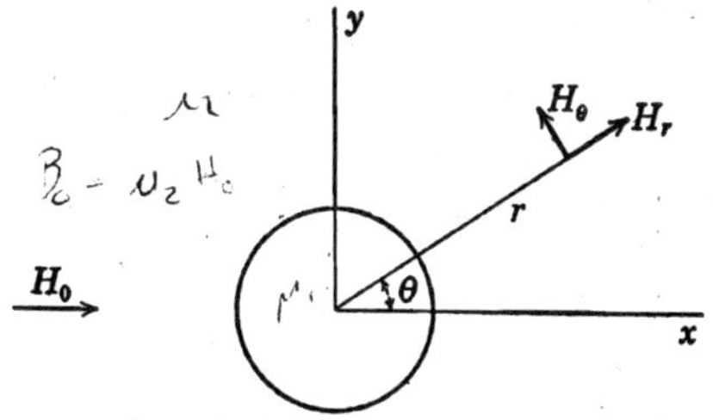

FIG. 43.—Cylinder in a uniform magnetostatic field.

current I, in part to the induced magnetization. Now clearly $\mathbf{B}_0$ can be derived from a vector potential directed along the z-axis. If $\mathbf{i}_3$ is the unit vector in the positive z-direction, then

$$(4) \qquad \mathbf{A}_0 = \mathbf{i}_3 B_0 y = \mathbf{i}_3 B_0 r \sin\theta.$$

Moreover $\mathbf{A}_1$ is also oriented along this same axis, for the magnetization can be represented by equivalent volume and surface currents. Certainly the induced magnetization lies in the transverse plane and consequently the equivalent currents $\mathbf{J}' = \nabla \times \mathbf{M}_1$ and

$$\mathbf{K}' = \mathbf{n} \times (\mathbf{M}_2 - \mathbf{M}_1)$$

of Sec. 4.10 are directed along the cylinder. Since the vector potential is always parallel to the current, $A_{1x} = A_{1y} = 0$. The expansion of (2) is now a simple matter. From (85), page 50, we have

$$(5) \qquad \frac{1}{r}\frac{\partial}{\partial r}\left(r\frac{\partial A}{\partial r}\right) + \frac{1}{r^2}\frac{\partial^2 A}{\partial \theta^2} = -\mu_1 J,$$

in which the subscript z has been dropped.

A general solution must first be found for the homogeneous equation obtained by putting the right-hand member of (5) equal to zero. The separation of such an equation was discussed in Sec. 3.21, and since the vector potential is a single-valued function of θ we write down at once

$$(6)\quad A' = \sum_{n=0}^{\infty} (a_n \cos n\theta + b_n \sin n\theta) r^n + \sum_{n=0}^{\infty} (c_n \cos n\theta + d_n \cos n\theta) r^{-n}.$$

Moreover the vector potential must be finite everywhere in space and consequently inside the cylinder the coefficients c_n and d_n are zero.

The general solution of (5) is obtained by the usual method of adding to (6) any particular solution of (5). Since $r^2 J$ depends only on r, we shall assume that this particular solution $\mathbf{A}''$ is independent of θ.

$$(7)\qquad \frac{d}{dr}\left(r \frac{d}{dr} A''\right) = -\mu_1 r J = -\frac{\mu_1 I}{\pi a^2} r,$$

$$(8)\qquad A'' = -\frac{\mu_1 I}{4\pi}\left(\frac{r}{a}\right)^2.$$

Consequently at all interior points of the cylinder the resultant vector potential is

$$(9)\qquad A^- = -\frac{\mu_1 I}{4\pi}\left(\frac{r}{a}\right)^2 + \sum_{n=0}^{\infty} (a_n \cos n\theta + b_n \sin n\theta) r^n.$$

Since the vector potential of the induced magnetization is regular at infinity, outside the wire we put $a_n = b_n = 0$ in (6). To this must be added the vector potential of the current I, which in Eq. (20), page 233, was shown to be $\mu_2 I/2\pi \ln 1/r$, and lastly the contribution of the external field itself from (4). The resultant vector potential at any point outside the cylinder is, therefore,

$$(10)\quad A^+ = B_0 r \sin\theta + \frac{\mu_2 I}{2\pi} \ln \frac{1}{r} + \sum_{n=0}^{\infty} (c_n \cos n\theta + d_n \sin n\theta) r^{-n}.$$

Next, the coefficients of the series expansions are determined from the condition that the vector potential and tangential components of $\mathbf{H}$ shall be continuous across the boundary $r = a$. From the expansion of the curl in cylindrical coordinates, page 51, we have

$$(11)\qquad B_r = \frac{1}{r}\frac{\partial A_z}{\partial \theta}, \qquad B_\theta = -\frac{\partial A_z}{\partial r},$$

so that the boundary conditions imposed on the vector potential are

$$(12)\qquad A^- = A^+, \qquad \frac{1}{\mu_2}\frac{\partial A^+}{\partial r} = \frac{1}{\mu_1}\frac{\partial A^-}{\partial r}, \qquad (r = a).$$

The result of equating coefficients of like terms from (9) and (10) is then

$$(13)\qquad \begin{aligned} a_0 &= 0, & c_0 &= (2\mu_2 \ln a - \mu_1)\frac{I}{4\pi}, \\ b_1 &= \frac{2\mu_1}{\mu_1 + \mu_2} B_0, & d_1 &= \frac{\mu_1 - \mu_2}{\mu_1 + \mu_2} a^2 B_0, \end{aligned}$$

and all other coefficients zero.

$$
(14)\qquad
\begin{aligned}
A^{+} &= \frac{\mu_2}{2\pi} I \ln \frac{a}{r} - \frac{\mu_1 I}{4\pi} + \left(r + \frac{\mu_1 - \mu_2}{\mu_1 + \mu_2}\frac{a^2}{r}\right) B_0 \sin\theta, && (r > a),\\
A^{-} &= -\frac{\mu_1 I}{4\pi}\left(\frac{r}{a}\right)^2 + \frac{2\mu_1}{\mu_1 + \mu_2} B_0 r \sin\theta, && (r < a).
\end{aligned}
$$

By means of (11) the field may be calculated.

$$
(15)\qquad
\begin{aligned}
H_r^{+} &= \left(1 + \frac{\mu_1 - \mu_2}{\mu_1 + \mu_2}\frac{a^2}{r^2}\right) H_0 \cos\theta,\\
H_\theta^{+} &= \frac{I}{2\pi r} - \left(1 - \frac{\mu_1 - \mu_2}{\mu_1 + \mu_2}\frac{a^2}{r^2}\right) H_0 \sin\theta,
\end{aligned}
\qquad (r > a),
$$

$$
(16)\qquad
\begin{aligned}
H_r^{-} &= \frac{2}{\mu_1 + \mu_2} B_0 \cos\theta,\\
H_\theta^{-} &= \frac{I}{2\pi a^2} r - \frac{2}{\mu_1 + \mu_2} B_0 \sin\theta.
\end{aligned}
\qquad (r < a),
$$

The nature of the field inside the cylinder is made somewhat clearer by a transformation to rectangular coordinates.

$$
(17)\qquad H_x = H_r \cos\theta - H_0 \sin\theta, \qquad H_y = H_r \sin\theta + H_\theta \cos\theta.
$$

$$
(18)\qquad H_x^{-} = \frac{2B_0}{\mu_1 + \mu_2} - \frac{I}{2\pi a^2} y, \qquad H_y^{-} = \frac{I}{2\pi a^2} x.
$$

The magnetization of the cylinder is given by the relation

$$
\mathbf{M} = \frac{\mu_1 - \mu_0}{\mu_0} \mathbf{H}^{-}.
$$

$$
(19)\qquad M_x = \frac{\mu_1 - \mu_0}{\mu_1 + \mu_2}\frac{2}{\mu_0} B_0 - \frac{\mu_1 - \mu_0}{\mu_0}\frac{I}{2\pi a^2} y, \qquad M_y = \frac{\mu_1 - \mu_0}{\mu_0}\frac{I}{2\pi a^2} x.
$$

Evidently the magnetization induced by the applied field $\mathbf{B}_0$ is in the direction of the positive x-axis. Upon this is superposed the magnetization induced by the current I.

4.21. Force Exerted on the Cylinder.—If the medium supporting the wire is fluid, or a nonmagnetic solid, the force per unit length exerted by the field can be computed from (15), page 155. Resolved into rectangular components, this becomes

$$
(20)\qquad
\begin{aligned}
F_x &= \int_0^{2\pi} \mu_2 \left(H_x H_r - \frac{1}{2} H^2 \cos\theta\right) r\, d\theta.\\
F_y &= \int_0^{2\pi} \mu_2 \left(H_y H_r - \frac{1}{2} H^2 \sin\theta\right) r\, d\theta.
\end{aligned}
$$

The rectangular components of the field outside the cylinder are first calculated from (15) and (17).

$$(21)\qquad \begin{aligned} H_x^+ &= H_0 + \frac{\mu_1 - \mu_2}{\mu_1 + \mu_2}\frac{a^2}{r^2} H_0 \cos 2\theta - \frac{I}{2\pi r}\sin\theta, \\ H_y^+ &= \frac{\mu_1 - \mu_2}{\mu_1 + \mu_2}\frac{a^2}{r^2} H_0 \sin 2\theta + \frac{I}{2\pi r}\cos\theta, \end{aligned}$$

while H^2 is equal to the sum of the squares of these two components.

$$(22)\qquad \begin{aligned} H^2 = H_0^2 &+ 2\frac{\mu_1 - \mu_2}{\mu_1 + \mu_2}\left(\frac{a}{r}\right)^2 H_0^2 \cos 2\theta + \left(\frac{\mu_1 - \mu_2}{\mu_1 + \mu_2}\right)^2\left(\frac{a}{r}\right)^4 H_0^2 \\ &+ \left[\frac{\mu_1 - \mu_2}{\mu_1 + \mu_2}\left(\frac{a}{r}\right)^2 - 1\right]\frac{IH_0}{\pi r}\sin\theta + \frac{I^2}{4\pi^2 r^2}. \end{aligned}$$

Upon introducing (21) and (22) into (20) and evaluating the integrals, one obtains the components of the total force exerted by the field upon the wire and the current it carries.

$$(23)\qquad F_x = 0, \qquad F_y = \mu_2 H_0 I.$$

That the induced magnetic moment of the wire contributes nothing to the resultant force follows from the uniformity of the applied field. If this field, on the other hand, were generated by the current in a neighboring conductor, the distribution of induced magnetization would no longer be symmetrical; a force would be exerted on the cylinder even in the absence of a current I.

Problems

1. An infinitely long, straight conductor is bounded externally by a circular cylinder of radius a and internally by a circular cylinder of radius b. The distance between centers is c, with $a > b + c$. The internal cylinder is hollow. The conductor carries a steady current I uniformly distributed over the cross section. Show that the field within the cylindrical hole is

$$H = \frac{cI}{2\pi(a^2 - b^2)}$$

and is directed transverse to the diameter joining the two centers.

2. Two straight, parallel wires of infinite length carry a direct current I in opposite directions. The conductivity σ of the wires is finite. The radius of each wire is a and the distance between centers is b.

a. Using a bipolar coordinate system find expressions for the electrostatic potential and the transverse and longitudinal components of electric field intensity at points inside and outside the conductors.

b. Find expressions for the corresponding components of magnetic field intensity.

c. Discuss the flow of energy in the field.

3. A circular loop of wire of radius a carrying a steady current I lies in the xy-plane with its center at the origin. A point in space is located by the cylindrical coordinates

r, ϕ, z, where $x = r\cos\phi$, $y = r\sin\phi$. Show that the vector potential at any point in the field is

$$A_\phi = \frac{a\mu I}{2\pi}\int_0^\pi \frac{\cos\alpha\, d\alpha}{(a^2 + r^2 + z^2 - 2ar\cos\alpha)^{\frac{1}{2}}}$$
$$= \frac{\mu}{\pi k} I \left(\frac{a}{r}\right)^{\frac{1}{2}} \left[\left(1 - \frac{1}{2}k^2\right)K - E\right],$$

where $k^2 = 4ar/[(a + r)^2 + z^2]$, and K and E are complete elliptic integrals of the first and second kinds.

Show that when $(r^2 + z^2)^{\frac{1}{2}} \gg a$, this reduces to Eq. (37), p. 237, for the vector potential of a magnetic dipole.

4. From the expression for the vector potential of a circular loop in Problem 3 show that the components of field intensity are

$$H_r = \frac{\mu I}{2\pi}\frac{z}{r[(a + r)^2 + z^2]^{\frac{1}{2}}}\left[-K + \frac{a^2 + r^2 + z^2}{(a - r)^2 + z^2}E\right],$$
$$H_z = \frac{\mu I}{2\pi}\frac{1}{[(a + r)^2 + z^2]^{\frac{1}{2}}}\left[K + \frac{a^2 - r^2 - z^2}{(a - r)^2 + z^2}E\right].$$

5. The field of two coaxial Helmholtz coils of radius a and separated by a distance a between centers is approximately uniform in a region near the axis and halfway between them. Assume that each coil has n turns, that the cross section of a coil is small relative to a, and that each coil carries the same current. From the results of Problem 4 write down expressions for the longitudinal and radial components of **H** at points on the axis and at any point in a plane normal to the axis and located midway between the coils. What is the field intensity at the mid-point on the axis? Find an expansion for the longitudinal component H_z in powers of z/a valid on the axis in the neighborhood of the mid-point, and corresponding expressions for H_r and H_z in powers of r/a valid over the transverse plane through the mid-point.

6. Two linear circuits C_1 and C_2 carry steady currents I_1 and I_2 respectively. Show that the magnetic energy of the system is

$$T_{12} = \frac{\mu}{4\pi} I_1 I_2 \oint_{C_1}\oint_{C_2} \frac{d\mathbf{s}_1 \cdot d\mathbf{s}_2}{r_{12}}$$

where $d\mathbf{s}_1$ and $d\mathbf{s}_2$ are vector elements of length along the contours and r_{12} the distance between these elements. The permeability μ is constant.

The *coefficient of mutual inductance* is defined by the relation

$$T_{12} = L_{12}I_1I_2, \qquad L_{12} = \frac{\mu}{4\pi}\oint_{C_1}\oint_{C_2}\frac{d\mathbf{s}_1 \cdot d\mathbf{s}_2}{r_{12}}.$$

7. Show that the mutual inductance of two circular, coaxial loops in a medium of constant permeability is

$$L_{12} = \mu\sqrt{ab}\left[\left(\frac{2}{k} - k\right)K - \frac{2}{k}E\right]$$

where a and b are the radii of the loops, c the distance between centers, K and E the complete elliptic integrals of the first and second kinds, and k is defined by

$$k^2 = \frac{4ab}{(a+b)^2 + c^2}.$$

When the separation c is very small compared to the radii and $a \simeq b$ show that this formula reduces to

$$L_{12} \simeq \mu a \left(\ln \frac{8a}{d} - 2 \right)$$

where $d = [(a-b)^2 + c^2]^{\frac{1}{2}}$.

8. The *coefficient of self-inductance* L_{11} of a circuit carrying a steady current I_1 is defined by the relation

$$T_1 = \tfrac{1}{2} L_{11} I_1^2$$

where T_1 is the magnetic energy of the circuit. Hence L_{11} can be calculated from either of the expressions

$$L_{11} = \frac{1}{I_1^2} \int \mathbf{J} \cdot \mathbf{A}\, dv = \frac{1}{I_1^2} \int \mu H^2\, dv$$

provided $\mathbf{B}$ is a linear function of $\mathbf{H}$. The volume integral in the first case is extended over the region occupied by current, in the second over the entire field. That portion of L_{11} associated with the energy of the field inside the conductor is called the *internal self-inductance* L'_{11}.

Show that the internal self-inductance of a long, straight conductor of constant permeability μ_1 is

$$L'_{11} = \frac{\mu_1}{8\pi} \qquad \text{henrys/meter.}$$

9. Show that the self-inductance of a circular loop of wire of radius R and cross-sectional radius r is

$$L = R\left[\mu_2 \left(\ln \frac{8R}{r} - 2 \right) + \frac{1}{4} \mu_1 \right]$$

where μ_1 is the permeability of the wire and μ_2 that of the external medium, both being assumed constant.

10. A toroidal coil is wound uniformly with a single layer of n turns on a surface generated by the revolution of a circle r meters in radius about an axis R meters from the center of the circle. Show that the self-inductance of the coil is

$$L = \mu n^2 (R - \sqrt{R^2 - r^2}) \qquad \text{henrys.}$$

11. A circular loop of wire is placed with its plane parallel to the plane face of a semi-infinite medium of constant permeability. Find the increase in the self-inductance of the loop due to the presence of the magnetic material.

Show that if the plane of the loop coincides with the surface of the magnetic material, the self-inductance is increased by a factor which is independent of the form of the loop.

12. Find an expression for the change in self-inductance of a circular loop of wire due to the presence of a second circular loop coaxial with the first.

13. An infinitely long, hollow cylinder of constant permeability μ_1 is placed in a fixed and uniform magnetic field whose direction is perpendicular to the generators of the cylinder. The potential of this initial field is $\phi_0^* = -H_0 r \cos\theta$, where r and θ are cylindrical coordinates in a transverse plane with the axis of the cylinder as an origin. The cross section of the cylinder is bounded by concentric circles of outer radius a and inner radius b. The permeability of the external and internal medium has the constant value μ_2. Show that the potential at any interior point is

$$\phi^* = -\left[1 - \frac{1 - \left(\frac{b}{a}\right)^2}{\left(\frac{\mu_1 + \mu_2}{\mu_1 - \mu_2}\right)^2 - \left(\frac{b}{a}\right)^2}\right] H_0 r \cos\theta, \qquad (r < b).$$

14. A hollow sphere of outer radius a and inner radius b and of constant permeabill ity μ_1 is placed in a fixed and uniform magnetic field $\mathbf{H}_0$. The external and internal medium has a constant permeability μ_2. Show that the internal field is uniform and is given by

$$\mathbf{H} = \left[1 - \frac{1 - \left(\frac{b}{a}\right)^3}{\frac{(\mu_1 + 2\mu_2)(2\mu_1 + \mu_2)}{2(\mu_1 - \mu_2)^2} - \left(\frac{b}{a}\right)^3}\right] \mathbf{H}_0, \qquad (r < b).$$

Discuss the relative effectiveness of a hollow sphere and a long hollow cylinder for magnetic shielding.

15. A magnetic field such as is observed on the earth's surface might be generated either by currents or magnetic matter within the earth, or by circulating currents above its surface. Actual measurement indicates that the field of external sources amounts to not more than a few per cent of the total field. Show how the contributions of external sources may be distinguished from those of internal sources by measurements of the horizontal and vertical components of magnetic field intensity on the earth's surface.

16. Assume that the earth's magnetic field is due to stationary currents or magnetic matter within the earth. The scalar potential at external points can then be represented as an expansion in spherical harmonics. Show that the coefficients of the first four harmonics can be determined by measurement of the components of magnetic field intensity at eight points on the surface.

17. A winding of fine wire is to be placed on a spheroidal surface so that the field of the coil will be identical with that resulting from a uniform magnetization of the spheroid in the direction of the major axis. How shall the winding be distributed?

18. A copper sphere of radius a carries a uniform charge distribution on its surface. The sphere is rotated about a diameter with constant angular velocity. Calculate the vector potential and magnetic field at points outside and inside the sphere.

19. A solid, uncharged, conducting sphere is rotated with constant angular velocity ω in a uniform magnetic field $\mathbf{B}$, the axis of rotation coinciding with the direction of the field. Find the volume and surface densities of charge and the electrostatic potential at points both inside and outside the sphere. Assume that the magnetic field of the rotating charges can be neglected.

20. The polarization $\mathbf{P}$ of a stationary, isotropic dielectric in an electrostatic field $\mathbf{E}$ is expressed by the formula

$$\mathbf{P} = (\epsilon - \epsilon_0)\mathbf{E}.$$

If each element of the dielectric is displaced with a velocity $\mathbf{v}$ in a magnetic field the polarization is

$$\mathbf{P} = (\epsilon - \epsilon_0)(\mathbf{E} + \mathbf{v} \times \mathbf{B}),$$

at least to terms of the first order in $1/c = \sqrt{\epsilon_0\mu_0}$.

A dielectric cylinder of radius a rotates about its axis with a constant angular velocity ω in a uniform magnetostatic field. The field is parallel to the axis of the cylinder. Find the polarization of the cylinder, the bound charge density appearing on the surface, and the electrostatic potential at points both inside and outside the cylinder.

21. A dielectric sphere of radius a in a uniform magnetostatic field is rotated with a constant angular velocity ω about an axis parallel to the direction of the field. Using the expression for polarization given in Problem 20, find the densities of bound volume and surface charge and the potential at points inside and outside the sphere.

22. Show that the force between two linear current elements is

$$d\mathbf{F} = \frac{\mu}{4\pi} I_1 I_2 \frac{d\mathbf{s}_2 \times (d\mathbf{s}_1 \times \mathbf{r}^0)}{r^2}$$
$$= \frac{\mu}{4\pi}\frac{I_1 I_2}{r^2}[(d\mathbf{s}_2 \cdot \mathbf{r}^0)\, d\mathbf{s}_1 - (d\mathbf{s}_1 \cdot d\mathbf{s}_2)\mathbf{r}^0],$$

where $\mathbf{r}^0$ is a unit vector directed along the line r joining $d\mathbf{s}_1$ and $d\mathbf{s}_2$ from $d\mathbf{s}_1$ toward $d\mathbf{s}_2$.

23. A long, straight wire carrying a steady current is embedded in a semi-infinite mass of soft iron of permeability μ at a distance d from the plane face. The wire is separated from the iron by an insulating layer of negligible thickness. Find expressions for the field at points inside and outside the iron.

24. A long, straight wire carrying a steady current is located in the air gap between two plane parallel walls of infinitely permeable iron. The wire is parallel to the walls and is assumed to be of infinitesimal cross section. Find the field intensity at any point in the air gap. Plot the force exerted on the conductor per unit length for a current of one ampere as a function of its distance from one wall.

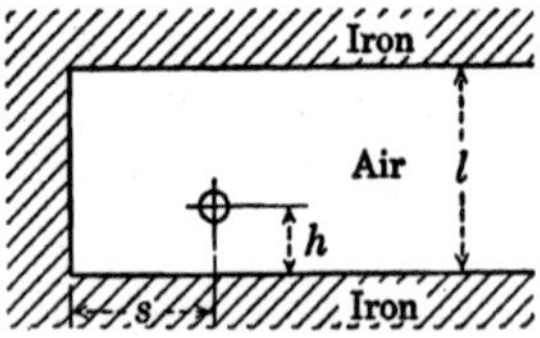

25. A long, straight conductor of infinitesimal cross section lies in an infinitely deep, parallel-sided slot in a mass of infinitely permeable iron. The thickness of the slot is l and the conductor is located by the parameters s and h as in the figure. Calculate the field intensity at points within the slot and the force on unit length of conductor per ampere. (*Cf.* Hague, "Electromagnetic Problems in Electrical Engineering," and Linder, *J. Am. Inst. Elec. Engrs.*, **46**, 614, 1927.)

26. An infinite elliptic cylinder of soft iron defined by the equation

$$b^2x^2 + a^2y^2 = a^2b^2$$

is placed in a fixed and uniform magnetic field whose direction is perpendicular to the generators of the cylinder. Find expressions for the magnetic scalar potential at internal and external points and calculate the force and torque exerted on the cylinder per unit length. Discuss the case of unit eccentricity in which the cylinder reduces to a thin slab.

27. A particle of mass m and charge q is projected into the field of a magnetic dipole.

a. Write the differential equations for the motion of the particle in a system of spherical coordinates whose origin coincides with the center of the dipole.

b. Show that the component of the angular momentum vector in the direction of the dipole axis is constant.

c. Discuss the trajectory in the case of a particle initially projected in the equatorial plane.

28. A small bar magnet is located near the plane face of a very large mass of soft iron whose permeability is μ. The magnet is located by its distance from the plane and the angle made by its axis with the perpendicular passing through its center. Considering only the dipole moment of the magnet, find the force and torque exerted on it by the induced magnetization in the iron.

29. Two small bar magnets whose dipole moments are respectively $\mathbf{m}_1$ and $\mathbf{m}_2$ are placed on a perfectly smooth table. The distance between their centers is large relative to the length of the magnets. At any instant the dipole axes make angles θ_1 and θ_2 with the line joining their centers.

a. Calculate the force exerted by $\mathbf{m}_1$ on $\mathbf{m}_2$ and the torque on $\mathbf{m}_2$ about a vertical axis through its center.

b. Calculate the force exerted by $\mathbf{m}_2$ on $\mathbf{m}_1$ and the torque on $\mathbf{m}_1$ about a vertical axis through its center.

c. Calculate the total angular momentum of the system about a fixed point in the plane. Is it constant?

30. Show that the force between two small bar magnets varies as the inverse fourth power of the distance between centers, whatever their orientation in space.

31. A magnetostatic field is produced by a distribution of magnetized matter. There are no currents at any point. Show that the integral

$$\frac{1}{2}\int \mathbf{B}\cdot\mathbf{H}\,dv = 0$$

if extended over the entire field.

32. If $\mathbf{J}(x, y, z)$ is the current density at any point in a region V bounded by a closed surface S, show that the magnetic field at any interior point x', y', z' is

$$\mathbf{B}(x', y', z') = \frac{\mu}{4\pi}\int_V \mathbf{J}\times\nabla\left(\frac{1}{r}\right)dv - \frac{1}{4\pi}\int_S\left[(\mathbf{n}\times\mathbf{B})\times\nabla\left(\frac{1}{r}\right) + (\mathbf{n}\cdot\mathbf{B})\nabla\left(\frac{1}{r}\right)\right]da,$$

where $r = \sqrt{(x'-x)^2 + (y'-y)^2 + (z'-z)^2}$, and that at exterior points the value of the surface integral is zero.

33. Determine the field of a magnetic quadrupole from the expansion of the vector potential given by Eq. (23), p. 234. Give a geometrical interpretation of the quadrupole moment in terms of infinitesimal linear currents.

CHAPTER V

PLANE WAVES IN UNBOUNDED, ISOTROPIC MEDIA

Every solution of Maxwell's equations which is finite, continuous, and single-valued at all points of a homogeneous, isotropic domain represents a possible electromagnetic field. Apart from the stationary fields investigated in the preceding chapters, the simplest solutions of the field equations are those that depend upon the time and a single space coordinate, and the factors characterizing the propagation of these elementary one-dimensional fields determine also in large part the propagation of the complex fields met with in practical problems. We shall study the properties of plane waves in unbounded, isotropic media without troubling ourselves for the moment as to the exact nature of the charge and current distribution that would be necessary to establish them.

PROPAGATION OF PLANE WAVES

5.1. Equations of a One-dimensional Field.—It will be assumed for the present that the medium is homogeneous as well as isotropic, and of unlimited extent. We shall suppose, furthermore, that the relations

$$\text{(1)} \qquad \mathbf{D} = \epsilon \mathbf{E}, \qquad \mathbf{B} = \mu \mathbf{H}, \qquad \mathbf{J} = \sigma \mathbf{E},$$

are linear so that the medium can be characterized electromagnetically by the three constants ϵ, μ, and σ. If the conductivity is other than zero any initial free-charge distribution in the medium must vanish spontaneously (Sec. 1.7). In the following, ρ will be put equal to zero in dielectrics as well as conductors. The Maxwell equations satisfied by the field vectors are then

$$\text{(I)} \quad \nabla \times \mathbf{E} + \mu \frac{\partial \mathbf{H}}{\partial t} = 0, \qquad \text{(III)} \quad \nabla \cdot \mathbf{H} = 0,$$

$$\text{(II)} \quad \nabla \times \mathbf{H} - \epsilon \frac{\partial \mathbf{E}}{\partial t} - \sigma \mathbf{E} = 0, \qquad \text{(IV)} \quad \nabla \cdot \mathbf{E} = 0.$$

We now look for solutions of this system which depend upon the time and upon distance measured along a single axis in space. This preferred direction need not coincide with a coordinate axis of the reference system. Let us suppose, therefore, that the field is a function of a coordinate ζ measured along a line whose direction is defined by the unit vector $\mathbf{n}$. The rectangular components n_x, n_y, n_z of this unit vector are obviously the direction cosines of the new coordinate axis ζ. Our assumption implies that at each instant the vectors **E** and **H** are constant in direction and magnitude over planes normal to **n**. These planes are defined by the equation

$$\mathbf{r} \cdot \mathbf{n} = \text{constant}, \tag{2}$$

where **r** is the radius vector drawn from the origin to any point in the plane as indicated by Fig. 44.

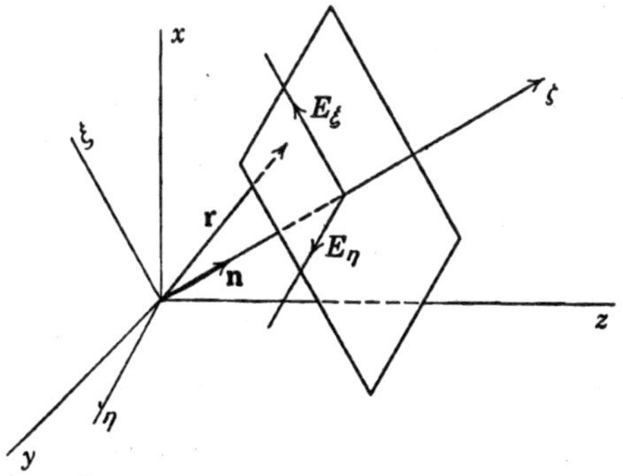

Fig. 44.—Homogeneous plane waves are propagated in a direction fixed by the unit vector **n**.

Since the fields are of the form

$$\mathbf{E} = \mathbf{E}(\zeta, t), \qquad \mathbf{H} = \mathbf{H}(\zeta, t), \tag{3}$$

the partial derivatives with respect to a set of rectangular coordinates may be expressed by

$$\frac{\partial}{\partial x} = n_x \frac{\partial}{\partial \zeta}, \qquad \frac{\partial}{\partial y} = n_y \frac{\partial}{\partial \zeta}, \qquad \frac{\partial}{\partial z} = n_z \frac{\partial}{\partial \zeta}. \tag{4}$$

From these we construct the operator ∇ and so obtain simplified expressions for the curl and divergence.

$$\nabla = \mathbf{i}\frac{\partial}{\partial x} + \mathbf{j}\frac{\partial}{\partial y} + \mathbf{k}\frac{\partial}{\partial z} = (\mathbf{i}n_x + \mathbf{j}n_y + \mathbf{k}n_z)\frac{\partial}{\partial \zeta} = \mathbf{n}\frac{\partial}{\partial \zeta}, \tag{5}$$

$$\nabla \times \mathbf{E} = \mathbf{n} \times \frac{\partial \mathbf{E}}{\partial \zeta} = \frac{\partial}{\partial \zeta}(\mathbf{n} \times \mathbf{E}), \tag{6}$$

$$\nabla \cdot \mathbf{E} = \mathbf{n} \cdot \frac{\partial \mathbf{E}}{\partial \zeta} = \frac{\partial}{\partial \zeta}(\mathbf{n} \cdot \mathbf{E}). \tag{7}$$

In virtue of these relations, the field equations assume the form

$$\text{(I)}\quad \mathbf{n} \times \frac{\partial \mathbf{E}}{\partial \zeta} + \mu \frac{\partial \mathbf{H}}{\partial t} = 0, \qquad \text{(III)}\quad \mathbf{n} \cdot \frac{\partial \mathbf{H}}{\partial \zeta} = 0,$$

$$\text{(II)}\quad \mathbf{n} \times \frac{\partial \mathbf{H}}{\partial \zeta} - \epsilon \frac{\partial \mathbf{E}}{\partial t} - \sigma \mathbf{E} = 0, \qquad \text{(IV)}\quad \mathbf{n} \cdot \frac{\partial \mathbf{E}}{\partial \zeta} = 0.$$

This system is now solved simultaneously by differentiating first (I) with respect to ζ and then multiplying vectorially by $\mathbf{n}$. (II) is next differentiated with respect to t. We have

$$\begin{aligned}\mathbf{n}\times\left(\mathbf{n}\times\frac{\partial^2\mathbf{E}}{\partial\zeta^2}\right)&=\mathbf{n}\left(\mathbf{n}\cdot\frac{\partial^2\mathbf{E}}{\partial\zeta^2}\right)-(\mathbf{n}\cdot\mathbf{n})\frac{\partial^2\mathbf{E}}{\partial\zeta^2}=-\frac{\partial^2\mathbf{E}}{\partial\zeta^2},\\ \mathbf{n}\times\frac{\partial^2\mathbf{H}}{\partial t\,\partial\zeta}&=\epsilon\frac{\partial^2\mathbf{E}}{\partial t^2}+\sigma\frac{\partial\mathbf{E}}{\partial t};\end{aligned}\tag{8}$$

upon elimination of the terms in $\mathbf{H}$, the vector $\mathbf{E}$ is found to satisfy the equation

$$\frac{\partial^2\mathbf{E}}{\partial\zeta^2}-\mu\epsilon\frac{\partial^2\mathbf{E}}{\partial t^2}-\mu\sigma\frac{\partial\mathbf{E}}{\partial t}=0.\tag{9}$$

A similar elimination of $\mathbf{E}$ leads to an identical equation for $\mathbf{H}$.

$$\frac{\partial^2\mathbf{H}}{\partial\zeta^2}-\mu\epsilon\frac{\partial^2\mathbf{H}}{\partial t^2}-\mu\sigma\frac{\partial\mathbf{H}}{\partial t}=0.\tag{10}$$

Since $\mathbf{n}\cdot\left(\mathbf{n}\times\frac{\partial\mathbf{E}}{\partial\zeta}\right)$ is identically zero, it follows from (I) that $\mathbf{n}\cdot\frac{\partial\mathbf{H}}{\partial t}=0$. Taken together with (III), this gives

$$\mathbf{n}\cdot\left(\frac{\partial\mathbf{H}}{\partial t}\,dt+\frac{\partial\mathbf{H}}{\partial\zeta}\,d\zeta\right)=\mathbf{n}\cdot d\mathbf{H}=0.\tag{11}$$

We are forced to conclude that a variation of the ζ-component of $\mathbf{H}$ with respect to either ζ or t is incompatible with the assumption of a field which is constant over planes normal to the ζ-axis at each instant. Equation (11) does admit the possibility of a static field in the ζ-direction but, since at present we are concerned only with variable fields, we shall put $\mathbf{H}_\zeta=0$.

Likewise, it follows from (II) that

$$\mathbf{n}\cdot\left(\epsilon\frac{\partial\mathbf{E}}{\partial t}+\sigma\mathbf{E}\right)=0;\tag{12}$$

this together with (IV) leads to

$$\mathbf{n}\cdot\left(\frac{\partial\mathbf{E}}{\partial t}\,dt+\frac{\sigma}{\epsilon}\,\mathbf{E}\,dt+\frac{\partial\mathbf{E}}{\partial\zeta}\,d\zeta\right)=\mathbf{n}\cdot\left(d\mathbf{E}+\frac{\sigma}{\epsilon}\,\mathbf{E}\,dt\right)=0.\tag{13}$$

The component of $\mathbf{E}$ normal to the family of planes, therefore, satisfies the condition

$$\frac{dE_\zeta}{dt}+\frac{\sigma}{\epsilon}\,E_\zeta=0,\tag{14}$$

which upon integration gives

$$E_\zeta=E_{0\zeta}e^{-\frac{t}{\tau}}\tag{15}$$

where $E_{0\zeta}$ is the longitudinal component of $\mathbf{E}$ at $t = 0$ and $\tau = \epsilon/\sigma$ is the relaxation time defined on page 15. If the conductivity is finite, the longitudinal component of $\mathbf{E}$ vanishes exponentially: a static electric field cannot be maintained in the interior of a conductor. According to (15) there may be a component E_ζ in a perfect dielectric medium, but this particular integral does not contain the time.

Equations (11) and (13) prove the *transversality* of the field. *The vectors* $\mathbf{E}$ *and* $\mathbf{H}$ *of every electromagnetic field subject to the condition* (3) *lie in planes normal to the axis of the coordinate* ζ.

Let us introduce a second system of rectangular coordinates ξ, η, ζ, whose origin coincides with that of the fixed system x, y, z and whose ζ-axis is oriented in the direction specified by $\mathbf{n}$. With respect to this new system the vector $\mathbf{E}$ has the components E_ξ, E_η, and $E_\zeta = 0$. Both E_ξ and E_η satisfy (9), which can be solved by separation of the variables. Let

$$E_\xi = f_1(\xi)f_2(t). \tag{16}$$

Then

$$\frac{1}{f_1}\frac{d^2f_1}{d\zeta^2} = \frac{\mu\epsilon}{f_2}\frac{d^2f_2}{dt^2} + \frac{\mu\sigma}{f_2}\frac{df_2}{dt} = -k^2, \tag{17}$$

where $-k^2$ is the separation constant. The general solution of the equation in f_1 is

$$f_1(\zeta) = Ae^{ik\zeta} + Be^{-ik\zeta}, \tag{18}$$

where A and B are complex constants; for f_2 we shall take the particular solution

$$f_2(t) = Ce^{-pt}. \tag{19}$$

Then p must satisfy the determinantal equation

$$p^2 - \frac{\sigma}{\epsilon}p + \frac{k^2}{\mu\epsilon} = 0. \tag{20}$$

There exists a fixed relation between p and the separation constant k^2; the value of either may be specified, whereupon the other is determined.

From (18) and (19) may be constructed a particular solution of the form

$$E_\xi = E_{1\xi}e^{ik\zeta - pt} + E_{2\xi}e^{-ik\zeta - pt}; \tag{21}$$

likewise, for the other rectangular component,

$$E_\eta = E_{1\eta}e^{ik\zeta - pt} + E_{2\eta}e^{-ik\zeta - pt}. \tag{22}$$

The four complex constants $E_{1\xi} \ldots E_{2\eta}$ are the components of two complex vector amplitudes lying in the $\xi\eta$ plane. Combining (21) and (22), one obtains

$$\mathbf{E} = \mathbf{E}_1 e^{ik\zeta - pt} + \mathbf{E}_2 e^{-ik\zeta - pt}. \tag{23}$$

This vector solution of (9) in turn is introduced into the field equations to obtain the associated magnetic vector **H**. Since **H** must have the

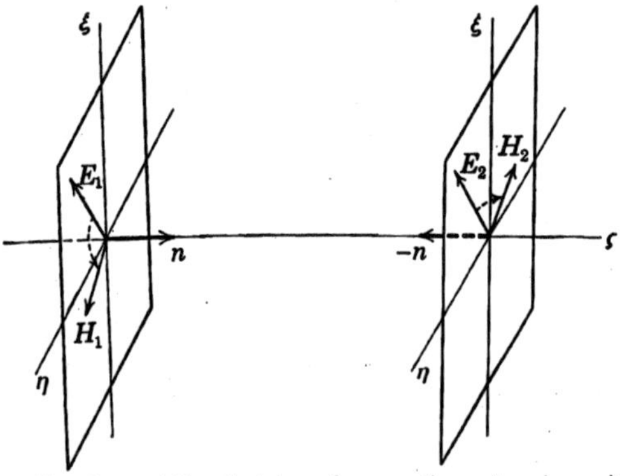

FIG. 45.—Relative directions of the electric and magnetic vectors in positive and negative waves.

same functional dependence on ζ and t, we can write

$$\mathbf{H} = \mathbf{H}_1 e^{ik\zeta - pt} + \mathbf{H}_2 e^{-ik\zeta - pt}, \tag{24}$$

and then determine the constants $\mathbf{H}_1$, $\mathbf{H}_2$ in terms of $\mathbf{E}_1$ and $\mathbf{E}_2$.

$$\frac{\partial \mathbf{E}}{\partial \zeta} = ik\mathbf{E}_1 e^{ik\zeta - pt} - ik\mathbf{E}_2 e^{-ik\zeta - pt}, \qquad \frac{\partial \mathbf{H}}{\partial t} = -p\mathbf{H}. \tag{25}$$

These derivatives are introduced into (I) and lead to

$$(ik\mathbf{n} \times \mathbf{E}_1 - p\mu\mathbf{H}_1)e^{ik\zeta - pt} - (ik\mathbf{n} \times \mathbf{E}_2 + p\mu\mathbf{H}_2)e^{-ik\zeta - pt} = 0. \tag{26}$$

The coefficients of the exponentials must vanish independently and, hence,

$$\mathbf{H}_1 = \frac{ik}{p\mu}\mathbf{n} \times \mathbf{E}_1, \qquad \mathbf{H}_2 = -\frac{ik}{p\mu}\mathbf{n} \times \mathbf{E}_2. \tag{27}$$

In terms of the rectangular components,

$$\begin{aligned} H_{1\xi} &= -\frac{ik}{p\mu}E_{1\eta}, & H_{1\eta} &= \frac{ik}{p\mu}E_{1\xi}, \\ H_{2\xi} &= \frac{ik}{p\mu}E_{2\eta}, & H_{2\eta} &= -\frac{ik}{p\mu}E_{2\xi}. \end{aligned} \tag{28}$$

Since $\mathbf{E} \cdot \mathbf{n} \times \mathbf{E} = 0$, it follows that $\mathbf{E} \cdot \mathbf{H} = 0$. *The electric and magnetic vectors of a field of the type defined by Eq.* (3) *are orthogonal to the direction* $\mathbf{n}$ *and to each other.* These mutual relations are illustrated in Fig. 45.

5.2. Plane Waves Harmonic in Time.—There are now two cases to be considered, dependent upon the choice of p and k, which it will be well to treat separately. Let us assume first that the field is harmonic in time, and that therefore p is a pure imaginary. The separation constant is then determined by (20).

$$p = i\omega, \qquad k^2 = \mu\epsilon\omega^2 + i\mu\sigma\omega. \tag{29}$$

In a conducting medium k^2, and hence k itself, is complex. *The sign of the root will be so chosen that the imaginary part of k is always positive.*

$$k = \alpha + i\beta. \tag{30}$$

The amplitudes $E_{1\xi} \ldots E_{2\eta}$ are also complex and will be written now in the form

$$\begin{aligned} E_{1\xi} &= a_1 e^{i\theta_1}, & E_{2\xi} &= a_2 e^{i\theta_2}, \\ E_{1\eta} &= b_1 e^{i\psi_1}, & E_{2\eta} &= b_2 e^{i\psi_2}, \end{aligned} \tag{31}$$

where the new constants $a_1 \ldots b_2$, $\theta_1 \ldots \psi_2$ are real. In virtue of these definitions we have for the ξ-component of the vector $\mathbf{E}$

$$E_\xi = a_1 e^{-\beta\zeta + i(\alpha\zeta - \omega t + \theta_1)} + a_2 e^{\beta\zeta - i(\alpha\zeta + \omega t - \theta_2)}. \tag{32}$$

Since (32) is the solution of a linear equation with real coefficients, both its real and imaginary parts must also be solutions. The real part of (32) is

$$E_\xi = a_1 e^{-\beta\zeta} \cos(\omega t - \alpha\zeta - \theta_1) + a_2 e^{\beta\zeta} \cos(\omega t + \alpha\zeta - \theta_2). \tag{33}$$

The phase angles θ_1 and θ_2 are arbitrary and consequently a choice of the imaginary part of (32) does not lead to a solution independent of (33).

In the same way the η-component of $\mathbf{E}$ is obtained from (22).

$$E_\eta = b_1 e^{-\beta\zeta} \cos(\omega t - \alpha\zeta - \psi_1) + b_2 e^{\beta\zeta} \cos(\omega t + \alpha\zeta - \psi_2). \tag{34}$$

The components of the associated magnetic field are found from (28). We now have

$$\frac{ik}{p\mu} = \frac{\alpha + i\beta}{\mu\omega} = \frac{\sqrt{\alpha^2 + \beta^2}}{\mu\omega} e^{i\gamma}, \qquad \gamma = \tan^{-1}\frac{\beta}{\alpha}. \tag{35}$$

The complex ξ-component of $\mathbf{H}$ is

$$H_\xi = H_{1\xi} e^{-\beta\zeta + i(\alpha\zeta - \omega t)} + H_{2\xi} e^{\beta\zeta - i(\alpha\zeta + \omega t)}; \tag{36}$$

upon substitution of the appropriate values of $H_{1\xi}$ and $H_{2\xi}$, we obtain for the real part

$$(37)\quad H_\xi = -b_1 \frac{\sqrt{\alpha^2+\beta^2}}{\mu\omega} e^{-\beta\zeta} \cos(\omega t - \alpha\zeta - \psi_1 - \gamma) + b_2 \frac{\sqrt{\alpha^2+\beta^2}}{\mu\omega} e^{\beta\zeta} \cos(\omega t + \alpha\zeta - \psi_2 - \gamma).$$

Similarly for the η-component,

$$(38)\quad H_\eta = a_1 \frac{\sqrt{\alpha^2+\beta^2}}{\mu\omega} e^{-\beta\zeta} \cos(\omega t - \alpha\zeta - \theta_1 - \gamma) - a_2 \frac{\sqrt{\alpha^2+\beta^2}}{\mu\omega} e^{\beta\zeta} \cos(\omega t + \alpha\zeta - \theta_2 - \gamma).$$

These solutions correspond obviously to plane waves propagated along the ζ-axis in both positive and negative directions. Consider first the most elementary case in which all amplitudes except a_1 are zero in a nonconducting medium. Then σ and consequently β are zero and the field is represented by the equations

$$(39)\quad E_\xi = a_1 \cos(\omega t - \alpha\zeta - \theta_1), \qquad H_\eta = \frac{\alpha}{\mu\omega} a_1 \cos(\omega t - \alpha\zeta - \theta_1).$$

This field is periodic in both space and time. The frequency is $\omega/2\pi = \nu$ and the period along the time axis is $2\pi/\omega = T$. The space period is called the wave length and is defined by the relation

$$(40)\qquad \alpha = \frac{2\pi}{\lambda}, \qquad \lambda = \frac{2\pi}{\alpha}.$$

The argument $\phi_1 = \omega t - \alpha\zeta - \theta_1$ of the periodic function is called the *phase* and the angle θ_1, which will be determined by initial conditions, is the *phase angle*. At each instant the vectors **E** and **H** are constant over the planes ζ = constant. Let us now choose a plane on which the phase has some given value at $t = 0$, and inquire how this plane must be displaced along the ζ-axis in order that its phase shall be invariant to a change in t. Since on any such plane the phase is constant, we have

$$(41)\quad \phi_1 = \omega t - \alpha\zeta - \theta_1 = \text{constant}, \qquad d\phi_1 = \omega\, dt - \alpha\, d\zeta = 0.$$

The surfaces of constant phase in a field satisfying (39) are, therefore, planes which displace themselves in the direction of the *positive* ζ-axis with a constant velocity

$$(42)\qquad v = \frac{d\zeta}{dt} = \frac{\omega}{\alpha}.$$

v is called the *phase velocity* of the wave. It represents simply the velocity of propagation of a phase or state and does not necessarily coincide with the velocity with which the energy of a wave or signal is propagated. In fact v may exceed the critical velocity c without violating in any way the relativity postulate.

In a nonconducting medium

$$(43)\qquad \alpha = \omega\sqrt{\mu\epsilon}, \qquad v = \frac{1}{\sqrt{\mu\epsilon}} = \frac{c}{\sqrt{\kappa_e\kappa_m}},$$

where c is the velocity of the wave in free space and κ_e, κ_m are the electric and magnetic specific inductive capacities. In optics the ratio $n = c/v = \alpha c/\omega$ is called the *index of refraction*. Since in all but ferromagnetic materials κ_m is very nearly unity, the index of refraction should be equal to the square root of κ_e. This result was first established by Maxwell and was the basis of his prediction that light is an electromagnetic phenomenon. Marked deviations from the expected values of n were observed by Maxwell, but these were later accounted for by the discovery that the inductive capacity does not necessarily maintain at high frequencies the value measured under static or quasi-static conditions. A functional dependence of n on the frequency results in a corresponding dependence of the phase velocity and leads to the phenomena known as dispersion.

At a given frequency the wave length is determined by the properties of the medium.

$$(44)\qquad \lambda = \frac{v}{\nu} = \frac{\lambda_0}{n},$$

where λ_0 is the wave length at the same frequency in free space. In all nonionized media $n > 1$, and consequently the phase velocity is decreased and the wave length is shortened.

Referring again to (39), we see that the relation of the magnetic to the electric vector is such that the cross product $\mathbf{E} \times \mathbf{H}$ is in the direction of propagation. The magnetic vector is propagated in the same direction with the same velocity, and in a nonconducting medium is exactly in phase with the electric vector. Their amplitudes differ by the factor

$$(45)\qquad \frac{\alpha}{\mu\omega} = \sqrt{\frac{\epsilon}{\mu}} = 2.654 \times 10^{-3}\sqrt{\frac{\kappa_e}{\kappa_m}}, \qquad (\sigma = 0).$$

If in Eqs. (33) to (38) all amplitudes except a_2 are put equal to zero, another particular solution is found which in nonconducting media reduces to

$$(46)\qquad E_\xi = a_2 \cos(\omega t + \alpha\zeta - \theta_2), \qquad H_\eta = a_2\sqrt{\frac{\epsilon}{\mu}} \cos(\omega t + \alpha\zeta - \theta_2).$$

This field differs from the preceding only in its direction of propagation. The phase is $\phi_2 = \omega t + \alpha\zeta - \theta_2$ and the surfaces of constant phase are propagated with the velocity $v = -\omega/\alpha$ in the direction of the *negative* ζ-axis.

The two particular solutions whose amplitudes are b_1 and b_2 represent a second pair of positive and negative waves, both characterized by electric vectors parallel to the η-axis.

Let us remove now the restriction to perfect dielectric media and examine the effect of a finite conductivity. It is apparent in the first place that both electric and magnetic vectors are attenuated exponentially in the direction of propagation. Waves traveling in the negative direction are multiplied by the factor $e^{+\beta\zeta}$, but since ζ decreases in the direction of propagation this too represents an attenuation. Not only does a conductivity of the medium lead to a damping of the wave, but it affects also its velocity The constants α and β can be calculated in terms of μ, ϵ, and σ by squaring (30) and equating real and imaginary parts respectively to the real and imaginary terms in (29).

$$\alpha^2 - \beta^2 = \mu\epsilon\omega^2, \qquad \alpha\beta = \frac{\mu\sigma\omega}{2}. \tag{47}$$

Upon solving these relations simultaneously, we obtain

$$\alpha = \omega\left[\frac{\mu\epsilon}{2}\left(\sqrt{1 + \frac{\sigma^2}{\epsilon^2\omega^2}} + 1\right)\right]^{1/2}, \tag{48}$$

$$\beta = \omega\left[\frac{\mu\epsilon}{2}\left(\sqrt{1 + \frac{\sigma^2}{\epsilon^2\omega^2}} - 1\right)\right]^{1/2}. \tag{49}$$

Ambiguities of sign that arise on extracting the first root are resolved by noting that α and β must be real.

The planes of constant phase are propagated with a velocity

$$v = c\left[\frac{\kappa_m\kappa_e}{2}\left(\sqrt{1 + \frac{\sigma^2}{\epsilon^2\omega^2}} + 1\right)\right]^{-1/2} \tag{50}$$

which *increases with frequency* so long as the constants κ_e, κ_m, and σ are independent of frequency. Attenuation of amplitudes in a surface of constant phase is determined by the *attenuation factor* β which also *increases with increasing frequency*. The complex factor k will be referred to as the *propagation constant*, and its real part α may be called the *phase constant*, although this term is applied also to the angles θ and ψ.

The effect of frequency and conductivity on the propagation of plane waves is most easily elucidated by consideration of two limiting cases. Examination of (48) and (49) shows that the behavior of the factors α and β is essentially determined by the quantity $\sigma^2/\epsilon^2\omega^2$. Now the total current density at any point in the medium is

$$\mathbf{J} = \sigma \mathbf{E} + \epsilon \frac{\partial \mathbf{E}}{\partial t} = (\sigma - i\omega\epsilon)\mathbf{E}, \tag{51}$$

whence it is apparent that $\sigma/\omega\epsilon$ is equal to the ratio of the densities of conduction current to displacement current.

Case I. $\dfrac{\sigma^2}{\epsilon^2\omega^2} \ll 1$. The displacement current is very much greater than the conduction current. This situation may arise either in a medium which is but slightly conducting, or in a relatively good conductor such as sea water through which is propagated a wave of very high frequency. Expansion of (48) and (49) in powers of $\sigma^2/\epsilon^2\omega^2$ gives

$$\alpha = \omega\sqrt{\mu\epsilon}\left(1 + \frac{1}{8}\frac{\sigma^2}{\epsilon^2\omega^2} + \cdots\right), \tag{52}$$

$$\beta = \frac{\sigma}{2}\sqrt{\frac{\mu}{\epsilon}} = 188.3\,\sigma\sqrt{\frac{\kappa_m}{\kappa_e}}. \tag{53}$$

One will remark that to this approximation the attenuation factor is independent of frequency. *In a given medium the attenuation approaches asymptotically a maximum defined by* (53) *as the frequency is increased.*

Case II. $\dfrac{\sigma^2}{\epsilon^2\omega^2} \gg 1$. The conduction current greatly predominates over the displacement current. This is invariably the case in metals, where σ is of the order of 10^7 mhos/meter. Not much is known about the inductive capacity of metals but there is no reason to believe that it assumes large values. Since the magnitude of ϵ is probably of the order of 10^{-11}, the displacement current could not possibly equal the conduction current at frequencies less than 10^{17}, lying in the domain of atomic phenomena to which the present obviously does not apply. For α and β we obtain the approximate formula

$$\alpha = \beta = \sqrt{\frac{\omega\mu\sigma}{2}} = 1.987 \times 10^{-3}\sqrt{\nu\sigma\kappa_m}. \tag{54}$$

An increase of frequency, permeability, or conductivity contributes in the same way to an increase in attenuation. The phase velocity also increases with frequency, but decreases with increasing σ or κ_m. Thus the higher order harmonics of a complex periodic wave are constantly advancing with respect to those of lower order.

The amplitudes of the electric and magnetic vectors of a plane wave are related by

$$|\mathbf{H}| = \frac{\sqrt{\alpha^2 + \beta^2}}{\mu\omega}|\mathbf{E}|. \tag{55}$$

$$\frac{\sqrt{\alpha^2 + \beta^2}}{\mu\omega} = \sqrt{\frac{\epsilon}{\mu}} \left(1 + \frac{\sigma^2}{\epsilon^2\omega^2}\right)^{1/4}. \tag{56}$$

In poorly conducting media expansion of (56) gives

$$\frac{\sqrt{\alpha^2 + \beta^2}}{\mu\omega} \simeq 2.654 \times 10^{-3} \sqrt{\frac{\kappa_e}{\kappa_m}} \left(1 + 0.807 \times 10^{20} \frac{\sigma^2}{\kappa_e^2\nu^2}\right), \tag{57}$$

whereas in good conductors,

$$\frac{\sqrt{\alpha^2 + \beta^2}}{\mu\omega} \simeq \sqrt{\frac{\sigma}{\mu\omega}} = 355.8 \sqrt{\frac{\sigma}{\nu\kappa_m}}, \qquad \left(\frac{\sigma^2}{\epsilon^2\omega^2} \gg 1\right). \tag{58}$$

In a perfect dielectric the electric and magnetic vectors oscillate in phase; if the medium is conducting, the magnetic vector lags by an angle γ.

$$\tan\gamma = \frac{\beta}{\alpha} = \left(\frac{\sqrt{1 + \frac{\sigma^2}{\epsilon^2\omega^2}} - 1}{\sqrt{1 + \frac{\sigma^2}{\epsilon^2\omega^2}} + 1}\right)^{1/2}. \tag{59}$$

If $\frac{\sigma^2}{\epsilon^2\omega^2} \gg 1$, this ratio reduces to unity; hence the magnetic vector of a plane wave penetrating a metal lags behind the electric vector by 45 deg.

5.3. Plane Waves Harmonic in Space.—The assumption that p in Eq. (20) is a pure imaginary leads to complex values of k and to electromagnetic fields which are simple harmonic functions of time. A field which at any given point on the ζ-axis is a known periodic function of t can be resolved by a Fourier analysis into harmonic components propagated along the ζ-axis as just described. The time variation at any other fixed point can then be found by recombining the components at the point in question. In place of the variation of the field with t at a certain point ζ, one may be given the distribution with respect to ζ at a specified value of t and asked to determine the field at any later instant. A harmonic analysis must then be made with respect to the variable ζ.

Let the separation constant k^2 be real. Then p is a complex quantity determined by

$$p = \frac{\sigma}{2\epsilon} \pm i \sqrt{\frac{k^2}{\mu\epsilon} - \frac{\sigma^2}{4\epsilon^2}}. \tag{60}$$

If for the moment we write $q = \sqrt{\frac{k^2}{\mu\epsilon} - \frac{\sigma^2}{4\epsilon^2}}$, the electric vector defined in (23) assumes the form

$$\mathbf{E} = \mathbf{E}_1 e^{-\frac{\sigma}{2\epsilon}t + i(k\zeta - qt)} + \mathbf{E}_2 e^{-\frac{\sigma}{2\epsilon}t - i(k\zeta + qt)}. \tag{61}$$

If now $\sigma^2/4\epsilon^2 < k^2/\mu\epsilon$, the quantity iq is a pure imaginary and the field may be interpreted as a plane wave propagated along the ζ-axis with a phase velocity

$$v = \frac{d\zeta}{dt} = \pm\frac{q}{k}. \tag{62}$$

The amplitude of oscillation at any point ζ decreases exponentially with t, the rate of decrease being determined essentially by the relaxation time $\tau = \epsilon/\sigma$.

When $\sigma^2/4\epsilon^2 > k^2/\mu\epsilon$, the quantity iq is real and there is no propagation in the sense considered heretofore. The field is periodic in ζ but decreases monotonically with the time. There is no displacement along the space axis of an initial wave form: the wave phenomenon has degenerated into *diffusion*.

5.4. Polarization.—Inasmuch as the properties of the positive and negative waves differ only in the direction of propagation, we shall confine our attention at present to the positive wave alone. Furthermore, since the attenuating effect of a finite conductivity in an isotropic, homogeneous medium enters as an exponential factor common to all field components, it plays no part in the polarization and will be neglected. Let us suppose then that the amplitudes and phase constants of the rectangular components of **E** have been specified and investigate the locus of $|\mathbf{E}| = \sqrt{E_\xi^2 + E_\eta^2}$ in the plane ζ = constant.

To determine this locus one must eliminate from the equations

$$E_\xi = a\cos(\phi + \theta), \qquad E_\eta = b\cos(\phi + \psi), \qquad E_\zeta = 0, \tag{63}$$

the variable component $\phi = \alpha\zeta - \omega t$ of the phase. To this end (63) is written in the form

$$\frac{E_\xi}{a} = \cos(\phi + \theta), \qquad \frac{E_\eta}{b} = \cos(\phi + \theta)\cos\delta - \sin(\phi + \theta)\sin\delta, \tag{64}$$

where $\delta = \psi - \theta$. Upon squaring, the periodic factor $\cos(\phi + \theta)$ can be eliminated and it is found that the rectangular components must satisfy the relation

$$\left(\frac{E_\xi}{a}\right)^2 - 2\frac{E_\xi}{a}\frac{E_\eta}{b}\cos\delta + \left(\frac{E_\eta}{b}\right)^2 = \sin^2\delta. \tag{65}$$

The discriminant of this quadratic form is negative,

$$\frac{4\cos^2\delta}{a^2b^2} - \frac{4}{a^2b^2} = -\frac{4}{a^2b^2}\sin^2\delta \leq 0, \tag{66}$$

and the locus of the vector whose components are E_ξ, E_η is, therefore, an ellipse in the $\xi\eta$-plane.

In this case the wave is said to be *elliptically polarized*. The points of contact made by the ellipse with the circumscribed rectangle are found from (65) to be $(\pm a, \pm b \cos \delta)$ and $(\pm a \cos \delta, \pm b)$. In general the principal axes of the ellipse fail to coincide with the coordinate axes ξ, η, but the two systems may be brought into coincidence by a rotation of the coordinate system about the ζ-axis through an angle ϑ defined by

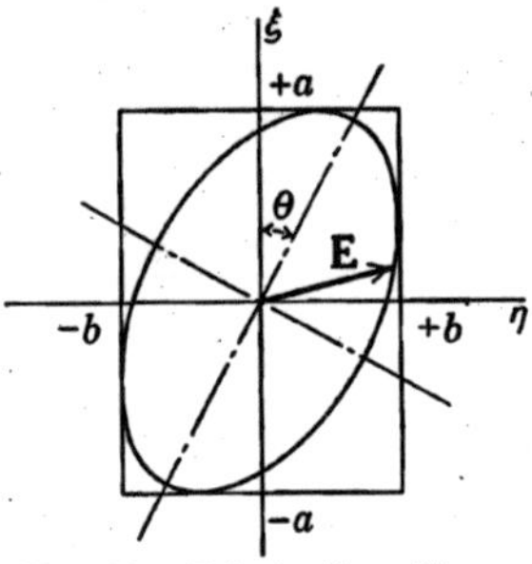

FIG. 46.—Polarization ellipse.

$$(67) \qquad \tan 2\vartheta = \frac{2ab}{a^2 - b^2} \cos \delta.$$

When the amplitudes of the rectangular components are equal and their phases differ by some odd integral multiple of $\pi/2$, the polarization ellipse degenerates into a circle.

$$(68) \qquad E_\xi^2 + E_\eta^2 = a^2, \left(a = b, \qquad \delta = \frac{m\pi}{2}, \qquad m = \pm 1, \pm 3, \cdots \right).$$

The wave is now said to be *circularly polarized*. Two cases are recognized according to the positive or negative rotation of the electric vector about the ζ-axis. It is customary to describe as right-handed circular polarization a clockwise rotation of **E** when viewed in a direction opposite to that of propagation, looking *towards* the source.

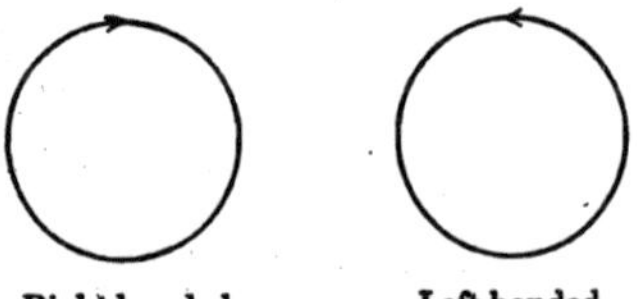

FIG. 47.—Circular polarization, the circles indicating the locus of **E**. The direction of propagation is normal to the page and toward the observer.

By far the most important of the special cases is that in which the polarization ellipse degenerates into a straight line. This occurs when $\delta = \pm m\pi$, where m is any integer. The locus of **E** in the $\xi\eta$-plane then reduces to a straight line making an angle ϑ with the ξ-axis defined by

$$(69) \qquad \tan \vartheta = \frac{E_\eta}{E_\xi} = (-1)^m \frac{b}{a}.$$

The wave is *linearly polarized*. The magnetic vector of a plane wave is at right angles to the electric vector and **H** oscillates, therefore, parallel to a line whose slope is $(-1)^m a/b$. It is customary to define the polarization in terms of **E** and to denote the line (69) as the axis of linear polarization. In optics, however, the orientation of the vectors is specified traditionally by the "plane of polarization," by which is meant the plane normal to **E** containing both **H** and the axis of propagation.

5.5. Energy Flow.—The rate at which energy traverses a surface in an electromagnetic field is measured by the Poynting vector, $\mathbf{S} = \mathbf{E} \times \mathbf{H}$ defined in Sec. 2.19. Since by (27) $\mathbf{H} = \pm \frac{ik}{p\mu} \mathbf{n} \times \mathbf{E}$, the sign depending upon the direction of propagation, it is apparent that the energy flow is normal to the planes of constant phase and in the direction of propagation.

To calculate the instantaneous value of the flow one must operate with the real parts of the complex wave functions. The mean value of the flow can be quickly determined, on the other hand, by constructing the complex Poynting vector $\mathbf{S}^* = \frac{1}{2}\mathbf{E} \times \tilde{\mathbf{H}}$ as in Sec. 2.20. For simplicity let us consider a plane wave linearly polarized along the x-axis and propagated in the direction of positive z. Then

$$(70) \qquad E_x = ae^{-\beta z + i(\alpha z - \omega t + \theta)}, \qquad \tilde{H}_y = a\,\frac{\sqrt{\alpha^2 + \beta^2}}{\mu\omega}\, e^{-\beta z - i(\alpha z - \omega t + \theta + \gamma)}.$$

The complex flow vector is, therefore,

$$(71) \qquad S_z^* = \frac{1}{2} E_x \tilde{H}_y = a^2 \frac{\sqrt{\alpha^2 + \beta^2}}{2\mu\omega}\, e^{-2\beta z - i\gamma},$$

whose real part represents the energy crossing on the average unit area of the xy-plane per second.

$$(72) \qquad \bar{S}_z = \frac{a^2}{2} \frac{\sqrt{\alpha^2 + \beta^2}}{\mu\omega}\, e^{-2\beta z} \cos\gamma.$$

The factor $\cos\gamma$ arising from the relative phase displacement between **E** and **H** may be expressed in terms of α and β.

$$(73) \qquad \tan\gamma = \frac{\beta}{\alpha}, \qquad \cos\gamma = \frac{\alpha}{\sqrt{\alpha^2 + \beta^2}}.$$

$$(74) \qquad \bar{S}_z = \frac{\alpha}{2\mu\omega}\, e^{-2\beta z} a^2.$$

According to Poynting's theorem the divergence of the mean flow vector measures the energy transformed per unit volume per second into heat. In the present instance

$$(75) \qquad \nabla \cdot \bar{\mathbf{S}} = \frac{\partial \bar{S}_z}{\partial z} = -\frac{\alpha\beta}{\mu\omega}\, a^2 e^{-2\beta z} = -\frac{\sigma}{2}\, a^2 e^{-2\beta z},$$

which in turn is obviously equal to the conductivity times the mean square value of the electric field vector **E**.

5.6. Impedance.—In a world which outwardly is all variety, it is comforting to discover occasional unity and tempting to speculate upon its significance. To the untutored mind a vibrating set of weights suspended from a network of springs appears to have little in common with the currents oscillating in a system of coils and condensers. But the electrical circuit may be so designed that its behavior, and the vibrations of the mechanical system, can be formulated by the same set of differential equations. Between the two there is a one-to-one correspondence. Current replaces velocity and voltage replaces force; and mass and the elastic property of the spring are represented by inductance and the capacity of a condenser. It would seem that the "absolute reality," if one dare think of such a thing, is an *inertia property*, of which *mass* and *inductance* are only representations or names.

Whatever the philosophic significance of mechanical, electrical, and chemical equivalences may be, the physicist has made good use of them to facilitate his own investigations. The technique developed in the last thirty years for the analysis of electrical circuits has been applied with success to mechanical systems which not long ago appeared too difficult to handle, and mechanical problems of the most complicated nature are represented by electrical analogues which can be investigated with ease in the laboratory. Not only the methods but the concepts of electrical circuits have been extended to other branches of physics. Certainly the most important among these is the concept of *impedance* relating voltage and current in both amplitude and phase. This idea has been applied in mechanics to express the ratio of force to velocity, and in hydrodynamics, notably in acoustics, to measure the ratio of pressure to flow.

The extension of the impedance concept to electromagnetic fields is not altogether new, but it has recently been revived and developed in a very interesting paper by Schelkunoff.[1] The impedance offered by a given medium to a wave of given type is closely related to the energy flow, but to bring out its complex nature we may best start with an analogy from a one-dimensional transmission line, as does Schelkunoff.

Let z measure length along an electric transmission line and let $V = V_0 e^{-i\omega t}$, $I = I_0 e^{-i\omega t}$ be respectively the voltage across and the current in the line at any point z. The quantities V_0, I_0 are functions of z alone. The resistance of the line per unit length is R and its inductance per unit length is L. There is a leakage across the line at each point represented by the conductance G and a shunt capacity C. The series impedance Z and the shunt admittance Y are, therefore,

$$Z = R - i\omega L, \qquad Y = G - i\omega C, \tag{76}$$

[1] SCHELKUNOFF, *Bell System Tech. J.*, **17**, 17, January, 1938.

while the voltage and current are found to satisfy the relations[1]

$$\frac{\partial V}{\partial z} = -ZI, \qquad \frac{\partial I}{\partial z} = -YV. \tag{77}$$

These equations are satisfied by two independent solutions which represent waves traveling respectively in the positive and negative directions.

$$\begin{aligned} I_1 &= A_1 e^{ikz - i\omega t}, & V_1 &= Z_0 I_1, \\ I_2 &= A_2 e^{-ikz - i\omega t}, & V_2 &= -Z_0 I_2, \end{aligned} \tag{78}$$

where

$$k = i\sqrt{YZ}, \qquad Z_0 = \sqrt{\frac{Z}{Y}}. \tag{79}$$

k is the propagation constant and Z_0 the *characteristic impedance* of the line.[2]

Consider now a plane electromagnetic wave propagated in a direction specified by a unit vector **n**. Distance in this direction will again be measured by the coordinate ζ and we shall suppose that the time enters only through the factor $e^{-i\omega t}$. Whereas voltage and current are scalar quantities, **E** and **H** are of course vectors. To establish a fixed convention for determining the algebraic sign, the field equations will be written in such a way as to connect the vector **E** with the vector **H** × **n** *which is parallel to* **E** *and directed in the same sense.*

$$\frac{\partial \mathbf{E}}{\partial \zeta} = i\omega\mu(\mathbf{H} \times \mathbf{n}), \qquad \frac{\partial}{\partial \zeta}(\mathbf{H} \times \mathbf{n}) = i(\omega\epsilon + i\sigma)\mathbf{E}. \tag{80}$$

By analogy with (77)

$$Z = -i\omega\mu, \qquad Y = -i(\omega\epsilon + i\sigma). \tag{81}$$

The propagation constant is

$$k = i\sqrt{YZ} = \sqrt{\omega^2\epsilon\mu + i\omega\mu\sigma} \tag{82}$$

as in (29), page 273, while the intrinsic impedance of the medium for plane waves is defined by Schelkunoff as the quantity

$$Z_0 = \sqrt{\frac{Z}{Y}} = \sqrt{\frac{\omega\mu}{\omega\epsilon + i\sigma}} = \frac{\mu\omega}{\sqrt{\alpha^2 + \beta^2}}\, e^{-i\gamma}. \tag{83}$$

[1] See for example, Guillemin, "Communication Networks," Vol. II, Chap. II, Wiley, 1935.

[2] These results differ in the algebraic sign of the imaginary component from those usually found in the literature on circuit theory due to our choice of $e^{-i\omega t}$ rather than $e^{i\omega t}$. Obviously this choice is wholly arbitrary. Although it is advantageous to use $e^{i\omega t}$ in circuit theory, we shall be concerned in wave theory primarily with the space rather than the time factor. The expansions in curvilinear coordinates to be carried out in the next chapter justify the choice of $e^{-i\omega t}$.

In free space this impedance reduces to

$$Z_0 = \sqrt{\frac{\mu_0}{\epsilon_0}} = 376.6 \text{ ohms.} \tag{84}$$

Assuming **n** to be in the direction of propagation so that the distinction between positive and negative waves is unnecessary, the relation between electric vectors becomes

$$\mathbf{n} \times \mathbf{E} = Z_0\mathbf{H}, \qquad \mathbf{E} = Z_0\mathbf{H} \times \mathbf{n}. \tag{85}$$

There is an intimate connection between the intrinsic impedance and the complex Poynting vector.

$$\mathbf{S}^* = \frac{1}{2}\mathbf{E} \times \tilde{\mathbf{H}} = \frac{1}{2}\mathbf{E} \times \frac{(\mathbf{n} \times \tilde{\mathbf{E}})}{\tilde{Z}_0} = \frac{1}{2\tilde{Z}_0}(\mathbf{E} \cdot \tilde{\mathbf{E}})\mathbf{n}, \tag{86}$$

and, consequently,

$$\tilde{Z}_0 = \frac{1}{2}\frac{E^2}{|\mathbf{S}^*|}. \tag{87}$$

GENERAL SOLUTIONS OF THE ONE-DIMENSIONAL WAVE EQUATION

In the course of this chapter we have investigated certain particular solutions of the field equations which depend on one space variable and the time. Owing to the linear character of the equations these particular solutions may be multiplied by arbitrary constants and summed to form a general solution, about which we now inquire. In virtue of the infinite set of constants at our disposal it must be possible to construct solutions that satisfy certain prescribed initial conditions. We shall show, for example, that the distribution of a field vector as a function of ζ may be specified at a stated instant $t = t_0$, and that the field is thereby uniquely determined at all subsequent times; or the variation of a field vector as a function of time may be prescribed over a single plane $\zeta = \zeta_0$ and the field thus determined at all other points of space and time. The means at our disposal for the investigation of general integrals of a partial differential equation of the second order fall principally into two classes: the methods of Fourier and Cauchy, and those of Riemann and Volterra. The latter have come to constitute in recent years the most fundamental approach to the theory of partial differential equations, and their application to the theory of wave propagation has been developed in a series of brilliant researches by Hadamard.[1] But although the method of characteristics, first proposed by Riemann, affords a deeper insight into the nature of the problem, it proves less well adapted as yet

[1] HADAMARD, "Leçons sur la propagation des ondes," A. Hermann, Paris, 1903; "Lectures on Cauchy's Problem," Yale University Press, 1923.

to the requirements of a practical solution than the methods of harmonic analysis which will occupy our attention in the present section.

5.7. Elements of Fourier Analysis.—For the convenience of the reader, who presumably is to some degree conversant with the subject, we shall set forth here without proof the more elementary facts concerning Fourier series and Fourier integrals.

Consider the trigonometric series

$$\frac{a_0}{2} + (a_1 \cos x + b_1 \sin x) + \cdots (a_n \cos nx + b_n \sin nx) + \cdots, \tag{1}$$

with known coefficients a_n and b_n, and let it be assumed that the series converges uniformly in the region $0 \leq x \leq 2\pi$. Then (1) converges uniformly for all values of x and represents a periodic function $f(x)$ with period 2π.

$$f(x + 2\pi) = f(x). \tag{2}$$

The coefficients of the series may now be expressed in terms of $f(x)$. In virtue of the assumed uniformity of convergence, (1) may be multiplied by either $\cos nx$ or $\sin nx$ and integrated term by term. In the domain $0 \leq x \leq 2\pi$, the trigonometric functions are orthogonal; that is to say,

$$\int_0^{2\pi} \cos nx \cos mx \, dx = 0, \qquad \int_0^{2\pi} \sin nx \sin mx \, dx = 0, \tag{3}$$
$$\int_0^{2\pi} \cos nx \sin mx \, dx = 0, \qquad (m \neq n).$$

Since

$$\int_0^{2\pi} \cos^2 nx \, dx = \int_0^{2\pi} \sin^2 mx \, dx = \pi, \tag{4}$$

it follows that

$$a_n = \frac{1}{\pi} \int_0^{2\pi} f(x) \cos nx \, dx, \tag{5}$$
$$(m = 0, 1, 2 \cdots),$$
$$b_n = \frac{1}{\pi} \int_0^{2\pi} f(x) \sin nx \, dx.$$

Inversely let us suppose that a function $f(x)$ is given. Then in a purely formal manner one may associate with $f(x)$ a "Fourier series" defined by

$$f(x) = \frac{a_0}{2} + \sum_{n=1}^{\infty} (a_n \cos nx + b_n \sin nx), \tag{6}$$

the coefficients to be determined from (5). The right-hand side of (6), however, will converge to a representation of $f(x)$ in the domain $0 \leq x \leq 2\pi$ only when the function $f(x)$ is subjected to certain conditions. The whole question of the convergence of Fourier series is an extremely delicate one, and even the continuity of $f(x)$ is not sufficient to ensure it. It turns out, fortunately, that under such circumstances the series can nevertheless be summed to represent accurately the function within the specified interval.[1] A statement of the least stringent conditions to be imposed on an otherwise arbitrary function in order that it may be expanded in a convergent or summable trigonometric series would necessitate a lengthy and difficult exposition; since we are concerned solely with functions occurring in physical investigations, we can content ourselves with certain *sufficient* (though not necessary) requirements. We shall ask only that in the interval $0 \leq x \leq 2\pi$ the function and its first derivative shall be piecewise continuous. A function $f(x)$ is said to be piecewise continuous in a given interval if it is continuous throughout that interval except at a finite number of points. If such a point be x_0, the function approaches the finite value $f(x_0 + 0)$ as x_0 is approached from the right, and $f(x_0 - 0)$ as it is approached from the left. At the discontinuity itself the value of the function is taken to be the arithmetic mean $\frac{f(x_0 + 0) + f(x_0 - 0)}{2}$.

The Fourier expansion is particularly well adapted to the representation of functions which cannot be expressed in a closed analytic form, but which are constituted of sections or pieces of analytic curves not necessarily joined at the ends. In view of the admissible discontinuities in the function or its derivative, the fact that two Fourier series represent the same function throughout a subinterval does not imply at all that they represent one and the same function outside that subinterval. It is thus essential that one distinguish between the *representation* of a function and the function itself. The expansion in a trigonometric series indicated on the right-hand side of Eq. (6) represents and is equivalent to the piecewise continuous function $f(x)$ within the region $0 \leq x \leq 2\pi$. Outside this domain the values assumed by the series are repeated periodically in x with a period 2π, whereas the function $f(x)$ may behave in any arbitrary manner. Only when $f(x)$ satisfies the additional relation (2) will it coincide with its Fourier representation over the entire domain $-\infty < x < \infty$.

If within a specified interval the Fourier series of $f(x)$ is known to be uniformly convergent, it may be integrated term by term and the series thus obtained placed equal to the integral of $f(x)$ between the same limits.

[1] WHITTAKER and WATSON, "Modern Analysis," 4th ed., Chap. IX, Cambridge University Press, London.

The uniform convergence of a Fourier series is in itself not sufficient to justify its differentiation term by term and the equating of the derived series to df/dx. If furthermore $f(x)$ is discontinuous at some point within an interval, its Fourier series certainly is not uniformly convergent everywhere within that interval, and consequently the Fourier expansions of its integral and of its derivative demand special attention.[1]

It is usually advantageous to replace (6) by an equivalent complex series of exponential terms of the form

$$f(x) = \sum_{n=-\infty}^{\infty} c_n e^{inx}, \tag{7}$$

whose coefficients are determined from

$$c_n = \frac{1}{2\pi} \int_0^{2\pi} f(\alpha) e^{-in\alpha}\, d\alpha, \quad (n = 0, \pm 1, \pm 2 \cdots). \tag{8}$$

Since

$$e^{inx} = \cos nx + i \sin nx, \tag{9}$$

it is clear that the complex coefficients c_n are related to the real coefficients a_n and b_n by the equations

$$\begin{aligned} 2c_n &= a_n - ib_n, && (n > 0), \\ 2c_0 &= a_0, && (n = 0), \\ 2c_n &= a_{-n} + ib_{-n}, && (n < 0). \end{aligned} \tag{10}$$

By an appropriate change of variable the Fourier expansion (7) may be modified to represent a function in the region $-l \le x \le l$.

$$f(x) = \sum_{n=-\infty}^{\infty} c_n e^{\frac{in\pi}{l}x}, \qquad c_n = \frac{1}{2l} \int_{-l}^{l} f(\alpha) e^{-\frac{in\pi}{l}\alpha}\, d\alpha. \tag{11}$$

Now this stretching of the basic interval or period from 2π to $2l$ suggests strongly the possibility of passing to the limit as the basic interval becomes infinite and thereby obtaining a Fourier representation of a nonperiodic function for all real values of x lying between $-\infty$ and $+\infty$. We shall assume that throughout the entire region $-\infty < x < \infty$, $f(x)$ and its first derivative are piecewise continuous, that at the discontinuities the value of the function is to be determined by the arithmetical mean, and further that the integral $\int_{-\infty}^{\infty} f(x)\, dx$ is absolutely convergent, or in other words, that the integral $\int_{-\infty}^{\infty} |f(x)|\, dx$ exists. Let $\pi/l = \Delta u$.

[1] These questions are thoroughly treated by Carslaw, "Introduction to the Theory of Fourier's Series and Integrals," Chap. VIII, Macmillan, 1921.

Then (11) may be written

(12) $$f(x) = \frac{1}{2\pi} \sum_{n=-\infty}^{\infty} \Delta u \int_{-l}^{l} f(\alpha) e^{in\,\Delta u(x-\alpha)}\,d\alpha.$$

On the other hand the definite integral $\int_{-\infty}^{\infty} \phi(u)\,du$ is defined as the limit of a sum as $\Delta u \to 0$:

(13) $$\int_{-\infty}^{\infty} \phi(u)\,du = \lim_{\Delta u \to 0} \sum_{n=-\infty}^{\infty} \phi(n\,\Delta u)\,\Delta u.$$

As l approaches infinity, Δu approaches zero and we may reasonably expect the limit of (12) to be

(14) $$f(x) = \frac{1}{2\pi} \int_{-\infty}^{\infty} du \int_{-\infty}^{\infty} f(\alpha) e^{iu(x-\alpha)}\,d\alpha.$$

This is the Fourier integral theorem according to which an arbitrary function, satisfying only conditions of piecewise continuity and the existence of $\int_{-\infty}^{\infty} |f(x)|\,dx$, may be expressed as a double integral.[1] It will be noted that the result has been obtained by a purely formal transition to the limit which makes the existence of the representation appear plausible but does not confirm it. A rigorous demonstration is beyond the scope of these introductory remarks.

If $f(x)$ is real, the imaginary part of the Fourier integral must vanish and (14) then reduces to

(15) $$f(x) = \frac{1}{\pi} \int_{0}^{\infty} du \int_{-\infty}^{\infty} f(\alpha) \cos u(x-\alpha)\,d\alpha.$$

Since

(16) $$\cos u(x-\alpha) = \cos ux \cos u\alpha + \sin ux \sin u\alpha,$$

it is apparent that the Fourier integral of a real, even function $f(-x) = f(x)$ is

(17) $$f(x) = \frac{2}{\pi} \int_{0}^{\infty} \cos ux \int_{0}^{\infty} f(\alpha) \cos u\alpha\,d\alpha\,du,$$

and that of a real, odd function $f(-x) = -f(x)$ is

(18) $$f(x) = \frac{2}{\pi} \int_{0}^{\infty} \sin ux \int_{0}^{\infty} f(\alpha) \sin u\alpha\,d\alpha\,du.$$

[1] As a general reference on the subject the reader may consult "Introduction to the Theory of Fourier Integrals," by E. C. Titchmarsh, Oxford University Press, 1937.

From these formulas we see that the Fourier integral of a function may be interpreted as a resolution into harmonic components of frequency $u/2\pi$ over the continuous spectrum of frequencies lying between zero and infinity. In (18), for example, one may consider the function

$$g(u) = \sqrt{\frac{2}{\pi}} \int_0^\infty f(\alpha) \sin u\alpha \, d\alpha \tag{19}$$

as the amplitude or *spectral density* of $f(x)$ in the frequency interval u to $u + du$. Then

$$f(x) = \sqrt{\frac{2}{\pi}} \int_0^\infty g(u) \sin ux \, du. \tag{20}$$

The relation between $f(x)$ and $g(u)$ is reciprocal. $g(u)$ is said to be the *Fourier transform* of $f(x)$, and $f(x)$ is likewise the transform of $g(u)$. In the more general case of Eq. (14) one may write for the spectral density of $f(x)$ the function

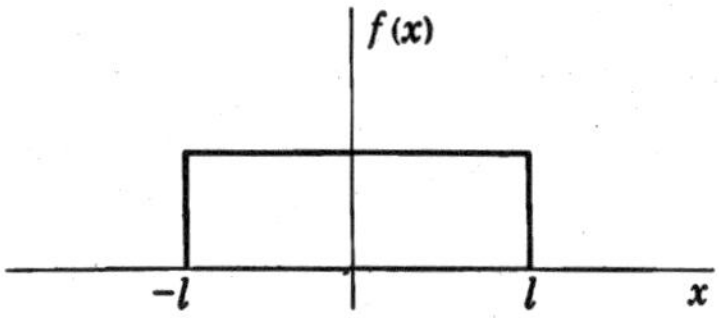

FIG. 48.—Step function.

$$g(u) = \frac{1}{\sqrt{2\pi}} \int_{-\infty}^{\infty} f(x) e^{-iux} \, dx, \tag{21}$$

and hence for $f(x)$ the reciprocal relation

$$f(x) = \frac{1}{\sqrt{2\pi}} \int_{-\infty}^{\infty} g(u) e^{iux} \, du. \tag{22}$$

An extensive table of Fourier transforms has been published by Campbell and Foster.[1]

The application of the Fourier integral may be illustrated by several brief examples of practical interest. Consider first the discontinuous *step function* defined by

$$\begin{aligned} f(x) &= 1, & &\text{when } |x| < l, \\ f(x) &= \tfrac{1}{2}, & &\text{when } |x| = l, \\ f(x) &= 0, & &\text{when } |x| > l. \end{aligned} \tag{23}$$

The function is real and even, so that we may employ (17). The transform is

$$g(u) = \sqrt{\frac{2}{\pi}} \int_0^l \cos ux \, dx = \sqrt{\frac{2}{\pi}} \frac{\sin ul}{u}, \tag{24}$$

and the Fourier integral

$$f(x) = \sqrt{\frac{2}{\pi}} \int_0^\infty g(u) \cos ux \, du = \frac{2}{\pi} \int_0^\infty \frac{\sin ul \cos ux}{u} \, du. \tag{25}$$

[1] CAMPBELL and FOSTER: "Fourier Integrals for Practical Applications," *Bell Telephone System Tech. Pub.*, Monograph B-584, 1931. Appeared in earlier form in *Bell System Tech. J.*, October, 1928, pp. 639–707.

Let us take next for $f(x)$ the "error function"

$$f(x) = e^{-\frac{a^2x^2}{2}}. \tag{26}$$

To find its Fourier transform we must calculate the integral

$$g(u) = \sqrt{\frac{2}{\pi}} \int_0^\infty e^{-\frac{a^2x^2}{2}} \cos ux \, dx, \tag{27}$$

which is clearly equal to the real part of the complex integral

$$g(u) = \sqrt{\frac{2}{\pi}} \int_0^\infty e^{-\frac{a^2x^2}{2}+iux} \, dx. \tag{28}$$

On completing the square, (28) may be written

$$\begin{aligned} g(u) &= \sqrt{\frac{2}{\pi}}\, e^{-\frac{u^2}{2a^2}} \int_0^\infty e^{-\frac{a^2}{2}\left(x-\frac{iu}{a^2}\right)^2} dx \\ &= \sqrt{\frac{2}{\pi}}\, e^{-\frac{u^2}{2a^2}} \int_{-\frac{iu}{a^2}}^\infty e^{-\frac{a^2\beta^2}{2}} d\beta. \end{aligned} \tag{29}$$

The path of integration follows the imaginary axis from $-iu/a^2$ to 0 and then runs along the real axis from 0 to ∞. The imaginary part of the integral arises solely from the interval $-iu/a^2 \leq \beta < 0$, and since we are concerned only with the real part the lower limit may be taken equal to zero. The result is a definite integral whose value is well known:

$$\int_0^\infty e^{-\frac{a^2\beta^2}{2}} d\beta = \frac{1}{a}\sqrt{\frac{\pi}{2}}; \tag{30}$$

so that the transform of $e^{-\frac{a^2x^2}{2}}$ is

$$g(u) = \frac{1}{a} e^{-\frac{u^2}{2a^2}}. \tag{31}$$

Then $f(x)$, which is reciprocally the transform of $g(u)$, is

$$f(x) = \sqrt{\frac{2}{\pi}} \int_0^\infty \frac{e^{-\frac{u^2}{2a^2}}}{a} \cos ux \, du = e^{-\frac{a^2x^2}{2}}. \tag{32}$$

In the particular case $a = 1$, Eq. (32) becomes a homogeneous integral equation satisfied by the function $e^{-\frac{x^2}{2}}$. The pair of functions defined by (31) and (32) have other properties more important for our present needs. Let us modify (31) slightly to form the function

$$S(x, a) = \frac{1}{\sqrt{2\pi}} \frac{e^{-\frac{x^2}{2a^2}}}{a}. \tag{33}$$

The area under this curve is equal to unity, whatever the value of the parameter a;

$$\int_{-\infty}^{\infty} S(x, a)\, dx = 1, \tag{34}$$

as follows directly from (30). Now let a become smaller and smaller. The breadth of the peak grows more and more narrow, while at the same time its height increases in such a manner in the neighborhood of $x = 0$ as to maintain the area constant, as indicated in Fig. 49. In the limit as $a \to 0$, the curve shrinks to the line $x = 0$ where it attains infinite amplitude. A singularity has been generated, an *impulse function* bounding unit area in the immediate neighborhood of $x = 0$. A unit

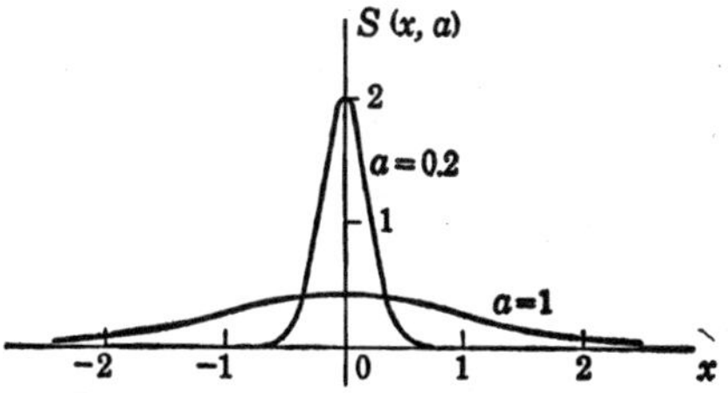

FIG. 49.—The impulse function $S(x, a)$.

impulse function which vanishes everywhere but at the point $x = x_0$ is represented by

$$S_0(x - x_0) = \lim_{a \to 0} \frac{1}{\sqrt{2\pi}} \frac{e^{-\frac{(x - x_0)^2}{2a^2}}}{a}. \tag{35}$$

Any arbitrary function $F(\xi)$, subject to the usual conditions of continuity, can now be expressed as an infinite integral.

$$F(\xi) = \int_{-\infty}^{\infty} S_0(\xi - x) F(x)\, dx. \tag{36}$$

It is apparent from (32) that the transform of the impulse function $S_0(x)$ is a straight line displaced by an amount $1/\sqrt{2\pi}$ from the horizontal axis.

As a final illustration of the Fourier integral theorem consider a harmonic wave train of finite duration. Such a pulse might result from closing and then reopening a switch connecting a circuit to an alternating current generator, or represent the emission of light from an atom in the course of an energy transition. Let

$$\begin{aligned} f(t) &= 0, && \text{when} \quad |t| > \frac{T}{2}, \\ f(t) &= \cos \omega_0 t, && \text{when} \quad |t| < \frac{T}{2}. \end{aligned} \tag{37}$$

To facilitate the integration we shall take $f(t)$ equal to the real part of $e^{i\omega_0 t}$ in the domain $|t| < \frac{T}{2}$ and use Eqs. (21) and (22).

$$g(\omega) = \frac{1}{\sqrt{2\pi}} \int_{-\frac{T}{2}}^{\frac{T}{2}} e^{i(\omega_0 - \omega)t}\, dt = \sqrt{\frac{2}{\pi}} \frac{\sin \frac{(\omega_0 - \omega)T}{2}}{\omega_0 - \omega}. \tag{38}$$

The Fourier integral of $f(t)$ corresponds now to the real part of (22), or

$$\begin{aligned} f(t) &= \frac{1}{\pi} \int_{-\infty}^{\infty} \frac{\sin \frac{(\omega_0 - \omega)T}{2}}{\omega_0 - \omega} \cos \omega t\, d\omega \\ &= \frac{1}{\pi} \int_0^{\infty} \left[\frac{\sin \frac{(\omega_0 - \omega)T}{2}}{\omega_0 - \omega} + \frac{\sin \frac{(\omega_0 + \omega)T}{2}}{\omega_0 + \omega} \right] \cos \omega t\, d\omega. \end{aligned} \tag{39}$$

Equation (39) may be interpreted as a spectral resolution of a function which during a finite interval T is sinusoidal with frequency ω_0. The amplitude of the disturbance in the neighborhood of any frequency ω is determined by the function

$$A(\omega) = \frac{1}{\pi} \frac{\sin \frac{(\omega_0 - \omega)T}{2}}{\omega_0 - \omega}. \tag{40}$$

The amplitude vanishes at the points

$$\omega = \omega_0 - \frac{2\pi n}{T}, \qquad (n = \pm 1, \pm 2 \cdots), \tag{41}$$

in the manner indicated by Fig. 50, and has its maximum value at $\omega = \omega_0$. As the duration T of the wave train increases, the envelope of the amplitude function is compressed horizontally until in the limit, as $T \to \infty$, the entire disturbance is confined to the line $\omega = \omega_0$ on the frequency spectrum. The simple harmonic variations that enter into so many of our discussions are mathematical ideals; the oscillations of natural systems, whether mechanical or electrical, are finite in duration, and the associated waves are periodic only in an approximate sense. We shall discover shortly that in the presence of a dispersive medium the entire character of the propagation may be governed by the duration of the wave train.

5.8. General Solution of the One-dimensional Wave Equation in a Nondissipative Medium.—To acquire some further facility in the use of the Fourier integral it will repay us to pause for a moment over the elementary problem of finding a general solution to the wave equation in a nonconducting medium. Let ψ represent either the x- or y-component of any electromagnetic vector. Then ψ satisfies

$$\frac{\partial^2\psi}{\partial z^2} - \frac{1}{v^2}\frac{\partial^2\psi}{\partial t^2} = 0, \qquad v^2 = \frac{1}{\mu\epsilon}. \tag{42}$$

According to Sec. 5.2, a particular solution of (42) is represented by

$$\psi = (Ae^{i\frac{\omega}{v}z} + Be^{-i\frac{\omega}{v}z})e^{-i\omega t}. \tag{43}$$

The coefficients A and B are arbitrary and may depend on the frequency ω; that is to say, we associate with each harmonic component an appropriate amplitude which may be indicated by writing $A(\omega)$ and $B(\omega)$.

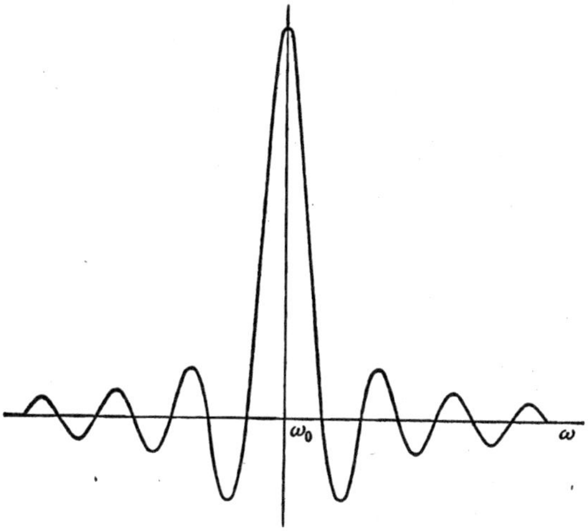

FIG. 50.—The amplitude function $A(\omega) = \frac{1}{\pi}\frac{\sin(\omega_0 - \omega)T/2}{\omega_0 - \omega}$.

Now the general solution of (42) is obtained by summing the particular solutions over a range of ω. In case ψ is to be a periodic function, the sum will extend over a discrete set of frequencies. In general the wave function is aperiodic in both space and time, and the frequency spectrum is consequently continuous.

$$\psi(z, t) = \int_{-\infty}^{\infty} [A(\omega)e^{i\frac{\omega}{v}z} + B(\omega)e^{-i\frac{\omega}{v}z}]\, e^{-i\omega t}\, d\omega. \tag{44}$$

Let us suppose that over the plane $z = 0$ the values of the function ψ and of its derivative in the direction of propagation are prescribed functions of time.

$$\psi(0, t) = f(t), \qquad \left(\frac{\partial\psi}{\partial z}\right)_{z=0} = F(t). \tag{45}$$

The problem is to find the coefficients $A(\omega)$ and $B(\omega)$ such that these two conditions are satisfied, and thereby to show that the prescription of the function and its derivative at a specified point in space (or time) is sufficient to determine $\psi(z, t)$ everywhere.

If we assume provisionally that the integral on the right of (44) is uniformly convergent, it may be differentiated under the sign of integration with respect to the parameter z, yielding

$$\frac{\partial\psi}{\partial z} = \frac{i}{v}\int_{-\infty}^{\infty} \omega[A(\omega)e^{i\frac{\omega}{v}z} - B(\omega)e^{-i\frac{\omega}{v}z}]e^{-i\omega t}\,d\omega. \tag{46}$$

On placing $z = 0$ in (44) and (46), we obtain

$$\begin{aligned} f(t) &= \int_{-\infty}^{\infty} [A(\omega) + B(\omega)]e^{-i\omega t}\,d\omega, \\ F(t) &= \frac{i}{v}\int_{-\infty}^{\infty} \omega[A(\omega) - B(\omega)]e^{-i\omega t}\,d\omega; \end{aligned} \tag{47}$$

upon comparison with (21) and (22) it is immediately evident that the coefficients of the factor $e^{-i\omega t}$ in the integrands of (47) are Fourier transforms.

$$\begin{aligned} A + B &= \frac{1}{2\pi}\int_{-\infty}^{\infty} f(t)e^{i\omega t}\,dt, \\ \frac{i\omega}{v}(A - B) &= \frac{1}{2\pi}\int_{-\infty}^{\infty} F(t)e^{i\omega t}\,dt. \end{aligned} \tag{48}$$

Solving these two relations simultaneously for A and B and substituting[1] α for t as a variable of integration, we obtain

$$\begin{aligned} A(\omega) &= \frac{1}{4\pi}\int_{-\infty}^{\infty} \left[f(\alpha) - \frac{iv}{\omega}F(\alpha)\right]e^{i\omega\alpha}\,d\alpha, \\ B(\omega) &= \frac{1}{4\pi}\int_{-\infty}^{\infty} \left[f(\alpha) + \frac{iv}{\omega}F(\alpha)\right]e^{i\omega\alpha}\,d\alpha, \end{aligned} \tag{49}$$

and on substitution into (44) there results:

$$\begin{aligned} \psi(z, t) = \frac{1}{4\pi}\int_{-\infty}^{\infty} d\omega \int_{-\infty}^{\infty} f(\alpha)[e^{i\omega\left(\alpha + \frac{z}{v} - t\right)} + e^{i\omega\left(\alpha - \frac{z}{v} - t\right)}]\,d\alpha \\ - \frac{iv}{4\pi}\int_{-\infty}^{\infty} d\omega \int_{-\infty}^{\infty} \frac{F(\alpha)}{\omega}(e^{i\frac{\omega}{v}z} - e^{-i\frac{\omega}{v}z})e^{i\omega(\alpha - t)}\,d\alpha. \end{aligned} \tag{50}$$

Now the first of these double integrals is the Fourier expansion of a function which on reference to (14) may be written down at once. In

[1] The variables α and β appearing in the next few pages obviously have no connection with the real and imaginary parts of $k = \alpha + i\beta$ defined on p. 273.

the second let us invert the order of integration.

$$(51)\quad \psi(z,t) = \frac{1}{2} f\left(t - \frac{z}{v}\right) + \frac{1}{2} f\left(t + \frac{z}{v}\right) + \frac{v}{2\pi}\int_{-\infty}^{\infty} F(\alpha)\,d\alpha \int_{-\infty}^{\infty} \sin\frac{\omega}{v} z\, e^{i\omega(\alpha-t)}\frac{d\omega}{\omega}.$$

The imaginary part of the last integral vanishes because

$$\frac{1}{\omega}\sin\frac{\omega}{v} z \sin\omega(\alpha - t)$$

integrates to an even function. Therefore,

$$(52)\quad \int_{-\infty}^{\infty} \sin\frac{\omega z}{v} e^{i\omega(\alpha-t)}\frac{d\omega}{\omega} = \int_{-\infty}^{\infty} \sin\frac{\omega z}{v}\cos\omega(\alpha - t)\frac{d\omega}{\omega}$$
$$= \int_0^{\infty}\left[\sin\omega\left(\alpha + \frac{z}{v} - t\right) - \sin\omega\left(\alpha - \frac{z}{v} - t\right)\right]\frac{d\omega}{\omega}.$$

By a slight modification of the conditions which defined the step function (23), it is easy to show that each of the last two integrals represents a function with a single discontinuity. In fact

$$(53)\quad \int_0^{\infty}\frac{\sin px}{x}\,dx = \begin{cases} \dfrac{\pi}{2}, & \text{when} \quad p > 0, \\ 0, & \text{when} \quad p = 0, \\ -\dfrac{\pi}{2}, & \text{when} \quad p < 0. \end{cases}$$

Hence (52) will vanish whenever the arguments $\alpha + \frac{z}{v} - t$ and $\alpha - \frac{z}{v} - t$ are of the same sign. Changes of sign occur at $\alpha = t - \frac{z}{v}$ and $\alpha = t + \frac{z}{v}$. The integral (52) has, therefore, a nonvanishing value only in the interval $t - \frac{z}{v} < \alpha < t + \frac{z}{v}$, where it is equal to π. The general solution of (42) is now established in terms of the prescribed initial conditions.

$$(54)\quad \psi(z,t) = \frac{1}{2} f\left(t - \frac{z}{v}\right) + \frac{1}{2} f\left(t + \frac{z}{v}\right) + \frac{v}{2}\int_{t-\frac{z}{v}}^{t+\frac{z}{v}} F(\alpha)\,d\alpha.$$

Let us define a function $h(\beta)$ by the relation

$$(55)\quad h(\beta) = -v\int_0^{\beta} F(\alpha)\,d\alpha.$$

Then

$$\frac{dh}{d\beta} = h'(\beta) = -vF(\beta). \tag{56}$$

In place of (54) one may write

$$\psi(z, t) = \frac{1}{2}f\left(t - \frac{z}{v}\right) + \frac{1}{2}f\left(t + \frac{z}{v}\right) + \frac{1}{2}h\left(t - \frac{z}{v}\right) - \frac{1}{2}h\left(t + \frac{z}{v}\right). \tag{57}$$

When $z = 0$, it is evident that $\psi(0, t)$ reduces to $f(t)$, and $(d\psi/dz)_{z=0}$ to $\frac{1}{v}h'$, or $F(t)$. It may also be verified by differentiation and substitution that (42) *is satisfied by any function whose argument is of the form* $t \pm \frac{z}{v}$. The replacement of the variable t in $f(t)$ by $t - \frac{z}{v}$ results in a displacement or propagation without distortion of the function to the right along the positive z-axis with a velocity v. The initial distribution therefore splits into two waves, the one traveling to the right and the other to the left. Upon these two waves are superposed the h waves, so chosen that at $z = 0$ they annul one another but such that the derivative of ψ assumes the prescribed value $F(t)$.

Prescription of the two functions $f(t)$ and $F(t)$ on the plane $z = 0$ is sufficient to determine completely the electromagnetic field. Let us suppose, for example, that the electric vector of the plane wave is polarized along the x-axis. The Maxwell equations are then

$$\frac{\partial E_x}{\partial z} + \mu\frac{\partial H_y}{\partial t} = 0, \qquad \frac{\partial H_y}{\partial z} + \epsilon\frac{\partial E_x}{\partial t} = 0. \tag{58}$$

Let E_x be represented by the function $\psi(z, t)$ in either (54) or (57); then on differentiating with respect to both t and z we find expressions for $\partial H_y/\partial t$ and $\partial H_y/\partial z$ with the aid of (58); from these in turn it is a simple matter to deduce the expression for H_y.

$$\begin{aligned} E_x &= \frac{1}{2}\left[f\left(t - \frac{z}{v}\right) + f\left(t + \frac{z}{v}\right) + h\left(t - \frac{z}{v}\right) - h\left(t + \frac{z}{v}\right)\right], \\ H_y &= \frac{1}{2}\sqrt{\frac{\epsilon}{\mu}}\left[f\left(t - \frac{z}{v}\right) - f\left(t + \frac{z}{v}\right) + h\left(t - \frac{z}{v}\right) + h\left(t + \frac{z}{v}\right)\right]. \end{aligned} \tag{59}$$

These fields are clearly such that over the plane $z = 0$,

$$\begin{aligned} E_x &= f(t), & \frac{\partial E_x}{\partial z} &= F(t), \\ H_y &= \sqrt{\frac{\epsilon}{\mu}}\,h(t) = -\frac{1}{\mu}\int_0^t F(t)\,dt, & \frac{\partial H_y}{\partial z} &= -\epsilon\frac{df}{dt}. \end{aligned} \qquad (z = 0) \tag{60}$$

The electromagnetic field is, therefore, determined by the specification, as independent functions of time on the plane $z = 0$, of any pair in (60) containing both $f(t)$ and $F(t)$. Thus the electric field and its normal derivative may be prescribed, or the electric and magnetic fields without restriction on the derivatives.

5.9. Dissipative Medium; Prescribed Distribution in Time.—We pass on now to the more difficult problem of finding an electromagnetic field in a conducting medium which reduces to certain prescribed functions of the time on a plane z = constant. Again we shall allow ψ to represent any rectangular component of an electromagnetic vector satisfying the equation

$$\frac{\partial^2\psi}{\partial z^2} - \mu\epsilon\frac{\partial^2\psi}{\partial t^2} - \mu\sigma\frac{\partial\psi}{\partial t} = 0. \tag{61}$$

It will be convenient in what follows to introduce two new constants.

$$a = \frac{1}{\sqrt{\mu\epsilon}}, \qquad b = \frac{\sigma}{2\epsilon},$$
$$k = \sqrt{\mu\epsilon\omega^2 + i\mu\sigma\omega} = \frac{1}{a}\sqrt{\omega^2 + 2b\omega i}. \tag{62}$$

Note that in a dissipative medium a differs from the phase velocity $v = c/n$ of a harmonic component. In this notation the wave equation assumes the form

$$\frac{\partial^2\psi}{\partial t^2} + 2b\frac{\partial\psi}{\partial t} - a^2\frac{\partial^2\psi}{\partial z^2} = 0. \tag{63}$$

Particular solutions of (63) that are harmonic in time were investigated in Sec. 5.2 and found to be

$$\psi = (Ae^{ikz} + Be^{-ikz})e^{-i\omega t}. \tag{64}$$

Assuming the coefficients A and B as well as the complex quantity k to be functions of the frequency, we construct a general solution by summation of the harmonic components.

$$\psi(z, t) = \int_{-\infty}^{\infty} [A(\omega)e^{ikz} + B(\omega)e^{-ikz}]e^{-i\omega t}\,d\omega. \tag{65}$$

Again, let us suppose that the wave function and its normal derivative are prescribed functions of time over the plane z = constant. For convenience the origin of the reference system is located within the plane so that the constant is zero.

$$\psi(0, t) = f(t), \qquad \left(\frac{\partial\psi}{\partial z}\right)_{z=0} = F(t). \tag{66}$$

In physical problems the character of the function $\psi(z, t)$ is such that the infinite integral converges uniformly. Assuming this convergence, it may be differentiated with respect to the parameter z.[1]

$$\frac{\partial \psi}{\partial z} = \int_{-\infty}^{\infty} ik[A(\omega)e^{ikz} - B(\omega)e^{-ikz}]e^{-i\omega t}\, d\omega. \tag{67}$$

On the initial plane, (65) and (67) reduce to

$$\begin{aligned} f(t) &= \int_{-\infty}^{\infty} [A(\omega) + B(\omega)]e^{-i\omega t}\, d\omega, \\ F(t) &= \int_{-\infty}^{\infty} ik[A(\omega) - B(\omega)]e^{-i\omega t}\, d\omega. \end{aligned} \tag{68}$$

Again the coefficients of $e^{-i\omega t}$ may be interpreted as the Fourier transforms of $f(t)$ and $F(t)$ respectively.

$$\begin{aligned} A + B &= \frac{1}{2\pi}\int_{-\infty}^{\infty} f(\alpha)e^{i\omega\alpha}\, d\alpha, \\ A - B &= -\frac{i}{2\pi k}\int_{-\infty}^{\infty} F(\alpha)e^{i\omega\alpha}\, d\alpha. \end{aligned} \tag{69}$$

Solving for A and B:

$$\begin{aligned} A(\omega) &= \frac{1}{4\pi}\int_{-\infty}^{\infty} \left[f(\alpha) - \frac{i}{k}F(\alpha)\right]e^{i\omega\alpha}\, d\alpha, \\ B(\omega) &= \frac{1}{4\pi}\int_{-\infty}^{\infty} \left[f(\alpha) + \frac{i}{k}F(\alpha)\right]e^{i\omega\alpha}\, d\alpha. \end{aligned} \tag{70}$$

When these values are reinserted into (65) one obtains, after inversion of the order of integration, the expression

$$\begin{aligned} \psi(z, t) = \frac{1}{2\pi}\int_{-\infty}^{\infty} d\alpha\, f(\alpha)\int_{-\infty}^{\infty} \cos kz\cdot e^{i\omega(\alpha - t)}\, d\omega \\ + \frac{1}{2\pi}\int_{-\infty}^{\infty} d\alpha\, F(\alpha)\int_{-\infty}^{\infty} \frac{\sin kz}{k}\, e^{i\omega(\alpha - t)}\, d\omega. \end{aligned} \tag{71}$$

Thus far the analysis has not deviated formally from that of the problem in a nondissipative medium. At this point, however, we encounter a difficulty due to the complexity of k and we shall have to digress momentarily to discuss the representation of the function $\sin kz/k$ as a definite integral. The basis of this representation is an integral due to Gegenbauer[2] which in the simpler case at hand reduces to

[1] See for example Carslaw, "Introduction to the Theory of Fourier's Series and Integrals," Chaps. IV and VI, Macmillan, 1921.

[2] WATSON, "A Treatise on the Theory of Bessel Functions," p. 379, Cambridge University Press, 1922. Equation (72) is a special case of (69), p. 411, derived in Sec. 7.7.

(72) $$\frac{\sin u}{u} = \frac{1}{2}\int_0^\pi J_0(u \sin\phi \sin\theta) e^{iu\cos\phi\cos\theta} \sin\theta \, d\theta.$$

$J_0(u \sin \phi \sin \theta)$ denotes the zero-order Bessel function of argument $u \sin \phi \sin \theta$.

Let us make the following substitutions:

(73) $$u = v\sqrt{(p+\lambda)(p+\mu)},$$

where λ and μ are real or complex constants.

(74) $$u\cos\phi = v\left(p + \frac{\lambda+\mu}{2}\right), \qquad u \sin\phi = i\frac{v}{2}(\lambda - \mu),$$
$$\cos\theta = \frac{\beta}{v}, \qquad \sin\theta = \sqrt{1 - \frac{\beta^2}{v^2}},$$
$$\sin\theta \, d\theta = -\frac{d\beta}{v}.$$

In terms of these new parameters, (72) becomes

(75) $$\frac{\sin v\sqrt{(p+\lambda)(p+\mu)}}{\sqrt{(p+\lambda)(p+\mu)}} = \frac{1}{2}\int_{-v}^{v} J_0\left(\frac{\lambda-\mu}{2}\sqrt{\beta^2 - v^2}\right) e^{i\beta\left(p+\frac{\lambda+\mu}{2}\right)} d\beta.$$

Observe that the function to the left of Eq. (75) is the Fourier transform of a function $g(\beta)$ defined as follows:[1]

(76) $$g(\beta) = 0, \text{ when } |\beta| > v,$$
$$g(\beta) = \sqrt{\frac{\pi}{2}} J_0\left(\frac{\lambda-\mu}{2}\sqrt{\beta^2 - v^2}\right) e^{i\beta\frac{(\lambda+\mu)}{2}}, \text{ when } |\beta| < v.$$

Finally, let us assign to the parameters in this general formula the particular values

$$v = \frac{z}{a}, \qquad p = \omega, \qquad \lambda = 2bi, \qquad \mu = 0.$$

The desired result is then

(77) $$\frac{\sin kz}{k} = \frac{a \sin \frac{z}{a}\sqrt{\omega^2 + 2b\omega i}}{\sqrt{\omega^2 + 2b\omega i}} = \frac{a}{2}\int_{-\frac{z}{a}}^{\frac{z}{a}} J_0\left(\frac{b}{a}\sqrt{z^2 - a^2\beta^2}\right) e^{i\beta\omega - b\beta} \, d\beta.$$

After this small excursion we return once more to Eq. (71). In the second integral on the right-hand side, which for convenience may be

[1] CAMPBELL and FOSTER, *loc. cit.*, No. 872.2.

denoted ψ_2, introduce (77) and then invert the order of integration.

$$\psi_2 = \frac{a}{2}\int_{-\frac{z}{a}}^{\frac{z}{a}} d\beta\, e^{-b\beta}\left[\frac{1}{2\pi}\int_{-\infty}^{\infty} d\omega \int_{-\infty}^{\infty} \phi(\alpha, \beta)e^{i\omega(\alpha+\beta-t)}\, d\alpha\right], \tag{78}$$

where

$$\phi(\alpha, \beta) = F(\alpha)J_0\left(\frac{b}{a}\sqrt{z^2 - a^2\beta^2}\right). \tag{79}$$

The double integral within the brackets is the Fourier expansion of the function $\phi(t - \beta, \beta)$; consequently,

$$\psi_2 = \frac{a}{2}\int_{-\frac{z}{a}}^{\frac{z}{a}} F(t - \beta)J_0\left(\frac{b}{a}\sqrt{z^2 - a^2\beta^2}\right)e^{-b\beta}\, d\beta, \tag{80}$$

or, upon a slight change of variable,

$$\psi_2 = \frac{a}{2}\int_{t-\frac{z}{a}}^{t+\frac{z}{a}} F(\beta)J_0\left(\frac{b}{a}\sqrt{z^2 - a^2(t-\beta)^2}\right)e^{-b(t-\beta)}\, d\beta. \tag{81}$$

To calculate the first integral to the right of Eq. (71), which shall be denoted ψ_1, we need only replace $F(\alpha)$ in ψ_2 by $f(\alpha)$ and differentiate partially with respect to z. Since the limits in (81) are functions of z, this differentiation must be effected according to the formula

$$\frac{\partial}{\partial z}\int_{x_1(z)}^{x_2(z)} f(x, z)\, dx = \int_{x_1}^{x_2} \frac{\partial f}{\partial z}\, dx + f(x_2, z)\frac{\partial x_2}{\partial z} - f(x_1 z)\frac{\partial x_1}{\partial z}. \tag{82}$$

$$\psi_1 = \frac{e^{\frac{b}{a}z}}{2} f\left(t + \frac{z}{a}\right) + \frac{e^{-\frac{b}{a}z}}{2} f\left(t - \frac{z}{a}\right) \tag{83}$$
$$+ \frac{a}{2}e^{-bt}\int_{-\frac{z}{a}}^{t+\frac{z}{a}} f(\beta)e^{b\beta}\frac{\partial}{\partial z}J_0\left(\frac{b}{a}\sqrt{z^2 - a^2(t-\beta)^2}\right)d\beta.$$

The wave function which in a conducting medium reduces on the plane $z = 0$ to $f(t)$ and whose normal derivative reduces to $F(t)$ is determined everywhere by the expression

$$\psi(z, t) = \frac{e^{\frac{b}{a}z}}{2} f\left(t + \frac{z}{a}\right) + \frac{e^{-\frac{b}{a}z}}{2} f\left(t - \frac{z}{a}\right) \tag{84}$$
$$+ \frac{a}{2}e^{-bt}\int_{t-\frac{z}{a}}^{t+\frac{z}{a}} f(\beta)e^{b\beta}\frac{\partial}{\partial z}J_0\left(\frac{b}{a}\sqrt{z^2 - a^2(t-\beta)^2}\right)d\beta$$
$$+ \frac{a}{2}e^{-bt}\int_{t-\frac{z}{a}}^{t+\frac{z}{a}} F(\beta)e^{b\beta}J_0\left(\frac{b}{a}\sqrt{z^2 - a^2(t-\beta)^2}\right)d\beta.$$

It is apparent from (84) that the character of the propagation is profoundly modified by the presence of even a slight conductivity. At $z = 0$ the integrals vanish due to the identity of the limits. The initial pulse $f(t)$ then splits into two waves of half the initial amplitude as in the nondissipative case. These partial waves are propagated to the right and to the left with a velocity a which is necessarily independent of frequency, and also independent of the conductivity. They are furthermore attenuated exponentially in the direction of propagation, as might have been anticipated. Note that the attenuation factor is that approached by a harmonic component in the limit as $\sigma^2/\epsilon^2\omega^2 \to \infty$.

$$\frac{b}{a} = \frac{\sigma}{2}\sqrt{\frac{\mu}{\epsilon}} = 188.3\,\sigma\sqrt{\frac{\kappa_m}{\kappa_e}}, \tag{85}$$

which is identical with (53), page 277. The initial function is no longer propagated without change of form, for the integrals represent contributions to the field persisting for an infinite time at points which have been traversed by the wave front. *The wave now leaves in its wake a residue or tail which subsides exponentially with the time.* The nature of these contributions will be clarified by the numerical example to be discussed in Sec. 5.11.

5.10. Dissipative Medium; Prescribed Distribution in Space.—The initial conditions are frequently prescribed in another fashion. Let us suppose that at the instant $t = 0$ the field ψ and its time derivative $\partial\psi/\partial t$ are specified as functions of the space coordinate z.

$$\psi(z, 0) = g(z), \qquad \left(\frac{\partial\psi}{\partial t}\right)_{t=0} = G(z). \tag{86}$$

A harmonic analysis is called for in space rather than in time and a general solution may be constructed from particular solutions of the type discussed in Sec. 5.3.

$$\psi = e^{-\frac{\sigma}{2\epsilon}t}(Ae^{iqt} + Be^{-iqt})e^{ikz}, \tag{87}$$

where

$$q = \sqrt{\frac{k^2}{\mu\epsilon} - \frac{\sigma^2}{4\epsilon^2}} = \sqrt{a^2k^2 - b^2}. \tag{88}$$

The constants a and b are defined as in Sec. 5.9 but k is now a real variable. To eliminate the common exponential factor it is convenient to take

$$\psi = u(z, t)e^{-bt}, \qquad \frac{\partial\psi}{\partial t} = \left(\frac{\partial u}{\partial t} - bu\right)e^{-bt}. \tag{89}$$

The general solution is to be constructed by assuming the amplitudes A and B to depend on k and then summing over all positive and negative values of k.

$$u(z, t) = \int_{-\infty}^{\infty} [A(k)e^{iqt} + B(k)e^{-iqt}]e^{ikz}\, dk. \tag{90}$$

For the derivative, we have

$$\frac{\partial u}{\partial t} = \int_{-\infty}^{\infty} iq[A(k)e^{iqt} - B(k)e^{-iqt}]e^{ikz}\, dk. \tag{91}$$

The procedure is now essentially the same as that described in the foregoing paragraph. At $t = 0$,

$$\begin{aligned} g(z) &= \int_{-\infty}^{\infty} [A(k) + B(k)]e^{ikz}\, dk, \\ G(z) &= \int_{-\infty}^{\infty} \{ip[A(k) - B(k)] - b[A(k) + B(k)]\}e^{ikz}\, dk, \end{aligned} \tag{92}$$

and the coefficients are readily determined by the Fourier transform theorem.

$$\begin{aligned} A(k) &= \frac{1}{4\pi q}\int_{-\infty}^{\infty} [(q - ib)g(\alpha) - iG(\alpha)]e^{-ik\alpha}\, d\alpha, \\ B(k) &= \frac{1}{4\pi q}\int_{-\infty}^{\infty} [(q + ib)g(\alpha) + iG(\alpha)]e^{-ik\alpha}\, d\alpha. \end{aligned} \tag{93}$$

Substitution of these values into (90) leads, after simple reductions, to

$$\begin{aligned} u(z, t) = \frac{1}{2\pi}\int_{-\infty}^{\infty} d\alpha\, g(\alpha) \int_{-\infty}^{\infty} \cos qt\, e^{ik(z-\alpha)}\, dk \\ + \frac{b}{2\pi}\int_{-\infty}^{\infty} d\alpha\, g(\alpha) \int_{-\infty}^{\infty} \frac{\sin qt}{q} e^{ik(z-\alpha)}\, dk \\ + \frac{1}{2\pi}\int_{-\infty}^{\infty} d\alpha\, G(\alpha) \int_{-\infty}^{\infty} \frac{\sin qt}{q} e^{ik(z-\alpha)}\, dk. \end{aligned} \tag{94}$$

In virtue of (75) the function $\dfrac{\sin qt}{q} = \dfrac{\sin at\sqrt{k^2 - \frac{b^2}{a^2}}}{a\sqrt{k^2 - \frac{b^2}{a^2}}}$ can be expressed as a definite integral. To the parameters in (75) are assigned the values

$$v = at, \qquad q = -k, \qquad \lambda = b/a, \qquad \mu = -b/a,$$

whence it follows that

$$(95)\qquad \frac{\sin at\sqrt{k^2-\dfrac{b^2}{a^2}}}{a\sqrt{k^2-\dfrac{b^2}{a^2}}} = \frac{1}{2a}\int_{-at}^{at} J_0\left(\frac{b}{a}\sqrt{\beta^2-a^2t^2}\right)e^{-ik\beta}\,d\beta.$$

Let us denote the three terms on the right-hand side of (94) by u_1, u_2, and u_3 respectively. Then

$$(96)\qquad u_3 = \frac{1}{2a}\int_{-at}^{at} d\beta\left[\frac{1}{2\pi}\int_{-\infty}^{\infty} dk\int_{-\infty}^{\infty}\phi(\alpha,\beta)e^{ik(z-\alpha-\beta)}\,d\alpha\right],$$

where

$$(97)\qquad \phi(\alpha,\beta) = G(\alpha)J_0\left(\frac{b}{a}\sqrt{\beta^2-a^2t^2}\right).$$

By the Fourier integral theorem, the double integral within the brackets represents the expansion of $\phi(z-\beta,\beta)$, and consequently

$$(98)\qquad u_3 = \frac{1}{2a}\int_{-at}^{at} G(z-\beta)J_0\left(\frac{b}{a}\sqrt{\beta^2-a^2t^2}\right)d\beta,$$

or, after a change of variable,

$$(99)\qquad u_3 = \frac{1}{2a}\int_{z-at}^{z+at} G(\beta)J_0\left(\frac{b}{a}\sqrt{(z-\beta)^2-a^2t^2}\right)d\beta.$$

In like manner, one obtains

$$(100)\qquad u_2 = \frac{b}{2a}\int_{z-at}^{z+at} g(\beta)J_0\left(\frac{b}{a}\sqrt{(z-\beta)^2-a^2t^2}\right)d\beta;$$

from this u_1 is derived by differentiation with respect to t.

$$(101)\qquad u_1 = \frac{1}{2}g(z+at)+\frac{1}{2}g(z-at) + \frac{1}{2a}\int_{z-at}^{z+at} g(\beta)\frac{\partial}{\partial t}J_0\left(\frac{b}{a}\sqrt{(z-\beta)^2-a^2t^2}\right)d\beta.$$

The wave function which in a conducting medium reduces at $t = 0$ to $g(z)$ and whose time derivative at the same instant is $G(z)$ is determined everywhere by the expression

$$(102)\qquad \psi(z,t) = e^{-bt}\left\{\frac{1}{2}g(z+at)+\frac{1}{2}g(z-at) + \frac{1}{2a}\int_{z-at}^{z+at}[bg(\beta)+G(\beta)]\,J_0\left(\frac{b}{a}\sqrt{(z-\beta)^2-a^2t^2}\right)d\beta + \frac{1}{2a}\int_{z-at}^{z+at} g(\beta)\frac{\partial}{\partial t}J_0\left(\frac{b}{a}\sqrt{(z-\beta)^2-a^2t^2}\right)d\beta\right\}.$$

In its essential characteristics this solution does not differ markedly from that derived under the assumptions of Sec. 5.9. The initial field distribution along the z-axis splits into two waves which are propagated to the right and left respectively. If the conductivity is other than zero, a residue or tail remains after the passage of the wave front, and the field subsides exponentially with the time.

5.11. Discussion of a Numerical Example.—The physical significance of these results can be most easily elucidated in terms of a specific example. Fresh water is a convenient medium for such a discussion, for at radio frequencies it occupies a position midway between the good conductors and the good dielectrics. Let us assume the following constants:

$$\kappa_m = 1, \qquad \kappa_e = 81, \qquad \sigma = 2 \times 10^{-4} \quad \text{mho/meter.}$$

From these values we calculate

$$a = \tfrac{1}{3} \times 10^8 \text{ meters/sec.}, \qquad b = 1.4 \times 10^5 \text{ sec.}^{-1},$$
$$\frac{b}{a} = 4.2 \times 10^{-3} \text{ meter}^{-1}.$$

It is usually easier to visualize a distribution and propagation along an axis in space than along one in time; for this reason we shall consider (102) first. Let us suppose that at $t = 0$ the field vanishes everywhere except within the region $-25 < z < 25$, where its amplitude is unity.

$$\begin{aligned} g(z) &= 1, && \text{when } |z| < 25, \\ g(z) &= 0, && \text{when } |z| > 25, \\ G(z) &= 0, && \text{for all values of } z. \end{aligned}$$

In Fig. 51a $\frac{1}{2}[g(z + at) + g(z - at)]$ is plotted as a function of z for various values of t, and in Fig. 51b the same as a function of t for various values of z. These two figures represent the propagation of the initial pulse in time and space in a medium of zero conductivity. The function $g(z - at)$ vanishes if

$$z < at - 25 \qquad \text{or} \qquad z > at + 25,$$

or, if

$$t < \frac{1}{a}(z - 25) \qquad \text{or} \qquad t > \frac{1}{a}(z + 25).$$

Likewise the negative wave $g(z + at)$ is zero at all points for which

$$z < -at - 25 \qquad \text{or} \qquad z > -at + 25,$$

or

$$t < -\frac{1}{a}(25 + z) \qquad \text{or} \qquad t > \frac{1}{a}(25 - z).$$

In Figs. 52*a* and 52*b* the same wave functions are multiplied by the attenuation factor e^{-bt}. Nothing has been stated regarding the origin

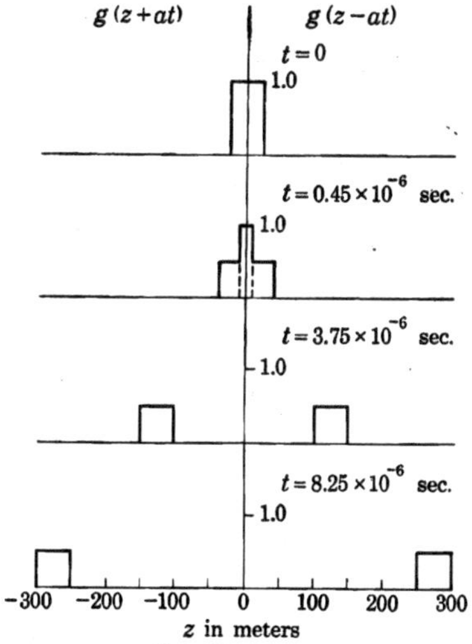

FIG. 51*a*.

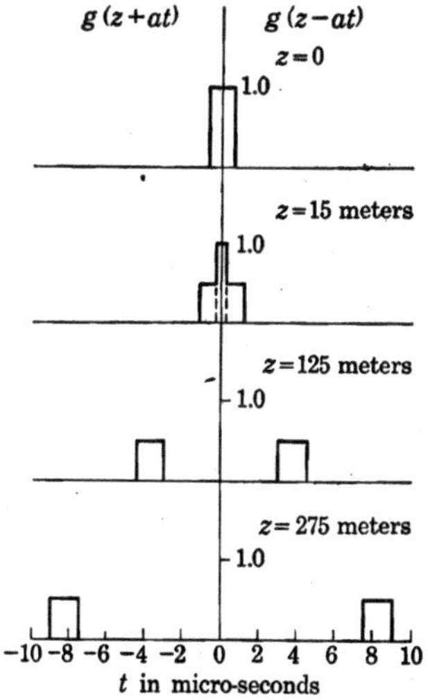

FIG. 51*b*.

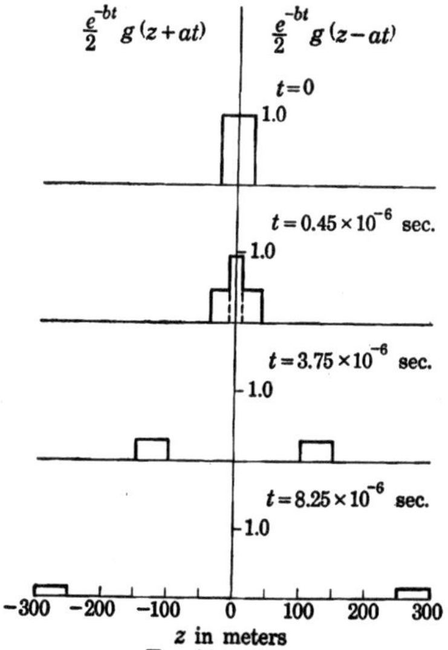

FIG. 52*a*.

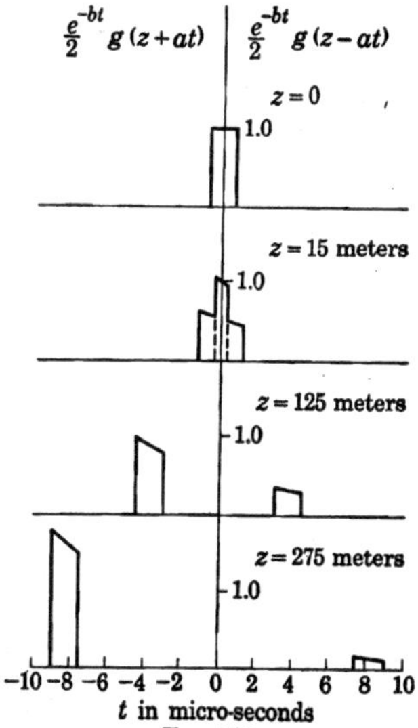

FIG. 52*b*.

of the field, but it is supposed that at some anterior time it was generated by a system of appropriately disposed sources from whence the field is propagated and attenuated in such a manner that at $t = 0$ it is distributed as shown at the top of Fig. 52a. In the present instance it is clear that the wave started as two pulses at some value of $t < -10 \times 10^{-6}$ sec., the one at a distant point on the positive z-axis, propagated to the left, and the other at a point on the negative z-axis, propagated to the right. In the neighborhood of $t = 0$, $z = 0$, they meet, pass through one another with consequent reinforcement, and then continue on their way. Their subsequent progress is indicated at $t = 0.45$, $t = 3.75$, and $t = 8.25$ microsec.. The succession of events occurring at a fixed point in space is illustrated in the series of Fig. 52b. Take, for example, $z = 275$ meters. At $t = -9$ microsec. the front of a negative wave arrives at this point, traveling toward smaller values of z. An interval of 1.5 sec. is now required for this wave to pass, in the course of which its amplitude is diminished exponentially. From $t = -7.5$ to $t = 7.5$ microsec. all is quiet; then arrives the front of the positive wave with greatly reduced amplitude, for it has passed through unit value at the origin and traveled on to the positive point of observation.

This simple history of a field must now be modified to take into account the contributions of the integral terms in Eq. (102). In most cases the evaluation of these terms necessitates numerical or mechanical processes of integration, but thanks to the simple functional form of $g(z)$ in the present instance the integration can be carried through analytically for sufficiently small values of z and t. From the theory of Bessel functions we take the following relations:

$$J_0(x) = 1 - \frac{x^2}{2^2} + \frac{x^4}{2^2 \cdot 4^2} - \frac{x^6}{2^2 \cdot 4^2 \cdot 6^2} + \cdots,$$

$$\frac{dJ_0(x)}{dx} = -J_1(x),$$

$$J_1(x) = \frac{x}{2} - \frac{x^3}{2^3 \cdot 2!} + \frac{x^5}{2^5 \cdot 2! \cdot 3!} - \cdots.$$

Introducing the appropriate value of x leads to

$$J_0\left(\frac{b}{a}\sqrt{(z-\beta)^2 - a^2t^2}\right) = 1 - \frac{1}{4}\left(\frac{b}{a}\right)^2 [(z-\beta)^2 - a^2t^2] + \frac{1}{64}\left(\frac{b}{a}\right)^4 [(z-\beta)^2 - a^2t^2]^2 - \cdots$$

$$\frac{\partial}{\partial t} J_0\left(\frac{b}{a}\sqrt{(z-\beta)^2 - a^2t^2}\right) = \frac{b^2t}{2}\left\{1 - \frac{1}{8}\left(\frac{b}{a}\right)^2 [(z-\beta)^2 - a^2t^2] + \cdots\right\}.$$

An investigation of the convergence of these series indicates that in the present instance no serious error will be committed if we limit ourselves to the first two terms, provided that z does not exceed 300 meters. The

Bessel functions will, therefore, be represented approximately by

$$J_0\left(\frac{b}{a}\sqrt{(z-\beta)^2 - a^2t^2}\right) = 1 + \frac{1}{4}b^2t^2 - \frac{1}{4}\left(\frac{b}{a}\right)^2 (z-\beta)^2,$$

$$\frac{\partial}{\partial t}J_0\left(\frac{b}{a}\sqrt{(z-\beta)^2 - a^2t^2}\right) = \frac{b^2t}{2}\left[1 + \frac{1}{8}b^2t^2 - \frac{1}{8}\left(\frac{b}{a}\right)^2 (z-\beta)^2\right].$$

These functions are now to be multiplied by $g(\beta)$ and integrated with respect to β. With regard to the limits we note the following. Suppose that t is positive. If $z - at > 25$ or $z + at < -25$, the resultant field $\psi(z, t)$ is zero. If $-at + 25 < z < at - 25$, the partial fields $g(z + at)$ and $g(z - at)$ are zero but the integrals do not vanish, and this no matter

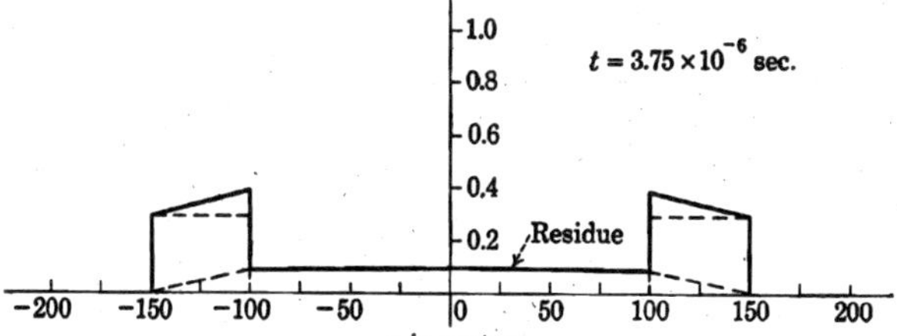

FIG. 53*a*.

how large t may become. *A residual field lingers in the region previously traversed by wave pulses.* If $z + at > 25$ and $z - at < -25$, the limits of integration may be fixed at ± 25, for beyond these values $g(\beta)$ is zero. Such limits are, therefore, appropriate for the entire region between the pulses, $-at + 25 < z < at - 25$.

The results of the integration are plotted in Figs. 53*a* and 53*b*. The effective field $\psi(z, t)$ is indicated by a heavy black outline; the partial

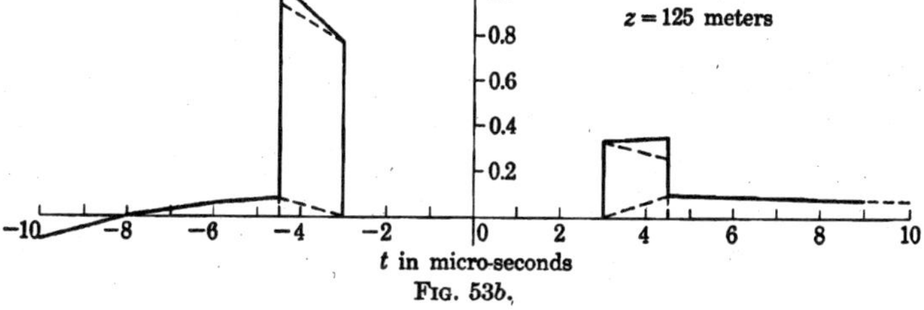

FIG. 53*b*.

fields expressed by the various terms of (102) are shown dotted. Figure 53*a* illustrates the distribution along the z-axis at the instant $t = 3.75$ microsec. We note the beginning of a deformation of the initially rectangular pulse, the top sloping forward and the residual tail trailing behind. As the wave progresses, the sharp front decreases in height, so that tail and pulse eventually merge into a rounded contour. In Fig. 53*b* observations are made at the fixed point $z = 125$ meters. At times anterior to $t = -3$ microsec. a field existed here, a field which was wiped out by the passage of the negative pulse. At $t = 3$ microsec. the positive wave arrives, followed by its tail which persists after the passage of the pulse.

The essential difference between the solutions represented in Eqs. (84) and (102) is due to an interchange of the roles of space and time. To illustrate the behavior of the field, let it be assumed that the initial time function $f(t)$ in Eq. (84) is a rectangular pulse defined by

$$\begin{aligned} f(t) &= 1, && \text{when } |t| < 1 \times 10^{-6} \text{ sec.,} \\ f(t) &= 0, && \text{when } |t| > 1 \times 10^{-6} \text{ sec.,} \\ F(t) &= 0, && \text{for all values of } t. \end{aligned}$$

The functions $\dfrac{e^{\frac{b}{a}z}}{2} f\left(t + \dfrac{z}{a}\right)$ and $\dfrac{e^{-\frac{b}{a}z}}{2} f\left(t - \dfrac{z}{a}\right)$ are plotted in Figs. 54*a* and 54*b* at various values of z and t. In virtue of the definition of $f(\beta)$ it is clear that the third term on the right-hand side of (84) contributes

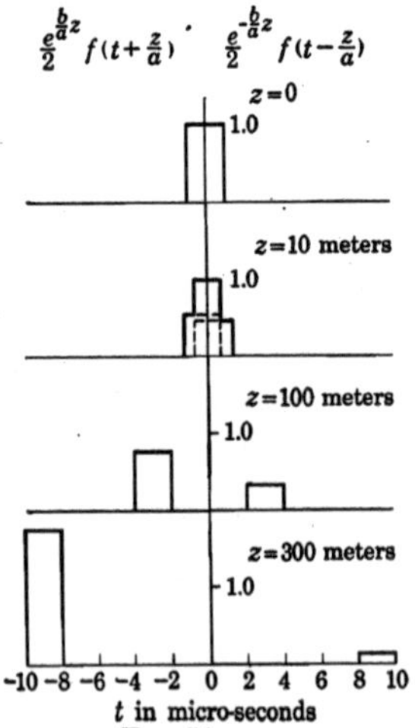

FIG. 54*a*.

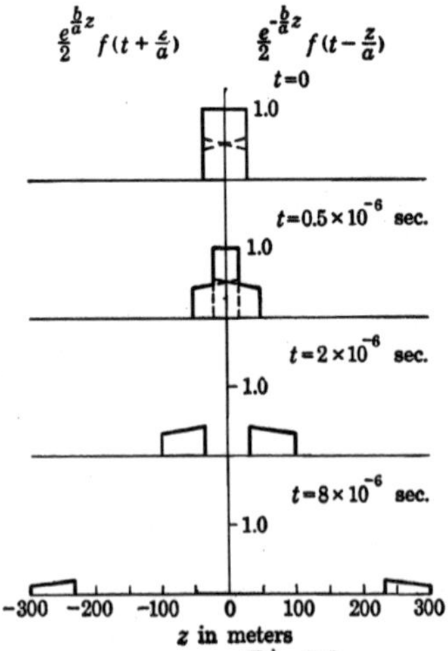

FIG. 54*b*.

nothing if $t - \dfrac{z}{a} > 1$ or $t + \dfrac{z}{a} < -1$. If $1 - \dfrac{z}{a} < t < \dfrac{z}{a} - 1$, the functions $f\left(t + \dfrac{z}{a}\right)$ and $f\left(t - \dfrac{z}{a}\right)$ vanish but the integral does not. In this interval, which lies between the pulses on the t-axis, the limits of integration are ± 1. In Figs. 55*a* and 55*b* are shown the deformations of an initially rectangular pulse at fixed points in time and space due to the tail of the wave.

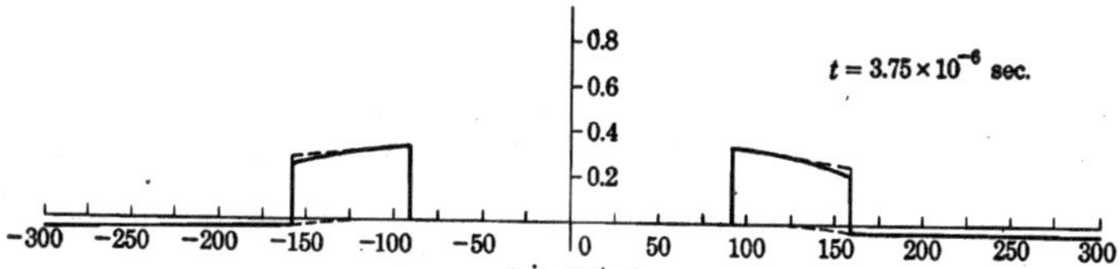

FIG. 55*a*.

When studying these figures it must be kept in mind that two *pre-existing* waves coming from sources at infinity have been assumed which superpose to give just the correct distribution at $t = 0$ or $z = 0$.

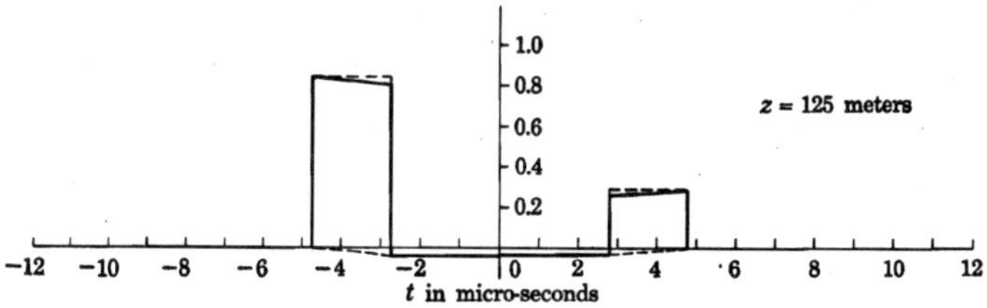

FIG. 55*b*.

A single wave whose amplitude is zero for all values of t and z less than zero can be generated only by a source at the origin constituting a singularity in the field—a problem we have not yet considered.

5.12. Elementary Theory of the Laplace Transformation.—Among the oldest and most important of the various methods devised for the solution of linear differential equations is that of the Laplace transformation. If in the equation

$$\frac{d^2w}{dz^2} + p(z)\frac{dw}{dz} + q(z)w = 0 \tag{103}$$

the dependent variable w is transformed by the relation

$$w(z) = \int u(t)e^{-zt}\,dt, \tag{104}$$

it will be found in many cases that $u(t)$ satisfies a differential equation which is simpler than (103), and in fact that $u(t)$ is frequently an elementary function whereas $w(z)$ cannot, in general, be expressed directly in terms of elementary transcendentals.[1] When the coefficients $p(z)$ and $q(z)$ are functions of the independent variable z, it is necessary to

[1] INCE, "Ordinary Differential Equations," Chaps. VIII and XVIII, Longmans, London, 1927.

choose the path of integration properly in the complex plane of t in order that (104) shall be a solution of (103); the independent solutions are then distinguished by the choice of path. The result is a representation of the particular solutions as contour integrals. If, however, the coefficients p and q are constant, the equation has no other singularities than the essential one at infinity and the paths of integration can be chosen to coincide with the real axis. The solutions are then expressed in terms of infinite integrals intimately related, if not equivalent, to a Fourier integral representation.

In recent years the application of the Laplace transformation to linear equations with constant coefficients has been considered with growing interest on the part of physicists and engineers, and its orderly and rigorous procedure appears to be rapidly replacing the quasi-empirical methods of the Heaviside operational calculus. Although the Laplace methods lead to no results unobtainable by direct application of Fourier integral analysis, they do offer certain definite advantages from the standpoint of convenience. They are particularly well adapted to the treatment of functions which do not vanish at $+\infty$ and which consequently fail to satisfy the condition of absolute convergence, and they lead in a simple and direct manner to the solution of an equation in terms of its initial conditions. On the other hand, they can be applied only to problems in which the field may be assumed to be zero for all negative values of the independent variable; in short, the Laplace integral is applicable to problems wherein all the future is of importance, but the past of no consequence.

We shall approach the theory of the Laplace transform from the standpoint of the Fourier integral and suggest rudimentary proofs for the more important theorems. Let us consider a function $f(t)$ which vanishes for all negative values of t. Then apart from a factor $1/\sqrt{2\pi}$ the Fourier transform of $f(t)$ is

$$F(\omega) = \int_0^\infty f(t)e^{-i\omega t}\,dt, \tag{105}$$

provided only that the integral $\int_0^\infty |f(t)|\,dt$ exists. In case $f(t)$ does not vanish properly at infinity the integral fails to converge, but under certain circumstances absolute convergence can be restored by introducing a factor $e^{-\gamma t}$. The Fourier transform of $e^{-\gamma t}f(t)$ is then

$$F(\gamma + i\omega) = \int_0^\infty f(t)e^{-(\gamma+i\omega)t}\,dt. \tag{106}$$

If there exists a real number γ, such that

$$\lim_{T\to\infty} \int_0^T |e^{-\gamma t}f(t)|\,dt < \infty, \tag{107}$$

then $f(t)$ will be said to be transformable. The lower bound γ_a of all the γ's which satisfy (107) is called the abscissa of absolute convergence. Suppose, for example, that $f(t) = e^{bt}$ when $t \geq 0$ and $f(t) = 0$ when $t < 0$. Then $f(t)$ is transformable in the sense of (106) and $\gamma > b = \gamma_a$. On the other hand a function $f(t) = e^{t^2}$ is not transformable, for there exists no real number γ leading to the convergence of (106).

The inverse transformation follows from the reciprocal properties of the Fourier transforms, Eqs. (21) and (22).

$$f(t)e^{-\gamma t} = \frac{1}{2\pi}\int_{-\infty}^{\infty} F(\gamma + i\omega)e^{it\omega}\,d\omega, \tag{108}$$

where $\gamma > \gamma_a$. The function defined by the integral on the right is equal to $f(t)e^{-\gamma t}$ only for positive values of t, and vanishes whenever $t < 0$. To give explicit expression to this fact we introduce the unit step function $u(t)$ which is zero when $t < 0$ and equal to unity when $t > 0$. The inverse transformation may thus be written

$$f(t) \cdot u(t) = \frac{1}{2\pi}\int_{-\infty}^{\infty} F(\gamma + i\omega)e^{(\gamma+i\omega)t}\,d\omega. \tag{109}$$

If $\gamma_a \leq 0$, one may pass to the limit as $\gamma \rightarrow 0$ after the integrations indicated in (106) or (109) have been effected and thus determine the Fourier transforms of functions not otherwise integrable. Such a step is unessential, and the Laplace transform contains this convergence factor implicitly. Let us introduce the complex variable $s = \gamma + i\omega$. Then the Laplace transform of $f(t)$, henceforth designated by the operator $L[f(t)]$, is

$$L[f(t)] = \int_0^{\infty} f(t)e^{-st}\,dt = F(s), \qquad Re(s) > \gamma_a, \tag{110}$$

where $Re(s)$ is the usual abbreviation for "the real part of s," and γ_a is determined in each case by the functional properties of $f(t)$. The inverse transformation is represented by an integral taken along a path in the complex plane of s.

$$L^{-1}[F(s)] = \frac{1}{2\pi i}\int_{\gamma-i\infty}^{\gamma+i\infty} F(s)e^{ts}\,ds = f(t) \cdot u(t), \tag{111}$$

where $\gamma > \gamma_a$. The Laplace transformation can be interpreted as a mapping of points lying on the positive real axis of t onto that portion of the complex plane of s which lies to the right of the abscissa γ_a. The domains of $f(t)$ and its transform $F(s)$ are indicated in Fig. 56.

From the linearity of the operators it follows that if $L[f_1(t)] = F_1(s)$, $L[f_2(t)] = F_2(s)$, then

$$L[f_1(t) \pm f_2(t)] = F_1(s) \pm F_2(s). \tag{112}$$

If, furthermore, a is any parameter independent of t and s, then

$$L[af(t)] = aF(s). \tag{113}$$

The application of the Laplace transform theory to the solution of differential equations is based on a theorem concerning the transform of a derivative. Assume that $f(t)$ and its derivative $\frac{d}{dt}f(t)$ are transformable in the sense of (107), and that $f(t)$ is continuous at $t = 0$. Then by partial integration

$$\int_0^\infty \frac{df}{dt} e^{-st}\, dt = f(t)e^{-st}\Big|_0^\infty - \int_0^\infty f(t)\frac{de^{-st}}{dt}\, dt;$$

hence, if $L[f(t)] = F(s)$, it follows that

$$L\left[\frac{df(t)}{dt}\right] = sF(s) - f(0). \tag{114}$$

The importance of this theorem lies in the fact that it introduces the initial value $f(0)$ of the function $f(t)$. Integrating partially a second time

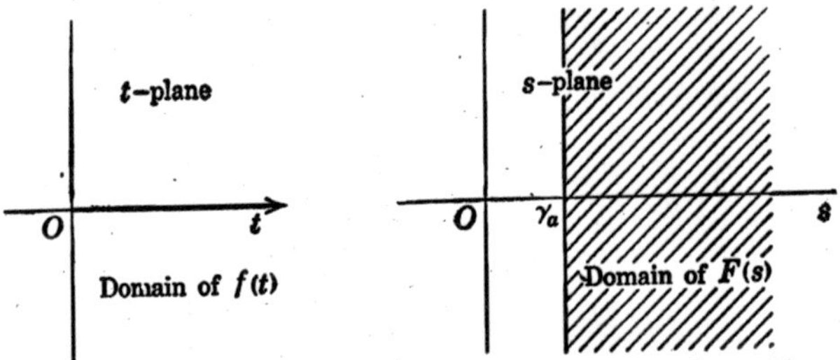

FIG. 56.—A Laplace transformation effects a mapping of points on the positive real axis of t onto the shaded portion of the s-plane.

leads to a theorem for the transform of a second derivative which involves not only the initial value $f(0)$ but also the value of the derivative $(df/dt)_{t=0}$.

$$L\left[\frac{d^2f(t)}{dt^2}\right] = s^2F(s) - sf(0) - \left(\frac{df}{dt}\right)_{t=0}. \tag{115}$$

Thus the transform of the dependent variable of a second-order differential equation is expressed in terms of the initial conditions. These results have no analogue in the theory of the Fourier transform.

The "Faltung" or "folding" theorem is a particular application of a well-known property of the Fourier integral.[1] Suppose again that

[1] See for example Bochner "Vorlesungen über Fouriersche Integrale," and Wiener, "The Fourier Integral," p. 45, Cambridge University Press, 1933.

$L[f_1(t)] = F_1(s)$, $L[f_2(t)] = F_2(s)$, and that we wish to determine the Laplace transform of the product function $f_1(t) \cdot f_2(t)$.

$$(116)\quad L[f_1(t) \cdot f_2(t)] = \int_0^\infty f_1(t) \cdot f_2(t) e^{-st}\, dt$$
$$= \int_0^\infty f_1(t) \cdot \left[\frac{1}{2\pi i} \int_{\gamma_2 - i\infty}^{\gamma_2 + i\infty} F_2(\sigma) e^{\sigma t}\, d\sigma \right] e^{-st}\, dt$$
$$= \frac{1}{2\pi i} \int_{\gamma_2 - i\infty}^{\gamma_2 + i\infty} d\sigma\, F_2(\sigma) \int_0^\infty f_1(t) e^{-(s-\sigma)t}\, dt,$$

and, hence, the first theorem:

$$(117)\quad L[f_1(t) \cdot f_2(t)] = \frac{1}{2\pi i} \int_{\gamma_2 - i\infty}^{\gamma_2 + i\infty} F_1(s - \sigma) \cdot F_2(\sigma)\, d\sigma,$$
$$\gamma_2 > \gamma_{a_2}, \qquad Re(s - \sigma) > \gamma_{a_1}.$$

Of more frequent use is a second Faltung theorem on the inverse transformation of a product of transforms. We wish to determine a function whose Laplace transform is $F_1(s) \cdot F_2(s)$.

$$(118)\quad L^{-1}[F_1(s) \cdot F_2(s)] = \frac{1}{2\pi i} \int_{\gamma - i\infty}^{\gamma + i\infty} F_1(s) \cdot F_2(s) e^{st}\, ds$$
$$= \frac{1}{2\pi i} \int_{\gamma - i\infty}^{\gamma + i\infty} F_1(s) \left[\int_0^\infty f_2(\tau) e^{-s\tau}\, d\tau \right] e^{st}\, ds$$
$$= \int_0^\infty d\tau\, f_2(\tau) \cdot \frac{1}{2\pi i} \int_{\gamma - i\infty}^{\gamma + i\infty} F_1(s) e^{s(t-\tau)}\, ds$$
$$= \int_0^\infty f_2(\tau) \cdot f_1(t - \tau) \cdot u(t - \tau)\, d\tau.$$

Since the unit function vanishes for negative values of the argument, the upper limit of τ must be t and hence

$$(119)\quad L^{-1}[F_1(s) \cdot F_2(s)] = \int_0^t f_1(t - \tau) \cdot f_2(\tau)\, d\tau,$$

or

$$(120)\quad L\left[\int_0^t f_1(t - \tau) \cdot f_2(\tau)\, d\tau \right] = F_1(s) \cdot F_2(s).$$

Geometrically, the product $F_1(s) \cdot F_2(s)$ represents the transform of an area constructed as follows: The function $f_1(\tau)$ is first folded over onto the negative half plane by replacing τ by $-\tau$, and then translated to the right by an amount t. The ordinate $f_1(t - \tau)$ is next multiplied by $f_2(\tau)\, d\tau$ and the area integrated from 0 to t.

A set of operations applied repeatedly in the course of our earlier discussion of the Fourier integral may be referred to as translations. Let a be a parameter independent of s and t.

$$L[e^{at}f(t)] = \int_0^\infty f(t)e^{-(s-a)t}\,dt = F(s-a). \tag{121}$$

Equation (121) is obviously valid for both positive and negative values of a. Next consider the transform of $f(t-a)$.

$$L[f(t-a)] = \int_0^\infty f(t-a)e^{-st}\,dt = e^{-as}\int_{-a}^\infty f(\tau)e^{-s\tau}\,d\tau. \tag{122}$$

Now it has been postulated throughout that the function $f(t)$ shall vanish for all negative values of the argument and hence the lower limit $-a$ may be replaced by zero. As a consequence, we obtain

$$L[f(t-a)] = e^{-as}F(s), \qquad \begin{array}{l} a > 0, \\ f(t) = 0 \end{array} \qquad \text{when } t < 0. \tag{123}$$

Likewise, we may write

$$L[f(t+a)] = e^{as}F(s), \qquad \begin{array}{l} a > 0, \\ f(t) = 0 \end{array} \qquad \text{when } t < a; \tag{124}$$

but one must note with care that this last result applies only to functions $f(t)$ which by definition vanish for $t < a$. A replacement of t by $t - a$ or $t + a$ represents a translation of $f(t)$ a distance a to the right or left respectively.

In addition to the foregoing fundamental theorems there exist a number of elementary but useful relations which may be deduced directly from (110). The following formulas are set down for convenience and their proof is left to the reader.

$$L\left[f\left(\frac{t}{a}\right)\right] = aF(as); \tag{125}$$

$$L[tf(t)] = -\frac{dF(s)}{ds}; \tag{126}$$

$$L\left[\frac{1}{t}f(t)\right] = \int_s^\infty F(\sigma)\,d\sigma; \tag{127}$$

$$L\left[\frac{df(t, a)}{da}\right] = \frac{dF(s, a)}{da}; \tag{128}$$

$$f(0) = \lim_{s\to\infty} sF(s); \tag{129}$$

$$\lim_{t\to\infty} f(t) = \lim_{s\to 0} sF(s). \tag{130}$$

Evaluation of the contour integrals that occur in the inverse transformations (111) is most easily effected by application of the theory of residues. The functions $f(t)$ with which one has to deal are analytic functions of the complex variable t except at a finite number of poles. It will be recalled that in the neighborhood of such a point, say $t = a$, the function may be expanded in a series of the form

$$f(t) = \frac{b_{-m}}{(t-a)^m} + \frac{b_{-m+1}}{(t-a)^{m-1}} + \cdots \frac{b_{-1}}{t-a} + \phi(t), \tag{131}$$

where $\phi(t)$ is analytic near and at a and, therefore, contains only positive powers of $t - a$. The singularity is said to be a pole of order m if m is a finite integer. If, however, an infinite number of negative powers is necessary for the representation of $f(t)$, the singularity is said to be essential. If now the function $f(t)$ is integrated about any closed contour C in the t-plane which encircles the pole a but no other singularity, it can be shown that all the terms in the expansion (131) vanish with the exception of the one whose coefficient is b_{-1}. The result is

$$\int_C f(t)\,dt = 2\pi i b_{-1}. \tag{132}$$

The coefficient b_{-1} is called the residue at the pole. In case the contour encloses a number of poles, the value of the integral is equal to $2\pi i$ times the sum of the residues. In particular, if $f(t)$ is analytic throughout the region bounded by the contour C, the integral vanishes. These results are a consequence of a fundamental theorem of function theory. If $f(t)$ is analytic on and within the contour C,—which is to say that it may be expanded in terms of positive powers of t only—then the value of $f(t)$ at any interior point a may be expressed as

$$f(a) = \frac{1}{2\pi i}\int_C \frac{f(t)}{t-a}\,dt. \tag{133}$$

This is known as Cauchy's theorem.

If the function $f(t)$ vanishes properly at infinity, an integral which is extended, as in (111), along an infinite line may be closed by a circuit of infinite radius and thus made subject to the theorems of the preceding paragraph. The restrictions that govern the application of the theory of residues to the evaluation of infinite integrals are specified in what is known as Jordan's lemma quoted as follows:

If $Q(z) \to 0$ uniformly with regard to $\arg z$ as $|z| \to \infty$ when $0 \leq \arg z \leq \pi$, and if $Q(z)$ is analytic when both $|z| > c$ (a constant) and $0 \leq \arg z \leq \pi$, then

$$\lim_{\rho\to\infty} \int_\Gamma e^{miz} Q(z)\,dz = 0,$$

where Γ is a semicircle of radius ρ above the real axis with center at the origin.[1]

[1] WHITTAKER and WATSON, *loc. cit.*, p. 115. In Chaps. V and VI of this reference the reader will find all that is essential for the application of function theory to the solution of physical problems.

As stated, this lemma applies to the evaluation of integrals along the real axis from $-\infty$ to $+\infty$, but it may be modified to include integrations along an imaginary axis; by appropriate changes the circuit may be transferred from the half plane lying above the real axis to that below, or from the right half plane to the left. In case $Q(z)$ has poles within the closed contour at distances $|z| < c$ from the origin, the value of the integral is equal not to zero but to the sum of the residues.

The bearing of these theorems on the theory of the Laplace transform may be illustrated by two elementary examples. Consider first the step function defined by

$$\begin{aligned} f(t) &= 0, \qquad \text{when } t < 0, \\ f(t) &= 1, \qquad \text{when } t > 0. \end{aligned} \tag{134}$$

It will be noted in passing that (134) does not fall within the province of functions admitting a Fourier analysis, for the integral $\int_0^\infty |f(t)|\, dt$ is nonconvergent. The Laplace transform of (134) is

$$L\,[f(t)] = \int_0^\infty 1 \cdot e^{-st}\, dt = \frac{1}{s} = F(s). \tag{135}$$

We verify this result by applying the inverse transformation (111).

$$L^{-1}\left(\frac{1}{s}\right) = \frac{1}{2\pi i}\int_{\gamma - i\infty}^{\gamma + i\infty} \frac{1}{s}\, e^{st}\, ds. \tag{136}$$

Now $F(s)$ has a pole of first order at $s = 0$. The abscissa of absolute convergence is zero, whence it is necessary that $Re(s) > 0$. The integration is to be extended along any line parallel to and to the right of the imaginary axis. The integrand of (136) satisfies the conditions of Jordan's lemma and the path of integration may be closed by an infinite circuit. If $t > 0$, the path may be deformed to the left so that $Re(s) \to -\infty$. The closed contour $\gamma - i\infty$ to $\gamma + i\infty$ to $-\infty + i\infty$ to $-\infty - i\infty$ to $\gamma - i\infty$ contains the pole at $s = 0$. The residue is unity, and hence by (132) the right-hand side of (136) is also unity for $t > 0$. Since $F(s)$ is only defined in the right half plane, a question may arise as to the justice of extending the integration to the left half plane. One must be careful to distinguish between a function and the representation of that function in a restricted domain. $1/s$ is a representation of $F(s)$ when $Re(s) > 0$, but in view of possible discontinuities in $F(s)$ the two functions need not coincide at all when $Re(s) < 0$. Now along the line $\gamma - i\infty$ to $\gamma + i\infty$ we are integrating the function $1/s$; for purposes of integration we may make use of any of its analytic properties. The analytic continuation into the left half plane is of the function $1/s$, and not of $F(s)$.

When $t < 0$, the vanishing of the integrand may be assured by deforming the path to the right. Within the closed contour $\gamma - i\infty$ to $\gamma + i\infty$ to $\infty + i\infty$ to $\infty - i\infty$ to $\gamma - i\infty$ there are no singularities and the right-hand side of (136) is zero.

We have verified that the unit step function may be expressed analytically by

$$u(t) = \frac{1}{2\pi i}\int_{\gamma - i\infty}^{\gamma + i\infty} \frac{1}{s} e^{st}\, ds, \qquad \gamma > 0. \tag{137}$$

It is interesting to apply the translation theorem (123) to this result, putting $a = z/v$.

$$L\left[u\left(t - \frac{z}{v}\right)\right] = e^{-\frac{z}{v}s} F(s) = \frac{1}{s} e^{-\frac{z}{v}s}, \tag{138}$$

or

$$u\left(t - \frac{z}{v}\right) = \frac{1}{2\pi i}\int_{\gamma - i\infty}^{\gamma + i\infty} \frac{1}{s} e^{s\left(t - \frac{z}{v}\right)}\, ds. \tag{139}$$

From the preceding discussion it is apparent that (139) vanishes when $t - \frac{z}{v} < 0$, and is equal to unity when $t - \frac{z}{v} > 0$. In Figs. 57a and 57b $u\left(t - \frac{z}{v}\right)$ is represented as a function of t and z respectively.

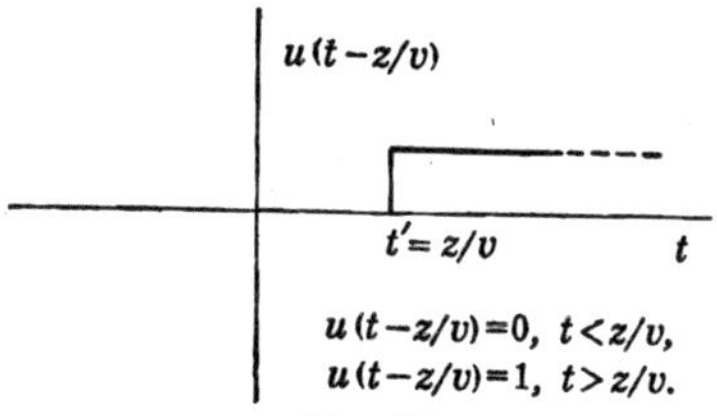

FIG. 57a.

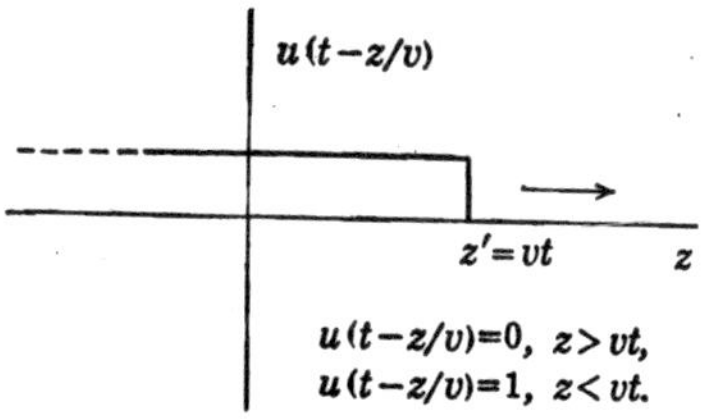

FIG. 57b.

As a second example let us imagine that a harmonic wave is switched on suddenly at $t = 0$ and continues indefinitely.

$$\begin{aligned} f(t) &= 0, && \text{when } t < 0,\\ f(t) &= e^{-i\omega t}, && \text{when } t > 0. \end{aligned} \tag{140}$$

$$L[f(t)] = \int_0^\infty e^{-i\omega t - st}\, dt = \frac{1}{s + i\omega}, \qquad Re(s) > 0. \tag{141}$$

As in the previous case we can verify (141) by applying the inverse transformation

$$L^{-1}\left[\frac{1}{s + i\omega}\right] = \frac{1}{2\pi i}\int_{\gamma - i\infty}^{\gamma + i\infty} \frac{e^{st}}{s + i\omega}\, ds, \qquad \omega \geq 0. \tag{142}$$

If $t > 0$, the path is again deformed to the left. The enclosed pole is located at $s = -i\omega$; on applying the Cauchy theorem we see that (142) is indeed equal to $e^{-i\omega t}$. On the other hand if $t < 0$, the path is deformed to the right and the integral vanishes. In general,

$$\frac{1}{2\pi i}\int_{\gamma-i\infty}^{\gamma+i\infty}\frac{e^{s\left(t-\frac{z}{v}\right)}}{s+i\omega}\,ds = \begin{cases} 0, & \text{when } t-\frac{z}{v}<0, \\ e^{-i\omega\left(t-\frac{z}{v}\right)}, & \text{when } t-\frac{z}{v}>0. \end{cases} \tag{143}$$

5.13. Application of the Laplace Transform to Maxwell's Equations.—Let $E_x = f(z, t)$ and $H_y = g(z, t)$ be the components of a plane electromagnetic field. The functions $f(z, t)$ and $g(z, t)$ are assumed to be transformable in the sense of (107) and to vanish for all negative values of t; moreover, they are related to one another by the field equations

$$\frac{\partial f}{\partial z} + \mu\frac{\partial g}{\partial t} = 0, \qquad \frac{\partial g}{\partial z} + \epsilon\frac{\partial f}{\partial t} + \sigma f = 0. \tag{144}$$

Treating z as a parameter, the Laplace transform of this system with respect to t proves to be

$$\begin{aligned} \frac{\partial F(z, s)}{\partial z} + \mu s G(z, s) - \mu g(z, 0) &= 0, \\ \frac{\partial G(z, s)}{\partial z} + \epsilon s F(z, s) + \sigma F(z, s) - \epsilon f(z, 0) &= 0. \end{aligned} \tag{145}$$

Upon elimination of $G(z, s)$ an ordinary, inhomogeneous equation for $F(z, s)$ is obtained in which s enters only as a parameter.

$$\frac{d^2F}{dz^2} - (\mu\epsilon s^2 + \mu\sigma s)F = \mu\left[\frac{\partial g(z, t)}{\partial z}\right]_{t=0} - \epsilon\mu s f(z, 0). \tag{146}$$

Since the derivative of $g(z, t)$ with respect to z at the instant $t = 0$ can be expressed in terms of $f(z, 0)$ and $(\partial f/\partial t)_{t=0}$ through (144), this last result is equivalent to

$$\frac{d^2F}{dz^2} - (\mu\epsilon s^2 + \mu\sigma s)F = -(\epsilon\mu s + \mu\sigma)f(z, 0) - \mu\epsilon\left(\frac{\partial f}{\partial t}\right)_{t=0}. \tag{147}$$

By $f(z, 0)$ one means strictly the limit of $f(z, t)$ as $t \to 0$. The functions

$$f_1(z) = \lim_{t\to 0} f(z, t), \qquad f_2(z) = \lim_{t\to 0}\frac{\partial f(z, t)}{\partial t}, \tag{148}$$

represent the *initial state of the field* and are assumed to be known. Upon writing $h^2 = \mu\epsilon s^2 + \mu\sigma s$, we obtain finally

$$\frac{d^2F}{dz^2} - h^2F = -\frac{h^2}{s}f_1(z) - \mu\epsilon f_2(z) = Z(z, s). \tag{149}$$

The general solution of (149) consists of a solution of the homogeneous equation

$$\frac{d^2F}{dz^2} - h^2F = 0 \tag{150}$$

containing two arbitrary constants, to which is added a particular solution of (149). One can verify easily enough that (149) is in fact satisfied by

$$F(z, s) = Ae^{hz} + Be^{-hz} + \frac{e^{hz}}{2h}\int e^{-hz}Z(z, s)\,dz - \frac{e^{-hz}}{2h}\int e^{hz}Z(z, s)\,dz, \tag{151}$$

in which h is that root of $\sqrt{h^2}$ which is positive when h^2 is real and positive. The constants A and B are to be determined such as to satisfy specified *boundary conditions* at the points $z = z_1$ and $z = z_2$. After A and B have been evaluated in terms of s, an inverse transformation leads from (151) back to a solution $f(z, t)$ of (144) satisfying both initial and boundary conditions.

To illustrate, consider first the elementary case in which the field is zero everywhere at the initial instant $t = 0$. Then $Z = 0$ and we are concerned only with the solution of the homogeneous equation. Let us suppose that at $z = 0$ there is a source switched on at the instant $t = 0$ which sends out a wave in the positive z-direction and whose intensity at the origin is $f(0, t) = f_3(t)$. Thus the second boundary condition at $z = z_2$ is replaced in the present example by the stipulation that the field shall be propagated to the right along the z-axis. The transform of the boundary value is represented by

$$F(0, s) = F_3(s) = L[f_3(t)]. \tag{152}$$

Since the wave is to travel to the right, the constant A must be placed equal to zero. This is clear when $\sigma = 0$, for then $h = s/a$, and by the translation theorem of page 314

$$F(z, s) = F_3(s)e^{-\frac{z}{a}s} = L\left[f_3\left(t - \frac{z}{a}\right)\right], \tag{153}$$

$$f(z, t) = f_3\left(t - \frac{z}{a}\right), \qquad (t \geq 0, \qquad z > 0). \tag{154}$$

If the conductivity is not equal to zero, the solution is more difficult. In this case we shall write $h^2 = \left(s\sqrt{\mu\epsilon} + \frac{1}{2}\sqrt{\frac{\mu}{\epsilon}}\,\sigma\right)^2 - \frac{\mu\sigma^2}{4\epsilon^2}$ and apply the integral[1]

$$\frac{e^{-hz}}{h} = \int_0^\infty J_0\left(\frac{i}{2}\sqrt{\frac{\mu}{\epsilon}}\,\sigma\beta\right) \frac{e^{-\left(\sqrt{\mu\epsilon}\,s+\frac{1}{2}\sqrt{\frac{\mu}{\epsilon}}\,\sigma\right)\sqrt{\beta^2+z^2}}}{\sqrt{\beta^2+z^2}}\,\beta\,d\beta. \tag{155}$$

Upon substitution of $\sqrt{\beta^2 + z^2} = at$, this reduces to

$$\frac{e^{-hz}}{h} = a\int_{\frac{z}{a}}^\infty e^{-bt} J_0\left(\frac{b}{a}\sqrt{z^2 - a^2t^2}\right) e^{-st}\,dt; \tag{156}$$

where $a = 1/\sqrt{\mu\epsilon}$, $b = \sigma/2\epsilon$, as defined on page 297. Finally, by differentiation we obtain

$$e^{-hz} = e^{-\frac{b}{a}z} \cdot e^{-s\frac{z}{a}} - a\int_{\frac{z}{a}}^\infty e^{-bt} \frac{\partial}{\partial z} J_0\left(\frac{b}{a}\sqrt{z^2 - a^2t^2}\right) e^{-st}\,dt, \tag{157}$$

or

$$e^{-hz} = e^{-\frac{b}{a}z} \cdot e^{-s\frac{z}{a}} - aL[\phi(z, t)], \tag{158}$$

where $\phi(z, t)$ is a function defined by

$$\phi(z, t) \begin{cases} = 0, & \text{when } 0 < t < \frac{z}{a}, \\ = e^{-bt} \frac{\partial}{\partial z} J_0\left(\frac{b}{a}\sqrt{z^2 - a^2t^2}\right), & \text{when } t \geq \frac{z}{a}. \end{cases} \tag{159}$$

With this transform at our disposal, we now write

$$F(z, s) = F_3(s)e^{-hz} = L[f_3(t)]e^{-\frac{b}{a}z} \cdot e^{-s\frac{z}{a}} - aL[f_3(t)] \cdot L[\phi]. \tag{160}$$

To this we apply the translation theorem and the Faltung theorem of page 313, followed by the inverse transformation of the complete equation.

$$f(z, t) = e^{-\frac{b}{a}z} f_3\left(t - \frac{z}{a}\right) - a\int_{\frac{z}{a}}^t f_3(t - \tau)e^{-b\tau} \frac{\partial}{\partial z} J_0\left(\frac{b}{a}\sqrt{z^2 - a^2\tau^2}\right) d\tau, \tag{161}$$

or, with a minor change of variable,

$$f(z, t) = e^{-\frac{b}{a}z} f_3\left(t - \frac{z}{a}\right) - ae^{-bt}\int_0^{t-\frac{z}{a}} f_3(\beta)e^{b\beta} \frac{\partial}{\partial z} J_0\left(\frac{b}{a}\sqrt{z^2 - a^2(t - \beta)^2}\right) d\beta. \tag{162}$$

[1] WATSON, *loc. cit.*, p. 416, (4).

This function, as the reader should verify, satisfies the wave equation in a dissipative medium and represents a wave traveling in the positive direction which at $z = 0$ reduces to $f_3(t)$ for all values of $t > 0$.

DISPERSION

5.14. Dispersion in Dielectrics.—A pulse or "signal" of any specified initial form may be constructed by superposition of harmonic wave trains of infinite length and duration. The velocities with which the constant-phase surfaces of these component waves are propagated have been shown to depend on the parameters ϵ, μ, and σ. In particular, if the medium is nonconducting and the quantities ϵ and μ are independent of the frequency of the applied field, the phase velocity proves to be constant and the signal is propagated without distortion. The presence of a conductivity, on the other hand, leads to a functional relation between the frequency and the phase velocity, as well as to attenuation. Consequently the harmonic components suffer relative displacements in phase in the direction of propagation and the signal arrives at a distant point in a modified and perhaps unrecognizable form. A medium in which the phase velocity is a function of the frequency is said to be *dispersive.*

At sufficiently high frequencies a substance may exhibit dispersive properties even when the conductivity σ due to free charges is wholly negligible. In dielectric media the phase velocity is related to the index of refraction n by $v = c/n$, where $n = \sqrt{\kappa_e\kappa_m}$. At frequencies less than 10^8 cycles/sec. the specific inductive capacities of most materials are substantially independent of the frequency, but they manifest a marked dependence on frequency within a range which often begins in the ultra-high-frequency radio region and extends into the infrared and beyond. Thus, while the refractive index of water at frequencies less than 10^8 is about 9, it fluctuates at frequencies in the neighborhood of 10^{10} cycles/sec. and eventually drops to 1.32 in the infrared. Apart from solutions or crystals of ferromagnetic salts, the dispersive action of a nonconductor can be attributed wholly to a dependence of κ_e on the frequency.

All modern theories of dispersion take into account the molecular constitution of matter and treat the molecules as dynamical systems possessing natural free periods which are excited by the incident field. A simple mechanical model which led to a strikingly successful dispersion formula was proposed by Maxwell and independently by Sellmeyer. A further advance was accomplished by Lorentz who extended the theory of the medium as a fine-grained assembly of molecular oscillators and who was able to account at least qualitatively for a large number of electrical and optical phenomena. According to Lorentz, however, these molecular systems obeyed the laws of classical dynamics; it is now known that they are in fact governed by the more stringent principles of quantum mechanics. Following upon the rapid advances of our knowledge of molecular and atomic structure revisions were made in the dispersion theory which at present may be considered to be in a very satisfactory state.

Both the classical and the quantum theories of dispersion undertake to calculate the displacement of charge from the center of gravity of an atomic system as a function of the frequency and intensity of the disturbing field. After a process of averaging over the atoms contained within an appropriately chosen volume element, one obtains an expression for the polarization of the medium; that is to say, the dipole moment per unit volume. The classical result corresponds closely in form to the quantum mechanical formula and leads in most cases to an adequate representation of the index of refraction as a function of frequency. We shall confine the discussion, therefore, to the case in which the electric polarization in the neighborhood of a resonance frequency can be expressed approximately by the real part of[1]

$$\mathbf{P} = \frac{a^2}{\bar{\omega}_0^2 - \omega^2 - i\omega g}\,\epsilon_0\mathbf{E}. \tag{1}$$

By the electric field intensity, we shall now understand the real part of the complex vector

$$\mathbf{E} = \mathbf{E}_0\, e^{-i\omega t}. \tag{2}$$

The constant a^2 is directly proportional to the number of oscillators per unit volume whose resonant frequency is ω_0. The constant $\bar{\omega}_0$ is related to this resonant frequency by

$$\bar{\omega}_0^2 = \omega_0^2 - \tfrac{1}{3}a^2, \tag{3}$$

such that $\bar{\omega}_0 \rightarrow \omega_0$ at sufficiently small densities of matter. The constant g takes account of dissipative, quasi-frictional forces introduced by collisions of the molecules. The constants $\bar{\omega}_0$ and g which characterize the molecules of a specific medium must be determined from experimental data.

At sufficiently low incident frequencies ω, the polarization $\mathbf{P}$ according to (1) approaches a constant value

$$\mathbf{P} = \frac{a^2}{\bar{\omega}_0^2}\,\epsilon_0\mathbf{E}, \tag{4}$$

[1] For the derivation of this result, see Lorentz, "Theory of Electrons," or any standard text on physical optics such as for example Born, "Optik," Springer, 1933, or Försterling. "Lehrbuch der Optik," Hirzel, 1928.

and, since the specific inductive capacity is related to the polarization by

$$\mathbf{P} = (\kappa - 1)\epsilon_0\mathbf{E}, \tag{5}$$

one may express κ in terms of the molecular constants.

$$\kappa = 1 + \frac{a^2}{\bar{\omega}_0^2}. \tag{6}$$

When, however, the incident frequency is increased, further neglect of the two remaining terms in the denominator becomes inadmissible. In that case we shall define by analogy a *complex inductive capacity* κ' through either of the equations

$$\mathbf{P} = (\kappa' - 1)\epsilon_0\mathbf{E}, \qquad \mathbf{D} = \kappa'\epsilon_0\mathbf{E}, \tag{7}$$

whence from (1) we obtain

$$\kappa' = 1 + \frac{a^2}{\bar{\omega}_0^2 - \omega^2 - i\omega g}. \tag{8}$$

In terms of this complex parameter the Maxwell equations in a medium whose magnetic permeability is μ_0 are

$$\nabla \times \mathbf{E} + \mu_0 \frac{\partial \mathbf{H}}{\partial t} = 0, \qquad \nabla \times \mathbf{H} - \epsilon_0\kappa' \frac{\partial \mathbf{E}}{\partial t} = 0, \tag{9}$$

as a consequence of which the rectangular components of the field vectors satisfy the wave equation

$$\nabla^2 \psi - \epsilon_0\mu_0\kappa' \frac{\partial^2\psi}{\partial t^2} = 0. \tag{10}$$

A plane wave solution of (10) is represented by

$$\psi = \psi_0 e^{ikz - i\omega t}, \tag{11}$$

where

$$k = \frac{\omega}{c}\sqrt{\kappa'} = \alpha + i\beta, \tag{12}$$

so that

$$\kappa' = \frac{c^2}{\omega^2}(\alpha + i\beta)^2. \tag{13}$$

The wave is propagated with a velocity $v = \omega/\alpha = c/n$, but α and the refractive index are now explicit functions of the frequency obtained by introducing (8) into (13).

In gases and vapors the density of polarized molecules is so low that κ' differs by a very small amount from unity. The constant a^2 is therefore small, so that $\bar{\omega}_0$ differs by a negligible amount from the natural frequency ω_0 and the root of κ' can be obtained by retaining the first two terms of a binomial expansion. Thus

$$\frac{c}{\omega}(\alpha + i\beta) \simeq 1 + \frac{1}{2}\frac{a^2}{\omega_0^2 - \omega^2 - i\omega g}. \tag{14}$$

When the impressed frequency ω is sufficiently low, the last two terms of the denominator may be neglected so that

$$\frac{\alpha c}{\omega} = n = 1 + \frac{a^2}{2\omega_0^2}. \tag{15}$$

The index of refraction and consequently the phase velocity in this case are independent of the frequency; there is no dispersion.

If the impressed frequency ω is appreciable with respect to the resonant frequency ω_0 but does not approach it too closely, the damping term may still be neglected. $|\omega_0^2 - \omega^2| \gg \omega g$,

$$\frac{\alpha c}{\omega} = n = 1 + \frac{1}{2}\frac{a^2}{\omega_0^2 - \omega^2}. \tag{16}$$

The attenuation factor is zero and the medium is transparent, but the refractive index and the phase velocity are functions of the frequency. If $\omega < \omega_0$, n will be greater than unity and *an increase in ω leads to an increase in n and a decrease in v.* If $\omega > \omega_0$, the refractive index is less than unity but an increase in ω still results in an increase in the numerical value of n. The dispersion in this case is said to be *normal.*

Finally let ω approach the resonance frequency ω_0. Upon resolving (14) into its real and imaginary parts one obtains

$$\begin{aligned} n &= \frac{\alpha c}{\omega} = 1 + \frac{a^2}{2}\frac{\omega_0^2 - \omega^2}{(\omega_0^2 - \omega^2)^2 + \omega^2 g^2}, \\ \beta &= \frac{a^2}{2c}\frac{g}{(\omega_0^2 - \omega^2)^2 + \omega^2 g^2}. \end{aligned} \tag{17}$$

In Fig. 58 are plotted the index of refraction and the absorption coefficient $\beta c/\omega$ of a gas as functions of frequency. The absorption coefficient exhibits a rather sharp maximum at ω_0, so that in this region the medium is opaque to the wave. As ω rises from values lying below ω_0, the index n reaches a peak at ω_1 and then falls off rapidly to a value less than unity at ω_2, whence it again increases with increasing ω, eventually approaching unity. When—as in the region $\omega_1\omega_2$—*an increase of frequency leads to a decrease in refractive index and an increase of phase velocity, the dispersion is said conventionally to be anomalous.* According to this definition dispersion introduced by the conductivity of the medium and discussed in the earlier sections of this chapter is anomalous. In fact it might almost be said that the "anomalous" behavior is more common than the "normal."

In the case of liquids and solids these results must be modified to some degree since the assumption of a small density of matter is no longer tenable, but the general form of the dispersion curves is not greatly affected. Discussion has been confined, moreover, to the behavior of the propagation constants in the neighborhood of a single resonance frequency. Actually a molecule is a complicated dynamical system possessing infinite series of natural frequencies, each affecting the reaction of the molecule to the incident field. The location of these natural periods cannot be determined by classical theory; by proper adjustment of constants to experimental data, an empirical dispersion formula can be set up, of which (8) is a typical term and which is found to satisfy the observed data over an extensive range of frequencies.

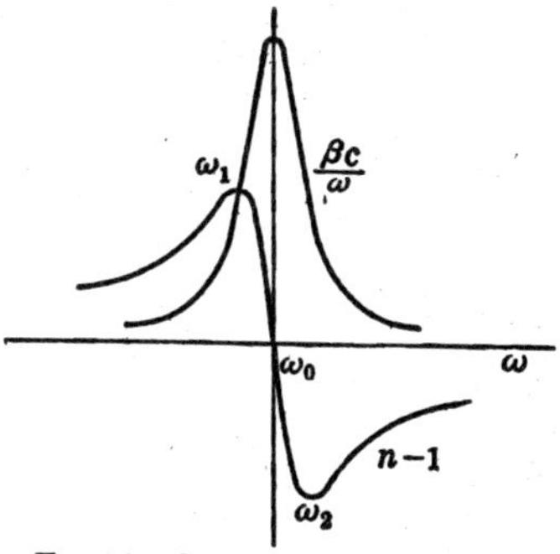

FIG. 58.—Dispersion and absorption curves in the neighborhood of a resonance frequency.

5.15. Dispersion in Metals.—A surprisingly accurate theory of the optical properties of metals may be deduced from the following rather crude model. One imagines the fixed, positive ions of a metallic conductor to constitute a region of constant electrostatic potential. Within this region a cloud or "gas" of freely circulating electrons has attained statistical equilibrium, so that on the average the force acting on any particular conduction electron is zero, and apart from normal fluctuations the total charge within any volume element is zero. The application of an external field **E** results in a general drift of the free electrons in the direction of the field. The motion is opposed by constantly recurring collisions at the lattice points occupied by the heavy ions; the consequent transfer of momentum from the drifting electrons to the lattice points results, on the one hand, in thermal vibrations of the ions, on the other in a damping of the electron motion. The exact nature of the local force exerted on a conduction electron is a somewhat delicate question, but in so far as one can ignore the contribution of bound electrons there appears to be justification for the assumption that the effective field intensity is equal to the macroscopic field **E** prevailing within the conductor.

Let **r** be the displacement from the neutral position of a free charge e whose mass is m. The charge is acted on by an external force $e\mathbf{E}$ and its motion is opposed by a force $-mg\frac{d\mathbf{r}}{dt}$ proportional to the velocity which accounts empirically for the dissipative effect of collisions. The equation for the average motion of e is, therefore,

$$m\frac{d^2\mathbf{r}}{dt^2} + mg\frac{d\mathbf{r}}{dt} = e\mathbf{E}_0 e^{-i\omega t}, \tag{18}$$

whose steady-state solution is

$$\mathbf{r} = \frac{1}{g - i\omega}\frac{ei}{\omega m}\mathbf{E}_0 e^{-i\omega t}. \tag{19}$$

If there are N free electrons in unit volume, the current density is

$$\mathbf{J} = Ne\frac{d\mathbf{r}}{dt} = \frac{N\dfrac{e^2}{m}}{g - i\omega}\mathbf{E}. \tag{20}$$

Then by analogy with the static conductivity defined by $\mathbf{J} = \sigma\mathbf{E}$, one is led to introduce a *complex conductivity* σ' which in virtue of (20) is

$$\sigma' = \frac{N\dfrac{e^2}{m}}{g - i\omega}. \tag{21}$$

The propagation constant and attenuation of a plane wave are now obtained in the usual way from the relation $k^2 = \mu\epsilon\omega^2 + i\sigma'\mu\omega$. The resonance frequencies of the bound electrons of metallic atoms are known to lie far in the violet or ultraviolet region of the spectrum, so that in the visible red, the infrared and certainly at radio frequencies the inductive capacity of a metal can be safely assumed equal to ϵ_0. Moreover if the conductor is nonferromagnetic, μ is also approximately equal to μ_0 and

$$k^2 = (\alpha + i\beta)^2 = \frac{\omega^2}{c^2}\left(1 + i\frac{\sigma'}{\omega\epsilon_0}\right). \tag{22}$$

Upon replacing σ' by (21) and separating into real and imaginary parts, one obtains

$$\begin{aligned} \alpha^2 - \beta^2 &= \frac{\omega^2}{c^2}\left(1 - \frac{Ne^2}{m\epsilon_0}\frac{1}{\omega^2 + g^2}\right), \\ 2\alpha\beta &= \frac{\omega^2}{c^2}\frac{Ne^2}{m\epsilon_0\omega}\frac{g}{\omega^2 + g^2}. \end{aligned} \tag{23}$$

At sufficiently low frequencies the inertia force $m\ddot{\mathbf{r}}$ in Eq. (18) is negligible with respect to the viscous force $mg\dot{\mathbf{r}}$. Within this range the conductivity σ' defined by (21) reduces to a real and constant value,

$$\sigma' = \frac{Ne^2}{mg}, \tag{24}$$

which must be identical with the static conductivity σ. In a series of researches of fundamental value, it was demonstrated by Hagen and Rubens[1] that the values of conductivity measured under static conditions may be employed without appreciable error well into the infrared. At wave lengths shorter than about 25×10^{-4} cm., however, the conductivity exhibits a pronounced dependence on frequency and the observed dispersion can be accounted for roughly by the relations set down in (23).

The sum and substance of these investigations is to show that dispersion formulas expressing the dependence of the parameters ϵ, μ, and σ on frequency permit the extension of classical electromagnetic theory far beyond the limits arbitrarily established in Chap. I.

5.16. Propagation in an Ionized Atmosphere.—Recent investigations of the propagation of radio waves in the ionized upper layers of the atmosphere have led to some very interesting problems. In these tenuous gases the mean free path of the electrons is exceedingly long and the damping factor consequently negligible, so that the apparent conductivity turns out to be purely imaginary,

$$\sigma' = i\frac{Ne^2}{m\omega}, \tag{25}$$

whence

$$n^2 = \frac{\alpha^2c^2}{\omega^2} = 1 - \frac{Ne^2}{m\epsilon_0\omega^2}, \qquad \beta = 0. \tag{26}$$

An electromagnetic wave is propagated in an electron atmosphere without attenuation and with a phase velocity *greater* than that of light in free space. At frequencies less than a certain critical value determined by the relation

$$\omega_0^2 = \frac{Ne^2}{m\epsilon_0} \tag{27}$$

the index of refraction becomes imaginary, and it can be shown that at the boundary of such a medium the wave will be totally reflected.

Actually the problem of propagation in the ionosphere is complicated by the presence of the earth's magnetic field. Let us suppose that a plane wave is propagated in the direction of the positive z-axis of a rectangular coordinate system. For the sake of example we shall assume only that there are N particles per unit volume of charge e and mass m in otherwise empty space. There is also present a static magnetic field that may be resolved into a component in the direction of propagation and a component transverse to this axis. Only the longi-

[1] *Ann. Physik*, **11**, 873, 1903. See Chap. IX, p. 508.

tudinal component, whose intensity will be denoted by H_0, leads to an effect of the first order.

The equations of motion of a particle are

$$(28)\qquad \begin{aligned} \ddot{x} &= \frac{e}{m} E_x + \frac{e\mu_0}{m} (\dot{y}H_0 - \dot{z}H_y), \\ \ddot{y} &= \frac{e}{m} E_y + \frac{e\mu_0}{m} (\dot{z}H_x - \dot{x}H_0), \\ \ddot{z} &= \frac{e\mu_0}{m} (\dot{x}H_y - \dot{y}H_x). \end{aligned}$$

Now the ratio of the force exerted by the magnetic vector of the traveling wave to that exerted by the electric vector is equal to the ratio of the velocity of the charge to the velocity of light. The terms involving H_x and H_y in (28) may, therefore, be neglected so that the equations of motion reduce to the simpler system

$$(29)\qquad \ddot{x} = \frac{e}{m} E_x + \frac{e\mu_0 H_0}{m} \dot{y}, \qquad \ddot{y} = \frac{e}{m} E_y - \frac{e\mu_0 H_0}{m} \dot{x}, \qquad \ddot{z} = 0.$$

Since the motion takes place entirely in the xy-plane and since the vectors of the incident wave are likewise confined to this plane, complex quantities can be employed to advantage. Let us write

$$(30)\qquad u = x + iy, \qquad E = E_x + iE_y, \qquad H = H_x + iH_y.$$

Then the equations of motion are expressed by the single complex equation

$$(31)\qquad \ddot{u} + i \frac{e\mu_0 H_0}{m} \dot{u} = \frac{e}{m} E;$$

while for the field equations one obtains

$$(32)\qquad \frac{\partial E}{\partial z} - i\mu_0 \frac{\partial H}{\partial t} = 0, \qquad \frac{\partial H}{\partial z} + i\epsilon_0 \frac{\partial E}{\partial t} = -iNe\dot{u},$$

in which the charge per unit volume times the average velocity has been introduced for current density.

A solution must now be found for this simultaneous system of *three* equations. Let us try

$$(33)\qquad E = Ae^{\pm i(hz - \omega t)}, \qquad H = Be^{\pm i(hz - \omega t)}, \qquad u = Ce^{\pm i(hz - \omega t)}.$$

These are harmonic waves traveling in the positive direction with a propagation constant h; whereas the algebraic sign of the exponent is immaterial in the absence of an applied field H_0, its choice in the present instance gives rise to two distinct solutions as we shall immediately see. For upon introducing (33) into (31) and (32) one obtains

$$
\begin{aligned}
-\frac{e}{m}A + \left(-\omega^2 \pm \frac{e\mu_0 H_0\omega}{m}\right)C &= 0, \\
hA + i\mu_0\omega B &= 0, \\
\epsilon_0\omega A + ihB + Ne\omega C &= 0.
\end{aligned}
\tag{34}
$$

In order that this homogeneous set of equations for the amplitudes A, B, C have other than a trivial solution, it is necessary that the determinant vanish.

$$
\begin{vmatrix}
-\frac{e}{m} & 0 & -\omega^2 \pm \frac{e\mu_0 H_0\omega}{m} \\
h & i\mu_0\omega & 0 \\
\epsilon_0\omega & ih & Ne\omega
\end{vmatrix} = 0.
\tag{35}
$$

Expansion of this determinant leads to an equation for the propagation constant h.

$$
\frac{c^2h^2}{\omega^2} = 1 - \frac{\frac{Ne^2}{m\epsilon_0}}{\omega^2 \mp \frac{e\mu_0 H_0\omega}{m}}.
\tag{36}
$$

The phase velocity $v = \omega/h$, and consequently (36) is equal to the square of the index of refraction n.

It turns out, therefore, that an electron atmosphere upon which is imposed a stationary magnetic field acts like an anisotropic, double-refracting crystal in that there are two modes of propagation with two distinct velocities. It is apparent, moreover, that by an appropriate choice of ω the index of refraction of the medium with respect to one of these waves becomes infinite.

$$
n_-^2 = \kappa_-' = 1 - \frac{\frac{Ne^2}{m\epsilon_0}}{\omega^2 - \frac{e\mu_0 H_0\omega}{m}}.
\tag{37}
$$

Likewise, at another frequency the index of refraction

$$
n_+^2 = \kappa_+' = 1 - \frac{\frac{Ne^2}{m\epsilon_0}}{\omega^2 + \frac{e\mu_0 H_0\omega}{m}}.
\tag{38}
$$

reduces to zero. The remarkable optical properties of the Kennelly-Heaviside layers can be largely accounted for by these formulas. A linearly polarized wave entering the medium is resolved into a right-handed and a left-handed circularly polarized component, the one propagated with the velocity v_+ and the other with the velocity v_-. Since the reflection occurring at the layer is determined by the index of refraction, the polarization of the wave returning to earth may be greatly altered and frequently contains a strong circularly polarized component which contributes to "fading" phenomena.[1]

VELOCITIES OF PROPAGATION

5.17. Group Velocity.—The concept of phase velocity applies only to fields which are periodic in space and which consequently represent wave trains of infinite duration. If a state of the medium is represented by the function $\psi(z, t)$, where

$$\psi(z, t) = Ae^{ikz-i\omega t}, \tag{1}$$

then the surfaces of constant state or phase are defined by

$$kz - \omega t = \text{constant}, \tag{2}$$

and these surfaces are propagated with the velocity

$$v = \frac{dz}{dt} = \frac{\omega}{k}. \tag{3}$$

A wave train of finite length, on the other hand, cannot be represented in the simple harmonic form (1), and the term phase velocity loses its precise significance. One speaks then rather loosely of the "velocity of light," or of the "wave-front velocity." The necessity of an exact formulation of the concept of wave velocity became apparent some years ago in connection with certain experiments fundamental to the theory of relativity, and the subject at present has an important bearing upon the problem of communication by means of short waves propagated along conductors. Since an electromagnetic field can never be completely localized in either space or time, there must be an essential arbitrariness about every definition of velocity. The slope and height of a wave front vary in the course of its progress, and the concept of a "center of gravity" grows vague as the pulse diffuses. It will probably be of more practical importance to the reader to know what certain common terms do *not* mean, than to attach to them a too precise physical significance.

[1] A more complete account of propagation in ionized media with bibliography will be found in an article by Mimno, *Rev. Mod. Phys.*, **9**, 1–43, 1937. See also the "Ergebnisse der exakten Naturwissenschaften," Vol. 17, Springer, 1938.

Next in importance to the idea of phase velocity is that of group velocity. Let us consider first the superposition of two harmonic waves which differ very slightly in frequency and wave number.

$$\psi_1 = \cos(kz - \omega t), \qquad (4)$$
$$\psi_2 = \cos[(k + \delta k)z - (\omega + \delta\omega)t].$$

The sum of these two is

$$\psi = \psi_1 + \psi_2 = 2\cos\frac{1}{2}(z\,\delta k - t\,\delta\omega)\cos\left[\left(k + \frac{\delta k}{2}\right)z - \left(\omega + \frac{\delta\omega}{2}\right)t\right], \qquad (5)$$

which is just the familiar expression for the phenomenon of "beats." The field oscillates at a frequency which differs negligibly from ω, while its effective amplitude

$$A = 2\cos\tfrac{1}{2}(z\,\delta k - t\,\delta\omega) \qquad (6)$$

varies slowly between the sum of the amplitudes of the component waves and zero. As a result of constructive and destructive interference the field distribution along both time and space axes appears as a series of

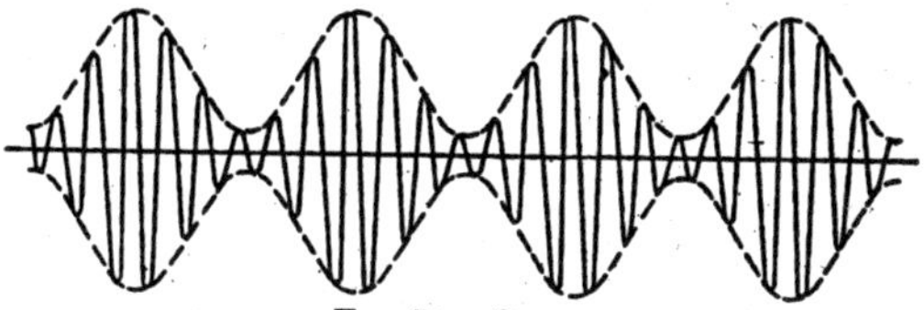

FIG. 59.—Beats.

periodically repeated "beats" or "groups" in the manner portrayed by Fig. 59. Now the surfaces over which the group amplitude A is constant are defined by the equation

$$z\,\delta k - t\,\delta\omega = \text{constant}, \qquad (7)$$

from which it follows that the groups themselves are propagated with the velocity

$$u = \frac{\delta\omega}{\delta k}. \qquad (8)$$

The group velocity is determined by the ratio of the difference of frequency to the difference of wave number. If the medium is nondispersive, $\delta k = \frac{1}{v}\,\delta\omega$ and, hence, the group velocity coincides with the phase velocity v. In a dispersive medium their values are distinct.

In the example just cited the superposition of two harmonic waves leads to a periodic series of groups. A single group or pulse of any desired shape may be constructed upon the theory of the Fourier integral by choosing the amplitude of the component harmonic waves as an appropriate function of the frequency or wave number and integrating.

$$\psi(z, t) = \int_{-\infty}^{\infty} A(k) e^{i(kz-\omega t)}\, dk. \tag{9}$$

The concept of group velocity applies only to such Fourier representations as are confined to a narrow band of the spectrum. If the amplitude function $A(k)$ is of negligible magnitude outside the region

$$k_0 - \delta k \le k \le k_0 + \delta k,$$

we may replace (9) by

$$\psi = \int_{k_0-\delta k}^{k_0+\delta k} A(k) e^{i(kz-\omega t)}\, dk, \tag{10}$$

which represents what is commonly called a "wave packet." For the present we shall assume that k is real and that ω is a known function of k. Within a sufficiently small interval $2\delta k$, $\omega(k)$ will deviate but slightly from its value at k_0 and may, therefore, be represented by the first two terms of a Taylor series.

$$\omega(k) = \omega(k_0) + \left(\frac{d\omega}{dk}\right)_{k=k_0} (k - k_0) + \cdots. \tag{11}$$

The condition that higher order terms in this expansion shall be negligible imposes the necessary restriction on δk.

$$kz - \omega t = k_0 z - \omega_0 t + (k - k_0)\left[z - \left(\frac{d\omega}{dk}\right)_{k=k_0} t\right] + \cdots. \tag{12}$$

A wave packet can now be represented by

$$\psi = \psi_0 e^{i(k_0 z - \omega_0 t)} \tag{13}$$

where ψ_0 is a mean amplitude defined by

$$\psi_0 = \int_{k_0-\delta k}^{k_0+\delta k} A(k) e^{i(k-k_0)\left[z - \left(\frac{d\omega}{dk}\right)_{k=k_0} t\right]}\, dk. \tag{14}$$

This amplitude is constant over surfaces defined by

$$z - \left(\frac{d\omega}{dk}\right)_{k=k_0} t = \text{constant}, \tag{15}$$

from which it is evident that the wave packet is propagated with the group velocity

$$u = \left(\frac{d\omega}{dk}\right)_{k_0}. \tag{16}$$

In case the medium is nondispersive, u coincides with the phase velocity v, but otherwise it is a function of the wave number k_0.

From the relations $k = 2\pi/\lambda$, $\omega = kv$, we may derive several equivalent expressions for the group velocity which are occasionally more convenient than (16).

$$u = \frac{d}{dk}(vk) = v + k\frac{dv}{dk} = v - \lambda\frac{dv}{d\lambda} = \frac{d\nu}{d\left(\frac{1}{\lambda}\right)} = \frac{1}{\frac{dk}{d\omega}}. \tag{17}$$

It is apparent from the manner in which the group velocity was defined that this concept is wholly precise only when the wave packet is composed of elementary waves lying within an infinitely narrow region of the spectrum. As the interval δk is increased, the spread in phase velocity of the harmonic components in a dispersive medium becomes more marked; the packet is deformed rapidly, and the group velocity as a velocity of the whole loses its physical significance. Note that a concentration of the field in space does not imply a corresponding concentration in the frequency or wave length spectrum, but rather the contrary. Consider, for example, the case of a harmonic wave train of finite length.

$$\begin{aligned} f(z) &= 0, & &\text{when } |z| > \frac{L}{2}, \\ f(z) &= \cos k_0 z, & &\text{when } |z| < \frac{L}{2}. \end{aligned} \tag{18}$$

The amplitude of the disturbance in the neighborhood of any wave number k is given by (40), page 292,

$$A(k) = \frac{1}{\pi}\frac{\sin\frac{(k_0 - k)}{2}L}{k_0 - k}, \tag{19}$$

and the form of this function is illustrated by Fig. 50. As the length L *increases*, the wave number interval δk within which $A(k)$ has an appreciable magnitude *decreases*. On the other hand a pulse produced by the abrupt switching on and off of a generator is diffused over the spectrum and cannot be represented even approximately by (10). This broad distribution over the spectrum is a common grief to every laboratory worker who must shield sensitive electrical apparatus from atmospheric disturbances or the surges resulting from the switching of heavy machinery.

5.18. Wave Front and Signal Velocities.—If the dispersion of the medium is normal and moderate, a pulse or wave packet may travel a great distance without appreciable diffusion; since the energy is presumed to be localized in the region occupied by the field, it is obvious that the velocity of energy propagation must equal at least approximately the group velocity. In a normally dispersive medium an increase of wave length results in an increase of phase velocity and hence by (17) the group velocity under these circumstances is always less than the phase velocity. If on the contrary the dispersion is anomalous—as is the case in conducting media—the derivative $dv/d\lambda$ is negative and the group velocity is *greater* than the phase velocity. There is in fact no lack of examples to show that u may exceed the velocity c. Since at one time it was generally believed that the group velocity was necessarily equivalent to the velocity of energy propagation, examples of this sort were proposed in the first years following Einstein's publication of the special theory of relativity as definite contradictions to the postulate that a signal can never be transmitted with a velocity greater than c. The objection was answered and the entire problem clarified in 1914 by a beautiful investigation conducted by Sommerfeld and Brillouin,[1] which may still be read with profit.

The particular problem considered by Sommerfeld is that of a signal which arrives or originates at the point $z = 0$ at the instant $t = 0$ and continues indefinitely thereafter as a harmonic oscillation of frequency ω. The signal is propagated to the right, into a dispersive medium, and we wish to know the time required to penetrate a given distance. We have seen that the Fourier integral of a nonterminating wave train does not converge, but this difficulty may be circumvented by deforming the path of integration into the complex domain of the variable ω. On the other hand, the theory of the Laplace transform was introduced for the very purpose of treating such functions; and indeed, it may be shown by a simple change of variable that the complex Fourier integral employed by Sommerfeld is exactly the transform defined in Eq. (143) of Sec. 5.12. It will facilitate the present discussion to modify the original analysis accordingly.

At $z = 0$ the signal is defined by the function

$$f(0, t) = \frac{1}{2\pi i}\int_{\gamma - i\infty}^{\gamma + i\infty} \frac{e^{st}}{s + i\omega}\,ds = \begin{cases} 0, & \text{when } t < 0, \\ e^{-i\omega t}, & \text{when } t > 0. \end{cases} \tag{20}$$

The representation of this field at any point z within the medium is obtained by extending (20) to a solution of the wave equation, which by (10), page 323, is

$$\frac{\partial^2 f}{\partial z^2} - \frac{\kappa'}{c^2}\frac{\partial^2 f}{\partial t^2} = 0. \tag{21}$$

[1] Sommerfeld, *Ann. Physik*, **44**, 177–202, 1914. Brillouin, *ibid.*, 203–240. A complete and more recent account of this work was published by Brillouin in the reports of the *Congrès International d'Électricité*, Vol. II, 1re Sec., Paris, 1932.

Equation (21) is satisfied by elementary functions of the form

$$exp\left(st - s\frac{\sqrt{\kappa'}}{c}z\right)$$

from which we construct our signal.

$$(22) \qquad f(z, t) = \frac{1}{2\pi i}\int_{\gamma-i\infty}^{\gamma+i\infty} \frac{e^{s\left(t-\sqrt{\kappa'}\frac{z}{c}\right)}}{s+i\omega}\,ds, \qquad \gamma > 0.$$

In a nondispersive medium, κ' is constant and (22) is identical with (143), page 318. If, however, the medium is dispersive, κ' is a function of s. We shall suppose that the dispersion can be represented by a formula such as (8), page 323, and upon replacing $-i\omega$ by s we obtain

$$(23) \qquad \sqrt{\kappa'} = \sqrt{1 + \frac{a^2}{s^2 + sg + \bar{\omega}_0^2}} = \beta(s).$$

To include metallic conductors we need only put $\bar{\omega}_0^2 = 0$. The wave function which for $t > 0$ reduces to a harmonic oscillation at $z = 0$ is, therefore,

$$(24) \qquad f(z, t) = \frac{1}{2\pi i}\int_{\gamma-i\infty}^{\gamma+i\infty} \frac{e^{s\left(t-\beta(s)\frac{z}{c}\right)}}{s+i\omega}\,ds, \qquad \gamma > 0.$$

As $s \to \pm i\infty$, $\beta(s) \to 1$. If now $\tau = t - \frac{z}{c} < 0$, the contour may be closed, according to Jordan's lemma, page 315, by a semicircle to the right of infinite radius. This path excludes all singularities of the integrand and consequently $f(z, t) = 0$. Thus we have proved that at a point z within the medium the field is zero as long as $t < z/c$, and hence that *the velocity of the wave front cannot exceed the constant* c.

If $\tau = t - \frac{z}{c} > 0$, the path may be closed only to the left. The singularities thus encircled occur at the pole $s = -i\omega$ and at the branch points of β. These last are located at the points where $\beta = 0$ and $\beta = \infty$. If $\beta(s)$ is written in the form

$$(25) \qquad \beta(s) = \sqrt{\frac{s^2 + sg + \bar{\omega}_0^2 + a^2}{s^2 + sg + \bar{\omega}_0^2}},$$

we see that

$$(26) \qquad \begin{aligned} \beta &= \infty, & \text{when } s &= -\frac{1}{2}g \pm \frac{i}{2}\sqrt{4\bar{\omega}_0^2 - g^2}, \\ \beta &= 0, & \text{when } s &= -\frac{1}{2}g \pm \frac{i}{2}\sqrt{4(\bar{\omega}_0^2 + a^2) - g^2}. \end{aligned}$$

The disposition of these singularities in the complex s-plane is indicated in Fig. 60*a*, where

$$(27)\quad a_{\pm} = -\frac{1}{2}g \pm \frac{i}{2}\sqrt{4\bar{\omega}_0^2 - g^2}, \qquad b_{\pm} = -\frac{1}{2}g \pm \frac{i}{2}\sqrt{4(\bar{\omega}_0^2 + a^2) - g}.$$

Now if one encircles a branch point in the plane of s, one returns to the initial value of β but with the opposite sign. This difficulty is obviated by introducing a "cut" or barrier along some line connecting a_+ and b_+, and another between a_- and b_-. Over this "cut" plane—which in the function-theoretical sense represents one sheet of the Riemann surface of $\beta(s)$—the function $\beta(s)$ is single-valued; for, since it is forbidden to traverse a barrier, any closed contour must encircle an even number of branch points. The path of integration in (24), which follows the line drawn from $\gamma - i\infty$ to $\gamma + i\infty$ and is then closed by an infinite semi-

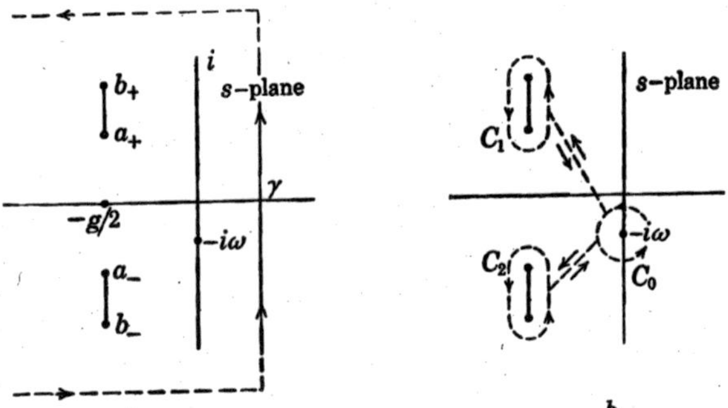

a b

FIGS. 60*a* and 60*b*.—Paths of integration in the s-plane.

circle to the left, may now be deformed in any manner on the "cut" plane without altering the value of the integral, provided only that in the process of deformation the contour does not sweep across the pole at $s = -i\omega$ or either of the two barriers. In particular the path may be shrunk to the form indicated in Fig. 60*b*. The contributions arising from a passage back and forth along the straight lines connecting C_0 and C_1 and C_0 and C_2 cancel one another and (24) reduces to three integrals about the closed contours C_0, C_1, and C_2.

The first of these three can be evaluated at once.

$$(28)\qquad f_0(z, t) = \frac{1}{2\pi i}\int_{C_0} \frac{e^{s\left(t - \beta(s)\frac{z}{a}\right)}}{s + i\omega}\, ds,$$

whence by Cauchy's theorem

$$(29)\qquad f_0(z, t) = e^{-i\omega\left[t - \beta(s)\frac{z}{c}\right]_{s = -i\omega}}.$$

By (25)

$$\beta(-i\omega) = \sqrt{\frac{\omega^2 + i\omega g - \bar{\omega}_0^2 - a^2}{\omega^2 + i\omega g - \bar{\omega}_0^2}} = n + ih; \tag{30}$$

hence,

$$f_0(z, t) = e^{-\frac{\omega h}{c} z} \cdot e^{i\left(\frac{\omega n}{c} z - \omega t\right)}. \tag{31}$$

The remaining two integrals which are looped about the barriers cannot be evaluated in any such simple fashion and they shall be designated by

$$f_{12}(z, t) = \frac{1}{2\pi i} \int_{C_1 + C_2} \frac{e^{s\left(t - \beta(s)\frac{z}{c}\right)}}{s + i\omega} \, ds. \tag{32}$$

Thus the resultant wave motion at any point within the medium can be represented by the sum of two terms,

$$f(z, t) = f_0(z, t) + f_{12}(z, t). \tag{33}$$

Physically these two components may be interpreted as forced and free vibrations of the charges that constitute the medium. The forced vibrations, defined by $f_0(z, t)$, are undamped in time and have the same frequency as the impinging wave train. The free vibrations $f_{12}(z, t)$ are damped in time as a result of the damping forces acting on the oscillating ions and their frequency is determined by the elastic binding forces. The course of the propagation into the medium can be traced as follows: Up to the instant $t = z/c$, all is quiet. Even when the phase velocity v is greater than c, no wave reaches z earlier than $t = z/c$. At $t = z/c$ the integral $f_{12}(z, t)$ first exhibits a value other than zero, indicating that the ions have been set into oscillation. If by the term "wave front" we understand the very first arrival of the disturbance, then *the wave front velocity is always equal to c, no matter what the medium.* It may be shown, however, that at this first instant $t = z/c$ the forced or steady-state term $f_0(z, t)$ just cancels the free or transient term $f_{12}(z, t)$, so that the process starts always from zero amplitude. The steady state is then gradually built up as the transient dies out, quite in the same way that the sudden application of an alternating e.m.f. to an electrical network results in a transient surge which is eventually replaced by a harmonic oscillation.

The arrival of the wave front and the rôle of the velocity v in adjusting the phase are illustrated in Fig. 61, which however shows only the steady-state term. The axis z/c is drawn normal to the axis of t. The line at 45 deg. then determines the wave-front velocity, for it passes through a point z at the instant z/c. The line whose slope is $\tan \theta = c/v$ determines the time $t = z/v$ of arrival at z of a wave whose velocity is v. Actually the phase velocity has nothing to do with the propagation; it gives only the arrangement of phases and, strictly speaking, this only in the case of infinite wave trains. The phase of the forced oscillation is measured from the intersection with the dotted line at $t = z/v$, and the phase at the wave front $t = z/c$ is adjusted to fit. If $v > c$, $\theta < 45$ deg. The phase of the steady state is again determined by the intersection with the line z/c at $t = z/v$, but the wave front itself arrives later.

The transition from vanishingly small amplitudes at the wave front to the relatively large values of the signal have been carefully examined

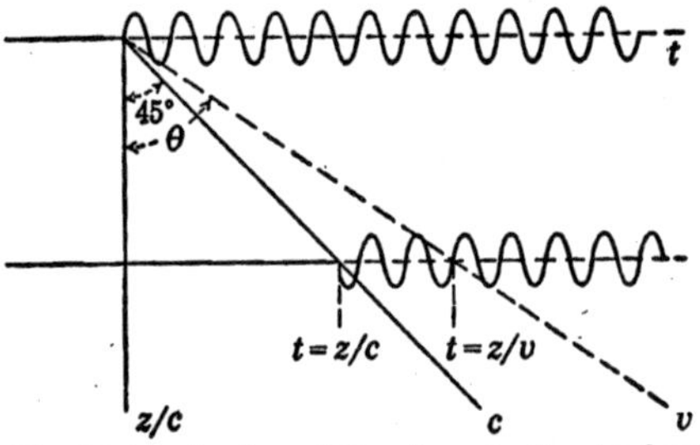

FIG. 61.—Determination of the phase in the steady state.

by Brillouin. This investigation is of a much more delicate nature and we must content ourselves here with a statement of conclusions which may be drawn after the evaluation of (32). According to Brillouin a signal is a train of oscillations starting at a certain instant. In the course of its journey the signal is deformed. The main body of the signal is

FIG. 62a.—Illustrating the variation in amplitude of the first precursor wave.

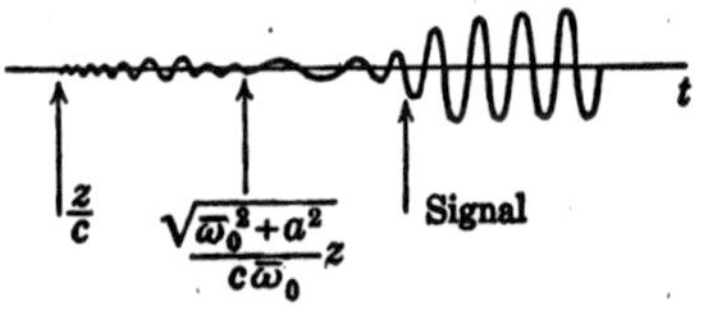

FIG. 62b.—Illustrating the arrival of the first and second precursors and the signal.

preceded by a first forerunner, or precursor, which in all media travels with the velocity c. This first precursor arrives with zero amplitude, and then grows slowly both in period and in amplitude, as indicated in Fig. 62*a*; the amplitude subsequently decreases while the period approaches the natural period of the electrons. There appears now a new phase of the disturbance which may be called the second precursor, traveling with the velocity $\dfrac{\bar{\omega}_0}{\sqrt{\bar{\omega}_0^2 + a^2}}\,c$. This velocity is obtained by assuming in the propagation factor $\beta(-i\omega)$ of Eqs. (30) and (32) that the impressed frequency ω is small compared to the atomic resonance frequency $\bar{\omega}_0$. The period of the second precursor, at first very large, decreases while the amplitude rises and then falls more or less in the manner of the first precursor.

With a sudden rise of amplitude the main body, or principal part, of the disturbance arrives, traveling with a velocity w which Brillouin defines as the signal velocity. An explicit and simple expression for w cannot be given and its definition is associated somewhat arbitrarily with the method employed to evaluate the integral (32). Physically its meaning is quite clear. In Fig. 62*b* are shown the two precursors and the final sudden rise to the steady state. When restricted to this third and last phase of the disturbance the term "signal" represents that portion of the wave which actuates a measuring device. Under the assumed conditions a measurement should, in fact, indicate a velocity

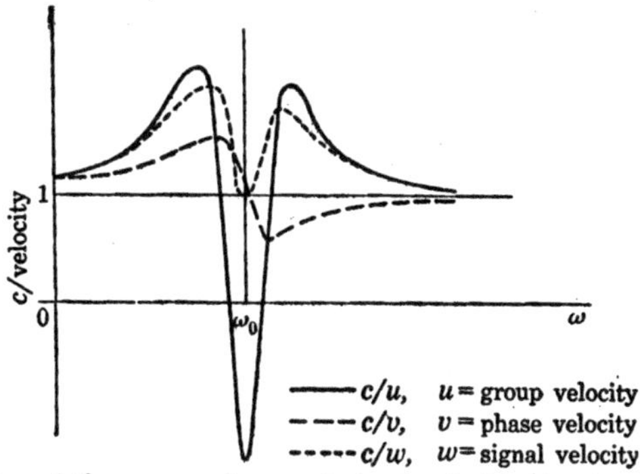

FIG. 63.—Behavior of the group, phase and signal velocities in the neighborhood of a resonance frequency.

of propagation approximately equal to w. It will be noted, however, that as the sensitivity of the detector is increased, the measured velocity also increases, until in the limit of infinite sensitivity we should record the arrival of the front of the first precursor which travels always with the velocity c. Qualitatively at least we may imagine the medium as a region of free space densely infested with electrons. An infinitesimal amount of energy penetrates the empty spaces, as through a sieve, traveling of course with the velocity c. Each successive layer of charges is excited into oscillation by the primary wave and reradiates energy both forward and backward. By reason of the inertia of the charges these secondary oscillations lag behind the primary wave in phase, and this constant retardation as the process progresses through successive layers results in a reduced velocity of the main body of the disturbance.

The *wave-front velocity* defined here is thus always equal to the constant c. *The phase velocity* v is associated only with steady states and may be either greater or less than c. The *group velocity* u differs from the phase velocity only in dispersive media. If the dispersion is normal, the group velocity is less than v; it is greater than the phase velocity when the dispersion is anomalous. In the neighborhood of an absorption band it may become infinite or even negative. The *signal velocity* w coincides with the group velocity in the region of normal dispersion, but deviates markedly from it whenever u behaves anomalously. The signal velocity is always less than c, but becomes somewhat difficult and arbitrary to define in the neighborhood of an absorption band. In Fig. 63 the velocities which characterize the propagation of a wave train in the neighborhood of an absorption band are plotted as ratios of c.

Problems

1. Let E_0 be the amplitude of the electric vector in volts per meter and $\bar{S}$ the mean energy flow in watts per square meter of a plane wave propagated in free space. Show that

$$\bar{S} = 1.327 \times 10^{-3} E_0^2 \text{ watt/meter}^2, \qquad E_0 = 27.45 \sqrt{\bar{S}} \text{ volts/meter.}$$

2. Let E_0 be the amplitude in volts per meter of the electric vector of a plane wave propagated in free space. Show that the amplitudes of the magnetic vectors H_0 and B_0 are related numerically to E_0 by

$$H_0 = 2.654 \times 10^{-3} E_0 \text{ ampere-turn/meter} = \tfrac{1}{3} \times 10^{-4} E_0 \text{ oersted,}$$
$$B_0 = \tfrac{1}{3} \times 10^{-8} E_0 \text{ weber/meter}^2 = \tfrac{1}{3} \times 10^{-4} E_0 \text{ gauss,}$$

where the oersted is the unrationalized c.g.s. electromagnetic unit. If E_0 is expressed in statvolts per centimeter, show that

$$H_0 = E_0 \text{ oersted}, \qquad B_0 = E_0 \text{ gauss.}$$

If a particle of charge q moves with a velocity v in the field of a plane electromagnetic wave, show that the ratio of the forces exerted by the magnetic and electric components is of the order v/c.

3. The theory of homogeneous, plane waves developed in Chap. V can be extended to inhomogeneous waves. If ψ is any rectangular component of an electromagnetic vector, we may take as a general definition of a plane wave the expression

$$\psi = \psi_0 e^{i\phi}, \qquad \phi = -\omega t + \mathbf{k} \cdot \mathbf{r},$$

in which ψ_0 is a complex amplitude and ϕ the complex phase. The propagation factor $\mathbf{k}$ is now *a complex vector* which we shall write as

$$\mathbf{k} = \mathbf{k}_1 + i\mathbf{k}_2 = \alpha\mathbf{n}_1 + i\beta\mathbf{n}_2,$$

where $\mathbf{n}_1$ and $\mathbf{n}_2$ are unit vectors. Show that α is the real phase constant; that the surfaces of constant real phase are planes normal to the axis $\mathbf{n}_1$, and the surfaces of constant amplitude are planes normal to $\mathbf{n}_2$; and that β is the factor measuring attenuation in the direction of most rapid change of amplitude. The true phase velocity v is in the direction $\mathbf{n}_1$.

Show that the operator ∇ can be replaced by $i\mathbf{k}$ and the field equations are

$$\mathbf{k} \times \mathbf{E} - \omega\mathbf{B} = 0, \qquad \mathbf{k} \cdot \mathbf{B} = 0,$$
$$\mathbf{k} \times \mathbf{H} + \omega\mathbf{D} = -i\mathbf{J}, \qquad \mathbf{k} \cdot \mathbf{D} = 0.$$

Show also that, if the medium is isotropic,

$$k^2 = k_1^2 - k_2^2 + 2i\mathbf{k}_1 \cdot \mathbf{k}_2 = \omega^2\mu\epsilon + i\omega\mu\sigma.$$

Let $\mathbf{k}_1 \cdot \mathbf{k}_2 = k_1k_2 \cos\theta$ and find expressions for k_1 and k_2 in terms of μ, ϵ, σ, and the angle θ.

4. Continuing Problem 3, assume that the electric vector is linearly polarized. Show that $\mathbf{E}$ is perpendicular to both $\mathbf{k}_1$ and $\mathbf{k}_2$ and that $\mathbf{H}$ lies in the plane of these two vectors. What is the locus of the magnetic vector $\mathbf{H}$? Inversely, show that if $\mathbf{H}$ is linearly polarized it is perpendicular to both $\mathbf{k}_1$ and $\mathbf{k}_2$, while $\mathbf{E}$ lies in the $\mathbf{k}_1$, $\mathbf{k}_2$-plane. These inhomogeneous plane waves are intimately related to the so-called "H waves" and "E waves," Sec. 9.18.

5. The theory of plane waves can be extended to homogeneous, *anisotropic* dielectrics. Let $\sigma = 0$, $\mu = \mu_0$, and the properties of the dielectric be specified by a tensor whose components are ϵ_{jk}. Choosing the principal axes as coordinate axes, we have

$$D_j = \epsilon_0\kappa_j E_j, \qquad (j = 1, 2, 3).$$

The field equations relating $\mathbf{E}$ and $\mathbf{D}$ are then

$$\mathbf{k} \times (\mathbf{k} \times \mathbf{E}) + \omega^2\mu_0\mathbf{D} = 0, \qquad \mathbf{k} \cdot \mathbf{D} = 0.$$

Let $\mathbf{k} = k\mathbf{n}$, where $\mathbf{n}$ is a unit real vector whose components along the principal axes are n_1, n_2, n_3. Let $v = \omega/k$ and $v_j = c/\sqrt{\kappa_j}$. The v_j are the *principal velocities.* Show that the components of $\mathbf{E}$ satisfy the homogeneous system

$$n_j \sum_i n_iE_i + \left(\frac{v^2}{v_j^2} - 1\right) E_j = 0, \qquad (j = 1, 2, 3),$$

and write down the equation which determines the phase velocity in any direction fixed by the vector $\mathbf{n}$. Show that this equation has three real roots, of which one is infinite and must be discarded. There are, therefore, two distinct modes of propagation whose phase velocities are v' and v'' in the direction $\mathbf{n}$. Correspondingly there are two types of linear oscillation in distinct directions, characterized by the vectors $\mathbf{E}'$, $\mathbf{D}'$ and $\mathbf{E}''$, $\mathbf{D}''$. Show that these are related so that

$$\mathbf{D}' \cdot \mathbf{E}'' = \mathbf{D}'' \cdot \mathbf{E}' = \mathbf{D}' \cdot \mathbf{D}'' = 0,$$
$$\mathbf{E}' \cdot \mathbf{E}'' \neq 0, \qquad \mathbf{E} \cdot \mathbf{n} \neq 0.$$

6. Show that in a homogeneous, anisotropic dielectric the phase velocity satisfies the Fresnel relation

$$\sum_{j=1}^{3} \frac{n_j^2}{v_j^2 - v^2} = 0,$$

where the n_j are the direction cosines of the wave normal $\mathbf{n}$ with respect to the principal axes, and the v_j are the phase velocities in the directions of the principal axes. (See Problem 5.) Show also that

$$v^2 = \sum_{j=1}^{3} \gamma_j^2 v_j^2,$$

where the γ_i are the direction cosines of the vector **D** with respect to the principal axes.

7. If in Problem 6 $v_1 > v_2 > v_3$, show that there are two directions in which v has only a single value. Find these directions in terms of v_1, v_2, v_3, and show that in each case the velocity is v_2. The directions defined thus are the *optic axes* of the medium which in this case is said to be *biaxial*. What are the necessary conditions in order that there be only one optic axis?

8. Show that for a plane wave in an anisotropic dielectric

$$\frac{1}{2}\mathbf{D}\cdot\mathbf{E} = \frac{1}{2\mu_0 v^2}[E^2 - (\mathbf{n}\cdot\mathbf{E})^2] = \frac{1}{2}\mu_0 H^2,$$

$$\mathbf{S} = \mathbf{E}\times\mathbf{H} = \frac{1}{\mu_0 v^2}[\mathbf{n}E^2 - (\mathbf{n}\cdot\mathbf{E})\mathbf{E}],$$

where v is the phase velocity defined in Problem 5.

A vectorial *velocity of energy propagation* **u** in the direction of **S** is defined by the equation

$$\mathbf{S} = h\mathbf{u}.$$

The velocity **u** is related to the phase velocity v by

$$v = \mathbf{n}\cdot\mathbf{u}.$$

Show that

$$h = \tfrac{1}{2}\mathbf{E}\cdot\mathbf{D} + \tfrac{1}{2}\mu_0 H^2$$

and, hence, that the energy flow is equal to the total energy density times the velocity **u**, with

$$\mathbf{u} = \frac{E^2\mathbf{n} - (\mathbf{n}\cdot\mathbf{E})\mathbf{E}}{E^2 - (\mathbf{n}\cdot\mathbf{E})^2}\,v,$$

or in magnitude

$$\left(\frac{v}{u}\right)^2 = 1 - \frac{(\mathbf{n}\cdot\mathbf{E})^2}{E^2}.$$

9. From the results of Problem 8 show that the magnitude of the velocity **u** of energy propagation in a plane wave can be found in terms of the constants v_1, v_2, v_3 from the equation

$$\sum_{j=1}^{3}\frac{l_j^2}{1 - \left(\dfrac{v_j}{u}\right)^2} = 1,$$

in which the l_j are direction cosines of the vector **u**.

Show also that there are, in general, two finite values of u corresponding to each direction. In what directions are these two roots identical? What is the relation of these preferred axes of u to the optic axes?

10. A nonconducting medium of infinite extent is isotropic but its specific inductive capacity $\kappa = \epsilon/\epsilon_0$ is a function of position. Show first that the electric field vector satisfies the equation

$$\nabla^2\mathbf{E} + k^2\mathbf{E} = -\nabla\left(\mathbf{E}\cdot\frac{\nabla\epsilon}{\epsilon}\right),$$

where $k^2 = \omega^2\epsilon\mu$ and the time enters only through the factor $\exp(-i\omega t)$.

Assume now that the spatial change of ϵ per wave length is small, so that $|\nabla\kappa|\,\lambda \ll 1$. Show that the right-hand term can be neglected and that the propagation is determined approximately by

$$\nabla^2\psi + k^2\psi = 0,$$

where ψ is a rectangular component of $\mathbf{E}$ or $\mathbf{H}$ and k is a slowly varying function of position. Next let

$$\psi = \psi_0\, e^{i\phi}, \qquad \phi = -\omega t + k_0 S,$$

where ϕ is the phase, $k_0 = \omega/c$, and S a function of position. Show that the phase function S is determined by the equation

$$(\nabla S)^2 + \frac{1}{ik_0}\nabla^2 S = \kappa,$$

where

$$(\nabla S)^2 = \left(\frac{\partial S}{\partial x}\right)^2 + \left(\frac{\partial S}{\partial y}\right)^2 + \left(\frac{\partial S}{\partial z}\right)^2.$$

Note that S is in general complex, even though κ is real, and that the amplitude of the wave is also a slowly varying function of position.

11. The transition from wave optics to geometrical optics can be based on the results of Problem 10. Show that as the wave number k becomes very large, the phase function is determined by the first-order, second-degree equation

$$(\nabla S)^2 = \left(\frac{k}{k_0}\right)^2 = \kappa,$$

while the amplitude satisfies

$$\nabla \ln \psi_0 \cdot \nabla S = -\tfrac{1}{2}\nabla^2 S.$$

Since S is now real, the wave fronts are represented by the family S = constant, and the wave normals or rays are given at each point by ∇S. The function S is called the "eikonal" and is identical with Hamilton's "characteristic function." Let the vector $\mathbf{n}$ represent the index of refraction $\sqrt{\kappa}$ times a unit vector in the direction of the wave normal. Then

$$\nabla S = \mathbf{n}, \qquad \mathbf{n}\cdot\nabla \ln \psi_0 = -\tfrac{1}{2}\nabla\cdot\mathbf{n}.$$

Note that these relations hold only in regions where the change in κ per wave length is small. Hence they fail in the neighborhood of sharp edges or of bodies whose dimensions are of the order of a wave length. In such cases the complete wave equation must be applied. (Sommerfeld and Runge, *Ann. Physik*, **35**, 290, 1911.)

Let $d\mathbf{r}$ be an element of length in the direction of a ray. Then

$$S = \int \mathbf{n} \cdot d\mathbf{r}.$$

S is a function of the coordinates x_0, y_0, z_0 of a fixed initial point and the coordinates x, y, z of the terminal point reached by the ray. Of all possible paths through these two points, the ray actually follows that which makes S a minimum, or

$$\delta S = \delta \int_{(x_0,y_0,z_0)}^{(x,y,z)} \mathbf{n} \cdot d\mathbf{r} = 0.$$

This is Fermat's principle. If the disturbance originates at x_0, y_0, z_0 at the instant t_0, and arrives at x, y, z at t, then

$$S(x_0, y_0, z_0; x, y, z) = t - t_0,$$

and the path of the ray is such that the time of arrival is a minimum.

12. Plane waves are set up in an inhomogeneous medium whose inductive capacity varies in the direction of propagation. Suppose this direction to be the z-axis and assume the field to be independent of x and y. The wave equation is then

$$\frac{\partial^2 \psi}{\partial z^2} + k_0^2 \kappa(z) \psi = 0,$$

where $k_0^2 = \omega^2 \epsilon_0 \mu_0$ and the time enters only in the factor $\exp(-i\omega t)$.

Suppose that $\kappa \to \kappa_1$ as $z \to -\infty$, and $\kappa \to \kappa_2$ as $z \to \infty$, where κ_1 and κ_2 are constants. It is possible to construct four particular solutions behaving asymptotically as follows:

$$\begin{aligned}
&\psi_1 \to \exp(ik_0 \sqrt{\kappa_1}\, z), && \text{as } z \to -\infty. \\
&\psi_2 \to \exp(-ik_0 \sqrt{\kappa_1}\, z), && \text{as } z \to -\infty. \\
&\psi_3 \to \exp(ik_0 \sqrt{\kappa_2}\, z), && \text{as } z \to \infty. \\
&\psi_4 \to \exp(-ik_0 \sqrt{\kappa_2}\, z), && \text{as } z \to \infty.
\end{aligned}$$

Thus ψ_1 and ψ_3 reduce to plane waves traveling in the positive direction; ψ_2 and ψ_4, to waves in the negative direction. Only two of these solutions are linearly independent and, hence, there must exist analytic connections of the form

$$\psi_3 = a_1\psi_1 + a_2\psi_2, \qquad \psi_4 = b_1\psi_1 + b_2\psi_2.$$

Mathematically this represents the *analytic continuation* of ψ_3 from one region to another, and physically the reflection coefficient is given by the ratio $r = a_2/a_1$.

Let

$$\kappa(z) = \kappa_1 + \frac{[(\kappa_2 - \kappa_1)(e^\zeta + 1) + \kappa_3]e^\zeta}{(e^\zeta + 1)^2},$$

where $\zeta = k_0 z/s$, κ_3 is a constant and s an additional parameter. Replace z by the new independent variable $u = \exp(\zeta)$ and show that the wave equation reduces to hypergeometric form. Express $\psi_1 \ldots \psi_4$ as hypergeometric functions and calculate the joining factors and the reflection coefficient. The form of $\kappa(z)$ is such as to admit a very wide choice in strata. (Epstein, *Proc. Nat. Acad. Sci.*, **16**, 627, 1930.)

13. Ten times the common logarithm of the ratio of the initial to the terminal energy flow measures the attenuation of a plane wave in *decibels*. Show that for a plane wave in a homogeneous, isotropic medium

$$\text{power loss} = 8.686\beta \qquad \text{db/meter},$$

where β is the attenuation factor defined in Eq. (49), page 276.

14. For the media whose constants are given below, plot the attenuation of a plane wave in decibels per meter against frequency from 0 to 10^7 cycles/second. Use semilog paper or plot against the logarithm of the frequency.

Material	Conductivity, mhos/meter	ϵ/ϵ_0
Sea water	4	81
Fresh water	10^{-3}	81
Wet earth	10^{-3}	10
Dry earth	10^{-5}	4

15. A function $f(x)$ is defined by

$$f(x) = e^{-\beta x}, \qquad x > 0,\ \beta > 0.$$
$$f(x) = e^{\beta x}, \qquad x < 0,\ \beta > 0.$$

Show that

$$f(x) = \frac{2}{\pi}\int_0^\infty \frac{\beta \cos ux}{\beta^2 + u^2}\,du.$$

16. Obtain a Fourier integral representation of the function $f(t)$ defined by

$$f(t) = 0, \qquad t < 0,$$
$$f(t) = e^{-bt}\cos \omega t, \qquad t > 0.$$

Plot the amplitudes of the harmonic components as a function of frequency (spectral density) and discuss the relation between the breadth of the peak and the logarithmic decrement.

17. The surges on transmission lines produced by lightning discharges are often simulated in electrical engineering practice by impulse generators developing a unidirectional impulse voltage of the form

$$V = V_0 \frac{\alpha}{\beta} \left[e^{-(\alpha-\beta)t} - e^{-(\alpha+\beta)t}\right],$$

where α and β are constants determined by the circuit parameters. This function represents a pulse or surge which rises sharply to a crest and then drops off more slowly in a long tail. The wave form can be characterized by two constants t_1 and t_2, where t_1 is the time required to reach the crest, and t_2 is the interval from zero voltage to that point on the tail at which the voltage is equal to half the crest voltage. A standard set of surges is represented by the constants

$$\begin{aligned} &(a)\ t_1 = 0.5, && t_2 = 5, \\ &(b)\ t_1 = 1.0, && t_2 = 10, \qquad \text{microseconds.} \\ &(c)\ t_1 = 1.5, && t_2 = 40, \end{aligned}$$

From these values one may determine the constants α and β.

Make a spectral analysis of the wave form and determine the frequencies at which the spectral density is a maximum for the three standard cases.

18. The "telegraphic equation"

$$\frac{\partial^2 u}{\partial t^2} = a^2 \frac{\partial^2 u}{\partial x^2} + b^2 u$$

has the initial conditions

$$u = f(x), \qquad \frac{\partial u}{\partial t} = g(x), \qquad \text{when } t = 0.$$

Show that

$$u = \tfrac{1}{2}[f(x + at) + f(x - at)] + \frac{1}{2a}\int_{-at}^{at} f(x + \beta)\frac{\partial}{\partial t} I_0\left[b\sqrt{t^2 - \left(\frac{\beta}{a}\right)^2}\right] d\beta$$
$$+ \frac{1}{2a}\int_{-at}^{at} g(x + \beta)\, I_0\left[b\sqrt{t^2 - \left(\frac{\beta}{a}\right)^2}\right] d\beta,$$

where

$$I_0(x) = \frac{1}{\pi}\int_{-\frac{\pi}{2}}^{\frac{\pi}{2}} \cos(x \cos\lambda)\, d\lambda.$$

This solution was obtained independently by Heaviside and by Poincaré.

19. An electromagnetic pulse is propagated in a homogeneous, isotropic medium whose constants are ϵ, μ, σ in the direction of the positive z-axis. At the instant $t = 0$ the form of the pulse is given by

$$f_1(z) = \lim_{t \to 0} f(z, t) = \frac{1}{\sqrt{2\pi}} \frac{e^{-\frac{z^2}{2\delta^2}}}{\delta},$$

where δ is a parameter. Find an expression for the pulse $f(z, t)$ at any subsequent instant.

20. Wherever the displacement current is negligible with respect to the conduction current ($\sigma/\epsilon\omega \gg 1$), the field satisfies approximately the equations

$$\nabla \times \mathbf{E} + \mu \frac{\partial \mathbf{H}}{\partial t} = 0, \qquad \nabla \times \mathbf{H} = \mathbf{J}.$$

Show that to the same approximation

$$\mathbf{E} = -\frac{\partial \mathbf{A}}{\partial t}, \qquad \mathbf{H} = \frac{1}{\mu}\nabla \times \mathbf{A},$$
$$\nabla^2 \mathbf{A} - \mu\sigma \frac{\partial \mathbf{A}}{\partial t} = 0, \qquad \nabla \cdot \mathbf{A} = 0,$$

and that the current density $\mathbf{J}$ satisfies

$$\nabla^2 \mathbf{J} - \mu\sigma \frac{\partial \mathbf{J}}{\partial t} = 0, \qquad \nabla \cdot \mathbf{J} = 0.$$

What boundary conditions apply to the current density vector? These equations govern the distribution of current in metallic conductors at all frequencies in the radio spectrum or below.

21. The equation obtained in Problem 20 for the current distribution in a metallic conductor is identical with that governing the diffusion of heat. Consider the one-dimensional case of a rectangular component J propagated along the z-axis

$$\frac{\partial^2 J}{\partial z^2} = \mu\sigma \frac{\partial J}{\partial t}.$$

Show that

$$J = \frac{1}{2}\sqrt{\frac{\mu\sigma}{\pi t}} \int_{-\infty}^{\infty} f(\alpha)\, e^{-\frac{\mu\sigma(\alpha - z)^2}{4t}}\, d\alpha$$

is a solution such that at $t = 0$, $J = f(z)$, and which is continuous with continuous derivatives for all values of z when $t > 0$.

22. In Problem 21 it was found that the equation

$$\nu \frac{\partial^2 \psi}{\partial z^2} = \frac{\partial \psi}{\partial t}$$

is satisfied by

$$\psi = \sqrt{\frac{1}{\pi\nu t}} \int_{-\infty}^{\infty} f(\alpha)\, e^{-\frac{(\alpha - z)^2}{4\nu t}}\, d\alpha.$$

From the theory of Fourier transforms show that this can be written in the form

$$\psi = \sqrt{\frac{1}{\pi\nu t}} \sum_{n=-\infty}^{\infty} \int_0^{2\pi} e^{-\frac{(z-\alpha+2n\pi)^2}{4\nu t}} f(\alpha)\, d\alpha.$$

Obviously the equation is also satisfied by

$$\psi = \frac{1}{\pi} \sum_{n=-\infty}^{\infty} \int_0^{2\pi} e^{i(z-\alpha)n - n^2\nu t} f(\alpha)\, d\alpha.$$

Compare these solutions and from the uniqueness theorem demonstrate the identity

$$\sum_{n=-\infty}^{\infty} e^{in(z-\alpha) - n^2\nu t} = \sqrt{\frac{\pi}{\nu t}} \sum_{n=-\infty}^{\infty} e^{-\frac{(z-\alpha+2n\pi)^2}{4\nu t}}.$$

This relation has been applied by Ewald to the theory of wave propagation in crystals.

23. Show that if radio waves are propagated in an electron atmosphere in the earth's magnetic field, a resonance phenomenon may be expected near a wave length of about 212 meters. Marked selective absorption is actually observed in this region. Assume $e = 1.60 \times 10^{-19}$ coulomb/electron, $m = 9 \times 10^{-31}$ kg./electron, and for the density of the magnetic field of the earth $B_0 = 0.5 \times 10^{-4}$ weber/meter^2.

24. A radio wave is propagated in an ionized atmosphere. Calculate and discuss the group velocity as a function of frequency, assuming for the propagation constant the expression given in Eq. (36), page 329. The concentration of electrons in certain regions of the ionosphere is presumed to be of the order of 10^{12} electrons/meter^3.

25. A plane, linearly polarized wave enters an electron atmosphere whose density is 10^{12} electrons/meter^3. A static magnetic field of intensity $B_0 = 0.5 \times 10^{-4}$ weber/meter^2 is applied in the direction of propagation. Obtain an expression representing the change in the state of polarization per wave length in the direction of propagation.

26. A radio wave is propagated in an ionized atmosphere in the presence of a fixed magnetic field. Find the propagation factor for the case in which the static field is perpendicular to the direction of propagation.

27. As a simple model of an atom one may assume a fixed, positive charge to which there is bound by a quasi-elastic force a negative charge e of mass m. Neglecting frictional forces, the equation of motion of such a system is

$$m\ddot{\mathbf{r}} + f\mathbf{r} = 0,$$

where $\mathbf{r}$ is the position vector of e and f is the binding constant. A static magnetic field is now applied. Show that there are two distinct frequencies of oscillation and that the corresponding motions represent circular vibrations in opposite directions in a plane perpendicular to the axis of the applied field. What is the frequency of rotation? This is the elementary theory of the Zeeman effect.

28. The force per unit charge exerted on a particle moving with a velocity $\mathbf{v}$ in a magnetic field $\mathbf{B}$ is $\mathbf{E}' = \mathbf{v} \times \mathbf{B}$. The motional electromotive force about a closed circuit is

$$V' = \oint \mathbf{E}' \cdot as = \int [\nabla \times (\mathbf{v} \times \mathbf{B})] \cdot \mathbf{n}\, da.$$

At any fixed point in space the rate of change of $\mathbf{B}$ with the time is $\partial\mathbf{B}/\partial t$. If by $\mathbf{E}$ we now understand the total force per unit charge in a moving body, then

$$\nabla \times \mathbf{E} = -\frac{\partial \mathbf{B}}{\partial t} + \nabla \times (\mathbf{v} \times \mathbf{B}).$$

Show that the right-hand side is the negative total derivative

$$\frac{d\mathbf{B}}{dt} = \frac{\partial \mathbf{B}}{\partial t} + (\mathbf{v} \cdot \nabla)\,\mathbf{B},$$

so that the Faraday law for a moving medium is

$$\nabla \times \mathbf{E} = -\frac{d\mathbf{B}}{dt}.$$

The displacement velocity is assumed to be small relative to the velocity of light.

CHAPTER VI

CYLINDRICAL WAVES

An electromagnetic field cannot, in general, be derived from a purely scalar function of space and time; as a consequence, the analysis of electromagnetic fields is inherently more difficult than the study of heat flow or the transmission of acoustic vibrations. The three-dimensional scalar wave equation is separable in 11 distinct coordinate systems,[1] but complete solutions of the *vector* wave equation in a form directly applicable to the solution of boundary-value problems are at present known only for certain separable systems of cylindrical coordinates and for spherical coordinates. It will be shown that in such systems an electromagnetic field can be resolved into two partial fields, each derivable from a purely scalar function satisfying the wave equation.

EQUATIONS OF A CYLINDRICAL FIELD

6.1. Representation by Hertz Vectors.—We shall suppose that one set of coordinate surfaces is formed by a family of cylinders whose elements are all parallel to the z-axis. Unless specifically stated, these cylindrical surfaces are not necessarily circular, or even closed. With respect to any surface of the family the unit vectors $\mathbf{i}_1$, $\mathbf{i}_2$, $\mathbf{i}_3$ are ordered as shown in Fig. 64. $\mathbf{i}_1$ is, therefore, normal to the cylinder, $\mathbf{i}_3$ tangent to it and directed along its elements, and $\mathbf{i}_2$ is tangent to the surface and perpendicular to $\mathbf{i}_1$ and $\mathbf{i}_3$. Position with respect to axes determined by the three unit vectors is measured by the coordinates u^1, u^2, z, and the infinitesimal line element is

$$(1) \qquad d\mathbf{s} = \mathbf{i}_1 h_1\, du^1 + \mathbf{i}_2 h_2\, du^2 + \mathbf{i}_3\, dz.$$

Now let us calculate the components of the electromagnetic field associated with a Hertz vector $\mathbf{\Pi}$ directed along the z-axis, so that $\Pi_1 = \Pi_2 = 0$, $\Pi_z \neq 0$. We shall assume in the present chapter that the medium is not only isotropic and homogeneous but also unbounded in extent. Then by (63) and (64), page 32, the electric and magnetic vectors of the field are given by

$$(2) \qquad \mathbf{E}^{(1)} = \nabla \times \nabla \times \mathbf{\Pi}, \qquad \mathbf{H}^{(1)} = \left(\epsilon \frac{\partial}{\partial t} + \sigma\right) \nabla \times \mathbf{\Pi};$$

[1] Eisenhart, *Annals of Math.*, **35**, 284, 1934.

since both Π_1 and Π_2 are zero, it is a simple matter to calculate from (81) and (85), page 49, the components of $\mathbf{E}^{(1)}$ and $\mathbf{H}^{(1)}$.

$$E_1^{(1)} = \frac{1}{h_1}\frac{\partial^2 \Pi_z}{\partial z\,\partial u^1}, \qquad E_2^{(1)} = \frac{1}{h_2}\frac{\partial^2 \Pi_z}{\partial z\,\partial u^2},$$

(3)

$$E_z^{(1)} = -\frac{1}{h_1 h_2}\left[\frac{\partial}{\partial u^1}\left(\frac{h_2}{h_1}\frac{\partial \Pi_z}{\partial u^1}\right) + \frac{\partial}{\partial u^2}\left(\frac{h_1}{h_2}\frac{\partial \Pi_z}{\partial u^2}\right)\right],$$

(4) $$H_1^{(1)} = \left(\epsilon\frac{\partial}{\partial t} + \sigma\right)\frac{1}{h_2}\frac{\partial \Pi_z}{\partial u^2}, \qquad H_2^{(1)} = -\left(\epsilon\frac{\partial}{\partial t} + \sigma\right)\frac{1}{h_1}\frac{\partial \Pi_z}{\partial u^1},$$

$$H_z^{(1)} = 0.$$

Thus we have derived from a *scalar* function $\Pi_z = \psi$, an electromagnetic field characterized by the absence of an axial or longitudinal component

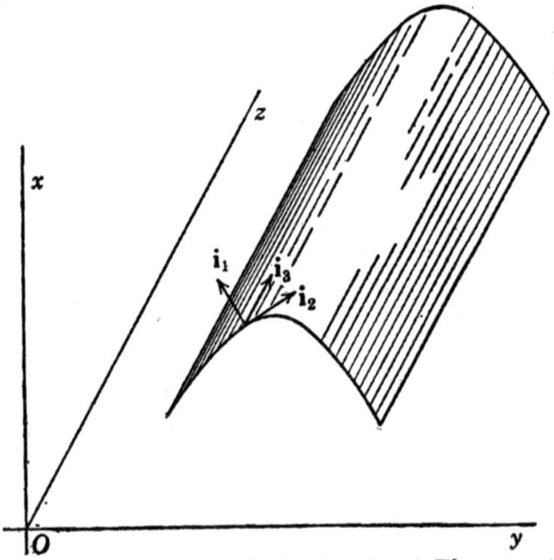

FIG. 64.—Relation of unit vectors to a cylindrical surface. The generators of the surface are parallel to the z-axis.

of the magnetic vector. Since $\mathbf{\Pi}$ is the electric polarization potential, this can be called a field of electric type (page 30), but in the present instance the term *transverse magnetic field* proposed recently by Schelkunoff[1] seems more apt.

Since Π_z is a rectangular component, it must satisfy the scalar wave equation,

(5) $$\nabla^2\psi - \mu\epsilon\frac{\partial^2\psi}{\partial t^2} - \mu\sigma\frac{\partial\psi}{\partial t} = 0,$$

[1] SCHELKUNOFF, Transmission Theory of Plane Electromagnetic Waves, *Proc. Inst. Radio Engrs.*, **25**, 1457–1492, November, 1937.

or by (82), page 49,

$$(6)\quad \frac{1}{h_1h_2}\frac{\partial}{\partial u^1}\left(\frac{h_2}{h_1}\frac{\partial\psi}{\partial u^1}\right)+\frac{1}{h_1h_2}\frac{\partial}{\partial u^2}\left(\frac{h_1}{h_2}\frac{\partial\psi}{\partial u^2}\right)+\frac{\partial^2\psi}{\partial z^2}-\mu\epsilon\frac{\partial^2\psi}{\partial t^2}-\mu\sigma\frac{\partial\psi}{\partial t}=0.$$

The elementary harmonic solutions of this equation are of the form

$$(7)\qquad \psi = f(u^1, u^2)e^{\pm ihz - i\omega t},$$

where $f(u^1, u^2)$ is a solution of

$$(8)\qquad \frac{1}{h_1h_2}\frac{\partial}{\partial u^1}\left(\frac{h_2}{h_1}\frac{\partial f}{\partial u^1}\right)+\frac{1}{h_1h_2}\frac{\partial}{\partial u^2}\left(\frac{h_1}{h_2}\frac{\partial f}{\partial u^2}\right)+(k^2-h^2)f=0.$$

In quite the same way a partial field can be derived from a second Hertz vector $\mathbf{\Pi}^*$ by the operations

$$(9)\qquad \mathbf{E}^{(2)} = -\mu\frac{\partial}{\partial t}\nabla\times\mathbf{\Pi}^*,\qquad \mathbf{H}^{(2)} = \nabla\times\nabla\times\mathbf{\Pi}^*;$$

if $\mathbf{\Pi}^*$ is directed along the z-axis, the components of these vectors are

$$(10)\qquad E_1^{(2)} = -\frac{\mu}{h_2}\frac{\partial^2\Pi_z^*}{\partial t\,\partial u^2},\qquad E_2^{(2)} = \frac{\mu}{h_1}\frac{\partial^2\Pi_z^*}{\partial t\,\partial u^1},\qquad E_z^{(2)} = 0,$$

$$(11)\qquad \begin{aligned} H_1^{(2)} &= \frac{1}{h_1}\frac{\partial^2\Pi_z^*}{\partial z\,\partial u^1},\qquad H_2^{(2)} = \frac{1}{h_2}\frac{\partial^2\Pi_z^*}{\partial z\,\partial u^2},\\ H_z^{(2)} &= -\frac{1}{h_1h_2}\left[\frac{\partial}{\partial u^1}\left(\frac{h_2}{h_1}\frac{\partial\Pi_z^*}{\partial u^1}\right)+\frac{\partial}{\partial u^2}\left(\frac{h_1}{h_2}\frac{\partial\Pi_z^*}{\partial u^2}\right)\right].\end{aligned}$$

The scalar function Π_z^* is a solution of (5), and the field of magnetic type, or *transverse electric*, derived from it is characterized by the absence of a longitudinal component of **E**.

The electromagnetic field obtained by superposing the partial fields derived from Π_z and Π_z^* is of such generality that one can satisfy a prescribed set of boundary conditions on any cylindrical surface whose generating elements are parallel to the z-axis; *i.e.*, on any coordinate surface defined by u^1 = constant—or prescribed conditions on either a surface of the family u^2 = constant or a plane z constant. The choice of these orthogonal families, however, is limited in practice to coordinate systems in which Eq. (8) is separable.

6.2. Scalar and Vector Potentials.—The transverse electric and transverse magnetic fields defined in the preceding paragraph have interesting properties which can be made clear by a study of the scalar and vector potentials. Consider first the transverse magnetic field in which

$H_z^{(1)} = 0.$

$$\mathbf{E}^{(1)} = -\nabla\phi - \frac{\partial \mathbf{A}}{\partial t}, \qquad \mathbf{B}^{(1)} = \nabla \times \mathbf{A}, \tag{12}$$

$$\phi = -\nabla\cdot\mathbf{\Pi}, \qquad \mathbf{A} = \mu\left(\epsilon\frac{\partial}{\partial t} + \sigma\right)\mathbf{\Pi}. \tag{13}$$

In the present instance $\psi = \Pi_z$ and, hence,

$$\phi = -\frac{\partial\psi}{\partial z}, \qquad A_z = \mu\epsilon\frac{\partial\psi}{\partial t} + \mu\sigma\psi, \qquad A_1 = A_2 = 0. \tag{14}$$

The components of $\mathbf{E}^{(1)}$ are then

$$\begin{aligned} E_1^{(1)} &= -\frac{1}{h_1}\frac{\partial\phi}{\partial u^1}, \qquad E_2^{(1)} = -\frac{1}{h_2}\frac{\partial\phi}{\partial u^2}, \\ E_z^{(1)} &= \frac{\partial^2\psi}{\partial z^2} - \mu\epsilon\frac{\partial^2\psi}{\partial t^2} - \mu\sigma\frac{\partial\psi}{\partial t}; \end{aligned} \tag{15}$$

for the components of $\mathbf{B}^{(1)}$, we obtain

$$B_1^{(1)} = \frac{1}{h_2}\frac{\partial A}{\partial u^2}, \qquad B_2^{(1)} = -\frac{1}{h_1}\frac{\partial A}{\partial u^1}, \qquad B_z^{(1)} = 0, \tag{16}$$

in which A without a subscript has been written for A_z.

Now it will be noted that in the plane z = constant the vector $\mathbf{E}^{(1)}$ is irrotational and therefore *in this transverse plane*, the line integral of $\mathbf{E}^{(1)}$ between any two points a and b is independent of the path joining them; for an element of length in the plane z = constant is expressed by $d\mathbf{s} = \mathbf{i}_1 h_1\,du^1 + \mathbf{i}_2 h_2\,du^2$ and, consequently,

$$\int_a^b \mathbf{E}^{(1)}\cdot d\mathbf{s} = -\int_a^b\left(\frac{\partial\phi}{\partial u^1}du^1 + \frac{\partial\phi}{\partial u^2}du^2\right) = \phi(a) - \phi(b). \tag{17}$$

The difference of potential, or voltage, between any two points in a transverse plane has a definite value at each instant, whatever the frequency and whatever the nature of the cylindrical coordinates.

Next, one will observe that the scalar function A plays the role of a stream or flow function for the vector $\mathbf{B}^{(1)}$. Let the curve joining the points a and b in Fig. 65 represent the trace of a cylindrical surface intersecting the plane z = constant. We shall calculate the flux of the vector $\mathbf{B}^{(1)}$ through the ribbonlike surface element bounded by the curve ab whose width in the z-direction is unity. If $\mathbf{n}$ is the unit normal to this surface and $\mathbf{i}_3$ a unit vector directed along the z-axis, we have

$$\mathbf{B}^{(1)}\cdot\mathbf{n}\,da = \mathbf{B}^{(1)}\cdot(\mathbf{i}_3 \times d\mathbf{s}) = \mathbf{i}_3\cdot(d\mathbf{s}\times\mathbf{B}^{(1)}), \tag{18}$$

where ds is again an element of length along the curve. Upon expanding (18) we obtain

$$\mathbf{B}^{(1)} \cdot \mathbf{n}\, da = h_1 B_2^{(1)}\, du^1 - h_2 B_1^{(1)}\, du^2 = -dA; \tag{19}$$

hence,

$$\int_a^b \mathbf{B}^{(1)} \cdot \mathbf{n}\, da = A(a) - A(b). \tag{20}$$

The magnetic flux through any unit strip of a cylindrical surface passing through two points in a plane z = constant is independent of the contour of the strip.

If the components of $\mathbf{E}^{(1)}$ and $\mathbf{B}^{(1)}$ are expressed in terms of the scalar function ψ, it is easy to show that

$$E_1^{(1)} B_1^{(1)} + E_2^{(1)} B_2^{(1)} = 0, \tag{21}$$

and consequently *the projection of the vector $\mathbf{E}^{(1)}$ on the plane z = constant is everywhere normal to* $\mathbf{B}^{(1)}$. Moreover, in the transverse plane the families of curves, ϕ = constant, A = constant, constitute an orthogonal set of equipotentials and streamlines. When the time variation is harmonic, it is clear from (14) and (7) that the equipotentials are the lines

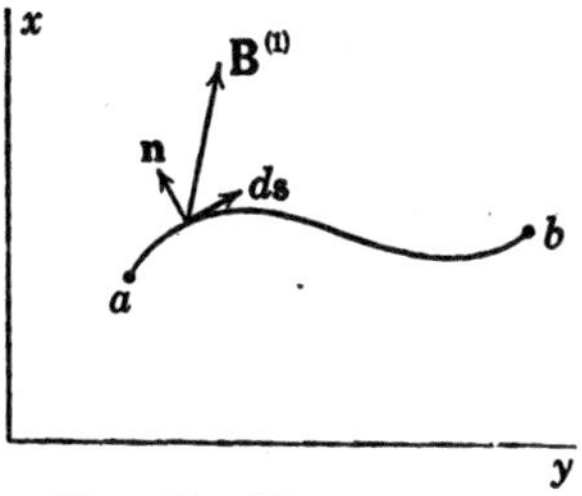

FIG. 65.—The curve a-b represents the intersection of a cylindrical surface with the xy-plane which contains also the vectors $\mathbf{n}$ and $\mathbf{B}^1$.

$$f(u^1, u^2) = \text{constant}, \tag{22}$$

in which $f(u^1, u^2)$ satisfies Eq. (8).

The transverse electric field has identical properties but with the roles of electric and magnetic vectors reversed. According to (35), page 27,

$$\mathbf{D}^{(2)} = -\nabla \times \mathbf{A}^*, \qquad \mathbf{H}^{(2)} = -\nabla\phi^* - \frac{\partial \mathbf{A}^*}{\partial t} - \frac{\sigma}{\epsilon}\mathbf{A}^*. \tag{23}$$

$$\phi^* = -\nabla \cdot \mathbf{\Pi}^*, \qquad \mathbf{A}^* = \mu\epsilon \frac{\partial \mathbf{\Pi}^*}{\partial t}. \tag{24}$$

If now we let $\psi = \Pi_z^*$, $\Pi_1^* = \Pi_2^* = 0$, then

$$\phi^* = -\frac{\partial \psi}{\partial z}, \qquad A_z^* = \mu\epsilon \frac{\partial \psi}{\partial t}, \qquad A_1^* = A_2^* = 0. \tag{25}$$

The components of the field vectors are, therefore,

$$D_1^{(2)} = -\frac{1}{h_2}\frac{\partial A^*}{\partial u^2}, \qquad D_2^{(2)} = \frac{1}{h_1}\frac{\partial A^*}{\partial u^1}, \qquad D_z^{(2)} = 0, \tag{26}$$

$$H_1^{(2)} = -\frac{1}{h_1}\frac{\partial \phi^*}{\partial u^1}, \qquad H_2^{(2)} = -\frac{1}{h_2}\frac{\partial \phi^*}{\partial u^2},$$

(27)
$$H_z^{(2)} = \frac{\partial^2\psi}{\partial z^2} - \mu\epsilon\frac{\partial^2\psi}{\partial t^2} - \mu\sigma\frac{\partial\psi}{\partial t}.$$

The projection of $\mathbf{H}^{(2)}$ on the transverse plane is irrotational and, consequently, the line integral representing the magnetomotive force between two points in this plane is independent of the path of integration.

$$\int_a^b \mathbf{H}^{(2)}\cdot d\mathbf{s} = \phi^*(a) - \phi^*(b). \tag{28}$$

Likewise the flux of the vector $\mathbf{D}^{(2)}$ crossing a strip of unit width as shown in Fig. 65 depends only on the terminal points.

$$\int_a^b \mathbf{D}^{(2)}\cdot\mathbf{n}\,da = \int_a^b dA^* = A^*(b) - A^*(a). \tag{29}$$

The projection of $\mathbf{H}^{(2)}$ on the plane z = constant is everywhere normal to $\mathbf{D}^{(2)}$; hence, the families ϕ^* = constant, A^* = constant are orthogonal. The field $\mathbf{D}^{(2)}$, $\mathbf{H}^{(2)}$ is in this sense conjugate to the field $\mathbf{E}^{(1)}$, $\mathbf{B}^{(1)}$.

6.3. Impedances of Harmonic Cylindrical Fields.—We shall suppose that the time variable enters only in the harmonic factor $e^{-i\omega t}$. Then the potentials and field components of a transverse magnetic field are

$$\phi = \mp ih\psi, \qquad A = -i(\mu\epsilon\omega + i\mu\sigma)\psi, \tag{30}$$

$$E_1^{(1)} = \pm\frac{ih}{h_1}\frac{\partial\psi}{\partial u^1}, \qquad E_2^{(1)} = \pm\frac{ih}{h_2}\frac{\partial\psi}{\partial u^2}, \qquad E_z^{(1)} = (k^2 - h^2)\psi, \tag{31}$$

$$H_1^{(1)} = -\frac{ik^2}{\mu\omega}\frac{1}{h_2}\frac{\partial\psi}{\partial u^2}, \qquad H_2^{(1)} = \frac{ik^2}{\omega\mu}\frac{1}{h_1}\frac{\partial\psi}{\partial u^1}, \qquad H_z^{(1)} = 0, \tag{32}$$

in which $k^2 = \mu\epsilon\omega^2 + i\mu\sigma\omega$. The upper sign applies to waves traveling in the positive direction along the z-axis, the lower sign to a negative direction of propagation.

A set of impedances relating the field vectors can now be defined on the basis of Sec. 5.6. It is apparent that the value of the impedance depends upon the direction in which it is measured. By definition,

$$Z_z^{(1)} = \frac{E_1^{(1)}}{H_2^{(1)}} = -\frac{E_2^{(1)}}{H_1^{(1)}} = \pm\frac{\omega\mu h}{k^2}. \tag{33}$$

The intrinsic impedance of a homogeneous, isotropic medium for plane waves is by (83), page 283,

$$Z_0 = \sqrt{\frac{\omega\mu}{\omega\epsilon + i\sigma}} = \frac{\omega\mu}{k}, \tag{34}$$

so that (33) can be expressed as

$$Z_z^{(1)} = \pm \frac{h}{k} Z_0. \tag{35}$$

Likewise impedances in the directions of the transverse axes can be defined by the relations

$$E_z^{(1)} = -Z_1^{(1)} H_2^{(1)}, \qquad E_z^{(1)} = Z_2^{(1)} H_1^{(1)}. \tag{36}$$

The correlation of the components of electric and magnetic vectors and the algebraic sign is obviously determined by the positive direction of the Poynting vector. Since $H_z^{(1)}$ is zero, there is no component of flow represented by the terms $E_2^{(1)} H_z^{(1)}$, $E_1^{(1)} H_z^{(1)}$, and the associated impedances consequently are infinite. Upon introducing the appropriate expressions for the field components into (36), we obtain

$$Z_1^{(1)} = \frac{h^2 - k^2}{ik} \frac{h_1 \psi}{\dfrac{\partial \psi}{\partial u^1}} Z_0, \qquad Z_2^{(1)} = \frac{h^2 - k^2}{ik} \frac{h_2 \psi}{\dfrac{\partial \psi}{\partial u^2}} Z_0. \tag{37}$$

The determination of corresponding impedances for the harmonic components of transverse electric fields needs no further explanation. The potentials and field vectors are in this case given by

$$\phi^* = \mp ih\psi, \qquad A^* = -i\mu\epsilon\omega\psi, \tag{38}$$

$$E_1^{(2)} = i\mu\omega \frac{1}{h_2} \frac{\partial \psi}{\partial u^2}, \qquad E_2^{(2)} = -i\mu\omega \frac{1}{h_1} \frac{\partial \psi}{\partial u^1}, \qquad E_z^{(2)} = 0, \tag{39}$$

$$H_1^{(2)} = \pm \frac{ih}{h_1} \frac{\partial \psi}{\partial u^1}, \qquad H_2^{(2)} = \pm \frac{ih}{h_2} \frac{\partial \psi}{\partial u^2}, \qquad H_z^{(2)} = (k^2 - h^2)\psi. \tag{40}$$

From these relations are calculated the various impedances.

$$\begin{aligned} E_1^{(2)} &= Z_z^{(2)} H_2^{(2)}, \qquad & E_2^{(2)} &= Z_z^{(2)} H_1^{(2)}, \\ E_2^{(2)} &= Z_1^{(2)} H_z^{(2)}, \qquad & E_1^{(2)} &= -Z_2^{(2)} H_z^{(2)}, \end{aligned} \tag{41}$$

$$Z_z^{(2)} = \pm \frac{\mu\omega}{h} = \pm \frac{k}{h} Z_0, \tag{42}$$

$$Z_1^{(2)} = \frac{ik}{h^2 - k^2} \frac{\dfrac{\partial \psi}{\partial u^1}}{h_1 \psi} Z_0, \qquad Z_2^{(2)} = \frac{ik}{h^2 - k^2} \frac{\dfrac{\partial \psi}{\partial u^2}}{h_2 \psi} Z_0. \tag{43}$$

There emerges from these results the rather curious set of relations:

$$Z_1^{(1)} Z_1^{(2)} = Z_2^{(1)} Z_2^{(2)} = Z_z^{(1)} Z_z^{(2)} = Z_0^{(2)}. \tag{44}$$

WAVE FUNCTIONS OF THE CIRCULAR CYLINDER

6.4. Elementary Waves.—By far the simplest of the separable cases is that in which the family u^1 = constant is represented by a set of coaxial circular cylinders. Then by 1, page 51,

$$u^1 = r, \qquad u^2 = \theta, \qquad h_1 = 1, \qquad h_2 = r, \tag{1}$$

and Eq. (8) of Sec. 6.1 reduces to

$$\frac{1}{r}\frac{\partial}{\partial r}\left(r\frac{\partial f}{\partial r}\right) + \frac{1}{r^2}\frac{\partial^2 f}{\partial \theta^2} + (k^2 - h^2)f = 0, \tag{2}$$

which is readily separated by writing $f(r, \theta)$ as a product

$$f = f_1(r)f_2(\theta), \tag{3}$$

wherein $f_1(r)$ and $f_2(\theta)$ are arbitrary solutions of the ordinary equations

$$r\frac{d}{dr}\left(r\frac{df_1}{dr}\right) + [(k^2 - h^2)r^2 - p^2]f_1 = 0, \tag{4}$$

$$\frac{d^2 f_2}{d\theta^2} + p^2 f_2 = 0. \tag{5}$$

The parameter p is, like h, a separation constant; its choice is governed by the physical requirement that at a fixed point in space the field must be single-valued. If, as in the present chapter, inhomogeneities and discontinuities of the medium are excluded, the field is necessarily periodic with respect to θ and the value of p is limited to the integers $n = 0, \pm 1, \pm 2, \ldots$. When, on the other hand, a field is represented by particular solutions of (4) and (5) in a sector of space bounded by the planes $\theta = \theta_1$, $\theta = \theta_2$, it is clear that nonintegral values must in general be assigned to p.

Equation (4) satisfied by the radial function $f_1(r)$ will be recognized as Bessel's equation. Its solutions can properly be called Bessel functions but, since this term is usually reserved for that particular solution $J_p(\sqrt{k^2 - h^2}\, r)$ which is finite on the axis $r = 0$, we shall adopt the name *circular cylinder function* to denote any particular solution of (4) and shall designate it by the letters $f_1 = Z_p(\sqrt{k^2 - h^2}\, r)$. The *order* of the function whose argument is $\sqrt{k^2 - h^2}\, r$ is p. Particular solutions of the wave equation (5), page 350, which are periodic in both t and θ, can therefore be constructed from elementary waves of the form

$$\psi_n = e^{in\theta}\, Z_n(\sqrt{k^2 - h^2}\, r)\, e^{\pm ihz - i\omega t}. \tag{6}$$

The propagation constant h is, in general, complex; consequently the field is not necessarily periodic along the z-axis. An explicit expression for h in terms of the frequency ω and the constants of the medium can be obtained only after the behavior of ψ over a cylinder $r =$ constant or on a plane $z =$ constant has been prescribed.

6.5. Properties of the Functions $Z_p(\rho)$.—Although it must be assumed that the reader is familiar with a considerable part of the theory of Bessel's equation, a review of the essential properties of its solutions will prove useful for reference.

If in Eq. (4) the independent variable be changed to $\rho = \sqrt{k^2 - h^2}\, r$, we find that $Z_p(\rho)$ satisfies

$$\frac{d^2Z_p}{d\rho^2} + \frac{1}{\rho}\frac{dZ_p}{d\rho} + \left(1 - \frac{p^2}{\rho^2}\right) Z_p = 0, \tag{7}$$

an equation characterized by a regular singularity at $\rho = 0$ and an essential singularity at $\rho = \infty$. The Bessel function $J_p(\rho)$, or cylinder function of the first kind, is a particular solution of (7) which is finite at $\rho = 0$. The Bessel function can, therefore, be expanded in a series of ascending powers of ρ and, since there are no singularities in the complex plane of ρ other than the points $\rho = 0$ and $\rho = \infty$, it is clear that this series must converge for all finite values of the argument. For any value of p, real or complex, and for both real and complex values of the argument ρ we have

$$J_p(\rho) = \sum_{m=0}^{\infty} \frac{(-1)^m}{m!\Gamma(p + m + 1)} \left(\frac{\rho}{2}\right)^{p+2m}. \tag{8}$$

When p in (7) is replaced by $-p$, the equation is unaltered; consequently, when p is not an integer, a second fundamental solution can be obtained from (8) by replacing p by $-p$. If, however, $p = n$ is an integer, $J_p(\rho)$ becomes a single-valued function of position. The gamma function $\Gamma(n + m + 1)$ is replaced by the factorial $(n + m)!$, so that

$$J_n(\rho) = \sum_{m=0}^{\infty} \frac{(-1)^m}{m!(n + m)!} \left(\frac{\rho}{2}\right)^{n+2m}, \qquad (n = 0, 1, 2, \cdots). \tag{9}$$

The function $J_{-n}(\rho)$ is now no longer independent of (9) but is related to it by

$$J_{-n}(\rho) = (-1)^n J_n(\rho), \tag{10}$$

so that one must resort to some other method in order to find a second solution.

The Bessel function of the second kind is defined by the relation

$$N_p(\rho) = \frac{1}{\sin p\pi} [J_p(\rho) \cos p\pi - J_{-p}(\rho)]. \tag{11}$$

This solution of (7) is independent of $J_p(\rho)$ for all values of p, but the right-hand side assumes the indeterminate form zero over zero when p

is an integer. It can be evaluated, however, in the usual way by differentiating numerator and denominator with respect to p and then passing to the limit as $p \to n$. The resultant expansion is a complicated affair[1] of which we shall give only the first term valid in the neighborhood of the origin.

$$(12) \qquad N_0(\rho) \simeq -\frac{2}{\pi} \ln \frac{2}{\gamma\rho}, \qquad N_n(\rho) \simeq \frac{-(n-1)!}{\pi}\left(\frac{2}{\rho}\right)^n, \qquad (n = 1, 2, \cdots),$$

where $\gamma = 1.78107$ and $|\rho| \ll 1$. The characterizing property of the functions of the second kind is their *singularity at the origin*. Since they become infinite at $\rho = 0$, they cannot be employed to represent fields that are physically finite in this neighborhood.

Further light is cast on the nature of the functions $J_p(\rho)$ and $N_p(\rho)$ when one examines their behavior for very large values of ρ. The expansions about the origin converge for all finite values of ρ, and $J_p(\rho)$ and $N_p(\rho)$ are analytic everywhere with respect to both p and ρ, the points $\rho = 0$ and $\rho = \infty$ excepted. However, when ρ is very large the convergence is so slow as to render the series useless for practical calculation and one seeks representations of these same functions in series of inverse powers of ρ. It can, in fact, be shown that Bessel's equation is satisfied *formally* by the expansions

$$(13) \qquad J_p(\rho) = \sqrt{\frac{2}{\pi\rho}}\,[P_p(\rho)\cos\phi - Q_p(\rho)\sin\phi],$$

$$(14) \qquad N_p(\rho) = \sqrt{\frac{2}{\pi\rho}}\,[P_p(\rho)\sin\phi + Q_p(\rho)\cos\phi],$$

$$(15) \qquad P_p(\rho) = 1 - \frac{(4p^2-1)(4p^2-9)}{2!(8\rho)^2} + \frac{(4p^2-1)(4p^2-9)(4p^2-25)(4p^2-49)}{4!(8\rho)^4} - \cdots,$$

$$(16) \qquad Q_p(\rho) = \frac{4p^2-1}{8\rho} - \frac{(4p^2-1)(4p^2-9)(4p^2-25)}{3!(8\rho)^3} + \cdots,$$

in which the phase angle ϕ is given by

$$(17) \qquad \phi = \rho - \left(p + \frac{1}{2}\right)\frac{\pi}{2}.$$

Now it turns out that these series diverge for *all* values of ρ, and consequently do not possess the exact analytical properties of the func-

[1] See WATSON, "A Treatise on the Theory of Bessel Functions," Chap. III, Cambridge University Press, 1922, and JAHNKE-EMDE, "Tables of Functions," p. 198, Teubner, 1933.

tions they are intended to represent. On the other hand, if ρ is large, the first few terms diminish rapidly in magnitude and in this sense the series are "semiconvergent." It can be shown that if the expansions are broken off near or before the point at which the successive terms begin to grow larger, they lead to an approximate value of the function and the error incurred can be estimated. The larger the value of ρ the more closely does the sum of the first few terms coincide with the true value of the function; for this reason such representations are said to be *asymptotic*.

In the present instance we note that when ρ is sufficiently large,

$$J_p(\rho) \simeq \sqrt{\frac{2}{\pi\rho}} \cos\left(\rho - \frac{2p+1}{4}\pi\right), \tag{18}$$

$$|\rho| \gg 1, \qquad |\rho| \gg |p|,$$

$$N_p(\rho) \simeq \sqrt{\frac{2}{\pi\rho}} \sin\left(\rho - \frac{2p+1}{4}\pi\right). \tag{19}$$

At very great distances from the origin the cylinder functions of the first and second kinds are related to each other as the cosine and sine functions, but are attenuated with increasing ρ due to the factor $1/\sqrt{\rho}$. They are the proper functions for the representation of *standing cylindrical waves*.

By analogy with the exponential functions one may construct a linear combination of the solutions $J_p(\rho)$ and $N_p(\rho)$ to obtain functions associated with *traveling waves*. The Bessel functions of the third kind, or *Hankel functions* as they are commonly known, are defined by the relations

$$H_p^{(1)}(\rho) = J_p(\rho) + iN_p(\rho), \tag{20}$$

$$H_p^{(2)}(\rho) = J_p(\rho) - iN_p(\rho). \tag{21}$$

From the preceding formulas we find readily that for very large ρ

$$H_p^{(1)}(\rho) \simeq \sqrt{\frac{2}{\pi\rho}}\, e^{i\left(\rho - \frac{2p+1}{4}\pi\right)}, \tag{22}$$

$$|\rho| \gg 1, \qquad |\rho| \gg |p|,$$

$$H_p^{(2)}(\rho) \simeq \sqrt{\frac{2}{\pi\rho}}\, e^{-i\left(\rho - \frac{2p+1}{4}\pi\right)}. \tag{23}$$

To the series expansions of the functions themselves we shall add for ready reference several of the more important recurrence relations.

$$Z_{p-1} + Z_{p+1} = \frac{2p}{\rho} Z_p. \tag{24}$$

(25) $$\frac{dZ_p}{d\rho} = \frac{1}{2} Z_{p-1} - \frac{1}{2} Z_{p+1}.$$

(26) $$\frac{d}{d\rho} [\rho^p Z_p(\rho)] = \rho^p Z_{p-1}.$$

(27) $$\frac{d}{d\rho} [\rho^{-p} Z_p(\rho)] = -\rho^{-p} Z_{p+1}.$$

6.6. The Field of Circularly Cylindrical Wave Functions.—Within a homogeneous, isotropic domain every electromagnetic field can be represented by linear combinations of the elementary wave functions

(28) $$\psi_{nhk} = e^{in\theta} J_n(\sqrt{k^2 - h^2}\, r)\, e^{\pm ihz - i\omega t},$$

(29) $$\psi_{nhk} = e^{in\theta} H_n^{(1)}(\sqrt{k^2 - h^2}\, r)\, e^{\pm ihz - i\omega t}.$$

Of these (28) alone applies to finite domains including the axis $r = 0$; at great distances from the source, (29) must be employed, since by (22) it reduces asymptotically to a wave traveling radially outward. Each elementary wave is identified by the parameter triplet n, h, k. When $n = 0$, the field is symmetric about the axis; when $h = 0$, the propagation is purely radial and the field strictly two-dimensional. Functions such as (28) and (29) can be said to represent *inhomogeneous plane waves.* The planes of constant phase are propagated along the z-axis with a velocity $v = \omega/\alpha$, where α is the real part of h, but the amplitudes over these planes are functions of r and θ. Such waves can be established only by sources located at finite distances from the origin of the reference system, or in media marked by discontinuities. The plane waves studied in the preceding chapter are in a strict sense *homogeneous,* since the planes of constant phase are likewise planes of constant amplitude. They can exist only in infinite, homogeneous media, generated by sources that are infinitely remote.

From the formulas developed in Sec. 6.3 one may calculate the impedances and the components of the field vectors in terms of a wave function ψ. We find

(30) $$Z_r^{(1)} = \frac{\omega\mu\sqrt{h^2 - k^2}}{k^2} \frac{Z_n(\rho)}{\dfrac{\partial Z_n(\rho)}{\partial \rho}}, \qquad Z_\theta^{(1)} = \frac{\omega\mu(k^2 - h^2)r}{nk^2},$$
$$Z_z^{(1)} = \pm \frac{\omega\mu h}{k^2}.$$

(31) $$E_r^{(1)} = \pm ih\frac{\partial \psi}{\partial r}, \qquad E_\theta^{(1)} = \pm\frac{ih}{r}\frac{\partial \psi}{\partial \theta}, \qquad E_z^{(1)} = (k^2 - h^2)\psi,$$

(32) $$H_r^{(1)} = -\frac{ik^2}{\mu\omega}\frac{1}{r}\frac{\partial \psi}{\partial \theta}, \qquad H_\theta^{(1)} = \frac{ik^2}{\mu\omega}\frac{\partial \psi}{\partial r}, \qquad H_z^{(1)} = 0.$$

Likewise for the transverse electric field,

$$Z_r^{(2)} = \frac{\omega\mu}{\sqrt{h^2 - k^2}} \frac{\dfrac{\partial Z_n(\rho)}{\partial \rho}}{Z_n(\rho)}, \qquad Z_\theta^{(2)} = \frac{\omega\mu n}{k^2 - h^2}\frac{1}{r}, \tag{33}$$

$$Z_z^{(2)} = \pm\frac{\mu\omega}{h}.$$

$$E_r^{(2)} = \frac{i\mu\omega}{r}\frac{\partial\psi}{\partial\theta}, \qquad E_\theta^{(2)} = -i\mu\omega\frac{\partial\psi}{\partial r}, \qquad E_z^{(2)} = 0, \tag{34}$$

$$H_r^{(2)} = \pm ih\frac{\partial\psi}{\partial r}, \qquad H_\theta^{(2)} = \pm\frac{ih}{r}\frac{\partial\psi}{\partial\theta}, \qquad H_z^{(2)} = (k^2 - h^2)\psi. \tag{35}$$

When initial conditions are prescribed over given plane or cylindrical surfaces, a solution is constructed by superposition of elementary wave functions. For fixed values of ω (or k) and h, one obtains for the resultant field in cylindrical coordinates the equations

$$\begin{aligned} E_r &= ih\sum_{n=-\infty}^{\infty} a_n\frac{\partial\psi_n}{\partial r} - \frac{\mu\omega}{r}\sum_{n=-\infty}^{\infty} nb_n\psi_n, \\ E_\theta &= -\frac{h}{r}\sum_{n=-\infty}^{\infty} na_n\psi_n - i\mu\omega\sum_{n=-\infty}^{\infty} b_n\frac{\partial\psi_n}{\partial r}, \\ E_z &= (k^2 - h^2)\sum_{n=-\infty}^{\infty} a_n\psi_n, \end{aligned} \tag{36}$$

$$\begin{aligned} H_r &= \frac{k^2}{\mu\omega}\frac{1}{r}\sum_{n=-\infty}^{\infty} na_n\psi_n + ih\sum_{n=-\infty}^{\infty} b_n\frac{\partial\psi_n}{\partial r}, \\ H_\theta &= \frac{ik^2}{\mu\omega}\sum_{n=-\infty}^{\infty} a_n\frac{\partial\psi_n}{\partial r} - \frac{h}{r}\sum_{n=-\infty}^{\infty} nb_n\psi_n, \\ H_z &= (k^2 - h^2)\sum_{n=-\infty}^{\infty} b_n\psi_n, \end{aligned} \tag{37}$$

where a_n and b_n are coefficients to be determined from initial conditions. The direction of propagation is positive or negative according to the sign of h.

INTEGRAL REPRESENTATIONS OF WAVE FUNCTIONS

6.7. Construction from Plane Wave Solutions.—If ζ measures distance along any axis whose direction with respect to a fixed reference system (x, y, z) is determined by a unit vector $\mathbf{n}$, then the most elementary type of plane wave can be represented according to Secs. 5.1 to 5.6 by the function

$$\psi = e^{ik\zeta - i\omega t}, \tag{1}$$

in which the constants k and ω are either real or complex. Let **R** be the radius vector drawn from the origin to a point of observation whose rectangular coordinates are x, y, z. The phase of the wave function at a given instant is then measured by

$$\zeta = \mathbf{n} \cdot \mathbf{R} = n_x x + n_y y + n_z z. \tag{2}$$

The direction cosines n_x, n_y, n_z of the vector **n** are best expressed in terms of the polar angles α and β shown in Fig. 66.

$$n_x = \sin\alpha\cos\beta, \qquad n_y = \sin\alpha\sin\beta, \qquad n_z = \cos\alpha; \tag{3}$$

hence,

$$\psi = e^{ik(x\sin\alpha\cos\beta + y\sin\alpha\sin\beta + z\cos\alpha) - i\omega t}. \tag{4}$$

As the parameters α and β are varied, the axis of propagation can be oriented at will. With each direction of propagation one associates an amplitude $g(\alpha, \beta)$ depending only on the angles α and β; since the field equations are in the present case linear, a solution can be constructed by superposing plane waves, all of the same frequency but traveling in various directions, each with its appropriate amplitude.[1]

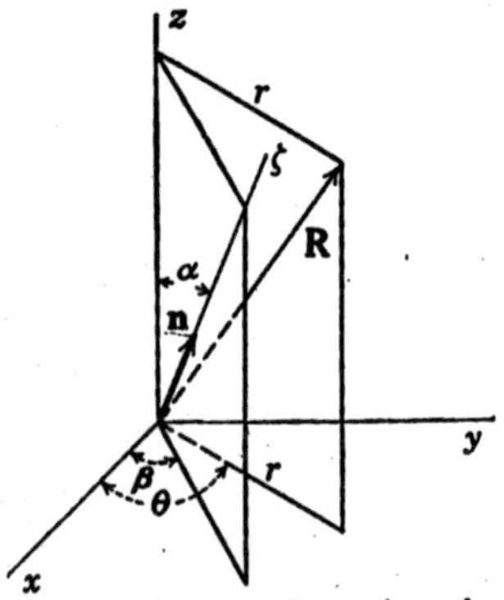

FIG. 66.—The phase of an elementary plane wave is measured along the ζ-axis whose direction is determined by the unit vector **n**. A fixed point of observation is located by the vector **R**.

$$\psi(x, y, z, t) = e^{-i\omega t}\int d\alpha \int d\beta\, g(\alpha, \beta)\, e^{ik(x\sin\alpha\cos\beta + y\sin\alpha\sin\beta + z\cos\alpha)}. \tag{5}$$

If the angles are real, the limits of integration for α are obviously 0 and π; β goes from 0 to 2π. But such a solution is mathematically by no means the most general, for (5) satisfies the wave equation for complex as well as real values of the parameters α and β and we shall discover shortly that complex angles must in fact be included if we are to represent arbitrary fields by such an integral.

When ω is real, the wave function defined by (5) is harmonic in time. To represent fields whose time variation over a specified coordinate

[1] The general theory of such solutions of the wave equation has been discussed by Whittaker. See WHITTAKER and WATSON, "Modern Analysis," 4th ed., Chap. XVIII, Cambridge University Press, 1927.

surface is more complex it will be necessary to sum or integrate (5) with respect to the parameter ω. We shall define a *vector propagation constant* $\mathbf{k} = k\mathbf{n}$ whose rectangular components are

$$(6) \qquad k_1 = k \sin\alpha \cos\beta, \qquad k_2 = k \sin\alpha \sin\beta, \qquad k_3 = k \cos\alpha,$$

so that the elementary plane wave function can be written

$$(7) \qquad \psi = e^{i\mathbf{k}\cdot\mathbf{R} - i\omega t}.$$

It follows upon introduction of (7) into the wave equation

$$(8) \qquad \frac{\partial^2\psi}{\partial x^2} + \frac{\partial^2\psi}{\partial y^2} + \frac{\partial^2\psi}{\partial z^2} - \mu\epsilon\frac{\partial^2\psi}{\partial t^2} - \mu\sigma\frac{\partial\psi}{\partial t} = 0,$$

that the components of $\mathbf{k}$ must satisfy the single relation

$$(9) \qquad k_1^2 + k_2^2 + k_3^2 = \mu\epsilon\omega^2 + i\mu\sigma\omega = k^2$$

and are otherwise completely arbitrary. Thus of the parameters k_1, k_2, k_3, ω, any three may be chosen arbitrarily whereupon the fourth is fixed by (9).

Let us suppose then that over the plane $z = 0$ the function ψ is prescribed. We shall say that $\psi(x, y, 0, t) = f(x, y, t)$. The desired solution is

(10)

$$\psi(x, y, z, t) = \left(\frac{1}{2\pi}\right)^{\frac{3}{2}} \int_{-\infty}^{\infty}\int_{-\infty}^{\infty}\int_{-\infty}^{\infty} g(k_1, k_2, \omega)\, e^{i(k_1x + k_2y + k_3z - \omega t)}\, dk_1\, dk_2\, d\omega,$$

in which k_1, k_2 and ω are real variables and k_3 is a complex quantity determined by

$$(11) \qquad k_3^2 = \omega^2\mu\epsilon + i\mu\sigma\omega - k_1^2 - k_2^2.$$

The amplitude function is to be such that

$$(12) \quad f(x, y, t) = \left(\frac{1}{2\pi}\right)^{\frac{3}{2}} \int_{-\infty}^{\infty}\int_{-\infty}^{\infty}\int_{-\infty}^{\infty} g(k_1, k_2, \omega)\, e^{i(k_1x + k_2y - \omega t)}\, dk_1\, dk_2\, d\omega.$$

If $f(x, y, t)$ and its first derivatives are piecewise continuous and absolutely integrable, then $g(k_1, k_2, \omega)$ is its Fourier transform and is given by

$$(13) \quad g(k_1, k_2, \omega) = \left(\frac{1}{2\pi}\right)^{\frac{3}{2}} \int_{-\infty}^{\infty}\int_{-\infty}^{\infty}\int_{-\infty}^{\infty} f(x, y, t)\, e^{-i(k_1x + k_2y - \omega t)}\, dx\, dy\, dt.$$

When $\sigma = 0$, each harmonic component is propagated along the z-axis with a velocity $v = \omega/k_3$, but since $k_3 = \sqrt{\omega^2\mu\epsilon - k_1^2 - k_2^2}$ is not a linear combination of ω, k_1, and k_2, it is apparent that the initial disturbance

$f(x, y, t)$ does not propagate itself without change of form even in the absence of a dissipative term. More precisely, there exists no general solution of (8) of the type $f\left(x, y, t - \frac{z}{v}\right)$.

Equally one might prescribe the function $\psi(x, y, z, t)$ throughout all space at an initial instant $t = 0$. Let us suppose, for example, that $\psi(x, y, z, 0) = f(x, y, z)$. The field is to be represented by the multiple integral

(14)
$$\psi(x, y, z, t) = \left(\frac{1}{2\pi}\right)^{\frac{3}{2}} \int_{-\infty}^{\infty} \int_{-\infty}^{\infty} \int_{-\infty}^{\infty} g(k_1, k_2, k_3)\, e^{i(k_1x + k_2y + k_3z - \omega t)}\, dk_1\, dk_2\, dk_3,$$

in which k_1, k_2, k_3 are real variables and ω is a complex quantity determined from (9).

(15)
$$i\omega = b \pm i\sqrt{a^2(k_1^2 + k_2^2 + k_3^2) - b^2},$$

where $a = 1/\sqrt{\mu\epsilon}$, $b = \sigma/2\epsilon$ as defined on page 297. The amplitude or weighting function $g(k_1, k_2, k_3)$ is to be such that

(16)
$$f(x, y, z) = \left(\frac{1}{2\pi}\right)^{\frac{3}{2}} \int_{-\infty}^{\infty} \int_{-\infty}^{\infty} \int_{-\infty}^{\infty} g(k_1, k_2, k_3)\, e^{i(k_1x + k_2y + k_3z)}\, dk_1\, dk_2\, dk_3.$$

If $f(x, y, z)$ has the necessary analytic properties, the Fourier transform exists and is given by

(17)
$$g(k_1, k_2, k_3) = \left(\frac{1}{2\pi}\right)^{\frac{3}{2}} \int_{-\infty}^{\infty} \int_{-\infty}^{\infty} \int_{-\infty}^{\infty} f(x, y, z)\, e^{-i(k_1x + k_2y + k_3z)}\, dx\, dy\, dz.$$

Only positive or outward waves have been considered in the foregoing and initial conditions have been imposed only upon the function ψ itself. If both ψ and a derivative with respect to one of the four variables are to satisfy specified conditions, it becomes necessary to include negative as well as positive waves according to the methods described for the one-dimensional case in Secs. 5.8, 5.9, and 5.10.

6.8. Integral Representations of the Functions $Z_n(\varrho)$.—In any system of cylindrical coordinates u^1, u^2, z, the wave equation (6), page 351, is satisfied by

(18)
$$\psi = f(u^1, u^2)\, e^{ihz - i\omega t},$$

where h and ω are real or complex constants. Following the notation of the preceding section $h = k_3 = k \cos \alpha$ and, since $k = \sqrt{\mu\epsilon\omega^2 + i\mu\sigma\omega}$, it follows that the angle α made by the directions of the component plane waves in (5) with the z-axis is also constant. In other words, the elementary cylindrical wave function (18) can be resolved into homogeneous plane waves whose directions form a circular cone about the z-axis, but the aperture of the cone is in the general case measured by a complex angle.

$$f(u^1, u^2) = \int g(\beta)\, e^{ik \sin \alpha (x \cos \beta + y \sin \beta)}\, d\beta, \tag{19}$$

in which x and y are to be expressed in terms of the cylindrical coordinates u^1, u^2.

In the coordinates of the circular cylinder we have $x = r \cos \theta$, $y = r \sin \theta$, and, hence,

$$x \cos \beta + y \sin \beta = r \cos (\theta - \beta). \tag{20}$$

In the notation of the preceding paragraphs we have also

$$kr \sin \alpha = r \sqrt{k^2 - h^2} = \rho, \tag{21}$$

so that

$$f(r, \theta) = \int g(\beta)\, e^{i\rho \cos (\theta - \beta)}\, d\beta. \tag{22}$$

We now change the variable of integration in (22) from β to $\phi = \beta - \theta$, and note that, since the equation for $f(r, \theta)$ is separable, it must be possible to express $g(\phi + \theta)$ as a product of two functions of one variable each.

$$g(\beta) = g(\phi + \theta) = g_1(\phi) g_2(\theta); \tag{23}$$

hence,

$$f(r, \theta) = f_1(r) f_2(\theta) = g_2(\theta) \int g_1(\phi)\, e^{i\rho \cos \phi}\, d\phi. \tag{24}$$

The angle function $g_2(\theta)$ must obviously be some linear combination of the exponentials $e^{ip\theta}$ and $e^{-ip\theta}$, and the amplitude or weighting factor $g_1(\phi)$ must be chosen so that the radial function $f_1(r)$ satisfies (4), page 356, which in terms of ρ is written

$$\rho^2 \frac{d^2 f_1}{d\rho^2} + \rho \frac{df_1}{d\rho} + (\rho^2 - p^2) f_1 = 0. \tag{25}$$

Let us substitute into (25) the integral

$$f_1(\rho) = \int g_1(\phi)\, e^{i\rho \cos \phi}\, d\phi. \tag{26}$$

Differentiating under the sign of integration, we have

(27)
$$\frac{df_1}{d\rho} = \int i\cos\phi\, g_1(\phi)\, e^{i\rho\cos\phi}\, d\phi, \qquad \frac{d^2f_1}{d\rho^2} = -\int \cos^2\phi\, g_1(\phi)\, e^{i\rho\cos\phi}\, d\phi;$$

hence from (25)

(28)
$$\int (\rho^2\sin^2\phi + i\rho\cos\phi - p^2) g_1(\phi)\, e^{i\rho\cos\phi}\, d\phi = 0.$$

This equation is next transformed by an integration by parts. Equation (28) is evidently equivalent to

(29)
$$\int \left[g_1(\phi) \frac{d^2 e^{i\rho\cos\phi}}{d\phi^2} + p^2 g_1(\phi)\, e^{i\rho\cos\phi} \right] d\phi = 0.$$

If P and Q are two functions of ϕ, then

(30)
$$Q\frac{d^2P}{d\phi^2} - P\frac{d^2Q}{d\phi^2} = \frac{d}{d\phi}\left(Q\frac{dP}{d\phi} - P\frac{dQ}{d\phi}\right);$$

this applied to (29) gives

(31)
$$\int \frac{d}{d\phi}\left[g_1(\phi)\frac{de^{i\rho\cos\phi}}{d\phi} - e^{i\rho\cos\phi}\frac{dg_1}{d\phi}\right] d\phi + \int \left(\frac{d^2g_1}{d\phi^2} + p^2 g_1\right) e^{i\rho\cos\phi}\, d\phi = 0.$$

The first of these two integrals can be made to vanish if the contour of integration C is so chosen that the differential has the same value at both the initial and terminal points; the second is zero if the integrand vanishes. Therefore (26) is a solution of the Bessel equation if $g_1(\phi)$ satisfies

(32)
$$\frac{d^2g_1}{d\phi^2} + p^2 g_1 = 0$$

and if the path C of integration is such that

(33)
$$\left(i\rho\sin\phi\, g_1(\phi) - \frac{dg_1}{d\phi}\right) e^{i\rho\cos\phi}\Big|_C = 0.$$

Equation (32) is evidently satisfied by $e^{ip\phi}$. However, we shall find that, if $f_1(\rho)$ is to be identical with the cylinder functions $Z_p(\rho)$ defined in Sec. 6.5, a constant factor $\frac{1}{\pi}e^{-ip\frac{\pi}{2}}$ must be added. Then by choosing

(34)
$$g_1(\phi) = \frac{1}{\pi}e^{ip\left(\phi - \frac{\pi}{2}\right)},$$

we obtain Sommerfeld's integral representation of the cylinder functions,[1]

$$(35)\qquad Z_p(\rho) = \frac{e^{-ip\frac{\pi}{2}}}{\pi}\int_C e^{i(\rho\cos\phi + p\phi)}\,d\phi,$$

the contour C to be such that

$$(36)\qquad (\rho\sin\phi + p)\,e^{i(\rho\cos\phi+p\phi)}\Big|_C = 0.$$

The distinction between the various particular solutions $J_p(\rho)$, $N_p(\rho)$, $H_p(\rho)$ is now reduced to a distinction in the contours of integration in the complex plane of ϕ. Consider first the most elementary case in which $p = n$, an integer. Then clearly, if the path of integration is extended along the real axis of ϕ from $-\pi$ to π, or any other segment of length 2π, the condition (36) is fulfilled and the definite integral

$$(37)\qquad J_n(\rho) = \frac{i^{-n}}{2\pi}\int_{-\pi}^{\pi} e^{i\rho\cos\phi + in\phi}\,d\phi$$

is a solution of the Bessel equation. That the function defined in (37) is actually identical with the particular solution (9), page 357, can be verified by expanding $e^{i\rho\cos\phi}$ in a series about the point $\rho = 0$ and integrating term by term.

In the general case when p is not an integer, or to obtain the independent solution, one must choose complex values of ϕ in order that (36) shall vanish at the terminal points. Let $\phi = \gamma + i\eta$. Then

$$(38)\qquad i\rho\cos\phi + ip\phi = \rho\sin\gamma\sinh\eta - p\eta + i\rho\cos\gamma\cosh\eta + ip\gamma.$$

If ρ is complex we shall assume that $\rho = a + ib$, where a is an essentially positive quantity. Then by an appropriate choice of γ and η the real part of (38) can be made infinitely negative, while $\exp(i\rho\cos\phi + ip\phi) \to 0$. This condition will be satisfied, for example, if we let $\eta \to +\infty$ and choose for γ either $-\pi/2$ or $+3\pi/2$; but the exponential also vanishes when $\eta \to -\infty$ provided $\gamma = +\pi/2$. In order that (35) shall represent a solution of the Bessel equation, it is only necessary that the contour C connect any two of these points.

Sommerfeld has chosen as a pair of fundamental solutions the two integrals

$$(39)\qquad H_p^{(1)}(\rho) = \frac{e^{-ip\frac{\pi}{2}}}{\pi}\int_{-\frac{\pi}{2}+i\infty}^{\frac{\pi}{2}-i\infty} e^{i\rho\cos\phi+ip\phi}\,d\phi,$$

$$(40)\qquad H_p^{(2)}(\rho) = \frac{e^{-ip\frac{\pi}{2}}}{\pi}\int_{\frac{\pi}{2}-i\infty}^{\frac{3\pi}{2}+i\infty} e^{i\rho\cos\phi+ip\phi}\,d\phi.$$

[1] SOMMERFELD, *Math. Ann.*, **47**, 335, 1896.

The contour C_1 followed by the integral (39) starts at $\eta = \infty, \gamma = -\pi/2$, crosses both real and imaginary axes at $\phi = 0$, and terminates eventually at $\eta = -\infty$, $\gamma = \pi/2$. The contour C_2 followed by (40) starts at the terminal point of C_1, crosses the real axis at $\gamma = \pi$, and comes to an end at $\eta = \infty$, $\gamma = 3\pi/2$. Both contours are illustrated in Fig. 67. The point at which the crossing of the real axis occurs is, of course, not essential to the definition. The contours may be deformed at will provided only they begin at an infinitely remote central point of one shaded area and terminate at a corresponding point in a second shaded area.

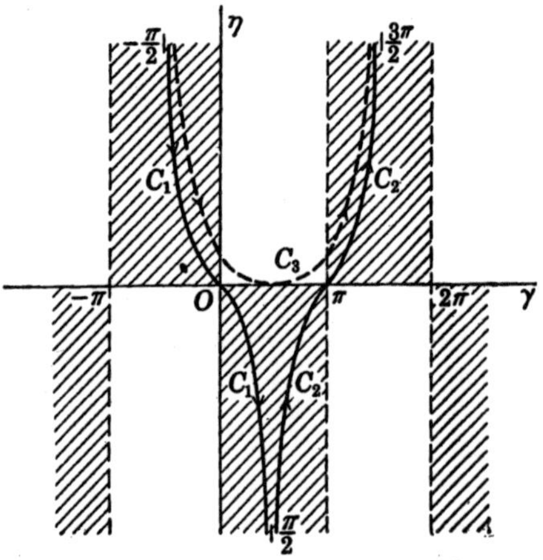

FIG. 67.—Contours of integration for the cylinder functions.

The advantages of carrying C_1 and C_2 across the real axis at the particular points $\gamma = 0$ and $\gamma = \pi$ become apparent when the integrals (39) and (40) are evaluated for very large values of ρ. For if the real part of ρ is very large, the factor $\exp(i\rho \cos \phi)$ becomes vanishingly small at all points of the shaded domains in Fig. 67 with the exception of the immediate neighborhood of the points $\eta = 0$, $\gamma = 0$, $\pm\pi$, $\pm 2\pi$, ... At these points the real part of $i\rho \cos \phi$, Eq. (38), is zero however large ρ; consequently, if C_1 and C_2 are drawn as in Fig. 67, the sole contribution to the contour integrals will be experienced in the neighborhood of the origin and the point $\eta = 0$, $\gamma = \pi$. Out of a shaded region in which the values of the integrand are vanishingly small, the contour C_1 leads over a steep "pass" or "saddle point" of high values at the origin, and then abruptly downwards into another shaded plane where the contributions to the integral are again of negligible amount. The contour C_2 encounters a similar saddle point at $\eta = 0$, $\gamma = \pi$. To confine the integration to as short a segment of the contour as possible, one must approach the pass by the line of steepest ascent, descending quickly from the top into the valley beyond by the line of steepest descent. This means in the present case that the contours C_1, C_2 must cross the axis at an angle of 45 deg. The behavior of the integrals (39) and (40) in the neighborhood of a saddle point has been used by Debye[1] to calculate the asymptotic expansions of the functions $H_p^{(1)}(\rho)$ and $H_p^{(2)}(\rho)$. When ρ is very large, both with respect to unity and to the order p, one obtains the relations (22) and (23) of page 359, thereby identifying the integrals (39) and (40) with the Hankel functions defined in Sec. 6.5. Debye considered also the case in which p was larger than the argument ρ, but his results have been improved and extended by more recent investigations.

A contour integral representation of $J_p(\rho)$, when p is not an integer, follows directly from the relation

$$J_p(\rho) = \tfrac{1}{2}[H_p^{(1)}(\rho) + H_p^{(2)}(\rho)]. \tag{41}$$

$$J_p(\rho) = \frac{e^{-ip\frac{\pi}{2}}}{2\pi} \int_{C_3} e^{i\rho\cos\phi + ip\phi}\, d\phi. \tag{42}$$

The contour C_3 shown in Fig. 67 represents a permissible deformation of $C_1 + C_2$.

6.9. Fourier-Bessel Integrals.—It has been shown how in cylindrical coordinates the two fundamental types of electromagnetic field can be derived from a scalar function ψ. In case the cylindrical coordinates are circular, ψ is in general a function obtained by superposition of elementary waves such as the particular solutions (28) and (29) of page 360. Our problem is now the following: at the instant $t = 0$ the value of $\psi(r, \theta, z, t)$ is prescribed over the plane $z = 0$; to determine ψ for all other values of z and t.

Let us suppose then that when $t = z = 0$, $\psi = f(r, \theta)$. We shall assume that $f(r, \theta)$ is a bounded, single-valued function of the variables and is, together with its first derivatives, piecewise continuous. Then $f(r, \theta)$ must be periodic in θ and can be expanded in a Fourier series whose coefficients are functions of r alone.

$$f(r, \theta) = \sum_{n=-\infty}^{\infty} f_n(r)\, e^{in\theta}, \qquad f_n(r) = \frac{1}{2\pi}\int_0^{2\pi} f(r, \theta)\, e^{-in\theta}\, d\theta. \tag{43}$$

If now $f(r, \theta)$ vanishes as $r \to \infty$ in such a way as to ensure the convergence of the integrals $\int_0^\infty |f_n(r)|\sqrt{r}\, dr$, then each coefficient $f_n(r)$ can be represented by a modified Fourier integral.[2]

[1] DEBYE, *Math. Ann.* **67**, 535, 1909. See also WATSON, *loc. cit.*, pp. 235*ff.*

[2] The Fourier-Bessel integral is established by more rigorous methods in Chap. XIV of Watson's "Bessel Functions."

Consider a function $f(x, y)$ of two variables admitting the Fourier integral representation

$$f(x, y) = \frac{1}{2\pi}\int_{-\infty}^{\infty}\int_{-\infty}^{\infty} g(k_1, k_2)\, e^{i(k_1x+k_2y)}\, dk_1\, dk_2. \tag{44}$$

A transformation to polar coordinates is now made both in coordinate space and in k space.

$$\begin{aligned} x &= r\cos\theta, & y &= r\sin\theta, \\ k_1 &= \lambda\cos\beta, & k_2 &= \lambda\sin\beta, \end{aligned} \tag{45}$$

so that in the notation of earlier paragraphs $\lambda = k\sin\alpha = \sqrt{k^2 - h^2}$. Then $k_1x + k_2y = \lambda r\cos(\beta - \theta)$ and (44), as a volume integral in k space, transforms to

$$f(r, \theta) = \frac{1}{2\pi}\int_0^{\infty} \lambda\, d\lambda \int_0^{2\pi} d\beta\, g(\lambda, \beta)\, e^{i\lambda r\cos(\beta-\theta)}. \tag{46}$$

The physical significance of this representation is worth noting. The function $\exp[i\lambda r\cos(\beta - \theta) - i\omega t]$ represents a plane wave whose propagation constant is λ, traveling in a direction which is normal to the z-axis and which makes an angle β with the x-axis. Each plane wave is multiplied by an amplitude factor $g(\lambda, \beta)$ and then summed first with respect to β from 0 to 2π and then with respect to the propagation constant, or space frequency λ.

The transform of $f(x, y)$ is

$$g(k_1, k_2) = \frac{1}{2\pi}\int_{-\infty}^{\infty}\int_{-\infty}^{\infty} f(\xi, \eta)\, e^{-i(k_1\xi+k_2\eta)}\, d\xi\, d\eta, \tag{47}$$

which when transformed to the polar coordinates $\xi = \rho\cos\mu$, $\eta = \rho\sin\mu$, leads to

$$g(\lambda, \beta) = \frac{1}{2\pi}\int_0^{\infty} \rho\, d\rho \int_0^{2\pi} d\mu\, f(\rho, \mu)\, e^{-i\lambda\rho\cos(\beta-\mu)}. \tag{48}$$

Suppose finally that $f(r, \theta) = f_n(r)\, e^{in\theta}$. Then

$$g(\lambda, \beta) = \frac{1}{2\pi}\int_0^{\infty} \rho\, d\rho\, f_n(\rho) \int_0^{2\pi} d\mu\, e^{-i\lambda\rho\cos(\beta-\mu)+in\mu}, \tag{49}$$

which upon a change of variable such as $\phi = \mu - \beta - \pi$ is clearly equal to

$$g(\lambda, \beta) = e^{in(\beta+\frac{1}{2}\pi)} \int_0^{\infty} f_n(\rho) J_n(\lambda\rho)\rho\, d\rho = g_n(\lambda)\, e^{in(\beta+\frac{1}{2}\pi)}. \tag{50}$$

Likewise (46) becomes now

$$f(r, \theta) = f_n(r)\, e^{in\theta} = \frac{1}{2\pi} \int_0^\infty \lambda\, d\lambda\, g_n(\lambda) \int_0^{2\pi} d\beta\, e^{i\lambda r \cos(\beta-\mu) + in(\beta + \frac{3}{2}\pi)}, \tag{51}$$

or upon placing $\phi = \beta - \theta$,

$$f_n(r)\, e^{in\theta} = e^{in\theta} \int_0^\infty g_n(\lambda) J_n(\lambda r) \lambda\, d\lambda. \tag{52}$$

Thus we obtain the pair of Fourier-Bessel transforms

$$f_n(r) = \int_0^\infty g_n(\lambda) J_n(\lambda r) \lambda\, d\lambda, \tag{53}$$

$$g_n(\lambda) = \int_0^\infty f_n(\rho) J_n(\lambda \rho) \rho\, d\rho. \tag{54}$$

The function

$$\psi = e^{-i\omega t} \sum_{n=-\infty}^{\infty} e^{in\theta} \int_0^\infty g_n(\lambda) J_n(\lambda r)\, e^{\pm i\sqrt{k^2-\lambda^2}\,z}\, \lambda\, d\lambda \tag{55}$$

is a solution of the wave equation in circular cylindrical coordinates, which at $t = 0$ reduces to $f(r, \theta)$ on the plane $z = 0$. But this obviously is not the most general solution satisfying these conditions, for we have still at our disposal the two parameters ω and k which are subject only to the relation $k^2 = \mu\epsilon\omega^2 + i\sigma\mu\omega$. If ω is real, harmonic wave functions such as (55) may be superposed to represent an arbitrary time variation of ψ over the plane $z = 0$. If both positive and negative waves are considered, it is permissible also to assign values to both ψ and $\partial\psi/\partial z$ at $z = 0$. The treatment of such a problem has been adequately illustrated in Chap. V.

6.10. Representation of a Plane Wave.—The Fourier-Bessel theorem offers a very simple means of representing an elementary plane wave in terms of cylindrical wave functions. Let the wave whose propagation constant is k travel in a direction defined by the unit vector $\mathbf{n}$, whose spherical polar angles with respect to a fixed reference system are α and β as in Fig. 66, page 362. Then

$$\psi = e^{ik \sin\alpha (x \cos\beta + y \sin\beta)} \cdot e^{ikz \cos\alpha - i\omega t}, \tag{56}$$

and our problem is to represent the function

$$f(x, y) = e^{ik \sin\alpha (x \cos\beta + y \sin\beta)} = e^{ikr \sin\alpha \cos(\beta - \theta)} \tag{57}$$

as a series of the form

$$f(r, \theta) = \sum_{n=-\infty}^{\infty} f_n(r)\, e^{in\theta}. \tag{58}$$

By (43) we have

$$f_n(r) = \frac{1}{2\pi} \int_0^{2\pi} e^{ikr \sin \alpha \cos (\beta - \theta) - in\theta}\, d\theta. \tag{59}$$

Upon replacing $\beta - \theta$ by ϕ, this becomes equal to

$$f_n(r) = e^{in\left(\frac{\pi}{2} - \beta\right)} J_n(kr \sin \alpha), \tag{60}$$

and we obtain thereby the useful expansion

$$e^{ikr \sin \alpha \cos (\beta - \theta)} = \sum_{n=-\infty}^{\infty} i^n J_n(kr \sin \alpha)\, e^{in(\beta - \theta)}. \tag{61}$$

Several other well-known series expansions are a direct consequence of this result. Thus, if we put $\rho = kr \sin \alpha$, $\theta - \beta = \phi - \frac{\pi}{2}$, (61) becomes

$$e^{i\rho \sin \phi} = \sum_{n=-\infty}^{\infty} J_n(\rho)\, e^{in\theta}, \tag{62}$$

which upon separation into real and imaginary parts leads to

$$\begin{aligned} \cos (\rho \sin \phi) &= \sum_{n=-\infty}^{\infty} J_n(\rho) \cos n\phi, \\ \sin (\rho \sin \phi) &= \sum_{n=-\infty}^{\infty} J_n(\rho) \sin n\phi. \end{aligned} \tag{63}$$

THE ADDITION THEOREM FOR CIRCULARLY CYLINDRICAL WAVES

6.11. Several important relations pertaining to a translation of the axis of propagation parallel to itself can be derived from the formulas of the foregoing section. In Fig. 68 O and O_1 denote the origins of two rectangular reference systems. The sheet of the figure coincides with the xy-plane of both systems and the axes x_1, y_1, z_1 through O_1 are respectively parallel to x, y, z. The function $J_n(\lambda r_1)\, e^{in\theta_1}$, when multiplied by $\exp(\pm ihz_1 - i\omega t)$, represents an elementary cylindrical wave referred to the z_1-axis. We wish to express this cylindrical wave in terms of a sum of cylindrical wave functions referred to the parallel z-axis through O.

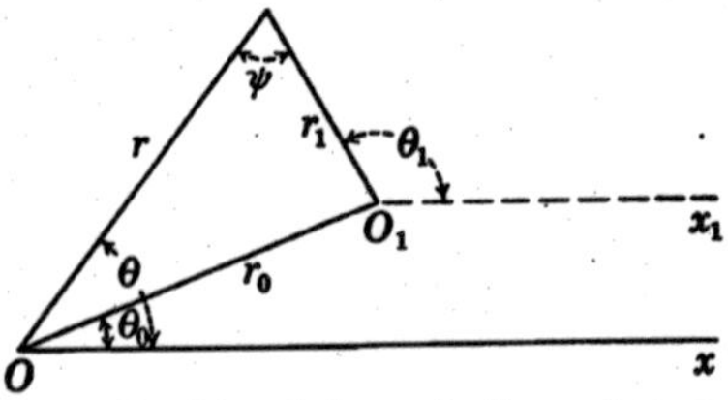

FIG. 68.—Translation of the reference system.

We write first

(1) $$J_n(\lambda r_1)\, e^{in\theta_1} = \frac{1}{2\pi}\int_{-\pi}^{\pi} e^{i\lambda r_1 \cos\phi + in\left(\phi+\theta_1-\frac{\pi}{2}\right)}\, d\phi.$$

From the figure it is clear that $\theta_1 = \theta + \psi$ and that

(2) $$\begin{aligned} r_1 \cos\psi &= r_1 \cos(\theta_1 - \theta) = r - r_0 \cos(\theta - \theta_0), \\ r_1 \sin\psi &= r_1 \sin(\theta_1 - \theta) = r_0 \sin(\theta - \theta_0). \end{aligned}$$

Moreover, in virtue of the periodic properties of the integrand, (1) is equivalent to

(3) $$J_n(\lambda r_1)\, e^{in\theta_1} = \frac{1}{2\pi}\int_{-\pi}^{\pi} e^{i\lambda r_1 \cos(\phi-\psi) + in\left(\phi-\psi+\theta_1-\frac{\pi}{2}\right)}\, d\phi.$$

It follows from (2) that

(4) $$r_1 \cos(\phi - \psi) = r\cos\phi - r_0 \cos(\phi + \theta - \theta_0);$$

hence,

(5) $$J_n(\lambda r_1)\, e^{in\theta_1} = \frac{1}{2\pi}\int_{-\pi}^{\pi} e^{i\lambda r\cos\phi - i\lambda r_0 \cos(\phi+\theta-\theta_0) + in\left(\phi+\theta-\frac{\pi}{2}\right)}\, d\phi.$$

By (61), page 372, we have

(6) $$e^{-i\lambda r_0 \cos(\phi+\theta-\theta_0)} = \sum_{m=-\infty}^{\infty} e^{-\frac{im\pi}{2}} J_m(\lambda r_0)\, e^{im(\phi+\theta-\theta_0)},$$

and in virtue of the uniformity of convergence one may interchange the order of summation and integration.

(7) $$\begin{aligned} J_n(\lambda r_1)\, e^{in\theta_1} &= \sum_{m=-\infty}^{\infty} J_m(\lambda r_0) \\ &\qquad \frac{1}{2\pi}\int_{-\pi}^{\pi} d\phi\, e^{i\lambda r\cos\phi + i(n+m)\left(\phi-\frac{\pi}{2}\right) + im(\theta-\theta_0) + in\theta} \\ &= \sum_{m=-\infty}^{\infty} J_m(\lambda r_0) J_{n+m}(\lambda r)\, e^{in\theta + im(\theta-\theta_0)}. \end{aligned}$$

Upon replacing θ_1 by $\theta + \psi$ this result becomes

(8) $$J_n(\lambda r_1)\, e^{in\psi} = \sum_{m=-\infty}^{\infty} J_m(\lambda r_0) J_{n+m}(\lambda r)\, e^{im(\theta-\theta_0)}.$$

An analogous expansion for the wave function $H_n^{(1)}(\lambda r_1)\, e^{in\theta_1}$ can be obtained by writing

(9) $$H_n^{(1)}(\lambda r_1)\, e^{in\psi} = \frac{1}{\pi}\int_{C_1} e^{i\lambda r_1 \cos(\phi-\psi) + in\left(\phi-\frac{\pi}{2}\right)}\, d\phi,$$

where C_1 is the contour described in Fig. 67, page 368, with a shift of amount ψ along the real axis to ensure the vanishing of the exponential at the terminals. Upon expanding the exponent by (2), we find

$$(10)\qquad H_n^{(1)}(\lambda r_1)\, e^{in\psi} = \frac{1}{\pi}\int_{C_1} d\phi\, e^{i\lambda r\cos\phi - i\lambda r_0\cos(\phi+\theta-\theta_0) + in\left(\phi - \frac{\pi}{2}\right)}.$$

If now $|r| > |r_0 \cos (\theta - \theta_0)|$, one can proceed exactly as above and is led to the expansion

$$(11)\qquad H_n^{(1)}(\lambda r_1)\, e^{in\psi} = \sum_{m=-\infty}^{\infty} J_m(\lambda r_0) H_{n+m}^{(1)}(\lambda r)\, e^{im(\theta-\theta_0)}.$$

If $r_0 = 0$, the two centers coincide; $J_m(0) = 0$ for all values of m except zero, and $J_0(0) = 1$. The angle ψ is zero and the right and left sides of (11) are clearly identical. The expansion can be verified also in the case when r and r_1 are very large, for then the Hankel functions can be replaced by their asymptotic representation, (22), page 359. The angle ψ is approximately zero and $r_1 \simeq r - r_0 \cos (\theta - \theta_0)$. The amplitude factor $\sqrt{2/\pi r_1}$ can be replaced by $\sqrt{2/\pi r}$ without appreciable error, but the term $r_0 \cos (\theta - \theta_0)$ must be retained in the phase. Then (11) reduces to

$$(12)\qquad e^{-i\lambda r_0 \cos(\theta-\theta_0)} = \sum_{m=-\infty}^{\infty} J_m(\lambda r_0)\, e^{-\frac{im\pi}{2} + im(\theta-\theta_0)},$$

and the right-hand side of (12) in turn is by (61), page 372, the correct expansion of the plane wave on the left. Indeed as r_1 becomes infinite, the expanding cylindrical wave function designated by (11) must become asymptotic to a plane wave.

When $|r| < |r_0 \cos (\theta - \theta_0)|$ the expansion (11) fails to converge and is replaced by

$$(13)\qquad H_n^{(1)}(\lambda r_1)\, e^{in\psi} = \sum_{m=-\infty}^{\infty} H_m^{(1)}(\lambda r_0) J_{n+m}(\lambda r)\, e^{im(\theta-\theta_0)},$$

which is finite at $r = 0$. At this point $J_{n+m}(\lambda r)$ is zero unless $m = -n$. Moreover $\psi \simeq \pi + \theta_0 - \theta$ and, since $H_{-n}^{(1)}(\rho) = e^{n\pi i} H_n^{(1)}(\rho)$, the right and left sides of (13) are clearly identical. In the other limiting case of r_0 very large, we write $r_1 = r_0 - r \cos (\theta - \theta_0)$ and find by means of the asymptotic representations of the Hankel functions that (13) approaches

$$(14)\qquad e^{-i\lambda r\cos(\theta-\theta_0)} = \sum_{m=-\infty}^{\infty} J_{n+m}(\lambda r)\, e^{-i(m+n)\frac{\pi}{2} + i(m+n)(\theta-\theta_0)},$$

a plane-wave function propagated from O_1 towards O along the line joining these two centers.

WAVE FUNCTIONS OF THE ELLIPTIC CYLINDER

6.12. Elementary Waves.—Circular wave functions are in a sense a degenerate form of elliptic wave functions obtained by placing the eccentricity of the cylinders equal to zero. The analysis of the field and the properties of the functions must inevitably prove more complex in elliptic than in circular coordinates, but the results are also fundamentally of greater interest.

From 3, page 52, we take

$$(1) \qquad u^1 = \xi, \qquad u^2 = \eta, \qquad \xi \geqq 1, \qquad -1 \leqq \eta \leqq 1,$$

$$(2) \qquad h_1 = c_0\sqrt{\frac{\xi^2 - \eta^2}{\xi^2 - 1}}, \qquad h_2 = c_0\sqrt{\frac{\xi^2 - \eta^2}{1 - \eta^2}};$$

upon introducing these into (8), page 351, we find that $f(\xi, \eta)$ must satisfy the equation

$$(3) \qquad \sqrt{\xi^2 - 1}\,\frac{\partial}{\partial\xi}\left(\sqrt{\xi^2 - 1}\,\frac{\partial f}{\partial\xi}\right) + \sqrt{1 - \eta^2}\,\frac{\partial}{\partial\eta}\left(\sqrt{1 - \eta^2}\,\frac{\partial f}{\partial\eta}\right) + c_0^2(k^2 - h^2)(\xi^2 - \eta^2)f = 0.$$

This in turn is readily separated by writing $f = f_1(\xi)f_2(\eta)$ and leads to

$$(4) \qquad (\xi^2 - 1)\frac{d^2f_1}{d\xi^2} + \xi\frac{df_1}{d\xi} + [c_0^2(k^2 - h^2)\xi^2 - b]f_1 = 0,$$

$$(5) \qquad (1 - \eta^2)\frac{d^2f_2}{d\eta^2} - \eta\frac{df_2}{d\eta} + [b - c_0^2(k^2 - h^2)\eta^2]f_2 = 0,$$

where b is an arbitrary constant of separation. Thus $f_1(\xi)$ and $f_2(\eta)$ *satisfy the same differential equation.* Equations (4) and (5) are in fact special cases derived from the associated Mathieu equation[1]

$$(6) \qquad (1 - z^2)w'' - 2(a + 1)zw' + (b - c^2z^2)w = 0$$

by putting the parameter $a = -\frac{1}{2}$. These equations are characterized by an irregular singularity at infinity and regular singularities at $z = \pm 1$. When $\xi \gg 1$, (4) goes over to a Bessel equation.

A certain simplification of (4) and (5) can be achieved by transformation of the independent variables. Let

$$(7) \qquad \xi = \cosh u, \qquad \eta = \cos v,$$

so that the transformation to rectangular coordinates is expressed by

$$(8) \qquad x = c_0 \cosh u \cos v, \qquad y = c_0 \sinh u \sin v.$$

[1] WHITTAKER and WATSON, *loc. cit.*, Chaps. X and XIX; INCE, "Ordinary Differential Equations," Chap. XX, Longmans, 1927.

Then in place of (4) and (5), we obtain

$$\frac{d^2f_1}{du^2} + (c_0^2\lambda^2 \cosh^2 u - b)f_1 = 0, \tag{9}$$

$$\frac{d^2f_2}{dv^2} + (b - c_0^2\lambda^2 \cos^2 v)f_2 = 0, \tag{10}$$

where as in the preceding sections $\lambda = \sqrt{k^2 - h^2}$. The distance between the focal points on the x-axis, Fig. 9, is $2c_0$ and the eccentricity of the confocal ellipses is $e = 1/\cosh u$. The transition to the circular case follows as the limit when $c_0 \to 0$ and $u \to \infty$. Then $c_0 \cosh u \to c_0 \sinh u \to r$, and v is clearly the angle made by the radius r with the x-axis. For this reason we shall refer to $f_1(u)$ as a radial function and $f_2(v)$ as an angular function. When $u = 0$, the eccentricity is one and the ellipse reduces to a line of length $2c$ joining the foci on the x-axis.

The Mathieu equations (9) and (10) have been studied by many writers. We shall consider first the angular functions $f_2(v)$. There are, of course, solutions of (10) whatever the value of the separation constant b. But the electromagnetic field is a single-valued function of position and hence, if the properties of the medium are homogeneous with respect to the variable v, it is necessary that $f_2(v)$ be a *periodic function of the angle* v. Now Eq. (10) admits periodic solutions only for certain *characteristic values* of the parameter b. These characteristic values form a denumerable set $b_1, b_2, \ldots, b_m, \ldots$. Their determination is a problem of some length, so we shall content ourselves here with a reference to tabulated results[1] and proceed directly to the definition of the functions satisfying (9) and (10) when b coincides with a characteristic value b_m.

For the proper values of b, Eq. (10) admits both even and odd periodic solutions. The denumerable set of characteristic values leading to even solutions may be designated as $b_m^{(e)}$ and the associated characteristic

[1] A very readable account of the theory of the Mathieu functions has been given by Whittaker and Watson, *loc. cit.*, Chap. XIX; further details with extensive references to the literature have been published by Strutt in a monograph entitled "Lamésche- Mathieusche- und verwandte Funktionen in Physik und Technik," in the collection "Ergebnisse der Mathematik," Springer, 1932. Tables of Mathieu functions and characteristic values have been published by Goldstein, *Trans. Cambridge Phil. Soc.*, **23**, 303–336, 1927. The functions Se_m and So_m defined in the text differ from the ce_m and se_m of these authors only by a proportionality factor. Whereas Goldstein chooses his coefficients such that the normalization factor is $\sqrt{\pi}$, we have found it advantageous to normalize in such a way that the even function and the derivative of the odd function have unit value at the pole $v = 0$. See Stratton, *Proc. Nat. Acad. Sci. U. S.*, **21**, 51–56, 316–321, 1935; and MORSE, *ibid.*, pp. 56–62. Extensive tables of the expansion coefficients D_n^m, the characteristic values b_m and the normalization factors have been computed by P. M. Morse and will appear shortly.

functions can then be represented by the cosine series

$$Se_m(c_0\lambda, \cos v) = \sum_n{}' D_n^m \cos nv, \quad (m = 1, 2, 3, \cdots), \tag{11}$$

where the primed summation is to be extended over even values of n if m is even and over odd values of n if m is odd. The recursion formula connecting the coefficients $D_n^m(c_0\lambda)$ can be found by introducing (11) into (10). All coefficients of the series are thus referred to an initial one which is arbitrary. It is advantageous to choose this initial coefficient in such a way that the function itself has unit value when $v = 0$, corresponding to $\eta = \cos v = 1$. To this end we impose upon the D_n^m the condition

$$\sum_n{}' D_n^m = 1, \qquad Se_m(c_0\lambda, 1) = 1. \tag{12}$$

The odd periodic solutions of (10) are associated with a second set of characteristic values to be designated by $b_m^{(o)}$. These functions can be represented by a sine series

$$So_m(c_0\lambda, \cos v) = \sum_n{}' F_n^m \sin nv \tag{13}$$

upon whose coefficients $F_n^m(c_0\lambda)$ we impose the condition $\sum_n{}' nF_n^m = 1$. Consequently the derivative of $So_m(c_0\lambda, \cos v)$ will have unit value at $v = 0$.

$$\left[\frac{d}{dv} So_m(c_0\lambda, \cos v)\right]_{v=0} = 1. \tag{14}$$

The characteristic functions Se_m and So_m constitute a complete orthogonal set. Let $b_i^{(e)}$ and $b_j^{(e)}$ be two characteristic values and Se_i, Se_j the associated functions. They satisfy the equations

$$\frac{d^2Se_i}{dv^2} + (b_i^{(e)} - c_0^2\lambda^2 \cos^2 v)Se_i = 0, \tag{15}$$

$$\frac{d^2Se_j}{dv^2} + (b_j^{(e)} - c_0^2\lambda^2 \cos^2 v)Se_j = 0. \tag{16}$$

Multiply (15) by Se_j, (16) by Se_i, and subtract one from the other.

$$\frac{d}{dv}\left(Se_j \frac{d}{dv} Se_i - Se_i \frac{d}{dv} Se_j\right) + (b_i^{(e)} - b_j^{(e)})Se_iSe_j = 0. \tag{17}$$

Upon integrating (17) from 0 to 2π and taking account of the periodicity of the functions, we obtain

$$\int_0^{2\pi} Se_i(c_0\lambda, \cos v)\, Se_j(c_0\lambda, \cos v)\, dv = \begin{cases} 0, & i \neq j, \\ N_i^{(e)}, & i = j. \end{cases} \tag{18}$$

The normalization factors $N_m^{(e)}$ can be computed from the series expansions of the functions.

In identical fashion one deduces that

$$\int_0^{2\pi} So_i(c_0\lambda, \cos v)\, So_j(c_0\lambda, \cos v)\, dv = \begin{matrix} 0, & i \neq j, \\ N_i^{(o)}, & i = j. \end{matrix} \tag{19}$$

and

$$\int_0^{2\pi} Se_i(c_0\lambda, \cos v)\, So_j(c_0\lambda, \cos v)\, dv = 0, \tag{20}$$

the last for $i = j$ as well as $i \neq j$.

With each characteristic value $b_m^{(e)}$ there is associated one and only one periodic solution Se_m. For this same number $b_m^{(e)}$ there must exist, however, an independent solution of (10). Since the second solution is nonperiodic it is unessential in physical problems so long as the medium is homogeneous with respect to the angle v. When, on the other hand, there are discontinuities in the properties of the medium across surfaces v = constant, the boundary conditions may require use of functions of the second kind.

We turn our attention now to the radial functions. It can be shown without great difficulty that Eq. (9) is satisfied by an expansion in Bessel functions whose coefficients differ only by a factor from those of (11) and (13). Thus, when the parameter b assumes one of the characteristic values $b_m^{(e)}$, an associated radial function is

$$Re_m^1(c_0\lambda, \xi) = \sqrt{\frac{\pi}{2}} \sum_n{}' i^{m-n} D_n^m J_n(c_0\lambda\xi), \tag{21}$$

where $\xi = \cosh u$, $i^{m-n} = \exp\left[i(m-n)\frac{\pi}{2}\right]$. Since all solutions of Bessel's equation satisfy the recurrence relations (24) and (25), page 359, the even radial functions of the second kind are defined in terms of the Bessel functions of the second kind by

$$Re_m^2(c_0\lambda, \xi) = \sqrt{\frac{\pi}{2}} \sum_n{}' i^{m-n} D_n^m N_n(c_0\lambda\xi). \tag{22}$$

The convergence of such series of functions is not easy to demonstrate but appears to be satisfactory in the present instance. The great advantage of these representations is that they lead us at once to asymptotic expressions for very large values of $c_0\lambda\xi$. By (18) and (19), page 359, we have

$$Re_m^1(c_0\lambda, \xi) \simeq \frac{1}{\sqrt{c_0\lambda\xi}} \cos\left(c_0\lambda\xi - \frac{2m+1}{4}\pi\right), \tag{23}$$

$$Re_m^2(c_0\lambda, \xi) \simeq \frac{1}{\sqrt{c_0\lambda\xi}} \sin\left(c_0\lambda\xi - \frac{2m+1}{4}\pi\right), \qquad c_0\lambda\xi \to \infty, \tag{24}$$

By analogy with the Hankel functions, we are led to form the linear combinations

$$(25) \qquad Re_m^3(c_0\lambda,\, \xi) = Re_m^1 + iRe_m^2 = \sqrt{\frac{\pi}{2}}\sum_n{}' i^{m-n} D_n^m H_n^{(1)}(c_0\lambda\xi),$$

$$(26) \qquad Re_m^4(c_0\lambda,\, \xi) = Re_m^1 - iRe_m^2 = \sqrt{\frac{\pi}{2}}\sum_n{}' i^{m-n} D_n^m H_n^{(2)}(c_0\lambda\xi),$$

whose asymptotic expansions are

$$(27) \qquad Re_m^3(c_0\lambda,\, \xi) \simeq \frac{1}{\sqrt{c_0\lambda\xi}}\, e^{i\left(c_0\lambda\xi - \frac{2m+1}{4}\pi\right)},$$

$$(28) \qquad Re_m^4(c_0\lambda,\, \xi) \simeq \frac{1}{\sqrt{c_0\lambda\xi}}\, e^{-i\left(c_0\lambda\xi - \frac{2m+1}{4}\pi\right)}. \qquad c_0\lambda\xi \to \infty,$$

The function Re_m^1 satisfies the same equation as the angle function Se_m; neither has singularities other than that at infinity. The two must, therefore, be proportional to one another.

$$(29) \qquad Se_m(c_0\lambda,\, \xi) = \sqrt{2\pi}\, l_m^{(e)} Re_m^1(c_0\lambda,\, \xi),$$

$$l_m^{(e)} = \begin{cases} \dfrac{1}{\pi}\displaystyle\sum_n{}' i^{n-m}\frac{D_n^m}{D_0^m}, & (m \text{ even}),\\[2ex] \dfrac{2}{c_0\lambda\pi}\displaystyle\sum_n{}' n i^{n-m}\frac{D_n^m}{D_1^m}, & (m \text{ odd}).\end{cases}$$

On the left, ξ replaces $\cos v$. Consequently when $u = 0$, $\xi = 1$, we have

$$(30) \qquad Re_m^1(c_0\lambda,\, 1) = \frac{1}{\sqrt{2\pi}}\frac{1}{l_m^{(e)}}, \qquad \left[\frac{d}{du} Re_m^1(c_0\lambda,\, \cosh u)\right]_{u=0} = 0.$$

A corresponding set of relations can be obtained for the odd radial functions. We define:

$$(31) \qquad Ro_m^1(c_0\lambda,\, \xi) = \frac{\sqrt{\xi^2-1}}{\xi}\sqrt{\frac{\pi}{2}}\sum_n{}' i^{n-m} n F_n^m J_n(c_0\lambda\xi),$$

$$(32) \qquad Ro_m^2(c_0\lambda,\, \xi) = \frac{\sqrt{\xi^2-1}}{\xi}\sqrt{\frac{\pi}{2}}\sum_n{}' i^{n-m} n F_n^m N_n(c_0\lambda\xi),$$

$$(33) \qquad Ro_m^3(c_0\lambda,\, \xi) = Ro_m^1 + iRo_m^2, \qquad Ro_m^4(c_0\lambda,\, \xi) = Ro_m^1 - iRo_m^2.$$

The asymptotic expansions of the odd functions are identical with those of the corresponding even functions. Finally,

$$(34) \qquad i\, So_m(c_0\lambda,\, \xi) = \sqrt{2\pi}\, l_m^{(o)} Ro_m^1(c_0\lambda,\, \xi),$$

$$l_m^{(o)} = \begin{cases} \dfrac{4}{\pi c_0^2\lambda^2}\displaystyle\sum_n{}' n\frac{F_n^m}{F_2^m} i^{n-m}, & (m \text{ even}),\\[2ex] \dfrac{2}{\pi c_0\lambda}\displaystyle\sum_n{}' \frac{F_n^m}{F_1^m} i^{n-m}, & (m \text{ odd}).\end{cases}$$

At $u = 0$, $\xi = 1$, we have

$$(35)\qquad Ro_m^1(c_0\lambda, 1) = 0, \qquad \left[\frac{d}{du} Ro_m^1(c_0\lambda, \cosh u)\right]_{u=0} = \frac{1}{\sqrt{2\pi}} \frac{1}{l_m^{(o)}}.$$

With these functions at our disposal we are now in a position to write down elementary wave functions for the elliptic cylinder. The even and odd wave functions, which are finite everywhere and in particular along the axis joining the foci, are constructed from radial functions of the first kind.

$$(36)\qquad \psi_{em}^1 = Se_m(c_0\lambda, \cos v)\, Re_m^1(c_0\lambda, \xi)\, e^{\pm i\sqrt{k^2-\lambda^2}z - i\omega t},$$

$$(37)\qquad \psi_{om}^1 = So_m(c_0\lambda, \cos v)\, Ro_m^1(c_0\lambda, \xi)\, e^{\pm i\sqrt{k^2-\lambda^2}z - i\omega t}.$$

When it is known that at great distances from the axis of the cylinders the field is traveling radially outward, the elementary wave functions will be constructed from radial functions of the third kind.

$$(38)\qquad \psi_{em}^3 = Se_m(c_0\lambda, \cos v)\, Re_m^3(c_0\lambda, \xi)\, e^{\pm i\sqrt{k^2-\lambda^2}z - i\omega t},$$

$$(39)\qquad \psi_{om}^3 = So_m(c_0\lambda, \cos v)\, Ro_m^3(c_0\lambda, \xi)\, e^{\pm i\sqrt{k^2-\lambda^2}z - i\omega t}.$$

The components of the electric and magnetic vectors can now be found by the rules set down in Sec. 6.3.

6.13. Integral Representations.—According to (19), page 365, we have in any system of cylindrical coordinates

$$(40)\qquad f(u^1, u^2) = \int g(\beta)\, e^{ik \sin\alpha(x\cos\beta + y\sin\beta)}\, d\beta.$$

If now the rectangular coordinates are replaced by elliptic coordinates through the transformation $x = c_0 \cosh u \cos v$, $y = c_0 \sinh u \sin v$, we can write (40) in the form

$$(41)\qquad f(u, v) = \int g(\beta)\, e^{i\lambda p}\, d\beta,$$

where as in the past $\lambda = \sqrt{k^2 - h^2} = k \sin\alpha$, and where

$$(42)\qquad p(u, v, \beta) = x\cos\beta + y\sin\beta = c_0(\cosh u \cos v \cos\beta + \sinh u \sin v \sin\beta).$$

This function $f(u, v)$ is to satisfy the equation

$$(43)\qquad \frac{\partial^2 f}{\partial u^2} + \frac{\partial^2 f}{\partial v^2} + c_0^2\lambda^2(\cosh^2 u - \cos^2 v)f = 0,$$

but since $f = f_1(u)f_2(v)$, it is clear that it must also satisfy

$$(44)\qquad \frac{\partial^2 f}{\partial u^2} + (c_0^2\lambda^2 \cosh^2 u - b)f = 0,$$

obtained by multiplying (9), page 376, by $f_2(v)$. Upon differentiating (41) with respect to u, we have

$$\frac{\partial^2 f}{\partial u^2} = \int g(\beta)\left[i\lambda \frac{\partial^2 p}{\partial u^2} - \lambda^2 \left(\frac{\partial p}{\partial u}\right)^2\right] e^{i\lambda p}\, d\beta; \tag{45}$$

if, therefore, (41) is to satisfy (44), it is necessary that

$$\int g(\beta)\left[i\lambda \frac{\partial^2 p}{\partial u^2} - \lambda^2 \left(\frac{\partial p}{\partial u}\right)^2 + c_0^2\lambda^2 \cosh^2 u - b\right] e^{i\lambda p}\, d\beta = 0. \tag{46}$$

Now p is a function of β as well as of u and it can be easily verified that

$$c_0^2 \cosh^2 u - \left(\frac{\partial p}{\partial u}\right)^2 = c_0^2 \cos^2 \beta + \left(\frac{\partial p}{\partial \beta}\right)^2, \tag{47}$$

$$\frac{\partial^2 p}{\partial u^2} = -\frac{\partial^2 p}{\partial \beta^2}. \tag{48}$$

Consequently (46) is equivalent to

$$\int g(\beta)\left[\frac{\partial^2 e^{i\lambda p}}{\partial \beta^2} + (b - c_0^2\lambda^2 \cos^2 \beta) e^{i\lambda p}\right] d\beta = 0. \tag{49}$$

Finally by (30), page 366,

$$g(\beta)\frac{\partial^2 e^{i\lambda p}}{\partial \beta^2} = e^{i\lambda p}\frac{d^2 g}{d\beta^2} + \frac{\partial}{\partial \beta}\left(g\frac{\partial e^{i\lambda p}}{\partial \beta} - e^{i\lambda p}\frac{\partial g}{\partial \beta}\right); \tag{50}$$

hence, (41) is a solution of (44) provided $g(\beta)$ satisfies

$$\frac{d^2 g}{d\beta^2} + (b - c_0^2\lambda^2 \cos^2 \beta) g = 0, \tag{51}$$

and the path C of integration is so chosen that

$$g\frac{\partial e^{i\lambda p}}{\partial \beta} - e^{i\lambda p}\frac{\partial g}{\partial \beta}\bigg|_C = 0. \tag{52}$$

It will be noted that the equation of the transform $g(\beta)$ is identical with that of the angular function $f_2(v)$, and in fact differs unessentially from that for $f_1(u)$ with which we set out. This property is common to the transforms of all solutions of equations belonging to the group defined by (6), page 375.

The results just derived are valid whatever the value of the separation constant b_m. If, however, we confine ourselves to the even and odd sets of characteristic values $b_m^{(e)}$ and $b_m^{(o)}$, it will be advantageous to choose for $g(\beta)$ the periodic solutions of (51) which have been designated by $Se_m(c_0\lambda, \cos\beta)$, $So_m(c_0\lambda, \cos\beta)$. Since $p(u, v, \beta)$ is also periodic in β with period 2π, it is apparent that (52) will be fulfilled when the path of integration is any section of the real axis of length 2π. An integral representation of $f(u, v)$ which holds when b belongs to the set $b_m^{(e)}$ is then

$$f_m^{(e)}(u, v) = \int_0^{2\pi} Se_m(c_0\lambda, \cos\beta)\, e^{i\lambda p(u,v,\beta)}\, d\beta. \tag{53}$$

From this an integral representing $f_1(u)$ can be found immediately by placing $v = 0$.

$$f_{1m}^{(e)}(u) = \int_0^{2\pi} Se_m(c_0\lambda, \cos\beta)\, e^{ic_0\lambda \cosh u \cos\beta}\, d\beta. \tag{54}$$

There remains the task of identifying the particular solution defined by (54). Upon introducing the cosine expansion of Se_m and abbreviating $c_0\lambda \cosh u = \rho$, we obtain

$$f_{1m}^{(e)}(u) = \sum_n{}' D_n^m \int_0^{2\pi} e^{i\rho\cos\beta} \cos n\beta\, d\beta = 2\pi \sum_n{}' i^n D_n^m J_n(\rho). \tag{55}$$

Next we note that $i^n = (-1)^n i^{-n}$, and remember that the summation extends over even values of n if m is even; over odd values only if m is odd. Therefore $(-1)^n = (-1)^m$, and in virtue of the definition (21), page 378, it follows that

$$Re_m^1(c_0\lambda, \xi) = \frac{i^{-m}}{\sqrt{8\pi}} \int_0^{2\pi} Se_m(c_0\lambda, \cos\beta)\, e^{ic_0\lambda\xi\cos\beta}\, d\beta, \tag{56}$$

where $\xi = \cosh u$. Moreover, $Se_m(c_0\lambda, \cos\beta)$ is proportional to $Re_m^1(c_0\lambda, \cos\beta)$, in which ξ has been replaced by $\cos\beta$, so that by (29), page 379, *the function $Re_m^1(c_0\lambda, \xi)$ satisfies an integral equation,*

$$Re_m^1(c_0\lambda, \xi) = i^{-m} \frac{l_m^{(e)}}{2} \int_0^{2\pi} Re_m^1(c_0\lambda, \cos\beta)\, e^{ic_0\lambda\xi\cos\beta}\, d\beta. \tag{57}$$

Still another representation of Re_m^1 can be found by expanding the integrand of (57). Thus

$$Re_m^1(c_0\lambda, \xi) = l_m^{(e)} \sqrt{\frac{\pi}{8}} \sum_n{}' i^{-n} D_n^m \int_0^{2\pi} J_n(c_0\lambda\cos\beta)\, e^{ic_0\lambda\xi\cos\beta}\, d\beta. \tag{58}$$

But

$$J_n(c_0\lambda\cos\beta) = \frac{i^{-n}}{\pi} \int_0^{\pi} \cos n\phi\, e^{ic_0\lambda\cos\beta\cos\phi}\, d\phi, \tag{59}$$

whence,

(60)
$$\int_0^{2\pi} J_n(c_0\lambda\cos\beta)\, e^{ic_0\lambda\xi\cos\beta}\, d\beta = i^{-n} \int_0^{2\pi} \cos n\phi\, J_0[c_0\lambda(\xi + \cos\phi)]\, d\phi.$$

Remembering once more that $(-1)^n = (-1)^m$, we obtain

$$(61)\quad Re_m^1(c_0\lambda, \xi) = l_m^{(e)}(-1)^m \sqrt{\frac{\pi}{8}} \int_0^{2\pi} J_0[c_0\lambda(\xi + \cos\phi)]\, Se_m(c_0\lambda, \cos\phi)\, d\phi.$$

Integral representations of the radial functions of the third and fourth kinds can be obtained by a variation in the choice of contour. To simplify matters a little, let us take $\cosh u = \xi$, $v = 0$, $\cos\beta = t$. Then the several radial functions are represented by integrals of the form

$$(62)\qquad f_{1m}^{(e)}(\xi) = \int_C Se_m(c_0\lambda, t)\, e^{ic_0\lambda\xi t}(1 - t^2)^{-\frac{1}{2}}\, dt,$$

where the contour C is such as to ensure the vanishing of the so-called "bilinear concomitant" (52), which now assumes the form

$$(63)\qquad (1 - t^2)^{\frac{1}{2}}\left[ic_0\lambda\xi Se_m(c_0\lambda, t) - \frac{d}{dt} Se_m(c_0\lambda, t)\right] e^{ic_0\lambda\xi t}\Big|_C = 0.$$

When t becomes very large, the asymptotic expansion of Se_m is

$$(64)\qquad Se_m(c_0\lambda, t) \simeq l_m^{(e)} \sqrt{\frac{2\pi}{c_0\lambda t}} \cos\left(c_0\lambda t - \frac{2m+1}{4}\pi\right);$$

consequently, the vanishing of (63) at infinity is governed by a factor $\exp ic_0\lambda(\xi - 1)t$. If, therefore, the real part of $c_0\lambda(\xi - 1)$ is greater than zero, the contour can begin or end at $t = i\infty$, while for the other limit one may choose either $+1$ or -1. As a result we have

$$(65)\qquad Re_m^3(c_0\lambda, \xi) = i^{-m}\sqrt{\frac{2}{\pi}} \int_{i\infty}^{1} Se_m(c_0\lambda, t)\, e^{ic_0\lambda\xi t}\, (1 - t^2)^{-\frac{1}{2}}\, dt,$$

$$(66)\qquad Re_m^4(c_0\lambda, \xi) = i^{-m}\sqrt{\frac{2}{\pi}} \int_{-1}^{i\infty} Se_m(c_0\lambda, t)\, e^{ic_0\lambda\xi t}\, (1 - t^2)^{-\frac{1}{2}}\, dt,$$

provided the real part of $[c_0\lambda(\xi - 1)] > 0$. Reverting to the complex β-plane, we find (65) equivalent to

$$(67)\qquad Re_m^3(c_0\lambda, \xi) = i^{-m}\sqrt{\frac{2}{\pi}} \int_0^{\frac{\pi}{2} - i\infty} Se_m(c_0\lambda, \cos\beta)\, e^{ic_0\lambda\xi\cos\beta}\, d\beta,$$

with a corresponding integral for $Re_m^4(c_0\lambda, \xi)$.

With the help of (67) one may deduce further integrals of the type (61). Thus, in place of (58) we write

$$(68)\quad Re_m^3(c_0\lambda, \xi) = l_m^{(e)}\sqrt{2\pi} \sum_n{}' i^{-n} D_n^m \int_0^{\frac{\pi}{2} - i\infty} J_n(c_0\lambda\cos\beta)\, e^{ic_0\lambda\xi\cos\beta}\, d\beta$$

$$= l_m^{(e)}\sqrt{\frac{2}{\pi}} \sum_n{}' (-1)^n D_n^m \int_0^{\pi} d\phi \cos n\phi \int_0^{\frac{\pi}{2} - i\infty} d\beta\, e^{ic_0\lambda(\xi + \cos\phi)\cos\beta}.$$

Now by (39), page 367, when $p = 0$, we have

$$(69) \qquad \int_0^{\frac{\pi}{2}-i\infty} e^{i\rho\cos\beta}\,d\beta = \frac{1}{2}\int_{-\frac{\pi}{2}+i\infty}^{\frac{\pi}{2}-i\infty} e^{i\rho\cos\beta}\,d\beta = \frac{\pi}{2}H_0^{(1)}(\rho),$$

provided the real part of $\rho > 0$; consequently,

$$(70) \quad Re_m^3(c_0\lambda, \xi) = l_m^{(e)}(-1)^m\sqrt{\frac{\pi}{8}}\int_0^{2\pi} H_0^{(1)}[c_0\lambda(\xi + \cos\phi)]\,Se_m(c_0\lambda, \cos\phi)\,d\phi$$

provided the real part of $[c_0\lambda(\xi - 1)] > 0$. A combination of (61) and (70) leads to

$$(71) \quad Re_m^2(c_0\lambda, \xi) = l_m^{(e)}(-1)^m\sqrt{\frac{\pi}{8}}\int_0^{2\pi} N_0[c_0\lambda(\xi + \cos\phi)]\,Se_m(c_0\lambda, \cos\phi)\,d\phi.$$

A clue to the formulation of integral representations of the odd functions is given by

$$(72) \qquad \int_0^{2\pi} e^{ic_0\lambda\xi\cos\beta}\sin n\beta\cos\beta\,d\beta = \frac{n}{ic_0\lambda\xi}\int_0^{2\pi} e^{ic_0\lambda\xi\cos\beta}\cos n\beta\,d\beta,$$

the result of an integration by parts. The proof of the relations

$$\begin{aligned}(73) \quad Ro_m^1(c_0\lambda, \xi) &= \frac{c_0\lambda}{\sqrt{8\pi}}\,i^{1-m}\sqrt{\xi^2 - 1}\int_0^{2\pi} So_m(c_0\lambda, \cos\beta)\,e^{ic_0\lambda\xi\cos\beta}\sin\beta\,d\beta \\ &= \frac{c_0\lambda}{2}\,l_m^{(o)}i^{-m}\sqrt{\xi^2 - 1}\int_0^{2\pi} Ro_m^1(c_0\lambda, \cos\beta)\,e^{ic_0\lambda\xi\cos\beta}\sin\beta\,d\beta \\ &= c_0\lambda l_m^{(o)}(-1)^m\sqrt{\frac{\pi}{8}}\sqrt{\xi^2 - 1}\int_0^{2\pi} J_1[c_0\lambda(\xi + \cos\phi)] \\ &\qquad\qquad So_m(c_0\lambda, \cos\phi)\,\frac{\sin\phi}{\xi + \cos\phi}\,d\phi,\end{aligned}$$

$$\begin{aligned}(74) \quad Ro_m^3(c_0\lambda, \xi) &= c_0\lambda l_m^{(o)}(-1)^m\sqrt{\frac{\pi}{8}}\sqrt{\xi^2 - 1}\int_0^{2\pi} H_1^{(1)}[c_0\lambda\,(\xi + \cos\phi)] \\ &\qquad\qquad So_m(c_0\lambda, \cos\phi)\,\frac{\sin\phi}{\xi + \cos\phi}\,d\phi,\end{aligned}$$

is left to the reader.

6.14. Expansion of Plane and Circular Waves.—The periodic functions $Se_m(c_0\lambda, \cos v)$ and $So_m(c_0\lambda, \cos v)$ constitute a complete, orthogonal set. Consequently, if a function $f(u, v)$ is periodic in v with period 2π and together with the first derivative $\partial f/\partial v$ is piecewise continuous with respect to v, it can be expanded in a series of the form

$$(75) \quad f(u, v) = \sum_{m=0}^{\infty} f_m^{(e)}(u)\,Se_m(c_0\lambda, \cos v) + \sum_{m=0}^{\infty} f_m^{(o)}(u)\,So_m(c_0\lambda, \cos\theta),$$

whose coefficients are determined from

$$
(76)\qquad
\begin{aligned}
f_m^{(e)}(u) &= \frac{1}{N_m^{(e)}} \int_0^{2\pi} f(u, v)\, Se_m(c_0\lambda, \cos v)\, dv, \\
f_m^{(o)}(u) &= \frac{1}{N_n^{(o)}} \int_0^{2\pi} f(u, v)\, So_m(c_0\lambda, \cos v)\, dv.
\end{aligned}
$$

This expansion may be applied to the representation of a plane wave whose propagation constant is k, traveling in a direction defined by the unit vector $\mathbf{n}$ whose spherical angles with respect to a fixed reference system are α and β as in Fig. 66, page 362.

$$
(77)\qquad \psi = e^{ik \sin\alpha(x\cos\beta + y\sin\beta)} \cdot e^{ikz\cos\alpha - i\omega t}.
$$

Upon expressing x and y in terms of u and v and putting $\lambda = k \sin \alpha$, we have

$$
(78)\qquad f(u, v, \beta) = e^{i\lambda p}, \quad p = c_0(\cosh u \cos v \cos \beta + \sinh u \sin v \sin \beta).
$$

Since there is complete symmetry in (78) between v and β, the expansion (75) can be written

$$
(79)\qquad e^{i\lambda p} = \sum_{m=0}^{\infty} a_m(u)\, Se_m(c_0\lambda, \cos\beta)\, Se_m(c_0\lambda, \cos v) + \sum_{m=0}^{\infty} b_m(u)\, So_m(c_0\lambda, \cos\beta)\, So_m(c_0\lambda, \cos v)
$$

in which the coefficients $a_m(u)$ and $b_m(u)$ depend on u alone. In virtue of (76), we have

$$
(80)\qquad a_m(u)\, N_m^{(e)} Se_m(c_0\lambda, \cos\beta) = \int_0^{2\pi} Se_m(c_0\lambda, \cos v)\, e^{i\lambda p}\, dv;
$$

since $a_m(u)$ is unaffected by the value of β, we may put $\beta = 0$, $Se_m(c_0\lambda, 1) = 1$, $p = c_0 \cosh u \cos v$. Then by (56),

$$
(81)\qquad a_m(u) = \frac{i^m \sqrt{8\pi}}{N_m^{(e)}} Re_m^1(c_0\lambda, \cosh u).
$$

Likewise

$$
(82)\qquad b_m(u)\, N_m^{(o)} So_m(c_0\lambda, \cos\beta) = \int_0^{2\pi} So_m(c_0\lambda, \cos v)\, e^{i\lambda p}\, dv.
$$

We now differentiate this last relation with respect to β and then put $\beta = 0$. In virtue of (14) the coefficient $b_m(u)$ proves to be

$$
(83)\qquad b_m(u) = \frac{i^m \sqrt{8\pi}}{N_m^{(o)}} Ro_m^1(c_0\lambda, \cosh u).
$$

The complete expansion for a plane wave of arbitrary direction is, therefore,

$$(84)\quad e^{i\lambda p} = \sqrt{8\pi}\sum_{m=0}^{\infty} i^m \left[\frac{1}{N_m^{(e)}} Re_m^1(c_0\lambda, \cosh u)\, Se_m(c_0\lambda, \cos v)\right.$$
$$\left. Se_m(c_0\lambda, \cos\beta) + \frac{1}{N_m^{(o)}} Ro_m^1(c_0\lambda, \cosh u)\, So_m(c_0\lambda, \cos v)\, So_m(c_0\lambda, \cos\beta)\right].$$

This last result enables us to write down at once the integral representations of those elementary elliptic wave functions that are finite on the axis. We need only multiply (84) by $Se_m(c_0\lambda, \cos\beta)$ and integrate over a period. From the orthogonal properties of the functions it follows that

$$(85)\quad Re_m^1(c_0\lambda, \cosh u)\, Se_m(c_0\lambda, \cos v) = \frac{i^{-m}}{\sqrt{8\pi}}\int_0^{2\pi} Se_m(c_0\lambda, \cos\beta)\, e^{i\lambda p}\, d\beta;$$

similarly, for the odd functions,

$$(86)\quad Ro_m^1(c_0\lambda, \cosh u)\, So_m(c_0\lambda, \cos v) = \frac{i^{-m}}{\sqrt{8\pi}}\int_0^{2\pi} So_m(c_0\lambda, \cos\beta)\, e^{i\lambda p}\, d\beta.$$

With the help of (67) it can be shown also that the elementary functions of elliptic waves traveling outward from the axis are represented by the contour integral

(87)

$$Re_m^3(c_0\lambda, \cosh u)\, Se_m(c_0\lambda, \cos v) = i^{-m}\sqrt{\frac{2}{\pi}}\int_0^{\frac{\pi}{2}-i\infty} Se_m(c_0\lambda, \cos\beta)\, e^{\pm i\lambda p}\, d\beta.$$

The upper sign in the exponential is to be chosen when $-\pi/2 < v < \pi/2$, the negative sign when $\pi/2 < v < 3\pi/2$. A similar relation can be deduced for the odd functions.

From the analysis of Sec. 6.8 it is easy to see that the even circular wave function $\cos n\theta\, J_n(\lambda r)$, which remains finite on the axis, can be represented by the integral

$$(88)\qquad 2\pi i^n \cos n\theta\, J_n(\lambda r) = \int_0^{2\pi} \cos n\beta\, e^{i\lambda p}\, d\beta,$$

where $p = x\cos\beta + y\sin\beta = r\cos(\beta - \theta)$. Upon multiplication by D_n^m and summation over even or odd values of n, one obtains

$$(89)\quad Re_m^1(c_0\lambda, \cosh u)\, Se_m(c_0\lambda, \cos v) = \sqrt{\frac{\pi}{2}}\sum_n{}' i^{n-m} D_n^m \cos n\theta\, J_n(\lambda r),$$

which expresses the even elliptic wave function as an expansion in circular wave functions. A similar expression can be found for the odd functions.

An addition theorem relating circular and elliptic waves referred to two parallel axes has been derived by Morse.[1]

Problems

1. Show that the equation

$$\nabla^2\psi - \mu\epsilon\frac{\partial^2\psi}{\partial t^2} - \sigma\mu\frac{\partial\psi}{\partial t} = 0$$

is satisfied by wave functions of the type

$$\psi = Ce^{i(\delta_1+\delta_2)}\cos(h_1x - \delta_1)\cos(h_2y - \delta_2)e^{ih_3z - i\omega t},$$
$$h_1^2 + h_2^2 + h_3^2 = k^2.$$

Let $\psi = \Pi_z$, $\Pi_x = \Pi_y = 0$, be the components of a Hertz vector oriented in the direction of propagation. Show that this leads to a transverse magnetic field whose electric components are

$$E_x = -ih_1h_3Ce^{i(\delta_1+\delta_2)}\sin(h_1x - \delta_1)\cos(h_2y - \delta_2)e^{ih_3z - i\omega t},$$
$$E_y = -ih_2h_3Ce^{i(\delta_1+\delta_2)}\cos(h_1x - \delta_1)\sin(h_2y - \delta_2)e^{ih_3z - i\omega t},$$
$$E_z = (k^2 - h_3^2)Ce^{i(\delta_1+\delta_2)}\cos(h_1x - \delta_1)\cos(h_2y - \delta_2)e^{ih_3z - i\omega t},$$

and whose impedances are

$$Z_x = -\frac{E_z}{H_y} = \frac{k^2 - h_3^2}{ik^2h_1}\,\omega\mu\cot(h_1x - \delta_1),$$
$$Z_y = \frac{E_z}{H_x} = \frac{k^2 - h_3^2}{ik^2h_2}\,\omega\mu\cot(h_2y - \delta_2),$$
$$Z_z = \frac{E_x}{H_y} = \frac{\omega\mu h_3}{k^2}.$$

Find the components of a corresponding transverse electric field by choosing $\psi = \Pi_z^*$, $\Pi_x^* = \Pi_y^* = 0$.

The factor $e^{i(\delta_1+\delta_2)}$ is introduced to include traveling as well as standing waves. Thus, if one chooses $\delta_1 = i\infty$, the field behaves as e^{ih_1x}.

2. The equation of a two-dimensional wave motion in a *nondissipative* medium is

$$\frac{\partial^2\psi}{\partial x^2} + \frac{\partial^2\psi}{\partial y^2} - \frac{1}{a^2}\frac{\partial^2\psi}{\partial t^2} = 0.$$

Show that

$$2\pi a\,\psi(x, y, t) = \frac{\partial}{\partial t}\int_0^{at}\int_0^{2\pi}\frac{g(x', y')}{\sqrt{a^2t^2 - r^2}}\,r\,dr\,d\theta + \int_0^{at}\int_0^{2\pi}\frac{G(x', y')}{\sqrt{a^2t^2 - r^2}}\,r\,dr\,d\theta,$$

where

$$x' - x = r\cos\theta, \qquad y' - y = r\sin\theta,$$

and

$$\psi = g(x, y), \qquad \frac{\partial\psi}{\partial t} = G(x, y), \qquad \text{when } t = 0.$$

This is the Poisson-Parseval formula. Note that at time t the wave function ψ is determined by the initial values g and G at all points on the circumference *and the interior* of a circle of radius at. Thus even in nondissipative media the disturbance

[1] MORSE, *loc. cit.*

persists after the passage of the wave front. This behavior is characteristic of wave motion in even-dimensional spaces.

3. Show that the wave equation

$$\frac{1}{r}\frac{\partial}{\partial r}\left(r\frac{\partial\psi}{\partial r}\right) + \frac{\partial^2\psi}{\partial z^2} - \frac{1}{c^2}\frac{\partial^2\psi}{\partial t^2} = 0$$

has a particular solution of the form

$$\psi = \frac{1}{2\pi}\int_0^{2\pi} F\left(z + ir\cos\alpha,\ t - \frac{r}{c}\sin\alpha\right) d\alpha,$$

which reduces to $\psi = F(z, t)$ on the axis $r = 0$, where F is an arbitrary function.

Show that another solution of the same equation is

$$\psi = \int_{-\infty}^{\infty} F\left(z - r\sinh\alpha,\quad t - \frac{r}{c}\cosh\alpha\right) d\alpha. \qquad \text{(Bateman)}$$

4. A special case of the solution obtained in Problem 3 is the symmetric, two-dimensional wave function.

$$\psi = \frac{1}{2\pi}\int_0^{\infty} F\left(t - \frac{r}{c}\cosh\alpha\right) d\alpha,$$

which represents a wave expanding about a uniform line source of strength $F(t)$ along the z-axis.

Show that if $F(t) = 0$ when $t < 0$, then

$$\psi = \frac{c}{2\pi}\int_{-\infty}^{t-\frac{r}{c}} \frac{F(\beta)}{\sqrt{c^2(t-\beta)^2 - r^2}}\, d\beta$$

which is zero everywhere as long as $t < r/c$.

If the source acts only for a finite time τ, so that $F(t) = 0$ except in the interval $0 < t < \tau$, show that the wave leaves a residue or "tail" whose form, when $t - \frac{r}{c} \gg \tau$, is determined by

$$\psi = \frac{c}{2\pi\sqrt{c^2t^2 - r^2}}\int_0^{\tau} F(\beta)\, d\beta.$$

(Lamb, "Hydrodynamics," 5th ed., p. 279, 1924. Cambridge University Press.)

5. Show that the equation

$$\frac{\partial^2\psi}{\partial x^2} + \frac{\partial^2\psi}{\partial y^2} + k^2\psi = 0$$

is satisfied by

$$\psi = \int_{r+r_0}^{\infty} \frac{\sin k(\xi + \alpha)\, d\xi}{\sqrt{\xi^2 - (x - x_0)^2 - (y - y_0)^2}}$$

where

$$r^2 = x^2 + y^2, \qquad r_0^2 = x_0^2 + y_0^2. \qquad \text{(Bateman)}$$

6. Show that the equation

$$\frac{\partial^2\psi}{\partial x^2} + \frac{\partial^2\psi}{\partial y^2} = \lambda^2\psi$$

is satisfied by

$$\psi = \int_0^\infty e^{-\frac{r^2}{4t}-\lambda^2 t}\left[f\left(\frac{x+iy}{t}\right) + F\left(\frac{x-iy}{t}\right)\right]\frac{dt}{t},$$

where $r^2 = x^2 + y^2$ and f and F are arbitrary functions. To what limitations is this solution subjected?

Obtain in this way the particular solution

$$\psi = \frac{1}{r}\, e^{-\lambda r}(x + iy)^{-\frac{1}{2}}. \qquad \text{(Bateman)}$$

7. The equation

$$a^2\left(\frac{\partial^2\psi}{\partial x^2} + \frac{\partial^2\psi}{\partial y^2}\right) - 2b\,\frac{\partial\psi}{\partial t} - \frac{\partial^2\psi}{\partial t^2} = 0$$

is reduced by the substitution

$$\psi = e^{-bt}u(x, y, t)$$

to

$$a^2\left(\frac{\partial^2 u}{\partial x^2} + \frac{\partial^2 u}{\partial y^2}\right) + b^2 u - \frac{\partial^2 u}{\partial t^2} = 0.$$

The initial conditions are

$$u = g(x, y), \qquad \frac{\partial u}{\partial t} = G(x, y), \qquad \text{when } t = 0.$$

Let r be the radius of a sphere drawn about the point of observation and α, β, γ the direction cosines of r with respect to the x, y, z axes. Show that a solution is represented by

$$u(x, y, t) = \frac{1}{2\pi}\frac{\partial}{\partial t}\int\int t\, g(x + at\alpha, y + at\beta)\cos(ibt\gamma)\, d\Omega$$
$$+ \frac{1}{2\pi}\int\int t\, G(x + at\alpha, y + at\beta)\cos(ibt\gamma)\, d\Omega,$$

(Boussinesq)

where $d\Omega$ is the solid angle subtended by an element of surface on the sphere.

8. Show that

$$J_p(\rho) = \frac{\rho^p}{2^p\Gamma(\frac{1}{2})\Gamma(p+\frac{1}{2})}\int_0^\pi \cos(\rho\cos\phi)\sin^{2p}\phi\, d\phi$$

is an integral representation of the Bessel function of the first kind valid for all values of the order p provided $Re(p) > -\frac{1}{2}$.

9. Derive the integral representations

$$N_0(z) = -\frac{2}{\pi}\int_0^\infty \cos(z\cosh\alpha)\, d\alpha, \qquad Im\, z > 0.$$

for the Bessel function of the second kind and zero order, and

$$H_p^{(2)}(z) = i^{p+1}\frac{2}{\pi}\int_0^\infty e^{-iz\cosh\alpha}\cosh p\alpha\, d\alpha, \qquad Im\, z < 0.$$

10. Show that the modified Bessel equation

$$z^2 \frac{d^2y}{dz^2} + z\frac{dy}{dz} - (z^2 + p^2)y = 0$$

is satisfied by the function

$$I_p(z) = \sum_{m=0}^{\infty} \frac{1}{m!\,\Gamma\,(p+m+1)} \left(\frac{z}{2}\right)^{p+2m}$$

where

$$I_p(z) = e^{-\frac{p\pi i}{2}} J_p(iz), \qquad \text{when } -\pi < \arg z < \frac{\pi}{2}.$$

The most commonly used independent solution is the one defined by

$$K_p(z) = \frac{\pi i}{2} e^{\frac{p\pi i}{2}} H_p^{(1)}(iz).$$

A complete set of recurrence relations for these functions are given by Watson, "Bessel Functions," page 79.

11. Demonstrate the following integral representations of the modified Bessel functions.

$$\begin{aligned} I_p(z) &= \frac{1}{\Gamma(\frac{1}{2})\Gamma(p+\frac{1}{2})}\left(\frac{z}{2}\right)^p \int_0^\pi \cosh\,(z\cos\alpha)\sin^{2p}\alpha\,d\alpha \\ &= \frac{1}{\Gamma(\frac{1}{2})\Gamma(p+\frac{1}{2})}\left(\frac{z}{2}\right)^p \int_{-1}^1 (1-t^2)^{p-\frac{1}{2}}\cosh\,(zt)\,dt \\ &= \frac{1}{\Gamma(\frac{1}{2})\Gamma(p+\frac{1}{2})}\left(\frac{z}{2}\right)^p \int_0^\pi e^{\pm z\cos\alpha}\sin^{2p}\alpha\,d\alpha, \end{aligned}$$

$$K_p(z) = \frac{\Gamma(\frac{1}{2})}{\Gamma(p+\frac{1}{2})}\left(\frac{z}{2}\right)^p \int_0^\infty e^{-z\cosh\alpha}\sinh^{2p}\alpha\,d\alpha.$$

It is assumed that $Re\left(p+\frac{1}{2}\right) > 0$ and $|\arg z| < \frac{\pi}{2}$.

12. Show that for very large values of the argument the asymptotic behavior of the modified Bessel functions is given by

$$I_p(z) \simeq \frac{e^z}{\sqrt{2\pi z}} \sum_{m=0}^{\infty} \frac{(-1)^m(p,m)}{(2z)^m} + \frac{e^{-z+(p+\frac{1}{2})\pi i}}{\sqrt{2\pi z}} \sum_{m=0}^{\infty} \frac{(p,m)}{(2z)^m},$$

provided $-\frac{\pi}{2} < \arg z < \frac{3}{2}\pi$,

$$K_p(z) \simeq \sqrt{\frac{\pi}{2z}}\, e^{-z} \sum_{m=0}^{\infty} \frac{(p,m)}{(2z)^m}, \qquad |\arg z| < \frac{3}{2}\pi,$$

where

$$(p, m) = \frac{\Gamma(p+m+\frac{1}{2})}{m!\,\Gamma(p-m+\frac{1}{2})} = \frac{(4p^2-1)(4p^2-3^2)\cdots[4p^2-(2m-1)^2]}{2^{2m}m!}.$$

Retaining only the first term,

$$I_p(z) \simeq \sqrt{\frac{1}{2\pi z}}\, e^z, \qquad K_p(z) \simeq \sqrt{\frac{\pi}{2z}}\, e^{-z}.$$

13. Prove that if $R = \sqrt{r^2 + z^2}$ is the distance between two points, the inverse distance is represented by

$$\frac{1}{R} = \frac{2}{\pi} \int_0^\infty \cos \beta z\, K_0(\beta r)\, d\beta,$$

where $K_0(\beta r)$ is the modified Bessel function defined in Problem 10.

14. As a special case of Problem 4, show that the equation

$$\frac{1}{r}\frac{\partial}{\partial r}\left(r \frac{\partial \psi}{\partial r}\right) - \frac{1}{c^2}\frac{\partial^2 \psi}{\partial t^2} = 0$$

is satisfied by

$$\psi(r, t) = K_0(ikr)\, e^{i\omega t},$$

where $k = \omega/c$ and K_0 is the modified Bessel function defined in Problem 10. This solution plays the same role in two-dimensional wave propagation as does the function e^{ikR}/R in three dimensions.

15. Show that the equation

$$\frac{1}{r}\frac{\partial}{\partial r}\left(r \frac{\partial \psi}{\partial r}\right) + \frac{1}{r^2}\frac{\partial^2 \psi}{\partial \theta^2} + \lambda^2 \psi = 0$$

is satisfied by

$$\psi = \frac{1}{n} J_0(\lambda r) + 2 \sum_{m=1}^{\infty} e^{i\frac{m}{n}\frac{\pi}{2}} J_{\frac{m}{n}}(\lambda r) \cos \frac{m}{n}\theta,$$

which has a period of $2n\pi$ in θ and hence is multivalued about the axis. Show that this solution can be represented by the integral

$$\psi = \frac{1}{2\pi n} \int_C \frac{e^{\frac{i\alpha}{n}}}{e^{\frac{i\alpha}{n}} - 1}\, e^{i\lambda r \cos(\theta - \alpha)}\, d\alpha$$

following a properly chosen contour C in the complex α-plane. This solution is everywhere finite and continuous. ψ behaves like a plane wave at infinity provided $|\theta| < \pi$, but vanishes (is regular) at infinity if θ lies between any pair of the limits

$$\pi < \theta < 3\pi, \qquad 3\pi < \theta < 5\pi, \quad \cdots \quad -3\pi < \theta < -\pi,$$

which define the 2nd, 3rd, . . . nth sheets of a Riemann surface. Such multiform wave functions have been applied by Sommerfeld to diffraction problems.

16. Derive explicit expressions for the radial and tangential impedances of elliptic cylindrical electromagnetic waves.

17. Show that the integral

$$\psi = \int_0^\infty e^{\pm iz\sqrt{k^2 - \lambda^2}} \frac{J_0(\lambda r)}{\sqrt{\lambda^2 - k^2}} f(\lambda)\lambda\, d\lambda$$

is a cylindrical wave function which is equal to e^{ikR}/R, $R^2 = r^2 + z^2$, when $f(\lambda) = 1$.

CHAPTER VII

SPHERICAL WAVES

In the preceding chapter it was shown how the analytic difficulties involved in the treatment of *vector* differential equations in curvilinear coordinates might be overcome in cylindrical systems by a resolution of the field into two partial fields, each derivable from a purely scalar function satisfying the wave equation. Fortunately this method is applicable also to spherical coordinates to which we now give our attention. The peculiar advantages of cylindrical and spherical systems are a consequence of the very simple character of their geometrical properties. A deeper insight into the nature of the problem and of the difficulties offered by curvilinear coordinates in general will be gained from a brief preliminary study of the vector wave equation.

THE VECTOR WAVE EQUATION

7.1. A Fundamental Set of Solutions.—Within any closed domain of a homogeneous, isotropic medium from which sources have been excluded, all vectors characterizing the electromagnetic field—the field vectors **E, B, D,** and **H,** the vector potential and the Hertzian vectors—satisfy one and the same differential equation. If **C** denotes any such vector, then

$$\nabla^2 \mathbf{C} - \mu\epsilon \frac{\partial^2 \mathbf{C}}{\partial t^2} - \mu\sigma \frac{\partial \mathbf{C}}{\partial t} = 0. \tag{1}$$

Because of the linearity of this wave equation, fields of arbitrary time variation can be constructed from harmonic solutions and there is no loss of generality in the assumption that the vector **C** contains the time only as a factor $e^{-i\omega t}$. By the operator ∇^2 acting on a vector one must understand $\nabla^2 \mathbf{C} = \nabla\nabla \cdot \mathbf{C} - \nabla \times \nabla \times \mathbf{C}$; therefore, in place of (1) we shall write

$$\nabla\nabla \cdot \mathbf{C} - \nabla \times \nabla \times \mathbf{C} + k^2 \mathbf{C} = 0, \tag{2}$$

where $k^2 = \epsilon\mu\omega^2 + i\sigma\mu\omega$ as usual.

Now the vector equation (2) can always be replaced by a simultaneous system of three scalar equations, but the solution of this system for any component of **C** is in most cases impractical. [*Cf.* (85), page 50.] It is only when **C** is resolved into its rectangular components that three independent equations are obtained and in this case

$$\nabla^2 C_j + k^2 C_j = 0, \qquad (j = x, y, z). \tag{3}$$

The operator ∇^2 can then be expressed in curvilinear as well as rectangular coordinates. Very little attention has been paid to the determination of independent *vector* solutions of (2), but the problem has been attacked recently by Hansen[1] in a series of interesting papers dealing with the radiation from antennas.

Let the scalar function ψ be a solution of the equation

$$\nabla^2\psi + k^2\psi = 0, \tag{4}$$

and let $\mathbf{a}$ be any *constant* vector of unit length. We now construct three independent vector solutions of (2) as follows:

$$\mathbf{L} = \nabla\psi, \qquad \mathbf{M} = \nabla \times \mathbf{a}\psi, \qquad \mathbf{N} = \frac{1}{k}\nabla \times \mathbf{M}. \tag{5}$$

If $\mathbf{C}$ is placed equal to $\mathbf{L}$, $\mathbf{M}$, or $\mathbf{N}$, one will verify that (2) is indeed satisfied identically by (5) subject to (4). Since $\mathbf{a}$ is a constant vector, it is clear that $\mathbf{M}$ can be written also as

$$\mathbf{M} = \mathbf{L} \times \mathbf{a} = \frac{1}{k}\nabla \times \mathbf{N}. \tag{6}$$

For one and the same generating function ψ the vector $\mathbf{M}$ is perpendicular to the vector $\mathbf{L}$, or

$$\mathbf{L} \cdot \mathbf{M} = 0. \tag{7}$$

The vector functions $\mathbf{L}$, $\mathbf{M}$, and $\mathbf{N}$ have certain notable properties that follow directly from their definitions. Thus

$$\nabla \times \mathbf{L} = 0, \qquad \nabla \cdot \mathbf{L} = \nabla^2\psi = -k^2\psi, \tag{8}$$

whereas $\mathbf{M}$ and $\mathbf{N}$ are solenoidal.

$$\nabla \cdot \mathbf{M} = 0, \qquad \nabla \cdot \mathbf{N} = 0. \tag{9}$$

The particular solutions of (4) which are finite, continuous, and single-valued in a given domain form a discrete set. For the moment we shall denote any one of these solutions by ψ_n. Associated with each characteristic function ψ_n are three vector solutions $\mathbf{L}_n$, $\mathbf{M}_n$, $\mathbf{N}_n$ of (2), no two of which are colinear. Presumably any arbitrary wave function can be represented as a linear combination of the characteristic vector functions; since the $\mathbf{L}_n$, $\mathbf{M}_n$, $\mathbf{N}_n$ possess certain orthogonal properties which we shall demonstrate in due course, the coefficients of the expansion can be deter-

[1] HANSEN, *Phys. Rev.*, **47**, 139–143, January, 1935; *Physics*, **7**, 460–465 December, 1936; *J. Applied Phys.*, **8**, 284–286, April, 1937.

mined. When the given function is purely solenoidal, the expansion is made in terms of $\mathbf{M}_n$ and $\mathbf{N}_n$ alone. If, however, the divergence of the function does not vanish, terms in $\mathbf{L}_n$ must be included.

The vectors **M** and **N** are obviously appropriate for the representation of the fields **E** and **H**, for each is proportional to the curl of the other. Thus, if the time enters only as a factor $e^{-i\omega t}$ and if the free-charge density is everywhere zero in a homogeneous, isotropic medium of conductivity σ, we have

$$\mathbf{E} = \frac{i\omega\mu}{k^2} \nabla \times \mathbf{H}, \qquad \mathbf{H} = \frac{1}{i\omega\mu} \nabla \times \mathbf{E}. \tag{10}$$

Suppose then that the vector potential can be represented by an expansion in characteristic vector functions of the form

$$\mathbf{A} = \frac{i}{\omega} \sum_n (a_n \mathbf{M}_n + b_n \mathbf{N}_n + c_n \mathbf{L}_n), \tag{11}$$

the coefficients a_n, b_n, c_n to be determined from the current distribution.

By (8), (10), and the relation $\mu\mathbf{H} = \nabla \times \mathbf{A}$, we find for the fields

$$\mathbf{E} = -\sum_n (a_n \mathbf{M}_n + b_n \mathbf{N}_n), \qquad \mathbf{H} = -\frac{k}{i\omega\mu} \sum_n (a_n \mathbf{N}_n + b_n \mathbf{M}_n). \tag{12}$$

The scalar potential ϕ plays no part in the calculation, but can, of course, be determined directly from (11). For by (27), page 27, $\nabla \cdot \mathbf{A} = \frac{i}{\omega} k^2 \phi$; hence, by (8)

$$\phi = -\sum_n c_n \psi_n. \tag{13}$$

Then $\nabla\phi = -\sum_n c_n \mathbf{L}_n$ and clearly the relation $\mathbf{E} = -\nabla\phi - \frac{\partial \mathbf{A}}{\partial t}$ leads again to (12). Finally, if we recall that under the specified conditions an electromagnetic field can be represented in terms of two Hertz vectors by the equations

$$\mathbf{E} = \nabla \times \nabla \times \mathbf{\Pi} + i\omega\mu \nabla \times \mathbf{\Pi}^*, \qquad \mathbf{H} = \frac{k^2}{i\omega\mu} \nabla \times \mathbf{\Pi} + \nabla \times \nabla \times \mathbf{\Pi}^*, \tag{14}$$

it is immediately apparent that

$$\mathbf{\Pi} = -\frac{1}{k} \sum_n b_n \psi_n \mathbf{a}, \qquad \mathbf{\Pi}^* = -\frac{1}{i\omega\mu} \sum_n a_n \psi_n \mathbf{a}, \tag{15}$$

where **a** is a constant vector.

Before applying these results to cylindrical and spherical systems, let us consider the elementary example of waves in rectangular coordinates.

A plane wave whose propagation vector is $\mathbf{k} = k\mathbf{n}$ is represented by

$$\psi = e^{i\mathbf{k}\cdot\mathbf{R}-i\omega t}, \tag{16}$$

where $\mathbf{R}$ is the radius vector drawn from a fixed origin. Then since $\mathbf{k}\cdot\mathbf{R} = k_x x + k_y y + k_z z$, it is easy to see that

$$\mathbf{L} = i\psi\mathbf{k}, \qquad \mathbf{M} = i\psi\mathbf{k}\times\mathbf{a}, \qquad \mathbf{N} = \frac{1}{k}\psi(\mathbf{k}\times\mathbf{a})\times\mathbf{k}. \tag{17}$$

In this particular case $\mathbf{L}\cdot\mathbf{M} = \mathbf{M}\cdot\mathbf{N} = \mathbf{N}\cdot\mathbf{L} = 0$; *all three vectors are mutually perpendicular and* $\mathbf{L}$ *is a purely longitudinal wave.* Now the polar angles determining the direction of $\mathbf{k}$ are α and β, as defined in Fig. 66, page 362, and general solutions of (2) can be constructed by integrating these plane-wave functions over all possible directions. If $g(\alpha, \beta)$ is a scalar amplitude or weighting factor, one may write for $\mathbf{L}$ in any coordinate system, subject to convergence requirements,

$$\mathbf{L} = ie^{-i\omega t}\int d\alpha \int d\beta\, g(\alpha, \beta)\mathbf{k}(\alpha, \beta)\, e^{i\mathbf{k}\cdot\mathbf{R}}; \tag{18}$$

likewise for $\mathbf{M}$ and $\mathbf{N}$

$$\mathbf{M} = ie^{-i\omega t}\int d\alpha \int d\beta\, g(\alpha, \beta)\mathbf{k}\times\mathbf{a}\, e^{i\mathbf{k}\cdot\mathbf{R}}, \tag{19}$$

$$\mathbf{N} = \frac{1}{k}e^{-i\omega t}\int d\alpha \int d\beta\, g(\alpha, \beta)(\mathbf{k}\times\mathbf{a})\times\mathbf{k}\, e^{i\mathbf{k}\cdot\mathbf{R}}. \tag{20}$$

7.2. Application to Cylindrical Coordinates.[1]—The scalar characteristic functions of the wave equation in cylindrical coordinates were set down in Eq. (6), page 356. There are certain disadvantages in the use of complex angle functions $\exp(in\theta)$ and it will prove simpler to deal with the two real functions $\cos n\theta$ and $\sin n\theta$ which may be denoted simply as even and odd. Cylindrical wave functions, constructed with Bessel functions of the first kind and hence finite on the axis, will be denoted by $\psi_n^{(1)}$, while the wave functions formed with $N_n(\lambda r)$ or $H_n^{(1)}(\lambda r)$ will be called $\psi_n^{(2)}$ and $\psi_n^{(3)}$ respectively. Thus

$$\begin{aligned}\psi_{en\lambda}^{(1)} &= \cos n\theta\, J_n(\lambda r)\, e^{ihz-i\omega t},\\ \psi_{on\lambda}^{(1)} &= \sin n\theta\, J_n(\lambda r)\, e^{ihz-i\omega t},\end{aligned} \tag{21}$$

$$\begin{aligned}\psi_{en\lambda}^{(3)} &= \cos n\theta\, H_n^{(1)}(\lambda r)\, e^{ihz-i\omega t},\\ \psi_{on\lambda}^{(3)} &= \sin n\theta\, H_n^{(1)}(\lambda r)\, e^{ihz-i\omega t},\end{aligned} \tag{22}$$

where $\lambda = \sqrt{k^2 - h^2} = k \sin\alpha$ as usual. Functions of the first kind are

[1] Compare the results of this section with Eqs. (36) and (37), p. 361.

represented by the definite integrals

$$
(23)\qquad
\begin{aligned}
\psi^{(1)}_{en\lambda} &= \frac{i^{-n}}{2\pi}\int_0^{2\pi} e^{i\lambda r\cos(\beta-\theta)}\cos n\beta\, d\beta\, e^{ihz-i\omega t},\\
\psi^{(1)}_{on\lambda} &= \frac{i^{-n}}{2\pi}\int_0^{2\pi} e^{i\lambda r\cos(\beta-\theta)}\sin n\beta\, d\beta\, e^{ihz-i\omega t},
\end{aligned}
$$

and representations of other kinds differ from these only in the choice of the contour of integration.

From (23) we now construct integral representations of the *vector* wave functions. Thus

$$
(24)\qquad \mathbf{L}^{(1)}_{en\lambda} = \nabla\psi^{(1)}_{en\lambda} = \frac{\partial\psi^{(1)}_{en\lambda}}{\partial r}\mathbf{i}_1 + \frac{1}{r}\frac{\partial\psi^{(1)}_{en\lambda}}{\partial\theta}\mathbf{i}_2 + \frac{\partial\psi^{(1)}_{en\lambda}}{\partial z}\mathbf{i}_3.
$$

Upon differentiating (23) with respect to r, θ, and z and noting that $k\sin\alpha\cos(\beta-\theta)$, $k\sin\alpha\sin(\beta-\theta)$, and $k\cos\alpha$ are the components of $\mathbf{k}$ respectively along the axes defined by the unit vectors $\mathbf{i}_1, \mathbf{i}_2, \mathbf{i}_3$, so that

$$
(25)\qquad \mathbf{k} = \mathbf{i}_1 k\sin\alpha\cos(\beta-\theta) + \mathbf{i}_2 k\sin\alpha\sin(\beta-\theta) + \mathbf{i}_3 k\cos\alpha,
$$

we find for the even function

$$
(26)\qquad \mathbf{L}^{(1)}_{en\lambda} = \frac{i^{1-n}}{2\pi}\int_0^{2\pi} \mathbf{k}e^{i\lambda r\cos(\beta-\theta)}\cos n\beta\, d\beta\, e^{ihz-i\omega t},
$$

which is of the form (18). The corresponding odd function is obtained by replacing $\cos n\beta$ by $\sin n\beta$ under the integral. When applying these characteristic functions to the expansion of an arbitrary vector wave, it proves convenient to split off the factor $\exp(ihz - i\omega t)$, and we shall define the vector counterparts of the scalar function $f(r, \theta)$ of Chap. VI by

$$
(27)\qquad \mathbf{L}_n = \mathbf{l}_n e^{ihz-i\omega t},\qquad \mathbf{M}_n = \mathbf{m}_n e^{ihz-i\omega t},\qquad \mathbf{N}_n = \mathbf{n}_n e^{ihz-i\omega t}.
$$

From (26) we have

$$
(28)\qquad \mathbf{l}^{(1)}_{\substack{e\\o}n\lambda} = \frac{i^{1-n}}{2\pi}\int_0^{2\pi} \mathbf{k}e^{i\lambda r\cos(\beta-\theta)}\begin{matrix}\cos\\ \sin\end{matrix}\, n\beta\, d\beta.
$$

In like fashion one finds integral representations of the independent solutions. For the constant vector $\mathbf{a}$ we choose the unit vector $\mathbf{i}_3$ directed along the z-axis. Then

$$
(29)\qquad \mathbf{M}^{(1)}_n = \nabla\psi^{(1)}_n \times \mathbf{i}_3 = \frac{1}{r}\frac{\partial\psi^{(1)}_n}{\partial\theta}\mathbf{i}_1 - \frac{\partial\psi^{(1)}_n}{\partial r}\mathbf{i}_2,
$$

whence it is easily deduced that

$$
(30)\qquad \mathbf{m}^{(1)}_{\substack{e\\o}n\lambda} = \frac{i^{1-n}}{2\pi}\int_0^{2\pi} (\mathbf{k}\times\mathbf{i}_3)e^{i\lambda r\cos(\beta-\theta)}\begin{matrix}\cos\\ \sin\end{matrix}\, n\beta\, d\beta;
$$

since $\nabla \times \mathbf{M}_n = k\mathbf{N}_n$, we have

$$(31) \qquad \mathbf{n}^{(1)}_{{}^e_on\lambda} = \frac{i^{-n}}{2\pi k} \int_0^{2\pi} (k^2\mathbf{i}_3 - h\mathbf{k}) e^{i\lambda r \cos(\beta - \theta)} \begin{matrix}\cos\\ \sin\end{matrix}\, n\beta\, d\beta.$$

From these integrals one can easily calculate the rectangular components of the wave functions. The vector functions of angle in the integrands are resolved into rectangular components which can then be combined with $\cos n\beta$ or $\sin n\beta$. The resulting scalar integrals for the components are evaluated by comparison with (23).

To expand an arbitrary vector function of r and θ in terms of the $\mathbf{l}_{{}^e_on\lambda}$, $\mathbf{m}_{{}^e_on\lambda}$, $\mathbf{n}_{{}^e_on\lambda}$, one must show that these functions are orthogonal. Let $\mathbf{i}_1, \mathbf{i}_2, \mathbf{i}_3$ be unit base vectors of a circularly cylindrical coordinate system, Fig. 7, page 51, and $Z_n(\lambda r)$ a cylindrical Bessel function of any kind. Then by (5) we obtain

$$(32) \quad \mathbf{l}_{{}^e_on\lambda} = \frac{\partial}{\partial r} Z_n(\lambda r) \begin{matrix}\cos\\ \sin\end{matrix}\, n\theta\, \mathbf{i}_1 \mp \frac{n}{r} Z_n(\lambda r) \begin{matrix}\sin\\ \cos\end{matrix}\, n\theta\, \mathbf{i}_2 + ih\, Z_n(\lambda r) \begin{matrix}\cos\\ \sin\end{matrix}\, n\theta\, \mathbf{i}_3,$$

$$(33) \qquad \mathbf{m}_{{}^e_on\lambda} = \mp \frac{n}{r} Z_n(\lambda r) \begin{matrix}\sin\\ \cos\end{matrix}\, n\theta\, \mathbf{i}_1 - \frac{\partial}{\partial r} Z_n(\lambda r) \begin{matrix}\cos\\ \sin\end{matrix}\, n\theta\, \mathbf{i}_2,$$

$$(34) \quad \mathbf{n}_{{}^e_on\lambda} = \frac{ih}{k}\frac{\partial}{\partial r} Z_n(\lambda r) \begin{matrix}\cos\\ \sin\end{matrix}\, n\theta\, \mathbf{i}_1 \mp \frac{ihn}{kr} Z_n(\lambda r) \begin{matrix}\sin\\ \cos\end{matrix}\, n\theta\, \mathbf{i}_2 + \frac{\lambda^2}{k} Z_n(\lambda r) \begin{matrix}\cos\\ \sin\end{matrix}\, n\theta\, \mathbf{i}_3.$$

Now it is immediately evident that the scalar product of any two functions integrated over θ from 0 to 2π must vanish if the two differ in the index n.

$$(35) \qquad \int_0^{2\pi} \mathbf{l}_{en\lambda} \cdot \mathbf{l}_{on\lambda}\, d\theta = \int_0^{2\pi} \mathbf{l}_{{}^e_on\lambda} \cdot \mathbf{l}_{{}^e_on'\lambda'}\, d\theta = 0, \qquad \text{if } n \neq n'.$$

Let us understand by $\bar{\mathbf{l}}$ the function $\mathbf{l}$ with the sign of ih reversed.[1] Then the normalizing factor is to be found from the integral

$$(36) \quad \int_0^{2\pi} \mathbf{l}_{{}^e_on\lambda} \cdot \bar{\mathbf{l}}_{{}^e_on\lambda'}\, d\theta = (1 + \delta)\pi \left[\frac{\partial}{\partial r} Z_n(\lambda r) \frac{\partial}{\partial r} Z_n(\lambda' r) + \frac{n^2}{r^2} Z_n(\lambda r)\, Z_n(\lambda' r) + hh' Z_n(\lambda r)\, Z_n(\lambda' r) \right],$$

where $\delta = 0$ unless $n \mp 0$, in which case $\delta = 1$. The first two terms on the right can be combined with the aid of the recurrence relations (24) and (25), page 359, giving

[1] This is not necessarily the conjugate, since λ and h may be complex quantities.

$$(37)\quad \int_0^{2\pi} \mathbf{l}_{\substack{e\\o}n\lambda} \cdot \mathbf{l}_{\substack{e\\o}n\lambda'}\, d\theta = (1+\delta)\pi \left\{ \frac{\lambda\lambda'}{2} [Z_{n-1}(\lambda r)\, Z_{n-1}(\lambda' r) + Z_{n+1}(\lambda r) Z_{n+1}(\lambda' r)] + hh' Z_n(\lambda r)\, Z_n(\lambda' r) \right\}.$$

The same recurrence relations lead us also to the integrals

$$(38)\quad \int_0^{2\pi} \mathbf{m}_{en\lambda} \cdot \mathbf{m}_{on\lambda}\, d\theta = \int_0^{2\pi} \mathbf{m}_{\substack{e\\o}n\lambda} \cdot \mathbf{m}_{\substack{e\\o}n'\lambda'}\, d\theta = 0, \qquad (n \neq n'),$$

$$(39)\quad \int_0^{2\pi} \mathbf{m}_{\substack{e\\o}n\lambda} \cdot \mathbf{m}_{\substack{e\\o}n\lambda'}\, d\theta = (1+\delta)\pi \frac{\lambda\lambda'}{2} [Z_{n-1}(\lambda r)\, Z_{n-1}(\lambda' r) + Z_{n+1}(\lambda r)\, Z_{n+1}(\lambda' r)],$$

$$(40)\quad \int_0^{2\pi} \mathbf{n}_{en\lambda} \cdot \mathbf{n}_{on\lambda}\, d\theta = \int_0^{2\pi} \mathbf{n}_{\substack{e\\o}n\lambda} \cdot \mathbf{n}_{\substack{e\\o}n'\lambda'}\, d\theta = 0, \qquad (n \neq n'),$$

$$(41)\quad \int_0^{2\pi} \mathbf{n}_{\substack{e\\o}n\lambda} \cdot \bar{\mathbf{n}}_{\substack{e\\o}n\lambda'}\, d\theta = (1+\delta) \frac{\pi}{k^2} \left\{ \frac{hh'\lambda\lambda'}{2} [Z_{n-1}(\lambda r)\, Z_{n-1}(\lambda' r) + Z_{n+1}(\lambda r)\, Z_{n+1}(\lambda' r)] + (\lambda\lambda')^2 Z_n(\lambda r)\, Z_n(\lambda' r) \right\}.$$

Furthermore

$$(42)\quad \int_0^{2\pi} \mathbf{l}_{\substack{e\\o}n\lambda} \cdot \mathbf{m}_{\substack{e\\o}n\lambda'}\, d\theta = \int_0^{2\pi} \mathbf{m}_{\substack{e\\o}n\lambda} \cdot \mathbf{n}_{\substack{e\\o}n\lambda'}\, d\theta = \int_0^{2\pi} \mathbf{l}_{\substack{e\\o}n\lambda} \cdot \bar{\mathbf{n}}_{\substack{o\\e}n\lambda'}\, d\theta = 0,$$

and by use of the recurrence relations once more

$$(43)\quad \int_0^{2\pi} \mathbf{l}_{\substack{e\\o}n\lambda} \cdot \mathbf{m}_{\substack{o\\e}n\lambda'}\, d\theta = \int_0^{2\pi} \mathbf{m}_{\substack{e\\o}n\lambda} \cdot \mathbf{n}_{\substack{o\\e}n\lambda'}\, d\theta = 0.$$

There remains only one other combination, and this, unfortunately, leads to a difficulty.

$$(44)\quad \int_0^{2\pi} \mathbf{l}_{\substack{e\\o}n\lambda} \cdot \bar{\mathbf{n}}_{\substack{e\\o}n\lambda'}\, d\theta = i(1+\delta) \frac{\pi}{k} \left\{ h\lambda'^2\, Z_n(\lambda r)\, Z_n(\lambda' r) - \frac{\lambda\lambda' h}{2} [Z_{n-1}(\lambda r)\, Z_{n-1}(\lambda' r) + Z_{n+1}(\lambda r)\, Z_{n+1}(\lambda' r)] \right\},$$

a quantity not identically zero. The set of vector functions $\mathbf{l}_{\substack{e\\o}n\lambda}$, $\mathbf{m}_{\substack{e\\o}n\lambda}$, $\mathbf{n}_{\substack{e\\o}n\lambda}$, therefore, just fail to be completely orthogonal with respect to θ because of (44). In many cases this is of no importance, for if a vector function is divergenceless—as are the electromagnetic field vectors in the absence of free charge—it can be expanded in terms of $\mathbf{m}_{\substack{e\\o}n\lambda}$, $\mathbf{n}_{\substack{e\\o}n\lambda}$ alone. A complete expansion of the vector potential, on the other hand, requires inclusion of the $\mathbf{l}_{\substack{e\\o}n\lambda}$.[1]

[1] The completeness of the set of vector functions has not been demonstrated but presumably follows from the completeness of the scalar set $\psi_{\substack{e\\o}n\lambda}$.

In the case of functions of the first kind the Fourier-Bessel theorem has been applied by Hansen to complete the orthogonality and to simplify the normalization factors. From (53) and (54), page 371, after an obvious interchange of variable, we have

$$f(\lambda) = \int_0^\infty r\,dr \int_0^\infty \lambda'\,d\lambda'\,f(\lambda')J_n(\lambda' r)J_n(\lambda r). \tag{45}$$

This is applied to (37), (39), (41), and (44) when $Z_n(\lambda r)$ is replaced by $J_n(\lambda r)$. For $f(\lambda')$ we shall have λ', $h' = \sqrt{k^2 - \lambda'^2}$, $h'\lambda'$ and the like. Now one must note that the validity of the Fourier-Bessel integral has been demonstrated only when $\int_0^\infty |f(\lambda)|\sqrt{\lambda}\,d\lambda$ exists, and that *this condition is not fulfilled in the present instance.* The artifices employed to ensure convergence involve mathematical difficulties beyond the scope of the present work. With certain reservations one may introduce an exponential convergence factor exactly as in the theory of the Laplace transform on page 310. By $\mathbf{l}^{(1)}_{\substack{e\\o}n\lambda}$ let us understand henceforth the *limit* approached by $\mathbf{l}^{(1)}_{\substack{e\\o}n\lambda}e^{-|s|\lambda}$ as $s \to 0$. Then

$$\lim_{s\to 0} \int_0^\infty r\,dr \int_0^\infty \lambda'\,d\lambda'\,[\lambda' e^{-|s|\lambda'}]J_n(\lambda' r)J_n(\lambda r) = \lambda. \tag{46}$$

Consequently the normalizing factors reduce to

$$\int_0^\infty \int_0^\infty \int_0^{2\pi} \mathbf{l}^{(1)}_{\substack{e\\o}n\lambda} \cdot \mathbf{l}^{(1)}_{\substack{e\\o}n\lambda'}\lambda'\,d\lambda' r\,dr\,d\theta = (1 + \delta)\pi k^2, \tag{47}$$

$$\int_0^\infty \int_0^\infty \int_0^{2\pi} \mathbf{m}^{(1)}_{\substack{e\\o}n\lambda} \cdot \mathbf{m}^{(1)}_{\substack{e\\o}n\lambda'}\lambda'\,d\lambda' r\,dr\,d\theta = (1 + \delta)\pi\lambda^2, \tag{48}$$

$$\int_0^\infty \int_0^\infty \int_0^{2\pi} \mathbf{n}^{(1)}_{\substack{e\\o}n\lambda} \cdot \bar{\mathbf{n}}^{(1)}_{\substack{e\\o}n\lambda'}\lambda'\,d\lambda' r\,dr\,d\theta = (1 + \delta)\pi\lambda^2, \tag{49}$$

while for the troublesome (44) we find[1]

$$\int_0^\infty \int_0^\infty \int_0^{2\pi} \mathbf{l}^{(1)}_{\substack{e\\o}n\lambda} \cdot \bar{\mathbf{n}}^{(1)}_{\substack{e\\o}n\lambda'}\lambda'\,d\lambda' r\,dr\,d\theta = 0. \tag{50}$$

THE SCALAR WAVE EQUATION IN SPHERICAL COORDINATES

7.3. Elementary Spherical Waves.—Since an arbitrary time variation of the field can be represented by Fourier analysis in terms of harmonic components, no essential loss of generality will be incurred hereafter by the assumption that

$$\psi = f(R, \theta, \phi)e^{-i\omega t}. \tag{1}$$

[1] The reader must note that the limit approached by (46) as $s \to 0$ is not necessarily equal to the value of the integral *at* $s = 0$, for this would imply continuity of the function defined by the integral in the neighborhood of $s = 0$. This point underlies the whole theory of the Laplace transform. *Cf.* Carslaw, "Introduction to the Theory of Fourier's Series and Integrals," 2d ed., p. 293, Macmillan, 1921.

In a homogeneous, isotropic medium the function $f(R, \theta, \phi)$ must satisfy

$$\nabla^2 f + k^2 f = 0, \tag{2}$$

which in spherical coordinates is expanded by (95), page 52, to

$$\frac{1}{R^2}\frac{\partial}{\partial R}\left(R^2 \frac{\partial f}{\partial R}\right) + \frac{1}{R^2 \sin\theta}\frac{\partial}{\partial \theta}\left(\sin\theta \frac{\partial f}{\partial \theta}\right) + \frac{1}{R^2 \sin^2\theta}\frac{\partial^2 f}{\partial \phi^2} + k^2 f = 0. \tag{3}$$

The equation is separable, so that upon placing $f = f_1(r)f_2(\theta)f_3(\phi)$ one finds

$$R^2 \frac{d^2 f_1}{dR^2} + 2R\frac{df_1}{dR} + (k^2R^2 - p^2)f_1 = 0, \tag{4}$$

$$\frac{1}{\sin\theta}\frac{d}{d\theta}\left(\sin\theta \frac{df_2}{d\theta}\right) + \left(p^2 - \frac{q^2}{\sin^2\theta}\right)f_2 = 0, \tag{5}$$

$$\frac{d^2 f_3}{d\phi^2} + q^2 f_3 = 0. \tag{6}$$

The parameters p and q are separation constants whose choice is governed by the physical requirement that at any fixed point in space the field must be single-valued. If the properties of the medium are independent of equatorial angle ϕ, it is necessary that $f_3(\phi)$ be a periodic function with period 2π and q is, therefore, restricted to the integers $m = 0, \pm 1, \pm 2, \ldots$. To determine p we first identify the solution f_2 as an associated Legendre function. Upon substitution of $\eta = \cos\theta$, Eq. (5) transforms to

$$(1 - \eta^2)\frac{d^2 f_2}{d\eta^2} - 2\eta\frac{df_2}{d\eta} + \left(p^2 - \frac{m^2}{1 - \eta^2}\right)f_2 = 0, \tag{7}$$

an equation characterized by regular singularities at the points $\eta = -1$, $\eta = +1$, $\eta = \infty$, and no others. Its solutions are, therefore, hypergeometric functions. Now when $m = 0$, (7) reduces to the Legendre equation. There are, of course, two independent solutions of the Legendre equation which may be expanded in ascending power series about the origin $\eta = 0$. These series do not, in general, converge for $\eta = \pm 1$. If, however, we choose $p^2 = n(n + 1)$, where $n = 0, 1, 2, \ldots$, then one of the series breaks off after a finite number of terms and has a finite value at the poles. These polynomial solutions satisfying the equation

$$(1 - \eta^2)\frac{d^2 v}{d\eta^2} - 2\eta\frac{dv}{d\eta} + n(n + 1)v = 0 \tag{8}$$

are known as Legendre polynomials and are designated by $P_n(\eta)$. If now we differentiate (8) m times with respect to η, we obtain

$$(1 - \eta^2)\frac{d^2 w}{d\eta^2} - 2(m + 1)\eta\frac{dw}{d\eta} + [n(n + 1) - m(m + 1)]w = 0, \tag{9}$$

where $w = d^m v/d\eta^m$, and upon making a last substitution of the dependent variable, $w = (1 - \eta^2)^{-\frac{m}{2}} f_2(\eta)$, we obtain (7) in the form

$$(10) \qquad (1 - \eta^2)\frac{d^2 f_2}{d\eta^2} - 2\eta\frac{df_2}{d\eta} + \left[n(n+1) - \frac{m^2}{1 - \eta^2}\right] f_2 = 0.$$

The solutions of (10) which are finite at the poles $\eta = \pm 1$ and which are, consequently, periodic in θ are the associated Legendre polynomials

$$(11) \qquad f_2(\eta) = P_n^m(\eta) = (1 - \eta^2)^{\frac{m}{2}} \frac{d^m P_n(\eta)}{d\eta^m}.$$

For each pair of integers there exists an independent solution $Q_n^m(\eta)$ which becomes infinite at $\eta = \pm 1$, and which consequently does not apply to physical fields in a complete spherical domain.

The definition of the associated Legendre polynomial as stated in (11) holds only when n and m are *positive* integers. The functions of negative index are related in a rather simple fashion to those of positive index, but we shall have no need of them here. To obviate any confusion on this matter we choose for particular solutions of (6) the *real* functions $\cos m\phi$, $\sin m\phi$, and restrict m and n to the positive integers and zero. It is clear from (11) that $P_n^m(\eta)$ vanishes when $m > n$, for $P_n(\eta)$ is a polynomial of nth degree. We have in fact

$$(12) \qquad P_n^m(\eta) = \frac{(1 - \eta^2)^{\frac{m}{2}}}{2^n n!} \frac{d^{n+m}(\eta^2 - 1)^n}{d\eta^{n+m}}.$$

The properties of the hypergeometric functions, of which the Legendre functions are the best known example, have been explored in great detail and it would be scarcely possible to give an adequate account of the various series and integral representations within the space at our disposal. We shall set down only the indispensable recurrence relations,

$$(13) \qquad \begin{aligned}
(n - m + 1)P_{n+1}^m - (2n + 1)\eta P_n^m + (n + m)P_{n-1}^m &= 0, \\
P_{n-1}^m &= \eta P_n^m - (n - m + 1)\sqrt{1 - \eta^2}\, P_n^{m-1}, \\
P_{n+1}^m &= \eta P_n^m + (n + m)\sqrt{1 - \eta^2}\, P_n^{m-1}, \\
\sqrt{1 - \eta^2}\, P_n^{m+1} &= (n + m + 1)\eta P_n^m - (n - m + 1)P_{n+1}^m, \\
\sqrt{1 - \eta^2}\, P_n^{m+1} &= 2m\eta P_n^m - (n + m)(n - m + 1)\sqrt{1 - \eta^2}\, P_n^{m-1}, \\
\sqrt{1 - \eta^2}\, P_n^m &= \frac{1}{2n + 1}(P_{n+1}^{m+1} - P_{n-1}^{m+1}), \\
\frac{m}{\sqrt{1 - \eta^2}} P_n^m &= \frac{1}{2}\eta[(n - m + 1)(n + m)P_n^{m-1} + P_n^{m+1}] \\
&\quad + m\sqrt{1 - \eta^2}\, P_n^m.
\end{aligned}$$

To these may be added the differential relations

$$
\begin{aligned}
&(1-\eta^2)\frac{dP_n^m}{d\eta} = (n+1)\eta P_n^m - (n-m+1)P_{n+1}^m,\\
(14)\quad &(1-\eta^2)\frac{dP_n^m}{d\eta} = (n+m)P_{n-1}^m - n\eta P_n^m,\\
&\frac{dP_n^m}{d\theta} = -\sqrt{1-\eta^2}\,\frac{dP_n^m}{d\eta}\\
&\qquad = \frac{1}{2}\left[(n-m+1)(n+m)P_n^{m-1} - P_n^{m+1}\right].
\end{aligned}
$$

It may be remarked in passing that these relations are satisfied also by functions of the second kind Q_n^m.

The functions $\cos m\phi\, P_n^m(\cos\theta)$ and $\sin m\phi\, P_n^m(\cos\theta)$ are periodic on the surface of a unit sphere and the indices m and n determine the number of nodal lines. Thus when $m = 0$, the field is independent of the equatorial angle ϕ. If n is also zero, the value of the function is everywhere constant on the surface of the sphere. When $n = 1$, there is a single nodal line at the equator $\theta = \pi/2$ along which the function is zero. When

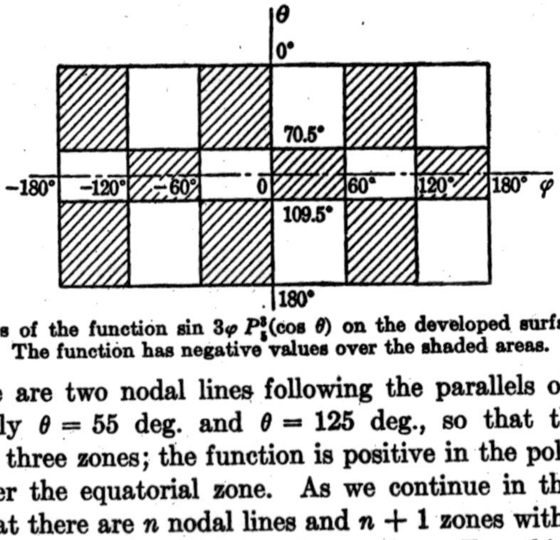

FIG. 69.—Nodes of the function $\sin 3\varphi\, P_5^3(\cos\theta)$ on the developed surface of a sphere. The function has negative values over the shaded areas.

$n = 2$, there are two nodal lines following the parallels of latitude at approximately $\theta = 55$ deg. and $\theta = 125$ deg., so that the sphere is divided into three zones; the function is positive in the polar zones and negative over the equatorial zone. As we continue in this way, it is apparent that there are n nodal lines and $n + 1$ zones within which the function is alternately positive and negative. For this reason the $P_n(\cos\theta)$ are often called *zonal harmonics*. If now m has a value other than zero, it will be observed upon examination of Appendix IV that the function is zero at the poles due to the factor $(1 - \eta^2)^{\frac{m}{2}}$, and that the number of nodal lines parallel to the equator is $n - m$. Moreover, the function vanishes along lines of longitude determined by the roots of $\cos m\phi$ and $\sin m\phi$. There are obviously m longitudinal nodes that intersect the nodal parallels of latitude orthogonally, thus dividing the surface of the sphere into rectangular domains, or *tesserae*, within which the function is alternately positive and negative. For this reason the functions $\cos m\phi\, P_n^m(\cos\theta)$ and $\sin m\phi\, P_n^m(\cos\theta)$ are sometimes called *tesseral harmonics* of nth degree and mth order. There are obviously $2n + 1$ tesseral harmonics of nth degree. This division into positive and negative domains is illustrated graphically in Fig. 69 for the function $\sin 3\phi\, P_5^3(\cos\theta)$.

If the tesseral harmonics are multiplied by a set of arbitrary constants and summed, one obtains the *spherical surface harmonics* of degree n with which we have already had something to do in Chap. III.

$$(15) \qquad Y_n(\theta, \phi) = \sum_{m=0}^{n} (a_{nm} \cos m\phi + b_{nm} \sin m\phi) P_n^m(\cos\theta).$$

The tesseral harmonics form a complete system of orthogonal functions on the surface of a sphere.[1] It is in fact easy to show through an integration of (12) by parts that

$$(16) \qquad \int_{-1}^{1} P_n^m(\eta) P_l^m(\eta)\, d\eta = 0, \qquad \int_{-1}^{1} P_n^m(\eta) P_n^l \frac{d\eta}{1-\eta^2} = 0,$$

when $n \neq l$ or $m \neq l$ respectively, and that

$$(17) \qquad \begin{aligned} \int_{-1}^{1} [P_n^m(\eta)]^2\, d\eta &= \frac{2}{2n+1} \frac{(n+m)!}{(n-m)!}, \\ \int_{-1}^{1} [P_n^m(\eta)]^2 \frac{d\eta}{1-\eta^2} &= \frac{1}{m} \frac{(n+m)!}{(n-m)!}. \end{aligned}$$

From these relations follows the fundamental theorem on the expansion of an arbitrary function in spherical surface harmonics: *Let $g(\theta, \phi)$ be an arbitrary function on the surface of a sphere which together with all its first and second derivatives is continuous. Then $g(\theta, \phi)$ can be represented by an absolutely convergent series of surface harmonics,*

$$(18) \qquad g(\theta, \phi) = \sum_{n=0}^{\infty} \left[a_{no} P_n(\cos\theta) + \sum_{m=1}^{n} (a_{nm} \cos m\phi + b_{nm} \sin m\phi) P_n^m(\cos\theta) \right],$$

whose coefficients are determined by

$$(19) \qquad \begin{aligned} a_{no} &= \frac{2n+1}{4\pi} \int_0^{2\pi} \int_0^{\pi} g(\theta, \phi)\, P_n(\cos\theta) \sin\theta\, d\theta\, d\phi, \\ a_{nm} &= \frac{2n+1}{2\pi} \frac{(n-m)!}{(n+m)!} \int_0^{2\pi} \int_0^{\pi} g(\theta, \phi)\, P_n^m(\cos\theta) \cos m\phi \sin\theta\, d\theta\, d\phi, \\ b_{nm} &= \frac{2n+1}{2\pi} \frac{(n-m)!}{(n+m)!} \int_0^{2\pi} \int_0^{\pi} g(\theta, \phi)\, P_n^m(\cos\theta) \sin m\phi \sin\theta\, d\theta\, d\phi. \end{aligned}$$

[1] The proof of this statement and of the expansion theorem which follows will be found in Courant-Hilbert, "Methoden der mathematischen Physik," 2d ed., Chap.

In the case of a function that depends on θ alone, the conditions determining the convergence of an expansion in Legendre polynomials are identical with those governing the convergence of a Fourier series. Under such circumstances it is sufficient that $g(\theta)$ and its first derivative be piecewise continuous in the interval $0 \leq \theta \leq 2\pi$. The convergence theory of the expansion (18) in surface harmonics, on the other hand, presents definitely greater difficulties and the extension of the theorem to discontinuous functions involves considerations beyond the scope of the present outline.

There remains the identification of the radial function $f_1(R)$ satisfying (4). If for f_1 we write $f_1 = (kR)^{-\frac{1}{2}}v(R)$, it is readily shown that $v(R)$ satisfies

$$R^2 \frac{d^2v}{dR^2} + R\frac{dv}{dR} + \left[k^2R^2 - \left(n + \frac{1}{2}\right)^2\right] v = 0, \tag{20}$$

and hence, by Sec. 6.5, is a cylinder function of half order.

$$f_1(R) = \frac{1}{\sqrt{kR}} Z_{n+\frac{1}{2}}(kR). \tag{21}$$

The characteristic, or elementary, wave functions which at all points on the surface of a sphere are finite and single-valued are, therefore,

$$f_{\substack{e\\o}mn} = \frac{1}{\sqrt{kR}} Z_{n+\frac{1}{2}}(kR)\, P_n^m(\cos\theta) \begin{matrix}\cos\\ \sin\end{matrix}\, m\phi. \tag{22}$$

As in the cylindrical case, we choose for $Z_{n+\frac{1}{2}}(kR)$ a Bessel function of the first kind within domains which include the origin, a function of the third kind wherever the field is to be represented as a traveling wave.

7.4. Properties of the Radial Functions.—Various notations have been employed at one time or another to designate the radial functions $(kR)^{-\frac{1}{2}}Z_{n+\frac{1}{2}}(kR)$ but none appears to have gained general acceptance.[1] What seems to be a logical proposal has been made recently by Morse[2] and will be adopted here. Accordingly we define the *spherical Bessel functions* by

$$\begin{aligned} z_n(\rho) &= \sqrt{\frac{\pi}{2\rho}}\, Z_{n+\frac{1}{2}}(\rho), & j_n(\rho) &= \sqrt{\frac{\pi}{2\rho}}\, J_{n+\frac{1}{2}}(\rho), \\ n_n(\rho) &= \sqrt{\frac{\pi}{2\rho}}\, N_{n+\frac{1}{2}}(\rho), & h_n^{(1)}(\rho) &= \sqrt{\frac{\pi}{2\rho}}\, H_{n+\frac{1}{2}}^{(1)}(\rho), \\ & & h_n^{(2)}(\rho) &= \sqrt{\frac{\pi}{2\rho}}\, H_{n+\frac{1}{2}}^{(2)}(\rho). \end{aligned} \tag{23}$$

VII, §5, 1931. For greater detail see Hobson, "The Theory of Spherical and Ellipsoidal Harmonics," Cambridge University Press, 1931.

[1] See WATSON, "Bessel Functions," p. 55, Cambridge University Press, 1922.

[2] MORSE: "Vibration and Sound," p. 246, McGraw-Hill, 1936.

Series expansions of these functions about the point $\rho = 0$ can be obtained directly from (8) and (11), page 357. If we recall that $\Gamma(z+1) = z\Gamma(z)$, $\Gamma(\frac{1}{2}) = \sqrt{\pi}$, and make use of the *duplication formula*

$$\Gamma\left(z + \frac{1}{2}\right) = \frac{\sqrt{\pi}\,\Gamma(2z)}{2^{2z-1}\Gamma(z)}, \tag{24}$$

we find for $j_n(\rho)$

$$j_n(\rho) = 2^n \rho^n \sum_{m=0}^{\infty} \frac{(-1)^m (n+m)!}{m!(2n+2m+1)!}\,\rho^{2m}, \tag{25}$$

whence it is apparent that $j_n(\rho)$ is an integral function. Likewise from the relation $N_{n+\frac{1}{2}}(\rho) = (-1)^{n-1} J_{-n-\frac{1}{2}}(\rho)$, we obtain for the function of the second kind

$$n_n(\rho) = -\frac{1}{2^n \rho^{n+1}} \sum_{m=0}^{\infty} \frac{\Gamma(2n-2m+1)}{m!\Gamma(n-m+1)}\,\rho^{2m}. \tag{26}$$

We turn next to representations when ρ is very large, and discover at once a notable property of the Bessel functions whose order is half an odd integer. On page 358 expansions in descending powers of ρ were given for the functions of arbitrary order p. Such series satisfy the Bessel equation formally, but are only semiconvergent. We observe now, however, that when $p = n + 1/2$, the series $P_{n+\frac{1}{2}}(\rho)$ and $Q_{n+\frac{1}{2}}(\rho)$ break off, so that there is no longer question of convergence. Consequently (13) and (14) of page 358 are *analytic* representations of $J_{n+\frac{1}{2}}(\rho)$ and $N_{n+\frac{1}{2}}(\rho)$; furthermore, it is apparent that *these functions of half order can be expressed in finite terms.* From (23) above and Eqs. (13) to (17) of page 358, we obtain

$$j_n(\rho) = \frac{1}{\rho}\left[P_{n+\frac{1}{2}}(\rho)\cos\left(\rho - \frac{n+1}{2}\pi\right) - Q_{n+\frac{1}{2}}(\rho)\sin\left(\rho - \frac{n+1}{2}\pi\right)\right], \tag{27}$$

$$n_n(\rho) = \frac{1}{\rho}\left[P_{n+\frac{1}{2}}(\rho)\sin\left(\rho - \frac{n+1}{2}\pi\right) + Q_{n+\frac{1}{2}}(\rho)\cos\left(\rho - \frac{n+1}{2}\pi\right)\right], \tag{28}$$

where

$$P_{n+\frac{1}{2}}(\rho) = 1 - \frac{n(n^2-1)(n+2)}{2^2\cdot 2!\,\rho^2} + \frac{n(n^2-1)(n^2-4)(n^2-9)(n+4)}{2^4\cdot 4!\,\rho^4} - \cdots, \tag{29}$$

$$Q_{n+\frac{1}{2}}(\rho) = \frac{n(n+1)}{2\cdot 1!\,\rho} - \frac{n(n^2-1)(n^2-4)(n+3)}{2^3\cdot 3!\,\rho^3} + \cdots. \tag{30}$$

For the functions of the third and the fourth kind this leads to

$$\begin{aligned} h_n^{(1)}(\rho) &= \frac{(-i)^{n+1}}{\rho}\,e^{i\rho}[P_{n+\frac{1}{2}}(\rho) + iQ_{n+\frac{1}{2}}(\rho)], \\ h_n^{(2)}(\rho) &= \frac{i^{n+1}}{\rho}\,e^{-i\rho}[P_{n+\frac{1}{2}}(\rho) - iQ_{n+\frac{1}{2}}(\rho)]. \end{aligned} \tag{31}$$

These series converge very rapidly so that for large values of ρ the first term alone gives the approximate value of the function. Thus, when $\rho \to \infty$,

$$(32)\qquad \begin{aligned} j_n(\rho) &\simeq \frac{1}{\rho}\cos\left(\rho - \frac{n+1}{2}\pi\right), & n_n(\rho) &\simeq \frac{1}{\rho}\sin\left(\rho - \frac{n+1}{2}\pi\right), \\ h_n^{(1)}(\rho) &\simeq \frac{1}{\rho}(-i)^{n+1}e^{i\rho}, & h_n^{(2)}(\rho) &\simeq \frac{1}{\rho}i^{n+1}e^{-i\rho}. \end{aligned}$$

The recurrence relations satisfied by the spherical Bessel function follow directly from Eqs. (24) to (27) of page 359.

$$(33)\qquad z_{n-1} + z_{n+1} = \frac{2n+1}{\rho} z_n,$$

$$(34)\qquad \frac{d}{d\rho} z_n(\rho) = \frac{1}{2n+1}[nz_{n-1} - (n+1)z_{n+1}],$$

$$(35)\qquad \frac{d}{d\rho}[\rho^{n+1}z_n(\rho)] = \rho^{n+1}z_{n-1}, \qquad \frac{d}{d\rho}[\rho^{-n}z_n(\rho)] = -\rho^{-n}z_{n+1}.$$

Having defined the radial functions, we can at last write down the general solutions of (3) as sums of elementary spherical wave functions. In case $f(R, \theta, \phi)$ is to be finite at the origin, we have

$$(36)\qquad f^{(1)}(R, \theta, \phi) = \sum_{n=0} j_n(kR)\left[a_{n0}P_n(\cos\theta) + \sum_{m=1}^{n}(a_{nm}\cos m\phi + b_{nm}\sin m\phi)P_n^m(\cos\theta)\right],$$

while a field whose surfaces of constant phase travel outward is represented by $f^{(3)}(R, \theta, \phi)$, obtained by replacing $j_n(kR)$ by $h_n^{(1)}(kR)$ in (36). A *spherically symmetric* solution results when all coefficients except a_{00} are zero. Then apart from an arbitrary factor, we have

$$(37)\qquad \begin{aligned} f_0^{(1)} &= \frac{\sin kR}{kR}, & f_0^{(2)} &= \frac{\cos kR}{kR}, \\ f_0^{(3)} &= \frac{1}{kR}e^{ikR}, & f_0^{(4)} &= \frac{1}{kR}e^{-ikR}. \end{aligned}$$

7.5. Addition Theorem for the Legendre Polynomials.—If $g(\theta, \phi)$ is any function satisfying the conditions of the expansion theorem (18), its value at the pole, $\theta = 0$, must be

$$(38)\qquad [g(\theta, \phi)]_{\theta=0} = \sum_{n=0}^{\infty} a_{n0} = \frac{1}{4\pi}\sum_{n=0}^{\infty}(2n+1)\int_0^{2\pi}\int_0^{\pi} g(\theta, \phi)P_n(\cos\theta) \sin\theta\, d\theta\, d\phi,$$

since $P_n(1) = 1$, $P_n^m(1) = 0$. In particular let us take for $g(\theta, \phi)$ any surface harmonic $Y_n(\theta, \phi)$ of degree n expressed by (15). Then in virtue of the orthogonality relations, the sum in (38) reduces to a single term and we obtain the formula

$$(39)\quad \int_0^{2\pi}\int_0^{\pi} Y_n(\theta, \phi)P_n(\cos\theta)\sin\theta\, d\theta\, d\phi = \frac{4\pi}{2n+1}[Y_n(\theta, \phi)]_{\theta=0}.$$

We shall apply this result to the problem of expressing a zonal harmonic in terms of a new axis of reference. Let P in Fig. 70 be a point on a sphere whose coordinates with respect to a fixed rectangular reference system are θ and ϕ. A second point Q has the coordinates α and β. The angle made by the axis OP with the axis OQ is γ. The zonal harmonics at P referred to the new polar axis OQ are of the form P_n $(\cos\gamma)$ and our problem, therefore, is to expand $P_n(\cos\gamma)$ in terms of the coordinates θ, ϕ, and α, β.

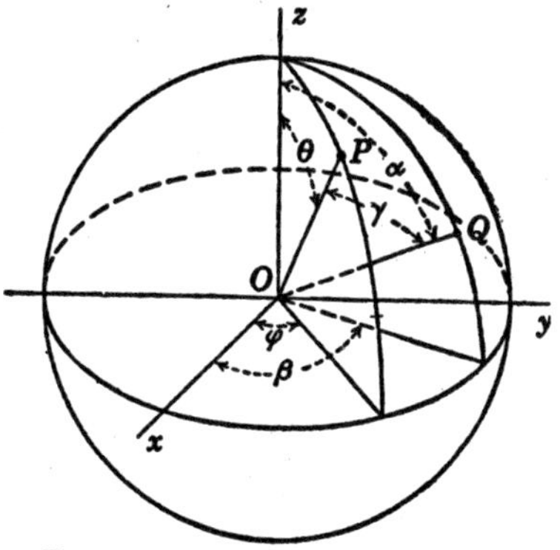

FIG. 70.—Rotation of the axis of reference from OP to OQ for a system of zonal harmonics.

It is apparent that on a unit sphere $\cos\gamma$ is the projection of the line OP on the axis OQ. If x, y, z are the coordinates of P, x', y', z' those of Q, then

$$(40)\quad \cos\gamma = xx' + yy' + zz' = \sin\theta\sin\alpha\cos(\phi - \beta) + \cos\theta\cos\alpha.$$

We now assume for $P_n(\cos\gamma)$ an expansion of the form

$$(41)\quad P_n(\cos\gamma) = \frac{c_0}{2}P_n(\cos\theta) + \sum_{m=1}^{n}(c_m\cos m\phi + d_m\sin m\phi)P_n^m(\cos\theta),$$

multiply both sides by $P_n^m(\cos\theta)\cos m\phi$, and integrate over the unit sphere. With the help once more of the orthogonality relations, we obtain

$$(42)\quad \int_0^{2\pi}\int_0^{\pi} P_n(\cos\gamma)\, P_n^m(\cos\theta)\cos m\phi\sin\theta\, d\theta\, d\phi$$
$$= \frac{2\pi}{2n+1}\frac{(n+m)!}{(n-m)!}c_m.$$

But by (39)

$$(43)\quad \int_0^{2\pi}\int_0^{\pi} P_n(\cos\gamma)P_n^m(\cos\theta)\cos m\phi\sin\theta\, d\theta\, d\phi$$
$$= \frac{4\pi}{2n+1}[P_n^m(\cos\theta)\cos m\phi]_{\gamma=0} = \frac{4\pi}{2n+1}P_n^m(\cos\alpha)\cos m\beta,$$

since when $\gamma = 0$, $\theta = \alpha$, $\phi = \beta$. Consequently,

$$c_m = 2\frac{(n-m)!}{(n+m)!}P_n^m(\cos\alpha)\cos m\beta. \tag{44}$$

Likewise,

$$d_m = 2\frac{(n-m)!}{(n+m)!}P_n^m(\cos\alpha)\sin m\beta. \tag{45}$$

The desired expansion, or *addition formula*, is therefore

$$P_n(\cos\gamma) = P_n(\cos\alpha)\,P_n(\cos\theta) + 2\sum_{m=1}^{n}\frac{(n-m)!}{(n+m)!}P_n^m(\cos\alpha)\,P_n^m(\cos\theta)\cos m(\phi-\beta). \tag{46}$$

This result leads to an alternative formulation of the expansion theorem stated on page 403. If $g(\theta, \phi)$ and its derivatives possess the necessary continuity on the surface of a sphere, its Laplace series (18) is

$$g(\theta, \phi) = \sum_{n=0}^{\infty} Y_n(\theta, \phi), \tag{47}$$

or by (19) and (46),

$$g(\theta, \phi) = \frac{1}{4\pi}\sum_{n=0}^{\infty}(2n+1)\int_0^{2\pi}\int_0^{\pi} g(\alpha, \beta)P_n(\cos\gamma)\sin\alpha\,d\alpha\,d\beta. \tag{48}$$

7.6. Expansion of Plane Waves.—It is now a relatively easy matter to find a representation of a plane wave propagated in an arbitrary direction in terms of elementary spherical waves about a fixed center. The direction and character of the plane wave will be determined by its propagation vector **k** whose rectangular components are

$$k_1 = k\sin\alpha\cos\beta, \qquad k_2 = k\sin\alpha\sin\beta, \qquad k_3 = k\cos\alpha, \tag{49}$$

while the coordinates of any arbitrary point of observation are

$$x = R\sin\theta\cos\phi, \qquad y = R\sin\theta\sin\phi, \qquad z = R\cos\theta. \tag{50}$$

Then the phase is given by

$$\mathbf{k}\cdot\mathbf{R} = kR[\sin\alpha\sin\theta\cos(\phi-\beta) + \cos\alpha\cos\theta] = kR\cos\gamma. \tag{51}$$

The function $f^{(1)}(R, \theta, \phi) = \exp(ikR\cos\gamma)$ is continuous and has continuous derivatives everywhere including the origin $R = 0$. It can, therefore, be expanded according to (36), page 406. Consider first an axis that coincides with the direction of the wave. Since the wave is symmetrical about this axis, we write

$$e^{ikR\cos\gamma} = \sum_{n=0}^{\infty} a_n j_n(kR)P_n(\cos\gamma). \tag{52}$$

The coefficients are determined in the usual manner by multiplying both sides by $P_n(\cos\gamma)\sin\gamma$ and integrating with respect to γ from 0 to π. Then by (17)

$$a_n j_n(kR) = \frac{2n+1}{2}\int_0^\pi e^{ikR\cos\gamma}\,P_n(\cos\gamma)\sin\gamma\,d\gamma. \tag{53}$$

To rid this relation of its dependence on R, differentiate both sides with respect to $\rho = kR$ and then place $\rho = 0$. From (25) it follows that

$$\left[\frac{d^n j_n(\rho)}{d\rho^n}\right]_{\rho=0} = \frac{2^n(n!)^2}{(2n+1)!}; \tag{54}$$

hence,

$$\frac{2^n(n!)^2}{(2n+1)!}\,a_n = \frac{2n+1}{2}\,i^n\int_0^\pi \cos^n\gamma\,P_n(\cos\gamma)\sin\gamma\,d\gamma. \tag{55}$$

The integral on the right of (55) is readily evaluated and we obtain $a_n = (2n+1)i^n$, or

$$e^{ikR\cos\gamma} = \sum_{n=0}^\infty i^n(2n+1)j_n(kR)\,P_n(\cos\gamma). \tag{56}$$

In case the z-axis of the coordinate system fails to coincide with the direction of the plane wave, one may use the addition theorem (46) to express (56) in terms of an arbitrary axis of reference. Then

$$\begin{aligned} e^{ikR\cos\gamma} = \sum_{n=0}^\infty i^n(2n+1)j_n(kR)\Bigg[P_n(\cos\alpha)\,P_n(\cos\theta) \\ + 2\sum_{m=1}^n \frac{(n-m)!}{(n+m)!}P_n^m(\cos\alpha)\,P_n^m(\cos\theta)\cos m(\phi-\beta)\Bigg]. \end{aligned} \tag{57}$$

7.7. Integral Representations.—We have found it convenient on various occasions to represent a wave function as a sum of plane waves with appropriate weighting or amplitude factors. The direction of each component wave is determined by the angles α and β. Integration is extended over all real directions in space, and may in certain cases include imaginary values of α and β as well. As in Sec. 6.7, we wish to find representations of the form

$$f(x, y, z) = \int d\alpha \int d\beta\, g(\alpha, \beta)\, e^{ik(x\sin\alpha\cos\beta + y\sin\alpha\sin\beta + z\cos\alpha)}, \tag{58}$$

or, when $x = R\sin\theta\cos\phi$, $y = R\sin\theta\sin\phi$, $z = R\cos\theta$,

$$f(R, \theta, \phi) = \int d\alpha \int d\beta\, g(\alpha, \beta)\, e^{ikR\cos\gamma}, \tag{59}$$

where α and β are the angles shown in both Figs. 66 and 70. In the case of cylindrical waves a fixed frequency and a fixed wave length along the z-axis lead to a constant value of α, so that the directions of the plane waves representing a cylindrical function $f(u^1, u^2)$ constitute a circular cone. Since for spherical waves there exists no such preferred axis, the integration must be extended to both α and β.

An integral representation of elementary spherical wave functions can be obtained directly from the expansion (57). If both sides of that equation are multiplied by $P_n^m(\cos\alpha)\cos m\beta$ or by $P_n^m(\cos\alpha)\sin m\beta$, one obtains, thanks to the orthogonality relations (19), the result

$$(60)\quad j_n(kR)\,P_n^m(\cos\theta)\,{\cos \atop \sin}\,m\phi = \frac{i^{-n}}{4\pi}\int_0^{2\pi}\int_0^{\pi} e^{ikR\cos\gamma}\,P_n^m(\cos\alpha)\,{\cos \atop \sin}\,m\beta\,\sin\alpha\,d\alpha\,d\beta.$$

Upon multiplying both sides by the arbitrary constants a_{nm}, b_{nm} and summing over m, this can be written also as

$$(61)\quad j_n(kR)Y_n(\theta,\phi) = \frac{i^{-n}}{4\pi}\int_0^{2\pi}\int_0^{\pi} e^{ikR\cos\gamma}Y_n(\alpha,\beta)\sin\alpha\,d\alpha\,d\beta.$$

From these general formulas may be derived a number of useful special cases. Thus by choosing $m = 0$, $\theta = 0$, we obtain a representation of the spherical Bessel function $j_n(kR)$.

$$(62)\qquad j_n(kR) = \frac{i^{-n}}{2}\int_0^{\pi} e^{ikR\cos\alpha}\,P_n(\cos\alpha)\sin\alpha\,d\alpha,$$

or

$$(63)\qquad j_n(kR) = \frac{i^{-n}}{2}\int_{-1}^{1} e^{ikR\eta}\,P_n(\eta)\,d\eta.$$

It is in fact easy to show that Eq. (4), page 400, is satisfied by the integral

$$(64)\qquad z_n(kR) = i^{-n}\int_C e^{ikR\eta}\,P_n(\eta)\,d\eta,$$

provided the contour C is such that the bilinear concomitant

$$(65)\qquad (1-\eta^2)\left(\frac{dP_n}{d\eta} - ikRP_n\right)e^{ikR\eta}\Big|_C = 0$$

vanishes at the limits. In place of $\eta = \pm 1$, we may choose a value that will cause the exponential to vanish. Thus if k is *real*, we take for the upper limit $\eta = i\infty$ and obtain the representations

$$(66)\qquad \begin{aligned} h_n^{(1)}(kR) &= i^{-n}\int_{i\infty}^{1} e^{ikR\eta}\,P_n(\eta)\,d\eta,\\ h_n^{(2)}(kR) &= i^{-n}\int_{-1}^{i\infty} e^{ikR\eta}\,P_n(\eta)\,d\eta. \end{aligned}$$

If on the other hand k is complex due to a conductivity of the medium, one may choose that root of k^2 whose imaginary part is positive and replace the limit $\eta = i\infty$ by $\eta = \infty$.

Again from (60) and (54), placing ϕ and R equal to zero, one obtains a representation of the function $P_n^m(\cos\theta)$:

$$(67)\quad P_n^m(\cos\theta) = \frac{(2n+1)!}{4\pi 2^n (n!)^2} \int_0^{2\pi}\int_0^{\pi} \cos^n \gamma\, P_n^m(\cos\alpha)\cos m\beta \sin\alpha\, d\alpha\, d\beta.$$

The integration with respect to α cannot be carried out easily with only the help of formulas from the present section at our disposal, but by other means it may be shown that (67) reduces to

$$(68)\quad P_n^m(\cos\theta) = \frac{(n+m)!}{2\pi n!} i^{-m} \int_0^{2\pi} (\cos\theta + i\sin\theta\cos\beta)^n \cos m\beta\, d\beta.$$

Finally, one will note that by making use of the integral representation (37), page 367, for a Bessel function, the right-hand side of (60) can be reduced to a single integral. When $\phi = 0$,

$$(69)\quad j_n(kR)\, P_n^m(\cos\theta) = \frac{i^{-n}}{2} \int_0^{\pi} e^{ikR\cos\alpha\cos\theta} J_m(kR\sin\theta\sin\alpha)\, P_n^m(\cos\alpha)\sin\alpha\, d\alpha.$$

7.8. A Fourier-Bessel Integral.—It will be assumed that the arbitrary function $f(x, y, z)$ and its first derivatives are piecewise continuous and that the integral of the absolute value of the function extended over all space exists. Subject to these conditions the Fourier integral of $f(x, y, z)$ is

$$(70)\quad f(x, y, z) = \left(\frac{1}{2\pi}\right)^{\frac{3}{2}} \int_{-\infty}^{\infty}\int_{-\infty}^{\infty}\int_{-\infty}^{\infty} g(k_1, k_2, k_3) e^{i(k_1x+k_2y+k_3z)}\, dk_1\, dk_2\, dk_3,$$

and its transform is

$$(71)\quad g(k_1, k_2, k_3) = \left(\frac{1}{2\pi}\right)^{\frac{3}{2}} \int_{-\infty}^{\infty}\int_{-\infty}^{\infty}\int_{-\infty}^{\infty} f(x, y, z) e^{-i(k_1x+k_2y+k_3z)}\, dx\, dy\, dz.$$

We now transform these integrals to spherical coordinates under the tacit assumption that a triple integral extended over an infinite cube can be replaced by an integral over a sphere of infinite radius. Proceeding as in Sec. 6.9 and noting that $k_1x + k_2y + k_3z = kR\cos\gamma$, we obtain

$$(72)\quad f(R, \theta, \phi) = \left(\frac{1}{2\pi}\right)^{\frac{3}{2}} \int_0^{\infty}\int_0^{\pi}\int_0^{2\pi} g(\alpha, \beta, k)\, e^{ikR\cos\gamma} k^2 \sin\alpha\, dk\, d\alpha\, d\beta,$$

$$(73)\quad g(\alpha, \beta, k) = \left(\frac{1}{2\pi}\right)^{\frac{3}{2}} \int_0^{\infty}\int_0^{\pi}\int_0^{2\pi} f(R, \theta, \phi)\, e^{-ikR\cos\gamma} R^2 \sin\theta\, dR\, d\theta\, d\phi.$$

Next we shall assume that $f(R, \theta, \phi) = f_n(R) Y_n(\theta, \phi)$ and impose upon the otherwise arbitrary function $f_n(R)$ the condition that it shall be piecewise continuous and have a piecewise continuous first derivative, and that the integral $\int_0^\infty |f_n(R)|\, dR$ shall exist. Then

$$(74)\quad g(\alpha, \beta, k) = \left(\frac{1}{2\pi}\right)^{\frac{3}{2}} \int_0^\infty R^2\, dR\, f_n(R) \int_0^\pi \int_0^{2\pi} Y_n(\theta, \phi) e^{-ikR\cos\gamma} \sin\theta\, d\theta\, d\phi,$$

which in virtue of (61) and the fact that $j_n(-kR) = (-1)^n j_n(kR)$ goes over into

$$(75)\qquad g(\alpha, \beta, k) = (-1)^n Y_n(\alpha, \beta) \sqrt{\frac{2}{\pi}} \int_0^\infty f_n(R)\, j_n(kR) R^2\, dR$$

or $g(\alpha, \beta, k) = (-1)^n Y_n(\alpha, \beta)\, g_n(k)$. Introducing this evaluation of the transform back into (72) and again interchanging the order of integration, we are led to

$$(76)\qquad f_n(R)\, Y_n(\theta, \phi) = \sqrt{\frac{2}{\pi}} \int_0^\infty g_n(k)\, j_n(kR) k^2\, dk\, Y_n(\theta, \phi),$$

and hence to the symmetrical pair of transforms

$$(77)\qquad f(R) = \sqrt{\frac{2}{\pi}} \int_0^\infty g(k)\, j_n(kR) k^2\, dk,$$

$$(78)\qquad g(k) = \sqrt{\frac{2}{\pi}} \int_0^\infty f(R)\, j_n(kR) R^2\, dR,$$

from which the subscript n has been dropped as having no longer any particular significance. In the special case $n = 0$, Eqs. (77) and (78) reduce to the ordinary Fourier integral (18), page 288. In case the chosen value of R coincides with a point of discontinuity in $f(R)$, we write

$$(79)\quad \frac{2}{\pi} \int_0^\infty k^2\, dk \int_0^\infty f(\rho)\, j_n(k\rho)\, j_n(kR)\, \rho^2\, d\rho = \frac{1}{2}[f(R+0) + f(R-0)].$$

7.9. Expansion of a Cylindrical Wave Function.—When calculating the radiation from oscillating current distributions it is sometimes necessary to express a cylindrical wave function in spherical coordinates. The integral representation of wave functions that are finite on the axis is

$$(80)\qquad {\cos \atop \sin} m\phi\, J_m(\lambda r)\, e^{ihz} = \frac{i^{-m}}{2\pi} \int_0^{2\pi} e^{i\lambda r \cos(\beta - \phi) + ihz} {\cos \atop \sin} m\beta\, d\beta.$$

Now

$$(81)\quad \mathbf{k} \cdot \mathbf{R} = \lambda r \cos(\beta - \phi) + ihz = kR[\sin\alpha \sin\theta \cos(\phi - \beta) + \cos\alpha \cos\theta],$$

and in (57), page 409, we have an expansion in the appropriate spherical wave functions of $\exp(i\mathbf{k}\cdot\mathbf{R})$. Upon integration with respect to β, we obtain

$$(82)\quad \begin{matrix}\cos\\ \sin\end{matrix} m\phi\, J_m(\lambda r)\, e^{ihz} = \sum_{n=0}^{\infty} i^{n-m}(2n+1)\frac{(n-m)!}{(n+m)!}\begin{matrix}\cos\\ \sin\end{matrix} m\phi\; P_n^m(\cos\alpha)\, P_n^m(\cos\theta)\, j_n(kR).$$

Since P_n^m vanishes if $m > n$ the first m terms of (82) are zero and the expansion can be written

$$(83)\quad J_m(\lambda r)\, e^{ihz} = \sum_{n=0}^{\infty} i^n(2n+2m+1)\frac{n!}{(n+2m)!} P_{n+m}^m(\cos\alpha)\; P_{n+m}^m(\cos\theta)\, j_{n+m}(kR).$$

When $\alpha = \pi/2$, $h = 0$, $\lambda = k$, and

$$(84)\quad P_{n+m}^m(0) = \begin{cases} 0, & \text{if } n \text{ is odd},\\ \dfrac{(n+2m-1)!}{2^{n+m-1}\left(\frac{n}{2}\right)!\left(\frac{n}{2}+m-1\right)!}, & \text{if } n \text{ is even}.\end{cases}$$

The result is an expansion of a cylindrical Bessel function in a series of spherical Bessel functions.

$$(85)\quad J_m(kr) = \sum_{l=0}^{\infty} \frac{(-1)^l}{2^{2l+m-1}}\,\frac{(4l+2m+1)}{2l+2m}\,\frac{(2l)!}{l!(l+m-1)!} P_{2l+m}^m(\cos\theta)\, j_{2l+m}(kR).$$

7.10. Addition Theorem for $z_0(kR)$.—Let $P(R_0, \theta, \phi)$ be a point of observation and $Q(R_1, \theta_1, \phi_1)$ the source point of a spherical wave. The coordinates R_0, θ, ϕ and R_1, θ_1, ϕ_1 are referred to a fixed coordinate system whose origin is at O. The distance from Q to P is, therefore,

$$R = \sqrt{R_0^2 + R_1^2 - 2R_0R_1\cos\gamma}$$

where $\cos\gamma = \cos\theta\cos\theta_1 + \sin\theta\sin\theta_1\cos(\phi - \phi_1)$. Our problem is to express a wave function referred to Q as a sum of spherical waves whose source is located at O. General treatment of this problem demands a lengthy calculation and its practical importance is insufficient to warrant discussion here. However, in the theory of radiation we shall have occasion to apply the special case in which the wave is spherically symmetric about the source Q. One finds then without great difficulty the expansions:

$$(86)\quad j_0(kR) = \frac{\sin kR}{kR} = \sum_{n=0}^{\infty}(2n+1)P_n(\cos\gamma)\, j_n(kR_0)\, j_n(kR_1),$$

$$(87)\quad h_0^{(1)}(kR) = \frac{e^{ikR}}{ikR} = \begin{cases} \displaystyle\sum_{n=0}^{\infty}(2n+1)P_n(\cos\gamma)\, j_n(kR_0)\, h_n^{(1)}(kR_1), & (R_0 < R_1), \\ \displaystyle\sum_{n=0}^{\infty}(2n+1)P_n(\cos\gamma)\, j_n(kR_1)\, h_n^{(1)}(kR_0), & (R_0 > R_1). \end{cases}$$

The proof is left to the reader.

THE VECTOR WAVE EQUATION IN SPHERICAL COORDINATES

7.11. Spherical Vector Wave Functions.—As in the case of cylindrical coordinates, one can deduce solutions of the vector wave equation in spherical coordinates directly from the characteristic functions of the corresponding scalar equation. Following the notation of the preceding section we shall put $\psi_{\substack{e\\o}mn} = f_{\substack{e\\o}mn} e^{-i\omega t}$, where $f_{\substack{e\\o}mn}$ is the characteristic solution

$$(1)\qquad f_{\substack{e\\o}mn} = \frac{\cos}{\sin} m\phi\, P_n^m(\cos\theta)\, z_n(kR).$$

According to Sec. 7.1, one solution of the vector wave equation

$$(2)\qquad \nabla\nabla\cdot\mathbf{C} - \nabla\times\nabla\times\mathbf{C} + k^2\mathbf{C} = 0$$

can be found simply by taking the gradient of (1). We define $\mathbf{L} = \nabla\psi$, and split off the time factor by writing $\mathbf{L} = \mathbf{l}e^{-i\omega t}$. Then by (95), page 52,

$$(3)\quad \mathbf{l}_{\substack{e\\o}mn} = \frac{\partial}{\partial R} z_n(kR)\, P_n^m(\cos\theta) \frac{\cos}{\sin} m\phi\, \mathbf{i}_1 + \frac{1}{R} z_n(kR) \frac{\partial}{\partial\theta} P_n^m(\cos\theta) \frac{\cos}{\sin} m\phi\, \mathbf{i}_2 \mp \frac{m}{R\sin\theta} z_n(kR)\, P_n^m(\cos\theta) \frac{\sin}{\cos} m\phi\, \mathbf{i}_3,$$

where $\mathbf{i}_1$, $\mathbf{i}_2$, $\mathbf{i}_3$ are the unit vectors defined in Fig. 8, page 52, for a spherical coordinate system.

To obtain the independent solutions **M** and **N** as described in Sec. 7.1, one should introduce a *fixed* vector **a**. Such a procedure in the present instance is of course permissible, but then **M** and **N** will be neither normal nor purely tangential over the entire surface of the sphere. If in place of **a** we use a radial vector $\mathbf{i}_1$, a vector function $\mathbf{L} \times \mathbf{i}_1$ is obtained which is tangent to the sphere; but $\mathbf{i}_1$ is *not* a constant vector and consequently it by no means follows à priori that it can be employed to generate an independent solution. However, we shall discover that a tangential solution **M** can in fact be constructed from a radial vector. Unfortunately the procedure fails in more general coordinate systems.

Let us try to find a solution of (2) in the form of a vector

$$\mathbf{M} = \nabla \times (\mathbf{i}_1 u(R)\psi) = \mathbf{L} \times \mathbf{i}_1 u(R),$$

where $u(R)$ is an unknown scalar function of R. Then if M_1, M_2, M_3 are the R, θ, and ϕ components of $\mathbf{M}$, we have

$$(4)\qquad M_1 = 0, \qquad M_2 = \frac{1}{R\sin\theta}\frac{\partial}{\partial\phi}(u\psi), \qquad M_3 = -\frac{1}{R}\frac{\partial}{\partial\phi}(u\psi).$$

The divergence of $\mathbf{M}$ is zero and we now expand the equation $\nabla \times \nabla \times \mathbf{M} - k^2\mathbf{M} = 0$ into R, θ, and ϕ components by (85), page 50. The R-component is identically zero, whatever $u(R)$. The conditions on the θ and ϕ components are satisfied provided $u(R)$ is such that

$$(5)\qquad \frac{\partial^2}{\partial R^2}(u\psi) + \frac{1}{R^2\sin\theta}\frac{\partial}{\partial\theta}\left[\sin\theta\frac{\partial}{\partial\theta}(u\psi)\right] + \frac{1}{R^2\sin^2\theta}\frac{\partial^2}{\partial\phi^2}(u\psi) + k^2u\psi = 0.$$

(The peculiar advantage of spherical coordinates over more general systems is that the conditions on both tangential components are identical.) If, therefore, we choose $u(R) = R$, Eq. (5) reduces to the required relation

$$(6)\qquad \nabla^2\psi + k^2\psi = 0.$$

Hence in spherical coordinates, (2) is satisfied by

$$(7)\qquad \mathbf{M} = \nabla \times \mathbf{R}\psi = \mathbf{L} \times \mathbf{R} = \frac{1}{k}\nabla \times \mathbf{N},$$

whose components are

$$(8)\qquad M_1 = 0, \qquad M_2 = \frac{1}{\sin\theta}\frac{\partial\psi}{\partial\phi}, \qquad M_3 = -\frac{\partial\psi}{\partial\theta}.$$

From the relation $k\mathbf{N} = \nabla \times \mathbf{M}$ the components of a third solution can be easily found.

$$(9)\qquad kN_1 = \frac{\partial^2(R\psi)}{\partial R^2} + k^2R\psi, \qquad N_2 = \frac{1}{kR}\frac{\partial^2(R\psi)}{\partial R\,\partial\theta}, \qquad N_3 = \frac{1}{kR\sin\theta}\frac{\partial^2(R\psi)}{\partial R\,\partial\phi}.$$

Since ψ must satisfy (4), page 400, the radial component reduces to the simpler form[1]

$$(10)\qquad N_1 = \frac{n(n+1)}{kR}\psi.$$

[1] The first thorough investigations of the electromagnetic problem of the sphere were made by Mie, *Ann. Physik*, **25**, 377, 1908 and Debye, *Ann. Physik*, **30**, 57, 1909. Both writers made use of a pair of potential functions leading directly to our vectors **M** and **N**. The connection between these solutions and a radial Hertzian vector has been pointed out by Sommerfeld in Riemann-Weber, "Differentialgleichungen der Physik," p. 497, 1927.

To obtain explicit expressions for the vector wave functions **M** and **N**, we need only carry out the differentiation of (1) as required by (8) and (9). The time factor is split off by writing $\mathbf{M} = \mathbf{m}e^{-i\omega t}$, $\mathbf{N} = \mathbf{n}e^{-i\omega t}$, and we find

$$(11)\quad \mathbf{m}_{\substack{e\\o}mn} = \mp \frac{m}{\sin\theta} z_n(kR)\, P_n^m(\cos\theta) \frac{\sin}{\cos} m\phi\, \mathbf{i}_2 - z_n(kR) \frac{\partial P_n^m}{\partial\theta} \frac{\cos}{\sin} m\phi\, \mathbf{i}_3,$$

$$(12)\quad \mathbf{n}_{\substack{e\\o}mn} = \frac{n(n+1)}{kR} z_n(kR)\, P_n^m(\cos\theta) \frac{\cos}{\sin} m\phi\, \mathbf{i}_1 + \frac{1}{kR}\frac{\partial}{\partial R}[R\, z_n(kR)] \frac{\partial}{\partial\theta} P_n^m(\cos\theta) \frac{\cos}{\sin} m\phi\, \mathbf{i}_2 \mp \frac{m}{kR\sin\theta}\frac{\partial}{\partial R}[R\, z_n(kR)]\, P_n^m(\cos\theta) \frac{\sin}{\cos} m\phi\, \mathbf{i}_3.$$

7.12. Integral Representations.—The wave functions **l**, **m**, and **n** can be represented as integrals of vector plane waves such as (18) to (20), page 395. For the functions of the first kind, which are finite at the origin, we have by (60), page 410,

$$(13)\quad f^{(1)}_{\substack{e\\o}mn} = \frac{i^{-n}}{4\pi} \int_0^{2\pi}\int_0^{\pi} e^{ikR\cos\gamma} P_n^m(\cos\alpha) \frac{\cos}{\sin} m\beta \sin\alpha\, d\alpha\, d\beta.$$

First, we calculate the components of the gradient; namely, $\partial f/\partial R$, $\frac{1}{R}\frac{\partial f}{\partial\theta}$, $\frac{1}{R\sin\theta}\frac{\partial f}{\partial\phi}$, differentiating under the sign of integration. Next we observe from an inspection of Fig. 70, page 407, that if the vector $\mathbf{k}(\alpha, \beta)$ is directed along the line OQ and $\mathbf{R}(\theta, \phi)$ along the line OP, then

$$(14)\quad \begin{aligned} \mathbf{k}\cdot\mathbf{i}_1 &= k\cos\gamma = k[\sin\alpha\sin\theta\cos(\phi-\beta) + \cos\alpha\cos\theta], \\ \mathbf{k}\cdot\mathbf{i}_2 &= k[\sin\alpha\cos\theta\cos(\phi-\beta) - \cos\alpha\sin\theta], \\ \mathbf{k}\cdot\mathbf{i}_3 &= -k\sin\alpha\sin(\phi-\beta), \end{aligned}$$

the unit spherical vectors $\mathbf{i}_1$, $\mathbf{i}_2$, $\mathbf{i}_3$ referring, of course, to the point of observation at P. Then it follows without further calculation that

$$(15)\quad \mathbf{l}^{(1)}_{\substack{e\\o}mn} = \frac{i^{1-n}}{4\pi} \int_0^{2\pi}\int_0^{\pi} \mathbf{k}(\alpha, \beta) e^{ikR\cos\gamma} P_n^m(\cos\alpha) \frac{\cos}{\sin} m\beta \sin\alpha\, d\alpha\, d\beta.$$

To find the corresponding representations of $\mathbf{m}^{(1)}$ and $\mathbf{n}^{(1)}$ we note that

$$(16)\quad \begin{aligned} \mathbf{R}\times\mathbf{k} &= kR\{\sin\alpha\sin(\phi-\beta)\mathbf{i}_2 + [\sin\alpha\cos\theta\cos(\phi-\beta) - \cos\alpha\sin\theta]\mathbf{i}_3\}, \\ (\mathbf{k}\times\mathbf{R})\times\mathbf{k} &= k^2R\{\sin^2\gamma\,\mathbf{i}_1 - \cos\gamma[\sin\alpha\cos\theta\cos(\phi-\beta) - \cos\alpha\sin\theta]\mathbf{i}_2 + \cos\gamma\sin\alpha\sin(\phi-\beta)\mathbf{i}_3\}, \end{aligned}$$

and, upon differentiating according to (8) and (9), obtain

$$(17)\quad \mathbf{m}^{(1)}_{\substack{e\\o}mn} = \frac{i^{1-n}}{4\pi} \int_0^{2\pi}\int_0^{\pi} \mathbf{k}\times\mathbf{R}\, e^{ikR\cos\gamma} P_n^m(\cos\alpha) \frac{\cos}{\sin} m\beta \sin\alpha\, d\alpha\, d\beta,$$

$$(18)\quad \mathbf{n}^{(1)}_{\substack{e\\o}mn} = \frac{i^{-n}}{4\pi k}\int_0^{2\pi}\int_0^{\pi} [(\mathbf{k}\times\mathbf{R})\times\mathbf{k} + 2i\mathbf{k}]$$

$$e^{ikR\cos\gamma}\, P_n^m(\cos\alpha)\,\frac{\cos}{\sin}\, m\beta \sin\alpha\, d\alpha\, d\beta.$$

To obtain the corresponding integral representations of functions of the third and fourth kinds the integration over α from 0 to π must be replaced by a contour in the complex domain as described on page 410.

Rectangular components of the vectors **l**, **m**, and **n** can be calculated without great difficulty with the help of these integral representations, although the resultant expressions are somewhat long and cumbersome. The vector integrands are resolved into their rectangular components and become explicit expressions of α, β, θ, ϕ. Thus, for example, $k_x = k \sin\alpha \cos\beta$. The recurrence relations for the spherical harmonics are then applied to eliminate these factors and to reduce the integrals to a form that can be evaluated by (13).

7.13. Orthogonality.—The scalar product of any even vector with any odd vector, or with any vector differing in index m, is obviously orthogonal and we need consider only such products as do not vanish when integrated over ϕ from 0 to 2π. Thus

$$(19)\quad \int_0^{2\pi} \mathbf{l}_{\substack{e\\o}mn}\cdot\mathbf{l}_{\substack{e\\o}mn'}\, d\phi = (1+\delta)\pi\left[P_n^m P_{n'}^m \frac{\partial}{\partial R} z_n(kR)\frac{\partial}{\partial R} z_{n'}(kR) + \left(\frac{\partial P_n^m}{\partial\theta}\frac{\partial P_{n'}^m}{\partial\theta} + \frac{m^2}{\sin^2\theta} P_n^m P_{n'}^m\right)\frac{1}{R^2} z_n(kR)\, z_{n'}(kR)\right],$$

where $\delta = 0$ if $m > 0$, $\delta = 1$ if $m = 0$. To reduce (19) we apply the formula

$$(20)\quad \int_0^{\pi}\left(\frac{dP_n^m}{d\theta}\frac{dP_{n'}^m}{d\theta} + \frac{m^2 P_n^m P_{n'}^m}{\sin^2\theta}\right)\sin\theta\, d\theta = \begin{cases} 0, & \text{when } n \neq n', \\ \dfrac{2}{2n+1}\dfrac{(n+m)!}{(n-m)!}\, n(n+1), & \text{when } n = n'. \end{cases}$$

Integration of (19) over θ gives, therefore, zero when $n \neq n'$, or, when $n = n'$,

$$(21)\quad \int_0^{2\pi}\int_0^{\pi} \mathbf{l}_{\substack{e\\o}mn}\cdot\mathbf{l}_{\substack{e\\o}mn} \sin\theta\, d\theta\, d\phi = (1+\delta)\frac{2\pi}{2n+1}\frac{(n+m)!}{(n-m)!}\left\{\left[\frac{\partial\, z_n(kR)}{\partial R}\right]^2 + \frac{n(n+1)}{R^2}[z_n(kR)]^2\right\}.$$

A final reduction follows from the recurrence relations (33) and (34), page 406, and we obtain the normalizing factor

$$(22)\quad \int_0^{2\pi}\int_0^{\pi} \mathbf{l}_{\substack{e\\o}mn} \cdot \mathbf{l}_{\substack{e\\o}mn} \sin\theta\, d\theta\, d\phi$$
$$= (1+\delta)\frac{2\pi}{(2n+1)^2}\frac{(n+m)!}{(n-m)!}k^2\{n[z_{n-1}(kR)]^2 + (n+1)[z_{n+1}(kR)]^2\}.$$

The same formulas lead directly to the integrals

$$(23)\quad \int_0^{2\pi}\int_0^{\pi} \mathbf{m}_{\substack{e\\o}mn} \cdot \mathbf{m}_{\substack{e\\o}mn} \sin\theta\, d\theta\, d\phi$$
$$= (1+\delta)\frac{2\pi}{2n+1}\frac{(n+m)!}{(n-m)!}n(n+1)[z_n(kR)]^2,$$

$$(24)\quad \int_0^{2\pi}\int_0^{\pi} \mathbf{n}_{\substack{e\\o}mn} \cdot \mathbf{n}_{\substack{e\\o}mn} \sin\theta\, d\theta\, d\phi$$
$$= (1+\delta)\frac{2\pi}{(2n+1)^2}\frac{(n+m)!}{(n-m)!}n(n+1)\{(n+1)[z_{n-1}(kR)]^2 + n[z_{n+1}(kR)]^2\},$$

while all products differing in the index n are zero.

Upon examining the cross products, we find first

$$(25)\quad \int_0^{2\pi} \mathbf{l}_{\substack{e\\o}mn} \cdot \mathbf{m}_{\substack{o\\e}mn'}\, d\phi = \pm(1+\delta)\frac{\pi}{R} z_n z_{n'} \frac{m}{\sin\theta}\frac{\partial}{\partial\theta}[P_n^m P_{n'}^m];$$

hence,

$$(26)\quad \int_0^{\pi}\int_0^{2\pi} \mathbf{l}_{\substack{e\\o}mn} \cdot \mathbf{m}_{\substack{o\\e}mn'} \sin\theta\, d\theta\, d\phi = 0$$

for arbitrary values of n and n'. Likewise

$$(27)\quad \int_0^{\pi}\int_0^{2\pi} \mathbf{m}_{\substack{e\\o}mn} \cdot \mathbf{n}_{\substack{o\\e}mn'} \sin\theta\, d\theta\, d\phi = 0.$$

The complete orthogonality of the system is spoiled, as in the cylindrical case, by the product $\mathbf{l}_{\substack{e\\o}mn} \cdot \mathbf{n}_{\substack{e\\o}mn'}$, whose integral over the sphere does not vanish if $n = n'$. In that case we find

$$(28)\quad \int_0^{\pi}\int_0^{2\pi} \mathbf{l}_{\substack{e\\o}mn} \cdot \mathbf{n}_{\substack{e\\o}mn} \sin\theta\, d\theta\, d\phi$$
$$= (1+\delta)\frac{2\pi}{(2n+1)^2}\frac{(n+m)!}{(n-m)!}n(n+1)k\{[z_{n-1}(kR)]^2 - [z_{n+1}(kR)]^2\}.$$

To complete the orthogonalization one might treat k' as a variable parameter and integrate over k' and R as in Sec. 7.2; such a procedure usually proves unnecessary.

7.14. Expansion of a Vector Plane Wave.—To study the diffraction of a plane wave of specified polarization by a spherical object one must first find an expansion of the incident vector wave in terms of the spherical wave functions $\mathbf{l}_{\substack{e\\o}mn}$, $\mathbf{m}_{\substack{e\\o}mn}$, $\mathbf{n}_{\substack{e\\o}mn}$.

Consider the vector function

$$\mathbf{f}(z) = \mathbf{a}\, e^{ikz} = \mathbf{a}\, e^{ikR\cos\theta}, \tag{29}$$

where $\mathbf{a}$ is an amplitude vector oriented arbitrarily with respect to a rectangular reference system. Let $\mathbf{a}$ be resolved into three unit *vectors* directed along the x-, y-, z-axis respectively. Then

$$\begin{aligned} \mathbf{a}_x &= \sin\theta\cos\phi\,\mathbf{i}_1 + \cos\theta\cos\phi\,\mathbf{i}_2 - \sin\phi\,\mathbf{i}_3, \\ \mathbf{a}_y &= \sin\theta\sin\phi\,\mathbf{i}_1 + \cos\theta\sin\phi\,\mathbf{i}_2 + \cos\phi\,\mathbf{i}_3, \\ \mathbf{a}_z &= \cos\theta\,\mathbf{i}_1 - \sin\theta\,\mathbf{i}_2, \end{aligned} \tag{30}$$

where $\mathbf{i}_1$, $\mathbf{i}_2$, $\mathbf{i}_3$ are again the unit vectors defined in Fig. (8), page 52, for a spherical coordinate system. Now the divergence of the vector functions $\mathbf{a}_x\exp(ikz)$ and $\mathbf{a}_y\exp(ikz)$ is zero and consequently they may be expanded in terms of the characteristic functions $\mathbf{m}$ and $\mathbf{n}$ alone. At $R = 0$ the field is finite and we shall, therefore, require functions of the first kind. It is apparent, moreover, that the dependence of (30) on ϕ limits us to $m = 1$; whence upon consideration of the odd and even properties of (11) and (12), we set up the expansion

$$\mathbf{a}_x e^{ikR\cos\theta} = \sum_{n=0}^{\infty} (a_n\mathbf{m}_{o1n}^{(1)} + b_n\mathbf{n}_{e1n}^{(1)}). \tag{31}$$

To determine the coefficients we apply the orthogonality relations of the preceding section.

$$\int_0^{\pi}\int_0^{2\pi} \mathbf{a}_x\cdot\mathbf{m}_{o1n}^{(1)}\, e^{ikR\cos\theta}\sin\theta\, d\theta\, d\phi = 2\pi i^n n(n+1)[j_n(kR)]^2, \tag{32}$$

whence by (23),

$$a_n = \frac{2n+1}{n(n+1)}\, i^n. \tag{33}$$

Likewise

$$\begin{aligned} \int_0^{\pi}\int_0^{2\pi} &\mathbf{a}_x\cdot\mathbf{n}_{e1n}^{(1)}\, e^{ikR\cos\theta}\sin\theta\, d\theta\, d\phi \\ &= -2\pi i^{n+1}\frac{n(n+1)}{2n+1}\{(n+1)[j_{n-1}(kR)]^2 + n[j_{n+1}(kR)]^2\}, \end{aligned} \tag{34}$$

which in virtue of (24) gives

$$b_n = -i^{n+1}\frac{2n+1}{n(n+1)}; \tag{35}$$

hence,

$$\mathbf{a}_x e^{ikz} = \sum_{n=0}^{\infty} i^n\frac{2n+1}{n(n+1)}[\mathbf{m}_{o1n}^{(1)} - i\,\mathbf{n}_{e1n}^{(1)}]. \tag{36}$$

The same process gives for the wave polarized in the y-direction the expansion

$$\mathbf{a}_y e^{ikz} = -\sum_{n=0}^{\infty} i^n \frac{2n+1}{n(n+1)} [\mathbf{m}_{e1n}^{(1)} + i\,\mathbf{n}_{o1n}^{(1)}]. \tag{37}$$

Since the *longitudinal* wave function $\mathbf{a}_z \exp(ikz)$ has a nonvanishing divergence, its expansion must involve the $\mathbf{l}$ functions. It turns out, in fact, that *only* this set is required and one finds without difficulty that

$$\mathbf{a}_z e^{ikz} = \frac{1}{k}\sum_{n=0}^{\infty} i^{n-1}(2n+1)\mathbf{l}_{eon}^{(1)}. \tag{38}$$

Problems

1. Show that the spherical Bessel functions satisfy

$$\int_{-\infty}^{\infty} j_n(\rho) j_m(\rho)\, d\rho = \begin{cases} 0, & \text{when } m \neq n, \\ \dfrac{\pi}{2}\dfrac{1}{2n+1}, & \text{when } n = m, \end{cases}$$

provided n and m are integers satisfying the conditions $n \geq 0$, $m > 0$. For the application of such integrals to the expansion of functions in Bessel series see Watson, "Bessel Functions," page 533.

2. Show that the cylindrical Bessel functions satisfy

$$\frac{d^n}{d(x^2)^n}[x^{-p}J_p(x)] = \left(-\frac{1}{2}\right)^n x^{-p-n}J_{p+n}(x),$$

$$\frac{d^n}{d(x^2)^n}[x^p J_p(x)] = \frac{x^{p-n}}{2^n} J_{p-n}(x),$$

and with the help of these formulas show that the spherical Bessel functions satisfy

$$j_n(x) = (-1)^n (2x)^n \frac{d^n}{d(x^2)^n}\left(\frac{\sin x}{x}\right) = x^n \left(-\frac{1}{x}\frac{d}{dx}\right)^n \left(\frac{\sin x}{x}\right),$$

$$n_n(x) = (-1)^{n-1}(2x)^n \frac{d^n}{d(x^2)^n}\left(\frac{\cos x}{x}\right) = -x^n \left(-\frac{1}{x}\frac{d}{dx}\right)^n \left(\frac{\cos x}{x}\right).$$

3. Show that if $R^2 = R_0^2 + R_1^2 - 2R_0R_1 \cos\gamma$,

$$\frac{1}{2}\int_{-1}^{1} \frac{\sin kR}{kR}\, d(\cos\gamma) = \frac{\sin kR_0}{kR_0}\frac{\sin kR_1}{kR_1},$$

$$\frac{1}{2}\int_{-1}^{1} \frac{\cos kR}{kR}\, d(\cos\gamma) = \begin{cases} \dfrac{\sin kR_0}{kR_0}\dfrac{\cos kR_1}{kR_1}, & R_0 \leq R_1, \\ \dfrac{\cos kR_0}{kR_0}\dfrac{\sin kR_1}{kR_1}, & R_0 \geq R_1. \end{cases}$$

4. Show that in the prolate spheroidal coordinates defined on page 56 the equation

$$\nabla^2\psi + k^2\psi = 0$$

assumes the form

$$\frac{\partial}{\partial \xi}\left[(\xi^2 - 1)\frac{\partial \psi}{\partial \xi}\right] + \frac{\partial}{\partial \eta}\left[(1 - \eta^2)\frac{\partial \psi}{\partial \eta}\right] + \left[\frac{1}{\xi^2 - 1} + \frac{1}{1 - \eta^2}\right]\frac{\partial^2 \psi}{\partial \phi^2} + k^2d^2(\xi^2 - \eta^2)\psi = 0.$$

Let $\psi = \psi_1(\xi)\psi_2(\eta)\psi_3(\phi)$, and show that the above separates into

$$\frac{d}{d\xi}\left[(\xi^2 - 1)\frac{d\psi_1}{d\xi}\right] + \left[k^2d^2\xi^2 - \frac{m^2}{\xi^2 - 1} - C\right]\psi_1 = 0,$$

$$\frac{d}{d\eta}\left[(1 - \eta^2)\frac{d\psi_2}{d\eta}\right] - \left[k^2d^2\eta^2 + \frac{m^2}{1 - \eta^2} - C\right]\psi_2 = 0,$$

$$\frac{d^2\psi_3}{d\phi^2} + m^2\psi_3 = 0,$$

where m and C are separation constants. Next show that a set of prolate spheroidal wave functions can be constructed from

$$\psi_{mn} = (\xi^2 - 1)^{\frac{m}{2}}(1 - \eta^2)^{\frac{m}{2}} Se^1_{mn}(c, \eta)Re_{mn}(c, \xi) \begin{matrix}\cos\\ \sin\end{matrix}\, m\phi$$

in which both the "angular functions" Se and the "radial functions" Re satisfy the equation

$$(1 - z^2)y'' - 2(m + 1)zy' + (b - c^2z^2)y = 0,$$

with m an integer or zero, $b = C - m$, and $c = kd = 2\pi d/\lambda$.

5. When the wave functions of Problem 4 apply to a complete spheroid one must choose $m = 0, 1, 2, \ldots$, and find the separation constant b such that the field is finite at the poles $\eta = \pm 1$. A discrete set of values can be found in the form

$$b_n = n(n + 1) + f_n(c), \qquad n = 0, 1, 2, \ldots.$$

Show that the functions Se and Re can be represented by the expansions

$$Se^1_{mn}(c, z) = \sum_s{}' d_s T^m_s(z), \qquad T^m_s(z) = (1 - z^2)^{-\frac{m}{2}} P^m_{s+m}(z)$$

where P^m_{s+m} is an associated Legendre function and the prime over the summation sign indicates that the sum is to be extended over even values of s if n is even and over odd values if n is odd.

$$Re^1_{mn}(c, z) = \frac{1}{2^m m!(cz)^m}\sum_s{}' i^{s-n}\frac{(2m + s)!}{s!} d_s j_{s+m}(cz)$$

where $j_{s+m}(cz)$ is a spherical Bessel function. The independent solution is obtained by replacing $j_{s+m}(cz)$ by $n_{s+m}(cz)$. Show that the coefficients satisfy the recursion formula

$$\frac{(s + 2m + 2)(s + 2m + 1)}{(2s + 2m + 5)(2s + 2m + 3)}c^2d_{s+2} + \frac{s(s - 1)}{(2s + 2m - 1)(2s + 2m - 3)}c^2d_{s-2} + \left[\frac{2s^2 + 2s(2m + 1) + 2m - 1}{(2s + 2m - 1)(2s + 2m + 3)}c^2 + s(s + 2m + 1) - b\right]d_s = 0$$

and from this relation give a method of determining the characteristic values b_n.

6. Prove that the spheroidal functions defined in Problem 5 satisfy the relations

$$Se^1_{mn} = i^n 2^{m+1} m!\, \lambda_n Re^1_n,$$
$$Se^1_{mn}(c, z) = \lambda_n \int_{-1}^{1} e^{icst}(1 - t^2)^m Se^1_{mn}(c, t)\, dt$$

where λ_n is one of a discrete set of characteristic values.

7. Write the equation $\nabla^2\psi + k^2\psi$ in oblate spheroidal coordinates and express the solutions in terms of the functions Se and Re defined in Problem 5.

8. A system of coordinates ξ, η, ϕ with rotational symmetry is defined by

$$ds^2 = h_1\, d\xi^2 + h_2\, d\eta^2 + r^2\, d\phi^2$$

where r is the perpendicular distance from the axis of rotation. If the field has the same symmetry as the coordinate system, its components are independent of ϕ. Show that the equations

$$\nabla \times \mathbf{E} - i\omega\mu\mathbf{H} = 0, \qquad \nabla \times \mathbf{H} + (i\omega\epsilon - \sigma)\mathbf{E} = 0$$

break up into two independent groups:

$$\text{(I)} \qquad \begin{aligned} &\frac{1}{rh_2}\frac{\partial}{\partial\eta}(rE_\phi) - i\omega\mu H_\xi = 0, \\ &\frac{1}{rh_1}\frac{\partial}{\partial\xi}(rE_\phi) + i\omega\mu H_\eta = 0, \\ \frac{1}{h_1h_2}&\left[\frac{\partial}{\partial\xi}(h_2H_\eta) - \frac{\partial}{\partial\eta}(h_1H_\xi)\right] + (i\omega\epsilon - \sigma)E_\phi = 0. \end{aligned}$$

$$\text{(II)} \qquad \begin{aligned} &\frac{1}{rh_2}\frac{\partial}{\partial\eta}(rH_\phi) + (i\omega\epsilon - \sigma)E_\xi = 0, \\ &\frac{1}{rh_1}\frac{\partial}{\partial\xi}(rH_\phi) - (i\omega\epsilon - \sigma)E_\eta = 0, \\ \frac{1}{h_1h_2}&\left[\frac{\partial}{\partial\xi}(h_2E_\eta) - \frac{\partial}{\partial\eta}(h_1E_\xi)\right] - i\omega\mu H_\phi = 0. \end{aligned}$$

Show that (II) is satisfied by a potential Q such that

$$Q = rH_\phi, \qquad E_\xi = \frac{i\omega\mu}{rh_2k^2}\frac{\partial Q}{\partial\eta}, \qquad E_\eta = -\frac{i\omega\mu}{rh_1k^2}\frac{\partial Q}{\partial\xi},$$
$$\frac{\partial}{\partial\xi}\left(\frac{h_2}{rh_1}\frac{\partial Q}{\partial\xi}\right) + \frac{\partial}{\partial\eta}\left(\frac{h_1}{rh_2}\frac{\partial Q}{\partial\eta}\right) + k^2\frac{h_1h_2}{r}Q = 0.$$

(I) can be integrated in the same manner. (Abraham)

9. Apply the method of Problem 8 to find a rotationally symmetric electromagnetic field in prolate spheroidal coordinates. In the case of electric oscillations directed along meridian lines show that Q satisfies

$$(\xi^2 - 1)\frac{\partial^2 Q}{\partial\xi^2} + (1 - \eta^2)\frac{\partial^2 Q}{\partial\eta^2} + d^2k^2(\xi^2 - \eta^2)Q = 0,$$

which separates into

$$(\xi^2 - 1)Q_1'' + (d^2k^2\xi^2 - C)Q_1 = 0,$$
$$(1 - \eta^2)Q_2'' + (-d^2k^2\eta^2 + C)Q_2 = 0,$$

where $Q = Q_1(\xi)Q_2(\eta)$. Note that both of these are a special case of the equation

$$(1 - z^2)y'' - 2(a + 1)zy' + (b - c^2z^2)y = 0$$

when $a = -1$. Show that the field is given by

$$E_\xi = \frac{i\omega\mu}{d^2k^2} \frac{Q_1}{\sqrt{(\xi^2 - 1)(\xi^2 - \eta^2)}} \frac{dQ_2}{d\eta},$$

$$E_\eta = -\frac{i\omega\mu}{d^2k^2} \frac{Q_2}{\sqrt{(1 - \eta^2)(\xi^2 - \eta^2)}} \frac{dQ_1}{d\xi}.$$

10. Two types of spherical electromagnetic waves have been discussed in this chapter: the transverse electric for which $E_R = 0$, and the transverse magnetic for which $H_R = 0$. Obtain expressions for the radial wave impedances as was done previously for cylindrical waves.

11. Prove the expansions (86) and (87) of Sec. 7.10 and show that when $\theta_1 = \pi$, $R_1 \to \infty$, Eq. (87) goes over asymptotically into the expansion of a plane wave.

12. Obtain expressions for the rectangular components of the vector spherical wave functions **l**, **m**, and **n** by the method suggested at the end of Sec. 7.12.

CHAPTER VIII

RADIATION

In the course of the past three chapters we have studied the propagation of electromagnetic fields with no concern for the manner in which they are established. We consider now the sources, and the fundamental problem of determining the intensity and structure of a field generated by a given distribution of charge and current.

THE INHOMOGENEOUS SCALAR WAVE EQUATION

8.1. Kirchhoff Method of Integration.—Mathematically the problem of relating a field to its source is that of integrating an inhomogeneous differential equation. Let ψ represent a scalar potential or any rectangular component of a field vector, and let $g(x, y, z, t)$ be the density of the source function. We shall assume that throughout a domain V the medium containing the source is homogeneous and isotropic, and that *its conductivity is zero.* The presence of conductivity introduces serious analytical difficulties which can be avoided in most practical applications of the theory. The effect of conducting bodies in the neighborhood of oscillating sources will be treated as a boundary-value problem in Chap. IX. Subject to these restrictions, the scalar function ψ satisfies the equation

$$\nabla^2\psi - \frac{1}{v^2}\frac{\partial^2\psi}{\partial t^2} = -g(x, y, z, t), \tag{1}$$

where $v = (\epsilon\mu)^{-\frac{1}{2}}$ is the phase velocity.

The Kirchhoff theory of integration is an extension to the wave equation of a method applied in Sec. 3.3 to Poisson's equation. Let V be a closed domain bounded by a regular surface S and let ϕ and ψ be any two scalar functions which with their first and second derivatives are continuous throughout V and on S. Then by (7), page 165,

$$\int_V (\phi\nabla^2\psi - \psi\nabla^2\phi)\,dv = \int_S \left(\phi\frac{\partial\psi}{\partial n} - \psi\frac{\partial\phi}{\partial n}\right) da, \tag{2}$$

where $\partial/\partial n$ denotes differentiation in the direction of the positive, or outward, normal. Let (x', y', z') be a fixed point of observation within V and

$$r = \sqrt{(x' - x)^2 + (y' - y)^2 + (z' - z)^2} \tag{3}$$

the distance from a variable point (x, y, z) within V or on S to the fixed point. For ϕ we shall choose the spherically symmetric solution

$$\phi = \frac{1}{r} f\left(t + \frac{r}{v}\right) \tag{4}$$

of the homogeneous equation

$$\nabla^2 \phi - \frac{1}{v^2} \frac{\partial^2 \phi}{\partial t^2} = 0, \tag{5}$$

where $f\left(t + \frac{r}{v}\right)$ is a completely arbitrary analytic function of the argument $t + \frac{r}{v}$.

As in the analogous problem of the stationary field, ϕ has a singularity at $r = 0$; this point must, therefore, be excluded from the domain V by a small sphere S_1 of radius r_1 drawn about (x', y', z') as a center. The volume V is now bounded externally by S and internally by S_1. Furthermore, if we denote the left side of (2) by I and make use of (1) and (5), we obtain

$$I = -\int_V \phi g \, dv + \frac{1}{v^2} \int_V \left(\phi \frac{\partial^2 \psi}{\partial t^2} - \psi \frac{\partial^2 \phi}{\partial t^2}\right) dv. \tag{6}$$

Or, since $\phi \frac{\partial^2 \psi}{\partial t^2} - \psi \frac{\partial^2 \phi}{\partial t^2} = \frac{\partial}{\partial t}\left(\phi \frac{\partial \psi}{\partial t} - \psi \frac{\partial \phi}{\partial t}\right)$,

$$I = -\int_V \phi g \, dv + \frac{1}{v^2} \frac{\partial}{\partial t} \int_V \left(\phi \frac{\partial \psi}{\partial t} - \psi \frac{\partial \phi}{\partial t}\right) dv. \tag{7}$$

Upon integrating over the entire duration of time this gives

$$\int_{-\infty}^{\infty} I \, dt = -\int_{-\infty}^{\infty} dt \int_V \phi g \, dv + \frac{1}{v^2} \int_V \left(\phi \frac{\partial \psi}{\partial t} - \psi \frac{\partial \phi}{\partial t}\right) dv \Bigg|_{t=-\infty}^{t=+\infty}. \tag{8}$$

The next task is to choose the function $f\left(t + \frac{r}{v}\right)$ in such a way that the last term of (8) vanishes. To this end let $t' = r/v$ and take for $f(t + t')$ the *impulse function*

$$f(t + t') = \frac{1}{\sqrt{2\pi}} \frac{e^{\frac{-(t+t')^2}{2\delta^2}}}{\delta} = S_0(t + t'), \tag{9}$$

defined by (35), page 291, with the property

$$\int_{-\infty}^{\infty} f(t + t') \, d(t + t') = 1. \tag{10}$$

To avoid any question as to the continuity of ϕ and its derivatives, we shall imagine δ to be exceedingly small but shall not pass at once to the limit $\delta = 0$. Thus defined, $S_0(t + t')$ vanishes everywhere outside an infinitesimal interval in the neighborhood of $t = -t'$. With this in mind it is apparent that if $F(t)$ is an arbitrary function of time, then

$$F(-t') = \int_{-\infty}^{\infty} S_0(t + t')F(t)\,dt, \tag{11}$$

and that $\partial f/\partial t$ approaches zero with vanishing δ for all values of t.

Since now ϕ and $\partial\phi/\partial t$ vanish at the limits $t = -\infty$ and $t = \infty$[1] for all finite values of r, it follows that the last term of (8) must likewise vanish and, consequently, that

$$\begin{aligned}\int_{-\infty}^{\infty} I\,dt &= -\int_V dv \int_{-\infty}^{\infty} S_0(t + t')\,\frac{g(x, y, z, t)}{r}\,dt \\ &= -\int_V \frac{1}{r}\,g(x, y, z, -t')\,dv.\end{aligned} \tag{12}$$

When equated to the right-hand side of (2), this leads to

$$\begin{aligned}\int_V \frac{1}{r}\,g\,dv + \int_{-\infty}^{\infty} dt \int_S \left(\phi\frac{\partial\psi}{\partial n} - \psi\frac{\partial\phi}{\partial n}\right) da \\ + \int_{-\infty}^{\infty} dt \int_{S_1} \left(\phi\frac{\partial\psi}{\partial n} - \psi\frac{\partial\phi}{\partial n}\right) da = 0.\end{aligned} \tag{13}$$

Over the surface S_1 we know that $\partial/\partial n = -\partial/\partial r$, so that

$$\begin{aligned}&\int_{S_1} \left(\phi\frac{\partial\psi}{\partial n} - \psi\frac{\partial\phi}{\partial n}\right) da \\ &= \left\{\frac{S_0\left(t + \frac{r_1}{v}\right)}{r_1^2} + \frac{\left[\frac{\partial}{\partial r}S_0\left(t + \frac{r}{v}\right)\right]_{r=r_1}}{r_1}\right\} \int_{S_1} \psi\,da \\ &\qquad - S_0\left(t + \frac{r_1}{v}\right) \int_{S_1} \left(\frac{\partial\psi}{\partial r}\right)_{r=r_1} da.\end{aligned} \tag{14}$$

Passing to the limit as $r_1 \to 0$, this reduces to

$$\int_{S_1} \left(\phi\frac{\partial\psi}{\partial n} - \psi\frac{\partial\phi}{\partial n}\right) da = -4\pi S_0(t)\,\psi(x', y', z', t), \tag{15}$$

and hence by (11),

$$\int_{-\infty}^{\infty} dt \int_{S_1} \left(\phi\frac{\partial\psi}{\partial n} - \psi\frac{\partial\phi}{\partial n}\right) da = -4\pi\psi(x', y', z', 0). \tag{16}$$

[1] Obviously these limits may now be drawn in to enclose an infinitesimal interval containing the instant $t = -r/v$.

Substituting this value into (13) gives

$$\psi(x', y', z', 0) = \frac{1}{4\pi}\int_V \frac{1}{r} g(x, y, z, -t')\, dv + \frac{1}{4\pi}\int_S \int_{-\infty}^{\infty} \left[\frac{1}{r} S_0(t + t') \frac{\partial\psi}{\partial n} - \psi \frac{\partial}{\partial n} \frac{S_0(t + t')}{r}\right] dt\, da. \tag{17}$$

To reduce these integrals to a simpler form we note that

$$\psi \frac{\partial}{\partial n}\left[\frac{S_0(t + t')}{r}\right] = S_0(t + t')\psi \frac{\partial}{\partial n}\left(\frac{1}{r}\right) + \frac{\psi}{r}\frac{\partial}{\partial n} S_0\, (t + t'), \tag{18}$$

and that

$$\frac{\partial}{\partial n} S_0\left(t + \frac{r}{v}\right) = \frac{\partial}{\partial u} S_0(u) \frac{\partial u}{\partial n} = \frac{1}{v}\frac{\partial r}{\partial n}\frac{\partial S_0}{\partial t}. \tag{19}$$

An integration by parts gives

$$\int_{-\infty}^{\infty} \frac{\psi}{r}\frac{\partial S_0(t + t')}{\partial n}\, dt = -\frac{1}{v}\frac{1}{r}\frac{\partial r}{\partial n}\int_{-\infty}^{\infty} S_0(t + t')\frac{\partial\psi}{\partial t}\, dt, \tag{20}$$

and application of (11) leads to

$$\psi(x', y', z', 0) = \frac{1}{4\pi}\int_V \frac{1}{r} g(x, y, z, -t')\, dv + \frac{1}{4\pi}\int_S \left[\frac{1}{r}\left(\frac{\partial\psi}{\partial n}\right)_{t=-t'} - \frac{\partial}{\partial n}\left(\frac{1}{r}\right)\psi_{t=-t'} + \frac{1}{vr}\frac{\partial r}{\partial n}\left(\frac{\partial\psi}{\partial t}\right)_{t=-t'}\right] da. \tag{21}$$

According to (21) the value of ψ at the point x', y', z' and the instant $t = 0$ is obtained by summing the contributions of elements whose phases have been retarded by an amount $t' = r/v$. But the location of the reference point on the time axis is purely arbitrary; consequently, the Kirchhoff formula becomes

$$\psi(x', y', z', t) = \frac{1}{4\pi}\int_V \frac{1}{r}\, [g]\, dv + \frac{1}{4\pi}\int_S \left[\frac{1}{r}\left[\frac{\partial\psi}{\partial n}\right] - \frac{\partial}{\partial n}\left(\frac{1}{r}\right)[\psi] + \frac{1}{vr}\frac{\partial r}{\partial n}\left[\frac{\partial\psi}{\partial t}\right]\right] da, \tag{22}$$

in which the symbol

$$[g] = g\left(x, y, z, t - \frac{r}{v}\right) \tag{23}$$

will denote a function with retarded phase.

The volume integral in (22) is a *particular solution* of the inhomogeneous wave equation representing physically the contribution to $\psi(x', y', z', t)$ of all sources contained within V. To this particular solution is added a *general solution* of the *homogeneous* equation (5) expressed as a surface integral extended over S and accounting for all sources located outside S. In case the values of ψ and its derivatives are known over S, the field is completely determined at all interior points, but it is no more permissible to assign these values arbitrarily than in the static case. We shall have occasion below to discuss this point further in connection with the Huygens principle.

The behavior of ψ at infinity is no longer obvious since r enters explicitly into $[g]$ through the retarded time variable $t - \frac{r}{v}$. The field, however, is propagated with a finite velocity; consequently, if all sources are located within a finite distance of some fixed point of reference and if they have been established within some finite period in the past, one may allow the surface S to recede beyond the first wave front. It lies then entirely within a region unreached by the disturbance at time t and within which ψ and its derivatives are zero. In this case

$$\psi(x', y', z', t) = \frac{1}{4\pi} \int_V \frac{1}{r} g\left(x, y, z, t - \frac{r}{v}\right) dv, \tag{24}$$

where V now represents the entire volume occupied by sources.

The reader has doubtless noted that the choice of the function $f\left(t + \frac{r}{v}\right)$ is highly arbitrary, since the homogeneous equation (5) admits also a solution $\frac{1}{r} f\left(t - \frac{r}{v}\right)$. This function leads obviously to an *advanced time*, implying that the quantity ψ can be observed *before* it has been generated by the source. The familiar chain of cause and effect is thus reversed and this alternative solution might be discarded as logically inconceivable. However the application of "logical" causality principles offers very insecure footing in matters such as these and we shall do better to restrict the theory to retarded action solely on the grounds that this solution alone conforms to the present physical data.

8.2. Retarded Potentials.—When (24) is applied to the scalar potential and to the rectangular components of the vector potential we obtain the formulas

$$\phi(x', t) = \frac{1}{4\pi\epsilon} \int_V \frac{1}{r} \rho(x, t^*) \, dv, \tag{25}$$

$$\mathbf{A}(x', t) = \frac{\mu}{4\pi} \int_V \frac{1}{r} \mathbf{J}(x, t^*) \, dv, \tag{26}$$

which enable one to calculate the field from a given distribution of charge and current. Here x' represents the coordinate triplet x', y', z', and x the triplet x, y, z, while the retarded time is expressed by $t^* = t - \frac{r}{v}$.

This statement, however, is subject to one condition. We have in fact only shown that (25) and (26) are solutions of the inhomogeneous wave equation. In order that they shall also represent potentials of an electromagnetic field it is necessary that they satisfy the relation

$$\nabla' \cdot \mathbf{A} + \mu\epsilon \frac{\partial \phi}{\partial t} = 0. \tag{27}$$

The prime attached to the operator ∇ denotes differentiation with respect to x', y', z' at the point of observation.

We shall show that this condition is fulfilled by (25) and (26) provided the densities of charge and current satisfy an equation of continuity. Since $\partial/\partial t = \partial/\partial t^*$ at a fixed point in space we have

$$\frac{\partial \phi}{\partial t} = \frac{1}{4\pi\epsilon} \int_V \frac{1}{r} \frac{\partial \rho(x, t^*)}{\partial t^*}\, dv, \tag{28}$$

$$\nabla' \cdot \mathbf{A} = \frac{\mu}{4\pi} \int_V \nabla' \cdot \frac{\mathbf{J}(x, t^*)}{r}\, dv = \frac{\mu}{4\pi} \int_V \left[\mathbf{J} \cdot \nabla' \left(\frac{1}{r}\right) + \frac{1}{r} \nabla' \cdot \mathbf{J}\right] dv. \tag{29}$$

Now when the operator ∇ is applied to any power of r, one may write $\nabla r^n = -\nabla' r^n$. The variables x' occur in $\mathbf{J}(x, t^*)$ only through $t^* = t - \frac{r}{v}$; consequently,

$$\nabla' \cdot \mathbf{J} = \frac{\partial \mathbf{J}}{\partial t^*} \cdot \nabla' t^* = -\frac{1}{v} \frac{\partial \mathbf{J}}{\partial t^*} \cdot \nabla' r = \frac{1}{v} \frac{\partial \mathbf{J}}{\partial t^*} \cdot \nabla r. \tag{30}$$

On the other hand

$$\nabla \cdot \mathbf{J} = -\frac{1}{v} \frac{\partial \mathbf{J}}{\partial t^*} \cdot \nabla r + (\nabla \cdot \mathbf{J})_{t^* = \text{constant}}, \tag{31}$$

so that

$$\nabla' \cdot \mathbf{J} = (\nabla \cdot \mathbf{J})_{t^* = \text{constant}} - \nabla \cdot \mathbf{J}. \tag{32}$$

Equation (29) can now be written

$$\nabla' \cdot \mathbf{A} = -\frac{\mu}{4\pi} \int_V \nabla \cdot \left(\frac{\mathbf{J}}{r}\right) dv + \frac{\mu}{4\pi} \int_V \frac{1}{r} (\nabla \cdot \mathbf{J})_{t^* = \text{constant}}\, dv. \tag{33}$$

By the divergence theorem

$$\int_V \nabla \cdot \left(\frac{\mathbf{J}}{r}\right) dv = \int_S \frac{\mathbf{J}}{r} \cdot \mathbf{n}\, da = 0, \tag{34}$$

since the closed surface S bounding V can be taken so large that $\mathbf{J}$ is zero everywhere over S. Then on combining (28) and (33), we find

$$\frac{1}{\mu} \nabla' \cdot \mathbf{A} + \epsilon \frac{\partial \phi}{\partial t} = \frac{1}{4\pi} \int_V \frac{1}{r} \left[(\nabla \cdot \mathbf{J})_{t^* = \text{constant}} + \frac{\partial \rho}{\partial t^*}\right] dv, \tag{35}$$

and this must vanish since the conservation of charge requires that

$$(\nabla \cdot \mathbf{J})_{t^*=\text{constant}} + \frac{\partial \rho}{\partial t^*} = 0 \tag{36}$$

at any point whose local time is t^*.

8.3. Retarded Hertz Vector.—By virtue of (36) it is sufficient to prescribe the current distribution in space and time. The charge density can then be calculated and the potentials determined by evaluating the integrals (25) and (26). However, in most cases it is advantageous to derive the field from a single source function by means of a Hertz vector.

Let us express the current and charge densities in terms of a single vector **P** defined by

$$\mathbf{J} = \frac{\partial \mathbf{P}}{\partial t}, \qquad \rho = -\nabla \cdot \mathbf{P}. \tag{37}$$

The equation of continuity is then satisfied identically and the field equations are

$$\text{(I)}\ \nabla \times \mathbf{E} + \mu \frac{\partial \mathbf{H}}{\partial t} = 0, \qquad \text{(III)}\ \nabla \cdot \mathbf{H} = 0,$$

$$\text{(II)}\ \nabla \times \mathbf{H} - \epsilon \frac{\partial \mathbf{E}}{\partial t} = \frac{\partial \mathbf{P}}{\partial t}, \qquad \text{(IV)}\ \nabla \cdot \mathbf{E} = -\frac{1}{\epsilon} \nabla \cdot \mathbf{P}.$$

As shown in Sec. 1.11, these equations are satisfied by

$$\mathbf{E} = \nabla\nabla \cdot \mathbf{\Pi} - \mu\epsilon \frac{\partial^2 \mathbf{\Pi}}{\partial t^2}, \qquad \mathbf{H} = \epsilon \nabla \times \frac{\partial \mathbf{\Pi}}{\partial t}, \tag{38}$$

where $\mathbf{\Pi}$ is any solution of

$$\nabla \times \nabla \times \mathbf{\Pi} - \nabla\nabla \cdot \mathbf{\Pi} + \mu\epsilon \frac{\partial^2 \mathbf{\Pi}}{\partial t^2} = \frac{1}{\epsilon} \mathbf{P}. \tag{39}$$

The rectangular components of $\mathbf{\Pi}$, therefore, satisfy the inhomogeneous wave equation

$$\nabla^2 \Pi_j - \mu\epsilon \frac{\partial^2 \Pi_j}{\partial t^2} = -\frac{1}{\epsilon} P_j, \qquad (j = 1, 2, 3). \tag{40}$$

It need scarcely be said that the vector **P** defined by (37) is *not* identical with the dielectric polarization defined first in Sec. 1.6. The vector **P** of the present section measures the polarization, or moment per unit volume, of the *free-charge* distribution and consequently is not equal to $\mathbf{D} - \epsilon_0 \mathbf{E}$.

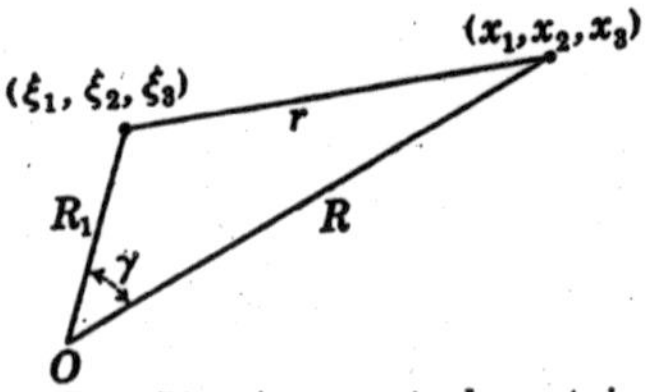

FIG. 71.—A current element is located at ξ_1, ξ_2, ξ_3, while x_1, x_2, x_3 locate a fixed point of observation.

To facilitate the calculations that follow a small change in notation will be made. The point of observation in Fig. 71 whose radial distance from a fixed origin O is R will now be located by the rectangular coordinates (x_1, x_2, x_3). The current element located a distance

R_1 from O has the coordinates (ξ_1, ξ_2, ξ_3) and its distance from the point of observation is r, so that

$$r = \sqrt{\sum_{j=1}^{3} (x_j - \xi_j)^2}. \tag{41}$$

We need consider only the harmonic components of the variation, from which general transient or steady-state solutions may be constructed. Let

$$\mathbf{J} = \mathbf{J}_0(\xi)\, e^{-i\omega t}, \qquad \rho = \rho_0(\xi)\, e^{-i\omega t}, \qquad \rho_0 = \frac{1}{i\omega} \nabla \cdot \mathbf{J}_0(\xi), \tag{42}$$

$$\mathbf{P} = \mathbf{P}_0(\xi)\, e^{-i\omega t}, \qquad \mathbf{P}_0 = \frac{i}{\omega} \mathbf{J}_0. \tag{43}$$

Applying (24) to (40) and noting that in the present case $k = \omega\sqrt{\mu\epsilon} = \omega/v$, we obtain the fundamental formula

$$\boxed{\mathbf{\Pi}(x, t) = \frac{e^{-i\omega t}}{4\pi\epsilon} \int_V \mathbf{P}_0(\xi) \frac{e^{ikr}}{r}\, dv.} \tag{44}$$

A MULTIPOLE EXPANSION

8.4. Definition of the Moments.—To evaluate the retarded Hertz integral one must frequently resort to some form of series expansion. The nature of this expansion will be governed by the frequency and the geometry of the current distribution. In the present section we shall consider an extension to variable fields of the theory of multipoles set forth in Secs. 3.8 to 3.12 and Secs. 4.4 to 4.7 for electrostatic and magnetostatic fields.

According to (87), page 414, when $R > R_1$, the integrand of (44) above can be expanded in the series

$$\frac{e^{ikr}}{r} = ik \sum_{n=0}^{\infty} (2n+1) P_n(\cos\gamma)\, j_n(kR_1)\, h_n^{(1)}(kR); \tag{1}$$

consequently, if R is greater than the radius of a sphere containing the entire source distribution,

$$\mathbf{\Pi}(x, t) = \frac{ik}{4\pi\epsilon} e^{-i\omega t} \sum_{n=0}^{\infty} (2n+1) h_n^{(1)}(kR) \int_V \mathbf{P}_0(\xi)\, j_n(kR_1)\, P_n(\cos\gamma)\, dv. \tag{2}$$

Suppose now that $\mathbf{P}_0(\xi)$ is expressed as a function of R_1, γ, ϕ, where ϕ is the equatorial angle about an axis drawn from O through the point of observation. Then the vector $\mathbf{P}_0(R_1, \gamma, \phi)$ can be resolved into scalar components and each component expanded in a series of spherical harmonics as in (36), page 406. In virtue of the ortnogonality of the functions $P_n(\cos\gamma)$, the integral on the right reduces essentially to an expansion coefficient times $[j_n(kR_1)]^2$. The success of this attack will be determined by the nature of the current distribution and the difficulties encountered in computing the expansion coefficients.

If the wave length of the oscillating field is many times greater than the largest dimension of the region occupied by current, the series (2) can be interpreted in terms of electric and magnetic multipoles located at the origin. We shall assume that for all values of R_1 within V

$$kR_1 = \frac{2\pi R_1}{\lambda} \ll 1, \tag{3}$$

where λ is the wave length in the medium defined by ϵ and μ. Consequently the function $j_n(kR_1)$ can be replaced by the first term of its series expansion. By (25), page 405,

$$j_n(kR_1) \simeq \frac{2^n n!}{(2n+1)!}(kR_1)^n. \tag{4}$$

The total field will be represented as the sum of partial fields by

$$\mathbf{\Pi}(x, t) = \sum_{n=0}^{\infty} \mathbf{\Pi}^{(n)}(x, t), \tag{5}$$

and, hence,

$$\mathbf{\Pi}^{(n)}(x, t) = \frac{i}{4\pi\epsilon}\frac{2^n n!}{(2n)!}k^{n+1}e^{-i\omega t}h_n^{(1)}(kR)\int_V \mathbf{P}_0(\xi)\,R_1^n P_n(\cos\gamma)\,dv. \tag{6}$$

Since $P_0(\cos\gamma) = 1$ and $h_0^{(1)}(kR) = -\frac{i}{kR}e^{ikR}$, we have at once for the first term.

$$\mathbf{\Pi}^{(0)} = \frac{1}{4\pi\epsilon}\mathbf{p}^{(1)}\frac{e^{ikR-i\omega t}}{R}, \qquad \mathbf{p}^{(1)} = \int_V \mathbf{P}_0(\xi)\,dv. \tag{7}$$

Let ψ be any scalar function of position. By the divergence theorem

$$\int_V \nabla\cdot(\psi\mathbf{P}_0)\,dv = \int_S \psi\mathbf{P}_0\cdot\mathbf{n}\,da, \tag{8}$$

and, hence,

$$\int_V \mathbf{P}_0\cdot\nabla\psi\,dv = -\int_V \psi\nabla\cdot\mathbf{P}_0 + \int_S \psi\mathbf{P}_0\cdot\mathbf{n}\,da. \tag{9}$$

If S encloses the entire source distribution, $\mathbf{P}_0\cdot\mathbf{n}$ will be zero over S and the surface integral will vanish. Choose now for ψ a rectangular component ξ_j of the radius vector $\mathbf{R}_1$. Then

$$\int_V P_{0j}\,dv = -\int \xi_j\nabla\cdot\mathbf{P}_0\,dv; \tag{10}$$

consequently

$$\mathbf{p}^{(1)} = \int_V \mathbf{P}_0\, dv = \int_V \mathbf{R}_1 \rho_0\, dv. \tag{11}$$

The partial field $\mathbf{\Pi}^{(0)}$ *is generated by an oscillating electric dipole whose moment* $\mathbf{p}^{(1)}$ *is defined exactly as in the electrostatic case, Eq.* (47), *page* 179.

To obtain the second term of (6) we note that $P_1(\cos\gamma) = \cos\gamma$, $h_1^{(1)}(kR) = -\frac{1}{kR}\left(1 + \frac{i}{kR}\right) e^{ikR}$. Hence,

$$\mathbf{\Pi}^{(1)} = -\frac{ik}{4\pi\epsilon}\left(\frac{1}{R} + \frac{i}{kR^2}\right) e^{ikR-i\omega t} \int_V \mathbf{P}_0(\xi)\, R_1 \cos\gamma\, dv. \tag{12}$$

To interpret the integral we first expand the integrand.

$$\begin{aligned} \mathbf{P}_0 R_1 \cos\gamma &= (\mathbf{R}_1 \cdot \mathbf{R}) \frac{\mathbf{P}_0}{R} \\ &= \frac{1}{2R}[(\mathbf{R}_1 \times \mathbf{P}_0) \times \mathbf{R} + \mathbf{P}_0(\mathbf{R}_1 \cdot \mathbf{R}) + \mathbf{R}_1(\mathbf{R} \cdot \mathbf{P}_0)]. \end{aligned} \tag{13}$$

The *magnetic dipole moment* was defined in (28), page 235, as

$$\mathbf{m}^{(1)} = \frac{1}{2}\int_V \mathbf{R}_1 \times \mathbf{J}_0\, dv. \tag{14}$$

Since $\mathbf{R}$ is independent of the variables ξ of integration, one has, therefore,

$$\frac{1}{2R}\int_V (\mathbf{R}_1 \times \mathbf{P}_0) \times \mathbf{R}\, dv = \frac{i}{\omega R}\mathbf{m}^{(1)} \times \mathbf{R}. \tag{15}$$

The *electric quadrupole moments* of a charge distribution were defined in (48), page 179, as the components of a tensor,

$$p_{ij} = \int_V \xi_i \xi_j \rho_0\, dv = -\int_V \xi_i \xi_j \nabla \cdot \mathbf{P}_0\, dv. \tag{16}$$

Upon putting $\psi = \xi_i \xi_j$ in (9) this gives

$$p_{ij} = \int (\xi_j \mathbf{P}_{0i} + \xi_i P_{0j})\, dv = p_{ji}. \tag{17}$$

Furthermore,

$$\mathbf{R}_1 \cdot \mathbf{R} = R \sum_{j=1}^{3} \xi_j \frac{\partial R}{\partial x_j}, \qquad \mathbf{P}_0 \cdot \mathbf{R} = R \sum_{j=1}^{3} P_{0j} \frac{\partial R}{\partial x_j}. \tag{18}$$

Consequently,

$$\begin{aligned} &\frac{1}{2R}\int_V [P_{0i}(\mathbf{R}_1 \cdot \mathbf{R}) + \xi_i(\mathbf{P}_0 \cdot \mathbf{R})]\, dv \\ &\qquad = \frac{1}{2}\sum_{j=1}^{3} \frac{\partial R}{\partial x_j}\int_V (\xi_j P_{0i} + \xi_i P_{0j})\, dv = \frac{1}{2}\sum_{j=1}^{3} p_{ij}\frac{\partial R}{\partial x_j} = \frac{1}{2}p_i^{(2)}. \end{aligned} \tag{19}$$

The quantities $p_i^{(2)}$ are the components of a vector

$$\mathbf{p}^{(2)} = {}^2\mathbf{p} \cdot \nabla R, \tag{20}$$

where ${}^2\mathbf{p}$ is the tensor whose components are p_{ij}. [*Cf.* Eqs. (23), (24), page 100.]

The partial field $\mathbf{\Pi}^{(1)}$ *represents the contributions of a magnetic dipole and an electric quadrupole.*

$$\mathbf{\Pi}^{(1)} = \frac{1}{4\pi}\sqrt{\frac{\mu}{\epsilon}}\left(\mathbf{m}^{(1)} \times \nabla R - \frac{i\omega}{2}\,\mathbf{p}^{(2)}\right)\left(\frac{1}{R} + \frac{i}{kR^2}\right)e^{ikR - i\omega t}. \tag{21}$$

Comparison of (21) with (7) shows that the two fields differ in the first place by a factor $\sqrt{\mu\epsilon}$, or approximately 3×10^8, and consequently the moments $\mathbf{m}^{(1)}$ and $\mathbf{p}^{(2)}$ must be very much larger than $\mathbf{p}^{(1)}$ if $\mathbf{\Pi}^{(1)}$ is to be of appreciable magnitude. On the other hand, it must be remembered that the electric moments of a *current* distribution are functions of the frequency as well as of the geometry [Eq. (43), page 431]. If $\mathbf{p}^{(2)}$ is expressed in terms of current rather than charge, we see that the magnitude of $\mathbf{\Pi}^{(1)}$ is independent of frequency, whereas the magnitude of $\mathbf{\Pi}^{(0)}$ diminishes as $1/\omega$ with increasing frequency.

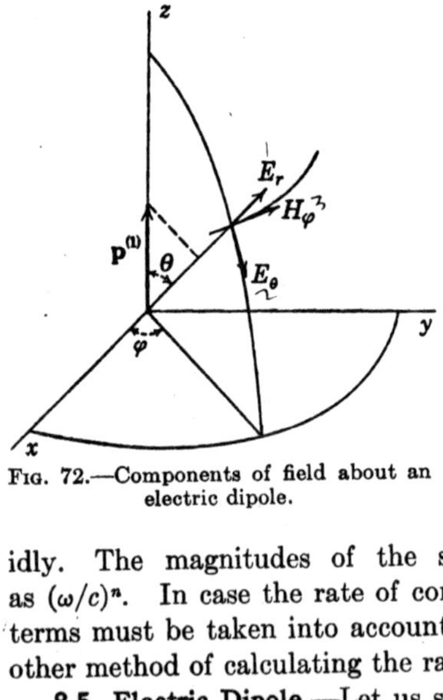

FIG. 72.—Components of field about an electric dipole.

The remaining terms of the series in $\mathbf{\Pi}^{(n)}$ represent the contributions of electric and magnetic multipoles of higher order. These must be expressed in terms of tensors and the labor of computation increases rapidly. The magnitudes of the successive terms progress essentially as $(\omega/c)^n$. In case the rate of convergence is so slow that higher order terms must be taken into account, one is usually forced to adopt some other method of calculating the radiation.

8.5. Electric Dipole.—Let us suppose that the charge distribution is such that only the electric dipole moment is of importance. This is the case, for example, of a current oscillating in a straight section of wire whose length is very small compared to that of the wave. The coordinate system is oriented so that the positive z-axis coincides with the dipole moment $\mathbf{p}^{(1)}$. The vector $\mathbf{p}^{(1)}$ can be resolved into its components in a spherical system as illustrated in Fig. 72.

$$\mathbf{p}^{(1)} = p^{(1)} \cos\theta\,\mathbf{i}_1 - p^{(1)} \sin\theta\,\mathbf{i}_2, \tag{22}$$

where $\mathbf{i}_1$, $\mathbf{i}_2$, $\mathbf{i}_3$ are the unit spherical vectors designated in Fig. 8, page 52. According to (7) the Hertz vector $\mathbf{\Pi}^{(0)}$ is parallel to $\mathbf{p}^{(1)}$ and can be resolved in the same manner. The field vectors are now calculated from the formulas (38), page 430, which in the case of a harmonic time factor become

$$\mathbf{E} = \nabla\nabla \cdot \mathbf{\Pi} + k^2\mathbf{\Pi}, \qquad \mathbf{H} = -i\omega\epsilon\nabla \times \mathbf{\Pi}. \tag{23}$$

The differential operators, expressed in spherical coordinates by (95), page 52, are applied to the vector

$$\mathbf{\Pi}^{(0)} = \Pi^{(0)} \cos\theta\, \mathbf{i}_1 - \Pi^{(0)} \sin\theta\, \mathbf{i}_2. \tag{24}$$

If $\mathbf{R}_0 \equiv \nabla R$ is a unit vector directed from the dipole towards the observer and $t^* = t - \frac{R}{v}$, then

$$\mathbf{E}^{(0)} = \frac{1}{4\pi\epsilon} e^{-i\omega t^*} \left\{\frac{1}{R^3} [3\mathbf{R}_0(\mathbf{R}_0 \cdot \mathbf{p}^{(1)}) - \mathbf{p}^{(1)}] - \frac{ik}{R^2} [3\mathbf{R}_0(\mathbf{R}_0 \cdot \mathbf{p}^{(1)}) - \mathbf{p}^{(1)}] - \frac{k^2}{R} [\mathbf{R}_0 \times (\mathbf{R}_0 \times \mathbf{p}^{(1)})]\right\}, \tag{25}$$

$$\mathbf{H}^{(0)} = -\frac{i\omega}{4\pi} e^{-i\omega t^*} \left(\frac{1}{R^2} - \frac{ik}{R}\right) \mathbf{p}^{(1)} \times \mathbf{R}_0. \tag{26}$$

The general structure of a dipole field is easily determined from these expressions. We note in the first place that the electric vector lies in a meridian plane through the axis of the dipole and the magnetic vector is perpendicular to this plane. The magnetic lines of force are coaxial circles about the dipole. At zero frequency all terms vanish with the exception of the first in $1/R^3$ in (25), which upon reference to (27), page 175, is seen to be identical with the field of an electrostatic dipole. Moreover, if we replace $\mathbf{p}^{(1)}$ by the linear current element $\frac{i}{\omega} I\, ds$ according to (43), page 431, it is apparent that the first term of (26) can be interpreted as an extension to variable fields of the Biot-Savart law, (14), page 232. The terms in (25) and (26) have been arranged in inverse powers of R and the ratio of the magnitudes of successive terms is kR, or $2\pi \frac{R}{\lambda}$. In the immediate neighborhood of the source the "static" and "induction" fields in $1/R^3$ and $1/R^2$ predominate, while at distances such that $R \gg \lambda$, or $kR \ll 1$, only the "radiation" field

$$\mathbf{E}^{(0)} \simeq -\frac{k^2}{4\pi\epsilon} \mathbf{R}_0 \times (\mathbf{R}_0 \times \mathbf{p}^{(1)}) \frac{e^{-i\omega t^*}}{R}, \qquad \mathbf{H}^{(0)} \simeq \frac{\omega k}{4\pi} \mathbf{R}_0 \times \mathbf{p}^{(1)} \frac{e^{-i\omega t^*}}{R} \tag{27}$$

need be taken into account. *At great distances from the source, measured in wave lengths, the field becomes transverse to the direction of propagation.*

If we let $R \to \infty$ and at the same time increase $\mathbf{p}^{(1)}$ so that the intensity of the field remains constant, we obtain in the limit the plane waves discussed in Chap. V.

This transverse property of the wave motion at great distances is common to the fields of all electromagnetic sources, but we must not overlook the fact that *in the vicinity of the source there is a longitudinal component in the direction of propagation.* The radiation field of an electric dipole, for example, vanishes along the axis of the dipole, but for sufficiently small values of R there is an appreciable longitudinal component of $\mathbf{E}^{(0)}$ of the form

$$E_z = \frac{2}{4\pi\epsilon}\left(\frac{1}{z} - \frac{ik}{z^2}\right)|p^{(1)}|e^{-i\omega t^*}. \tag{28}$$

This component, and not the radiation field, determines the distribution of current in a vertical antenna above the earth.

Let us calculate next the energy flow within the field. For this purpose we shall make use of the complex Poynting vector derived in Sec. 2.20. If $\bar{\mathbf{S}}$ is the mean value of the energy crossing unit area in the direction of the positive normal per second, then

$$\bar{\mathbf{S}} = Re(\mathbf{S}^*), \qquad \mathbf{S}^* = \tfrac{1}{2}\mathbf{E} \times \tilde{\mathbf{H}}. \tag{29}$$

To calculate the complex Poynting vector $\mathbf{S}^*$, we first write (25) and (26) in component form.

$$\begin{aligned} E_R^{(0)} &= \frac{1}{2\pi\epsilon}\left(\frac{1}{R^3} - \frac{ik}{R^2}\right)\cos\theta\,|p^{(1)}|e^{-i\omega t^*}, \\ E_\theta^{(0)} &= \frac{1}{4\pi\epsilon}\left(\frac{1}{R^3} - \frac{ik}{R^2} - \frac{k^2}{R}\right)\sin\theta\,|p^{(1)}|e^{-i\omega t^*}, \\ H_\phi^{(0)} &= -\frac{i\omega}{4\pi}\left(\frac{1}{R^2} - \frac{ik}{R}\right)\sin\theta\,|p^{(1)}|e^{-i\omega t^*}. \end{aligned} \tag{30}$$

From these we construct

$$\mathbf{S}^* = \tfrac{1}{2}E_\theta \tilde{H}_\phi\,\mathbf{i}_1 - \tfrac{1}{2}E_R\tilde{H}_\phi\,\mathbf{i}_2, \tag{31}$$

and obtain

$$\begin{aligned} \mathbf{S}^* = \frac{\omega k^3}{32\pi^2\epsilon}\frac{\sin^2\theta}{R^2}|p^{(1)}|^2\,\mathbf{i}_1 + \frac{i\omega}{32\pi^2\epsilon}\frac{\sin^2\theta}{R^5}|p^{(1)}|^2\,\mathbf{i}_1 \\ - \frac{i\omega}{16\pi^2\epsilon}\left(\frac{1}{R^5} - \frac{k^2}{R^3}\right)\sin\theta\cos\theta\,|p^{(1)}|^2\,\mathbf{i}_2. \end{aligned} \tag{32}$$

Next the normal component of $\mathbf{S}^*$ is integrated over the surface of a sphere of radius R.

$$\int_0^{2\pi}\int_0^{\pi} S_R^* R^2 \sin\theta\,d\theta\,d\phi = \frac{\omega^4}{12\pi}\mu\sqrt{\epsilon\mu}\,|p^{(1)}|^2 + \frac{i\omega}{12\pi\epsilon}\frac{1}{R^3}|p^{(1)}|^2. \tag{33}$$

Now by (29), page 137, the real part of (33) gives the mean outward flow of energy through a spherical surface enclosing the oscillating dipole. Since this real part is independent of the radius of the sphere, it must be the same through all concentric spheres and we discover that *the system is losing energy at a constant rate*

$$W^{(0)} = \frac{\omega^4}{12\pi}\mu\sqrt{\epsilon\mu}\,|p^{(1)}|^2 \qquad \text{watts.} \tag{34}$$

The quantity W measures the *radiation loss* from the dipole. In the expansion of an electromagnetic field, only those terms that vanish at infinity as R^{-1} give rise to radiation. In the present instance there is, of course, an instantaneous energy flow associated with the other terms of (25) and (26), but the time average of this flow over a complete cycle is zero. The terms in R^{-2} and R^{-3} account for the energy stored in the field, energy which periodically flows out from the source and returns to it, without ever being lost from the system.

In free space $\mu_0 = 4\pi \times 10^{-7}$, $\sqrt{\mu_0\epsilon_0} = \frac{1}{3} \times 10^{-8}$, and hence

$$W^{(0)} = \tfrac{1}{9} \times 10^{-15}\,\omega^4|p^{(1)}|^2 = 173 \times 10^{-15}\,\nu^4|p^{(1)}|^2 \quad \text{watts} \tag{35}$$

where ν is the frequency. The radiation increases very rapidly with increasing frequency and decreasing wave length.

8.6. Magnetic Dipole.—The current distribution may be such that only the magnetic dipole moment $\mathbf{m}^{(1)}$ is of importance. At sufficiently low frequencies this is certainly the case for any closed loop of wire. Then

$$\mathbf{\Pi}^{(1)} = \frac{1}{4\pi}\sqrt{\frac{\mu}{\epsilon}}\,(\mathbf{m}^{(1)} \times \mathbf{R}_0)\left(\frac{1}{R} + \frac{i}{kR^2}\right)e^{-i\omega t*}. \tag{36}$$

The components of the field calculated from (23) are

$$\begin{aligned} E_\phi^{(1)} &= \frac{k^2}{4\pi}\sqrt{\frac{\mu}{\epsilon}}\left(\frac{1}{R} + \frac{i}{kR^2}\right)\sin\theta\,|m^{(1)}|e^{-i\omega t*}, \\ H_R^{(1)} &= \frac{1}{2\pi}\left(\frac{1}{R^3} - \frac{ik}{R^2}\right)\cos\theta\,|m^{(1)}|e^{-i\omega t*}, \\ H_\theta^{(1)} &= \frac{1}{4\pi}\left(\frac{1}{R^3} - \frac{ik}{R^2} - \frac{k^2}{R}\right)\sin\theta\,|m^{(1)}|e^{-i\omega t*}. \end{aligned} \tag{37}$$

The structure of this field is identical with that of the electric dipole but the roles of $\mathbf{E}$ and $\mathbf{H}$ are reversed. The vector $\mathbf{H}^{(1)}$ lies in a meridian plane through the dipole and has a radial as well as a transverse component. The lines of $\mathbf{E}^{(1)}$ are concentric circles about the dipole axis. When $k = 0$, (37) reduces to the expressions derived in Sec. 4.5 for the static field of a fixed magnetic dipole.

To calculate the radiation we construct the complex Poynting vector as in (31) and integrate its radial component over a sphere of radius R.

Only the terms of (37) in $1/R$ contribute to the result and we obtain

$$W^{(1)} = \frac{k^4}{12\pi}\sqrt{\frac{\mu}{\epsilon}}\,|m^{(1)}|^2 \qquad \text{watts.} \tag{38}$$

In free space $\sqrt{\mu_0/\epsilon_0} = 376.6$ ohms, in which case

$$W^{(1)} = 10k^4|m^{(1)}|^2 \qquad \text{watts.} \tag{39}$$

For the sake of example consider an alternating current in a single circular loop of radius R_1 meters. The maximum value of the current is I_0 and the wave length is assumed to be very much greater than R_1. The phase of the current is then essentially the same at all points of the circuit, whence it follows from (43), page 431, and (11), page 433, that the electric dipole moment is negligible. The magnetic moment $|m^{(1)}| = \pi R_1^2 I_0$; consequently, the energy radiated per second is

$$W = 160\pi^6\left(\frac{R_1}{\lambda}\right)^4 I_0^2. \tag{40}$$

The *radiation resistance* $\mathcal{R}$ is defined as the equivalent ohmic resistance dissipating energy at an equal rate. Since I_0 represents the maximum current amplitude,

$$W = \tfrac{1}{2}\mathcal{R}I_0^2 \tag{41}$$

and, hence, in the present instance

$$\mathcal{R} = 320\pi^6\left(\frac{R_1}{\lambda}\right)^4 = 3.075 \times 10^5\left(\frac{R_1}{\lambda}\right)^4 \qquad \text{ohms.} \tag{42}$$

This result is restricted in the first place by the condition that the second term in the expansion of $j_1(kR_1)$ shall be negligible with respect to the first.

$$j_1(kR_1) = \frac{kR_1}{3}\left(1 - \frac{(kR_1)^2}{10} + \cdots\right), \tag{43}$$

and hence $j_1(kR_1)$ can be approximated by the first term if $kR_1 < 1$. It is in fact a general rule that these functions can be represented by their first term as long as their argument is not greater than their order. If then $\lambda = 10\,R_1$, we find the radiation resistance to be about 30 ohms. As λ decreases, $\mathcal{R}$ increases very rapidly, but the electric dipole and quadrupole moments now begin to play a part. On the other hand if the wave length increases, it is clear that radiation soon becomes entirely negligible. Currents whose wave length is long compared to the largest dimension of the metallic circuit give rise to no appreciable radiation and are said to be *quasi-stationary*.

RADIATION THEORY OF LINEAR ANTENNA SYSTEMS

8.7. Radiation Field of a Single Linear Oscillator.—The rigorous determination of the current distribution in a conductor is a boundary-value problem. A steady-state solution of the field equations must be found, corresponding to a given impressed e.m.f., which satisfies the necessary conditions of finiteness at interior points of the conductor and continuity of the tangential components of field intensity across its surface. This problem has never been completely solved for a finite length of a cylindrical conductor, but a good idea of what takes place may be gained from a study of a conducting, prolate spheroid whose eccentricity differs but little from unity. Such an investigation was made first by Abraham,[1] and a number of papers have dealt more recently with the same subject. All this work confirms what one would naturally expect: the current distribution along an isolated straight wire is essentially harmonic and the ratio of antenna length to wave length is half an integer, as long as the cross-sectional dimensions are very small with respect to the length. A much greater source of error than the finite diameter of the conductor is the damping due to radiation. As the antenna current flows outward from the feeding point towards the ends, energy is dissipated in heat and radiation. Consequently the assumption of a constant amplitude at the current loops along the wire is incorrect. Superimposed on a standing antenna current wave is a damped traveling wave supplying the losses. The effect of this modified distribution on the radiation pattern may be considerable.

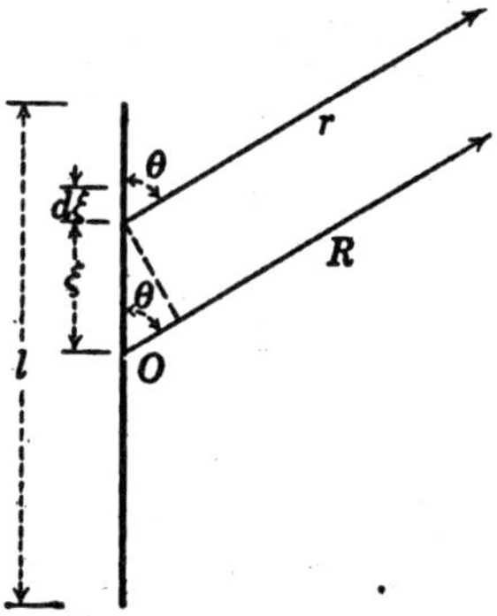

FIG. 73.—Single linear oscillator.

The current distribution is also affected by proximity to the ground or other objects. The trend in contemporary antenna design is towards systems of linear antennas supported some distance above the earth and in this case the effect of the earth on current distribution can usually be neglected. When, on the other hand, a structural steel tower is used as an antenna, the capacity to ground may have an important influence on the distribution.

In the present section we shall consider the radiating properties of linear antennas completely isolated in free space. The manner in which the radiation is reflected and refracted by the earth will be taken up in the following chapter. We shall assume that the current has been measured at a number of points along each conductor and that this measured distribution can be represented by one or more harmonic standing or traveling waves.

Let us consider first a single straight wire of length l as shown in Fig. 73. The current at any point ξ measured from the center of the

[1] ABRAHAM, *Ann. Physik*, **66**, 435, 1898.

wire is $I = u(\xi)\exp(-i\omega t)$. Then by (44), page 431, the Hertz vector of the distribution is

$$\mathbf{\Pi}(x, t) = \frac{i}{4\pi\omega\epsilon_0} e^{-i\omega t} \int_{-\frac{l}{2}}^{\frac{l}{2}} u(\xi) \frac{e^{ikr}}{r} d\xi, \tag{1}$$

where $d\xi$ is an element of length along the conductor. The *radiation field* of an element $d\xi$ is by (27), page 435,

$$\begin{aligned} dE_\theta &= -\frac{i\omega\mu_0}{4\pi} \frac{\sin\theta}{r} e^{ikr-i\omega t} u(\xi)\, d\xi, \\ dH_\phi &= -\frac{i\omega}{4\pi c} \frac{\sin\theta}{r} e^{ikr-i\omega t} u(\xi)\, d\xi, \end{aligned} \tag{2}$$

so that upon integrating the contributions of all elements along the length of the wire we obtain

$$E_\theta = \sqrt{\frac{\mu_0}{\epsilon_0}} H_\phi = -\frac{i\omega\mu_0}{4\pi} e^{-i\omega t} \int_{-\frac{l}{2}}^{\frac{l}{2}} \frac{\sin\theta}{r} e^{ikr} u(\xi)\, d\xi. \tag{3}$$

Now if $r \gg l$ one may consider $\sin\theta/r$ to be a *slowly varying* function of ξ. This means that the radius vectors from current elements to the point of observation are essentially parallel, so that $\sin\theta/r$ may be replaced by $\sin\theta/R$ and removed from under the sign of integration. This approximation is the more nearly exact the larger R, and since the radiation is the same through all concentric spheres about the origin O, we may take R as large as we please. The phase, however, must be considered with more care. From the figure it is clear that the time required for a wave emitted by the element at ξ to reach a distant point is less than that required for a wave originating at O by the amount $\xi\cos\theta/c$, and consequently their phase eventually differs by $k\xi\cos\theta$, since $k = 2\pi/\lambda$. It follows that, as $R \to \infty$, the field intensities approach the limit

$$E_\theta = \sqrt{\frac{\mu_0}{\epsilon_0}} H_\phi = -\frac{i\omega\mu_0}{4\pi} \frac{\sin\theta}{R} e^{ikR-i\omega t} \int_{-\frac{l}{2}}^{\frac{l}{2}} e^{-ik\xi\cos\theta} u(\xi)\, d\xi. \tag{4}$$

A sinusoidal current distribution is represented by

$$u(\xi) = I_0 \sin(k\xi - \alpha), \tag{5}$$

and the condition that the current shall be zero at the ends $\xi = \pm l/2$ is satisfied by

$$u(\xi) = I_0 \sin m\pi\left(\frac{\xi}{l} + \frac{1}{2}\right), \qquad k = \frac{m\pi}{l}. \tag{6}$$

The integer m is equal to the number of half wave lengths contained in the length l, and I_0 is the current amplitude at a loop position. The integral (4) is readily evaluated and leads to

$$(7)\quad E_\theta = -\frac{i}{4\pi}\sqrt{\frac{\mu_0}{\epsilon_0}}\, I_0\,[e^{\frac{im\pi}{2}\cos\theta} - (-1)^m e^{-\frac{im\pi}{2}\cos\theta}]\,\frac{1}{R\sin\theta}\, e^{ikR-i\omega t}.$$

The numerical factor is

$$(8)\qquad \frac{1}{2\pi}\sqrt{\frac{\mu_0}{\epsilon_0}} = 59.92 \simeq 60;$$

hence,

$$(9)\qquad E_\theta = -i60I_0\,\frac{\cos\left(\frac{m\pi}{2}\cos\theta\right)}{\sin\theta}\,\frac{e^{ikR-i\omega t}}{R},\qquad \text{when } m \text{ is odd},$$

or

$$(10)\qquad E_\theta = 60I_0\,\frac{\sin\left(\frac{m\pi}{2}\cos\theta\right)}{\sin\theta}\,\frac{e^{ikR-i\omega t}}{R},\qquad \text{when } m \text{ is even}.$$

The expressions do not depend on the length of the radiator, but only upon the mode of excitation. *The field is that of a dipole located at the center whose amplitude is modified by a phase factor $F(\theta)$.*

The intensity of radiation in any given direction is again expressed in terms of the complex Poynting vector which is radial and whose magnitude is $S^* = \frac{1}{2}E_\theta\tilde{H}_\phi$.

$$(11)\qquad \begin{aligned} S^* &= \frac{30}{2\pi}\frac{I_0^2}{R^2}\,\frac{\cos^2\left(\frac{m\pi}{2}\cos\theta\right)}{\sin^2\theta},\qquad (m = 1, 3, 5 \cdots),\\ S^* &= \frac{30}{2\pi}\frac{I_0^2}{R^2}\,\frac{\sin^2\left(\frac{m\pi}{2}\cos\theta\right)}{\sin^2\theta},\qquad (m = 2, 4, 6, \cdots). \end{aligned}$$

From the location of the zeros and maxima of these functions one can gain a fair idea of the radiation pattern. When m is odd, the zeros of S^* occur at

$$(12)\qquad \cos\theta = \frac{1}{m}, \frac{3}{m}, \cdots \frac{m}{m},$$

while the maxima are found from the relation

$$(13)\qquad \frac{m\pi}{2}\tan\theta\sin\theta = \operatorname{cotan}\left(\frac{m\pi}{2}\cos\theta\right).$$

Likewise, when m is even, the zeros occur at

$$(14)\qquad \cos\theta = 0, \frac{2}{m}, \frac{4}{m}, \cdots, \frac{m}{m},$$

and the maxima correspond to angles satisfying

$$(15)\qquad \frac{m\pi}{2}\tan\theta\sin\theta = \tan\left(\frac{m\pi}{2}\cos\theta\right).$$

The radiation patterns of the first four modes are shown in Fig. 74. In each case the current distribution is indicated by a dotted curve along the radiator, which is broken up into oscillating current sections alternately in phase and 180 deg. out of phase. The interference action of these current sections gives rise to lobes in the radiation pattern equal in number to that of sections m. Between lobes is a radiation node.

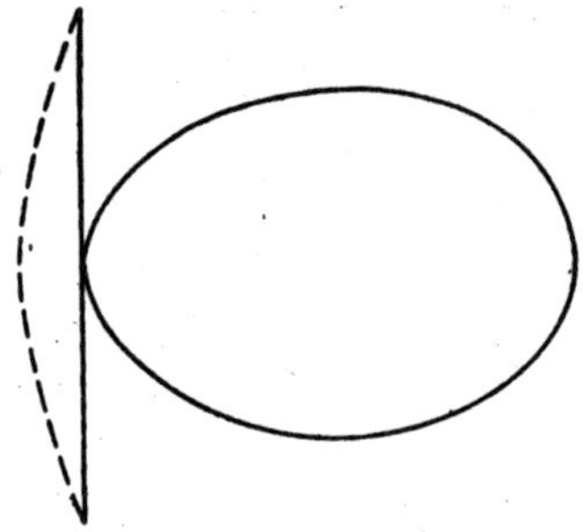

FIG. 74a. $m = 1$.

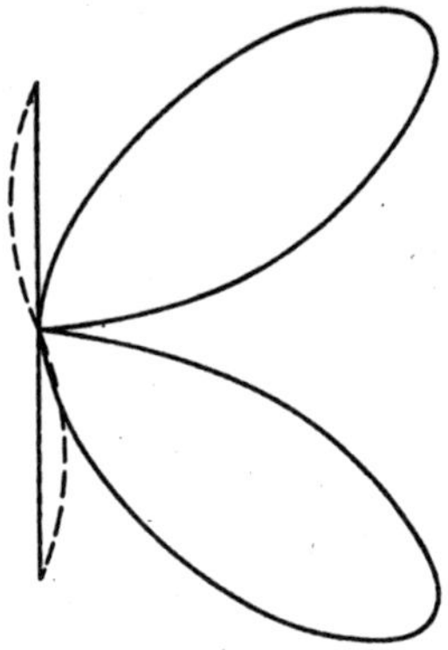

FIG. 74b. $m = 2$.

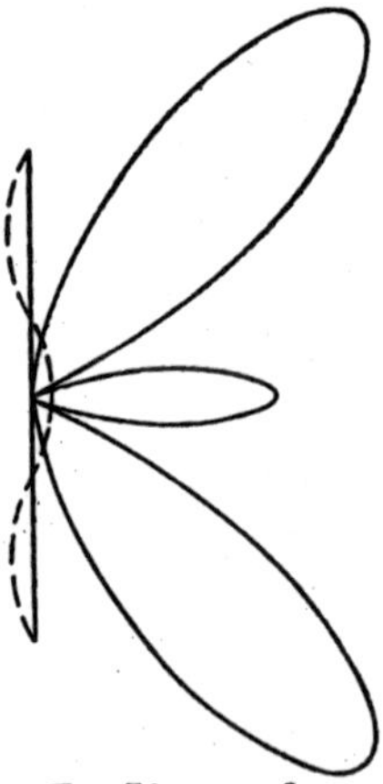

FIG. 74c. $m = 3$.

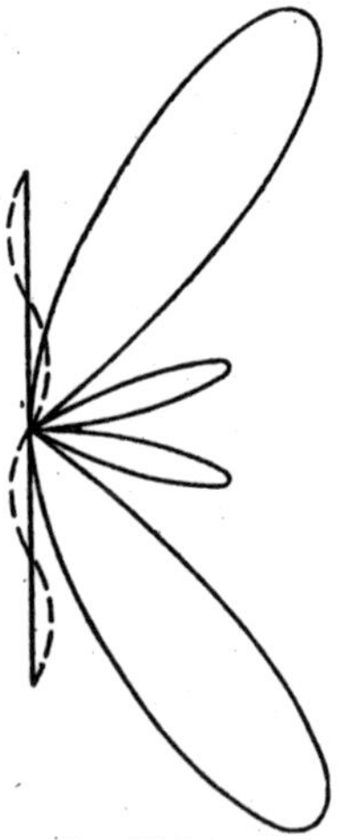

FIG. 74d. $m = 4$.

FIG. 74.—Radiation patterns of a linear oscillator excited in the modes $m = 1, 2, 3, 4$.

The radiation intensity is greatest along the axes of the lobes lying nearest the radiator and as $m \to \infty$ the radiation is directed entirely along the wire. *Energy can be propagated along an infinite conductor but there is no outward radial flow.* We shall see later that a Joule heat loss within the conductor gives rise to a compensating inward flow.

These theoretical patterns are modified in practice by the damping due to Joule heat and radiation losses. As a consequence, the successive current loops diminish in amplitude as one proceeds along the antenna outward from the point of feeding. The radiation lobes in or near the equatorial plane $\theta = \pi/2$ are strengthened at the expense of the outer ones. This effect becomes more pronounced as the resistance of the antenna wire is increased. It is illustrated by Fig. 75. The curves were obtained by Bergmann[1] from measurements on linear oscillators excited in the mode $m = 3$ and constructed of bronze and two sizes of steel wire respectively. The ratio of the current maximum at the central loop to that at either of the outer loops was of the order of 1.4:1. When the

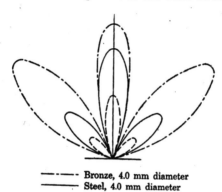

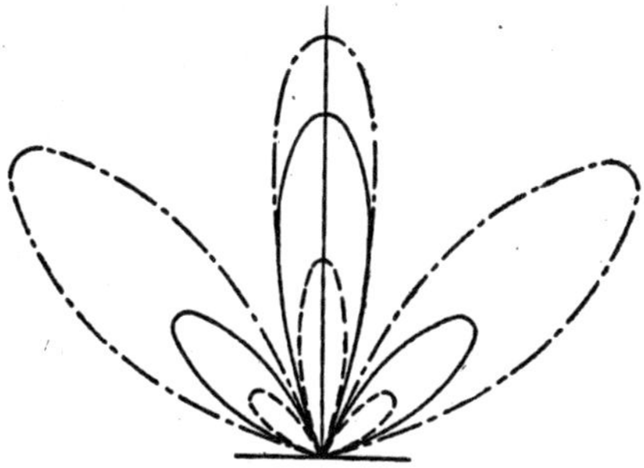

—·—·— Bronze, 4.0 mm diameter
——— Steel, 4.0 mm diameter
- - - - - Steel, 0.5 mm diameter

FIG. 75.—Effect of antenna resistance on the radiation according to measurements by Bergmann.

current amplitudes were weighted in this ratio, a calculation of the radiation gave results in close agreement with the observed values.

The power dissipated in radiation is obtained by integration of S^* over the surface of a sphere of very large radius R drawn about the center O. When m is odd, we have

$$(16)\qquad W = \int_0^{2\pi}\int_0^{\pi} S^* R^2 \sin\theta \, d\theta \, d\phi = 30 I_0^2 \int_0^{\pi} \frac{\cos^2\left(\frac{m\pi}{2}\cos\theta\right)}{\sin\theta} d\theta.$$

To evaluate this integral, we replace the variable θ by $u = \cos\theta$ between the limits $+1$ and -1. Then, since

$$(17)\qquad \frac{1}{1-u^2} = \frac{1}{2}\left(\frac{1}{1+u} + \frac{1}{1-u}\right),$$

(16) transforms to

$$(18)\qquad W = 15 I_0^2 \int_{-1}^{1} \frac{1+\cos m\pi u}{1+u} du.$$

[1] BERGMANN, *Ann. Physik*, **82**, 504, 1927.

Now let $1 + u = v/m\pi$, which leads to

$$W = 15I_0^2 \int_0^{2m\pi} \frac{1 - \cos v}{v}\, dv, \tag{19}$$

or

$$W = 15I_0^2[\ln 2m\pi\gamma - \text{Ci}\, 2m\pi], \tag{20}$$

where $\gamma = 1.7811$ and $\text{Ci}\cdot x$ is the cosine integral

$$\text{Ci}\, x = -\int_x^{\infty} \frac{\cos v}{v}\, dv, \tag{21}$$

for which values have been tabulated in Jahnke-Emde, "Tables of Functions."

An identical result is obtained when m is even. In either case the radiation resistance is

$$\mathcal{R} = 30(\ln 2m\pi\gamma - \text{Ci}\, 2m\pi), \quad (m = 1, 2, 3, \cdots), \tag{22}$$

or, since $\ln 2\pi\gamma = 2.4151$,

$$\mathcal{R} = 72.45 + 30 \ln m - 30\text{Ci}\, 2\pi m \qquad \text{ohms.} \tag{23}$$

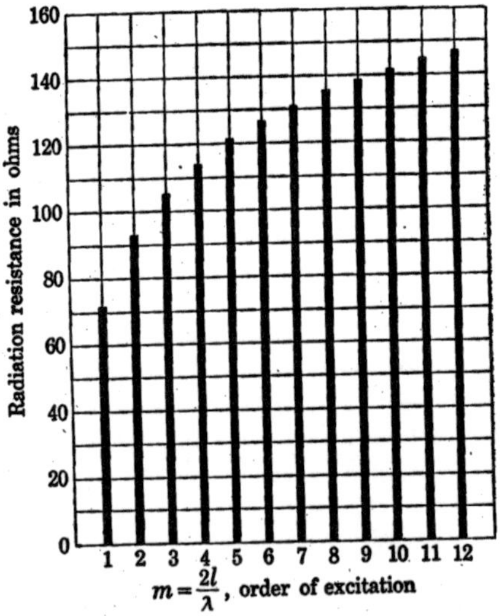

FIG. 76.—Radiation resistance of a linear antenna at resonance.

If $m > 2$, the term in Ci $2\pi m$ is negligible. The relation of radiation resistance to the order of excitation is shown graphically in Fig. 76. Equation (23) is valid only at resonance, when the quantity $2l/\lambda = m$ is an integer. At intermediate wave lengths the radiation resistance follows a curve which oscillates above and below the resonance values indicated by Fig. 76. The exact form of this curve depends on the nature of the excitation.[1]

[1] See, for example, an interesting paper by Labus, Hochfrequenztech. u. Elektroak., **41**, 17, 1933.

8.8. Radiation Due to Traveling Waves.—The current distribution considered in the preceding paragraph constitutes a standing wave with nodes located at the end points. By a proper termination of the wire, a part or all of the reflected wave may be suppressed with the result that the current is propagated along the conductor as a traveling wave. Such, for example, is the case of a transmission line, where in order to secure the most efficient transfer of energy every effort is made to avoid standing waves.

To calculate the radiation losses of a single-wire antenna of finite length[1] with no reflected current component, let us assume for the current distribution the function

$$u(\xi) = I_0 e^{ik\xi}. \tag{24}$$

If, as in the previous case, the antenna is fed at the central point, the radiation field intensities are

$$E_\theta = \sqrt{\frac{\mu_0}{\epsilon_0}}\, H_\phi = -\frac{i\omega\mu_0}{4\pi} I_0 \frac{\sin\theta}{R} e^{ikR - i\omega t} \int_{-\frac{l}{2}}^{\frac{l}{2}} e^{ik(1-\cos\theta)\xi}\, d\xi, \tag{25}$$

which, when evaluated, gives

$$E_\theta = -i60 I_0 \sin\theta \frac{\sin\left[\frac{kl}{2}(1 - \cos\theta)\right]}{1 - \cos\theta} \frac{e^{ikR - i\omega t}}{R}. \tag{26}$$

The radiant energy flow is determined by the function

$$S^* = \frac{1}{2} E_\theta \tilde{H}_\phi = \frac{30 I_0^2}{2\pi} \frac{\sin^2\theta \sin^2\left[\frac{kl}{2}(1 - \cos\theta)\right]}{(1 - \cos\theta)} \frac{1}{R^2}. \tag{27}$$

The power dissipated in radiation is, therefore, equal to

$$W = 30 I_0^2 \int_0^\pi \frac{\sin^2\theta \sin^2\left[\frac{kl}{2}(1 - \cos\theta)\right]}{(1 - \cos\theta)^2} \sin\theta\, d\theta. \tag{28}$$

To integrate we put $u = 1 - \cos\theta$, $du = \sin\theta\, d\theta$, and reduce (28) to

$$W = 30 I_0^2 \int_0^2 \frac{(2 - u)\sin^2\frac{klu}{2}}{u}\, du. \tag{29}$$

Omitting the elementary intervening steps, one finds for W the expression

$$W = 30 I_0^2 \left(1.415 + \ln\frac{2l}{\lambda} - \mathrm{Ci}\,\frac{4\pi l}{\lambda} + \frac{\sin\frac{4\pi l}{\lambda}}{\frac{4\pi l}{\lambda}}\right), \tag{30}$$

[1] The problem of a line with parallel return was first treated correctly by Manneback, *J. Am. Inst. Elec. Engrs.*, **42**, 95, 1923.

and for the radiation resistance

$$\mathcal{R} = 84.9 + 60\ln\frac{2l}{\lambda} + 60\frac{\sin\frac{4\pi l}{\lambda}}{\frac{4\pi l}{\lambda}} - 60\mathrm{Ci}\,\frac{4\pi l}{\lambda}. \tag{31}$$

It appears that for equal wave lengths and the same antenna the radiation resistance of a traveling wave is greater than that of a standing wave.

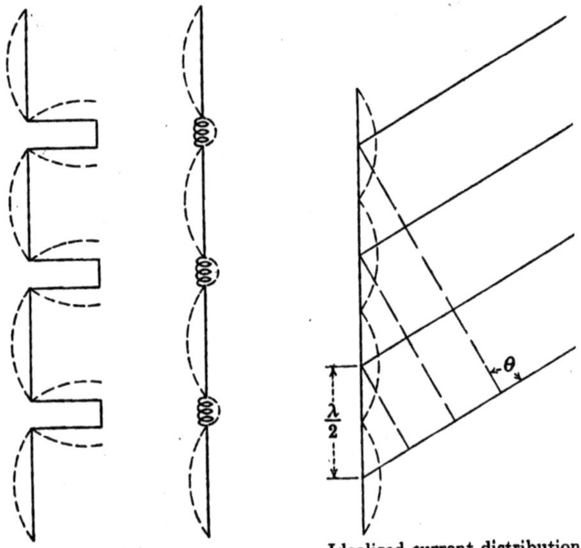

Marconi-Franklin antennas   Idealized current distribution

FIG. 77.

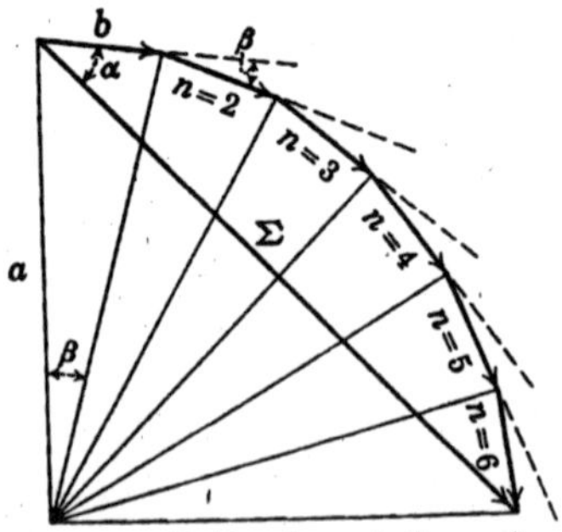

FIG. 78.—Evaluation of the series $\sum_{m=0}^{n} b\, e^{-im\beta}$.

8.9. Suppression of Alternate Phases. The directional properties of a single-wire antenna can be accentuated by suppression of alternate current loops. This is accomplished in the Marconi-Franklin type antenna by proper loading with inductance at half-wave-length intervals along the line. The alternate sections of opposite phase are shortened to such an extent as to render them negligible as radiators and the antenna is then equivalent to a colinear set of half-wave oscillators placed end to end and all in phase. This is shown schematically in Fig. 77.

The contribution of each oscillating segment to the radiation field at a distant point arrives with a phase advance of $\beta = \omega d/c$ over its next lower neighbor. Consequently the resultant field is

$$(32) \qquad E_\theta = -60iI_0 \frac{\cos\left(\frac{\pi}{2}\cos\theta\right)}{\sin\theta} \frac{e^{ikR-i\omega t}}{R} \sum_{m=0}^{n} e^{-im\beta},$$

where n is the number of current loops. Now the quantity $b\, e^{-im\beta}$ is a vector in a complex plane rotated through an angle $-m\beta$ from the real axis. The sum indicated in (32) can, therefore, be evaluated by simple geometry. The vector diagram is drawn in Fig. 78, and the sum is clearly the chord of a regular polygon. Following the notation of the figure we have

$$b = 2a \sin\frac{\beta}{2}, \qquad \sum = 2a \sin\frac{n\beta}{2},$$

whence $\sum = \dfrac{b \sin\frac{n\beta}{2}}{\sin\frac{\beta}{2}}$ at an angle $\alpha = -(n-1)\dfrac{\beta}{2}$ with the real axis.

Since in the present instance $b = 1$, we obtain

$$(33) \qquad \sum_{m=0}^{n} e^{-im\beta} = \frac{\sin\frac{n\beta}{2}}{\sin\frac{\beta}{2}} e^{-i(n-1)\frac{\beta}{2}}.$$

For β we write $\omega d/c = \pi \cos\theta$, since $d = \dfrac{\lambda}{2}\cos\theta$, and so obtain

$$(34) \qquad E_\theta = 60iI_0 \frac{\cos\left(\frac{\pi}{2}\cos\theta\right)}{\sin\theta} \frac{\sin\left(\frac{n\pi}{2}\cos\theta\right)}{\sin\left(\frac{\pi}{2}\cos\theta\right)} \frac{e^{ikR-i\omega t-i(n-1)\frac{\pi}{2}\cos\theta}}{R}.$$

The radiation intensity is

$$(35) \qquad S^* = \frac{30\, I_0^2}{2\pi}\left[\frac{\cos\left(\frac{\pi}{2}\cos\theta\right)}{\sin\theta} \frac{\sin\left(\frac{n\pi}{2}\cos\theta\right)}{\sin\left(\frac{\pi}{2}\cos\theta\right)}\right]^2 \frac{1}{R^2},$$

and the total power radiated

$$(36) \qquad W = 30I_0^2 \int_0^\pi \frac{\cos^2\left(\frac{\pi}{2}\cos\theta\right)}{\sin\theta} \frac{\sin^2\left(\frac{n\pi}{2}\cos\theta\right)}{\sin^2\left(\frac{\pi}{2}\cos\theta\right)}\, d\theta.$$

This integral has been evaluated by Bontsch-Bruewitsch,[1] who obtains

$$(37)\quad W = (-1)^{n-1} 30\, I_0^2 \left[\overset{n\pi}{\underset{0}{S}} - 4\overset{(n-1)\pi}{\underset{0}{S}} + 8\overset{(n-2)\pi}{\underset{0}{S}} - 12\overset{(n-3)\pi}{\underset{0}{S}} + \cdots \pm 4(n-1)\overset{\pi}{\underset{0}{S}} \right],$$

where

$$(38)\qquad \overset{m\pi}{\underset{0}{S}} = \int_0^{m\pi} \frac{\sin^2 u}{u}\, du = \frac{1}{2}(\ln 2m\pi\gamma - \mathrm{Ci}\, 2m\pi).$$

The radiation resistance is, therefore,

$$(39)\qquad \mathcal{R} = (-1)^{n-1}\, 60 \left(\overset{n\pi}{\underset{0}{S}} - 4\overset{(n-1)\pi}{\underset{0}{S}} + 8\overset{(n-2)\pi}{\underset{0}{S}} - \cdots \right).$$

Figure 79*a* shows the concentration of radiant energy in the equatorial plane for the case of three oscillating segments; Fig. 79*b* is a plot of radiation resistance which shows that a marked increase of radiating efficiency is achieved by the suppression of alternate interfering phases.

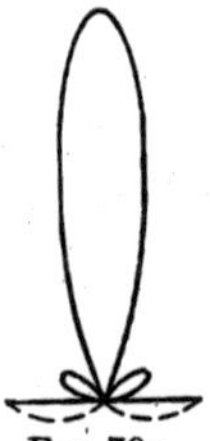

FIG. 79*a*.

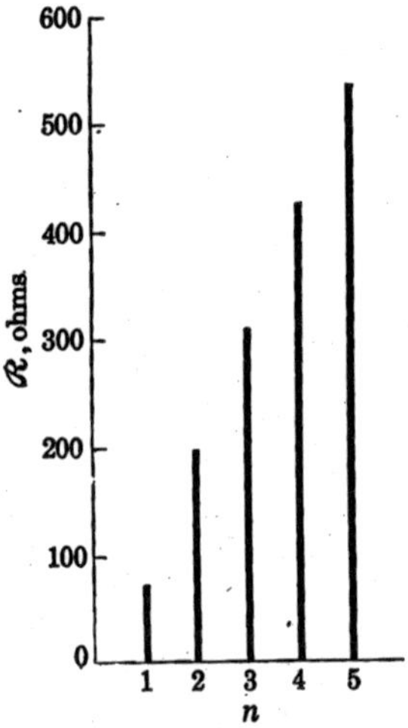

FIG. 79*b*.

FIG. 79*a*.—Radiation pattern about antenna whose current distribution is indicated by the dotted curve.

FIG. 79*b*.—Radiation resistance of *n* half-wave segments oscillating in phase.

8.10. Directional Arrays.—The single-wire antenna is directive only in the sense that the radiation is concentrated in cones of revolution about the antenna as an axis. A colinear set of half-wave oscillators excited in phase confines the radiant energy to a thin, circular disklike region such as that of Fig. 79*a*. If now one wishes to introduce a preferred direction in the equatorial plane itself, or any other form of axial asymmetry, the single-wire radiator must be replaced by a group or array of half-wave antennas suitably spaced and excited in the proper phase. In practice the spacing between centers is usually regular, so that the array constitutes a lattice structure. There is in fact a certain parallel between the problems of x-ray crystallography and the design of short-wave antenna systems. From the x-ray diffraction pattern the physicist endeavors to locate the centers of the diffracting atomic dipoles and hence determine the structure of the crystal; the radio engineer must choose the lattice spacing and phases of the current dipoles so as to

[1] BONTSCH-BRUEWITSCH, *Ann. Physik*, **81**, 437, 1926.

obtain a prescribed radiation pattern. In both cases the distribution of radiation is characterized by a phase or form factor dependent upon the polar angles θ and ϕ, and the methods which have been developed for the analysis of crystal structure can be applied in large part to antenna design.

To simplify matters we shall assume that all radiators of the array are parallel. The antennas are excited at the central point by means of

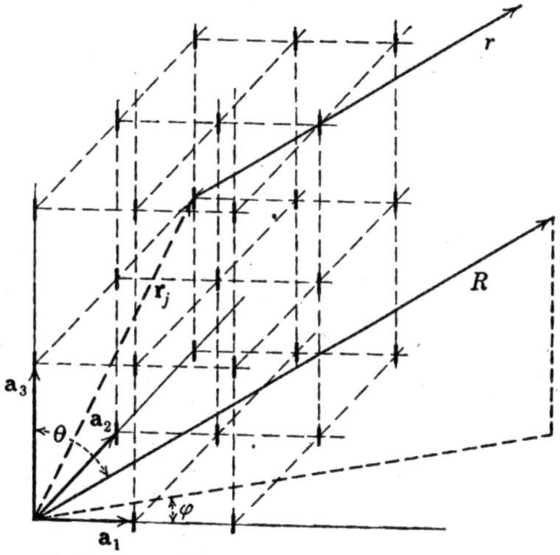

FIG. 80.—A system of half-wave oscillators arrayed in a regular lattice structure defined by the base vectors $\mathbf{a}_1$, $\mathbf{a}_2$, $\mathbf{a}_3$.

transmission lines which can be designed to radiate a negligible amount of energy. Moreover the phase velocity of propagation along these lines can be adjusted to values either less or greater than c; consequently, the phase relations between the members of the array can be prescribed practically at will.[1]

The phase factor of a single half-wave antenna will be denoted by F_0,

$$F_0(\theta) = \frac{\cos\left(\frac{\pi}{2}\cos\theta\right)}{\sin\theta}. \tag{40}$$

The centers of the oscillators are spaced regularly in a lattice structure whose constants are fixed by the three "base vectors" $\mathbf{a}_1$, $\mathbf{a}_2$, $\mathbf{a}_3$, as shown

[1] The technical aspects of this problem have been discussed by many recent writers. One may consult: Carter, Hansell and Lindenblad, *Proc. Inst. Radio Engrs.*, **19**, 1773, 1931; Wilmotte and McPetrie, *J. Inst. Elec. Eng. (London)*, **66**, 949, 1928; Hund, "Phenomena in High-frequency Systems," McGraw-Hill, 1936; Hollman, "Physik und Technik der ultrakurzen Wellen," Vol. II, Springer, 1936.

in Fig. 80. The jth oscillator is located in the network by a vector

$$\mathbf{r}_j = j_1\mathbf{a}_1 + j_2\mathbf{a}_2 + j_3\mathbf{a}_3, \tag{41}$$

where j_1, j_2, j_3 are three whole numbers not excluding zero. The phase of the jth oscillator with respect to the phase of the oscillator at the origin is β_j,

$$\beta_j = j_1\alpha_1 + j_2\alpha_2 + j_3\alpha_3. \tag{42}$$

Thus α_1 is the phase of any oscillator relative to its nearest neighbor on the axis defined by the base vector $\mathbf{a}_1$, and so on for the others. The radiation field of the jth oscillator is, therefore,

$$E_j = -i60I_{0j}F_0(\theta)\frac{e^{ikR-i\omega t-ik\mathbf{R}_0\cdot\mathbf{r}_j-i\beta_j}}{R} \tag{43}$$

where $\mathbf{R}_0$ is a unit vector in the direction of the radius vector $\mathbf{R}$. The resultant field intensity is obtained by summing over the entire array.

$$E = -i60F_0\frac{e^{ikR-i\omega t}}{R}\sum_j I_{0j}\,e^{-i(k\mathbf{R}_0\cdot\mathbf{r}_j+\beta_j)}. \tag{44}$$

By a proper adjustment of amplitude, spacing, and phase in (44) a radiation pattern of almost any desired form can be obtained. We shall consider only the most common arrangement in which all current amplitudes are the same and the base vectors $\mathbf{a}_1$, $\mathbf{a}_2$, $\mathbf{a}_3$ define a rectangular structure whose sides are parallel to the x-, y-, and z-axes. In each row parallel to the x-axis there are n_1 radiators; in each row parallel to the y-axis there are n_2; and in each column parallel to the z-axis there are n_3, so that the lattice is completely filled.

$$\mathbf{R}_0\cdot\mathbf{r}_j = j_1a_1\sin\theta\cos\phi + j_2a_2\sin\theta\sin\phi + j_3a_3\cos\theta \tag{45}$$

in which a_1, a_2, a_3 are the spacings along the x-, y-, z-axes respectively. The field intensity at any point in the radiation field is

$$E = -i60I_0F_0f_1f_2f_3\frac{e^{ikR-i\omega t}}{R}, \tag{46}$$

whose complex phase factors are

$$\begin{aligned} f_1 &= \sum_{j_1=0}^{n_1} e^{-ij_1(ka_1\sin\theta\cos\phi+\alpha_1)}, \\ f_2 &= \sum_{j_2=0}^{n_2} e^{-ij_2(ka_2\sin\theta\sin\phi+\alpha_2)}, \\ f_3 &= \sum_{j_3=0}^{n_3} e^{-ij_3(ka_3\cos\theta+\alpha_3)}. \end{aligned} \tag{47}$$

If we let

(48) $\gamma_1 = ka_1 \sin\theta\cos\phi + \alpha_1, \qquad \gamma_2 = ka_2 \sin\theta\sin\phi + \alpha_2,$
$\gamma_3 = ka_3 \cos\theta + \alpha_3,$

then (47) can be written

$$f_s = \sum_{j_s=0}^{n_s} e^{-ij_s\gamma_s} = \frac{\sin\dfrac{n_s\gamma_s}{2}}{\sin\dfrac{\gamma_s}{2}} e^{-i(n_s-1)\frac{\gamma_s}{2}}, \qquad \begin{matrix}(s = 1, 2, 3)\\ (n_s \geqq 1),\end{matrix} \tag{49}$$

by (33). The complex factors are eliminated from the radiation intensity and we get

$$S^* = \frac{30}{2\pi}\frac{I_0^2}{R^2}(F_0F_1F_2F_3)^2, \tag{50}$$

where

$$F_s = \frac{\sin\dfrac{n_s\gamma_s}{2}}{\sin\dfrac{\gamma_s}{2}}, \qquad (s = 1, 2, 3). \tag{51}$$

All the radiation formulas derived above are special cases of this result. For example, the single wire of Sec. 8.7 excited in an upper mode constitutes a linear distribution of half-wave oscillators with $a_1 = a_2 = 0$, $a_3 = \lambda/2$, $\alpha_3 = \pi$. Then

$$F_1 = F_2 = 1, \qquad F_3 = \frac{\sin\dfrac{n_3\pi}{2}(\cos\theta + 1)}{\cos\left(\dfrac{\pi}{2}\cos\theta\right)}, \tag{52}$$

which leads directly to (11) on page 441. If the colinear oscillators are in phase, as described in Sec. 8.9, we take $a_1 = a_2 = 0$, $a_3 = \lambda/2$, $\alpha_3 = 0$, and obtain (35).

Consider next a row of vertical half-wave oscillators spaced along the *horizontal* x-axis. We shall assume them to be excited in phase and spaced a half wave length apart. $a_1 = \lambda/2$, $a_2 = a_3 = 0$, $\alpha_i = 0$. Then

$$S^* = \frac{30}{2\pi}\frac{I_0^2}{R^2}\left[\frac{\cos\left(\dfrac{\pi}{2}\cos\theta\right)}{\sin\theta}\;\frac{\sin\left(\dfrac{n_1\pi}{2}\sin\theta\cos\phi\right)}{\sin\left(\dfrac{\pi}{2}\sin\theta\cos\phi\right)}\right]^2, \tag{53}$$

which now depends on the equatorial angle ϕ. The zeros occur at those angles for which the quantity $n_1\pi/2 \sin\theta\cos\phi$ has any one of the values $\pi, 2\pi, \ldots$. In the equatorial plane $\theta = \pi/2$ and there are radiation nodes along the lines determined by

$$\cos\phi = \pm\frac{2}{n_1}, \pm\frac{4}{n_1} \cdots \pm\frac{n_1}{n_1}, \tag{54}$$

or

$$\cos\phi = \pm\frac{2}{n_1}, \pm\frac{4}{n_1} \cdots \pm\frac{n_1 - 1}{n_1},$$

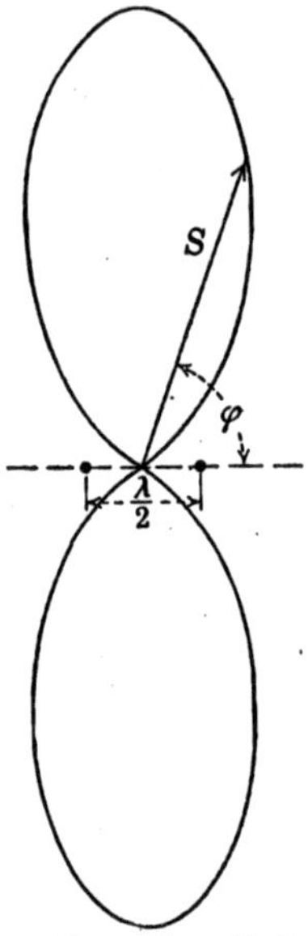

FIG. 81a.—Radiation pattern in the equatorial plane of two half-wave oscillators operating in phase and separated by a half wave length.

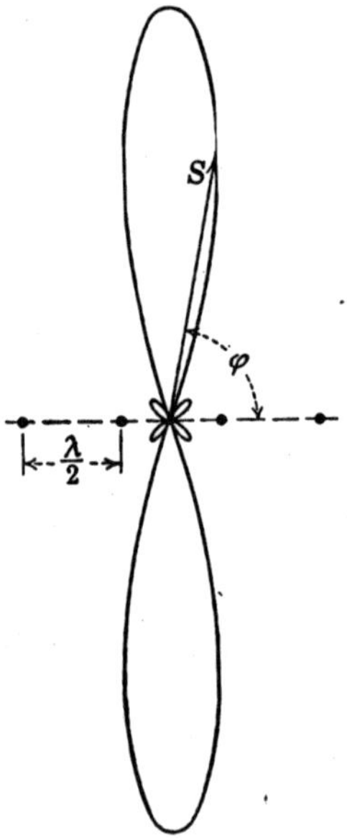

FIG. 81b.—Radiation pattern in the equatorial plane of four half-wave oscillators operating in phase and with half-wave-length separation.

as n_1 is even or odd. The principal maximum occurs at $\phi = \pi/2$, so that this array gives rise to a broadside emission of energy. Subsidiary maxima are determined by the condition $\partial F_1/\partial\phi = 0$, which in the present instance leads to the relation

$$n_1 \tan\left(\frac{\pi}{2}\sin\theta\cos\phi\right) = \tan\left(\frac{n_1\pi}{2}\sin\theta\cos\phi\right). \tag{55}$$

When $\theta = \phi = \pi/2$, $F_0F_1 = n_1$, so that *the radiation intensity in the direction of the principal maximum is n_1^2 times the maximum intensity of a single oscillator.* On the other hand, each oscillator receives only the n_1th part of the total power delivered by the source (assuming equipartition), and hence *the net gain in radiation intensity in the preferred direction is given by the factor n_1, the number of oscillators.*

The neighboring zeros lying on either side of the principal maximum in the equatorial plane are fixed by the relation $\cos\phi = \pm 2/n_1$. Hence the larger n_1, the narrower the beam. The beam is evidently confined to a region bounded by the angles ϕ_1 and ϕ_2, where $\cos\phi_2 - \cos\phi_1 = 4/n_1$,

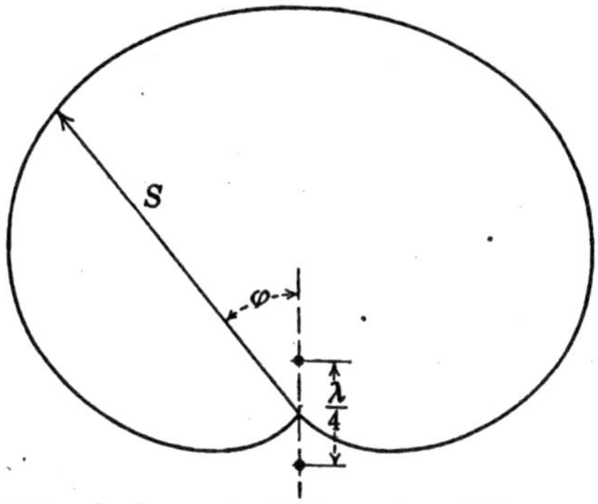

FIG. 82.—Radiation pattern in the equatorial plane of two half-wave oscillators 90 deg. out of phase and with quarter-wave spacing.

lying on either side of $\phi = \pi/2$. The radiation patterns in an equatorial plane have been plotted in Figs. 81*a* and 81*b* for arrays containing two and four oscillators.

Lastly, we shall apply (50) to two antennas spaced a quarter wave length apart and 90 deg. out of phase. Let $a_1 = \lambda/4$, $a_2 = a_3 = 0$, $\alpha_1 = -\pi/2$.

$$(56)\quad S^* = \frac{120}{2\pi}\frac{I_0^2}{R^2}\left[\frac{\cos\left(\frac{\pi}{2}\cos\theta\right)}{\sin\theta}\cos\left(\frac{\pi}{4}(\sin\theta\cos\phi - 1)\right)\right]^2.$$

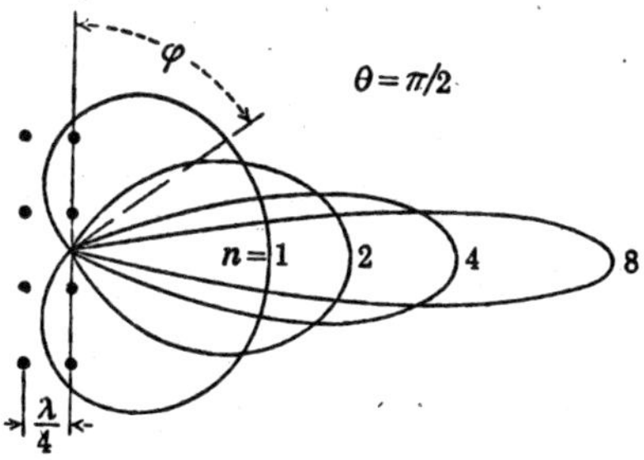

FIG. 83.—Radiation patterns in the equatorial plane for a horizontal row of vertical half-wave oscillators backed by a row of reflectors, illustrating the concentration of radiation in a forward beam by increasing the number of oscillators. (*Hollmann.*)

The radiation pattern in the equatorial plane is shown in Fig. 82. The energy is thrown forward in a broad beam in the direction of the positive x-axis. Radiation to the rear is reduced to a very small amount. In many commercial installations the rearward antenna is not connected directly to the source. Its excitation is then due to inductive coupling with the forward oscillator and its action is that of a reflector. In this case, however, the spacing is not quite $\lambda/4$ and a small adjustment must also be made in the length of the reflector.

A needlelike beam in space can be obtained by a combination of the simple arrays described above. Concentration of the radiation in a meridian plane is accomplished by means of a horizontal row of vertical oscillators. Extending the length of these vertical lines into a series of half-wave radiators excited in phase results in a concentration in the equatorial plane. Finally, this two-dimensional array is backed up by a corresponding set of reflectors at an approximate distance of $\lambda/4$. In Fig. 83 the radiation diagram in the plane $\theta = \pi/2$ of a horizontal row of vertical oscillators backed by reflectors is drawn for several values of n_2, the number in a row. In Fig. 84 the same diagram is drawn for a vertical set in the meridian plane $\phi = \pi/2$. In both cases the effectiveness of an increase in n is apparent. The greatest improvement naturally occurs when n is small, in which case the addition of another oscillator makes a very considerable difference. The relative gain decreases with increasing numbers and a point is soon reached at which any further addition to the array is economically unwarranted.

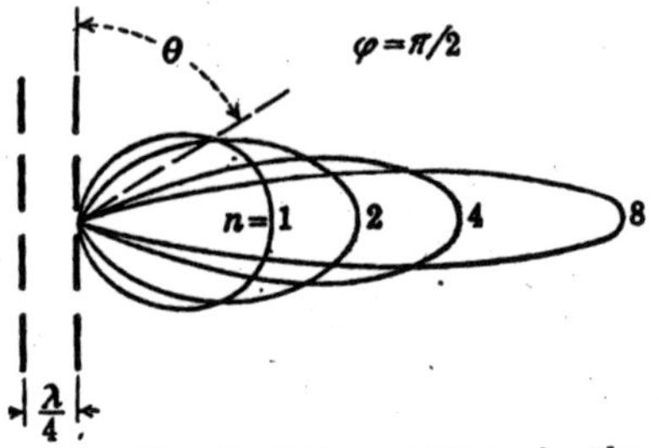

FIG. 84.—Radiation patterns in the meridian plane for a vertical set of half-wave oscillators backed by reflectors. (*Hollmann.*)

8.11. Exact Calculation of the Field of a Linear Oscillator.—The method of Poynting, which has thus far been employed to calculate the radiation resistance, is based entirely on a determination of the outward flow of energy at great distances from the center of the radiating system. Now if energy is constantly lost from the system it must also be supplied by the source; consequently, work must be done on the antenna currents at a constant rate. There must be a component of the e.m.f. parallel to the wire at its surface which is 180 deg. out of phase with the antenna current. A knowledge of the phase relations between current and e.m.f. *along the conductor* must lead to the same value of radiation resistance. In fact it can give us more; for this resistance is but the real part of a complex radiation impedance which must eventually be determined if the radiator is to be treated as a circuit element.

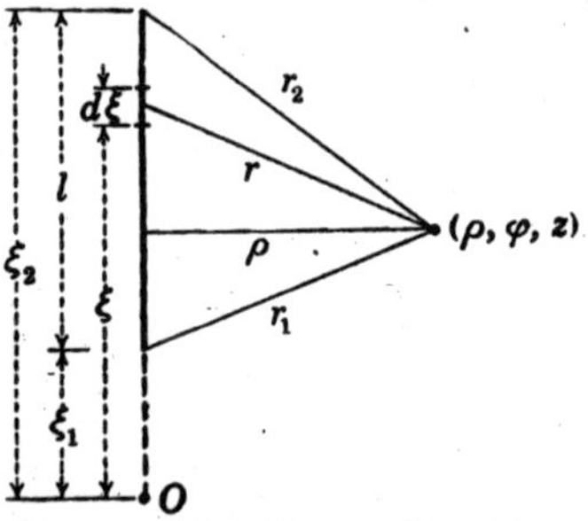

FIG. 85.—Coordinates of the linear oscillator discussed in Sec. 8.11.

It is apparent that the results of the preceding paragraphs cannot be applied to this alternative method, for they are valid only at great distances from the source. It is essential that we have an exact expression for the field in the immediate neighborhood of each conductor. The demonstrations by Brillouin,[1] Pistalkors,[2] and Bechmann[3] that an exact expression in closed form can in fact be found were an extremely important contribution to antenna theory.

We shall consider again a single linear conductor coinciding with a section of the z-axis, but the center of the conductor need no longer be located at the origin. A point on the conductor is fixed by its coordinate ξ, and its length is $l = \xi_2 - \xi_1$, as shown in Fig. 85. The current distribution is sinusoidal, but the current need not be zero at the ends if terminal loading is taken into account. Such loading can be accomplished through the capacitative effect of metal plates, by grounding through a lumped inductance or resistance, or by any of a variety of other devices. The current is, therefore, again

$$I = I_0 \sin (k\xi - \alpha)e^{-i\omega t} \tag{57}$$

and the Hertz vector of the distribution is

$$\Pi_z(x, t) = \frac{iI_0}{4\pi\epsilon_0\omega} e^{-i\omega t} \int_{\xi_1}^{\xi_2} \sin (k\xi - \alpha) \frac{e^{ikr}}{r} d\xi, \tag{58}$$

which can be written also in the form

$$\Pi_z(x, t) = \frac{I_0 e^{-i\omega t}}{8\pi\epsilon_0\omega}\left[e^{-i\alpha} \int_{\xi_1}^{\xi_2} \frac{e^{ik(r+\xi)}}{r} d\xi - e^{i\alpha} \int_{\xi_1}^{\xi_2} \frac{e^{ik(r-\xi)}}{r} d\xi\right]. \tag{59}$$

The cylindrical coordinates of an arbitrary point in the field are ρ, ϕ, z, and the distance from any current element to this point is $r = \sqrt{\rho^2 + (z - \xi)^2}$. Now let

$$u = r + \xi - z, \qquad du = \frac{u}{r} d\xi. \tag{60}$$

Then

$$\int_{\xi_1}^{\xi_2} \frac{e^{ik(r+\xi)}}{r} d\xi = e^{ikz} \int_{u_1}^{u_2} \frac{e^{iku}}{u} du, \tag{61}$$

where $u_1 = r_1 + \xi_1 - z$, $u_2 = r_2 + \xi_2 - z$, and r_1, r_2 are the distances from the ends of the wire to the point of observation. Likewise, if

$$v = r - \xi + z, \qquad dv = -\frac{v}{r} d\xi, \tag{62}$$

$$\int_{\xi_1}^{\xi_2} \frac{e^{ik(r-\xi)}}{r} d\xi = -e^{-ikz} \int_{v_1}^{v_2} \frac{e^{ikv}}{v} dv, \tag{63}$$

where $v_1 = r_1 - \xi_1 + z$, $v_2 = r_2 - \xi_2 + z$. The Hertz vector is, therefore,

$$\Pi_z = \frac{I_0 e^{-i\omega t}}{8\pi\epsilon_0\omega}\left[e^{i(kz-\alpha)} \int_{u_1}^{u_2} \frac{e^{iku}}{u} du + e^{-i(kz-\alpha)} \int_{v_1}^{v_2} \frac{e^{ikv}}{v} dv\right]. \tag{64}$$

[1] BRILLOUIN, *Radioélectricité*, April, 1922.
[2] PISTALKORS, *Proc. Inst. Radio Engrs.*, **17**, 562, 1929.
[3] BECHMANN, *ibid.*, **19**, 461, 1471, 1931.

A simple transformation reduces these exponential integrals to a form for which tabulated values are available. We note first that

$$\int_{u_1}^{u_2} \frac{e^{iku}}{u}\,du = \int_{u_1}^{\infty} \frac{e^{iku}}{u}\,du - \int_{u_2}^{\infty} \frac{e^{iku}}{u}\,du. \tag{65}$$

By definition the cosine and sine integrals are

$$\mathrm{Ci}\,ku_1 = -\int_{u_1}^{\infty} \frac{\cos ku}{u}\,du, \qquad \mathrm{Si}\,ku_1 = \int_0^{u_1} \frac{\sin ku}{u}\,du; \tag{66}$$

since $\mathrm{Si}\,\infty = \pi/2$, we have

$$\int_{u_1}^{\infty} \frac{e^{iku}}{u}\,du = -\mathrm{Ci}\,ku_1 - i\,\mathrm{Si}\,ku_1 + \frac{\pi i}{2}, \tag{67}$$

or

$$\int_{u_1}^{u_2} \frac{e^{iku}}{u}\,du = \mathrm{Ci}\,ku\Big|_{u_1}^{u_2} + i\,\mathrm{Si}\,ku\Big|_{u_1}^{u_2}. \tag{68}$$

Tables of these functions will be found in Jahnke-Emde.

To calculate the field we apply the formulas

$$\mathbf{E} = \nabla\nabla\cdot\mathbf{\Pi} + k^2\mathbf{\Pi}, \qquad \mathbf{H} = -i\omega\epsilon\nabla\times\mathbf{\Pi}, \tag{69}$$

which in cylindrical coordinates reduce to

$$\begin{aligned} E_z &= \frac{\partial^2\Pi_z}{\partial z^2} + k^2\Pi_z, \qquad E_\rho = \frac{\partial^2\Pi_z}{\partial\rho\,\partial z}, \qquad E_\phi = 0, \\ H_z &= H_\rho = 0, \qquad H_\phi = i\omega\epsilon\frac{\partial\Pi_z}{\partial\rho}. \end{aligned} \tag{70}$$

Since u_1, u_2, etc. in (64) depend upon both z and ρ, the integrals must be differentiated with respect to their limits. For example,

$$\begin{aligned} \frac{\partial}{\partial z}\int_{u_1}^{u_2} \frac{e^{iku}}{u}\,du &= \frac{e^{iku_2}}{u_2}\frac{\partial u_2}{\partial z} - \frac{e^{iku_1}}{u_1}\frac{\partial u_1}{\partial z}, \\ e^{ikz}\frac{\partial}{\partial z}\int_{u_1}^{u_2} \frac{e^{iku}}{u}\,du &= \frac{e^{ik(r_1+\xi_1)}}{r_1} - \frac{e^{ik(r_2+\xi_2)}}{r_2}. \end{aligned} \tag{71}$$

Similarly,

$$e^{ikz}\frac{\partial}{\partial\rho}\int_{u_1}^{u_2} \frac{e^{iku}}{u}\,du = \frac{\rho}{r_2-(z-\xi_2)}\frac{e^{ik(r_2+\xi_2)}}{r_2} - \frac{\rho}{r_1-(z-\xi_1)}\frac{e^{ik(r_1+\xi_1)}}{r_1}. \tag{72}$$

After differentiating in this manner and combining terms one obtains for the components of field intensity

$$E_z = -\frac{iI_0e^{-i\omega t}}{4\pi\epsilon_0\omega}\left[k\frac{e^{ikr_2}}{r_2}\cos(k\xi_2-\alpha) - k\frac{e^{ikr_1}}{r_1}\cos(k_{\xi 1}-\alpha) + \frac{\partial}{\partial z}\left(\frac{e^{ikr_2}}{r_2}\right)\sin(k\xi_2-\alpha) - \frac{\partial}{\partial z}\left(\frac{e^{ikr_1}}{r_1}\right)\sin(k\xi_1-\alpha)\right], \tag{73}$$

$$(74)\quad E_\rho = \frac{iI_0 e^{-i\omega t}}{4\pi\epsilon_0\omega}\left\{\frac{k(z-\xi_2)}{\rho}\frac{e^{ikr_2}}{r_2}\cos(k\xi_2-\alpha) - \frac{k(z-\xi_1)}{\rho}\frac{e^{ikr_1}}{r_1}\cos(k\xi_1-\alpha) + \frac{1}{\rho}\frac{\partial}{\partial z}\left[(z-\xi_2)\frac{e^{ikr_2}}{r_2}\right]\sin(k\xi_2-\alpha) - \frac{1}{\rho}\frac{\partial}{\partial z}\left[(z-\xi_1)\frac{e^{ikr_1}}{r_1}\right]\sin(k\xi_1-\alpha)\right\},$$

$$(75)\quad H_\phi = \frac{iI_0e^{-i\omega t}}{4\pi}\left[\frac{e^{ikr_2}}{\rho}\cos(k\xi_2-\alpha) - \frac{e^{ikr_1}}{\rho}\cos(k\xi_1-\alpha) + \frac{i(z-\xi_2)}{\rho}\frac{e^{ikr_2}}{r_2}\sin(k\xi_2-\alpha) - \frac{i(z-\xi_1)}{\rho}\frac{e^{ikr_1}}{r_1}\sin(k\xi_1-\alpha)\right].$$

In the case of an isolated wire whose ends are free the current will be zero at the points $\xi = \xi_1$, $\xi = \xi_2$. Then $k = \pi m/l$, where m is again the number of half wave lengths along the wire. Since

$$\sin\left(\frac{\pi m\xi_1}{l} - \alpha\right) = \sin\left(\frac{\pi m\xi_2}{l} - \alpha\right) = 0,$$

we place $\alpha = \frac{\pi m\xi_1}{l} = \pi m\left(1 - \frac{\xi_2}{l}\right)$. The field intensities then reduce to the comparatively simple expressions

$$(76)\quad \begin{aligned} E_z &= -i30I_0e^{-i\omega t}\left[(-1)^m\frac{e^{ikr_2}}{r_2} - \frac{e^{ikr_1}}{r_1}\right],\\ E_\rho &= i30I_0e^{-i\omega t}\left[(-1)^m\frac{(z-\xi_2)}{\rho}\frac{e^{ikr_2}}{r_2} - \frac{(z-\xi_1)}{\rho}\frac{e^{ikr_1}}{r_1}\right],\\ H_\phi &= \frac{iI_0}{4\pi}e^{-i\omega t}\frac{1}{\rho}\left[(-1)^m e^{ikr_2} - e^{ikr_1}\right]. \end{aligned}$$

8.12. Radiation Resistance by the E.M.F. Method.—The expressions for the field derived in the preceding paragraph are valid and exact at all points of space, including the immediate neighborhood of the current. The occurrence of singularities at the end points $r_1 = 0$, $r_2 = 0$, results from the assumption of a line distribution of current. In practice the finite cross section of the wire ensures finite values of the field intensity, although relatively large intensities must be expected near these terminals. At any other point along the conductor the phase relation between current and the parallel component of **E** can be determined from (73) or (76); consequently, the work necessary to maintain the current against the reactive forces of the field can be calculated.

In Sec. 2.19 it was shown that for *real* vectors **E, H,** and **J** and a current distribution in free space the divergence of the Poynting vector satisfies the relation

$$(77)\quad \nabla\cdot\mathbf{S} = -\mathbf{E}\cdot\mathbf{J} - \frac{\partial}{\partial t}\left(\frac{\epsilon_0}{2}E^2 + \frac{\mu_0}{2}H^2\right).$$

If the field is periodic, the average of the time rate of change of stored energy is zero, so that

$$\overline{\nabla \cdot \mathbf{S}} = \nabla \cdot \overline{\mathbf{S}} = -\overline{\mathbf{E} \cdot \mathbf{J}}. \tag{78}$$

If now in place of real vectors we deal with *complex* field intensities and currents, we have by Sec. 2.20

$$\begin{aligned} \overline{\mathbf{S}} &= \tfrac{1}{2}Re(\mathbf{E} \times \tilde{\mathbf{H}}) = Re\mathbf{S}^*, \\ \overline{Re\mathbf{E} \cdot Re\mathbf{J}} &= \tfrac{1}{2}Re(\mathbf{E} \cdot \tilde{\mathbf{J}}), \end{aligned} \tag{79}$$

the sign ~ indicating conjugate values. The total average power radiated from a system of currents is, therefore,

$$W = \int_S \overline{\mathbf{S}} \cdot \mathbf{n}\, da = -\frac{1}{2} Re \int_V \mathbf{E} \cdot \tilde{\mathbf{J}}\, dv. \tag{80}$$

The radiation can be calculated either by integrating the normal component of the Poynting vector over a closed surface including all the sources, or by integrating the power expended per unit volume over the current distribution.

Suppose next that there are n linear conductors carrying current in otherwise empty space. Without any great loss of generality it may be assumed that the conductors of the system are parallel to one another and to the z-axis. The current in the jth conductor is I_j, a complex quantity. The z-component of the electric field at the jth conductor due to the ith current is E_{ji} Then by (80) the total radiation from the system is

$$W = -\frac{1}{2} Re \sum_{j=1}^{n} \sum_{i=1}^{n} \int_{\xi_{1j}}^{\xi_{2j}} E_{ji}\tilde{I}_j \, d\xi_j, \tag{81}$$

the integrations to be extended the length of each conductor. This is the rate at which the induced e.m.f.'s do work on the currents. Let us write

$$I_j = I_{0j} \sin (k\xi_j - \alpha_j)e^{-i\omega t}, \qquad E_{ji} = I_{0i}U_{ji}e^{-i\omega t}. \tag{82}$$

The amplitudes I_{0j} and I_{0i} are in general complex, and the function U_{ji} is determined by (73). Then

$$W = \frac{1}{2} Re \sum_{j=1}^{n} \sum_{i=1}^{n} I_{0i}\tilde{I}_{0j}Z_{ji}, \tag{83}$$

where

$$Z_{ji} = -\int_{\xi_{1j}}^{\xi_{2j}} U_{ji} \sin (k\xi_j - \alpha_j)\, d\xi_j. \tag{84}$$

The quantities Z_{ij} are coupling coefficients or, in a certain sense, transfer impedances. The real part of the coefficient Z_{ii} represents the radiation resistance of the ith conductor when all other conductors are removed.

A very neat method of calculating the radiation has been suggested by Bechmann.[1] If the current distribution is sinusoidal as in (82), one may write

$$\frac{d^2\tilde{I}_j}{d\xi_j^2} = -k^2\tilde{I}_j. \tag{85}$$

The field intensity E_{ji} satisfies the relation

$$E_{ji} = \frac{\partial^2 \Pi_i}{\partial \xi_j^2} + k^2\Pi_i. \tag{86}$$

Moreover,

$$\tilde{I}_j \frac{\partial^2 \Pi_i}{\partial \xi_j^2} = \Pi_i \frac{d^2\tilde{I}_j}{\partial \xi_j^2} + \frac{\partial}{\partial \xi_j}\left(\tilde{I}_j \frac{\partial \Pi_i}{\partial \xi_j} - \Pi_i \frac{\partial \tilde{I}_j}{\partial \xi_j}\right), \tag{87}$$

and hence

$$W = \frac{1}{2} Re \sum_{j=1}^{n} \sum_{i=1}^{n} \left[\Pi_i \frac{\partial \tilde{I}_j}{\partial \xi_j} - \tilde{I}_j \frac{\partial \Pi_i}{\partial \xi_j}\right]_{\xi_{1j}}^{\xi_{2j}}. \tag{88}$$

From (64) and (68) we have

$$\Pi_i = \frac{I_{0i}\, e^{-i\omega t}}{8\pi\epsilon_0\omega}\left[e^{i(k\xi_j - \alpha)}\left[\text{Ci}\; ku + i\;\text{Si}\; ku\right]_{u_{1i}}^{u_{2i}} + e^{-i(k\xi_j - \alpha)}\left[\text{Ci}\; kv + i\;\text{Si}\; kv\right]_{v_{1i}}^{v_{2i}}\right], \tag{89}$$

where, in the present case $u = r_{ij} + \xi_i - \xi_j$, $v = r_{ij} - \xi_i + \xi_j$. The coordinate ξ_i refers to points on the antenna whose current is I_i, while ξ_j is a point on the jth conductor. The subscripts 1 and 2 denote as before the coordinates of the lower and upper ends respectively of an antenna. *No further integration is necessary to find the radiation from a system of linear conductors.* Since the current amplitudes I_{0i} and I_{0j} are factors in (88), an integrated expression for the impedance Z_{ji} can be found.

In the case of an unloaded antenna the current I_j is zero at the terminals ξ_{1j} and ξ_{2j}. Then $k\xi_{1j} - \alpha_j = 0$, $k\xi_{2j} - \alpha_j = m_j\pi$, and we obtain

$$W = \frac{1}{2} Re \sum_{j=1}^{n} \sum_{i=1}^{n} k\tilde{I}_{0j}[(-1)^{m_j}\Pi_i(\xi_{2j}) - \Pi_i(\xi_{1j})]. \tag{90}$$

A difficulty arises when this formula is applied to colinear conductors. Consider, for example, the radiation from a single linear oscillator as in Sec. 8.7. This case is the limit of two parallel wires whose separation approaches zero. If we take the terminals of the two conductors to be coincident, then

[1] BECHMANN, *loc. cit.*, p. 1471.

when $\xi_i = \xi_1$, $u_{1i} = 0 + \xi_1 - \xi_1 = 0$, $u_{2i} = l + \xi_2 - \xi_1 = 2l$,
$v_{1i} = 0 - \xi_1 + \xi_1 = 0$, $v_{2i} = l - \xi_2 + \xi_1 = 0$,
and when $\xi_i = \xi_2$, $u_{1i} = l + \xi_1 - \xi_2 = 0$, $u_{2i} = 0 + \xi_2 - \xi_2 = 0$,
$v_{1i} = l - \xi_1 + \xi_2 = 2l$, $v_{2i} = 0 - \xi_2 + \xi_2 = 0.$

By (89),

$$\Pi_i(\xi_2) = \frac{I_0 e^{-i\omega t}}{8\pi\epsilon_0\omega}(-1)^m \Big[\operatorname{Ci} kv + i\operatorname{Si} kv\Big]_{v=2l}^{v=0},$$

$$\Pi_1(\xi_1) = \frac{I_0 e^{-i\omega t}}{8\pi\epsilon_0\omega}\Big[\operatorname{Ci} ku + i\operatorname{Si} ku\Big]_{u=0}^{u=2l}.$$

But $k = \pi m/l = 2\pi/\lambda$, $\frac{1}{2\pi}\sqrt{\mu_0/\epsilon_0} = 60$; hence by (90)

$$W = -15 I_0^2 \operatorname{Ci} \frac{\pi m u}{l}\Big|_0^{2l}.$$

This result fails, however, since Ci 0 is infinite. The difficulty can be avoided by calculating first the radiation of two parallel wires separated by a small but *finite* distance ρ. Upon passing to the limit $\rho = 0$, the infinite terms drop out[1] and one obtains, as in Sec. 8.7,

$$W = 15 I_0^2(\ln 2m\pi\gamma - \operatorname{Ci} 2m\pi). \tag{91}$$

THE KIRCHHOFF-HUYGENS PRINCIPLE

8.13. Scalar Wave Functions.—Let V be any region within a homogeneous, isotropic medium bounded by a closed, regular surface S, and let $\psi(x, y, z)$ be any solution of

$$\nabla^2\psi + k^2\psi = 0 \tag{1}$$

which is continuous and has continuous first derivatives within V and on S. Then it follows directly from Green's theorem, (7), page 165, that the value of ψ at any interior point x', y', z' can be expressed as an integral of ψ and its normal derivative over S.

$$\psi(x', y', z') = \frac{1}{4\pi}\int_S \left[\frac{\partial\psi}{\partial n}\frac{e^{ikr}}{r} - \psi\frac{\partial}{\partial n}\left(\frac{e^{ikr}}{r}\right)\right] da, \tag{2}$$

where as usual

$$r = \sqrt{(x' - x)^2 + (y' - y)^2 + (z' - z)^2} \tag{3}$$

is the distance from the variable point x, y, z on S to the fixed interior point x', y', z'. Equation (2) is in fact the special case of (22), page 427, for which all sources are located outside S and the time factor $\exp(-i\omega t)$ has been split off.

[1] Details are given by Bechmann, *loc. cit.*, p. 1477.

The function defined by (2) is continuous and has continuous derivatives at all interior points, but exhibits discontinuities as the point x', y', z' traverses the surface S. The transition of the function

$$f(x', y', z') = \frac{1}{4\pi}\int_S \frac{\psi}{r}\frac{\partial e^{ikr}}{\partial n}\,da \tag{4}$$

across S was shown to be continuous in Sec. 3.15; consequently, the discontinuities of (2) are identical with those discussed in Sec. 3.17 in connection with the static potential. If x', y', z' is *any* point, either interior or exterior, the function defined by the integral

$$u(x', y', z') = \frac{1}{4\pi}\int_S \left[\frac{\partial\psi}{\partial n}\frac{e^{ikr}}{r} - \psi\frac{\partial}{\partial n}\left(\frac{e^{ikr}}{r}\right)\right] da \tag{5}$$

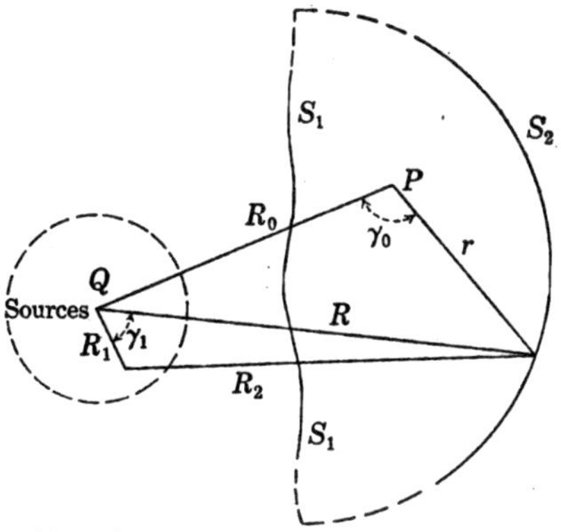

FIG. 86.—The surface S_1 may be closed at infinity.

is at all interior points identical with the solution ψ whose values have been specified over S, and at all exterior points is zero.[1]

As in the static case, too, the values of ψ and its derivatives at all interior points are uniquely determined by the values of ψ alone on S (Dirichlet problem), or by $\partial\psi/\partial n$ alone (Neumann problem). The values of *both* ψ and $\partial\psi/\partial n$, therefore, cannot be specified arbitrarily over S if u and ψ are to be identical at interior points. The function u defined by (5) satisfies (1) and is regular within V whatever the choice of ψ and $\partial\psi/\partial n$ over S, but the values assumed by u and $\partial u/\partial n$ on S will in general differ from those assigned to ψ and $\partial\psi/\partial n$.

All elements of the surface S which are infinitely remote from both the source and the point of observation contribute nothing to the value

[1] The analytic properties of the Kirchhoff wave functions are discussed in detail by Poincaré, "Théorie mathématique de la lumière," Vol. II, Chap. VII. See also the very useful treatise by Pockels, "Über die partielle Differentialgleichung $\Delta u + k^2 u = 0$," Teubner, 1891, and "Lectures on Cauchy's Problem" by Hadamard, Yale University Press, 1923.

of the integral (2). In Fig. 86 all elements of the source are located within a finite distance R_1 of a fixed origin Q. The surface S is represented as an infinite surface S_1 closed by a spherical surface S_2 of very large radius r about P as a center. The point of observation $P(x', y', z')$ lies within S and all elements of the source are exterior to S. If R, the distance from Q to an element of area on S_2, is very much larger than the largest value of R_1, then by (87), page 414, and (32), page 406,

$$\frac{e^{ikR_2}}{R_2} \simeq \frac{e^{ikR}}{R} \sum_{n=0}^{\infty} (-i)^n (2n+1) P_n(\cos\gamma_1)\, j_n(kR_1) = \frac{e^{ikR}}{R} \sum_{n=0}^{\infty} A_n(\gamma_1, R_1), \tag{6}$$

$$\frac{e^{ikR}}{R} \simeq \frac{e^{ikr}}{r} \sum_{n=0}^{\infty} B_n(\gamma_0, R_0); \tag{7}$$

hence,

$$\psi = \frac{e^{ikR_2}}{R_2} \simeq \frac{e^{ikr}}{r} f(\gamma_0, \gamma_1, R_0, R_1). \tag{8}$$

Over S_2

$$\begin{aligned} \frac{\partial\psi}{\partial n}\frac{e^{ikr}}{r} &= \frac{\partial\psi}{\partial r}\frac{e^{ikr}}{r} \simeq \left(ik - \frac{1}{r}\right)\left(\frac{e^{ikr}}{r}\right)^2 f(\gamma_0, \gamma, R_0, R_1), \\ \psi\frac{\partial}{\partial n}\left(\frac{e^{ikr}}{r}\right) &\simeq \left(ik - \frac{1}{r}\right)\left(\frac{e^{ikr}}{r}\right)^2 f(\gamma_0, \gamma, R_0, R_1). \end{aligned} \tag{9}$$

The terms in r^{-1}, r^{-2}, and r^{-3} in the expanded integrands of (2), therefore, cancel one another, while the remaining terms are of the order r^{-n}, $n > 3$, and consequently contribute nothing to the integral over S_2 as $r \to \infty$.

Let us suppose now that the surface S represents an opaque screen separating the source from the observer. In virtue of the theorem just proved it can be assumed that open surfaces of infinite extent are closed at infinity. If a small opening S_1 is made in the screen, the field will penetrate to some extent into the region occupied by the observer. The problem is to determine the intensity and distribution of this diffracted field. Clearly if the values of ψ and $\partial\psi/\partial n$ were known over the opening S_1 and over the observer's side of the screen, the diffracted field could be calculated by evaluating (2). These values are not known, but to obtain an approximate solution one may assume tentatively with Kirchhoff that:

(a) On the inner surface of the screen

$$\psi = 0, \qquad \frac{\partial\psi}{\partial n} = 0;$$

(b) Over the surface S_1 of the slit or opening the field is identical with that of the unperturbed incident wave.

Nearly all calculations of the diffraction patterns of slits and gratings made since Kirchhoff's time have been based on these assumptions.[1] It is easy to point out fundamental analytic errors involved in this procedure. In the first place, the assumption that ψ and $\partial\psi/\partial n$ are zero over the inside of the screen implies a discontinuity about the contour C_1 which bounds the opening S_1, whereas Green's theorem is valid only for functions which are continuous everywhere on the complete surface S. This difficulty cannot be obviated by the simple expedient of replacing the contour of discontinuity by a thin region of rapid but continuous transition. If ψ and $\partial\psi/\partial n$ are zero over any finite part of S, they are zero at all points of the space enclosed by S.[2] In the second place, an electromagnetic field cannot in general be represented by a single scalar wave function. It is characterized by a set of scalar functions which represent rectangular components of the electric and magnetic vectors. Each of these scalar functions satisfies (1) and its value at an interior point x', y', z' is, therefore, expressed by (2) in terms of its values over the boundary S. But these components at an interior point must not only satisfy the wave equation, they must also be solutions of the Maxwell field equations. The real problem is not the integration of a wave equation, either scalar or vector, but of a simultaneous system of first-order vector equations relating the vectors **E** and **H**.

In spite of such objections the classical Kirchhoff theory leads to satisfactory solutions of many of the diffraction problems of physical optics. This success is due primarily to the fact that the ratio of wave length to the largest dimension of the opening in optical problems is small. As a consequence the diffracted radiation is thrown largely forward in the direction of the incident ray and the assumption of zero intensity on the shadow side of the screen is approximately justified. Measurements are usually of intensity and do not take account of the polarization. As the wave length is increased, the diffraction pattern broadens. A computation of ψ from (2) now leads to intensities directly behind the screen which are by no means zero, contrary to the initial assumption. Various writers have suggested that this might be con-

[1] On the classical theory of diffraction see, for example, Planck, "Einführung in die theoretische Optik," Chap. IV, Hirzel, Leipzig, 1927, or Born, "Optik," Chap. IV, Springer, Berlin, 1933. To obtain a clearer understanding of the physical significance as well as the shortcomings of Kirchhoff's method the reader should consult the papers by Rubinowicz, *Ann. Physik*, **53**, 257, 1917, and Kottler, *ibid.*, **70**, 405, 1923. Since the writing of this and the following sections a treatment of the subject has appeared in book form: "The Mathematical Theory of Huygens' Principle," by Baker and Copson, Oxford, 1939.

[2] A direct consequence of Green's theorem. See POCKELS, *loc. cit.*, p. 212.

sidered as the first of a series of successive approximations, the values of ψ and $\partial\psi/\partial n$ over S obtained from the first calculation to be used as the boundary conditions in the second. There is, however, no proof that the process converges and the difficulties of integration make it impractical.

Recent advances in the technique of generating ultrahigh-frequency radio waves have stimulated interest in a number of problems previously of little practical importance. A natural consequence of this trend towards short waves is the application of methods of physical optics to the calculation of intensity and distribution of electromagnetic radiation from hollow tubes, horns, or small openings in cavity resonators. In radio practice, however, the length of the wave is commonly of the same order as the dimensions of the opening, and the polarization of the diffracted radiation is easily observed. It is hardly to be expected that the Kirchhoff formula (2) can be relied upon under these circumstances.

8.14. Direct Integration of the Field Equations.[1]—The problem of expressing the vectors **E** and **H** at an interior point in terms of the values of **E** and **H** over an enclosing surface has been discussed by a number of writers.[2] A simple and direct proof of the desired result can be obtained by applying the vector analogue of Green's theorem to the field equations. In Sec. 4.14 it was shown that if **P** and **Q** are two vector functions of position with the proper continuity, then

$$(10)\quad \int_V (\mathbf{Q}\cdot\nabla\times\nabla\times\mathbf{P} - \mathbf{P}\cdot\nabla\times\nabla\times\mathbf{Q})\,dv = \int_S (\mathbf{P}\times\nabla\times\mathbf{Q} - \mathbf{Q}\times\nabla\times\mathbf{P})\cdot\mathbf{n}\,da,$$

where S is as usual a regular surface bounding the volume V.

Let us assume that the field vectors contain the time only as a factor $\exp(-i\omega t)$ and write the field equations in the form

$$\text{(I)}\quad \nabla\times\mathbf{E} - i\omega\mu\mathbf{H} = -\mathbf{J}^*, \qquad \text{(III)}\quad \nabla\cdot\mathbf{H} = \frac{1}{\mu}\rho^*,$$

$$\text{(II)}\quad \nabla\times\mathbf{H} + i\omega\epsilon\mathbf{E} = \mathbf{J}, \qquad \text{(IV)}\quad \nabla\cdot\mathbf{E} = \frac{1}{\epsilon}\rho.$$

The medium is assumed to be homogeneous and isotropic, and of zero conductivity. The quantities $\mathbf{J}^*$ and ρ^* are fictitious densities of "magnetic current" and "magnetic charge," which to the best of our

[1] Sections 8.14 and 8.15 were written in collaboration with Dr. L. J. Chu.

[2] LOVE, *Phil. Trans.*, (A) **197**, 1901; LARMOR, *Lond. Math. Soc. Proc.*, **1**, 1, 1903; IGNATOWSKY, *Ann. Physik*, **23**, 875, 1907, and **25**, 99, 1908; TONOLO, *Annali di Mat.*, **3**, 17, 1910; MACDONALD, *Proc. Lond. Math. Soc.*, **10**, 91, 1911 and *Phil. Trans.*, (A) **212**, 295, 1912; TEDONE, *Linc. Rendi.*, (5) **1**, 286, 1917; KOTTLER, *Ann. Physik*, **71**, 457, 1923; SCHELKUNOFF, *Bell. System Tech. J.*, **15**, 92, 1936; BAKER and COPSON, *loc. cit.*, Chap. III.

knowledge have no physical existence. However, we shall have occasion shortly to assume arbitrary discontinuities in both **E** and **H**, discontinuities which are in fact physically impossible, but which *would* be generated by surface distributions of magnetic current or charge were they to exist. Currents and charges of both types are related by the equations of continuity,

$$\text{(V)}' \qquad \nabla \cdot \mathbf{J} - i\omega\rho = 0, \qquad \nabla \cdot \mathbf{J}^* - i\omega\rho^* = 0.$$

The vectors **E** and **H** satisfy

$$\text{(11)} \qquad \nabla \times \nabla \times \mathbf{E} - k^2\mathbf{E} = i\omega\mu\mathbf{J} - \nabla \times \mathbf{J}^*,$$

$$\text{(12)} \qquad \nabla \times \nabla \times \mathbf{H} - k^2\mathbf{H} = i\omega\epsilon\mathbf{J}^* + \nabla \times \mathbf{J},$$

with $k^2 = \omega^2\epsilon\mu$.

In (10) let $\mathbf{P} = \mathbf{E}$, $\mathbf{Q} = \phi\mathbf{a}$, where **a** is a unit vector in an arbitrary direction and $\phi = e^{ikr}/r$. Distance r is measured from the element at x, y, z to the point of observation at x', y', z' and is defined by (3). We have identically

$$\text{(13)} \qquad \nabla \times \mathbf{Q} = \nabla\phi \times \mathbf{a}, \qquad \nabla \times \nabla \times \mathbf{Q} = \mathbf{a}k^2\phi + \nabla(\mathbf{a} \cdot \nabla\phi),$$
$$\nabla \times \nabla \times \mathbf{P} = k^2\mathbf{E} + i\omega\mu\mathbf{J} - \nabla \times \mathbf{J}^*.$$

Following the procedure of Sec. 4.15, it is easily shown that **a** is a factor common to all the terms of (10) and, since **a** is arbitrary, it follows that

$$\text{(14)} \qquad \int_V \left(i\omega\mu\mathbf{J}\phi - \nabla \times \mathbf{J}^*\phi + \frac{1}{\epsilon}\rho\nabla\phi \right) dv$$
$$= \int_S [i\omega\mu(\mathbf{n} \times \mathbf{H})\phi + (\mathbf{n} \times \mathbf{E}) \times \nabla\phi + (\mathbf{n} \cdot \mathbf{E})\nabla\phi - \mathbf{n} \times \mathbf{J}^*\phi]\, da.$$

The identity

$$\text{(15)} \qquad \int_V \nabla \times \mathbf{J}^*\phi\, dv = \int_S \mathbf{n} \times \mathbf{J}^*\phi\, da + \int_V \mathbf{J}^* \times \nabla\phi\, dv$$

reduces this to

$$\text{(16)} \qquad \int_V \left(i\omega\mu\mathbf{J}\phi - \mathbf{J}^* \times \nabla\phi + \frac{1}{\epsilon}\rho\nabla\phi \right) dv$$
$$= \int_S [i\omega\mu(\mathbf{n} \times \mathbf{H}) + (\mathbf{n} \times \mathbf{E}) \times \nabla\phi + (\mathbf{n} \cdot \mathbf{E})\nabla\phi]\, da.$$

The exclusion of the singularity at $r = 0$ was described in Sec. 4.15. A sphere of radius r_1 is circumscribed about the point x', y', z', its normal directed out of V and consequently radially *toward* the center.

$$\text{(17)} \qquad \nabla\phi = \left(\frac{1}{r} - ik\right)\frac{e^{ikr}}{r}\mathbf{r}_0,$$

and on the sphere $\mathbf{n} = \mathbf{r}_0$. The area of the sphere vanishes with the radius as $4\pi r_1^2$ and, since

$$(\mathbf{n} \times \mathbf{E}) \times \mathbf{n} + (\mathbf{n} \cdot \mathbf{E})\mathbf{n} = \mathbf{E}, \tag{18}$$

the contribution of the spherical surface to the right-hand side (16) reduces to $4\pi\mathbf{E}(x', y', z')$. The value of $\mathbf{E}$ at any interior point of V is, therefore,

$$\begin{aligned}\mathbf{E}(x', y', z') = \frac{1}{4\pi}\int_V \left(i\omega\mu\mathbf{J}\phi - \mathbf{J}^* \times \nabla\phi + \frac{1}{\epsilon}\rho\,\nabla\phi\right) dv \\ - \frac{1}{4\pi}\int_S [i\omega\mu(\mathbf{n} \times \mathbf{H})\phi + (\mathbf{n} \times \mathbf{E}) \times \nabla\phi + (\mathbf{n} \cdot \mathbf{E})\nabla\phi]\, da.\end{aligned} \tag{19}$$

An obvious interchange of vectors leads to the corresponding expression for $\mathbf{H}$,

$$\begin{aligned}\mathbf{H}(x', y', z') = \frac{1}{4\pi}\int_V \left(i\omega\epsilon\mathbf{J}^*\phi + \mathbf{J} \times \nabla\phi + \frac{1}{\mu}\rho^*\nabla\phi\right) dv \\ + \frac{1}{4\pi}\int_S [i\omega\epsilon(\mathbf{n} \times \mathbf{E})\phi - (\mathbf{n} \times \mathbf{H}) \times \nabla\phi - (\mathbf{n} \cdot \mathbf{H})\nabla\phi]\, da.\end{aligned} \tag{20}$$

This last is the extension of (23), page 254, to the dynamic field.

If all currents and charges can be enclosed within a sphere of finite radius, the field is regular at infinity and either side of S may be chosen as its "interior."

In the earlier sections of this chapter the calculation of the fields of specified distributions of charge was discussed in terms of vector and scalar potentials and of Hertzian vectors. We have now shown that $\mathbf{E}$ and $\mathbf{H}$ may be calculated directly without the intervention of potentials. The surface integrals of (19) and (20) represent the contributions of sources located outside S. If S recedes to infinity, it may be assumed that these contributions vanish. Discarding the densities of magnetic charges and currents, one obtains the useful formulas

$$\begin{aligned}\mathbf{E}(x', y', z') &= \frac{1}{4\pi}\int_V \left[i\omega\mu\mathbf{J}\,\frac{e^{ikr}}{r} + \frac{1}{\epsilon}\rho\,\nabla\left(\frac{e^{ikr}}{r}\right)\right] dv, \\ \mathbf{H}(x', y', z') &= \frac{1}{4\pi}\int_V \mathbf{J} \times \nabla\left(\frac{e^{ikr}}{r}\right) dv.\end{aligned} \tag{21}$$

Since the current distribution is assumed to be known, the charge density can be determined from the equation of continuity.

In Sec. 9.2 it will be shown that an electromagnetic field within a bounded domain is completely determined by specification of the tangential components of $\mathbf{E}$ or $\mathbf{H}$ on the surface and the initial distribution of the field throughout the enclosed volume. It follows that when $\mathbf{n} \times \mathbf{E}$ and $\mathbf{n} \times \mathbf{H}$ in (19) and (20) have been fixed, the choice of $\mathbf{n} \cdot \mathbf{E}$

and $\mathbf{n} \cdot \mathbf{H}$ is no longer arbitrary. The selection must be consistent with the conditions imposed on a field satisfying Maxwell's equations. The same limitations on the choice of ψ and $\partial\psi/\partial n$ were pointed out in Sec. 8.13. The dependence of the normal component of $\mathbf{E}$ upon the tangential component of $\mathbf{H}$ is equivalent to that of ρ upon $\mathbf{J}$.

Let us suppose for the moment that the charge and current distributions in (19) are confined to a thin layer at the surface S. As the depth of the layer diminishes, the densities may be increased so that in the limit the volume densities are replaced in the usual way by surface densities. If the region V contains no charge or current within its interior or on its boundary S, the field at an interior point is

$$(22)\qquad \mathbf{E}(x',y',z') = -\frac{1}{4\pi}\int_S [i\omega\mu(\mathbf{n}\times\mathbf{H})\phi + (\mathbf{n}\times\mathbf{E})\times\nabla\phi + (\mathbf{n}\cdot\mathbf{E})\nabla\phi]\,da.$$

It is now clear that this is exactly the field that would be produced by a distribution of electric current over S with surface density $\mathbf{K}$, a distribution of magnetic current of density $\mathbf{K}^*$, and a surface electric charge of density η, where

$$(23)\qquad \mathbf{K} = -\mathbf{n}\times\mathbf{H},\qquad \mathbf{K}^* = \mathbf{n}\times\mathbf{E},\qquad \eta = -\epsilon\,\mathbf{n}\cdot\mathbf{E}.$$

The values of $\mathbf{E}$ and $\mathbf{H}$ in (23) are those just *inside* the surface S.

The function $\mathbf{E}(x', y', z')$ defined by (22) is discontinuous across S. It can be shown as above that discontinuities associated with $\phi = e^{ikr}/r$ are identical with those of the stationary regime, $\phi = 1/r$. Then by Sec. 3.15 the integral

$$(24)\qquad \mathbf{E}_3(x', y', z') = \frac{1}{4\pi\epsilon}\int_S \eta\,\nabla\phi\,da$$

suffers a discontinuity on transition through S equal to $\mathbf{n}\cdot\Delta\mathbf{E}_3 = \eta/\epsilon$, where $\Delta\mathbf{E}_3$ is the difference of the values outside and inside. The third term of (22), therefore, does not affect the transition of the tangential component but reduces the normal component of $\mathbf{E}$ to zero. Likewise by Sec. 4.13 the discontinuity of

$$(25)\qquad \mathbf{E}_2(x', y', z') = \frac{1}{4\pi}\int_S \mathbf{K}^*\times\nabla\phi\,da$$

is specified by $\mathbf{n}\times\Delta\mathbf{E}_2 = \mathbf{K}^*$, so that the second term in (22) reduces the tangential component of $\mathbf{E}$ to zero without affecting the normal component. The first term in (22) is continuous across S, but has discontinuous derivatives. The curl of $\mathbf{E}$ at x', y', z' is

$$(26)\qquad \nabla'\times\mathbf{E}(x', y', z') = -\frac{i\omega\mu}{4\pi}\int_S (\mathbf{n}\times\mathbf{H})\times\nabla\phi\,da$$

which is of the type (25). *The vector* **E** *and the tangential component of its curl are zero on the positive side of S;* **E** *is, therefore, zero at all external points.* The same analysis applies to **H**.

8.15. Discontinuous Surface Distributions.—The results of the preceding section hold only if the vectors **E** and **H** are continuous and have continuous first derivatives at all points of S. They cannot, therefore, be applied directly to the problem of diffraction at a slit. To obtain the required extension of (22) to such cases, consider the closed surface S (surfaces closed at infinity are included) to be divided into two zones S_1 and S_2 by a closed contour C lying on S, as in Fig. 87. The vectors **E** and **H** and their first derivatives are continuous over S_1 and satisfy the field equations. The same is true over S_2. However, the components of **E** and **H** which are tangential to the surface are now subject to a discontinuous change in passing across C from one zone to the other. The occurrence of such discontinuities can be reconciled with the field equations only by the further assumption of a line distribution of charges or currents about the contour C. This line distribution of sources contributes to the field, and only when it is taken into account do the resultant expressions for **E** and **H** satisfy Maxwell's equations.

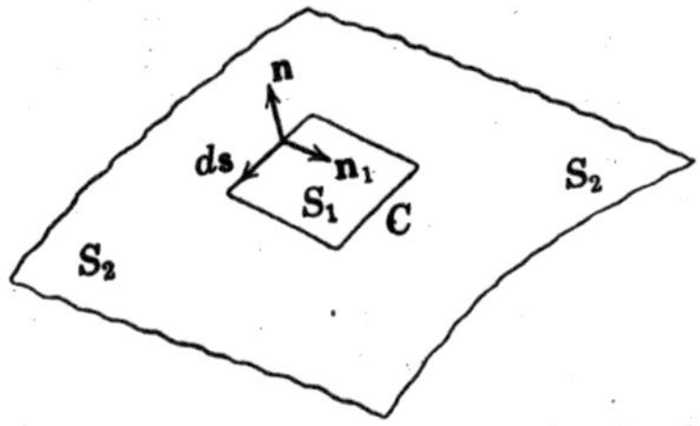

FIG. 87.—C is a contour on the closed surface S dividing it into the two parts S_1 and S_2.

A method of determining a contour distribution consistent with the requirements of the problem was proposed by Kottler.[1] It has been shown that the field at an interior point is identical with that produced by the surface currents and charges specified in (23). A discontinuity in the tangential components of **E** and **H** in passing on the surface from zone S_1 to zone S_2 implies therefore an abrupt change in the surface current density. The termination of a line of current, in turn, can be accounted for according to the equation of continuity by an accumulation of charge on the contour. Let $d\mathbf{s}$ be an element of length along the contour in the positive direction as determined by the positive normal $\mathbf{n}$, Fig. 87. Let $\mathbf{n}_1$ be a unit vector lying in the surface, normal to both $\mathbf{n}$ and the contour element $d\mathbf{s}$, and directed into zone (1). The line densities of electric and magnetic charge will be designated by σ and σ^*. Then Eqs. (V), when applied to surface currents, become

$$(27) \qquad \mathbf{n}_1 \cdot (\mathbf{K}_1 - \mathbf{K}_2) = i\omega\sigma, \qquad \mathbf{n}_1 \cdot (\mathbf{K}_1^* - \mathbf{K}_2^*) = i\omega\sigma^*,$$

[1] KOTTLER, *Ann. Physik*, 71, 457, 1923.

and hence by (23)

$$
\begin{aligned}
i\omega\sigma &= \mathbf{n}_1 \cdot (\mathbf{n} \times \mathbf{H}_2 - \mathbf{n} \times \mathbf{H}_1) = (\mathbf{H}_2 - \mathbf{H}_1) \cdot (\mathbf{n}_1 \times \mathbf{n}), \\
i\omega\sigma^* &= \mathbf{n}_1 \cdot (\mathbf{n} \times \mathbf{E}_1 - \mathbf{n} \times \mathbf{E}_2) = -(\mathbf{E}_2 - \mathbf{E}_1) \cdot (\mathbf{n}_1 \times \mathbf{n}).
\end{aligned} \tag{28}
$$

The vector $\mathbf{n}_1 \times \mathbf{n}$ is in the direction of ds. If S_2 represents an opaque screen over which $\mathbf{E}_2 = \mathbf{H}_2 = 0$, the field at any point on the shadow side is

$$
\begin{aligned}
\mathbf{E}(x', y', z') = &-\frac{1}{i\omega\epsilon}\frac{1}{4\pi}\oint_C \nabla\phi\, \mathbf{H}_1 \cdot d\mathbf{s} \\
&- \frac{1}{4\pi}\int_{S_1} [i\omega\mu(\mathbf{n} \times \mathbf{H}_1)\phi + (\mathbf{n} \times \mathbf{E}_1) \times \nabla\phi + (\mathbf{n} \cdot \mathbf{E}_1)\nabla\phi]\, da,
\end{aligned} \tag{29}
$$

which can be shown to be identical with

$$
\begin{aligned}
4\pi\mathbf{E}(x', y', z') = &-\frac{1}{i\omega\epsilon}\oint_C \nabla\phi\, \mathbf{H}_1 \cdot d\mathbf{s} + \oint_C \phi\, \mathbf{E}_1 \times d\mathbf{s} \\
&- \int_{S_1}\left(\mathbf{E}_1\frac{\partial\phi}{\partial n} - \phi\frac{\partial\mathbf{E}_1}{\partial n}\right) da.
\end{aligned} \tag{30}
$$

For the magnetic field one obtains

$$
\begin{aligned}
4\pi\mathbf{H}(x', y', z') = &\frac{1}{i\omega\mu}\oint_C \phi\mathbf{E}_1 \cdot d\mathbf{s} \\
&+ \int_{S_1} [i\omega\epsilon(\mathbf{n} \times \mathbf{E}_1)\phi - (\mathbf{n} \times \mathbf{H}_1) \times \nabla\phi - (\mathbf{n} \cdot \mathbf{H}_1)\,\nabla\phi]\, da \\
= &\frac{1}{i\omega\mu}\oint_C \nabla\phi\, \mathbf{E}_1 \cdot d\mathbf{s} + \oint_C \phi\, \mathbf{H}_1 \times d\mathbf{s} - \int_{S_1}\left(\mathbf{H}_1\frac{\partial\phi}{\partial n} - \phi\frac{\partial\mathbf{H}_1}{\partial n}\right) da.
\end{aligned} \tag{31}
$$

It remains to be shown that the fields expressed by these integrals are in fact divergenceless and satisfy (I) and (II). Consider first the divergence of (29) at a point x', y', z'.

$$
\begin{aligned}
\nabla' \cdot \mathbf{E}(x', y', z') = &\frac{1}{i\omega\epsilon}\frac{1}{4\pi}\oint_C \nabla^2\phi\, \mathbf{H}_1 \cdot d\mathbf{s} \\
&+ \frac{1}{4\pi}\int_S [i\omega\mu(\mathbf{n} \times \mathbf{H}_1) \cdot \nabla\phi + (\mathbf{n} \cdot \mathbf{E}_1)\nabla^2\phi]\, da \\
= &-\frac{k^2}{i\omega\epsilon}\frac{1}{4\pi}\oint_C \phi\, \mathbf{H}_1 \cdot d\mathbf{s} - \frac{k^2}{4\pi}\int_{S_1} (\mathbf{n} \cdot \mathbf{E}_1)\phi\, da \\
&+ \frac{i\omega\mu}{4\pi}\int_{S_1} (\mathbf{n} \times \mathbf{H}_1) \cdot \nabla\phi\, da,
\end{aligned} \tag{32}
$$

taking into account the relation $\nabla' = -\nabla$ when applied to ϕ or its derivatives. Now

$$
\int_S (\mathbf{n} \times \mathbf{H}) \cdot \nabla\phi\, da = \int_S \phi\nabla \times \mathbf{H} \cdot \mathbf{n}\, da - \oint_C \phi\mathbf{H} \cdot d\mathbf{s}. \tag{33}
$$

The line integral resulting from this transformation is zero when S is closed, but otherwise just cancels the contour integral in (32). Inversely, *only the presence of the contour integral in* (29) *leads to a zero divergence and waves which are transverse at great distances from the opening* S_1. From (II) and the relation $k^2/i\omega\epsilon = -i\omega\mu$, follows immediately the result

$$\nabla' \cdot \mathbf{E}(x', y', z') = 0. \tag{34}$$

An identical proof holds for $\mathbf{H}(x', y', z')$.

Finally, it will be shown that (29) and (31) satisfy (I) and (II).

$$\nabla' \times \mathbf{H}(x', y', z') = -\frac{1}{4\pi}\int_{S_1}[(\mathbf{n}\times\mathbf{H}_1)\cdot\nabla\nabla\phi + k^2\phi(\mathbf{n}\times\mathbf{H}_1) - i\omega\epsilon(\mathbf{n}\times\mathbf{E}_1)\times\nabla\phi]\,da, \tag{35}$$

since the curl of the gradient is identically zero. Furthermore,

$$\begin{aligned}\int_{S_1}(\mathbf{n}\times\mathbf{H}_1)\cdot\nabla\nabla\phi\,da &= -\int_{S_1}\mathbf{n}\cdot(\nabla_{\nabla\phi}\times\mathbf{H}_1)\nabla\phi\,da\\ &= \int_{S_1}(\mathbf{n}\cdot\nabla\times\mathbf{H}_1)\nabla\phi\,da - \int_{S_1}\mathbf{n}\times\nabla\cdot(\mathbf{H}_1\nabla\phi)\,da\\ &= -\oint_C \nabla\phi\,\mathbf{H}_1\cdot d\mathbf{s} - i\omega\epsilon\int_{S_1}(\mathbf{n}\cdot\mathbf{E}_1)\nabla\phi\,da,\end{aligned} \tag{36}$$

where the operator $\nabla_{\nabla\phi}$ acts on $\nabla\phi$ only. The last integral takes account of the fact that the field equations are by hypothesis satisfied on S_1. Then

$$\begin{aligned}\nabla'\times\mathbf{H}(x', y', z') &= \frac{1}{4\pi}\oint_C\nabla\phi\,\mathbf{H}_1\cdot d\mathbf{s} + \frac{i\omega\epsilon}{4\pi}\int_{S_1}[i\omega\mu(\mathbf{n}\times\mathbf{H}_1)\phi\\ &\qquad + (\mathbf{n}\times\mathbf{E}_1)\times\nabla\phi + (\mathbf{n}\cdot\mathbf{E}_1)\,\nabla\phi]\,da\\ &= -i\omega\epsilon\,\mathbf{E}(x', y', z').\end{aligned} \tag{37}$$

The validity of (I) is established in the same manner.[1]

FOUR-DIMENSIONAL FORMULATION OF THE RADIATION PROBLEM

8.16. Integration of the Wave Equation.—In Sec. 1.21 it was shown how by means of tensors the field equations could be written in an exceedingly concise form. If the imaginary distance $x_4 = ict$ be introduced as a coordinate in a four-dimensional manifold, the equations of a variable field are in fact *formally* identical with those which govern the static regime, and the methods which were applied to the integration of Poisson's equation can be extended directly to the more general case. The four-

[1] These formulas have been applied by L. J. Chu and the author to the calculation of diffraction by a rectangular slit in a perfectly conducting screen. The results compare favorably with those obtained by Morse and Rubenstein, *Phys. Rev.*, **54**, 895, 1938, who solved a two-dimensional slit problem rigorously by introducing coordinates of a hyperbolic cylinder. *Phys. Rev.*, **56**, 99. 1939. See also the treatment of such problems by Schelkunoff, *ibid.*, p. 308.

dimensional theory is more abstract than the methods described earlier in this chapter, and consequently has not been applied to the practical problem of calculating antenna radiation. Quite apart from its formal elegance, however, the four-dimensional treatment sometimes leads in the most direct manner to very useful results. This is particularly true of the rather difficult problem of calculating the field of an isolated charge moving in an arbitrary way.

The discussion will be confined to the case of charges and currents in *free space*. As in (82), page 73, the vector and scalar potentials can be represented by a single four-vector whose rectangular components are

(1) $$\Phi_1 = A_x, \qquad \Phi_2 = A_y, \qquad \Phi_3 = A_z, \qquad \Phi_4 = \frac{i}{c}\phi.$$

Likewise the current and charge densities are represented by a four-vector whose components are

(2) $$J_1 = J_x, \qquad J_2 = J_y, \qquad J_3 = J_z, \qquad J_4 = ic\rho.$$

The four-potential satisfies the relations

(3) $$\sum_{k=1}^{4} \frac{\partial^2 \Phi_j}{\partial x_k^2} = -\mu_0 J_j, \qquad \sum_{k=1}^{4} \frac{\partial \Phi_k}{\partial x_k} = 0, \qquad (j = 1, 2, 3, 4).$$

According to the notation of (35), page 64, $\Box$ is a symbolic four-vector whose components are $\partial/\partial x_k$. Then (3) can be expressed concisely as

(4) $$\Box^2 \mathbf{\Phi} = -\mu_0 \mathbf{J}, \qquad \Box \cdot \mathbf{\Phi} = 0.$$

Let V be a region of a four-dimensional space bounded by a three-dimensional "surface" S. $\mathbf{n}$ is the unit outward normal to S. Then in four as in three dimensions

(5) $$\int_V \Box \cdot \mathbf{\Phi}\, dv = \int_S \mathbf{\Phi} \cdot \mathbf{n}\, da.$$

If u and w are two scalar functions of the four coordinates x_1, x_2, x_3, x_4 which together with their first and second derivatives are continuous throughout V and on S, then

(6) $$\int_V \Box \cdot (u \Box w)\, dv = \int_S u \Box w \cdot \mathbf{n}\, da,$$

(7) $$\int_V \Box u \cdot \Box w\, dv + \int_V u \Box^2 w\, dv = \int_S u \frac{\partial w}{\partial n}\, da.$$

Thus, one obtains a four-dimensional analogue of Green's theorem,

(8) $$\int_V (w \Box^2 u - u \Box^2 w)\, dv = \int_S \left(w \frac{\partial u}{\partial n} - u \frac{\partial w}{\partial n} \right) da.$$

Let x'_j be the coordinates of a fixed point of observation within S, and x_j those of any variable point in V or on S. The distance from x_j to x'_j is

$$R = \sqrt{\sum_{j=1}^{4} (x'_j - x_j)^2}. \tag{9}$$

It can be verified by direct differentiation that the equation

$$\Box^2 w = 0 \tag{10}$$

is satisfied by

$$w = \frac{1}{R^2} \tag{11}$$

at all points exclusive of the singularity $R = 0$, which will be excised in the usual fashion by a sphere S_1 of radius R_1. If u is identified with any component Φ_j of the four-potential, there follows

$$\int_V \frac{\Box^2\Phi_j}{R^2}\, dv = \int_{S_1+S_2} \left[\frac{1}{R^2}\frac{\partial\Phi_j}{\partial n} - \Phi_j \frac{\partial}{\partial n}\left(\frac{1}{R^2}\right)\right] da. \tag{12}$$

Over S_1 we have

$$\frac{\partial\Phi_j}{\partial n} = -\frac{\partial\Phi_j}{\partial R}, \qquad \left[\frac{\partial}{\partial n}\left(\frac{1}{R^2}\right)\right]_{R=R_1} = \frac{2}{R_1^3}. \tag{13}$$

To determine the area of the hyperspherical surface S_1 polar coordinates are introduced,

$$\begin{aligned} x_1 &= R\cos\theta_1, & x_2 &= R\sin\theta_1\cos\theta_2, \\ x_3 &= R\sin\theta_1\sin\theta_2\cos\phi, & x_4 &= R\sin\theta_1\sin\theta_2\sin\phi, \end{aligned} \tag{14}$$

which satisfy

$$x_1^2 + x_2^2 + x_3^2 + x_4^2 = R^2. \tag{15}$$

The scale factors h_i are calculated as in (70), page 48, and are found to be

$$h_1 = 1, \qquad h_2 = R, \qquad h_3 = R\sin\theta_1, \qquad h_4 = R\sin\theta_1\sin\theta_2. \tag{16}$$

The element of volume is

$$dv = h_1h_2h_3h_4\, dR\, d\theta_1\, d\theta_2\, d\phi = R^3\sin^2\theta_1\sin\theta_2\, dR\, d\theta_1\, d\theta_2\, d\phi; \tag{17}$$

hence, the area of the hypersphere is

$$R^3\int_0^{\pi}\int_0^{\pi}\int_0^{2\pi} \sin^2\theta_1\sin\theta_2\, d\theta_1\, d\theta_2\, d\phi = 2\pi^2R^3. \tag{18}$$

Therefore,

$$\int_{S_1}\left[\frac{1}{R^2}\frac{\partial\Phi_j}{\partial n} - \Phi_j\frac{\partial}{\partial n}\left(\frac{1}{R^2}\right)\right] da \to -4\pi^2\Phi_j(x'), \qquad \text{as } R_1 \to 0. \tag{19}$$

Upon replacing $\Box^2\Phi_j$ by $-\mu_0 J_j$ and recombining the components Φ_j into a four-vector, one obtains for $\mathbf{\Phi}$ at the point x_j' the expression

$$(20) \qquad \mathbf{\Phi}(x') = \frac{\mu_0}{4\pi^2}\int_V \frac{\mathbf{J}(x)}{R^2}\,dv + \frac{1}{4\pi^2}\int_S \left[\frac{1}{R^2}\frac{\partial \mathbf{\Phi}}{\partial n} - \mathbf{\Phi}\frac{\partial}{\partial n}\left(\frac{1}{R^2}\right)\right] da.$$

To find the field vectors one must compute the components of the tensor ${}^2\mathbf{F}$ at the point x_j'. By (79), page 72,

$$(21) \qquad F_{jk} = \frac{\partial \Phi_k}{\partial x_j'} - \frac{\partial \Phi_j}{\partial x_k'}, \qquad (j, k = 1, 2, 3, 4).$$

Let us suppose for the moment that all sources are contained within V. Then the integral over S contributes nothing in (20) to the value of $\mathbf{\Phi}$ and in this case

$$(22) \qquad F_{jk}(x') = \frac{\mu_0}{4\pi^2}\int_V \left[J_k \frac{\partial}{\partial x_j'}\left(\frac{1}{R^2}\right) - J_j \frac{\partial}{\partial x_k'}\left(\frac{1}{R^2}\right)\right] dv$$
$$= \frac{\mu_0}{2\pi^2}\int_V \frac{(\mathbf{J}\times\mathbf{R})_{jk}}{R^4}\,dv,$$

where $(\mathbf{J}\times\mathbf{R})_{jk} = J_j R_k - J_k R_j$, and where $\mathbf{R}$ is the radius vector drawn from x to x'. The cross product $\mathbf{J}\times\mathbf{R}$ is a six-vector or antisymmetric tensor whose components are defined as in (62), page 69. In concise form,

$$(23) \qquad {}^2\mathbf{F}(x') = \frac{\mu_0}{2\pi^2}\int_V \frac{\mathbf{J}\times\mathbf{R}}{R^4}\,dv.$$

8.17. Field of a Moving Point Charge.—An isolated charge q moves with an arbitrary velocity $\mathbf{v}$ in free space. It will be assumed that the distance from the charge to the observer is such that the charge can be represented by a geometrical point. The motion of the charge along its trajectory is specified by expressing its coordinates as functions of t,

$$(24) \qquad x_1 = f_1(t), \qquad x_2 = f_2(t), \qquad x_3 = f_3(t), \qquad x_4 = f_4(t) = ict.$$

One must note that here t is not the observer's time but time as measured on the charge. Then

$$(25) \qquad \int_{-\infty}^{\infty}\int_{-\infty}^{\infty}\int_{-\infty}^{\infty} J_j\,dx_1\,dx_2\,dx_3 = q\frac{dx_j}{dt} = q\frac{df_j}{dt}, \qquad (j = 1, 2, 3, 4),$$

and by (20)

$$(26) \qquad \Phi_j = \frac{\mu_0}{4\pi^2}q\int_{-\infty}^{\infty}\frac{df_j}{dt}\frac{1}{R^2}\,dx_4.$$

The origin on the time axis can be shifted at will without loss of generality. It will be convenient to assume that the observer's time

$t' = 0$ at the instant of observation. Then $x_4' = 0$ and

$$R^2 = r^2 + x_4^2, \tag{27}$$

where r is the distance in three-dimensional configuration space from the charge to the observer. The poles of the integrand in (26) are located at

$$x_4 = -ir, \qquad x_4 = +ir. \tag{28}$$

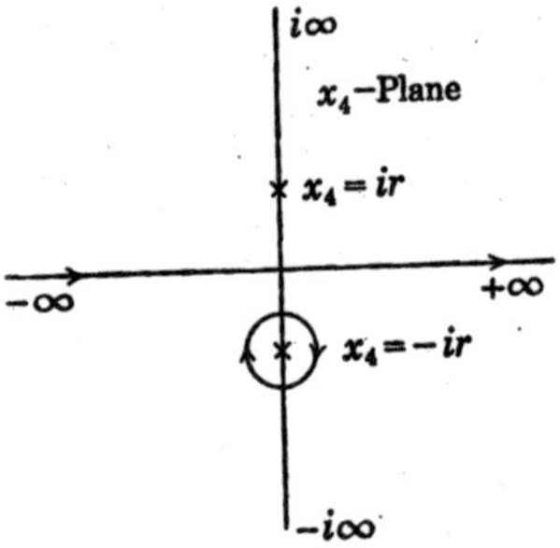

FIG. 88.—The contour of integration is reduced to a small circle about the pole at $x_4 = -ir$.

Now the data of the problem specify the coordinates of the trajectory for real values t which are less than t', and in the present instance less, therefore, than zero. Thus the $f_j(t)$ are given only for values of x_4 on the negative imaginary axis. However, the assumption will be made that the $f_j(t)$ are analytic functions of $x_4 = ict$ and that they can be continued analytically over the entire complex x_4-plane. Since the only singularities of (26) are the poles at (28), it follows from Jordan's lemma, page 315, that the contour of integration can be deformed from the real axis to a small circle about the pole $x_4 = -ir$ in the direction indicated by Fig. 88, and the integral then evaluated by the method of residues, page 315. Expansion of R^2 about the point $x_4 = -ir$ gives

$$R^2 = \left(\frac{dR^2}{dx_4}\right)_{R^2=0} (x_2 + ir) + \cdots. \tag{29}$$

$$\frac{dR^2}{dx_4} = \frac{2i}{c} \sum_{j=1}^{3} (x_j' - x_j) \frac{dx_j}{dt} + 2x_4 = \frac{2i}{c} \mathbf{v} \cdot \mathbf{r} + 2x_4; \tag{30}$$

hence,

$$\left(\frac{dR^2}{dx_4}\right)_{R^2=0} = \frac{2i}{c} \mathbf{v} \cdot \mathbf{r} - 2ir = 2ir\left(\frac{v_r}{c} - 1\right), \tag{31}$$

where v_r is the component of charge velocity in the direction of the radius vector $\mathbf{r}$ drawn from charge to observer.

The integral (26) now has the form

$$\Phi_j = \frac{\mu_0 q}{4\pi^2} \frac{i}{2r\left(1 - \dfrac{v_r}{c}\right)} \oint \frac{df_j}{dt} \frac{1}{x_4 + ir}\, dx_4. \tag{32}$$

The contour is traced in the clockwise direction; consequently, the integral in (32) is by Cauchy's theorem equal to $-2\pi i \dfrac{df_j}{dt}$.

$$\Phi_j = \frac{\mu_0}{4\pi} \frac{q}{r\left(1 - \frac{v_r}{c}\right)} \frac{df_j}{dt}, \tag{33}$$

or, in terms of vector and scalar potentials,

$$\mathbf{A} = \frac{\mu_0}{4\pi} \frac{q\mathbf{v}}{r\left(1 - \frac{v_r}{c}\right)}, \qquad \phi = \frac{1}{4\pi\epsilon_0} \frac{q}{r\left(1 - \frac{v_r}{c}\right)}. \tag{34}$$

These are the formulas of Liénard and Wiechert. The values of r and $\mathbf{v}$ are those associated with q at an instant preceding the observation by the interval r/c.

Rather than differentiate these expressions to obtain the field vectors, it is most direct to apply the same procedure to (23). This has been done by Sommerfeld[1] who obtains formulas, the derivation of which by older methods led to great complications. It is convenient to resolve the expressions for the field into two parts: a *velocity field* which contains no terms involving the acceleration $\dot{\mathbf{v}}$, and an *acceleration field* which vanishes as $\dot{\mathbf{v}}$ goes to zero. For the velocity field one finds

$$\begin{aligned} \mathbf{E} &= \frac{q}{4\pi\epsilon_0} \frac{\gamma^3}{r^2} \left(\mathbf{r}_0 - \frac{\mathbf{v}}{c}\right)\left(1 - \frac{v^2}{c^2}\right), \\ \mathbf{H} &= \frac{q}{4\pi} \frac{\gamma^3}{r^2} (\mathbf{v} \times \mathbf{r}_0)\left(1 - \frac{v^2}{c^2}\right), \end{aligned} \tag{35}$$

where

$$\gamma = \frac{1}{1 - \frac{\mathbf{v} \cdot \mathbf{r}_0}{c}} \tag{36}$$

and $\mathbf{r}_0$ is a unit vector in the direction of $\mathbf{r}$ from charge to observer at the instant $t' - \frac{r}{c}$. These formulas can be obtained also by applying the Lorentz transformation (111), page 79, to the field of a static charge and noting that (111) refers to the observer's time.

The acceleration field is

$$\begin{aligned} \mathbf{E} &= \frac{q}{4\pi\epsilon_0} \frac{\gamma^2}{rc^2} \left[\gamma\left(\mathbf{r}_0 - \frac{\mathbf{v}}{c}\right)(\mathbf{r}_0 \cdot \dot{\mathbf{v}}) - \dot{\mathbf{v}}\right], \\ \mathbf{H} &= \frac{q}{4\pi} \frac{\gamma^2}{rc} \left[\gamma\left(\frac{\mathbf{v}}{c} \times \mathbf{r}_0\right)(\mathbf{r}_0 \cdot \dot{\mathbf{v}}) + \dot{\mathbf{v}} \times \mathbf{r}_0\right]. \end{aligned} \tag{37}$$

[1] SOMMERFELD, in Riemann-Weber, "Partiellen Differentialgleichungen der mathematischen Physik," p. 786, 8th ed., 1935. See also, ABRAHAM, "Theorie der Elektrizität," Vol. II, pp. 74*ff*., 5th ed., 1923, and FRENKEL, "Lehrbuch der Elektrodynamik," Vol. I, Chap. VI.

Note first that *both* velocity and acceleration fields satisfy the condition

$$\mathbf{H} = \sqrt{\frac{\epsilon_0}{\mu_0}}\,\mathbf{r}_0 \times \mathbf{E}. \tag{38}$$

The magnetic vector is always perpendicular to the radius vector drawn from the effective position of the charge, by which one means the position at the instant $t' - \frac{r}{c}\cdot$ The electric vector of the velocity field is not transverse, but in the acceleration field

$$\mathbf{E} = \sqrt{\frac{\mu_0}{\epsilon_0}}\,\mathbf{H} \times \mathbf{r}_0, \qquad \mathbf{E} \cdot \mathbf{r}_0 = \mathbf{H} \cdot \mathbf{r}_0 = 0. \tag{39}$$

The velocity field decreases as $1/r^2$, while the acceleration field diminishes only as $1/r$. *At great distances the acceleration field predominates; it is purely transverse and it alone gives rise to radiation.*

In case the velocity of the charge is very much less than that of light, the equations (37) for the field at large distances become approximately

$$\begin{aligned} \mathbf{E} &\simeq \frac{q}{4\pi\epsilon_0}\frac{1}{rc^2}\,\mathbf{r}_0 \times (\mathbf{r}_0 \times \dot{\mathbf{v}}), \\ \mathbf{H} &\simeq \frac{q}{4\pi}\frac{1}{rc}\,\dot{\mathbf{v}} \times \mathbf{r}_0, \end{aligned} \tag{40}$$

which remind one of (27), page 435, for the radiation field of a dipole.

The results of this section have been based on the assumption that the motion and trajectory of the charge are known, just as earlier the assumption was made that the current distribution can be specified. In neither case is it exact to treat the problem of the motion or distribution apart from that of the radiation. Let us suppose, for example, that a charge is projected into a known magnetic field. If the radiation is ignored, the mechanical force exerted on the charge is $q\mathbf{v} \times \mathbf{B}$, from which the motion can be calculated by the methods of classical mechanics. In case the velocities are large, a correction can also be made for the relativistic change in mass. Now the effect of this force is to accelerate the charge in a direction transverse to its motion and, consequently, to introduce an energy loss through radiation. The dissipation of energy through radiation is not accounted for by the force $q\mathbf{v} \times \mathbf{B}$. An additional force, the *radiation reaction,* which can be compared roughly to friction, must be included. The radiation reaction in turn affects the trajectory, whence it is obvious that the exact solution can be found only by introducing the total effective force from the outset—a very much more difficult problem. Fortunately the radiation reaction is in most cases exceedingly small, so that a satisfactory approximation for the motion can be obtained by ignoring it entirely.

Problems

1. Show that the r.m.s. field intensity at large distances from a half-wave linear oscillator is given by the formula

$$E = \frac{7\sqrt{W}}{R} \frac{\cos\left(\frac{\pi}{2}\cos\theta\right)}{\sin\theta} \qquad \text{volts/meter,}$$

where W is the radiated power in watts, R the distance from the oscillator in meters, and θ the angle made by the radius vector $\mathbf{R}$ with the axis of the oscillator.

Show that the r.m.s. field intensity at large distances from an electric dipole is

$$E = 6.7 \frac{\sin\theta}{R}\sqrt{W} \qquad \text{volts/meter,}$$

provided the wave length is very large relative to the length of the dipole.

2. Let ψ be a solution of the equation

$$\frac{\partial^2\psi}{\partial x^2} + \frac{\partial^2\psi}{\partial y^2} + k^2\psi = 0,$$

which together with its first and second partial derivatives is continuous within and on a closed contour C in the xy-plane, and let

$$u(x', y') = \frac{1}{4}\int_C \left[H_0^{(1)}(kr)\frac{\partial\psi}{\partial n} - \psi\frac{\partial}{\partial n}H_0^{(1)}(kr)\right] ds,$$

where $r = \sqrt{(x'-x)^2 + (y'-y)^2}$, n denotes the outward normal to C, and $H_0^{(1)}(kr)$ is a Hankel function. Show that, if the fixed point x', y' lies outside the contour C, then $u = 0$, while $u = \psi$ if it lies within.

The theorem holds also when $H_0^{(1)}(kr)$ is replaced by the function $N(kr)$. This is a two-dimensional analogue of the Helmholtz formula (5), page 461, derived by Weber, *Math. Ann.*, **1**, 1–36, 1869.

3. Discuss the analogue of the Kirchhoff formula in two dimensions. A proof based on the Weber formula of Problem 2 has been given by Volterra, *Acta Math.*, **18**, 161, 1894. (Very interesting work on the propagation of waves in a two-dimensional space has recently been done by M. Riesz. See the discussion by Baker and Copson, "Huygens' Principle," Cambridge University Press, p. 54, 1939.)

4. Let x', y' be any fixed point within a plane two-dimensional domain bounded by a closed curve C, and let ψ be a solution of

$$\frac{\partial^2\psi}{\partial x^2} + \frac{\partial^2\psi}{\partial y^2} + k^2\psi = -\eta(x, y)$$

which is continuous and has continuous first derivatives on C and within the enclosed area S. Show that

$$2\pi\psi(x', y') = \int_C \left[\frac{\partial\psi}{\partial n}K_0(ikr) - \psi\frac{\partial}{\partial n}K_0(ikr)\right] ds + \frac{1}{2\pi}\int_S \eta(x, y)\,K_0(ikr)\,da,$$

where $r^2 = (x'-x)^2 + (y'-y)^2$ and $K_0(ikr)$ is a modified Bessel function discussed in Problem 10, Chap. VI. The normal n is drawn outward from the contour.

5. Let the current distribution throughout a volume V be specified by the function $\mathbf{J}(x)e^{-i\omega t}$, where x stands for the three coordinates x, y, z. Show that the electric field intensity due to the distribution can be represented by the integral

$$\mathbf{E}(x', t) = \frac{ie^{-i\omega t}}{4\pi\omega\epsilon}\int_V [(\mathbf{J}\cdot\nabla)\nabla + k^2\mathbf{J}]\frac{e^{ikr}}{r}\,dv,$$

where x' stands for the three coordinates of the point of observation, $k = \omega\sqrt{\mu\epsilon}$, and

$$r^2 = (x' - x)^2 + (y' - y)^2 + (z' - z)^2.$$

6. When the theorem of Problem 5 is applied to a perfect conductor one obtains

$$\mathbf{E}(x', t) = \frac{ie^{-i\omega t}}{4\pi\omega\epsilon}\int_S [(\mathbf{K}\cdot\nabla)\nabla + k^2\mathbf{K}]\frac{e^{ikr}}{r}\,da$$

where $\mathbf{K}$ is the surface current density at a point x, y, z on the conductor S. Let S now be the surface of a linear conductor. Assume that the curvature of the wire is continuous and that at all points the cross section is small in comparison with the radius of curvature and with the wave length. Show that the field of such a linear conductor is given by

$$\mathbf{E}(x', t) = \frac{i}{4\pi\omega\epsilon}e^{-i\omega t}\nabla'\left\{I\frac{e^{ikr}}{r}\Bigg|_{s=s_2}^{s=s_1} + \int_C \frac{\partial I}{\partial s}\frac{e^{ikr}}{r}\,ds\right\} + \frac{i\omega\mu}{4\pi}\int_C I\frac{e^{ikr}}{r}\,ds,$$

where ∇' is applied at the point of observation and the integrals are extended along the contour C of the wire between points s_1 and s_2.

7. Apply the formula of Problem 6 to obtain the field of a linear oscillator of length l and compare with the results of Sec. 8.11.

8. In case the linear circuit of Problem 6 is closed, $s_1 = s_2$ and the integrated term is zero. On the surface of the conductor the tangential component of $\mathbf{E}$ is approximately

$$E_s \simeq \frac{ie^{-i\omega t}}{4\pi\omega\epsilon}\oint_C\left[\frac{\partial^2 I}{\partial s^2} + k^2 I\right]\frac{e^{ikr}}{r}\,ds = 0,$$

whence the current I must be of the form

$$I = A_m\,e^{-ik_m s}, \qquad k_m = \frac{2\pi m}{l},$$

where l is the length of the circuit and m is an integer. Show that the field of such a closed, oscillating loop is given by

$$\mathbf{E}(x', t) = \frac{1}{4\pi}\sqrt{\frac{\mu}{\epsilon}}\sum_{m=1}^{\infty}A_m e^{-i\omega_m t}\left\{\nabla'\oint_C\frac{e^{ik_m(r-s)}}{r}\,ds + ik_m\oint_C\frac{e^{ik_m(r-s)}}{r}\,ds\right\}.$$

The expression is exact in the limit of vanishing cross section and is approximately correct if the cross section is small in comparison with the wave length and the radius of curvature at each point of the circuit. Its application to Hertzian oscillators, such as were used in the early days of wireless telegraphy, was discussed in an Adams Prize essay by Macdonald in 1902 entitled "Electric Waves."

9. A semi-infinite linear conductor carrying a current of frequency $\omega/2\pi$ coincides with the negative z-axis of a coordinate system. Find expressions for the components of electric and magnetic field intensity and calculate the total energy crossing a plane transverse to the conductor.

10. Compute the radiation resistance of a linear half-wave oscillator by the e.m.f. method described in Sec. 8.12.

11. An isolated, linear, half-wave antenna radiates 50 kw. at a wave length of 15 meters. Plot the parallel and perpendicular components of electric field intensity along the antenna on the assumption of a harmonic current distribution and a No. 4 wire.

12. The odd and even functions

$$F^{(o)} = \frac{\sin(\alpha\cos\theta)}{\sin\theta}, \qquad F^{(e)} = \frac{\cos(\alpha\cos\theta)}{\sin\theta},$$

occur frequently in the theory of linear oscillators. Let $\eta = \cos\theta$ and show that these function are solutions of

$$(1 - \eta^2)F'' - 2\eta F' + \left[\alpha^2(1 - \eta^2) - \frac{1}{1 - \eta^2}\right]F = 0.$$

Put $F = \sqrt{1 - \eta^2}\,G$, and show that G satisfies the associated Mathieu equation (6), page 375,

$$(1 - \eta^2)w'' - 2(a + 1)\eta w' + (b - c^2\eta^2)w = 0$$

for the particular case $a = 1$, $b = \alpha^2 - 1$, $c = \alpha$.

Although $F^{(e)}$ and $F^{(o)}$ are periodic in θ for all values of α, note that they remain finite at the poles only for certain characteristic values. For the even function, these values belong to the discrete set $\alpha_m^{(e)} = \frac{2m + 1}{2}\pi$; for the odd function, they form the set $\alpha_m^{(o)} = m\pi$, where m is an integer.

Demonstrate the orthogonality of the functions expressed by

$$\int_{-1}^{1} F_m^{(e)}F_n^{(e)}(1 - \eta^2)d\eta = \int_{-1}^{1} F_m^{(o)}F_n^{(o)}(1 - \eta^2)d\eta = 0, \qquad m \neq n,$$

$$\int_{-1}^{1} F_m^{(e)}F_n^{(o)}(1 - \eta^2)d\eta = 0, \qquad \begin{matrix} m = n, \\ m \neq n \end{matrix}$$

the subscripts referring to characteristic values of α, and show that the functions are normal. Discuss the relation of these functions to the radiation field of an arbitrary current distribution on a linear antenna.

13. An electromagnetic source is located at a point P_1, and another operating at the same frequency is located at P_2. The intervening medium is isotropic but not necessarily homogeneous. All relations between field vectors are linear, and the time variation is harmonic. Let the field vectors due to the source at P_1 be $\mathbf{E}_1$ and $\mathbf{H}_1$, those due to the source at P_2 be $\mathbf{E}_2$ and $\mathbf{H}_2$. Show that wherever these vectors are continuous and finite, they satisfy the symmetrical relation

$$\nabla \cdot (\mathbf{E}_1 \times \mathbf{H}_2 - \mathbf{E}_2 \times \mathbf{H}_1) = 0.$$

This result is due to Lorentz and has been developed into a number of reciprocity theorems of fundamental importance for radio communication.

14. On page 434 it was shown that the Hertz vector of an electric quadrupole is

$$\mathbf{\Pi} = -\frac{i\omega}{8\pi}\,\mathbf{p}^{(2)}\sqrt{\frac{\mu}{\epsilon}}\left(\frac{1}{R} + \frac{i}{kR^2}\right)e^{ikR-i\omega t},$$

$$p_i^{(2)} = \sum_{j=1}^{3} p_{ij}\frac{\partial R}{\partial x_j}, \qquad p_{ij} = \int_V \xi_i\xi_j\rho_0\,dv,$$

$$R^2 = x_1^2 + x_2^2 + x_3^2,$$

where $\rho_0(\xi_1, \xi_2, \xi_3)$ is the density of charge. Let p_R, p_θ, p_ϕ be the spherical components of the vector $\mathbf{p}^{(2)}$. Show that the radiation field of the quadrupole, in the region $kR \gg 1$, is given by

$$E_\theta \simeq \sqrt{\frac{\mu}{\epsilon}}\,H_\phi, \qquad E_\phi \simeq -\sqrt{\frac{\mu}{\epsilon}}\,H_\theta,$$

$$H_\theta \simeq \frac{i\omega^3\mu\epsilon}{8\pi}\,p_\phi\,\frac{e^{ikR-i\omega t}}{R}, \qquad H_\phi \simeq -\frac{i\omega^3\mu\epsilon}{8\pi}\,p_\theta\,\frac{e^{ikR-i\omega t}}{R},$$

while the radial components vanish as $1/R^2$. Show that the mean radiation intensity is

$$S_R \simeq \frac{\omega^6\mu^2\epsilon^2}{128\pi^2}\sqrt{\frac{\mu}{\epsilon}}\,(p_\theta^2 + p_\phi^2)\,\frac{1}{R^2}.$$

15. Show that the components p_θ and p_ϕ of the vector $\mathbf{p}^{(2)}$ in the preceding problem are related to the components p_{ij} of the quadrupole tensor by

$$p_\theta = \tfrac{1}{2}\sin 2\theta\,(p_{11}\cos^2\phi + p_{22}\sin^2\phi - p_{33} + p_{12}\sin 2\phi) + \cos 2\theta\,(p_{13}\cos\phi + p_{23}\sin\phi),$$

$$p_\phi = \tfrac{1}{2}\sin\theta\sin 2\phi\,(p_{22} - p_{11}) + p_{12}\sin\theta\cos 2\phi + p_{23}\cos\theta\cos\phi - p_{31}\cos\theta\sin\theta.$$

Rotate the coördinate system to coincide with the principal axes of the quadrupole tensor. Then $p_{12} = p_{23} = p_{31} = 0$ and the above reduces to

$$p_\theta = \sin 2\theta\,(Q_1 - Q_2\cos 2\phi), \qquad p_\phi = 2\sin\theta\sin 2\phi\,Q_2,$$

where

$$Q_1 = \tfrac{1}{4}(p_{11} + p_{22} - 2p_{33}), \qquad Q_2 = \tfrac{1}{4}(p_{22} - p_{11}).$$

Show now that the total quadrupole radiation is

$$W = \frac{\omega^6\mu^2\epsilon^2}{60\pi}\sqrt{\frac{\mu}{\epsilon}}\,(Q_1^2 + 3Q_2^2) \qquad \text{watts.}$$

If the medium about the quadrupole is air, $\mu\epsilon = 1/c^2$, and $\frac{1}{2\pi}\sqrt{\frac{\mu}{\epsilon}} = 60$, so that

$$W = \frac{2^7\pi^6c^2}{\lambda^6}\,(Q_1^2 + 3Q_2^2).$$

16. A charge $-e$ is located at each end of a line rotating with constant angular velocity about a perpendicular axis through its center. A charge $+2e$ is fixed at the center. The dipole moment of the configuration is zero. Calculate the components of the quadrupole moment and find the total radiation. (Van Vleck.)

17. Two fixed dipoles are located in a plane, their axes parallel but their moments directed in opposite sense. The dipoles rotate with constant angular velocity about a parallel axis located halfway between them. Calculate the components of the quadrupole moment and find the total radiation. (Van Vleck.)

18. A positive and negative charge are bound together by quasielastic forces to form a harmonic oscillator. Show that the radiation losses can be accounted for by a frictional force which, however, is proportional to the rate of change of acceleration rather than to the rate of change of displacement.

19. A positive point charge oscillates with very small amplitude about a fixed negative charge of equal magnitude. Show that the formulas for the field of an oscillating dipole can be obtained directly from Eqs. (35) and (37) on page 475 which apply to an accelerated point charge.

CHAPTER IX

BOUNDARY-VALUE PROBLEMS

It can now be assumed that the reader is familiar with the principles that govern the generation of electromagnetic waves and the manner in which they are propagated in a limitless region of homogeneous, isotropic space. The "boundless region" is, of course, simply an abstraction. In point of fact the most interesting electromagnetic phenomena are those induced by surfaces of discontinuity or rapid change in the physical properties of the medium. These boundary phenomena are roughly of three types. Suppose that a wave, propagated in one medium, is incident upon a surface of discontinuity marking the boundary of another. In the first case the linear dimensions of the surface measured in wave lengths are very large. A fraction of the incident energy is *reflected* at the surface and the remainder transmitted into the second medium. The direction of propagation is in general modified and this bending of the transmitted rays is referred to as *refraction*. The laws that govern the reflection and refraction of electromagnetic waves at surfaces of infinite extent are relatively simple. If, however, any or all the dimensions of the surface of discontinuity are of the order of the wave length, the difficulties of a mathematical discussion are vastly increased. The perturbation of the primary field under these circumstances is referred to as *diffraction*. In both cases the secondary field of induced charges and polarization is excited by a primary wave of independent origin. Both are *inhomogeneous boundary-value problems* as defined first for the static field on page 195.

The third case referred to above is the *homogeneous problem*. A conducting body is embedded in a dielectric medium. Charge is displaced from the equilibrium distribution on the surface and then released. The resulting oscillations of charge are accompanied by oscillations in the surrounding field. This field can in every case be represented as a superposition of *characteristic wave functions* whose form is determined by the configuration of the body and whose relative amplitudes are fixed by the initial conditions. Associated with each characteristic function is a *characteristic number* that determines the frequency of that particular oscillation. The oscillations are damped, partly due to the finite conductivity of the body and partly as a result of energy dissipated in radiation. However, the positions of conductor and dielectric relative to the surface of separation can be interchanged. Electromagnetic oscillations then take place within a dielectric cavity bounded by conducting walls. If the conductivity is infinite, there is neither heat loss nor radiation and the oscillations are undamped at all frequencies.

GENERAL THEOREMS

9.1. Boundary Conditions.—A conventional proof of the conditions satisfied at a boundary by normal and tangential components of the field vectors was presented in Sec. 1.13. In Chaps. III and IV the relation of these boundary conditions to the discontinuous properties of certain integrals was established for the stationary regime; on the basis of Secs. 8.13 and 8.14, the same procedure can be extended to the variable field. We shall forego this analysis but shall give some further attention to the important case in which the conductivity of one of the two media approaches infinity.

At the interface of two media the transition of the tangential component of **E** and the normal component of **D** is expressed by

$$\mathbf{n} \times (\mathbf{E}_2 - \mathbf{E}_1) = 0, \qquad \mathbf{n} \cdot (\mathbf{D}_2 - \mathbf{D}_1) = \delta, \tag{1}$$

where by previous convention **n** is the unit normal directed from the medium (1) into (2) and δ denotes the surface charge to avoid confusion with $\omega = 2\pi\nu$. The flow of charge across or to the boundary must also satisfy the equation of continuity in case either or both the conductivities are finite and not zero.

$$\mathbf{n} \cdot (\mathbf{J}_2 - \mathbf{J}_1) = -\frac{\partial \delta}{\partial t}. \tag{2}$$

Suppose now that the time enters only as a common factor $\exp(-i\omega t)$ and that apart from the boundary the two media are homogeneous and isotropic. Then (1) and (2) together give

$$\begin{aligned} \epsilon_2 E_{2n} - \epsilon_1 E_{1n} &= \delta, \\ \sigma_2 E_{2n} - \sigma_1 E_{1n} &= i\omega\delta. \end{aligned} \tag{3}$$

Let us see first under what circumstances the surface charge can be zero. If $\delta = 0$, the determinant of (3) must vanish and hence

$$\sigma_1\epsilon_2 - \sigma_2\epsilon_1 = 0. \tag{4}$$

If δ is not zero (and this is the usual case), it may be eliminated from (3) and we obtain as an *alternative boundary condition on the normal component of* **E**:

$$\mu_1 k_2^2 E_{2n} - \mu_2 k_1^2 E_{1n} = 0. \tag{5}$$

If either conductivity is infinite, (5) becomes indeterminate; but from (3)

$$E_{1n} = \frac{i\omega\epsilon_2 - \sigma_2}{\epsilon_1\sigma_2 - \epsilon_2\sigma_1}\,\delta, \qquad E_{2n} = \frac{i\omega\epsilon_1 - \sigma_1}{\epsilon_1\sigma_2 - \epsilon_2\sigma_1}\,\delta, \tag{6}$$

and in the limit as $\sigma_1 \to \infty$,

$$(7) \qquad E_{1n} \to 0, \qquad E_{2n} \to \frac{1}{\epsilon_2}\,\delta, \qquad J_{1n} = \sigma_1 E_{1n} \to \frac{\sigma_2 - i\omega\epsilon_2}{\epsilon_2}\,\delta.$$

From the field equations, moreover,

$$(8) \qquad \mathbf{E}_1 = \frac{1}{\sigma_1 - i\omega\epsilon_1}\,\nabla \times \mathbf{H}_1, \qquad \mathbf{H}_1 = \frac{1}{i\omega\mu_1}\,\nabla \times \mathbf{E}_1,$$

so that, if by hypothesis the field intensities are bounded, both $\mathbf{E}_1$ and $\mathbf{H}_1$ approach zero at all *interior* points of (1) as $\sigma_1 \to \infty$. The transition of the tangential component of $\mathbf{E}$ is continuous across the boundary and, consequently, in the case of an infinitely conducting medium

$$(9) \qquad \mathbf{n} \times \mathbf{E}_2 = \mathbf{n} \times \mathbf{E}_1 = 0.$$

Consider now the behavior of the magnetic field at the boundary. In general

$$(10) \qquad \mathbf{n} \cdot (\mathbf{B}_2 - \mathbf{B}_1) = 0, \qquad \mathbf{n} \times (\mathbf{H}_2 - \mathbf{H}_1) = \mathbf{K},$$

where the surface current density $\mathbf{K}$ is zero unless the conductivity of medium at the boundary is infinite. It has just been shown that $\mathbf{H}_1$ vanishes if σ_1 becomes infinite and in this case

$$(11) \qquad \mathbf{n} \cdot \mathbf{B}_2 = 0, \qquad \mathbf{n} \times \mathbf{H}_2 = \mathbf{K}.$$

A further useful boundary condition on the magnetic field in the case of perfect conductivity can be derived as follows.[1] Let the boundary surface S defined by the equation,

$$(12) \qquad \zeta = f_1(x, y, z) = \text{constant},$$

coincide with a coordinate surface in an orthogonal system of curvilinear coordinates ξ, η, ζ as in Sec. 1.16. If $\sigma_1 \to \infty$, the tangential component of $\mathbf{E}_2$ and the tangential component of $\nabla \times \mathbf{H}_2$, which is proportional to it, approach zero. Therefore by (80), page 49,

$$(13) \qquad \frac{1}{h_\eta h_\zeta}\left[\frac{\partial}{\partial \eta}(h_\zeta H_\zeta) - \frac{\partial}{\partial \zeta}(h_\eta H_\eta)\right]\mathbf{i}_\xi + \frac{1}{h_\zeta h_\xi}\left[\frac{\partial}{\partial \zeta}(h_\xi H_\xi) - \frac{\partial}{\partial \xi}(h_\zeta H_\zeta)\right]\mathbf{i}_\eta \to 0.$$

The normal coordinate is ζ. Since σ_1 is infinite, H_ζ is zero; hence, since the coefficients of the unit vectors must vanish independently, we have just outside the conductor

$$(14) \qquad \frac{\partial}{\partial \zeta}(h_\eta H_\eta) = 0, \qquad \frac{\partial}{\partial \zeta}(h_\xi H_\xi) = 0.$$

The quantities $h_\eta H_\eta$ and $h_\xi H_\xi$ are covariant components (page 48) of the vector $\mathbf{H}_2$ tangent to the surface. The boundary condition is, therefore: *the normal derivatives of the covariant components of magnetic field*

[1] The author is indebted to E. H. Smith for the proof.

tangent to the boundary vanish as the conductivity on one side becomes infinite.

Under the head of "boundary conditions" we shall consider also *the behavior of the field at infinity.* In this respect the variable field differs notably from the static. If in the stationary regime all sources are located within a finite distance of the origin, the field intensities vanish at infinity such that lim $R^2\mathbf{E}$ and lim $R^2\mathbf{H}$ are bounded as $R \to \infty$, where R is the radial distance from the origin. The scalar potential satisfies $\nabla^2\phi = 0$ and lim $R\phi$ is bounded as $R \to \infty$. Every function ϕ which is regular at infinity and which satisfies Laplace's equation at *all* points of space (no sources) is necessarily zero. On the other hand, if *variable* sources are located within a finite distance of the origin, the field intensities vanish such that lim $R\mathbf{E}$ and lim $R\mathbf{H}$ are bounded as $R \to \infty$. The scalar potential and rectangular components of the vector potential *and* the field intensities all satisfy the equation,

$$\nabla^2\psi + k^2\psi = 0, \tag{15}$$

at points where the source density is zero. Moreover there exist functions ψ satisfying (15) throughout all space and vanishing at infinity which are *not* everywhere zero.

A solution of Laplace's equation is uniquely determined by the sources of the potential and the condition that it shall be regular at infinity. These two conditions are not sufficient to determine uniquely the wave function ψ, for (15) admits the possibility of *convergent* as well as *divergent* waves. This question has been investigated by Sommerfeld[1] in connection with the Green's function of (15) for spaces of infinite extent. To conditions that are analogous to those of the static problem must be added a "radiation condition." The problem is formulated as follows: The density $g(x, y, z)$ of the source distribution is specified, and these sources are assumed to lie entirely within a domain of finite extent. Then ψ is uniquely determined if:

(*a*) at all points exterior to a closed surface S (which can, if necessary, be resolved into a number of separate closed surfaces), ψ satisfies

$$\nabla^2\psi + k^2\psi = -g; \tag{16}$$

(*b*) ψ satisfies homogeneous boundary conditions over S of the type $\alpha\psi + \beta \dfrac{\partial\psi}{\partial n} = 0$, where α and β are constants or specified functions of position;

(*c*) ψ vanishes in such a way that lim $R\psi$ is bounded as $R \to \infty$, a condition we shall again refer to as regularity at infinity;

[1] SOMMERFELD, *Jahresber. deut. math. Ver.*, **21**, 326, 1912.

(*d*) ψ satisfies the radiation condition

$$\lim_{R\to\infty} R\left(\frac{\partial\psi}{\partial R} - ik\psi\right) = 0, \tag{17}$$

which ensures that at great distances from the source the field represents a *divergent traveling wave.* It has been tacitly assumed that the time enters explicitly in the factor $\exp(-i\omega t)$.

The significance of (17) may be made clear by applying Green's theorem to (16) within a region bounded internally by S and externally by a surface S_0. Then, as in Secs. 8.1 and 8.13,

$$\psi(x', y', z') = \frac{1}{4\pi}\int_V g\frac{e^{ikR}}{R}\,dv + \frac{1}{4\pi}\int_S\left[\frac{\partial\psi}{\partial n}\frac{e^{ikR}}{R} - \psi\frac{\partial}{\partial n}\left(\frac{e^{ikR}}{R}\right)\right]da + \frac{1}{4\pi}\int_{S_0}\left[\frac{\partial\psi}{\partial n}\frac{e^{ikR}}{R} - \psi\frac{\partial}{\partial n}\left(\frac{e^{ikR}}{R}\right)\right]da. \tag{18}$$

The first two terms to the right of (18) represent traveling waves diverging from the source. The third term, however, expresses the sum of all waves traveling inwards from the elements of S_0 and must, therefore, vanish as S_0 recedes to infinity. If this third term be designated by U, we have

$$4\pi U = \int_{S_0}\left(\frac{\partial\psi}{\partial n} - ik\psi\frac{\partial R}{\partial n}\right)\frac{e^{ikR}}{R}\,da + \int_{S_0}\psi\,e^{ikR}\,d\Omega, \tag{19}$$

where $d\Omega$ is an element of solid angle. The second integral in (19) is extended over the finite domain 4π and vanishes as $R \to \infty$ provided ψ is regular at infinity as prescribed in (*c*) above. In order that U shall vanish, it is sufficient that

$$\lim_{R\to\infty} R\left(\frac{\partial\psi}{\partial n} - ik\psi\frac{\partial R}{\partial n}\right) = 0. \tag{20}$$

If at great distances S_0 is replaced by a sphere of radius R, (20) is identical with (17).[1]

9.2. Uniqueness of Solution.—Let V be a region of space bounded internally by the surface S and externally by S_0. The surface S can be resolved, if the case demands, into a number of distinct closed surfaces S_i as in Fig. 18, page 108. Then V is multiply connected (page 226) and the surfaces S_i represent the boundaries of various foreign objects in the field. It will be assumed for the moment that the properties of V are isotropic but the parameters μ, ϵ, and σ can be arbitrary functions of position. Now let $\mathbf{E}_1$, $\mathbf{H}_1$, and $\mathbf{E}_2$, $\mathbf{H}_2$ be two solutions of the field equations which at the instant $t = 0$ are identical at all points of V. We wish to find the minimum number of conditions to be imposed on the

[1] An alternative proof was given in Sec. 8.13, p. 461.

components of the field vectors at the boundaries S and S_0 in order that the two solutions shall remain identical at all times $t > 0$.

In virtue of the linearity of the field equations (we exclude ferromagnetic materials) the *difference field* $\mathbf{E} = \mathbf{E}_2 - \mathbf{E}_1$ and $\mathbf{H} = \mathbf{H}_2 - \mathbf{H}_1$ is also a solution. One may assume without loss of generality that the sources of the fields lie entirely outside the region V, since it was shown in the preceding chapter that the field is uniquely determined when the charges and currents are prescribed. Or, if sources appear within V, one must specify that the distribution and rate of working of the electromotive forces are in both cases the same. Then within V, by Poynting's theorem, page 132, the difference field satisfies

$$(21)\qquad \frac{\partial}{\partial t}\int_V \left(\frac{\epsilon}{2}E^2 + \frac{\mu}{2}H^2\right) dv + \int_V \frac{1}{\sigma} J^2\, dv = -\int_{S+S_0} (\mathbf{E} \times \mathbf{H}) \cdot \mathbf{n}\, da.$$

In order that the right-hand member of (21) shall vanish, it is only necessary that *either* the tangential components of $\mathbf{E}_1$ and $\mathbf{E}_2$, *or* the tangential components of $\mathbf{H}_1$ and $\mathbf{H}_2$ be identical for all values of $t > 0$; for then either $\mathbf{n} \times \mathbf{E} = 0$ or $\mathbf{n} \times \mathbf{H} = 0$ and $\mathbf{E} \times \mathbf{H}$ has no normal component over the boundaries. In that case we have

$$(22)\qquad \frac{\partial}{\partial t}\int_V \left(\frac{\epsilon}{2}E^2 + \frac{\mu}{2}H^2\right) dv = -\int_V \frac{1}{\sigma} J^2\, dv.$$

The right-hand member of (22) is always equal to or less than zero. The energy integral on the left is essentially positive or zero and vanishes at $t = 0$. Hence (22) can only be satisfied by $\mathbf{E} = \mathbf{E}_2 - \mathbf{E}_1 = 0$, $\mathbf{H} = \mathbf{H}_2 - \mathbf{H}_1 = 0$ for all values of $t > 0$, as was to be shown. *An electromagnetic field is uniquely determined within a bounded region V at all times $t > 0$ by the initial values of electric and magnetic vectors throughout V, and the values of the tangential component of the electric vector (or of the magnetic vector) over the boundaries for $t \geq 0$.*

If S_0 recedes to infinity, V is externally unbounded. To ensure the vanishing of the integral of a Poynting vector over an infinitely remote surface, it is only necessary to assume that the medium has a conductivity, however slight. If the field was initially established in the finite past, the difficulty may also be circumvented by the assumption that S_0 lies beyond the zone reached at time t by a field propagated with a finite velocity c. The theorem just proved does not take full account of this finiteness of propagation. We have established that the values of $\mathbf{E}$ and $\mathbf{H}$ are uniquely determined throughout V at time t by a tangential boundary condition and the initial values *everywhere in* V. Physically it is obvious, however, that this is more information than should be necessary. The field is propagated with a finite velocity and, consequently, only those elements of V whose distance from the point of observation is $\leq (t - t_0)c$ need be taken into account. The classical uniqueness proof given above has been extended in this sense by Rubinowicz.[1]

The theorem applies also to anisotropic bodies. Electric and magnetic energies are then positive definite forms (*cf.* page 141), which is to say that they are positive or zero for all values of the variables. The proof remains unmodified.

There is one aspect of the theorem which at first thought may be puzzling. We have seen that the field is determined by either the values of $\mathbf{n} \times \mathbf{E}$ *or* $\mathbf{n} \times \mathbf{H}$ on the boundary, yet in the problems to be discussed shortly we find it necessary to apply *both* boundary conditions,

$$\mathbf{n} \times (\mathbf{E}_2 - \mathbf{E}_1) = 0, \qquad \mathbf{n} \times (\mathbf{H}_2 - \mathbf{H}_1) = 0, \tag{23}$$

where here $\mathbf{E}_2$ and $\mathbf{E}_1$ denote values of field intensity on either side of the boundary. The reason for the apparent discrepancy is, of course, that $\mathbf{n} \times \mathbf{E}$ and $\mathbf{n} \times \mathbf{H}$ refer to the tangential components of the *resultant* field on one side of the surface. These are in general unknown and it is the object of a boundary-value problem to find them. Equations (23) simply specify the transition of the field across a surface of discontinuity and the two together enable us to continue analytically a given primary field from one region into another. Having thus determined the total field, the uniqueness theorem shows that there is no other possible solution. If, however, one side of a boundary is infinitely conducting, we do know the tangential component of the total field, for in this case $\mathbf{n} \times \mathbf{E} = 0$, and a single-boundary condition is sufficient for the solution of the problem.

9.3. Electrodynamic Similitude.—Since Newton's time the principle of similitude and the theory of models have had a most important influence on the development of applied mechanics. This is particularly true of ship and airplane design which is governed very largely by the data obtained from small models in towing tanks and wind tunnels. It is customary to express the conditions of an experiment in terms of certain dimensionless quantities such as the Reynolds number. Thus the results of a single measurement of a given model can be applied to a series of objects of identical form and differing only in scale, provided the viscosity of the fluid and the velocity of flow are varied in such a way as to keep the Reynolds number constant.

Similar principles prove very helpful in the design of electromagnetic apparatus. We shall write first the field equations in a dimensionless form. In a homogeneous, isotropic conductor

$$\nabla \times \mathbf{E} + \mu \frac{\partial \mathbf{H}}{\partial t} = 0, \qquad \nabla \times \mathbf{H} - \epsilon \frac{\partial \mathbf{E}}{\partial t} - \sigma \mathbf{E} = 0. \tag{24}$$

[1] RUBINOWICZ, *Physik. Z.*, **27**, 707, 1926.

Now let

$$\begin{gathered}\mathbf{E} = e\mathsf{E}, \qquad \mathbf{H} = h\mathsf{H}, \\ \epsilon = \epsilon_0\kappa_e, \qquad \mu = \mu_0\kappa_m, \qquad \sigma = \sigma_0 s, \\ \text{length} = l_0 L, \qquad \text{time} = t_0 T,\end{gathered} \tag{25}$$

where E, H, κ_e, κ_m, s, L, and T are the dimensionless measure numbers of the field variables in a system for which the unit quantities are e, h, ϵ_0, μ_0, l_0, σ_0, and t_0. The measure numbers satisfy the equations

$$\begin{gathered}\nabla \times \mathsf{E} + \alpha\kappa_m \frac{\partial \mathsf{H}}{\partial T} = 0, \\ \nabla \times \mathsf{H} - \beta\kappa_e \frac{\partial \mathsf{E}}{\partial T} - \gamma s \mathsf{E} = 0,\end{gathered} \tag{26}$$

where

$$\alpha = \frac{\mu_0 l_0}{t_0}\frac{h}{e}, \qquad \beta = \frac{\epsilon_0 l_0}{t_0}\frac{e}{h}, \qquad \gamma = \sigma_0 l_0 \frac{e}{h}, \tag{27}$$

are dimensionless constants. Upon eliminating the common ratio e/h, one obtains

$$\mu_0\epsilon_0 \left(\frac{l_0}{t_0}\right)^2 = \text{constant}, \qquad \mu_0\sigma_0 \frac{l_0^2}{t_0} = \text{constant}. \tag{28}$$

From the first of these it follows at once that the product $\mu_0\epsilon_0$ must have the dimensions of an inverse velocity squared. This is, no doubt, the most fundamental approach to the problem of units and dimensions.[1]

In order that two electromagnetic boundary-value problems be similar, it is necessary and sufficient that the coefficients $\alpha\kappa_m$, $\beta\kappa_e$, and γs be identical in both. For t_0 let us take for example the period τ of the field, and for l_0 any length that characterizes one of a family of bodies which differ only in scale. Thus l_0 may be the radius of a set of concentric spheres, or the major axis of a set of ellipsoids. The condition of similitude requires that the *two* characteristic parameters C_1 and C_2 in

$$\mu\epsilon\left(\frac{l_0}{\tau}\right)^2 = C_1, \qquad \mu\sigma\frac{l_0^2}{\tau} = C_2, \tag{29}$$

be invariant to a change of scale. Suppose that the characteristic length l_0 is halved. Then both C_1 and C_2 remain unchanged if the permeability μ at every point of the field is quadrupled. This is an awkward remedy from a practical standpoint, but it is the only way in which the initial state can be simulated through the adjustment of a single parameter. If μ and ϵ are left as they were, constancy of C_1 results also from halving the period, or doubling the frequency, but this alone does not take care of C_2. In order that the half-scale model shall exactly reproduce the

[1] *Cf.* Secs. 1.8. 4.8, and 4.9.

full-scale conditions, it is necessary also that the conductivity be *doubled* at every point.

This principle can be illustrated by reference to the type of high-frequency radio generator that maintains standing electromagnetic waves in a cavity resonator bounded by metal walls. The frequency of oscillation is determined essentially by the dimensions of the cavity, while the losses depend largely on the conductivity of the walls. If the dimensions are halved, the frequency will approximately be doubled; if the conductivity of the walls is unchanged, it is entirely possible that the resulting increase of the *relative* loss will pass the critical value, so that the half-scale apparatus fails to oscillate.

REFLECTION AND REFRACTION AT A PLANE SURFACE

9.4. Snell's Laws.—Two homogeneous, isotropic media have as a common boundary the plane S, and are otherwise infinite in extent. The unit vector $\mathbf{n}$ is normal to the plane S and directed from the region

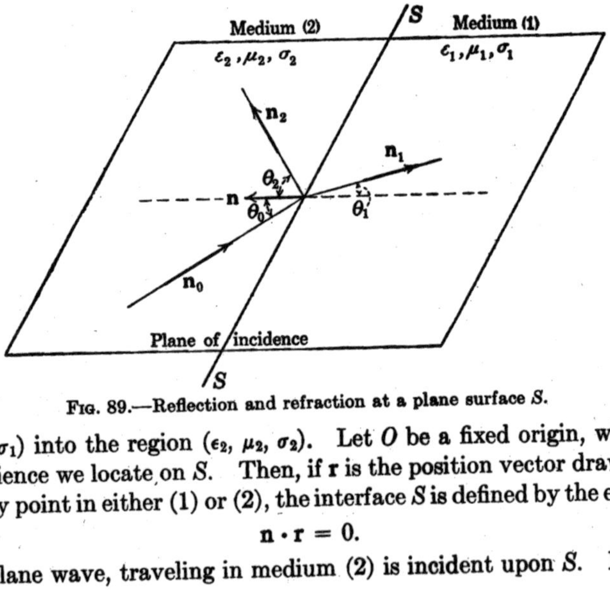

FIG. 89.—Reflection and refraction at a plane surface S.

$(\epsilon_1, \mu_1, \sigma_1)$ into the region $(\epsilon_2, \mu_2, \sigma_2)$. Let O be a fixed origin, which for convenience we locate on S. Then, if $\mathbf{r}$ is the position vector drawn from O to any point in either (1) or (2), the interface S is defined by the equation

$$\mathbf{n} \cdot \mathbf{r} = 0. \tag{1}$$

A plane wave, traveling in medium (2) is incident upon S. By (27), page 272,

$$\mathbf{E}_i = \mathbf{E}_0\, e^{ik_2\mathbf{n}_0\cdot\mathbf{r} - i\omega t}, \qquad \mathbf{H}_i = \frac{k_2}{\omega\mu_2}\, \mathbf{n}_0 \times \mathbf{E}_i, \tag{2}$$

where $\mathbf{E}_0$ is the complex amplitude of the incident wave and $\mathbf{n}_0$ a unit vector which fixes its direction of propagation,[1] as in Fig. 89. The plane defined by the pair of vectors $\mathbf{n}$ and $\mathbf{n}_0$ is called the *plane of incidence*.

[1] It is apparent that so far as the present problem is concerned it would be neater

The continuation of the primary field into medium (1) is determined by the boundary conditions at S. To satisfy these boundary conditions, a reflected or secondary field must be postulated in (2). Physically it is clear that the primary field induces an oscillatory motion of free and bound charge in the neighborhood of S, which in turn radiates a secondary field back into (2) as well as forward into (1). We shall make the tentative assumption that both transmitted and reflected waves are plane, and write

$$(3) \qquad \begin{aligned} \mathbf{E}_t &= \mathbf{E}_1 e^{ik_1\mathbf{n}_1\cdot\mathbf{r} - i\omega t}, \qquad & \mathbf{H}_t &= \frac{k_1}{\omega\mu_1}\mathbf{n}_1 \times \mathbf{E}_t, \\ \mathbf{E}_r &= \mathbf{E}_2 e^{ik_2\mathbf{n}_2\cdot\mathbf{r} - i\omega t}, \qquad & \mathbf{H}_r &= \frac{k_2}{\omega\mu_2}\mathbf{n}_2 \times \mathbf{E}_r, \end{aligned}$$

where the unit vectors $\mathbf{n}_1$ and $\mathbf{n}_2$ are in the directions of propagation of transmitted and reflected waves respectively, and $\mathbf{E}_1$ and $\mathbf{E}_2$ are complex amplitudes, all as yet undetermined. By hypothesis, $\mathbf{E}_1$, $\mathbf{E}_2$, and $\mathbf{E}_0$ are independent of the coordinates and, consequently, if the tangential components of the resultant field vectors are to be continuous across S, it is necessary that the arguments of the exponential factors in (2) and (3) be identical over the surface $\mathbf{n}\cdot\mathbf{r} = 0$. But

$$(4) \qquad \mathbf{r} = (\mathbf{n}\cdot\mathbf{r})\mathbf{n} - \mathbf{n} \times (\mathbf{n} \times \mathbf{r});$$

hence, at any point on the interface $\mathbf{r} = -\mathbf{n} \times (\mathbf{n} \times \mathbf{r})$. Therefore,

$$(5) \qquad \begin{aligned} k_2\mathbf{n}_0 \cdot \mathbf{n} \times (\mathbf{n} \times \mathbf{r}) &= k_2\mathbf{n}_2 \cdot \mathbf{n} \times (\mathbf{n} \times \mathbf{r}), \\ k_2\mathbf{n}_0 \cdot \mathbf{n} \times (\mathbf{n} \times \mathbf{r}) &= k_1\mathbf{n}_1 \cdot \mathbf{n} \times (\mathbf{n} \times \mathbf{r}), \end{aligned}$$

or, since $\mathbf{n}_0 \cdot \mathbf{n} \times (\mathbf{n} \times \mathbf{r}) = (\mathbf{n}_0 \times \mathbf{n}) \cdot (\mathbf{n} \times \mathbf{r})$,

$$(6) \qquad \begin{aligned} (\mathbf{n}_0 \times \mathbf{n} - \mathbf{n}_2 \times \mathbf{n}) \cdot \mathbf{n} \times \mathbf{r} &= 0, \\ (k_2\mathbf{n}_0 \times \mathbf{n} - k_1\mathbf{n}_1 \times \mathbf{n}) \cdot \mathbf{n} \times \mathbf{r} &= 0. \end{aligned}$$

From these two relations it follows that $\mathbf{n}$, $\mathbf{n}_0$, $\mathbf{n}_1$, and $\mathbf{n}_2$ are all coplanar. *The planes of constant phase of both transmitted and reflected waves are normal to the plane of incidence.* From the first of (6) also

$$(7) \qquad \sin\theta_2 = \sin(\pi - \theta_0) = \sin\theta_0,$$

whence *the angle of incidence* θ_0 *is equal to the angle of reflection* θ_2. From the second of (6)

$$(8) \qquad k_2 \sin\theta_0 = k_1 \sin\theta_1.$$

Equations (7) and (8) express *Snell's laws* of reflection and refraction.

---

from a notational point of view to let the incident wave travel from (1) into (2). Shortly, however, the plane S will be replaced by a closed surface S bounding a complete body. By previous convention $\mathbf{n}$ is directed outward from a closed surface and from medium (1) into (2). The choice of $\mathbf{n}_0$ as above will facilitate the comparison of formulas from the present with those of later sections.

9.5 Fresnel's Equations.—The boundary conditions will now be applied to determine the relation between the amplitudes $\mathbf{E}_0$, $\mathbf{E}_1$ and $\mathbf{E}_2$. At all points on S

$$\mathbf{n} \times (\mathbf{E}_0 + \mathbf{E}_2) = \mathbf{n} \times \mathbf{E}_1, \qquad \mathbf{n} \times (\mathbf{H}_0 + \mathbf{H}_2) = \mathbf{n} \times \mathbf{H}_1. \tag{9}$$

By virtue of (2) and (3) the second of these two can be expressed in terms of the electric vectors.

$$\mathbf{n} \times (\mathbf{n}_0 \times \mathbf{E}_0 + \mathbf{n}_2 \times \mathbf{E}_2) \frac{k_2}{\mu_2} = \mathbf{n} \times (\mathbf{n}_1 \times \mathbf{E}_1) \frac{k_1}{\mu_1}. \tag{10}$$

Expansion of (10) leads to such terms as

$$\mathbf{n} \times (\mathbf{n}_0 \times \mathbf{E}_0) = (\mathbf{n} \cdot \mathbf{E}_0)\mathbf{n}_0 - (\mathbf{n} \cdot \mathbf{n}_0)\mathbf{E}_0. \tag{11}$$

The orientation of the primary vector $\mathbf{E}_0$ is quite arbitrary but can always be resolved into a component normal to the plane of incidence and consequently tangent to S, and a second component lying in the plane of incidence. (*Cf.* Sec. 5.4, page 279.) The analysis is greatly simplified by a separate treatment of these two components of the incident wave.

Case I. $\mathbf{E}_0$ *Normal to the Plane of Incidence.*—Then

$$\mathbf{n} \cdot \mathbf{E}_0 = \mathbf{n}_0 \cdot \mathbf{E}_0 = 0.$$

Since the media are isotropic, the induced electric vectors of the transmitted and reflected waves must be parallel to $\mathbf{E}_0$ and hence also normal to the plane of incidence, so that $\mathbf{n} \cdot \mathbf{E}_1 = \mathbf{n} \cdot \mathbf{E}_2 = 0$. From Fig. 89

$$\begin{aligned} \mathbf{n} \cdot \mathbf{n}_0 &= \cos(\pi - \theta_0) = -\cos\theta_0, \\ \mathbf{n} \cdot \mathbf{n}_1 &= \cos(\pi - \theta_1) = -\cos\theta_1, \\ \mathbf{n} \cdot \mathbf{n}_2 &= \cos\theta_2. \end{aligned} \tag{12}$$

Upon multiplying the first of Eqs. (9) vectorially by $\mathbf{n}$ and making use of (11) and (12), we find that the amplitudes must satisfy

$$\begin{aligned} \mathbf{E}_0 + \mathbf{E}_2 &= \mathbf{E}_1, \\ \cos\theta_0\mathbf{E}_0 - \cos\theta_2\mathbf{E}_2 &= \frac{\mu_2 k_1}{\mu_1 k_2}\cos\theta_1\mathbf{E}_1. \end{aligned} \tag{13}$$

The relative directions of electric and magnetic vectors for this case are shown in Fig. 90. Solving (13) for $\mathbf{E}_1$ and $\mathbf{E}_2$ in terms of the primary amplitude $\mathbf{E}_0$ leads to

$$\begin{aligned} \mathbf{E}_1 &= \frac{\mu_1 k_2(\cos\theta_2 + \cos\theta_0)}{\mu_1 k_2\cos\theta_2 + \mu_2 k_1\cos\theta_1}\mathbf{E}_0, \\ \mathbf{E}_2 &= \frac{\mu_1 k_2\cos\theta_0 - \mu_2 k_1\cos\theta_1}{\mu_1 k_2\cos\theta_2 + \mu_2 k_1\cos\theta_1}\mathbf{E}_0. \end{aligned} \qquad (\text{when } \mathbf{n} \cdot \mathbf{E}_0 = 0), \tag{14}$$

These relations are not quite so simple as they appear at first sight, for θ_1 is complex if either (1) or (2) is conducting and *may* be complex even if both media are dielectric. By (7) and (8)

(15) $$\cos\theta_2 = \cos\theta_0, \qquad k_1\cos\theta_1 = \sqrt{k_1^2 - k_2^2\sin^2\theta_0}.$$

The angles of reflection and refraction can be eliminated from (14) and we obtain as an alternative form the relations

(16) $$\begin{aligned} \mathbf{E}_1 &= \frac{2\mu_1 k_2\cos\theta_0}{\mu_1 k_2\cos\theta_0 + \mu_2\sqrt{k_1^2 - k_2^2\sin^2\theta_0}}\,\mathbf{E}_0, \qquad (\text{when } \mathbf{n}\cdot\mathbf{E}_0 = 0),\\ \mathbf{E}_2 &= \frac{\mu_1 k_2\cos\theta_0 - \mu_2\sqrt{k_1^2 - k_2^2\sin^2\theta_0}}{\mu_1 k_2\cos\theta_0 + \mu_2\sqrt{k_1^2 - k_2^2\sin^2\theta_0}}\,\mathbf{E}_0. \end{aligned}$$

Complex values of the coefficients of $\mathbf{E}_0$ imply that the amplitudes $\mathbf{E}_1$ and $\mathbf{E}_2$ themselves are complex and that the transmitted and reflected waves differ in phase from the incident wave.

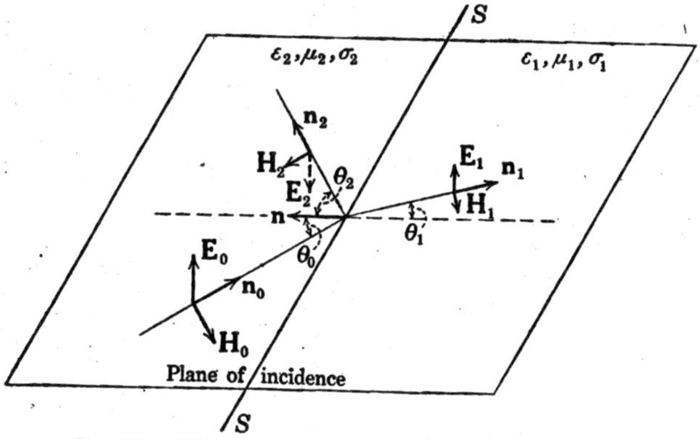

FIG. 90.—Polarization normal to the plane of incidence.

Case II. $\mathbf{E}_0$ *in the Plane of Incidence.*—The magnetic vectors are then normal to the plane of incidence and parallel to S.

$$\mathbf{n}\cdot\mathbf{H}_0 = \mathbf{n}\cdot\mathbf{H}_1 = \mathbf{n}\cdot\mathbf{H}_2.$$

From (2) and (3) we have

(17) $$\begin{aligned} \mathbf{E}_0 &= -\frac{\omega\mu_2}{k_2}\mathbf{n}_0\times\mathbf{H}_0, \qquad \mathbf{E}_1 = -\frac{\omega\mu_1}{k_1}\mathbf{n}_1\times\mathbf{H}_1,\\ \mathbf{E}_2 &= -\frac{\omega\mu_2}{k_2}\mathbf{n}_2\times\mathbf{H}_2, \end{aligned}$$

which when substituted into (9) give

(18) $$\begin{aligned} \cos\theta_0\,\mathbf{H}_0 - \cos\theta_2\,\mathbf{H}_2 &= \frac{\mu_1 k_2}{\mu_2 k_1}\cos\theta_1\,\mathbf{H}_1,\\ \mathbf{H}_0 + \mathbf{H}_2 &= \mathbf{H}_1, \end{aligned}$$

as the conditions at the boundary. The relative directions of electric and magnetic vectors for this case are shown in Fig. 91. Solution of (18) leads to

$$\text{(19)}\qquad \begin{aligned} \mathbf{H}_1 &= \frac{\mu_2 k_1(\cos\theta_2 + \cos\theta_0)}{\mu_2 k_1 \cos\theta_2 + \mu_1 k_2 \cos\theta_1}\mathbf{H}_0, \\ &\qquad\qquad (\text{when } \mathbf{n}\cdot\mathbf{H}_0 = 0), \\ \mathbf{H}_2 &= \frac{\mu_2 k_1 \cos\theta_0 - \mu_1 k_2 \cos\theta_1}{\mu_2 k_1 \cos\theta_0 + \mu_1 k_2 \cos\theta_1}\mathbf{H}_0, \end{aligned}$$

or, upon elimination of θ_1 and θ_2 by (15),

$$\text{(20)}\qquad \begin{aligned} \mathbf{H}_1 &= \frac{2\mu_2 k_1^2 \cos\theta_0}{\mu_2 k_1^2 \cos\theta_0 + \mu_1 k_2\sqrt{k_1^2 - k_2^2\sin^2\theta_0}}\mathbf{H}_0, \\ &\qquad\qquad (\text{when } \mathbf{n}\cdot\mathbf{H}_0 = 0), \\ \mathbf{H}_2 &= \frac{\mu_2 k_1^2 \cos\theta_0 - \mu_1 k_2\sqrt{k_1^2 - k_2^2\sin^2\theta_0}}{\mu_2 k_1^2 \cos\theta_0 + \mu_1 k_2\sqrt{k_1^2 - k_2^2\sin^2\theta_0}}\mathbf{H}_0. \end{aligned}$$

When the incidence is normal, $\theta_0 = 0$, the two cases cannot be distinguished and the amplitudes of transmitted and reflected waves reduce to

$$\text{(21)}\qquad \mathbf{E}_1 = \frac{2\mu_1 k_2}{\mu_1 k_2 + \mu_2 k_1}\mathbf{E}_0, \qquad \mathbf{E}_2 = \frac{\mu_1 k_2 - \mu_2 k_1}{\mu_1 k_2 + \mu_2 k_1}\mathbf{E}_0.$$

The relations expressed by Eqs. (14) and (19) were first derived in a slightly less general form by Fresnel in 1823 from the dynamical properties of a hypothetical elastic ether.

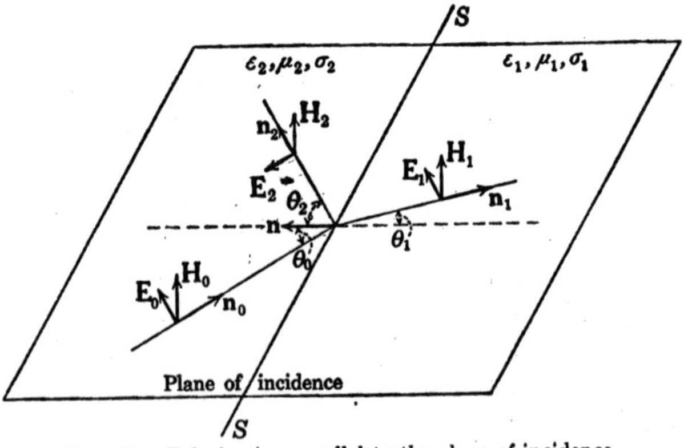

FIG. 91.—Polarization parallel to the plane of incidence.

9.6. Dielectric Media.—We shall study first the case in which the two conductivities σ_1 and σ_2 are zero, so that both media are perfectly transparent. The permeabilities will differ by a negligible amount from μ_0 and Snell's law can now be written

$$\frac{\sin\theta_1}{\sin\theta_0} = \frac{\sin\theta_1}{\sin\theta_2} = \sqrt{\frac{\epsilon_2}{\epsilon_1}} = \frac{v_1}{v_2} = n_{12}, \tag{22}$$

where v_1 and v_2 are the phase velocities and n_{12} is the *relative* index of refraction of the two media. If $\epsilon_1 > \epsilon_2$ it follows that $n_{12} < 1$ and there corresponds to every angle of incidence θ_0 a real angle of refraction θ_1. If, however, $\epsilon_1 < \epsilon_2$, as is the case when a wave emerges from a liquid or solid dielectric into air, then θ_1 is real only in the range for which $n_{12}\sin\theta_0 \leq 1$. The phenomena of *total reflection* occur when $n_{12}\sin\theta_0 > 1$. We shall exclude this possibility for the moment and consider the Fresnel laws for real angles between the limits 0 and $\pi/2$.

When $\mathbf{E}$ is normal to the plane of incidence we now obtain from (14)

$$\begin{aligned} \mathbf{E}_1 &= \frac{2\cos\theta_0\sin\theta_1}{\sin(\theta_1+\theta_0)}\mathbf{E}_0, \\ \mathbf{E}_2 &= \frac{\sin(\theta_1-\theta_0)}{\sin(\theta_1+\theta_0)}\mathbf{E}_0, \end{aligned} \qquad (\mathbf{n}\cdot\mathbf{E}=0), \tag{23}$$

and for the components of $\mathbf{E}$ lying in the plane of incidence from (19)

$$\begin{aligned} \mathbf{n}_1\times\mathbf{E}_1 &= \frac{2\cos\theta_0\sin\theta_1}{\sin(\theta_0+\theta_1)\cos(\theta_0-\theta_1)}\mathbf{n}_0\times\mathbf{E}_0, \\ \mathbf{n}_2\times\mathbf{E}_2 &= \frac{\tan(\theta_0-\theta_1)}{\tan(\theta_0+\theta_1)}\mathbf{n}_0\times\mathbf{E}_0. \end{aligned} \qquad (\mathbf{n}\cdot\mathbf{H}_0=0), \tag{24}$$

Since the coefficients of $\mathbf{E}_0$ in (23) and (24) are real, the reflected and transmitted waves are either in phase with the incident wave, or out of phase by 180 deg. It is apparent that the phase of the transmitted wave is in both cases identical with that of the incident wave. The phase of the reflected wave, however, will depend on the relative magnitudes of θ_0 and θ_1. Thus if $\epsilon_1 > \epsilon_2$, then $\theta_1 < \theta_0$, so that $\mathbf{E}_2$ is opposed in direction to $\mathbf{E}_0$ in (23) and therefore differs from it in phase by 180 deg. Under the same circumstances $\tan(\theta_0-\theta_1)$ is positive, but the denominator $\tan(\theta_0+\theta_1)$ becomes negative if $\theta_0+\theta_1 > \pi/2$ and the phase shifts accordingly.

The mean energy flow is given by the real part of the complex Poynting vector. In optics this quantity is usually referred to as the "intensity" of light, but the term is ambiguous since it is also applied to the amplitudes of the fields. In the present case

$$\begin{aligned} \overline{\mathbf{S}}_i &= \frac{1}{2}\mathbf{E}_i\times\tilde{\mathbf{H}}_i = \frac{\sqrt{\epsilon_2}}{2}\mathbf{E}_0\times(\mathbf{n}_0\times\mathbf{E}_0) = \frac{\sqrt{\epsilon_2}}{2}E_0^2\,\mathbf{n}_0, \\ \overline{\mathbf{S}}_t &= \frac{\sqrt{\epsilon_1}}{2}\mathbf{E}_1^2\,\mathbf{n}_1, \qquad \overline{\mathbf{S}}_r = \frac{\sqrt{\epsilon_2}}{2}E_2^2\,\mathbf{n}_2. \end{aligned} \tag{25}$$

Now the primary energy which is incident per second on a *unit* area of the dielectric interface is not $\overline{\mathbf{S}}_i$ but the normal component of this flow vector, or $\mathbf{n} \cdot \overline{\mathbf{S}}_i = -\frac{\sqrt{\epsilon_2}}{2} E_0^2 \cos \theta_0$. Likewise the energies leaving a unit area of the boundary by reflection and transmission are

$$\mathbf{n} \cdot \overline{\mathbf{S}}_r = \frac{\sqrt{\epsilon_2}}{2} E_2^2 \cos \theta_0, \qquad \mathbf{n} \cdot \overline{\mathbf{S}}_t = -\frac{\sqrt{\epsilon_1}}{2} E_1^2 \cos \theta_1. \tag{26}$$

According to the energy principle the normal component of energy flow across the interface must be continuous,

$$\mathbf{n} \cdot (\overline{\mathbf{S}}_i + \overline{\mathbf{S}}_r) = \mathbf{n} \cdot \overline{\mathbf{S}}_t, \tag{27}$$

or

$$\sqrt{\epsilon_2}\, E_0^2 \cos \theta_0 = \sqrt{\epsilon_1}\, E_1^2 \cos \theta_1 + \sqrt{\epsilon_2}\, E_2^2 \cos \theta_0. \tag{28}$$

It can be easily verified that the Fresnel formulas (23) and (24) satisfy this condition. The *reflection* and *transmission coefficients* are defined by the ratios

$$R = \frac{\mathbf{n} \cdot \overline{\mathbf{S}}_r}{\mathbf{n} \cdot \overline{\mathbf{S}}_i} = \frac{E_2^2}{E_0^2}, \qquad T = \frac{\mathbf{n} \cdot \overline{\mathbf{S}}_t}{\mathbf{n} \cdot \overline{\mathbf{S}}_i} = \sqrt{\frac{\epsilon_1}{\epsilon_2}} \frac{\cos \theta_1}{\cos \theta_0} \frac{E_1^2}{E_0^2}, \tag{29}$$

$$R + T = 1.$$

In case $\mathbf{E}_0$ is normal to the plane of incidence, these coefficients are

$$R_\perp = \frac{\sin^2 (\theta_1 - \theta_0)}{\sin^2 (\theta_1 + \theta_0)}, \qquad T_\perp = \frac{\sin 2\theta_0 \sin 2\theta_1}{\sin^2 (\theta_1 + \theta_0)}, \tag{30}$$

and in the case of $\mathbf{E}_0$ lying in the plane of incidence

$$R_\parallel = \frac{\tan^2 (\theta_0 - \theta_1)}{\tan^2 (\theta_0 + \theta_1)}, \qquad T_\parallel = \frac{\sin 2\theta_0 \sin 2\theta_1}{\sin^2 (\theta_0 + \theta_1) \cos^2 (\theta_0 - \theta_1)}. \tag{31}$$

If the incidence is normal, $\theta_0 = \theta_1 = 0$, and it follows from (21) and (22) that

$$\begin{aligned} R &= \left(\frac{\sqrt{\epsilon_2} - \sqrt{\epsilon_1}}{\sqrt{\epsilon_2} + \sqrt{\epsilon_1}}\right)^2 = \left(\frac{n_{12} - 1}{n_{12} + 1}\right)^2, \\ T &= \frac{4\sqrt{\epsilon_1 \epsilon_2}}{(\sqrt{\epsilon_2} + \sqrt{\epsilon_1})^2} = \frac{4n_{12}}{(n_{12} + 1)^2}. \end{aligned} \tag{32}$$

There is only one condition under which a reflection coefficient is zero. As $\theta_0 + \theta_1 \to \pi/2$, the $\tan (\theta_0 + \theta_1) \to \infty$ and in this case $R_\parallel \to 0$. The reflected and transmitted rays are then normal to one another ($\mathbf{n}_1 \cdot \mathbf{n}_2 = 0$) and $\sin \theta_1 = \sin \left(\frac{\pi}{2} - \theta_0\right) = \cos \theta_0$; it follows from (22) that

$$\tan \theta_0 = \sqrt{\frac{\epsilon_1}{\epsilon_2}} = n_{21}. \tag{33}$$

The angle that satisfies (33) is known as the *polarizing*, or *Brewster angle*. A wave incident upon a plane surface can be resolved, as we have seen, into two components, one polarized in a direction normal and the other parallel to the plane of incidence. The reflection coefficients of the two types differ and, consequently, the polarization of the reflected wave depends upon the angle of incidence. In particular, if incidence occurs at the polarizing angle, the reflected wave is polarized entirely in the direction normal to the plane of incidence. Use is sometimes made of this principle in optics to polarize natural light, although in practice it is less efficient than methods based on double-refracting prisms.

The optical indices relative to air are usually of the order of $n_{21} = 1.5$; at radio frequencies they may be very much larger with a corresponding increase in the polarizing angle. Thus in the case of water, n_{21} increases from about 1.33 to 9 at radio frequencies and the polarizing angle from 53 to 84.6 deg., which is not far removed from grazing incidence. Irregularities of radio transmission over water can doubtless be attributed on certain occasions to this cause. One will note also that at the polarizing angle $T_{\parallel}$ is unity and that this is the only condition under which all the energy of the primary wave can enter the second medium without loss by reflection at the surface.

9.7. Total Reflection.—We return now to the case excluded from Sec. 9.6 of transmission from medium (2) into a medium (1) whose index of refraction is less than that of (2). The formulas of Sec. 9.6 are valid whatever the relative values of ϵ_1 and ϵ_2, but if θ_0 is such that

$$n_{12} \sin \theta_0 = \sqrt{\frac{\epsilon_2}{\epsilon_1}} \sin \theta_0 > 1,$$

they can be satisfied only by complex values of θ_1. Physically a complex angle of refraction implies a shift of phase and the appearance of an attenuation factor.

Let us suppose, then, that $\sin \theta_1 > 1$. The cosine is a pure imaginary.

$$\cos \theta_1 = \frac{i}{\sqrt{\epsilon_1}} \sqrt{\epsilon_2 \sin^2 \theta_0 - \epsilon_1} = i n_{12} \sqrt{\sin^2 \theta_0 - n_{21}^2}. \tag{34}$$

The radical has two roots whose choice will be governed always by the condition that the field shall never become infinite. To simplify matters a bit, the reflecting surface S will be made to coincide with the plane $x = 0$, as in Fig. 92. All points of medium (1) correspond to negative values of x. The phase of the transmitted wave is, therefore,

$$k_1\mathbf{n}_1 \cdot \mathbf{r} = \omega\sqrt{\epsilon_1\mu_1}\,(-x\cos\theta_1 + z\sin\theta_1) \tag{35}$$
$$= \omega\sqrt{\epsilon_2\mu_2}\,(-ix\sqrt{\sin^2\theta_0 - n_{21}^2} + z\sin\theta_0),$$

where it is assumed that $\mu_1 = \mu_2 = \mu_0$, and the field intensity of the transmitted wave is

$$\mathbf{E}_t = \mathbf{E}_1\, e^{\beta_1 x + i\alpha z - i\omega t}, \qquad (x < 0), \tag{36}$$

where

$$\alpha = \omega\sqrt{\epsilon_2\mu_2}\sin\theta_0, \qquad \omega\sqrt{\epsilon_2\mu_2} = \frac{\omega}{v_2} = \frac{2\pi}{\lambda_2}, \tag{37}$$
$$\beta_1 = \omega\sqrt{\epsilon_2\mu_2}\sqrt{\sin^2\theta_0 - n_{21}^2}.$$

The field defined by (36) vanishes exponentially as $x \to -\infty$, which confirms the choice of the positive root in (34).

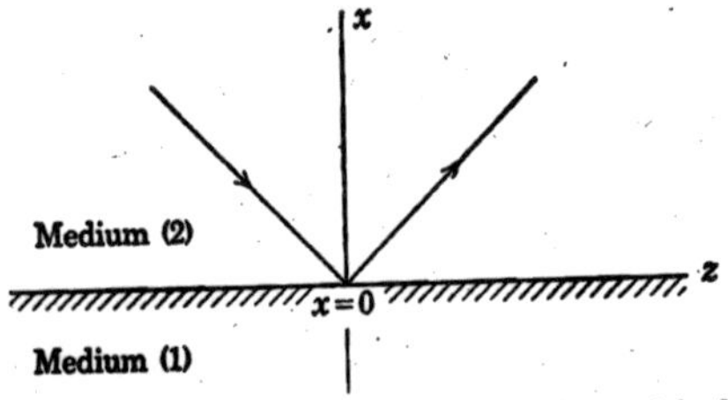

FIG. 92.—Total reflection from a surface coinciding with the yz-plane.

The amplitudes of reflected and transmitted fields are next determined from the Fresnel laws after elimination of θ_1. From (14) we obtain

$$\mathbf{E}_{1\perp} = \frac{2\cos\theta_0}{\cos\theta_0 + i\sqrt{\sin^2\theta_0 - n_{21}^2}}\,\mathbf{E}_{0\perp}, \tag{38}$$
$$\mathbf{E}_{2\perp} = \frac{\cos\theta_0 - i\sqrt{\sin^2\theta_0 - n_{21}^2}}{\cos\theta_0 + i\sqrt{\sin^2\theta_0 - n_{21}^2}}\,\mathbf{E}_{0\perp},$$

and from (19) and (3), for polarization in the plane of incidence,

$$\mathbf{n}_1 \times \mathbf{E}_{1\parallel} = \frac{2n_{21}\cos\theta_0}{n_{21}^2\cos\theta_0 + i\sqrt{\sin^2\theta_0 - n_{21}^2}}\,\mathbf{n}_0 \times \mathbf{E}_{0\parallel}, \tag{39}$$
$$\mathbf{n}_2 \times \mathbf{E}_{2\parallel} = \frac{n_{21}^2\cos\theta_0 - i\sqrt{\sin^2\theta_0 - n_{21}^2}}{n_{21}^2\cos\theta_0 + i\sqrt{\sin^2\theta_0 - n_{21}^2}}\,\mathbf{n}_0 \times \mathbf{E}_{0\parallel}.$$

Since the coefficients of $\mathbf{E}_0$ are complex, it is apparent that the transmitted and reflected waves are no longer in phase at the surface with the incident wave. The reflection coefficient according to (29) is $R = \mathbf{E}_2 \cdot \tilde{\mathbf{E}}_2/E_0^2$, and it follows at once from (38) and (39) that

$$R_\perp = R_\parallel = 1, \qquad T_\perp = T_\parallel = 0. \tag{40}$$

The intensity of energy flow in the reflected wave is exactly equal to the intensity in the incident wave; *there is no average flow into the medium of lesser refractive index.* The *field* intensity in medium (1), however, is by no means zero. There is, in fact, an instantaneous normal component of energy flow across the surface whose time average is zero; the time average of the flow within (1) parallel to the surface (*i.e.*, in the direction of z) does not vanish. This latter component is unattenuated in the direction of propagation but falls off very rapidly as the distance from S increases due to the factor exp $\beta_1 x$. The surfaces of constant phase in the transmitted wave are the planes z = constant, which are normal to surfaces of constant amplitude, x = constant. It is clear that $\mathbf{E}_{t\|}$ and $\mathbf{H}_{t\|}$ have components along the z-axis, which is the direction of propagation within (1). *The excitation within medium* (1) *in the case of total reflection is a nontransverse wave* (*it may be either transverse electric or transverse magnetic, page* 350) *confined to the immediate neighborhood of the surface.*

This analysis gives no clue as to how the energy initially entered (1), for it is based on assumptions of a steady state and of surfaces and wave fronts of infinite extent. Actually, the incident wave is bounded in both time and space. The total reflection of a beam of light of finite cross section has been treated by Picht[1] who showed that the average flow normal to the surface is in this case not strictly zero. Any change causing a fluctuation in the energy flow of the transmitted wave destroys the totality of reflection.

Figures showing the course of the magnetic lines of force and the lines of energy flow in total reflection have been published by Eichenwald and by Schaefer and Gross.[2]

Finally, let us examine the relative phases of the reflected waves. In (38) and (39) write

$$E_{2\perp} = e^{-i\delta\perp} E_{0\perp}, \qquad E_{2\|} = e^{-i\delta\|} E_{0\|}, \qquad |e^{-i\delta\perp}| = |e^{-i\delta\|}| = 1. \tag{41}$$

Then since

$$\frac{a - ib}{a + ib} = e^{-i\delta}, \qquad \tan\frac{\delta}{2} = \frac{b}{a}, \tag{42}$$

we have

$$\tan\frac{\delta_\perp}{2} = \frac{\sqrt{\sin^2\theta_0 - n_{21}^2}}{\cos\theta_0}, \qquad \tan\frac{\delta_\|}{2} = \frac{\sqrt{\sin^2\theta_0 - n_{21}^2}}{n_{21}^2\cos\theta_0}. \tag{43}$$

Suppose that the incident wave is linearly polarized in a direction that is neither parallel nor normal to the plane of incidence. We then resolve it into components and discover that the resultant reflected wave is

[1] PICHT, *Ann. Physik*, **3**, 433, 1929. See also NOETHER, *ibid.*, **11**, 141, 1931.

[2] SCHAEFER and GROSS, Untersuchungen über die Totalreflexion, *Ann. Physik*, **32**, 648, 1910.

formed by the superposition of two harmonic oscillations at right angles to one another and differing in phase by an amount

$$\delta = \delta_{\parallel} - \delta_{\perp}. \tag{44}$$

The reflected wave is elliptically polarized, since by (43) *the relative phase difference* δ *is not, in general, zero.*

$$\tan\frac{\delta}{2} = \frac{\tan\frac{\delta_{\parallel}}{2} - \tan\frac{\delta_{\perp}}{2}}{1 + \tan\frac{\delta_{\parallel}}{2}\tan\frac{\delta_{\perp}}{2}} = \frac{\cos\theta_0\sqrt{\sin^2\theta_0 - n_{21}^2}}{\sin^2\theta_0}. \tag{45}$$

The phase shifts $\delta_{\perp}$ and $\delta_{\parallel}$ both vanish at the polarizing angle $\theta_0 = \sin^{-1} n_{21}$, and their difference δ is also zero at grazing incidence, $\theta_0 = \pi/2$. The relative phase δ attains a maximum between these limits at an angle found by differentiating (45) with respect to θ_0 and equating to zero. The maximum occurs when

$$\sin^2\theta_0 = \frac{2n_{21}^2}{1 + n_{21}^2}, \tag{46}$$

which, when substituted into (45), gives

$$\tan\frac{\delta_{\max}}{2} = \frac{1 - n_{21}^2}{2n_{21}}. \tag{47}$$

This property of the totally reflected wave was used by Fresnel to produce circularly polarized light. It is first necessary that the incident wave be linearly polarized in a direction making an angle of 45 deg. with the normal to the plane of incidence. The amplitudes $E_{2\perp}$ and $E_{2\parallel}$ are then equal in magnitude. The relative index n_{21} and the angle of incidence θ_0 are next adjusted so that $\delta = \pi/2$, or $\tan \delta/2 = 1$. According to (47) this condition can be satisfied only if $1 - n_{21}^2 \geq 2n_{21}$, or $n_{21} < 0.414$, or $n_{12} > 2.41$. In the visible spectrum such a minimum value of the index of refraction is larger than occurs in any common transparent substance. To overcome this difficulty, Fresnel caused the ray of light to be totally reflected twice between the inner surfaces of a glass parallelepiped of proper angle. In the radio spectrum, on the other hand, the index of refraction may assume very much larger values. Thus, in the case of a surface formed by water and air, $n_{12} = 9$, $n_{21} = 0.11$. The condition $\tan \delta/2 = 1$ is then satisfied by either of the angles $\theta_0 = 6.5$ deg. or $\theta_0 = 44.6$ deg. The latter figure has been confirmed by measurements at a wave length of 250 cm.[1]

9.8. Refraction in a Conducting Medium.—The phenomena of reflection and refraction are modified to a striking degree by the presence of a

[1] BERGMANN, Die Erzeugung zirkular polarisierter elektrischer Wellen durch einmalige Totalreflexion, *Physik. Z.*, **33**, 582, 1932.

conductivity in either medium. The laws of Snell and Fresnel are still valid in a purely formal way but, as in the case of total reflection, the complex range of the angle θ_1 leads to a very different physical interpretation.

Let us suppose that medium (2) is still a perfect dielectric but that the refracting medium (1) is now conducting. The propagation constants are defined by

$$\text{(48)} \qquad \begin{aligned} k_1^2 &= \omega^2\epsilon_1\mu_1 + i\omega\sigma_1\mu_1, & k_2^2 &= \omega^2\epsilon_2\mu_2, \\ k_1 &= \alpha_1 + i\beta_1, & k_2 &= \alpha_2 = \omega\sqrt{\epsilon_2\mu_2}, \end{aligned}$$

where α_1 and β_1 are expressed in terms of ϵ_1, μ_1, and σ_1 by (48) and (49), page 276. By Snell's law we have

$$\text{(49)} \qquad \sin\theta_1 = \frac{k_2}{k_1}\sin\theta_0 = \frac{\alpha_2}{\alpha_1^2 + \beta_1^2}(\alpha_1 - i\beta_1)\sin\theta_0,$$

which it is convenient to write as

$$\text{(50)} \qquad \sin\theta_1 = (a - ib)\sin\theta_0.$$

The complex cosine is then

$$\text{(51)} \qquad \cos\theta_1 = \sqrt{1 - (a^2 - b^2 - 2abi)\sin^2\theta_0} = \rho e^{i\gamma}.$$

The magnitude ρ and phase γ are found by squaring (51) and equating real and imaginary parts on either side.

$$\text{(52)} \qquad \begin{aligned} \rho^2\cos 2\gamma &= \rho^2(2\cos^2\gamma - 1) = 1 - (a^2 - b^2)\sin^2\theta_0, \\ \rho^2\sin 2\gamma &= 2\rho^2\sin\gamma\cos\gamma = 2ab\sin^2\theta_0. \end{aligned}$$

The phase of the refracted wave is, as in (35) and Fig. 92,

$$\text{(53)} \qquad \begin{aligned} k_1\mathbf{n}_1\cdot\mathbf{r} &= (\alpha_1 + i\beta_1)(-x\cos\theta_1 + z\sin\theta_1) \\ &= -x\rho(\alpha_1\cos\gamma - \beta_1\sin\gamma) - ix\rho(\beta_1\cos\gamma + \alpha_1\sin\gamma) \\ &\qquad + z(a\alpha_1 + b\beta_1)\sin\theta_0 + iz(a\beta_1 - b\alpha_1)\sin\theta_0. \end{aligned}$$

From (49) and (50) it is readily seen that $(a\alpha_1 + b\beta_1)\sin\theta_0 = \alpha_2\sin\theta_0$ and $(a\beta_1 - b\alpha_1) = 0$. Within the conducting medium the transmitted wave is represented, therefore, by

$$\text{(54)} \qquad \mathbf{E}_t = \mathbf{E}_1\, e^{px + i(-qx + \alpha_2 z\sin\theta_0 - \omega t)}, \qquad (x < 0),$$

where

$$\text{(55)} \qquad \begin{aligned} p &= \rho(\beta_1\cos\gamma + \alpha_1\sin\gamma) \\ q &= \rho(\alpha_1\cos\gamma - \beta_1\sin\gamma). \end{aligned}$$

Note that the *surfaces of constant amplitude are the planes* $px =$ *constant, the surfaces of constant phase are the planes* $-qx + \alpha_2 \sin\theta_0\, z =$ *constant, and these two families do not, in general, coincide.* Within (1) the field is represented by a system of *inhomogeneous plane waves*, as in the case of total reflection. The planes of constant amplitude are parallel to the reflecting surface S; the direction of propagation is determined by the normal to the planes of constant phase. The angle ψ made by this wave normal with the normal to the boundary plane (in the present instance the negative x-axis) is the true angle of refraction and is defined by

$$-x \cos \psi + z \sin \psi = \text{constant}, \tag{56}$$

or

$$\cos \psi = \frac{q}{\sqrt{q^2 + \alpha_2^2 \sin^2 \theta_0}}, \qquad \sin \psi = \frac{\alpha_2 \sin \theta_0}{\sqrt{q^2 + \alpha_2^2 \sin^2 \theta_0}}. \tag{57}$$

The relation of the planes of constant amplitude to the planes of constant phase is illustrated in Fig. 93.

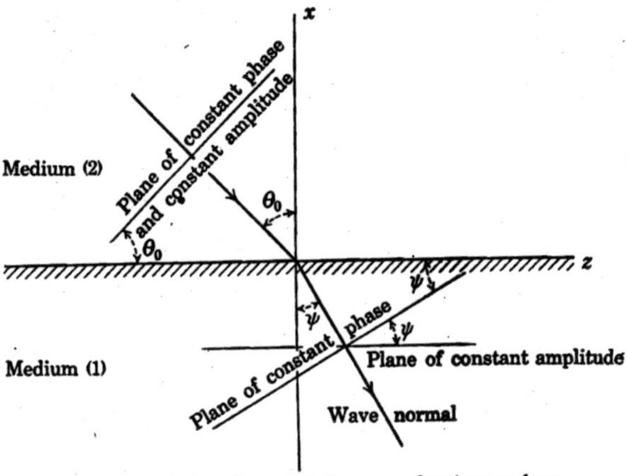

FIG. 93.—Refraction at a plane, conducting surface.

A modified Snell's law for real angles is expressed by (57).

$$n(\theta_0) = \frac{\sin \theta_0}{\sin \psi} = \frac{1}{\alpha_2} \sqrt{\alpha_2^2 \sin^2 \theta_0 + q^2}. \tag{58}$$

The quantity $n(\theta_0)$ is a real index of refraction which now depends on the angle of incidence, a notable deviation from the law of refraction in nonabsorbing media. The phase velocity, defined as the velocity of propagation of the planes of constant phase, is

$$v_1(\theta_0) = \frac{\omega}{\sqrt{q^2 + \alpha_2^2 \sin^2 \theta_0}} = \frac{v_2}{n(\theta_0)}. \tag{59}$$

Not only does this velocity depend on the angle of incidence, but as in the case of total reflection there are also components of field intensity in the direction of propagation. The field within the conductor is not strictly transverse.

The calculation of p, q, and n in terms of the constants of the media and the angle of incidence is a tedious but elementary task. From (52) and (55), together with the definition of a and b implicit in (49), one obtains the following relations:

$$
\begin{aligned}
p^2(\theta_0) &= \tfrac{1}{2}[-\alpha_1^2 + \beta_1^2 + \alpha_2^2 \sin^2 \theta_0 \\
&\qquad + \sqrt{4\alpha_1^2\beta_1^2 + (\alpha_1^2 - \beta_1^2 - \alpha_2^2 \sin^2 \theta_0)^2}], \\
q^2(\theta_0) &= \tfrac{1}{2}[\alpha_1^2 - \beta_1^2 - \alpha_2^2 \sin^2 \theta_0 \\
&\qquad + \sqrt{4\alpha_1^2\beta_1^2 + (\alpha_1^2 - \beta_1^2 - \alpha_2^2 \sin^2 \theta_0)^2}], \\
\alpha_2^2 n^2(\theta_0) &= \tfrac{1}{2}[\alpha_1^2 - \beta_1^2 + \alpha_2^2 \sin^2 \theta_0 \\
&\qquad + \sqrt{4\alpha_1^2\beta_1^2 + (\alpha_1^2 - \beta_1^2 - \alpha_2^2 \sin^2 \theta_0)^2}].
\end{aligned} \tag{60}
$$

From these a subsidiary set of relations, known as Ketteler's equations, can easily be derived.

$$
\begin{aligned}
\alpha_2^2 n^2(\theta_0) - p^2(\theta_0) &= \alpha_1^2 - \beta_1^2, \\
p(\theta_0) q(\theta_0) &= \alpha_1 \beta_1, \\
\alpha_2 n(\theta_0) p(\theta_0) &= \frac{\alpha_1 \beta_1}{\cos \psi}.
\end{aligned} \tag{61}
$$

The *functional* relation of $n(\theta_0)$ to the angle of incidence and the constants α_1 and β_1 expressed by (60) has been confirmed by measurements in the visible region of the light spectrum.[1] No direct connection exists, however, between the observed values of α_1 and β_1 at optical frequencies and the static or quasi-static values of the parameters ϵ_1, μ_1, and σ_1. In fact α_1 and β_1 can assume values at optical frequencies which at radio frequencies are possible only in densely ionized media. Thus Shea found for copper $\alpha_1/\alpha_2 = 0.48$, a value *less than unity*, and $\beta_1/\alpha_2 = 2.61$, instead of the exceedingly large value one might anticipate for such a good conductor. In this case the apparent phase velocity within the metal is greater than that of light. The anomalous behavior of these parameters at optical frequencies gives rise to some very interesting phenomena in the domain of metal optics, which lie beyond the limits we have imposed upon the present theory.[2]

Although there appear to be no experimental data available in support of Eqs. (60) at radio frequencies, there is every reason to believe them exact. We shall discuss only the case in which the conduction current in the medium is very much greater than the displacement current. Let $\eta = \sigma/\omega\epsilon$. Then the assumption is that $\eta_1^2 \gg 1$, and under these circumstances it will be recalled (page 277) that

[1] SHEA, *Wied. Ann.*, **47**, 271, 1892; WILSEY, *Phys. Rev.*, **8**, 391, 1916.

[2] An excellent account of the optical problems of reflection and refraction is given by König in his chapter on the electromagnetic theory of light in the "Handbuch der Physik," Vol. XX, pp. 197–253, Springer, 1928.

(62)
$$\alpha_1 \simeq \beta_1 \simeq \sqrt{\frac{\omega\mu_1\sigma_1}{2}}.$$

From (60),

(63)
$$p^2 \simeq \frac{\omega^2\epsilon_2\mu_2}{2}\left(\sin^2\theta_0 + \sqrt{\frac{\mu_1^2\sigma_1^2}{\omega^2\mu_2^2\epsilon_2^2} + \sin^4\theta_0}\right),$$

but, since η_1^2 is assumed to be very much larger than the maximum value of $\sin^2\theta_0$, we obtain the approximate formulas

(64)
$$p \simeq q \simeq \sqrt{\frac{\omega\mu_1\sigma_1}{2}}, \qquad n \simeq \sqrt{\frac{\mu_1\sigma_1}{2\omega\mu_2\epsilon_2}};$$

hence, $n \to \infty$ as $\sigma_1 \to \infty$ or $\omega \to 0$. At the same time

(65)
$$\sin\psi \simeq \sqrt{\frac{2\omega\mu_2\epsilon_2}{\mu_1\sigma_1}}\sin\theta_0$$

and $\psi \to 0$. *As the conductivity increases or the frequency decreases, the planes of constant phase align themselves parallel to the planes of constant amplitude and the propagation is into the conductor in a direction normal to the surface.*

In the case of copper, $\sigma_1 = 5.82 \times 10^7$ mhos/meter, and it is obvious that ψ differs from zero by an imperceptible amount. Whatever the angle of incidence, the transmitted wave travels in the direction of the normal. The factor

(66)
$$\delta = \sqrt{\frac{2}{\omega\mu_1\sigma_1}}$$

measures the depth of penetration. It is characteristic of all skin-effect phenomena and gives the distance within the conductor of a point at which the amplitude of the electric vector is equal to $1/e = 0.3679$ of its value at the surface. In this distance the phase lags 180 deg. Since the value of δ measures the effectiveness of a material for shielding purposes, it is interesting to know its order of magnitude. The table below gives the values of δ in the case of copper for several frequencies. They are obtained from the approximate formula $\delta = 6.6\nu^{-\frac{1}{2}}$ cm.

ν, cycles/sec.	δ, cm.
60	0.85
10^3	0.21
10^6	0.007

The depth of penetration is decreased by an increase in permeability, but this is usually offset by the poor conductivity of many highly permeable magnetic materials.

The angle ψ may be approximately zero for materials of much lower conductivity than the metals. In the case of sea water,

$$\sigma_1 = 3 \text{ mhos/meter}, \qquad \epsilon_1 = 81\epsilon_0,$$

and we find

$$\eta_1 \simeq \frac{6.7}{\nu} \times 10^8, \qquad n = \frac{1.64}{\sqrt{\nu}} \times 10^5, \qquad \delta = \frac{290}{\sqrt{\nu}} \text{ meters.} \tag{67}$$

At $\nu = 10^6$ we see that $n = 164$, $\psi < 0.35$ deg., and $\delta = 29$ cm. The approximation is just valid at $\nu = 10^8$, for then $\eta_1^2 = 45 \gg 1$, and in this case $n = 16.4$, $\psi < 3.5$ deg., $\delta = 2.9$ cm.

9.9. Reflection at a Conducting Surface.—We shall examine next the phase and amplitude of the wave reflected at the plane interface of a dielectric and a conductor. By Snell's law

$$k_1 \cos\theta_1 = \sqrt{k_1^2 - k_2^2 \sin^2\theta_0}, \tag{68}$$

and from (60) and (61) it follows that

$$\begin{aligned} pq &= \alpha_1\beta_1, \\ q^2 - p^2 &= \alpha_1^2 - \beta_1^2 - \alpha_2^2 \sin^2\theta_0, \\ q^2 + p^2 &= [4\alpha_1^2\beta_1^2 + (\alpha_1^2 - \beta_1^2 - \alpha_2^2 \sin^2\theta_0)^2]^{\frac{1}{2}}. \end{aligned} \tag{69}$$

It can easily be verified from these relations that

$$k_1 \cos\theta_1 = \sqrt{q^2 + p^2}\, e^{\frac{i\varphi}{2}}, \tag{70}$$

where

$$\tan\frac{\phi}{2} = \frac{p}{q}, \qquad \cos\frac{\phi}{2} = \frac{q}{\sqrt{q^2 + p^2}}, \qquad \sin\frac{\phi}{2} = \frac{p}{\sqrt{q^2 + p^2}}. \tag{71}$$

Upon substituting (70) into the Fresnel equation (14) for the reflected component of the electric vector polarized normal to the plane of incidence, one obtains

$$\mathbf{E}_{2\perp} = \frac{\mu_1\alpha_2 \cos\theta_0 - \mu_2\sqrt{q^2 + p^2}\, e^{\frac{i\varphi}{2}}}{\mu_1\alpha_2 \cos\theta_0 + \mu_2\sqrt{q^2 + p^2}\, e^{\frac{i\varphi}{2}}}\, \mathbf{E}_{0\perp}. \tag{72}$$

The fraction must be rationalized to give amplitude and phase. Then

$$\mathbf{E}_{2\perp} = \rho_\perp e^{-i\delta_\perp}\, \mathbf{E}_{0\perp}, \tag{73}$$

and after a relatively simple calculation we find

$$\begin{aligned} \rho_\perp{}^2 &= \frac{(\mu_2 q - \mu_1\alpha_2 \cos\theta_0)^2 + \mu_2^2 p^2}{(\mu_2 q + \mu_1\alpha_2 \cos\theta_0)^2 + \mu_2^2 p^2}, \\ \tan\delta_\perp &= \frac{2\mu_1\mu_2\alpha_2 p \cos\theta_0}{\mu_1^2\alpha_2^2 \cos^2\theta_0 - \mu_2^2(q^2 + p^2)}. \end{aligned} \tag{74}$$

The other component of polarization is found in the same way but the computation is considerably more tedious. According to (20), the second Fresnel equation is

$$\text{(75)}\quad \mathbf{H}_{2\perp} = \frac{\mu_2 \cos\theta_0\,(\alpha_1^2 - \beta_1^2 + 2\alpha_1\beta_1 i) - \mu_1\alpha_2\sqrt{q^2+p^2}\,e^{\frac{i\varphi}{2}}}{\mu_2 \cos\theta_0\,(\alpha_1^2 - \beta_1^2 + 2\alpha_1\beta_1 i) + \mu_1\alpha_2\sqrt{q^2+p^2}\,e^{\frac{i\varphi}{2}}}\,\mathbf{H}_{0\perp}.$$

The lengthy process of rationalizing this expression may be skipped over and the result set down at once. We abbreviate

$$\text{(76)}\quad \mathbf{n}_2 \times \mathbf{E}_{2\|} = \rho_\| e^{-i\delta_\|}\,\mathbf{n}_0 \times \mathbf{E}_{0\|},$$

and find

$$\text{(77)}\quad \begin{aligned} \rho_\|^2 &= \frac{[\mu_2(\alpha_1^2 - \beta_1^2)\cos\theta_0 - \mu_1\alpha_2 q]^2 + [2\mu_2\alpha_1\beta_1\cos\theta_0 - \mu_1\alpha_2 p]^2}{[\mu_2(\alpha_1^2 - \beta_1^2)\cos\theta_0 + \mu_1\alpha_2 q]^2 + [2\mu_2\alpha_1\beta_1\cos\theta_0 + \mu_1\alpha_2 p]^2}, \\ \tan\delta_\| &= \frac{2\mu_1\mu_2\alpha_2 p(q^2 + p^2 - \alpha_2^2\sin^2\theta_0)\cos\theta_0}{\mu_1^2\alpha_2^2(q^2+p^2) - \mu_2^2(\alpha_1^2+\beta_1^2)^2\cos^2\theta_0}. \end{aligned}$$

If the conductor is nonmagnetic, so that $\mu_1 = \mu_2$, the expression for the amplitude $\rho_\|$ factors into the form deduced by Pfeiffer[1] in the course of his optical studies:

$$\text{(78)}\quad \rho_\|^2 = \frac{(q - \alpha_2\cos\theta_0)^2 + p^2}{(q + \alpha_2\cos\theta_0)^2 + p^2}\,\frac{(q - \alpha_2\sin\theta_0\tan\theta_0)^2 + p^2}{(q + \alpha_2\sin\theta_0\tan\theta_0)^2 + p^2}.$$

As in the case of total reflection, the two components of polarization are reflected from an absorbing surface according to different laws. Consequently *an incident wave which is linearly polarized, but whose direction of polarization is neither normal nor parallel to the plane of incidence, will be reflected with elliptic polarization.* The polarization is determined by the ratio

$$\text{(79)}\quad \frac{\rho_\|}{\rho_\perp}\,e^{i(\delta_\| - \delta_\perp)} = \rho e^{i\delta}.$$

Only in the case of nonmagnetic materials do the expressions for ρ and δ reduce to a relatively simple form. If $\mu_1 = \mu_2$, one finds after another laborious calculation that

$$\text{(80)}\quad \begin{aligned} \rho^2 &= \frac{(q - \alpha_2\sin\theta_0\tan\theta_0)^2 + p^2}{(q + \alpha_2\sin\theta_0\tan\theta_0)^2 + p^2}, \\ \tan\delta &= \frac{2\alpha_2\,p\sin\theta_0\tan\theta_0}{\alpha_2^2\sin^2\theta_0\tan^2\theta_0 - (q^2+p^2)}. \end{aligned}$$

The reflection coefficients are again defined as the ratio of the energy flows in the incident and reflected waves. Thus

$$\text{(81)}\quad R_\perp = \rho_\perp^2, \qquad R_\| = \rho_\|^2.$$

In the case of normal incidence $R_\perp = R_\| = R$,

$$\text{(82)}\quad R = \frac{(\mu_2\alpha_1 - \mu_1\alpha_2)^2 + \mu_2^2\beta_1^2}{(\mu_2\alpha_1 + \mu_1\alpha_2)^2 + \mu_2^2\beta_1^2}.$$

[1] PFEIFFER, Diss. Giessen, 1912; KÖNIG, *loc. cit.*, p. 242. *Cf.* also WILSEY, *loc. cit.*, p. 393.

The degree of polarization is commonly measured by the ratio

$$s = \frac{|E_{2\perp}|^2 - |E_{2\parallel}|^2}{|E_{2\perp}|^2 + |E_{2\parallel}|^2} = \frac{1 - \rho^2 g^2}{1 + \rho^2 g^2}, \tag{83}$$

where $g^2 = E_{0\parallel}^2/E_{0\perp}^2$, the ratio of the incident intensities.

When one considers that the quantities q and p are functions of the angle of incidence as well as of all the parameters of the medium [Eqs. (60)], the complexity of what appeared at first to be the simplest of problems—the reflection of a plane wave from a plane, absorbing surface—is truly amazing. The formulas are far too involved to be understood from a casual inspection, and the nature of the reflection phenomena becomes apparent only when one treats certain limiting cases. Of these, one of the most important for electrical theory is that in which

Case I. $\eta_1^2 = \dfrac{\sigma_1^2}{\omega^2 \epsilon_1^2} \gg 1.$ Then

$$\alpha_1 \simeq \beta_1 \simeq q \simeq p \simeq \sqrt{\frac{\omega\mu_1\sigma_1}{2}}, \tag{84}$$

as was shown on page 504, and

$$\rho_\perp{}^2 \simeq \frac{\left(1 - \dfrac{\mu_1\alpha_2}{\mu_2\alpha_1}\cos\theta_0\right)^2 + 1}{\left(1 + \dfrac{\mu_1\alpha_2}{\mu_2\alpha_1}\cos\theta_0\right)^2 + 1}$$

$$\tan\delta_\perp \simeq \frac{\mu_1\alpha_2}{\mu_2\alpha_1}\,\frac{2\cos\theta_0}{\left(\dfrac{\mu_1\alpha_2}{\mu_2\alpha_1}\right)^2\cos^2\theta_0 - 1}. \tag{85}$$

Since $\mu_1/\mu_2 = \mu_1/\mu_0 = \kappa_{m1}$ is the magnetic permeability of the conductor and $\epsilon_2/\epsilon_0 = \kappa_{e1}$ is the specific inductive capacity of the dielectric, we may write

$$x = \frac{\mu_1\alpha_2}{\mu_2\alpha_1} = \sqrt{\frac{2\omega\mu_1\epsilon_2}{\mu_2\sigma_1}} = 2.11 \times 10^{-4}\sqrt{\frac{\nu\kappa_{m1}\kappa_{e2}}{\sigma_1}}, \tag{86}$$

where ν is the frequency. In the case of all metallic conductors $\mu_1\alpha_2/\mu_2\alpha_1 \ll 1$, and (85) reduces further to

$$\rho_\perp{}^2 \simeq 1 - 2x\cos\theta_0,$$
$$\tan\delta_\perp \simeq -2x\cos\theta_0. \tag{87}$$

This approximation is obtained by applying the binomial theorem to (85), and it is valid as long as the square of $\mu_1\alpha_2/\mu_2\alpha_1$ can be neglected with respect to its first power. Likewise in the case of polarization parallel to the plane of incidence

$$\rho_{\|}^2 = \frac{2\cos^2\theta_0 - 2x\cos\theta_0 + x^2}{2\cos^2\theta_0 + 2x\cos\theta_0 + x^2},$$
$$\tan\delta_{\|} = \frac{2x\cos\theta_0}{x^2 - \cos^2\theta_0}. \tag{88}$$

Here the square term in x has been retained since at a certain value of θ_0 the numerator of $\rho_{\|}^2$ becomes very small. This critical angle is obviously the analogue of the Brewster angle discussed above. The minimum value of $\rho_{\|}^2$ corresponds to an angle of incidence satisfying the relation $\sqrt{2}\cos\theta_0 = x$. Consequently if the minimum is sharply defined the ratio μ_1/σ_1 can be determined directly from the observed reflecting properties of the material.

It is clear from these results that the reflection coefficients of the metals are practically independent of the angle of incidence and differ by a negligible amount from unity even at the highest radio frequencies. Thus at an air-copper surface, $\kappa_{m1} = 1$, $\kappa_{e2} = 1$, $\sigma_1 = 5.82 \times 10^7$ mhos/meter, so that $x = 2.77 \times 10^{-8}\sqrt{\nu}$. In the case of iron the conductivity is about one-tenth as large and the permeability[1] may be of the order of 10^3, but x is still exceedingly small, even at wave lengths of a few centimeters. The formula

$$R = 1 - 4.22 \times 10^{-4}\sqrt{\frac{\nu\kappa_{m1}\kappa_{e2}}{\sigma_1}}, \tag{89}$$

to which (87) and (88) reduce at normal incidence, was verified in the infrared region of the spectrum by Hagen and Rubens.[2]

The results of their measurements are shown in the following table at $\lambda = 12\mu$ and $\lambda = 25.5\mu$, where $1\mu = 10^{-4}$ cm. The theoretical values with which they are compared were calculated from the electrostatic values of the conductivities. At wave lengths less than 12μ the devia-

Metal	$1 - R$, % at $\lambda = 12\mu$		$1 - R$, % at $\lambda = 25.5\mu$	
	Observed	Calculated	Observed	Calculated
Ag	1.15	1.3	1.13	1.15
Cu	1.6	1.4	1.17	1.27
Al	—	—	1.97	1.60
Au	2.1	1.6	1.56	1.39
Pt	3.5	3.5	2.82	2.96
Ni	4.1	3.6	3.20	3.16
Sn	—	—	3.27	3.23
Hg	—	—	7.66	7.55

[1] At ultrahigh frequencies the permeability may be very much smaller. Compare the measurements by Glathart, *Phys. Rev.*, **55**, 833, 1939.

[2] HAGEN and RUBENS, *Ann. Physik*, **11**, 873, 1903.

tions become more marked and at optical wave lengths there is no correspondence. These figures are extremely significant. *They imply that as long as we are concerned only with metals, the macroscopic electromagnetic theory is valid for all wave lengths greater than about* 10^{-3} *cm., or frequencies less than about* 3×10^{13} *cycles/sec.*

The reflection of radio waves from the surface of the sea or land gives results of another order. Consider first the case of an air-sea surface, for which $\kappa_{m1} = \kappa_{e2} = 1$, $\sigma_1 = 3$ mhos/meter, and $x = 1.22 \times 10^{-4}\sqrt{\nu}$. Since the approximations (87) and (88) are valid only when $x \ll 1$, we are limited now to frequencies less than 10^6 cycles/sec. Let us take, for example, a 600-kc. broadcast, frequency, $\lambda = 500$ meters. Then $x = 0.094$. In Fig. 94 are plotted the reflection coefficients $R_\perp$ and $R_\parallel$ of waves polarized normal and parallel to the plane of incidence as functions of the angle of incidence. One will note that $R_\perp > R_\parallel$, and that at 86.2 deg. the coefficient $R_\parallel$ reaches its minimum. Hence, if the electric vector of the incident wave is linearly polarized in a direction that is neither normal nor parallel to the plane of incidence, *the wave reflected from the surface of the sea will be elliptically polarized.* Moreover, if the transmitting antenna is vertical, the wave will be polarized in the plane of incidence; thus, if it meets the sea at a grazing angle, it may be almost completely absorbed.

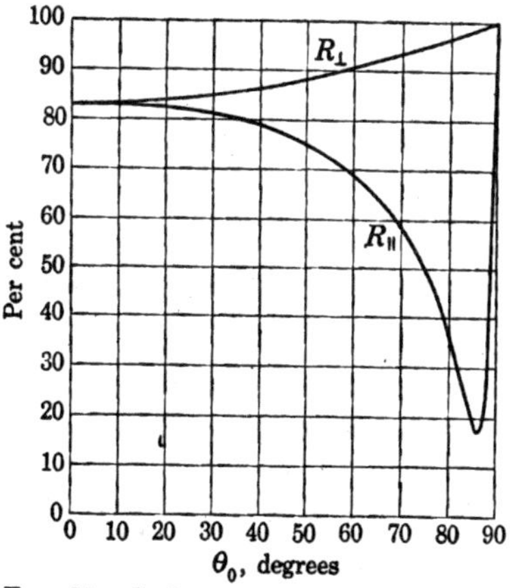

FIG. 94.—Reflection coefficients of plane radio waves from the surface of the sea.

If now the frequency is increased, or if the wave is incident upon fresh water or earth for which the conductivity is usually of the order of 10^{-4} or less, the approximations (87) and (88) no longer hold but the character of the reflection phenomena is not greatly modified. We have then to consider

Case II. $\eta_1^2 = \dfrac{\sigma_1^2}{\omega^2\epsilon_1^2} \ll 1$. By (52) and (53), page 277,

$$\alpha_1 \simeq \frac{\omega}{c}\sqrt{\kappa_{e1}}, \qquad \beta_1 \simeq 188.3\,\frac{\sigma_1}{\sqrt{\kappa_{e1}}}. \tag{90}$$

If the frequency is sufficiently high $\alpha_1 \gg \beta_1$, Eqs. (60) above reduce to

$$p^2 \simeq 0, \qquad q^2 \simeq \alpha_1^2 - \alpha_2^2 \sin^2 \theta_0. \tag{91}$$

Then

$$(92)\qquad \begin{aligned} \rho_\perp &\simeq \frac{(\kappa_{e1} - \kappa_{e2}\sin^2\theta_0)^{\frac{1}{2}} - \sqrt{\kappa_{e2}}\cos\theta_0}{(\kappa_{e1} - \kappa_{e2}\sin^2\theta_0)^{\frac{1}{2}} + \sqrt{\kappa_{e2}}\cos\theta_0}, \\ \rho_\parallel &\simeq \frac{\kappa_{e1}\cos\theta_0 - \sqrt{\kappa_{e2}}\,(\kappa_{e1} - \kappa_{e2}\sin^2\theta_0)^{\frac{1}{2}}}{\kappa_{e1}\cos\theta_0 + \sqrt{\kappa_{e2}}\,(\kappa_{e1} - \kappa_{e2}\sin^2\theta_0)^{\frac{1}{2}}}. \end{aligned}$$

To this approximation, only the dielectric properties of the medium enter and (92) is identical with the results obtained in Sec. 9.6. The next order would introduce a small correction in β_1.[1]

Let us take for example a fresh-water surface. $\kappa_{e2} = 1$, $\kappa_{e1} = 81$, $\sigma_1 \simeq 2 \times 10^{-4}$ mho/meter. Then

$$\eta_1 = \frac{4.45 \times 10^4}{\nu}, \qquad \alpha_1 \simeq 1.88 \times 10^{-7}\nu, \qquad \beta_1 \simeq 4.18 \times 10^{-3}.$$

The approximation (92) holds, therefore, if $\nu > 10^6$. The constants for earth and rock vary widely, but we may take as typical the case $\kappa_{e1} = 6$, $\sigma_1 \simeq 10^{-5}$. Then

$$\eta_1 = \frac{3 \times 10^4}{\nu}, \qquad \alpha_1 \simeq 5.1 \times 10^{-8}\nu, \qquad \beta_1 \simeq 7.7 \times 10^{-4},$$

so that again (92) is valid for $\nu > 10^6$. *At frequencies greater than 10^6 cycles/sec., fresh water and dry earth or rock act as dielectrics in so far as their reflecting properties are concerned.* The transmitted wave is of course rapidly attenuated due to the factor β_1. The reflection coefficients for these two cases have been plotted in Fig. 95 against angle of incidence. Such curves have been verified experimentally by Pfannenberg.[2]

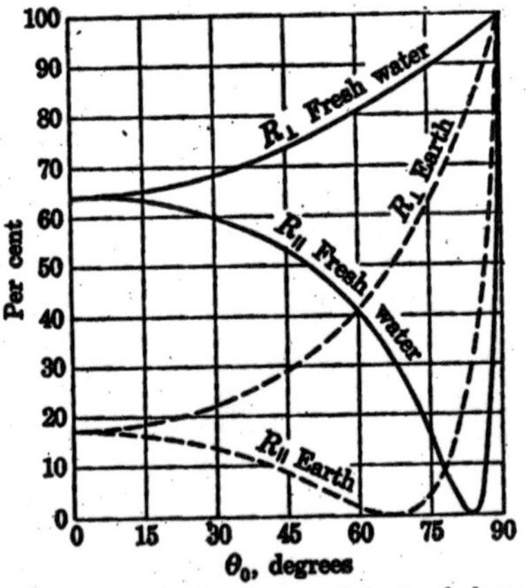

FIG. 95.—Reflection coefficients of plane radio waves from earth and fresh-water surfaces at frequencies greater than one megacycle per second.

Although in the case of metals the formulas hold for frequencies extending into the infrared, this is not true for dielectric or poorly conducting materials. At wave lengths of the order of a few centimeters most dielectrics exhibit a marked increase in absorption owing to causes other than electrical conductivity. Thus at a wave length of 2.8 cm., the value of β for distilled water computed on the basis of an electrical conductivity $\sigma = 2 \times 10^{-4}$ mho/meter is 4.2×10^{-3}, while

[1] An account of various approximation formulas useful in optics is given by König, *loc. cit.*, p. 246.

[2] PFANNENBERG, *Z. Physik*, **37**, 758, 1926.

the measured value[1] is 494 (m.k.s. units, distance in meters!). This is an extreme case corresponding to dipole resonance of the water molecule, but it is obvious that under such circumstances the approximations (92) are invalid. The general formulas may always be applied, even in the optical spectrum, provided the values of α and β are those determined by measurements at a specific frequency.

PLANE SHEETS

9.10. Reflection and Transmission Coefficients.—We shall consider only the case of normal incidence, which is of considerable practical

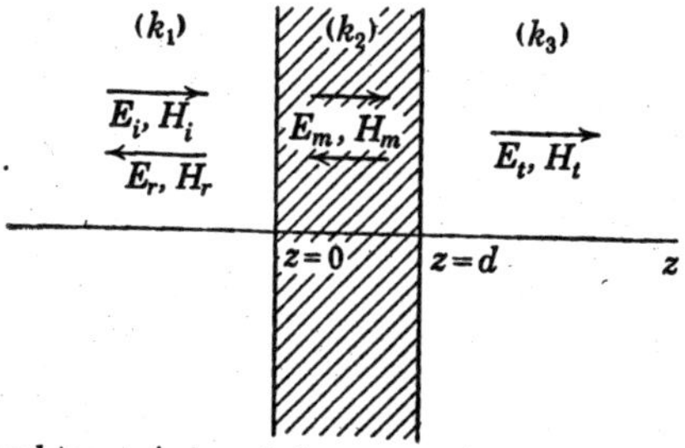

FIG. 96.—Reflection and transmission of plane waves by a plane sheet at normal incidence.

interest. Three arbitrary homogeneous media characterized by the propagation factors k_1, k_2, k_3 are separated by plane boundaries as shown in Fig. 96. We need deal only with the magnitudes of the vectors and write for the incident and reflected waves in medium (1):

$$E_i = E_0\, e^{ik_1z - i\omega t}, \qquad H_i = \frac{k_1}{\omega\mu_1} E_i, \tag{1}$$

$$E_r = E_1\, e^{-ik_1z - i\omega t}, \qquad H_r = -\frac{k_1}{\omega\mu_1} E_r. \tag{2}$$

The field within the sheet (2) must be expressed in terms of both positive and negative waves.

$$\begin{aligned} E_m &= (E_2^+ e^{ik_2z} + E_2^- e^{-ik_2z})e^{-i\omega t}, \\ H_m &= \frac{k_2}{\omega\mu_2}(E_2^+ e^{ik_2z} - E_2^- e^{-ik_2z})e^{-i\omega t}, \end{aligned} \tag{3}$$

while the transmitted wave is

$$E_t = E_3\, e^{ik_3z - i\omega t}, \qquad H_t = \frac{k_3}{\omega\mu_3} E_t. \tag{4}$$

It will be convenient to employ here the intrinsic impedance concept introduced in Sec. 5.6, page 282. For a plane wave in a homogeneous, isotropic medium

$$E_j = \pm Z_j H_j, \qquad Z_j = \frac{\omega\mu_j}{k_j}. \tag{5}$$

[1] BÄZ, *Physik. Z.*, **40**, 394, 1939.

We shall now define the impedance ratios

(6) $$Z_{jk} = \frac{Z_j}{Z_k} = \frac{\mu_j k_k}{\mu_k k_j},$$

and note that

(7) $$Z_{jk}Z_{kj} = 1, \qquad Z_{ij}Z_{jk} = Z_{ik}.$$

In terms of the impedance ratios the boundary conditions lead to four relations between the five amplitudes.

(8) $$\begin{aligned} E_0 + E_1 &= E_2^+ + E_2^-, \\ E_0 - E_1 &= Z_{12}(E_2^+ - E_2^-), \\ E_2^+ e^{ik_2d} + E_2^- e^{-ik_2d} &= E_3 e^{ik_3d}, \\ E_2^+ e^{ik_2d} - E_2^- e^{-ik_2d} &= Z_{23}E_3 e^{ik_3d}. \end{aligned}$$

Upon solving for the amplitudes of the reflected and transmitted waves, one finds

(9) $$\begin{aligned} E_1 &= \frac{(1 - Z_{12})(1 + Z_{23}) + (1 + Z_{12})(1 - Z_{23})e^{2ik_2d}}{(1 + Z_{12})(1 + Z_{23}) + (1 - Z_{12})(1 - Z_{23})e^{2ik_2d}} E_0, \\ E_3 &= \frac{4e^{-ik_3d}E_0}{(1 - Z_{12})(1 - Z_{23})e^{ik_2d} + (1 + Z_{12})(1 + Z_{23})e^{-ik_2d}}. \end{aligned}$$

We shall now define the complex ratios

(10) $$r_{jk} = \frac{1 - Z_{jk}}{1 + Z_{jk}} = \frac{Z_k - Z_j}{Z_k + Z_j} = -r_{kj}.$$

The physical significance of these ratios is apparent when one notes that

(11) $$r_{jk} = \frac{\mu_k k_j - \mu_j k_k}{\mu_k k_j + \mu_j k_k},$$

(12) $$|r_{jk}|^2 = \frac{(\mu_k\alpha_j - \mu_j\alpha_k)^2 + (\mu_k\beta_j - \mu_j\beta_k)^2}{(\mu_k\alpha_j + \mu_j\alpha_k)^2 + (\mu_k\beta_j + \mu_j\beta_k)^2} = R_{jk}.$$

The R_{jk} are the reflection coefficients at normal incidence for the plane interface dividing two semi-infinite media [Eq. (82), page 506]; the quantity r_{jk} is the complex ratio of the amplitudes of reflected and incident waves. The reflection coefficient at an interface is zero when the intrinsic impedances of the adjoining media are equal.

In terms of these r_{jk}, (9) reduces to

(13) $$\begin{aligned} \frac{E_1}{E_0} &= \frac{r_{12} + r_{23}\, e^{2ik_2d}}{1 + r_{12}r_{23}\, e^{2ik_2d}}, \\ \frac{E_3}{E_0} &= \frac{1}{(1 + Z_{12})(1 + Z_{23})} \frac{4e^{i(k_2 - k_3)d}}{1 + r_{12}r_{23}\, e^{2ik_2d}}. \end{aligned}$$

The reflection and transmission coefficients of the sheet are equal to the squares of the absolute values of these ratios. We write

$$
(14)\qquad
\begin{aligned}
Z_{jk} &= \frac{\mu_j}{\mu_k}\sqrt{\frac{\alpha_k^2+\beta_k^2}{\alpha_j^2+\beta_j^2}}\,e^{i\gamma_{jk}}\\
\tan\gamma_{jk} &= \frac{\alpha_j\beta_k-\alpha_k\beta_j}{\alpha_j\alpha_k+\beta_j\beta_k},
\end{aligned}
$$

$$
(15)\qquad
\begin{aligned}
r_{jk} &= |r_{jk}|e^{i\delta_{jk}},\\
\tan\delta_{jk} &= \frac{2\mu_j\mu_k(\alpha_k\beta_j-\alpha_j\beta_k)}{\mu_k^2(\alpha_j^2+\beta_j^2)-\mu_j^2(\alpha_k^2+\beta_k^2)}.
\end{aligned}
$$

One will note that $\gamma_{jk} = -\gamma_{kj}$, but $\delta_{jk} = +\delta_{kj}$. Upon introducing these phase angles into (13), and then multiplying each ratio by its conjugate value, we obtain

(16)

$$
R = \left|\frac{E_1}{E_0}\right|^2 = \frac{R_{12}+2\sqrt{R_{12}R_{23}}\,e^{-2\beta_2 d}\cos(\delta_{23}-\delta_{12}+2\alpha_2 d)+R_{23}\,e^{-4\beta_2 d}}{1+2\sqrt{R_{12}R_{23}}\,e^{-2\beta_2 d}\cos(\delta_{12}+\delta_{23}+2\alpha_2 d)+R_{12}R_{23}\,e^{-4\beta_2 d}},
$$

$$
(17)\qquad T = \left|\frac{E_3}{E_0}\right|^2
$$

$$
= \frac{\mu_2^2\mu_3^2(\alpha_1^2+\beta_1^2)(\alpha_2^2+\beta_2^2)}{[(\mu_2\alpha_1+\mu_1\alpha_2)^2+(\mu_2\beta_1+\mu_1\beta_2)^2][(\mu_3\alpha_2+\mu_2\alpha_3)^2+(\mu_3\beta_2+\mu_2\beta_3)^2]}
\times\frac{16e^{-2(\beta_2-\beta_3)d}}{1+2\sqrt{R_{12}R_{23}}\,e^{-2\beta_2 d}\cos(\delta_{12}+\delta_{23}+2\alpha_2 d)+R_{12}R_{23}\,e^{-4\beta_2 d}}.
$$

Equation (17) can be reduced to the form

$$
(18)\qquad T
= \frac{\mu_3}{\mu_1}\,\frac{\alpha_1^2+\beta_1^2}{\alpha_1\alpha_3+\beta_1\beta_3}\,\frac{[(1-R_{12})(1-R_{23})-4\sqrt{R_{12}R_{23}}\sin\delta_{12}\sin\delta_{23}]e^{-2(\beta_2-\beta_3)d}}{1+2\sqrt{R_{12}R_{23}}\,e^{-2\beta_2 d}\cos(\delta_{12}+\delta_{23}+2\alpha_2 d)+R_{12}R_{23}\,e^{-4\beta_2 d}}
$$

by applying the formulas

$$
(19)\qquad \sqrt{R_{jk}}\sin\delta_{jk} = \frac{2\mu_j\mu_k(\alpha_k\beta_j-\alpha_j\beta_k)}{(\mu_k\alpha_j+\mu_j\alpha_k)^2+(\mu_k\beta_j+\mu_j\beta_k)^2},
$$

$$
(20)\qquad 1-R_{jk} = \frac{4\mu_j\mu_k(\alpha_j\alpha_k+\beta_j\beta_k)}{(\mu_k\alpha_j+\mu_j\alpha_k)^2+(\mu_k\beta_j+\mu_j\beta_k)^2}.
$$

9.11. Application to Dielectric Media.—The interference phenomena represented by (16) and (18) have frequently been applied to the measurement of the optical constants of materials in the infrared and electrical as well as the visible regions of the spectrum.[1] To make clear the

[1] See, for example, Drude, "Lehrbuch der Optik," Chap. II, 1900. The method has been used in the infrared by Czerny and his students. Czerny, *Z. Physik*, **65**, 600, 1930; Barnes and Czerny, *Phys. Rev.*, **38**, 338, 1931; Cartwright and Czerny, *Z. Physik*, **85**, 269, 1933; Woltersdorff, *ibid.*, **91**, 230, 1934. Recently the same

physical behavior of the reflection coefficient as the thickness of the sheet is varied, we shall treat the simple case of a thin dielectric sheet separating two semi-infinite dielectric media as in Fig. 96. The inductive capacities of the three media are ϵ_1, ϵ_2, ϵ_3. $\beta_1 = \beta_2 = \beta_3 = 0$, and consequently $\delta_{12} = \delta_{23} = 0$. Then

$$R_{jk} = \left(\frac{\alpha_j - \alpha_k}{\alpha_j + \alpha_k}\right)^2 = \left(\frac{\sqrt{\epsilon_j} - \sqrt{\epsilon_k}}{\sqrt{\epsilon_j} + \sqrt{\epsilon_k}}\right)^2 = r_{jk}^2, \tag{21}$$

$$R = \frac{(r_{12} + r_{23})^2 - 4r_{12}r_{23}\sin^2 \alpha_2 d}{(1 + r_{12}r_{23})^2 - 4r_{12}r_{23}\sin^2 \alpha_2 d}, \tag{22}$$

and $\alpha_2 = 2\pi/\lambda_2$, where λ_2 *is the wave length within the sheet.* Now if ϵ_2 lies between ϵ_1 and ϵ_3 in value, the product $r_{12}r_{23}$ is positive and R oscillates

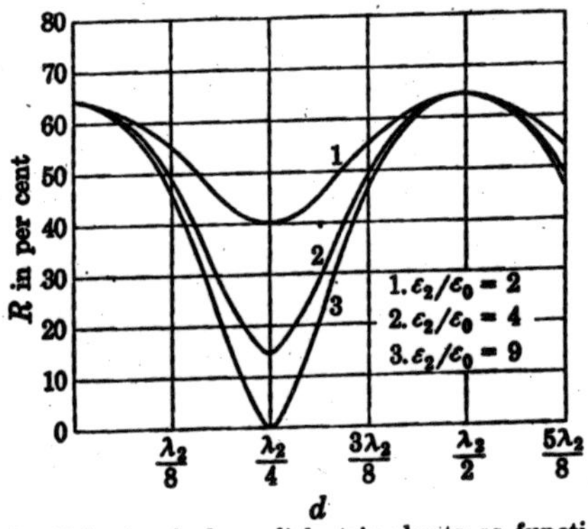

FIG. 97.—Reflection coefficients of plane dielectric sheets as functions of thickness and dielectric constant.

as a function of d. The minimum value of R occurs at $d = \lambda_2/4$, and is zero when $r_{12} = r_{23}$; *i.e.*, when ϵ_2 is the geometric mean of ϵ_1 and ϵ_3, or $\epsilon_2 = \sqrt{\epsilon_1\epsilon_3}$. Under these conditions all the energy is transmitted from medium (1) into medium (2), none is reflected at the surface. *The introduction of a quarter-wave length sheet of the proper dielectric constant accomplishes the same purpose as the matching of impedance at the junction of two electrical transmission lines.* In Fig. 97 curves are plotted which show the effect of a dielectric sheet at the interface of air and water.

This principle has been applied recently to the manufacture of optically "invisible" glass. The reflection of light from a glass surface is reduced almost to zero by coating with a film of the proper index of refraction.

method has been applied to ultrahigh-frequency radio waves. See for example Bäz, *Phys. Z.*, **40**, 394, 1939, and Pfister and Roth, *Hochfreq. und Elektroak.*, **51**, 156, 1938, who have extended the formula for the reflection coefficient to arbitrary angles of incidence.

9.12. Absorbing Layers.—If the medium on either side of the sheet is the same, Eqs. (16) and (18) reduce to

$$R = \frac{R_{12}[(1 - e^{-2\beta_2 d})^2 + 4e^{-2\beta_2 d}\sin^2 \alpha_2 d]}{(1 - R_{12}\,e^{-2\beta_2 d})^2 + 4R_{12}\,e^{-2\beta_2 d}\sin^2 (\delta_{12} + \alpha_2 d)}, \tag{23}$$

$$T = \frac{[(1 - R_{12})^2 + 4R_{12}\sin^2 \delta_{12}]e^{-2(\beta_2 - \beta_1)d}}{(1 - R_{12}\,e^{-2\beta_2 d})^2 + 4R_{12}\,e^{-2\beta_2 d}\sin^2 (\delta_{12} + \alpha_2 d)}. \tag{24}$$

A further simplification is achieved by defining

$$s_{12} = -\tfrac{1}{2}\ln R_{12}, \tag{25}$$

which when introduced into (23) and (24) leads to

$$R = \frac{\sin^2 \alpha_2 d + \sinh^2 \beta_2 d}{\sin^2 (\alpha_2 d + \delta_{12}) + \sinh^2 (\beta_2 d + s_{12})}, \tag{26}$$

$$T = \frac{(\sin^2 \delta_{12} + \sinh^2 s_{12})e^{\beta_1 d}}{\sin^2 (\alpha_2 d + \delta_{12}) + \sinh^2 (\beta_2 d + s_{12})}. \tag{27}$$

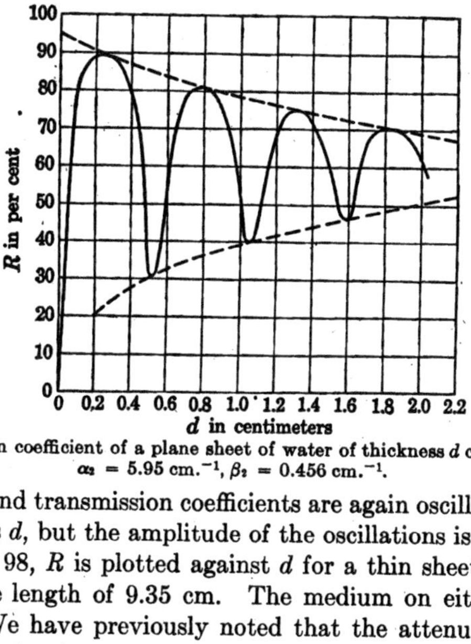

FIG. 98.—Reflection coefficient of a plane sheet of water of thickness d cm., $\lambda_0 = 9.35$ cm., $\alpha_2 = 5.95$ cm.$^{-1}$, $\beta_2 = 0.456$ cm.$^{-1}$.

The reflection and transmission coefficients are again oscillating functions of the thickness d, but the amplitude of the oscillations is no longer constant. In Fig. 98, R is plotted against d for a thin sheet of water at a free-space wave length of 9.35 cm. The medium on either side of the sheet is air. We have previously noted that the attenuation factor of water in this region is very much larger than can be accounted for by the electrical conductivity alone. The observed values are $\alpha_2 = 595$ meters^{-1}, $\beta_2 = 45.6$ meters^{-1}. Within the sheet the wave length is $\lambda_2 = \alpha_0\lambda_0/\alpha_2 = 1.05$ cm., and the minima occur approximately at half-wave-length intervals. In general, the smaller the attenuation factor, the more nearly do the minima coincide with the half-wave-length points. In the case of commercial insulators α_2 is at least 10^4 times as large as β_2, so that λ_2, and hence α_2, can be determined directly from the observed oscillations of the reflection coefficient. The numerical values of the minima are determined by the losses, and a curve such as that of Fig. 98 provides one of the best means of finding β_2 at ultrahigh frequencies.

In the case of a metal sheet the attenuation is so great, and the reflection coefficient so nearly unity, that transmission, through even the thinnest sheets, is wholly negligible. The transmission coefficient through a copper sheet 1μ thick (10^{-4} cm.), at 10^8 cycles/sec. for example, is less than 10^{-6}. However, this calculation gives a wholly erroneous impression of the effectiveness of shielding, for it applies only to plane surfaces of infinite extent. The reflection losses from surfaces whose radius of curvature is small compared to the wave length is by no means always large,[1] and the only significant general criterion is the value of the attenuation factor β_2, or its reciprocal, the skin depth, defined on page 504.

SURFACE WAVES

9.13. Complex Angles of Incidence.—The possibilities of the Fresnel equations are far from exhausted by the discussion of the preceding sections, for it has been confined exclusively to real angles of incidence. These equations represent the formal solution of a boundary-value problem which analytically is valid for complex as well as real angles. Complex angles of refraction already have occurred in connection with total reflection and reflection from conducting surfaces. There the associated plane waves proved to be inhomogeneous: the planes of constant amplitude fail to coincide with the planes of constant phase. Thus one may reasonably anticipate that a complex angle of incidence will be associated with an inhomogeneous primary wave. We shall not consider at this point how such waves are generated, but recall that in Secs. 6.8 and 7.7 it was shown that cylindrical and spherical waves of the most general type can be represented as a superposition of plane waves whose directional cosines include complex as well as real values.

If the angle of incidence is real, the primary wave gives rise in general to a reflected wave at the surface of discontinuity. Only in the case of a plane wave polarized in the plane of incidence and meeting the interface of two semi-infinite *dielectrics* at the Brewster angle (page 497) does there occur an exception to this rule. There is always a reflected wave if one of the media is conducting. If, however, complex angles of incidence are admitted, the reflection coefficients can be made to vanish whatever the nature of the media.

We shall consider in some detail the case of a wave whose electric vector is parallel to the plane of incidence. Letting $r_{\|}$ represent as in

[1] A point well illustrated in the case of cylindrical shields by Schelkunoff, *Bell System Tech. J.*, **13**, 532, 1934.

(15), page 513, the ratio of the reflected to the incident magnetic vector, the Fresnel equation (19), page 494, gives for this case

$$r_{\parallel} = \frac{\cos\theta_0 - Z_{12}\cos\theta_1}{\cos\theta_0 + Z_{12}\cos\theta_1}, \tag{1}$$

where

$$Z_{12} = \frac{\mu_1 k_2}{\mu_2 k_1}, \qquad Z_{21} = \frac{1}{Z_{12}}, \tag{2}$$

as defined on page 512. In order that there shall be no reflected wave the numerator of (1) must vanish. This condition, together with Snell's law, leads to the two relations

$$\cos\theta_1 = Z_{21}\cos\theta_0, \qquad \sin\theta_1 = \frac{\mu_2}{\mu_1} Z_{12}\sin\theta_0, \tag{3}$$

which determine the angles of incidence and refraction. Upon eliminating θ_1 one obtains

$$\sin^2\theta_0 = \frac{1 - Z_{12}^2}{1 - \left(\frac{\mu_2}{\mu_1}\right)^2 Z_{12}^4}, \qquad \cos^2\theta_0 = Z_{12}^2 \frac{1 - \left(\frac{\mu_2}{\mu_1}\right)^2 Z_{12}^2}{1 - \left(\frac{\mu_2}{\mu_1}\right)^2 Z_{12}^4}. \tag{4}$$

To simplify matters we shall assume that the interface of the two media coincides with the surface $x = 0$. The x-axis is normal to the

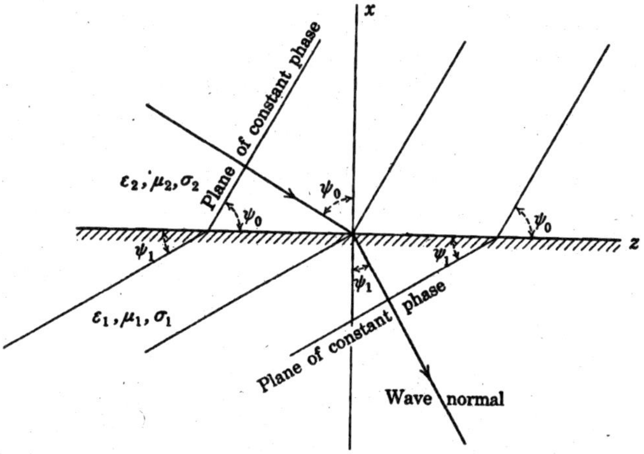

FIG. 99.—Refraction of the planes of constant phase characteristic of a surface wave.

surface and is positive in the direction leading from medium (1) into medium (2), as shown in Fig. 99. The plane of incidence is parallel to the xz-plane; consequently, the field has only the one magnetic component H_y. The electric vector lies in the plane of incidence and has in general both the components E_x and E_z. *The field is transverse magnetic.*[1] Since the magnetic vector is tangent to the surface of discontinuity, its transition from one medium to the other is continuous; hence, by Sec. 9.4

$$(5) \qquad \begin{aligned} H_y &= Ce^{ik_2(-x\cos\theta_0+z\sin\theta_0)-i\omega t}, \qquad &(x>0),\\ H_y &= Ce^{ik_1(-x\cos\theta_1+z\sin\theta_1)-i\omega t}, \qquad &(x<0),\end{aligned}$$

where C is an arbitrary constant.

Next we shall let

$$(6) \qquad h = k_2 \sin\theta_0 = k_1 \sin\theta_1,$$

and define the real and imaginary parts of the factors appearing in the exponents of (5) as follows:

$$(7) \qquad \begin{aligned} h &= \alpha + i\beta,\\ ik_2\cos\theta_0 &= -i\sqrt{k_2^2 - h^2} = p_2 + iq_2,\\ ik_1\cos\theta_1 &= +i\sqrt{k_1^2 - h^2} = -p_1 + iq_1.\end{aligned}$$

The signs of the radicals have been so chosen that the field will vanish for infinite values of $\pm x$. Then in terms of (7),

$$(8) \qquad \begin{aligned} H_y &= C\Psi_2, \qquad &(x>0),\\ H_y &= C\Psi_1, \qquad &(x<0),\end{aligned}$$

where

$$(9) \qquad \begin{aligned} \Psi_1 &= \exp[+p_1x - \beta z + i(-q_1x + \alpha z) - i\omega t],\\ \Psi_2 &= \exp[-p_2x - \beta z + i(-q_2x + \alpha z) - i\omega t].\end{aligned}$$

Let us consider first the nature of the incident wave. We note that the planes of constant phase are defined by

$$(10) \qquad -q_2x + \alpha z = \text{constant},$$

while the planes of constant amplitude satisfy

$$(11) \qquad -p_2x - \beta z = \text{constant}.$$

The planes of constant phase and constant amplitude are not coincident. The real angle of incidence ψ_0 is the angle made by the normal to the planes of constant phase with the positive x-axis. It satisfies the relation

$$(12) \qquad x\cos\psi_0 + z\sin\psi_0 = \text{constant},$$

whence

$$(13) \qquad \cos\psi_0 = \frac{q_2}{\sqrt{\alpha^2 + q_2^2}}, \qquad \sin\psi_0 = \frac{\alpha}{\sqrt{\alpha^2 + q_2^2}}.$$

The planes of constant phase are propagated in the direction of their positive normal with a phase velocity

$$(14) \qquad v_2 = \frac{\omega}{\sqrt{\alpha^2 + q_2^2}},$$

[1] See Sec. 6.1, pp. 349*ff*.

a velocity which is *less* than the normal velocity of a homogeneous plane wave in an infinite medium of the same character. On the other hand, the velocity parallel to the z-axis of a given point on a phase plane may be *greater* than the velocity in free space. The reason is clear from Fig. 100. The wave length λ_2 is the distance between two planes of corresponding phase measured in the direction of the normal.

$$\lambda_2 = \frac{2\pi}{\sqrt{\alpha^2 + q_2^2}}. \tag{15}$$

The distances between the same two planes measured along the x- and z-axes, respectively, are

$$\lambda_{2x} = \frac{\lambda_2}{\cos\psi_0} = \frac{2\pi}{q_2}, \qquad \lambda_{2z} = \frac{\lambda_2}{\sin\psi_0} = \frac{2\pi}{\alpha}. \tag{16}$$

The phase point must travel these distances in the course of one period so that the apparent velocities are

$$v_{2x} = -\frac{\omega}{q_2}, \qquad v_{2z} = \frac{\omega}{\alpha}, \tag{17}$$

and these quantities have values which lie between v_2 and infinity.

The normal to the planes of constant amplitude makes an angle ψ_0' with the x-axis, satisfying the relation

$$-x\cos\psi_0' + z\sin\psi_0' = \text{constant}, \tag{18}$$

whence

$$\tan\psi_0' = -\frac{\beta}{p_2}. \tag{19}$$

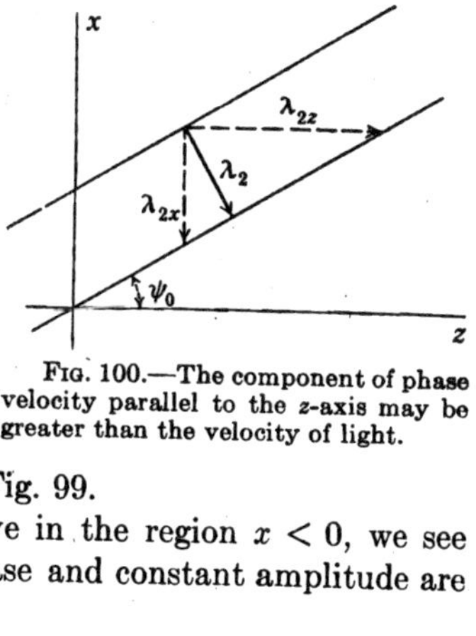

FIG. 100.—The component of phase velocity parallel to the z-axis may be greater than the velocity of light.

Planes of constant amplitude, which in the case of large conductivity are approximately at right angles to planes of constant phase, have not been drawn in Fig. 99.

Turning next to the transmitted wave in the region $x < 0$, we see from (9) that the planes of constant phase and constant amplitude are defined respectively by

$$-q_1x + \alpha z = \text{constant} \tag{20}$$

and

$$p_1x - \beta z = \text{constant}, \tag{21}$$

whose normals make angles ψ_1 and ψ_1' with the negative x-axis defined by

$$\tan\psi_1 = -\frac{\alpha}{q_1}, \qquad \tan\psi_1' = \frac{\beta}{p_1}. \tag{22}$$

The phase velocity is

$$v_1 = \frac{\omega}{\sqrt{\alpha^2 + q_1^2}}. \tag{23}$$

The components of the electric vector are best calculated from the field equation

$$\mathbf{E} = \frac{i\omega\mu}{k^2} \nabla \times \mathbf{H}, \tag{24}$$

which resolved into rectangular components gives

$$E_x = -\frac{i\omega\mu}{k^2}\frac{\partial H_y}{\partial z}, \qquad E_z = \frac{i\omega\mu}{k^2}\frac{\partial H_y}{\partial x}. \tag{25}$$

From (8) and (9), it follows that

$$\begin{aligned} E_x &= \frac{\omega\mu_2}{k_2^2}(\alpha + i\beta)C\Psi_2, \\ E_z &= -\frac{i\omega\mu_2}{k_2^2}(p_2 + iq_2)C\Psi_2, \end{aligned} \qquad (x > 0), \tag{26}$$

$$\begin{aligned} E_x &= \frac{\omega\mu_1}{k_1^2}(\alpha + i\beta)C\Psi_1, \\ E_z &= \frac{i\omega\mu_1}{k_1^2}(p_1 - iq_1)C\Psi_1. \end{aligned} \qquad (x < 0), \tag{27}$$

Physically, the field described by (8), (26), and (27) is associated with a current in the z-direction whose distribution with respect to y is uniform. Such a field is transverse magnetic, and the longitudinal component of electric field accounts for the losses in either medium, since E_zH_y is the component of energy flow normal to the interface. The transverse component E_x, on the other hand, gives rise to an energy flow parallel to the plane. The whole problem is comparable to the limiting case of a current propagated along a wire whose radius becomes infinite. The other symmetry, corresponding to the condition $r_\perp = 0$, leads to transverse electric waves and may be interpreted as the limiting problem of propagation along a solenoid of very large radius. In this case the current is transverse to the elements of the cylindrical surface.

9.14. Skin Effect.—If both media are perfect dielectrics, the angle θ_0 is real in the case of $r_\parallel = 0$ and is identical with the Brewster polarizing angle. Of much greater practical interest is the case in which medium (1) is a conductor and medium (2) a dielectric. One may then assume that $|Z_{12}|^2 \ll 1$. According to (4) and (6)

$$h^2 = k_2^2 \frac{1 - Z_{12}^2}{1 - \left(\dfrac{\mu_2}{\mu_1}\right)^2 Z_{12}^4}, \tag{28}$$

which can now be expanded in powers of Z_{12}.

$$h = k_2\left[1 - \frac{1}{2}Z_{12}^2 + \frac{1}{2}\left(\frac{\mu_2^2}{\mu_1^2} - \frac{1}{8}\right)Z_{12}^4 - \cdots\right], \tag{29}$$

of which we shall retain in the present approximation only the square term. Since by hypothesis the conductivity of (2) is zero, we may write

$$Z_{12}^2 = \frac{\mu_1^2}{\mu_2^2}\frac{\epsilon_2\mu_2\omega^2}{\epsilon_1\mu_1\omega^2 + i\sigma_1\mu_1\omega} = \frac{\mu_1\epsilon_2}{\mu_2\epsilon_1}\frac{1 - i\eta_1}{1 + \eta_1^2}, \tag{30}$$

where as above $\eta_1 = \sigma_1/\epsilon_1\omega$. Then

$$h \simeq k_2\left(1 - \frac{1}{2}\frac{\epsilon_2\mu_1}{\epsilon_1\mu_2}\frac{1}{1 + \eta_1^2} + \frac{i}{2}\frac{\epsilon_2\mu_1}{\epsilon_1\mu_2}\frac{\eta_1}{1 + \eta_1^2}\right). \tag{31}$$

Note that α, the real part of this expression, is less than k_2 and that, consequently, the velocity in the z-direction of a point on a plane of constant phase is greater than that of a free wave in the dielectric. We saw above that this is only an apparent velocity and that the true phase velocity is in fact less than that fixed by k_2.

We now make the further assumption that $\eta_1^2 \gg 1$, and that $1/\eta_1^2$ can, therefore, be neglected with respect to $1/\eta_1$. To this approximation

$$\alpha \simeq \omega\sqrt{\epsilon_2\mu_2}, \qquad \beta \simeq \frac{\omega^2\epsilon_2\mu_1}{2\sigma_1}\sqrt{\frac{\epsilon_2}{\mu_2}}. \tag{32}$$

Thus the velocity in the z-direction is essentially the characteristic phase velocity of the medium and the attenuation approaches zero with increasing conductivity.

To the same approximation

$$k_2^2 - h^2 \simeq \frac{\epsilon_2\mu_1}{\epsilon_1\mu_2}\frac{k_2^2}{1 + \eta_1^2}(1 - i\eta_1), \tag{33}$$

which by (7) leads to

$$q_2 \simeq p_2 \simeq \omega\epsilon_2\sqrt{\frac{\mu_1\omega}{2\sigma_1}}. \tag{34}$$

Likewise,

$$k_1^2 - h^2 \simeq \omega^2(\epsilon_1\mu_1 - \epsilon_2\mu_2) + i\omega\mu_1\sigma_1, \tag{35}$$

$$q_1 \simeq p_1 \simeq \sqrt{\frac{\omega\mu_1\sigma_1}{2}}. \tag{36}$$

From these results one will note, first, that

$$\frac{\alpha}{q_2} \simeq \frac{v_2}{\beta} \simeq \sqrt{\frac{2\mu_2\sigma_1}{\mu_1\epsilon_2\omega}}, \tag{37}$$

whence it appears from (13) and (19) that the *planes of constant amplitude in the incident wave are normal to the planes of constant phase.* As $\sigma_1 \to \infty$, $\tan \psi_0 \to \infty$ and $\psi_0 \to \pi/2$. *As the conductivity increases, the planes of constant phase in the dielectric become very nearly normal to the surface of the conductor.*

Within the conductor, the corresponding relations are determined by (22).

$$\tan \psi_1 \simeq -\sqrt{\frac{2\omega\epsilon_2\mu_2}{\mu_1\sigma_1}}, \qquad \tan \psi_1' \simeq \sqrt{\frac{\mu_1\omega^3\epsilon_2^3}{\mu_2\sigma_1^3}}. \tag{38}$$

As $\sigma_1 \to \infty$, both ψ_1 and ψ_1' approach zero. *The planes of constant amplitude and constant phase within the conductor are very nearly parallel to the interface.* At the same time the wave length becomes very small, for

$$\lambda_1 = \frac{v_1}{\nu} \simeq 2\pi \sqrt{\frac{2}{\omega\mu_1\sigma_1}}. \tag{39}$$

It will be recalled that in the case of metals σ_1 is of the order of 10^7 mhos/meter. The deviation of ψ_0 from $\pi/2$ and of ψ_1 from zero is then completely negligible, and the wave length in meters within the metal is

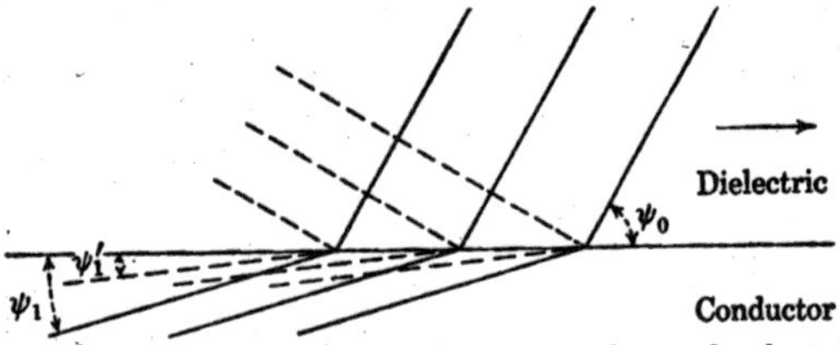

FIG. 101.—The solid lines represent planes of constant phase, the dotted lines planes of constant amplitude.

in magnitude of the order $\nu^{-\frac{1}{2}}$. Even in the case of sea water, at a frequency of 1 megacycle/sec., the forward tilt of the phase planes above the surface amounts to only some 10′ of arc. The forward tilt of a radio wave propagated over dry earth may be more marked. The relation of the planes of constant phase and constant amplitude is then somewhat as illustrated in Fig. 101. It must be remembered, however, that in such a case the assumption $\eta_1^2 \gg 1$ is not necessarily justified.

Upon introducing the approximations (32), (34), and (36) into (26) and (27), we obtain for the field components the expressions

$$E_x \simeq \sqrt{\frac{\mu_2}{\epsilon_2}}\, C\Psi_2, \qquad E_z \simeq \sqrt{\frac{\mu_1\omega}{\sigma_1}}\, e^{-i\frac{\pi}{4}}\, C\Psi_2, \qquad (x > 0), \tag{40}$$

$$E_x \simeq \frac{\omega\sqrt{\epsilon_2\mu_2}}{\sigma_1}\, e^{-i\frac{\pi}{2}}\, C\Psi_1, \qquad E_z \simeq \sqrt{\frac{\mu_1\omega}{\sigma_1}}\, e^{-i\frac{\pi}{4}}\, C\Psi_1, \qquad (x < 0). \tag{41}$$

Note that the tangential component E_z is continuous at the boundary

$x = 0$ and equal in magnitude to $(\mu_1\omega/\sigma_1)^{\frac{1}{2}}$ times the magnitude of the tangential component of magnetic field. As $\sigma_1 \to \infty$, the longitudinal electric field vanishes.

The normal and tangential components of **E** differ in phase by 45 deg. (A more exact calculation shows this angle to be less than 45 deg. in the case of poor conductors.) Consequently *the locus of the resultant electric vector at any point is an ellipse.* At a point within the dielectric just above the surface, the ellipse is inscribed within a rectangle whose sides are $2a_2 = 2(\mu_2/\epsilon_2)^{\frac{1}{2}}$, $2b_2 = 2(\mu_1\omega/\sigma_1)^{\frac{1}{2}}$, $b_2 \ll a_2$, as shown in

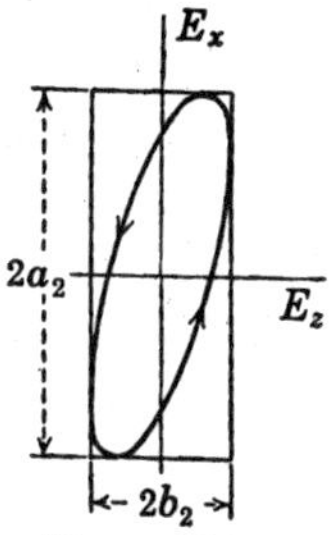

FIG. 102a.—Locus of the electric vector at a point in the dielectric.

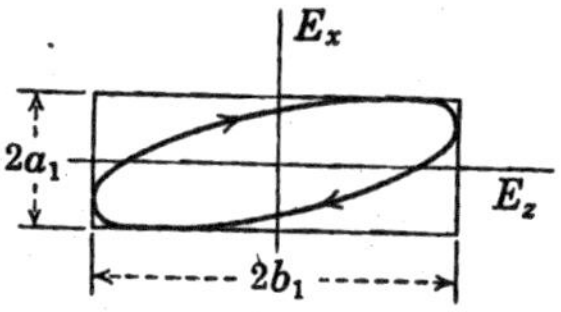

FIG. 102b.—Locus of the electric vector at a point in the conductor.

Fig. 102*a*. Since the particular value of z is of no importance, the coordinates of the ellipse are

$$E_x = a_2 \cos \omega t, \qquad E_z = b_2 \cos\left(\omega t + \frac{\pi}{4}\right). \tag{42}$$

The electric vector in the dielectric rotates in a counterclockwise direction. The ellipse is identical in form at other points of the dielectric but reduced in size by reason of the attenuation factor.

Within the conductor the electric vector describes an ellipse bounded by a rectangle whose sides are $2a_1 = 2\omega(\epsilon_2\mu_2)^{\frac{1}{2}}/\sigma_1$, $2b_1 = 2(\mu_1\omega/\sigma_1)^{\frac{1}{2}}$, $b_1 \gg a_1$. The coordinates of the ellipse are

$$E_x = a_1 \cos\left(\omega t + \frac{\pi}{2}\right), \qquad E_z = b_1 \cos\left(\omega t + \frac{\pi}{4}\right). \tag{43}$$

E_z is now retarded with respect to E_x and, consequently, *within the conductor the electric vector rotates in a clockwise direction.*

The depth of penetration within the conductor is determined by the damping factor p_1, whose reciprocal is exactly the factor δ defined on page 504. The current density in the z-direction is $J_z = \sigma_1 E_z$, and the total current passing through an infinite strip 1 meter in width parallel to the xy-plane in the conductor is

$$I = \sigma_1 \int_0^1 \int_{-\infty}^0 E_z \, dx \, dy = Ce^{-\beta z + i(\alpha z - \omega t)}. \tag{44}$$

On the other hand, the current through such a strip of depth d is

$$\Delta I = I(1 - e^{-p_1 d + i q_1 d}). \tag{45}$$

Since p_1 is very large in the case of good conductors, the current is confined almost entirely to a thin "skin" in the neighborhood of the surface. One will note, moreover, that the phase of a current filament varies with its depth. This effect may be expressed in terms of an internal self-inductance.[1]

PROPAGATION ALONG A CIRCULAR CYLINDER

9.15. Natural Modes.—A circular cylinder of radius a and infinite length is embedded in an infinite homogeneous medium. The propagation constant of the cylinder is k_1, that of the external medium k_2. No restriction is imposed as yet upon the conductivity of either. The components of the field in circular coordinates were derived in Sec. 6.6. This field must be finite at the center and, consequently, the wave functions within the wire will be constructed from Bessel functions of the first kind. Outside the cylinder the Hankel functions $H_n^{(1)}$ ensure the proper behavior at infinity. According to (36) and (37), page 361, we have at all interior points, $r < a$,

$$
\begin{aligned}
E_r^i &= \sum_{n=-\infty}^{\infty} \left[\frac{ih}{\lambda_1} J_n'(\lambda_1 r)\, a_n^i - \frac{\mu_1 \omega n}{\lambda_1^2 r} J_n(\lambda_1 r)\, b_n^i \right] F_n, \\
E_\theta^i &= - \sum_{n=-\infty}^{\infty} \left[\frac{nh}{\lambda_1^2 r} J_n(\lambda_1 r)\, a_n^i + \frac{i\mu_1 \omega}{\lambda_1} J_n'(\lambda_1 r)\, b_n^i \right] F_n, \qquad (1) \\
E_z^i &= \sum_{n=-\infty}^{\infty} [J_n(\lambda_1 r)\, a_n^i] F_n,
\end{aligned}
$$

$$
\begin{aligned}
H_r^i &= \sum_{n=-\infty}^{\infty} \left[\frac{nk_1^2}{\mu_1 \omega \lambda_1^2 r} J_n(\lambda_1 r)\, a_n^i + \frac{ih}{\lambda_1} J_n'(\lambda_1 r)\, b_n^i \right] F_n, \\
H_\theta^i &= \sum_{n=-\infty}^{\infty} \left[\frac{ik_1^2}{\mu_1 \omega \lambda_1} J_n'(\lambda_1 r)\, a_n^i - \frac{nh}{\lambda_1^2 r} J_n(\lambda_1 r)\, b_n^i \right] F_n, \qquad (2) \\
H_z^i &= \sum_{n=-\infty}^{\infty} [J_n(\lambda_1 r)\, b_n^i] F_n.
\end{aligned}
$$

[1] See the treatment by Sommerfeld in Riemann-Weber, "Differentialgleichungen der Physik," Vol. II, pp. 507–511, 7th ed., 1927.

At all exterior points, $r > a$,

$$
\begin{aligned}
E_r^e &= \sum_{n=-\infty}^{\infty} \left[\frac{ih}{\lambda_2} H_n^{(1)\prime}(\lambda_2 r)\, a_n^e - \frac{\mu_2 \omega n}{\lambda_2^2 r} H_n^{(1)}(\lambda_2 r)\, b_n^e \right] F_n, \\
E_\theta^e &= - \sum_{n=-\infty}^{\infty} \left[\frac{nh}{\lambda_2^2 r} H_n^{(1)}(\lambda_2 r)\, a_n^e + \frac{i\mu_2 \omega}{\lambda_2} H_n^{(1)\prime}(\lambda_2 r)\, b_n^e \right] F_n, \qquad (3) \\
E_z^e &= \sum_{n=-\infty}^{\infty} [H_n^{(1)}(\lambda_2 r)\, a_n^e] F_n,
\end{aligned}
$$

$$
\begin{aligned}
H_r^e &= \sum_{n=-\infty}^{\infty} \left[\frac{nk_2^2}{\mu_2 \omega \lambda_2^2 r} H_n^{(1)}(\lambda_2 r)\, a_n^e + \frac{ih}{\lambda_2} H_n^{(1)\prime}(\lambda_2 r)\, b_n^e \right] F_n, \\
H_\theta^e &= \sum_{n=-\infty}^{\infty} \left[\frac{ik_2^2}{\mu_2 \omega \lambda_2} H_n^{(1)\prime}(\lambda_2 r)\, a_n^e - \frac{nh}{\lambda_2^2 r} H_n^{(1)}(\lambda_2 r)\, b_n^e \right] F_n, \qquad (4) \\
H_z^e &= \sum_{n=-\infty}^{\infty} [H_n^{(1)}(\lambda_2 r)\, b_n^e] F_n.
\end{aligned}
$$

In these relations

$$
\begin{aligned}
\lambda_1^2 = k_1^2 - h^2, \qquad \lambda_2^2 = k_2^2 - h^2, \\
F_n = \exp(in\theta + ihz - i\omega t),
\end{aligned} \qquad (5)
$$

and the prime above a cylinder function denotes differentiation with respect to the argument λr.

The coefficients of the expansions and the propagation factor h are as yet undetermined. However, at the boundary $r = a$, the tangential components of the field are continuous, and this condition imposes a relation between the coefficients. From the continuity of the tangential components of **E**, we obtain

$$
\begin{aligned}
\frac{nh}{u^2} J_n(u)\, a_n^i + \frac{i\mu_1 \omega}{u} J_n'(u)\, b_n^i &= \frac{nh}{v^2} H_n^{(1)}(v)\, a_n^e + \frac{i\mu_2 \omega}{v} H_n^{(1)\prime}(v)\, b_n^e, \\
J_n(u)\, a_n^i &= H_n^{(1)}(v)\, a_n^e,
\end{aligned} \qquad (6)
$$

while from the tangential components of **H**

$$
\begin{aligned}
\frac{ik_1^2}{\mu_1 \omega u} J_n'(u)\, a_n^i - \frac{nh}{u^2} J_n(u)\, b_n^i &= \frac{ik_2^2}{\mu_2 \omega v} H_n^{(1)\prime}(v)\, a_n^e - \frac{nh}{v^2} H_n^{(1)}(v)\, b_n^e, \\
J_n(u)\, b_n^i &= H_n^{(1)}(v)\, b_n^e,
\end{aligned} \qquad (7)
$$

where

$$
u = \lambda_1 a, \qquad v = \lambda_2 a. \qquad (8)
$$

Equations (7) and (8) constitute a *homogeneous* system of linear relations satisfied by the four coefficients a_n^i, b_n^i, a_n^e, b_n^e. The system admits a nontrivial solution only in case its determinant is zero. *The propagation factor h is determined by the condition that the determinant of* (6) *and* (7) *shall vanish.* Expansion of this determinant leads to the transcendental equation

$$(9)\qquad \left[\frac{\mu_1}{u}\frac{J_n'(u)}{J_n(u)} - \frac{\mu_2}{v}\frac{H_n^{(1)\prime}(v)}{H_n^{(1)}(v)}\right]\left[\frac{k_1^2}{\mu_1 u}\frac{J_n'(u)}{J_n(u)} - \frac{k_2^2}{\mu_2 v}\frac{H_n^{(1)\prime}(v)}{H_n^{(1)}(v)}\right] = n^2h^2\left(\frac{1}{v^2} - \frac{1}{u^2}\right)^2,$$

whose roots are the allowed values of the propagation factor h and so determine the characteristic or natural modes of propagation. These roots are discrete and form a twofold infinity. The Bessel functions are transcendental; thus, for each value of n there is a denumerable infinity of roots, any one of which can be denoted by the subscript m. Any root of (9) can then be designated by h_{nm}.

No general method can be stated for finding the roots of (9), but fortunately it is possible to make approximations in practical problems involving metallic conductors which greatly simplify the solution. One will observe from (6) that *if the conductivity is finite, a superposition of electric and magnetic type waves is necessary to satisfy the boundary conditions, except in the symmetrical case* $n = 0$. The field is not transverse, with respect to either the electric or the magnetic vector. If, however, the conductivity of either medium is infinite, the boundary conditions can be fulfilled by either a transverse magnetic or a transverse electric field. In this case the problem is completely determined by either the conditions

$$(10)\qquad E_\theta = 0, \qquad E_z = 0, \qquad (r = a),$$

or, alternatively, by

$$(11)\qquad \frac{\partial}{\partial r}(rH_\theta) = 0, \qquad \frac{\partial}{\partial r}H_z = 0, \qquad (r = a),$$

taken from (14), page 484, noting that the tangential components of $\mathbf{H}$ do *not* vanish. In case the cylinder is a perfect conductor embedded in a dielectric, the allowed transverse magnetic and transverse electric modes are determined by roots respectively of the equations

$$(12)\qquad \frac{H_n^{(1)}(v)}{H_n^{(1)\prime}(v)} = 0, \qquad \frac{H_n^{(1)\prime}(v)}{H_n^{(1)}(v)} = 0.$$

The amplitudes a_n^e and b_n^e are now quite independent and are determined solely by the nature of the excitation. In the inverse case of a dielectric cylinder bounded externally by a perfect conductor, the transverse magnetic and transverse electric modes are determined by the roots respectively of

$$(13)\qquad \frac{J_n(u)}{J_n'(u)} = 0, \qquad \frac{J_n'(u)}{J_n(u)} = 0.$$

In practice the conductivity of a metal, while not infinite, is yet so very large that the cylinder functions in such a region can be replaced by their asymptotic representations. It is then found that in each possible mode, either a transverse electric or transverse magnetic wave predominates. A small component of opposite type which spoils the transversality may be considered a correction term taking account of the finiteness of the conductivity.

If only the symmetrical mode $n = 0$ is excited, the field is independent of θ and the coefficients a_0, b_0 are independent of one another, whatever the conductivity. From (6) it is apparent that in this case the transverse magnetic wave satisfies

$$\begin{aligned} J_0(u)\, a_0^i &= H_0^{(1)}(v)\, a_0^e, \\ \frac{k_1^2}{\mu_1 u} J_1(u)\, a_0^i &= \frac{k_2^2}{\mu_2 v} H_1^{(1)}(v)\, a_0^e, \end{aligned} \tag{14}$$

at the boundary, while for the transverse electric wave

$$\begin{aligned} \frac{\mu_1}{u} J_1(u)\, b_0^i &= \frac{\mu_2}{v} H_1^{(1)}(v)\, b_0^e, \\ J_0(u)\, b_0^i &= H_0^{(1)}(v)\, b_0^e. \end{aligned} \tag{15}$$

The allowed transverse magnetic modes are then determined by the roots h_{0m} of the transcendental equation

$$\frac{k_1^2}{\mu_1 u} \frac{J_1(u)}{J_0(u)} = \frac{k_2^2}{\mu_2 v} \frac{H_1^{(1)}(v)}{H_0^{(1)}(v)}, \tag{16}$$

while the transverse electric modes are found from

$$\frac{\mu_1}{u} \frac{J_1(u)}{J_0(u)} = \frac{\mu_2}{v} \frac{H_1^{(1)}(v)}{H_0^{(1)}(v)}. \tag{17}$$

9.16. Conductor Embedded in a Dielectric.—We shall discuss only the case of waves that are predominantly transverse magnetic, corresponding to the practically important problem of an axial current in a long, straight wire.[1] The axially symmetric solution, $n = 0$, shall be considered first. It will then be shown that the asymmetric modes, if excited, are immediately damped out and, consequently, never play a part in the propagation of current along a solid conductor.

Suppose, first, that the conductivity of the cylinder is infinite. According to (12) the propagation factor is then determined by

$$\frac{H_0^{(1)}(v)}{H_1^{(1)}(v)} = 0. \tag{18}$$

[1] In the case of transverse electric waves the lines of current are circles concentric with the axis of the conductor. This solution may, therefore, be applied to the study of propagation along an idealized solenoid. *Cf.* Sommerfeld, *Ann. Physik*, **15**, 673, 1904. If the conductor is solid the transverse electric modes are damped out.

If v is very small, $H_0^{(1)}(v)$ can be represented approximately by

$$H_0^{(1)}(v) \simeq \frac{2i}{\pi} \ln \frac{\gamma v}{2i}, \qquad \frac{dH_0^{(1)}(v)}{dv} \simeq \frac{2i}{\pi v}, \tag{19}$$

where $\gamma = 1.781$. These expressions are obtained by evaluating the integral (39), page 367, in the neighborhood of the origin. The Hankel function is evidently multivalued, and as in the case of the logarithm we use only its *principal value.* This principal value can be defined most easily in terms of the asymptotic expression

$$H_p^{(1)}(v) \simeq \sqrt{\frac{2}{\pi v}}\, e^{i\left(v - \frac{2p+1}{4}\pi\right)} \tag{20}$$

which holds when $v \gg 1$, $p \ll v$. If one puts $v = |v|e^{i\phi}$, then the principal value of $H_p^{(1)}(v)$ lies in the domain $-\pi/2 < \phi < 3\pi/2$. *The principal value or branch of $H_p^{(1)}(v)$ vanishes at all infinite points of the positive-imaginary half plane.* It is only with the roots of the principal branch that we are concerned, and it can be shown that the principal branches of $H_0^{(1)}(v)$ and $H_1^{(1)}(v)$ have none.[1] Consequently, the ratio

$$\frac{H_0^{(1)}(v)}{H_1^{(1)}(v)} \simeq -v \ln \frac{\gamma v}{2i} = 0 \tag{21}$$

has only the solution $v = 0$, and hence $h = k_2$. *If the cylinder is infinitely conducting, the field is propagated in the axial direction with a velocity determined solely by the external medium.* If this is free space, the waves travel along the cylinder without attenuation and with the velocity of light.

In case the conductivity of the cylinder, while not infinite, is very large, as of a metal, h differs from k_2 but there will be at least one mode for which the difference is small, so that $v = a\sqrt{k_2^2 - h^2}$ is a small quantity. The approximation (21) still applies to the right side of (16). On the left side we note that $|k_1| \gg k_2$; since $k_2 \simeq |h|$, $u \simeq k_1 a \gg 1$. The Bessel functions $J_0(u)$ and $J_1(u)$ can, therefore, be replaced by their asymptotic representations. The imaginary part of u is positive. One will remember that $k_1 = \alpha_1 + i\beta_1$ and that, if σ_1 is very large, $\alpha_1 \simeq \beta_1$. Then in virtue of (18), page 359,

$$\lim_{\beta_1 \to \infty} \frac{J_1(u)}{J_0(u)} = \lim_{\beta_1 \to \infty} \tan\left[a\left(\alpha_1 + i\beta_1 - \frac{\pi}{4a}\right)\right] = +i. \tag{22}$$

Equation (16) now assumes the form

$$v^2 \ln \frac{\gamma v}{2i} = i\frac{\mu_1}{\mu_2}\frac{k_2^2 a}{k_1}, \tag{23}$$

whose right-hand member is independent of h.

[1] SOMMERFELD, in Riemann-Weber, *loc. cit.*, p. 461.

This equation for h can be solved easily by successive approximations. The method proposed by Sommerfeld[1] goes as follows. Let

$$\xi = \left(\frac{\gamma v}{2i}\right)^2, \qquad \eta = -\frac{i\gamma^2}{2}\frac{\mu_1}{\mu_2}\frac{k_2^2 a}{k_1}, \tag{24}$$

$$\xi \ln \xi = \eta.$$

The quantity η is known and ξ is to be determined. Now the logarithm of ξ is a slowly varying function with respect to ξ itself. Hence, if the nth approximation has been found, the $(n + 1)$th is obtained from

$$\xi_{n+1} \ln \xi_n = \eta, \tag{25}$$

or

$$\xi = \cfrac{\eta}{\ln \cfrac{\eta}{\ln \cfrac{\eta}{\ln \cfrac{\eta}{\ln \cfrac{\eta}{\cdots}}}}} \tag{26}$$

The convergence can be demonstrated for complex as well as real values of η. Roots of (24) which lie near $\xi = 1$ or which are associated with other branches than the principal of the logarithm are rejected. It turns out that $\ln \xi$ is a negative number of the order -20, and the convergence is hastened by assuming this value for $\ln \xi_0$. From (24), with suitable approximations,

$$\eta \simeq (1 + i)1.75 \times 10^{-13} a\nu \sqrt{\frac{\kappa_{1m}\kappa_{2e}\nu}{\sigma_1}}, \tag{27}$$

where ν is the frequency and a is expressed in meters.

To illustrate, the following numerical example is quoted from Sommerfeld.[2] The wire is of copper, 1 mm. in radius. The frequency is 10^9 cycles/sec. and the outside medium is air. Then (27) gives

$$\eta = -(1 + i)7.25 \times 10^{-7}.$$

Assuming $\ln \xi_0 = -20$, we obtain for the first approximation

$$\xi_1 = (1 + i)3.6 \times 10^{-8},$$

$$\ln \xi_1 = -16.8 + i\frac{\pi}{4} = -(16.8 - 0.78i).$$

[1] *Göttinger Nachr.*, 1899. Described in Riemann-Weber, *loc. cit.*, p. 529.
[2] RIEMANN-WEBER, *loc. cit.*, p. 530.

Similarly

$$\xi_2 = \frac{\eta}{\ln \xi_1} = (4.1 + 4.5i)10^{-8},$$
$$\ln \xi_2 = -(16.7 - 0.83i),$$
$$\xi_3 = \frac{\eta}{\ln \xi_2} = (4.2 + 4.5i)10^{-8}.$$

ξ_3 differs but slightly from ξ_2 and the series may be broken off here. The smallness of ξ justifies the initial approximations for the Bessel and Hankel functions. Returning now to (24), we calculate the value of v, and hence of the propagation factor h, for $a = 1$ mm.

$$h \simeq k_2[1 + (6.0 + 6.4i)10^{-5}] = \alpha + i\beta.$$

The phase velocity along the wire is

$$V \simeq c(1 - 6.0 \times 10^{-5}),$$

while the distance the wave must travel to be reduced to $1/e$ of its initial amplitude is $z = 770$ meters. Although the velocity differs an inappreciable amount from that of free space, the damping at this high frequency and for this size of wire (about No. 12, Brown and Sharpe gauge) is relatively high from the point of view of a communication line.

If the radius a is extremely small, the quantity $u \simeq ak_1$ may also be small in spite of a large conductivity. To illustrate this point, Sommerfeld has worked out the case of a platinum wire, conductivity one-eighth that of copper, $a = 2\mu = 2 \times 10^{-6}$ meter and $\nu = 3 \times 10^8$ cycles/sec. One then finds that $u \simeq (1 + i)0.19$, and consequently the Bessel functions $J_0(u)$ and $J_1(u)$ cannot be replaced by their asymptotic expressions. In the case cited the value of u is so small that one can put approximately $J_0(u) = 1$, $J_1(u) = u/2$, and proceed according to the same method outlined above. One obtains

$$h \simeq 0.083 + 0.061\, i.$$

The phase velocity proves to be only 76 per cent of its free-space value and the attenuation is so large that the amplitude is reduced to $1/e$ of its initial value in a distance $z = 16$ cm.

Thus far we have considered only one root of the determinantal equation (16), the root which in the case of high conductivity makes $h \simeq k_2$ and leads to what will be termed the *principal wave*. There are other roots, however, for which $v = a\sqrt{k_2^2 - h^2}$ is very large. With the help of the asymptotic expressions for the Hankel functions, (22), page 359, we may now approximate (16) by

$$v \simeq -i\,\frac{\mu_2}{\mu_1}\frac{k_1^2}{k_2^2}\frac{1}{u}\frac{J_0(u)}{J_1(u)}. \tag{28}$$

If the left side of (28) is very large, the roots are given to the first approximation by $J_1(u) = 0$. But the roots of $J_1(u)$ lie on the real axis of the u-plane. Consequently the imaginary part of h^2 must be equal to the imaginary part of k_1^2. This means that in the case of metallic conductors the attenuation factor β of $h = \alpha + i\beta$ is of the order $2 \times 10^{-3} \sqrt{\nu \sigma_1 \kappa_{1m}}$, and hence *the complementary waves corresponding to the other roots of* (16) *are damped out at once.*

The roots of (17), which give the modes of axially symmetric transverse electric waves can be investigated according to the same procedure. However, since there is no "magnetic conductivity," there is also no principal wave characterized by low attenuation. All axially symmetric transverse electric waves in a solid metal cylinder are damped out too quickly to be observable.

In the asymmetric case we have seen that the field is a linear combination of transverse magnetic and transverse electric modes. The existence of a principal wave with small attenuation implies that $h \simeq k_2$, $v \ll 1$. Since n is now different from zero, this would lead to extremely large values of the right-hand member of (9), a condition that could be fulfilled only by nearly infinite values of the conductivity σ_1. It turns out that even with a conductivity as high as that of copper, the imaginary part of the propagation factor h is still very large and the attenuation such as to preclude the existence of asymmetric waves at any finite distance from the source.

The results of this study may be summed up as follows. There are an infinite number of modes of propagation along a solid conducting cylinder. The amplitudes of these modes are the coefficients a_n and b_n, whose values are determined by the initial excitation—the nature of the source. Of all these modes, however, only one, the principal mode, a symmetric, transverse magnetic wave, possesses a relatively low attenuation. All others are damped out within a centimeter or two from the source, even at frequencies as low as 60 cycles/sec. We shall see in the next section that the field of the principal wave within the conductor tends to concentrate in a thin skin near the surface. At high frequencies the center of the cylinder is essentially free of field and current, with a correspondingly small conversion of energy into heat. On the other hand the field and current distribution within the cylinder in the case of complementary waves is nearly uniform. The "skin effect" is then external, and the conversion into heat correspondingly large. As the conductivity of the cylinder is decreased, the attenuation of the principal wave increases, while that of the complementary waves is diminished until finally, in the limit of a perfect dielectric cylinder, we find the principal wave to have vanished and the complementary waves propagated without attenuation.

9.17. Further Discussion of the Principal Wave.—We have seen that in the case of a solid conducting cylinder only the principal wave plays an important role. Our next task is to put the results of the preceding section in the form best adapted to numerical discussion. The first step is to express the coefficients a_0^i and a_0^e in terms of the total conduction current in the cylinder, since this is usually the quantity most easily measured. The current density parallel to the axis at any point within the cylinder is $J_z = \sigma_1 E_z^i$, and the total current, therefore,

$$I = \int_0^a \int_0^{2\pi} J_z r \, d\theta \, dr = 2\pi a_0^i \sigma_1 \int_0^a J_0(\lambda_1 r) \, r \, dr \, e^{ihz - i\omega t}. \tag{29}$$

By (26), page 360,

$$\int_0^a J_0(\lambda_1 r) \, r \, dr = \frac{r}{\lambda_1} J_1(\lambda_1 r) \Big|_0^a = \frac{a}{\lambda_1} J_1(\lambda_1 a), \tag{30}$$

so that

$$I = \frac{2\pi a \sigma_1}{\lambda_1} a_0^i J_1(\lambda_1 a) \, e^{ihz - i\omega t}. \tag{31}$$

The amplitude of the total current is defined by

$$I = I_0 \, e^{ihz - i\omega t}, \tag{32}$$

and hence

$$a_0^i = \frac{\lambda_1 I_0}{2\pi a \sigma_1 J_1(\lambda_1 a)}. \tag{33}$$

Likewise by (14),

$$a_0^e = \frac{\lambda_2}{2\pi a \sigma_1} \frac{\mu_2 k_1^2}{\mu_1 k_2^2} \frac{I_0}{H_1^{(1)}(\lambda_2 a)}. \tag{34}$$

The longitudinal and transverse impedances were defined on page 354 by the relations

$$E_r = Z_z H_\theta, \qquad E_z = -Z_r H_\theta. \tag{35}$$

Then from (30), page 360, or directly from Eqs. (1) to (4), in Sec. 9.15, one obtains

$$Z_z^i = \frac{\omega \mu_1 h}{k_1^2}, \qquad Z_r^i = \frac{\omega \mu_1 \lambda_1}{i k_1^2} \frac{J_0(\lambda_1 r)}{J_1(\lambda_1 r)}, \qquad (r < a), \tag{36}$$

$$Z_z^e = \frac{\omega \mu_2 h}{k_2^2}, \qquad Z_r^e = \frac{\omega \mu_2 \lambda_2}{i k_2^2} \frac{H_0^{(1)}(\lambda_2 r)}{H_1^{(1)}(\lambda_2 r)}, \qquad (r > a). \tag{37}$$

It will be noted in passing that by (16),

$$Z_r^i = Z_r^e \qquad \text{at} \qquad r = a. \tag{38}$$

The continuity of the tangential components of **E** *and* **H** *at the boundary is ensured by matching the radial impedances at* $r = a$.

The *surface impedance* Z_s is defined as the ratio of longitudinal e.m.f. measured at the surface of the conductor to the total current.

$$E_z = Z_s I \qquad \text{at} \qquad r = a, \tag{39}$$

whence by (1) and (33),

$$Z_s = \frac{\lambda_1}{2\pi a \sigma_1} \frac{J_0(\lambda_1 a)}{J_1(\lambda_1 a)}. \tag{40}$$

Finally, upon introducing the expressions for a_0^i and a_0^e into (2) and (4), one obtains

$$\begin{aligned} H_\theta^i &= -\frac{ik_1^2 I}{2\pi a \omega \mu_1 \sigma_1} \frac{J_1(\lambda_1 r)}{J_1(\lambda_1 a)}, \\ H_\theta^e &= -\frac{ik_1^2 I}{2\pi a \omega \mu_1 \sigma_1} \frac{H_1^{(1)}(\lambda_2 r)}{H_1^{(1)}(\lambda_2 a)}. \end{aligned} \tag{41}$$

From these relations and the various impedance functions the components of the field can be determined at any point.

The approximation $\sigma_1/\epsilon_1\omega \gg 1$ is valid for all metallic conductors. In all practical cases it can be assumed, therefore, that

$$\begin{aligned} &k_1^2 \simeq i\omega\mu_1\sigma_1, \qquad && k_1 \simeq \sqrt{i\omega\mu_1\sigma_1}, \\ &|k_1| \gg k_2 \simeq h, && \lambda_1 \simeq k_1. \end{aligned} \tag{42}$$

Thus one obtains for the longitudinal impedances

$$Z_z^i \simeq -\frac{i\omega\sqrt{\mu_2\epsilon_2}}{\sigma_1}, \qquad Z_z^e \simeq \sqrt{\frac{\mu_2}{\epsilon_2}}. \tag{43}$$

Outside the cylinder the longitudinal impedance approaches the intrinsic impedance of the dielectric medium. Inside the cylinder it is a pure imaginary; the mean axial flow of energy within the cylinder is, therefore, zero. For

$$\bar{S}_z^i = -\tfrac{1}{2}Re(E_r^i \bar{H}_\theta^i) = -\tfrac{1}{2}Re(Z_z^i |H_\theta^i|^2) \simeq 0. \tag{44}$$

The transport of energy along the cylinder takes place entirely in the external dielectric. The internal energy surges back and forth and supplies the Joule heat losses.

The transverse impedance at an internal point reduces to

$$Z_r^i \simeq -\sqrt{\frac{\omega\mu_1}{\sigma_1}} \frac{J_0(r\sqrt{\omega\mu_1\sigma_1 i})}{J_1(r\sqrt{\omega\mu_1\sigma_1 i})} e^{\frac{i\pi}{4}}. \tag{45}$$

If $r\sqrt{\omega\mu_1\sigma_1} \to \infty$, the ratio $J_0(\lambda_1 r)/J_1(\lambda_1 r) \to -i$, as in (22) above. In this limit

$$Z_r^i \to -\sqrt{\frac{\omega\mu_1}{\sigma_1}} e^{-\frac{i\pi}{4}}, \tag{46}$$

or exactly the relation obtained in (40), page 522, for a surface wave propagated over an infinite plane. The conclusions to be drawn from this result are of the greatest practical importance. If by r we understand a radius of curvature of any highly conducting surface, it is plausible to assume that when $r\sqrt{\omega\mu_1\sigma_1} \gg 1$, one can approximate

$$E_3 \simeq \sqrt{\frac{\omega\mu_1}{\sigma_1}}\, e^{-\frac{i\pi}{4}} H_2, \tag{47}$$

where E_3 and H_2 are components tangent to the surface and orthogonal to each other. To the same approximation it can be assumed that the intensity of the magnetic field at the surface of a metallic conductor differs negligibly from the intensity at the same point were the conductivity infinite. Consequently a problem involving metallic guides

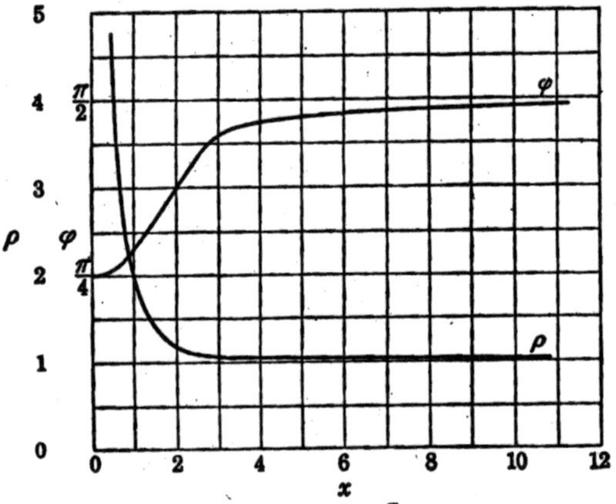

FIG. 103.—Plot of $\dfrac{J_0(x\sqrt{i})}{J_1(x\sqrt{i})} = \rho e^{-i\phi}$

can be solved first on the assumption of perfect conductivity. An approximation to the proper value of tangential component of **E** is next found from (47), and the losses of the system then determined by calculating the energy flow into the metal and assuming that this is converted entirely into heat. Examples of this procedure will be given later.

In Fig. 103 are plotted the magnitude and phase angle of the ratio

$$\frac{J_0(x\sqrt{i})}{J_1(x\sqrt{i})} = \rho e^{-i\phi} \tag{48}$$

against $x = r\sqrt{\omega\mu_1\sigma_1}$. The values are taken from the tables of Jahnke and Emde.[1] It is apparent that even for values of x as low as $x = 3$, the deviation from the asymptotic value $-i$ is very small. In m.k.s. units

$$\sqrt{\omega\mu_1\sigma_1} = 2.81 \times 10^{-3}\sqrt{\nu\kappa_{1m}\sigma_1}, \qquad \sqrt{\frac{\omega\mu_1}{\sigma_1}} = 2.81 \times 10^{-3}\sqrt{\frac{\nu\kappa_{1m}}{\sigma_1}}, \tag{49}$$

[1] "Tables of Functions," 2d ed., pp. 316–317, 1933.

and the conductivity of metals is of the order of 10^7 mhos/meter. The radius of the wire must be expressed in meters. Consequently at low frequencies x may be very much less than one unless the radius is relatively large. Take, for example, the case of a No. 12-gauge copper wire, $r = 1.026$ mm. Then $x = 0.022\sqrt{\nu}$. At $\nu = 60$ cycles/sec., $x = 0.17$. At $\nu = 10^5$ cycles/sec., $x = 6.9$, and from the curves of Fig. 103 it is clear that for this and all higher frequencies the field near the surface of the wire is very nearly the same as that in the neighborhood of an infinite plane.

The external radial impedance (37), with the help of (21), reduces to

$$Z_r^e \simeq -\sqrt{\frac{\omega\mu_1}{\sigma_1}}\, e^{-\frac{\pi i}{4}} \frac{r}{a} + i\omega\mu_2\left(1 - \frac{h^2}{k_2^2}\right) r \ln \frac{r}{a}. \tag{50}$$

Inside the magnetic field is given approximately by

$$H_\theta^i \simeq \frac{I}{2\pi a} \frac{J_1(r\sqrt{\omega\mu_1\sigma_1 i})}{J_1(a\sqrt{\omega\mu_1\sigma_1 i})}, \tag{51}$$

while outside, if r is not too large,

$$H_\theta^e \simeq \frac{I}{2\pi r}. \tag{52}$$

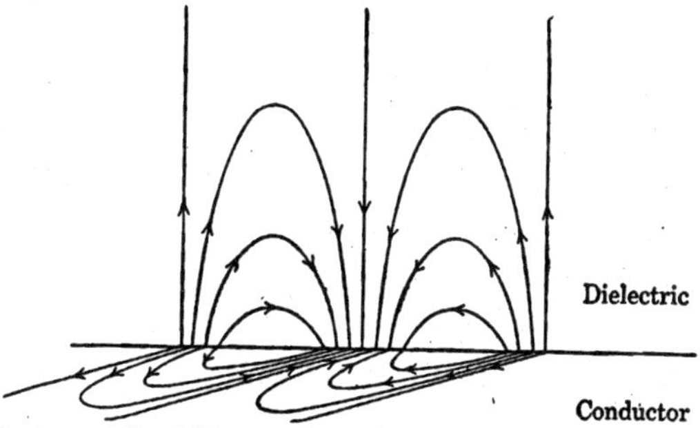

FIG. 104.—Structure of the field near the surface separating conductor and dielectric.

The structure of the field can be visualized by noting that $|Z_z^i| \ll |Z_r^i|$, $|Z_z^e| \gg |Z_r^e|$. Consequently $|E_r^i| \ll |E_z^i|$, $|E_r^e| \gg |E_z^e|$. Inside the cylinder the lines of **E** run almost parallel to the axis, trailing behind the wave; outside they leave the surface almost radially, with a very slight forward slant. The charge on the surface of the cylinder is propagated as a wave in the axial direction. Thus there are alternate bands of positive and negative surface charge moving with the field. Lines of **E** emerge from a positive band and return to a neighboring negative band. A few terminate at infinity. The structure is pictured roughly in Fig. 104.

The current distribution within the cylinder is expressed by the quantity

$$U = \frac{J_0(k_1 r)}{J_0(k_1 a)} \simeq \frac{J_0(x\sqrt{i})}{J_0(x_0\sqrt{i})}, \tag{53}$$

which is the ratio of current density at any internal point r to the current density at the surface. Here $x = r\sqrt{\omega\mu_1\sigma_1}$, as before, and $x_0 = a\sqrt{\omega\mu_1\sigma_1}$. If $x \ll 1$, U may be calculated from the power series representations of the Bessel function; if $x \gg 1$, the asymptotic expansions can be used. Tables of this ratio are given in Jahnke and Emde and also appear in many handbooks and texts on electrical engineering. In Fig. 105 the ratio of the current densities is plotted as a function of the radius for several values of the parameter x_0. For sufficiently large values of x_0 and x the asymptotic expression for U becomes

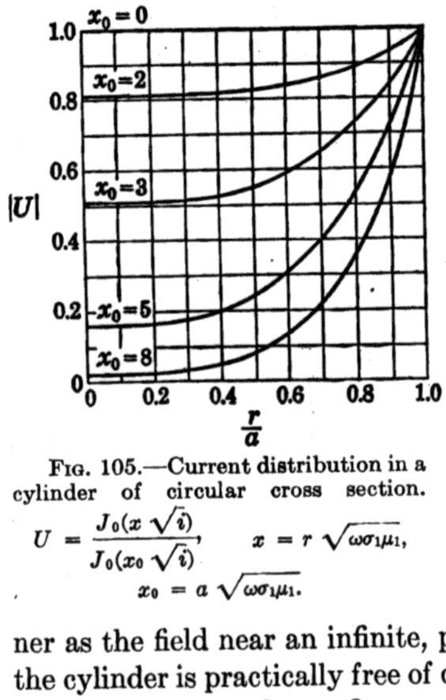

FIG. 105.—Current distribution in a cylinder of circular cross section.
$U = \frac{J_0(x\sqrt{i})}{J_0(x_0\sqrt{i})}$, $x = r\sqrt{\omega\sigma_1\mu_1}$, $x_0 = a\sqrt{\omega\sigma_1\mu_1}$.

$$U \simeq \sqrt{\frac{a}{r}}\, e^{\frac{(i-1)(a-r)}{\delta}}, \tag{54}$$

where $\delta = \sqrt{2/\omega\mu_1\sigma_1}$ is the *depth of penetration* first defined on page 504. The magnitude of the current density ratio is, therefore,

$$|U| \simeq \sqrt{\frac{a}{r}}\, e^{-\frac{a-r}{\delta}}. \tag{55}$$

The current density decreases exponentially from the surface inwards and, to this approximation, in the same manner as the field near an infinite, plane conductor. If $a \gg \delta$, the center of the cylinder is practically free of current; if $a < \delta$, the current layers interfere with one another. *One may assume for conductors of arbitrary cross section that the field and current distributions near the surface differ negligibly from those near the surface of an infinite plane provided the radius of curvature is very much greater than the skin depth.*

In conclusion we shall derive the classical expression for the alternating-current resistance of the cylinder. According to (39) the surface impedance Z_s is the ratio of the longitudinal component of electric field at a point on the surface of the wire to the total current. The circuit equation for a unit length of line is

$$E_z = RI + L\frac{dI}{dt} = (R - i\omega L)I, \tag{56}$$

where R is the a.c. resistance and L the internal inductance per unit length. Then

$$Z_s = R - i\omega L = \frac{\lambda_1}{2\pi a\sigma_1}\frac{J_0(\lambda_1 a)}{J_1(\lambda_1 a)}, \tag{57}$$

and hence

$$\frac{R - i\omega L}{R_0} \simeq \frac{x_0}{2}\frac{J_0(x_0\sqrt{i})}{J_1(x_0\sqrt{i})}e^{\frac{i\pi}{4}}, \tag{58}$$

where again $x_0 = a\sqrt{\omega\mu_1\sigma_1}$, and $R_0 = 1/\pi a^2\sigma_1$ is the direct-current resistance per unit length. The magnitude and phase of the Bessel function ratio can be taken directly from Fig. 103. If $x_0 \ll 1$, we obtain

$$\frac{R - i\omega L}{R_0} = 1 - \frac{1}{2}\left(\frac{x_0}{2}\right)^2 i + \frac{1}{12}\left(\frac{x_0}{2}\right)^4 + \cdots, \tag{59}$$

or

$$\frac{R}{R_0} \simeq 1 + \frac{1}{12}\left(\frac{x_0}{2}\right)^4, \qquad \frac{\omega L}{R_0} \simeq \frac{1}{2}\left(\frac{x_0}{2}\right)^2. \tag{60}$$

If $x_0 \gg 1$, as in the case of a well-developed skin effect, the asymptotic expansion of (58) leads to the Rayleigh formula

$$\frac{R - i\omega L}{R_0} \simeq \frac{1 - i}{2}\frac{x_0}{\sqrt{2}} = \frac{1}{2}(1 - i)\frac{a}{\delta}. \tag{61}$$

9.18. Waves in Hollow Pipes.—An interesting application of the general theory developed in Sec. 9.15 has been made to the propagation of waves along dielectric cylinders bounded externally by a medium of different inductive capacity or by a metallic sheath. The propagation of electromagnetic waves through tubes was discussed first in detail by Rayleigh,[1] while the theory of transmission along dielectric wires was investigated some years later by Hondros and Debye[2] and confirmed experimentally by Schriever.[3] Until recently no great practical importance was attached to this work because of the difficulty in generating waves of sufficiently high frequency. Progress in the development of short-wave oscillators has largely overcome this obstacle and the possibilities of such a mode of guiding waves were examined independently in 1936 by workers at the Bell Laboratories[4] and by Barrow.[5] Since then numerous contributions to the subject have appeared.

At extremely high frequencies the losses in the best available commercial dielectrics are considerable; for this reason there is obvious

[1] RAYLEIGH, *Phil. Mag.*, **43**, 125, 1897.

[2] HONDROS and DEBYE, *Ann. Physik*, **32**, 465, 1910.

[3] ZAHN, *Ann. Physik*, **49**, 907, 1916; SCHRIEVER, *Ann. Physik*, **63**, 645, 1920.

[4] SOUTHWORTH, *Bell System Tech. J.*, **15**, 284, 1936; CARSON, MEAD, and SCHELKUNOFF, *ibid.*, p. 310.

[5] BARROW, *Proc. Inst. Radio Engrs.*, **24**, 1298, 1936.

advantage to a wave guide from which all dielectric supports within the field have been eliminated. This can be accomplished by propagating the wave within a hollow cylinder whose walls are highly conducting. A further advantage is due to the absence in such a system of stray field outside the metal cylinder. At the extreme frequencies that prove necessary, even a very thin sheath acts as a perfect shield. We may assume, therefore, that the interior of the cylinder is a perfect dielectric, while the exterior is an infinite, conducting medium. The allowed modes of propagation are in general given by (9). If the conductivity of the external cylinder is finite, there will always be some small degree of coupling between the transverse magnetic and transverse electric modes except in the axially symmetric case. Practically, however, the conductivity of a metal is so high that this effect can usually be neglected and each mode treated as belonging to one type or the other. A wave that is essentially transverse magnetic is now commonly called an E wave (longitudinal component of **E**); a wave that is essentially transverse electric is called an H wave (longitudinal component of **H**).

We shall consider first the nondissipative case in which the conductivity of the external medium is infinite. The allowed modes of E and H waves are given by the roots of (13) and at once we come upon a fundamental difference with respect to the earlier problem of a solid metal cylinder. It was noted above that the principal branches of the Hankel functions $H_0^{(1)}(v)$ and $H_1^{(1)}(v)$ have no zeros and consequently their ratio vanishes only at $v = 0$. This led to the principal wave, for which $h \simeq k_2$. When the conductivity of the wire was finite, certain complementary waves could be excited but these disappeared immediately because of their high attenuation. Now, on the other hand, the situation is just reversed. The functions $J_n(u)$ and $J_n'(u)$ have an infinity of real zeros so long as n is real, and the zeros of $J_n(u)$ and $J_{n+1}(u)$ can be shown to be interlaced (to separate one another).[1] Therefore $J_n(u)$ and $J_n'(u)$ cannot vanish simultaneously and the modes of E waves are determined by

$$J_n(u) = 0, \tag{62}$$

while the modes of H waves are found from the roots of

$$J_n'(u) = 0, \tag{63}$$

where $u = \lambda_1 a = a\sqrt{k_1^2 - h^2}$. Now $u = 0$ is *not* a root of $J_0(u) = 0$. It does satisfy $J_n(u) = 0$ for $n > 0$, but in this case the field itself vanishes, as is apparent from Eqs. (1), page 524. *Consequently there is no principal wave in hollow pipes.* If such guides are to be useful for com-

[1] WATSON, "Theory of Bessel Functions," Chap. XV, Cambridge University Press, 1922.

munication purposes, it must be shown that the attenuation of the modes allowed by (62) and (63), corresponding to the complementary modes of the previous analysis, is relatively small.

For each order n of (62) there is a denumerable set of roots, any one of which will be designated u_{nm}. The propagation factor h_{nm} corresponding to the indicated mode of E wave is, therefore,

$$h_{nm} = \sqrt{k_1^2 - \left(\frac{u_{nm}}{a}\right)^2}. \tag{64}$$

If k_1 is real and greater than u_{nm}/a, then h_{nm} is real and the wave is propagated without attenuation. If $k_1 < u_{nm}/a$, h_{nm} is a pure imaginary and the attenuation is such that the mode is immediately suppressed. In this sense the tube acts as a high-pass wave filter. Cutoff takes place at the critical frequency ν_{nm}. In a dielectric $k_1 = a\sqrt{\epsilon_1\mu_1}$; hence,

$$\nu_{nm} = \frac{u_{nm}}{2\pi a\sqrt{\epsilon_1\mu_1}}. \tag{65}$$

The same condition can be expressed in terms of a critical wave length λ_{nm}.

$$\lambda_{nm} = \frac{2\pi a}{u_{nm}}, \tag{66}$$

with the understanding that λ_{nm} is the length of a wave of frequency ν_{nm} propagated in an *unbounded* medium k_1. If the frequency is less than ν_{nm}, or the free wave length greater than λ_{nm}, the mode is suppressed. The first few roots of (62) are found to be

$$\begin{array}{lll} u_{01} = 2.405, & u_{02} = 5.520, & u_{03} = 8.654, \\ u_{11} = 3.832, & u_{12} = 7.016, & \ldots\ldots\ldots, \\ u_{21} = 5.136, & \ldots\ldots\ldots, & \ldots\ldots\ldots \end{array} \tag{67}$$

The longest possible free wave length corresponds to the mode u_{01}, and is

$$\lambda_{01} = 2.61a. \tag{68}$$

Thus the limit is very roughly the diameter of the tube, so that a hollow-pipe wave guide can be of practical utility only in the centimeter wave band.

The allowed modes of the transverse electric, or H waves, are determined in the same manner. Designating the roots of (63) by u'_{nm}, we find

$$\begin{array}{lll} u'_{01} = 3.832, & u'_{02} = 7.016, & \ldots, \\ u'_{11} = 1.84, & u'_{12} = 5.33, & \ldots, \end{array} \tag{69}$$

The formulas for the propagation constant and critical frequency are identical with (64) and (65). It appears that of all possible modes, the

H_{11} component of the H wave has the longest critical free wave length, but in practice this advantage is outweighed by the fact that it also suffers a more rapid attenuation than some of the others.

The character of the propagation in the neighborhood of the critical frequencies can be made clear by graphs of phase and group velocities. We write the propagation factor of any allowed mode in the form

$$h_{nm} = k_1 \sqrt{1 - \left(\frac{\nu_{nm}}{\nu}\right)^2}. \tag{70}$$

Then the phase velocity is

$$v_{nm} = \frac{\omega}{h_{nm}} = \frac{v_1}{\sqrt{1 - \left(\frac{\nu_{nm}}{\nu}\right)^2}}, \tag{71}$$

where $v_1 = 1/\sqrt{\epsilon_1\mu_1}$ is the phase velocity in an unbounded medium whose propagation constant is k_1, while the group velocity, according to (17), page 333, is

$$V_{nm} = \frac{1}{\frac{dh_{nm}}{d\omega}} = v_1 \sqrt{1 - \left(\frac{\nu_{nm}}{\nu}\right)^2}. \tag{72}$$

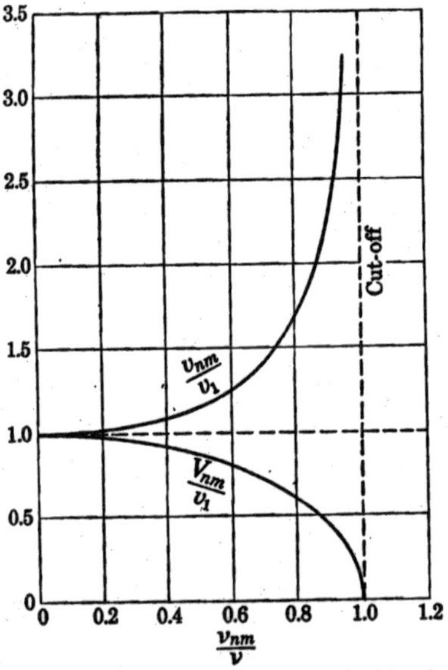

FIG. 106.—Phase and group velocities in a hollow pipe. $v_1 = 1/\sqrt{\epsilon_1\mu_1}$.

Phase and group velocities are related by

$$V_{nm}v_{nm} = v_1^2. \tag{73}$$

The relation of phase and group velocities to the impressed frequency is shown in Fig. 106. One will note that the phase velocity in the tube is always greater than that in the free medium. It increases with decreasing frequency and approaches infinity near the cutoff frequency ν_{nm}. The group velocity decreases with decreasing frequency and is zero at $\nu = \nu_{nm}$. The dispersion is normal and becomes very large in the neighborhood of ν_{nm}.

The components of the field within the tube can be written down directly from (1) and (2), page 524. In virtue of (70) and (65), we have

$$\lambda_1 = k_1 \frac{\nu_{nm}}{\nu} = \frac{u_{nm}}{a}. \tag{74}$$

For any mode of E-wave type, the field is

$$
(75)\quad \left.\begin{aligned}
E_r &= ia_{nm}\sqrt{\left(\frac{\nu}{\nu_{nm}}\right)^2 - 1}\; J_n'\left(\frac{r}{a}u_{nm}\right)\begin{matrix}\sin\\ \cos\end{matrix}\, n\theta \\
E_\theta &= \pm\, ia_{nm}\frac{na}{ru_{nm}}\sqrt{\left(\frac{\nu}{\nu_{nm}}\right)^2 - 1}\; J_n\left(\frac{r}{a}u_{nm}\right)\begin{matrix}\cos\\ \sin\end{matrix}\, n\theta \\
E_z &= a_{nm}\, J_n\left(\frac{r}{a}u_{nm}\right)\begin{matrix}\cos\\ \sin\end{matrix}\, n\theta
\end{aligned}\right\} e^{ih_{nm}z - i\omega t},
$$

$$
(76)\quad \left.\begin{aligned}
H_r &= \mp\, ia_{nm}\sqrt{\frac{\epsilon_1}{\mu_1}}\frac{n\nu}{\nu_{nm}}\frac{a}{ru_{nm}}\, J_n\left(\frac{r}{a}u_{nm}\right)\begin{matrix}\cos\\ \sin\end{matrix}\, n\theta \\
H_\theta &= ia_{nm}\sqrt{\frac{\epsilon_1}{\mu_1}}\frac{\nu}{\nu_{nm}}\, J_n'\left(\frac{r}{a}u_{nm}\right)\begin{matrix}\sin\\ \cos\end{matrix}\, n\theta
\end{aligned}\right\} e^{ih_{nm}z - i\omega t}.
$$

For any mode of H-wave type, one finds

$$
(77)\quad \left.\begin{aligned}
E_r &= \pm\, ib_{nm}\sqrt{\frac{\mu_1}{\epsilon_1}}\frac{n\nu}{\nu_{nm}'}\frac{a}{ru_{nm}'}\, J_n\left(\frac{r}{a}u_{nm}'\right)\begin{matrix}\cos\\ \sin\end{matrix}\, n\theta \\
E_\theta &= -ib_{nm}\sqrt{\frac{\mu_1}{\epsilon_1}}\frac{\nu}{\nu_{nm}'}\, J_n'\left(\frac{r}{a}u_{nm}'\right)\begin{matrix}\sin\\ \cos\end{matrix}\, n\theta
\end{aligned}\right\} e^{ih'_{nm}z - i\omega t},
$$

$$
(78)\quad \left.\begin{aligned}
H_r &= ib_{nm}\sqrt{\left(\frac{\nu}{\nu_{nm}'}\right)^2 - 1}\; J_n'\left(\frac{r}{a}u_{nm}'\right)\begin{matrix}\sin\\ \cos\end{matrix}\, n\theta \\
H_\theta &= \pm\, ib_{nm}\frac{na}{ru_{nm}'}\sqrt{\left(\frac{\nu}{\nu_{nm}'}\right)^2 - 1}\; J_n\left(\frac{r}{a}u_{nm}'\right)\begin{matrix}\cos\\ \sin\end{matrix}\, n\theta \\
H_z &= b_{nm}\, J_n\left(\frac{r}{a}u_{nm}'\right)\begin{matrix}\sin\\ \cos\end{matrix}\, n\theta
\end{aligned}\right\} e^{ih'_{nm}z - i\omega t}.
$$

The arbitrary amplitudes a_{nm} and b_{nm} will be determined by the initial excitation. There are theorems analogous to those of Fourier analysis which govern the representation of a function $f(r/a)$ in terms of Bessel functions. Suppose that $f(t)$ is a function defined arbitrarily in the interval $0 \leq t \leq 1$, and that $\int_0^1 |t^{\frac{1}{2}}f(t)|\, dt$ exists. Let

$$
(79)\qquad a_{nm} = \frac{2}{J_{n+1}^2(u_{nm})}\int_0^1 tf(t)J_n(tu_{nm})\, dt.
$$

Let $x = r/a$ be any internal point of an interval (x_1, x_2) such that $0 < x_1 < x_2 < 1$ and such that $f(t)$ has limited total fluctuation in (x_1, x_2). Then the Fourier-Bessel series

$$
(80)\qquad \sum_{m=1}^{\infty} a_{nm}\, J_n(xu_{nm}) = \frac{1}{2}\left[f(x + 0) + f(x - 0)\right]
$$

converges and its sum is the right-hand side of (80).[1] Of the modes which are excited, only those will be propagated whose propagation factors h_{nm} are real.

It may be noted in passing that at the critical frequency $\nu = \nu_{nm}$, the energy flow takes place entirely in the transverse plane. There is no flow parallel to the axis of the cylinder. The longitudinal impedance of the E waves is

$$Z_z = \frac{E_r}{H_\theta} = \sqrt{1 - \left(\frac{\nu_{nm}}{\nu}\right)^2} \sqrt{\frac{\mu_1}{\epsilon_1}}, \tag{81}$$

which approaches zero as $\nu \to \nu_{nm}$, while for the H waves

$$Z_z = \frac{1}{\sqrt{1 - \left(\frac{\nu_{nm}}{\nu}\right)^2}} \sqrt{\frac{\mu_1}{\epsilon_1}}. \tag{82}$$

Let us calculate next the total energy flow in the axial direction. The time average of this quantity is

$$\bar{W}_z = \int_0^a \int_0^{2\pi} \bar{S}_z \, r \, d\theta \, dr, \tag{83}$$

where

$$\bar{S}_z = \tfrac{1}{2} Re(E_r \tilde{H}_\theta - E_\theta \tilde{H}_r). \tag{84}$$

For the E waves

$$\bar{S}_z = \frac{|a_{nm}|^2}{2} \sqrt{\frac{\epsilon_1}{\mu_1}} \sqrt{1 - \left(\frac{\nu_{nm}}{\nu}\right)^2} \left(\frac{\nu}{\nu_{nm}}\right)^2 \left\{ \left[J_n'\left(\frac{r}{a} u_{nm}\right)\right]^2 \begin{matrix}\sin^2\\ \cos^2\end{matrix} n\theta + \left(\frac{na}{r u_{nm}}\right)^2 \left[J_n\left(\frac{r}{a} u_{nm}\right)\right]^2 \begin{matrix}\cos^2\\ \sin^2\end{matrix} n\theta \right\}. \tag{85}$$

The $\sin^2 n\theta$ and $\cos^2 n\theta$ factors reduce to π after integration. Let $r = ax$. Then in virtue of the recurrence relations (24) and (25) on page 359 and the condition $J_n(u_{nm}) = 0$, we obtain

$$\int_0^a \left\{ \left[J_n'\left(\frac{r}{a} u_{nm}\right)\right]^2 + \left(\frac{na}{r u_{nm}}\right)^2 \left[J_n\left(\frac{r}{a} u_{nm}\right)\right]^2 \right\} r \, dr = \frac{a^2}{2} \int_0^1 [J_{n-1}^2(x u_{nm}) + J_{n+1}^2(x u_{nm})] \, x \, dx = \frac{a^2}{2} J_{n+1}^2(u_{nm}), \tag{86}$$

and, therefore,

$$\bar{W}_z = \frac{\pi}{4} |a_{nm}|^2 a^2 \sqrt{\frac{\epsilon_1}{\mu_1}} \sqrt{1 - \left(\frac{\nu_{nm}}{\nu}\right)^2} \left(\frac{\nu}{\nu_{nm}}\right)^2 J_{n+1}^2(u_{nm}). \tag{87}$$

[1] WATSON, *loc. cit.*, Chap. XVIII.

Likewise for the H waves,

$$(88)\quad \bar{S}_z = \frac{1}{2}|b_{nm}|^2\sqrt{\frac{\mu_1}{\epsilon_1}}\sqrt{1-\left(\frac{\nu'_{nm}}{\nu}\right)^2}\left(\frac{\nu}{\nu'_{nm}}\right)^2\left\{\left[J'_n\left(\frac{r}{a}u'_{nm}\right)\right]^2 \frac{\sin^2}{\cos^2} n\theta + \left(\frac{na}{ru'_{nm}}\right)^2\left[J_n\left(\frac{r}{a}u'_{nm}\right)\right]^2\frac{\cos^2}{\sin^2} n\theta\right\}.$$

By using the same recurrence formulas and the relation $J'_n(u'_{nm}) = 0$, this integrates to give for the total energy flow

$$(89)\quad \bar{W}_z = \frac{\pi}{4}|b_{nm}|^2a^2\sqrt{\frac{\mu_1}{\epsilon_1}}\sqrt{1-\left(\frac{\nu'_{nm}}{\nu}\right)^2}\left(\frac{\nu}{\nu'_{nm}}\right)^2\left[1-\left(\frac{n}{u'_{nm}}\right)^2\right]J_n^2(u'_{nm}).$$

There remains the question of dissipation. If the conductivity of the walls is finite, the propagation factors h_{nm} have a small imaginary part which gives rise to an attenuation of the field as it travels along the axis of the tube. These complex propagation constants must now be determined from Eq. (9). The calculation is simplified to some extent by the fact that the Hankel functions may be replaced by their asymptotic representations, owing to the large conductivity of the outer metal cylinder; the roots of (9) will differ, therefore, by a very small amount from the values obtained from (62) and (63) for the case of infinite conductivity. They can be determined without great difficulty by expanding the functions $J_n(u)$ and $J'_n(u)$ about the points u_{nm} and u'_{nm}.

It will be more instructive at this point to calculate the attenuation by the approximate method outlined on page 534. If the conductivity is infinite, the tangential components E_z and E_θ are zero at $r = a$; if the cylinder is metal, the skin depth δ will certainly be very much less than the radius a. Then to a very good approximation, at $r = a$

$$(90)\qquad E_z \simeq \sqrt{\frac{\omega\mu_2}{\sigma_2}}\,H_\theta\,e^{-\frac{i\pi}{4}},\qquad E_\theta \simeq -\sqrt{\frac{\omega\mu_2}{\sigma_2}}\,H_z\,e^{-\frac{i\pi}{4}}.$$

Thus we can compute the radial flow of energy into the metal; since this outflow is at the expense of the flow down the tube, we can find the attenuation. The propagation factor is written in the form

$$(91)\qquad h_{nm} = \alpha_{nm} + i\beta_{nm}.$$

The mean radial flow per unit length *at the surface of the cylinder* is

$$(92)\qquad \bar{W}_r = \int_0^{2\pi}\bar{S}_r a\,d\theta,$$

where

$$(93)\qquad \bar{S}_r = \tfrac{1}{2}Re(E_\theta\tilde{H}_z - E_z\tilde{H}_\theta).$$

For the E waves

$$\bar{W}_r \simeq -\frac{\pi}{2}|a_{nm}|^2 a\sqrt{\frac{\omega\mu_2}{2\sigma_2}\frac{\epsilon_1}{\mu_1}}\left(\frac{\nu}{\nu_{nm}}\right)^2 J_{n+1}^2(u_{nm})\, e^{-2\beta_{nm}z}, \tag{94}$$

while for the transverse electric, or H waves, one obtains

$$\bar{W}_r \simeq -\frac{\pi}{2}|b_{nm}|^2 a\sqrt{\frac{\omega\mu_2}{2\sigma_2}}\left[1-\left(\frac{n}{u'_{nm}}\right)^2 + \left(\frac{n\nu}{u'_{nm}\nu'_{nm}}\right)^2\right] J_n^2(u'_{nm})\, e^{-2\beta'_{nm}z}. \tag{95}$$

To the same approximation the axial flow is still given by (87) and (89) when these expressions are multiplied by an attenuation factor $\exp(-2\beta_{nm}z)$ or $\exp(-2\beta'_{nm}z)$. The rate at which $\bar{W}_z$ decreases along the axis is

$$\frac{d\bar{W}_z}{dz} = -2\beta_{nm}\bar{W}_z = \bar{W}_r. \tag{96}$$

Consequently the attenuation factors of the E waves are given by

$$\beta_{nm} = -\frac{1}{2}\frac{\bar{W}_r}{\bar{W}_z} \simeq \frac{1}{a}\sqrt{\frac{\pi\epsilon_1\mu_2\nu}{\mu_1\sigma_2}}\,\frac{1}{\sqrt{1-\left(\frac{\nu_{nm}}{\nu}\right)^2}}; \tag{97}$$

for the H waves,

$$\beta'_{nm} \simeq \frac{1}{a}\sqrt{\frac{\pi\epsilon_1\mu_2\nu}{\mu_1\sigma_2}}\,\frac{1}{\sqrt{1-\left(\frac{\nu'_{nm}}{\nu}\right)^2}}\left[\left(\frac{\nu'_{nm}}{\nu}\right)^2 + \frac{\left(\frac{n}{u'_{nm}}\right)^2}{1-\left(\frac{n}{u'_{nm}}\right)^2}\right]. \tag{98}$$

These results are valid only when ν is greater than the critical frequency.

If either a or ν approaches infinity, the attenuation factors of the E waves approach the limit β_0, where

$$\beta_0 = \frac{1}{a}\sqrt{\frac{\pi\epsilon_1\mu_2\nu}{\mu_1\sigma_2}}, \tag{99}$$

which increases in the normal manner as the square root of the frequency. In Fig. 107 the attenuation ratios β_{nm}/β_0 and β'_{nm}/β_0 are plotted as functions of ν_{01}/ν for the two lowest modes of E and H waves. ν_{01} is the critical frequency of the fundamental E_{01} wave. In the neighborhood of the critical frequencies the attenuation increases with great rapidity. Most notable is the behavior of the attenuation curve of the H_{01} wave, which decreases with increasing frequency and approaches zero as $\nu \to \infty$. A further investigation shows, however, that this anomalous behavior is not common to all hollow tubes, but is peculiar to certain symmetrical cross sections, such as the circle and square. The stability of the lower modes with respect to small changes in cross section has been discussed

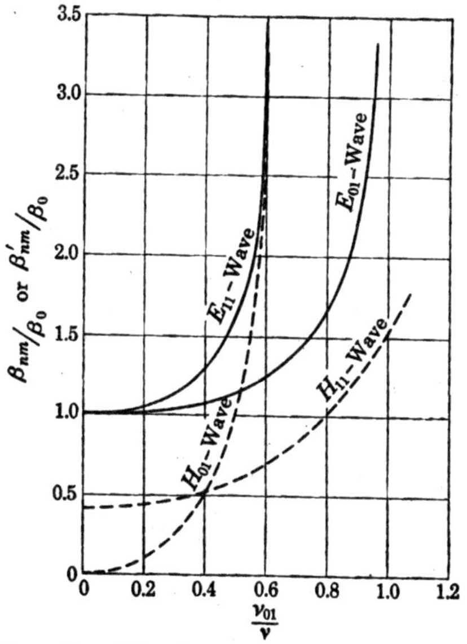

FIG. 107.—Attenuation ratios of E and H waves. ν_{01} = critical frequency of E_{01} wave.

by Brillouin[1] and by Chu[2] in connection with the propagation of waves in hollow tubes of elliptic and rectangular cross section.

COAXIAL LINES

9.19. Propagation Constant.—The most serious disadvantage of the hollow-tube transmission line is the absence of a principal mode of oscillation and the consequent limitation to wave lengths which are of the order of the diameter of the tube or less. This difficulty can be overcome by introducing an axial conductor. A principal mode then exists which allows propagation at all frequencies in the space bounded by two concentric cylinders. The coaxial line shares with the hollow tube the advantage of confining the field to its interior and thus eliminating interference with external circuits at high frequencies. Its losses, on the other hand, tend to be larger due to skin effect in the central conductor and imperfect insulation at the points of support.

In practice the outer cylinder is, of course, of finite thickness. At low frequencies the skin depth may exceed this thickness, in which case the field penetrates to the outside. The technical problem of "cross talk"

[1] BRILLOUIN, *Electrical Communication*, **16**, 350, 1938.

[2] CHU, *J. Applied Phys.*, **9**, 583, 1938. See also CHU and BARROW, *Proc. Inst. Radio Engrs.*, **26**, 1520, 1938.

between telephone cables and of interference caused by currents induced in communication lines through shielded power cables is a very serious one. However, an exhaustive discussion of such problems has no place here, for it is our purpose merely to illustrate the application of electromagnetic theory. Coaxial lines now play a very important part in radio and telephone communication at frequencies greater than 100 kc./sec. and the analysis of this section will be restricted to frequencies of this order.

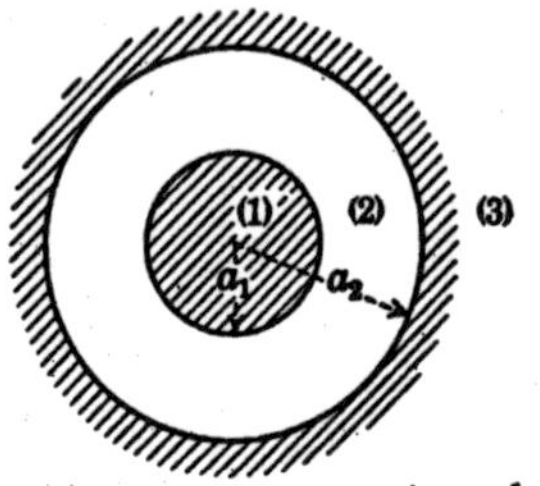

FIG. 108.—Cross section of a coaxial line.

We assume, therefore, that the skin depth is very much less than the thickness of the outer conducting cylinder, which can hence be assumed to have an infinite outer radius. In Fig. 108 the concentric cylinders (1) and (3) are conductors, while the intermediate region (2) is a dielectric to which, however, it may be necessary to ascribe a small conductivity. The inner and outer radii are a_1 and a_2. Only the symmetric, transverse magnetic modes (E waves) need be considered. Asymmetric modes can be excited, but unless the wave length is of the same order as the difference $a_2 - a_1$, their propagation constants will be pure imaginaries and they will be immediately damped out. The components of the field are taken directly from (36) and (37), page 361, for the case $n = 0$. Let I be the total current in the central conductor and I_0 its amplitude.

$$I = I_0 e^{ihz - i\omega t}. \tag{1}$$

Then exactly as in Sec. 9.17, we have for the central conductor, $r < a_1$,

$$H_\theta^{(1)} = -\frac{ik_1^2 I}{2\pi a_1 \omega\mu_1\sigma_1} \frac{J_1(\lambda_1 r)}{J_1(\lambda_1 a)}. \tag{2}$$

The longitudinal and radial impedances are defined by the relations

$$E_r = Z_z H_\theta, \qquad E_z = -Z_r H_\theta. \tag{3}$$

For the region (1) we have, as in Sec. 9.17,

$$Z_z^{(1)} = \frac{\omega\mu_1 h}{k_1^2}, \qquad Z_r^{(1)} = \frac{\omega\mu_1\lambda_1}{ik_1^2} \frac{J_0(\lambda_1 r)}{J_1(\lambda_1 r)}. \tag{4}$$

To construct a solution appropriate to the intermediate region (2) which can be fitted at both internal and external boundaries, two independent Bessel functions must be used. Which two is not important, since every solution can be represented as a linear combination of any pair of independent solutions. If the radii a_1 and a_2 are very large, the region (2) is bounded effectively by two plane surfaces. To represent a solution under these circumstances one would naturally use sine and cosine functions. Now it will be recalled (page 359) that if the argument ρ is very large, the function $\sqrt{\rho}\, J_0(\rho) \to \sqrt{\frac{2}{\pi}} \cos\left(\rho - \frac{\pi}{4}\right)$, while the function of the second kind $\sqrt{\rho}\, N_0(\rho) \to \sqrt{\frac{2}{\pi}} \sin\left(\rho - \frac{\pi}{4}\right)$. Consequently we shall write for the region $a_1 < r < a_2$,

$$H_\theta^{(2)} = -\frac{ik_2^2}{\omega\mu_2\lambda_2}[A\, J_1(\lambda_2 r) + B\, N_1(\lambda_2 r)]e^{ihz - i\omega t}, \tag{5}$$

in which A and B are undetermined coefficients. The components of $\mathbf{E}^{(2)}$ are then obtained from (3) and the relations

$$Z_z^{(2)} = \frac{\omega\mu_2 h}{k_2^2}, \qquad Z_r^{(2)} = \frac{\omega\mu_2\lambda_2}{ik_2^2}\frac{AJ_0(\lambda_2 r) + BN_0(\lambda_2 r)}{AJ_1(\lambda_2 r) + BN_1(\lambda_2 r)}. \tag{6}$$

In the outer conductor the proper behavior of the field at infinity is ensured by choosing a Hankel function, as in Sec. 9.15. If $r > a_2$,

$$H_\theta^{(3)} = -\frac{ik_3^2}{\omega\mu_3\lambda_3} CH_1^{(1)}(\lambda_3 r)\, e^{ihz - i\omega t}, \tag{7}$$

$$Z_z^{(3)} = \frac{\omega\mu_3 h}{k_3^2}, \qquad Z_r^{(3)} = \frac{\omega\mu_3}{ik_3^2}\frac{H_0^{(1)}(\lambda_3 r)}{H_1^{(1)}(\lambda_3 r)}. \tag{8}$$

The coefficients A, B, and C are next determined from the boundary conditions. At both $r = a_1$ and $r = a_2$ the z-components of $\mathbf{E}$ and the θ-components of $\mathbf{H}$ are continuous. These two conditions are equivalent to the continuity of the radial impedance functions. Hence

$$\begin{aligned} Z_r^{(1)} &= Z_r^{(2)}, && \text{at } r = a_1, \\ Z_r^{(2)} &= Z_r^{(3)}, && \text{at } r = a_2, \end{aligned} \tag{9}$$

or

$$\begin{aligned} \frac{\mu_1\lambda_1}{k_1^2}\frac{J_0(\lambda_1 a_1)}{J_1(\lambda_1 a_1)} &= \frac{\mu_2\lambda_2}{k_2^2}\frac{AJ_0(\lambda_2 a_1) + BN_0(\lambda_2 a_1)}{AJ_1(\lambda_2 a_1) + BN_1(\lambda_2 a_1)}, \\ \frac{\mu_3\lambda_3}{k_3^2}\frac{H_0^{(1)}(\lambda_3 a_2)}{H_1^{(1)}(\lambda_3 a_2)} &= \frac{\mu_2\lambda_2}{k_2^2}\frac{AJ_0(\lambda_2 a_2) + BN_0(\lambda_2 a_2)}{AJ_1(\lambda_2 a_2) + BN_1(\lambda_2 a_2)}. \end{aligned} \tag{10}$$

These are homogeneous equations in A and B and yield the ratio of A to B, but not their independent values. From (10) there follows

$$-\frac{A}{B} = \frac{N_0(\lambda_2 a_1) - \dfrac{\mu_1\lambda_1 k_2^2}{\mu_2\lambda_2 k_1^2}\dfrac{J_0(\lambda_1 a_1)}{J_1(\lambda_1 a_1)} N_1(\lambda_2 a_1)}{J_0(\lambda_2 a_1) - \dfrac{\mu_1\lambda_1 k_2^2}{\mu_2\lambda_2 k_1^2}\dfrac{J_0(\lambda_1 a_1)}{J_1(\lambda_1 a_1)} J_1(\lambda_2 a_1)} = \frac{N_0(\lambda_2 a_2) - \dfrac{\mu_3\lambda_3 k_2^2}{\mu_2\lambda_2 k_3^2}\dfrac{H_0^{(1)}(\lambda_3 a_2)}{H_1^{(1)}(\lambda_3 a_2)} N_1(\lambda_2 a_2)}{J_0(\lambda_2 a_2) - \dfrac{\mu_3\lambda_3 k_2^2}{\mu_2\lambda_2 k_3^2}\dfrac{H_0^{(1)}(\lambda_3 a_2)}{H_1^{(1)}(\lambda_3 a_2)} J_1(\lambda_2 a_2)}. \tag{11}$$

The roots of this determinantal equation are the characteristic values h_{0m}, ($m = 1, 2, 3 \ldots$), which fix the allowed symmetric modes of propagation.

9.20. Infinite Conductivity.—The complexity of (11) is such that one must resort to approximate methods of determining the roots. Let us consider first the relatively simple case in which the inner and outer cylinders are perfect conductors. We recall that $\lambda_1^2 = k_1^2 - h^2$, $\lambda_3^2 = k_3^2 - h^2$. The imaginary part of h must remain finite, otherwise the wave would not be propagated. Hence, as the conductivity approaches infinity, $\lambda_1 \to k_1$, $\lambda_3 \to k_3$; at the same time the absolute values of both k_1 and k_2 approach infinity. Then (11) reduces to

$$-\frac{A}{B} = \frac{N_0(\lambda_2 a_1)}{J_0(\lambda_2 a_1)} = \frac{N_0(\lambda_2 a_2)}{J_0(\lambda_2 a_2)}. \tag{12}$$

If λ_2 is very small, $J_0(\lambda_2 a_1) \simeq J_0(\lambda_2 a_2) \simeq 1$; for the functions N_0 we use the approximation (12), page 358. It is then apparent that the principal root of (12) corresponds to $\lambda_2 = 0$, or

$$h = k_2. \tag{13}$$

There is a principal wave whose propagation constant is that of the medium bounded by the two infinitely conducting cylinders.

There are also complementary waves. If λ_2 is very large, the asymptotic representations (18) and (19), page 359, are valid and (12) takes the form

$$\sin \lambda_2(a_2 - a_1) \simeq 0, \tag{14}$$

whose roots are

$$h_{0m} \simeq \sqrt{k_2^2 - \left(\frac{m\pi}{a_2 - a}\right)^2}, \qquad (m = 1, 2, 3, \cdots). \tag{15}$$

If k_2 is real and greater than $\frac{m\pi}{a_2 - a_1}$, then h_{0m} is real and the wave is propagated without attenuation. If, however, k_2 is less than $\frac{m\pi}{a_2 - a_1}$, then h_{0m} is a pure imaginary and the attenuation is such that the wave is immediately suppressed. All this is comparable to the hollow tube discussed in Sec. 9.18. The critical frequencies occur at

$$\nu_{0m} \simeq \frac{m\pi}{2\pi(a_2 - a_1)\sqrt{\epsilon_2\mu_2}}, \tag{16}$$

and critical wave lengths at

$$\lambda_{0m} = \frac{2(a_2 - a_1)}{m}. \tag{17}$$

In communication practice the distance $a_2 - a_1$ usually amounts to a few centimeters at the most and complementary waves are completely ruled out. The same considerations apply to the asymmetric modes, so that henceforth we shall deal only with the principal wave.

In (5) put $A = 0$ and note that as $\lambda_2 \to 0$, $N_1(\lambda_2 r) \to -2/\pi\lambda_2 r$. Then for sufficiently small values of λ_2 the magnetic field between the conductors is equal to

$$H_\theta^{(2)} = -\frac{2ik_2^2}{\pi\omega\mu_2\lambda_2^2}\frac{B}{r}e^{ik_2z-i\omega t} \tag{18}$$

The major part of the conduction current follows the surface of the inner and outer cylinders. If the intervening dielectric has some conductivity, there will also be a transverse component of current from one cylinder to the other, with a resultant decrease in current amplitude along the line. The longitudinal current carried by the inner cylinder can be found directly from Ampère's law. It is $2\pi a_1 H_\theta^{(2)}$, if $H_\theta^{(2)}$ is measured at $r = a_1$. Hence

$$B = \frac{i\omega\mu_2\lambda_2^2}{4k_2^2}I_0, \tag{19}$$

where I_0 is the conduction-current amplitude along the inner conductor.

$$H_\theta^{(2)} = \frac{I_0}{2\pi r}e^{ik_2z-i\omega t} = \frac{I}{2\pi r}, \tag{20}$$

$$Z_z^{(2)} = \frac{\omega\mu_2}{k_2}, \qquad Z_r^{(2)} = 0, \tag{21}$$

$$E_r^{(2)} = \frac{\omega\mu_2}{2\pi k_2}\frac{I}{r}, \qquad E_z^{(2)} = 0. \tag{22}$$

The *transverse voltage* at any point along the line is by definition

$$V = \int_{a_1}^{a_2} E_r^{(2)}\,dr = \frac{\omega\mu_2}{2\pi k_2}\ln\frac{a_2}{a_1}I, \tag{23}$$

while the ratio of the transverse voltage to the current in an infinite line is called its *characteristic impedance* Z_c.

$$Z_c = \frac{Z_0}{2\pi}\ln\frac{a_2}{a_1}, \tag{24}$$

where $Z_0 = \omega\mu_2/k_2$ is the intrinsic impedance of a medium whose propagation constant is k_2, as derived in (83), page 283.

Consider for a moment the distributed parameter line represented schematically in Fig. 109. L is the series inductance; R, the series resistance; C, the shunt capacity; and G, the shunt conductance; all per unit length of line and uniformly distributed. The equations governing the distribution of current and transverse voltage along the line are easily shown to be

$$\frac{\partial V}{\partial z} = -L\frac{\partial I}{\partial t} - RI = -(R - i\omega L)I, \tag{25}$$

$$\frac{\partial I}{\partial z} = -C\frac{\partial V}{\partial t} - GV = -(G - i\omega C)V. \tag{26}$$

The quantity $Z = R - i\omega L$ is the series impedance; $Y = G - i\omega C$ is the

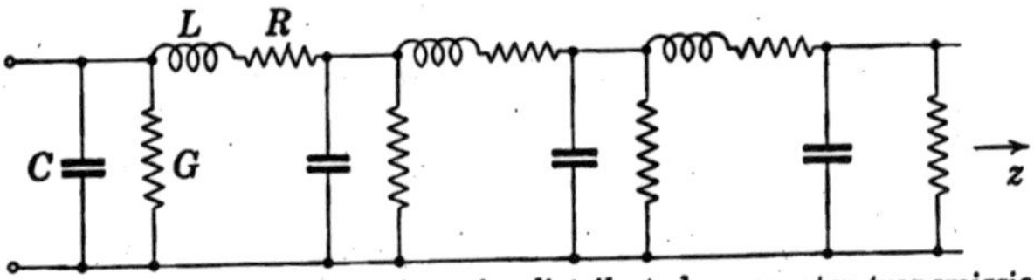

FIG. 109.—Schematic representation of a distributed parameter transmission line.

shunt admittance of the line. Upon solving (25) and (26) simultaneously, one obtains the so-called "telegrapher's equation,"

$$\frac{\partial^2 V}{\partial z^2} - LC\frac{\partial^2 V}{\partial t^2} - (LG + RC)\frac{\partial V}{\partial t} - RGV = 0, \tag{27}$$

or

$$\frac{\partial^2 V}{\partial z^2} = ZYV, \tag{28}$$

with an identical equation for the current. The propagation constant is $i\sqrt{YZ}$ and the characteristic impedance $\sqrt{Z/Y}$ as in Sec. 5.6.

If now we differentiate (23) with respect to z, the result is

$$\frac{\partial V}{\partial z} = \frac{i\omega\mu_2}{2\pi}\ln\frac{a_2}{a_1}I. \tag{29}$$

In the present instance the series resistance is zero; consequently, it follows that the series inductance per unit length of the coaxial line is

$$L = \frac{\mu_2}{2\pi}\ln\frac{a_2}{a_1} \qquad \text{henrys/meter.} \tag{30}$$

The propagation constant is k_2, from which the shunt admittance is found to be

$$Y = \frac{2\pi k_2^2}{i\omega\mu_2}\frac{1}{\ln\frac{a_2}{a_1}} = \frac{2\pi}{\ln\frac{a_2}{a_1}}(\sigma_2 - i\omega\epsilon_2). \tag{31}$$

The shunt conductance and shunt capacity are, therefore,

$$G = \frac{2\pi\sigma_2}{\ln\frac{a_2}{a_1}} \text{ mhos/meter}, \qquad C = \frac{2\pi\epsilon_2}{\ln\frac{a_2}{a_1}} \text{ farads/meter.} \tag{32}$$

It will be observed that L and C satisfy the fundamental relation

$$LC = \epsilon_2\mu_2. \tag{33}$$

9.21. Finite Conductivity.—In practice the cylindrical guides are of metal whose conductivity, although not infinite, is very large. To a very good approximation the distribution of the field between the coaxial cylinders is the same as in the case of perfect conductors, but the propagation factor must be modified to take into account the additional attenuation. To simplify the problem it will be assumed that both inner and outer cylinders are of the same material, as is ordinarily the case. Then

$$k_1^2 = k_3^2 \simeq i\omega\mu_1\sigma_1, \qquad \lambda_1 = \lambda_3 \simeq k_1. \tag{34}$$

Since λ_2, although no longer zero, must still be very small, the approximations (12), page 358, are valid.

$$N_0(\lambda_2 a_1) \simeq -\frac{2}{\pi}\ln\frac{2}{\gamma\lambda_2 a_1}, \qquad N_1(\lambda_2 a_1) \simeq -\frac{2}{\pi}\frac{1}{\lambda_2 a_1}, \qquad \gamma = 1.781. \tag{35}$$

Then with reference to (11) we have

$$\frac{\mu_1\lambda_1 k_2^2}{\mu_2\lambda_2 k_1^2} N_1(\lambda_2 a_1) \simeq -\frac{2}{\pi}\frac{k_2^2}{\mu_2\lambda_2^2}\sqrt{\frac{\mu_1}{i\omega\sigma_1}}\frac{1}{a_1}, \tag{36}$$

$$\frac{\mu_1\lambda_1 k_2^2}{\mu_2\lambda_2 k_1^2} J_1(\lambda_2 a_1) \simeq \frac{k_2^2}{2\mu_2}\sqrt{\frac{\mu_1}{i\omega\sigma_1}}\, a_1. \tag{37}$$

We have admitted the possibility of a small conductivity in the intermediate dielectric k_2, but this may be neglected in estimating orders of magnitude. Thus assuming for the moment that $k_2 \simeq \omega/c$, we have for copper

$$\frac{k_2^2}{2\mu_2}\sqrt{\frac{\mu_1}{\omega\sigma_1}} \simeq 10^{-17}\nu^{3/2}. \tag{38}$$

Remembering that a_1 and a_2 are measured in meters and hence are also small quantities, it is apparent that the denominators of the determinantal equation (11) can be replaced by unity and, therefore,

$$-\frac{A}{B}\frac{\pi}{2} \simeq -\ln\frac{2}{\gamma\lambda_2 a_1} + \frac{k_2^2}{\mu_2\lambda_2^2}\sqrt{\frac{\mu_1}{i\omega\sigma_1}}\frac{1}{a_1}\frac{J_0(\lambda_1 a_1)}{J_1(\lambda_1 a_1)}$$
$$\simeq -\ln\frac{2}{\gamma\lambda_2 a_2} + \frac{k_2^2}{\mu_2\lambda_2^2}\sqrt{\frac{\mu_1}{i\omega\sigma_1}}\frac{1}{a_2}\frac{H_0^{(1)}(\lambda_1 a_2)}{H_1^{(1)}(\lambda_1 a_2)}, \tag{39}$$

whence we obtain

$$\lambda_2^2 \simeq \frac{1}{\ln \frac{a_2}{a_1}} \frac{k_2^2}{\mu_2} \sqrt{\frac{\mu_1}{i\omega\sigma_1}} \left[\frac{1}{a_1} \frac{J_0(x_1\sqrt{i})}{J_1(x_1\sqrt{i})} - \frac{1}{a_2} \frac{H_0^{(1)}(x_2\sqrt{i})}{H_1^{(1)}(x_2\sqrt{i})} \right], \tag{40}$$

$$x_1 = a_1\sqrt{\omega\mu_1\sigma_1}, \qquad x_2 = a_2\sqrt{\omega\mu_1\sigma_1}. \tag{41}$$

But $\lambda_2^2 = k_2^2 - h^2$ and so finally

$$\begin{aligned} h &= k_2 - \frac{k_2}{2\mu_2} \sqrt{\frac{\mu_1}{i\omega\sigma_1}} \frac{1}{\ln \frac{a_2}{a_1}} \left[\frac{1}{a_1} \frac{J_0(x_1\sqrt{i})}{J_1(x_1\sqrt{i})} - \frac{1}{a_2} \frac{H_0^{(1)}(x_2\sqrt{i})}{H_1^{(1)}(x_2\sqrt{i})} \right] \\ &= k_2(1 - \Delta k). \end{aligned} \tag{42}$$

The approximate value of the ratio A/B as expressed in (39) can now be introduced into (5) to determine the magnetic field intensity at any point in the dielectric. One obtains

$$H_\theta^{(2)} \simeq \frac{2}{\pi} \frac{ik_2^2}{\omega\mu_2\lambda_2^2} B \left[\frac{1}{r} + \frac{k_2^2}{2\mu_2} \sqrt{\frac{\mu_1}{i\omega\sigma_1}} \frac{J_0(x_1\sqrt{i})}{J_1(x_1\sqrt{i})} \frac{r}{a_1} \right] e^{ihz - i\omega t}. \tag{43}$$

A logarithmic term has been dropped as obviously negligible. We assume that the conductivity σ_2 of the dielectric is very small, if not zero. Then the second term of (43) increases essentially with the three-halves power of the frequency. An easy computation will show as in (38) that unless a_1 is very small, or r very large, this second term is entirely negligible with respect to the first, even at frequencies as high as 10^{10} cycles/sec. The constant B is determined by the condition $2\pi a_1 H_\theta^{(2)} = I$ at the surface of the inner conductor, so that very nearly

$$H_\theta^{(2)} \simeq \frac{I}{2\pi r}, \tag{44}$$

as in the previous case of perfect conductivity. Physically this means that the conduction and displacement currents in the dielectric are negligible with respect to the conduction current carried by the inner conductor.

The radial electric field intensity is

$$E_r^{(2)} = \frac{\omega\mu_2 h}{k_2^2} H_\theta^{(2)} \simeq \frac{\omega\mu_2}{k_2}(1 - \Delta k)\frac{I}{2\pi r}, \tag{45}$$

and the transverse voltage

$$V = \int_{a_1}^{a_2} E_r^{(2)}\, dr = \frac{\omega\mu_2}{2\pi k_2}(1 - \Delta k) \ln \frac{a_2}{a_1} I. \tag{46}$$

This gives for the characteristic impedance

$$Z_c = \frac{\omega\mu_2}{2\pi k_2}(1 - \Delta k) \ln \frac{a_2}{a_1}; \tag{47}$$

the series impedance Z and shunt admittance Y of the line follow directly from the relations

$$h^2 = -YZ, \qquad Z_c^2 = \frac{Z}{Y}, \qquad Z = -ihZ_c, \qquad Y = -i\frac{h}{Z_c}. \tag{48}$$

The approximate value of the shunt admittance—and in practical cases it is an exceedingly good approximation—is

$$Y \simeq \frac{2\pi k_2^2}{i\omega\mu_2}\frac{1}{\ln\frac{a_2}{a_1}} = \frac{2\pi}{\ln\frac{a_2}{a_1}}(\sigma_2 - i\omega\epsilon_2), \tag{49}$$

or exactly the same expression as was obtained in the case of infinitely conducting guides. For the series impedance, on the other hand, we obtain

$$Z \simeq -\frac{i\omega\mu_2}{2\pi}(1 - \Delta k)^2 \ln\frac{a_2}{a_1} = R - i\omega L. \tag{50}$$

The behavior of the line from an engineering standpoint is now completely determined by the Eqs. (25) and (26).

If $x_1 \gg 1$, as is usually the case, one can replace the cylinder functions by their asymptotic values

$$\frac{J_0(x_1\sqrt{i})}{J_1(x_1\sqrt{i})} \simeq -i, \qquad \frac{H_0^{(1)}(x_2\sqrt{i})}{H_1^{(1)}(x_2\sqrt{i})} \simeq +i. \tag{51}$$

Then

$$\Delta k \simeq -\frac{(1+i)}{2\sqrt{2}}\frac{1}{\mu_2}\sqrt{\frac{\mu_1}{\omega\sigma_1}}\left(\frac{1}{a_1} + \frac{1}{a_2}\right)\frac{1}{\ln\frac{a_2}{a_1}}. \tag{52}$$

This correction term is small, so that $(\Delta k)^2$ can be discarded. The series resistance and inductance are now

$$\begin{aligned} R &\simeq \frac{1}{\sqrt{4\pi}}\sqrt{\frac{\nu\mu_1}{\sigma_1}}\left(\frac{1}{a_1} + \frac{1}{a_2}\right) && \text{ohms/meter}, \\ L &\simeq \frac{\mu_2}{2\pi}\left[\ln\frac{a_2}{a_1} + \frac{1}{\sqrt{4\pi}}\frac{1}{\mu_2}\sqrt{\frac{\mu_1}{\nu\sigma_1}}\left(\frac{1}{a_1} + \frac{1}{a_2}\right)\right] && \text{henrys/meter}. \end{aligned} \tag{53}$$

For copper guides, $\sigma_1 = 5.8 \times 10^7$ mhos/meter, and we obtain

$$\begin{aligned} R &\simeq 0.416 \times 10^{-7}\sqrt{\nu}\left(\frac{1}{a_1} + \frac{1}{a_2}\right) && \text{ohms/meter}, \\ L &\simeq 2 \times 10^{-7}\left[\ln\frac{a_2}{a_1} + \frac{1}{30.2}\sqrt{\frac{1}{\nu}}\left(\frac{1}{a_1} + \frac{1}{a_2}\right)\right] && \text{henrys/meter}. \end{aligned} \tag{54}$$

The attenuation is calculated from the propagation factor

$$h = k_2(1 - \Delta k) = \alpha + i\beta.$$

At very high frequencies the losses in the dielectric spacers rise rapidly, but when ν is less than a megacycle the shunt conductance of a well-constructed line can often be neglected. If we assume that $k_2 = \omega\sqrt{\epsilon_2\mu_2}$, the attenuation factor reduces to

$$(55) \qquad \beta \simeq \frac{1}{2\sqrt{2}}\sqrt{\frac{\omega\mu_1\epsilon_2}{\sigma_1\mu_2}}\left(\frac{1}{a_1} + \frac{1}{a_2}\right)\frac{1}{\ln\dfrac{a_2}{a_1}}.$$

In engineering practice it is customary to define the attenuation in terms of the logarithm of the power ratio. *Ten times the logarithm to the base 10 of the ratio of output power to input power expresses the attenuation in decibels.* Since the power ratio is $\exp(-2\beta z)$, the attenuation per unit length of line is

$$(56) \qquad N_{db} = -20\beta \log_{10} e = -8.686\,\beta \quad \text{decibels/meter.}$$

OSCILLATIONS OF A SPHERE

9.22. Natural Modes.—The problems of the plane and the cylinder treated previously in this chapter concern infinite surfaces which serve as guides to traveling waves. The sphere, on the other hand, is an example of a body bounded by a closed surface within which there can be set up a system of standing waves.

We shall suppose that the sphere, of radius a and characterized electrically by the constant k_1, is embedded in an infinite homogeneous medium k_2. The field will be represented in terms of the spherical, vector wave functions discussed in Secs. 7.11 to 7.14. According to (12), page 394,

$$(1) \quad \mathbf{E} = -\sum_n (a_n\mathbf{M}_n + b_n\mathbf{N}_n), \qquad \mathbf{H} = -\sum_n \frac{k}{i\omega\mu}(a_n\mathbf{N}_n + b_n\mathbf{M}_n).$$

Here the subscript n stands for all indices, of which there are in the present case three. The functions may be odd or even, and the indices m and n will determine the number of nodes with respect to the spherical angles ϕ and θ. It will be recalled that the radial component of every function $M_{\substack{e\\o}mn}$ is zero. Hence, if the coefficients a_n are all zero, only the b_n being excited, the field has a radial component of $\mathbf{E}$ but the magnetic vector is always perpendicular to the radius vector. There must be, therefore, a distribution of electric charge on the surface of the sphere. *The oscillations whose amplitudes are represented by the coefficients b_n are of* *electric type.* They have in the past been referred to also as *transverse magnetic*, or *E waves.* If, on the other hand, only the a_n are excited, the field is such as would be produced by oscillating magnetic charges on the surface of the sphere and the field is said to be of *magnetic type.* These oscillations may also be said to be *transverse electric*, or *H waves.*

Let us consider first an oscillation of magnetic type.

$$\text{(2)} \qquad \begin{aligned} \mathbf{E} &= -(A_{emn}\mathbf{M}_{emn} + A_{omn}\mathbf{M}_{omn}), \\ \mathbf{H} &= -\frac{k}{i\omega\mu}(A_{emn}\mathbf{N}_{emn} + A_{omn}\mathbf{N}_{omn}). \end{aligned}$$

The coefficients $A_{\substack{e\\o}mn}$ are as yet arbitrary and the frequency ω will shortly be determined by the boundary conditions. The functions $\mathbf{M}_{\substack{e\\o}mn}$ and $\mathbf{N}_{\substack{e\\o}mn}$ must now be chosen such that the field is finite at the origin and is regular at infinity. By (11) and (12), page 416, we have when $R < a$:

$$\text{(3)} \qquad \begin{aligned} E_R^i &= 0, \\ E_\theta{}^i &= -\frac{1}{\sin\theta}\frac{\partial Y_{mn}^i}{\partial\phi}\, j_n(k_1R)\, e^{-i\omega t}, \\ E_\phi{}^i &= \frac{\partial Y_{mn}^i}{\partial\theta}\, j_n(k_1R)\, e^{-i\omega t}, \end{aligned}$$

$$\text{(4)} \qquad \begin{aligned} H_R^i &= -\frac{n(n+1)}{i\omega\mu_1}\, Y_{mn}^i\, \frac{1}{R}\, j_n(k_1R)\, e^{-i\omega t}, \\ H_\theta{}^i &= -\frac{1}{i\omega\mu_1}\frac{\partial Y_{mn}^i}{\partial\theta}\frac{1}{R}\,[k_1R\, j_n(k_1R)]' e^{-i\omega t}, \\ H_\phi{}^i &= -\frac{1}{i\omega\mu_1}\frac{1}{\sin\theta}\frac{\partial Y_{mn}^i}{\partial\phi}\frac{1}{R}\,[k_1R\, j_n(k_1R)]' e^{-i\omega t}, \end{aligned}$$

where Y_{mn}^i is the tesseral harmonic

$$\text{(5)} \qquad Y_{mn}^i = (A_{emn}^i \cos m\phi + A_{omn}^i \sin m\phi) P_n^m(\cos\theta),$$

and the prime in (4) denotes differentiation with respect to the argument k_1R.

The external field in the region $R > a$ is obtained by replacing in (3) and (4) Y_{mn}^i by Y_{mn}^e,

$$\text{(6)} \qquad Y_{mn}^e = (A_{emn}^e \cos m\phi + A_{omn}^e \sin m\phi) P_n^m(\cos\phi),$$

k_1 and μ_1 by k_2 and μ_2, and $j_n(k_1R)$ by $h_n^{(1)}(k_2R)$.

At the surface of the sphere the continuity of the tangential components leads to the relations

$$\text{(7)} \qquad \begin{aligned} E_\theta{}^i &= E_\theta{}^e, & E_\phi{}^i &= E_\phi{}^e, \\ H_\theta{}^i &= H_\theta{}^e, & H_\phi{}^i &= H_\phi{}^e, \end{aligned} \qquad (r = a),$$

all four reducing to the conditions

$$(8)\qquad \begin{aligned} A^i_{emn} j_n(k_1a) &= A^e_{emn} h^{(1)}_n(k_2a), \\ \frac{1}{\mu_1} A^i_{emn}[k_1a\, j_n(k_1a)]' &= \frac{1}{\mu_2} A^e_{emn}[k_2a\, h^{(1)}_n(k_2a)]', \end{aligned}$$

with an identical pair relating the odd coefficients A^i_{omn} and A^e_{omn}. Let now

$$(9)\qquad \rho = k_2a, \qquad k_1 = Nk_2, \qquad k_1a = N\rho.$$

Then the boundary conditions (8) are satisfied only by a discrete set of *characteristic values* ρ_{ns} which are the roots of the transcendental equation

$$(10)\qquad \frac{[N\rho\, j_n(N\rho)]'}{\mu_1\, j_n(N\rho)} = \frac{[\rho\, h^{(1)}_n(\rho)]'}{\mu_2\, h^{(1)}_n(\rho)}.$$

Since the allowed values of ρ belong to a discrete set, it follows that there is a corresponding set of natural frequencies, or modes of oscillation. According to (9),

$$(11)\qquad \rho^2_{ns} = (\omega^2_{ns}\epsilon_2\mu_2 + i\omega_{ns}\mu_2\sigma_2)a^2;$$

hence, the natural frequencies of the magnetic modes are

$$(12)\qquad \omega_{ns} = \sqrt{\frac{\epsilon_2\mu_2\rho^2_{ns}}{a^2} - \frac{\mu_2^2\sigma_2^2}{4}} - i\,\frac{\mu_2\sigma_2}{2}$$

As in every characteristic-value problem, the amplitudes of the allowed modes are determined by the initial distribution of the field. Suppose that within the sphere at the instant $t = 0$ the radial component of $\mathbf{H}^i$ is a specified function of the form $f_1(\theta, \phi)f_2(R)$. We note first that the natural frequencies ω_{ns} are independent of the distribution in ϕ. The coefficients $A^i_{\substack{e\\o}mn}$ are uniquely determined by $f(\theta, \phi)$ as in (18), page 403. The series

$$(13)\qquad H^i_R = -\sum_{n=1}^{\infty} \frac{n(n+1)}{i\omega_{ns}\mu_1} Y^i_n \frac{j_n(k_1R)}{R} e^{-i\omega_{ns}t},$$

where $Y^i_n = \sum_{m=0}^{\infty} Y^i_{mn}$, then depends in the desired manner on θ and ϕ at $t = 0$. But there are an infinite number of such series, one for each root s of (10). Each of these series can be multiplied by another coefficient A^i_s and the summation extended over s.

$$(14)\qquad H^i_R = -\sum_{s=0}^{\infty}\sum_{n=1}^{\infty} A^i_s \frac{n(n+1)}{i\omega_{ns}\mu_1} Y^i_n \frac{j_n(k_1R)}{R} e^{-i\omega_{ns}t},$$

in which the A_s^i are to be determined so that (14) matches $f_2(R)$ at $t = 0$. The representation of arbitrary functions in Fourier-Bessel series is treated by Watson,[1] and the theory can be adapted to the spherical Bessel functions $j_n(k_1R)$. Finally, we note that before summing one can multiply the particular solutions by k_1R and then differentiate with respect to the same variable. No change is necessary in the coefficients $A^i_{\substack{e\\o}mn}$, but the summation over s is modified to fit the conditions imposed on either $H_\theta{}^i$ or $H_\phi{}^i$. A field has now been determined which fulfills the specified conditions within the sphere at $t = 0$. The factor $\exp(-i\omega_{ns}t)$ fixes the behavior at all subsequent times, and in virtue of the uniqueness theorem of Sec. 9.2 the field is also determined at all external points.

The oscillations of electric type are quite independent of the magnetic modes, but the fields and natural frequencies are determined in exactly the same manner. In place of (2) we have in this case

$$\text{(15)}\qquad \begin{aligned} \mathbf{E} &= -(B_{emn}\mathbf{N}_{emn} + B_{omn}\mathbf{N}_{omn}),\\ \mathbf{H} &= -\frac{k}{i\omega\mu}(B_{emn}\mathbf{M}_{emn} + B_{omn}\mathbf{M}_{omn}), \end{aligned}$$

and the field at any internal point $R < a$ is

$$\text{(16)}\qquad \begin{aligned} E_R^i &= -n(n+1)Y^i_{mn}\frac{j_n(k_1R)}{k_1R}e^{-i\omega t},\\ E_\theta{}^i &= -\frac{\partial Y^i_{mn}}{\partial\theta}\frac{1}{k_1R}[k_1R\,j_n(k_1R)]'e^{-i\omega t},\\ E_\phi{}^i &= -\frac{1}{\sin\theta}\frac{\partial Y^i_{mn}}{\partial\phi}\frac{1}{k_1R}[k_1R\,j_n(k_1R)]'e^{-i\omega t}, \end{aligned}$$

$$\text{(17)}\qquad \begin{aligned} H_R^i &= 0,\\ H_\theta{}^i &= -\frac{k_1}{i\omega\mu_1}\frac{1}{\sin\theta}\frac{\partial Y^i_{mn}}{\partial\phi}j_n(k_1R)\,e^{-i\omega t},\\ H_\phi{}^i &= \frac{k_1}{i\omega\mu_1}\frac{\partial Y^i_{mn}}{\partial\theta}j_n(k_1R)\,e^{-i\omega t}. \end{aligned}$$

The amplitudes of the internal and external fields are related by the boundary conditions

$$\text{(18)}\qquad \begin{aligned} \frac{k_1}{\mu_1}B^i_{\substack{e\\o}mn}j_n(k_1a) &= \frac{k_2}{\mu_2}B^e_{\substack{e\\o}mn}h_n^{(1)}(k_2a),\\ \frac{1}{k_1}B^i_{\substack{e\\o}mn}[k_1a\,j_n(k_1a)]' &= \frac{1}{k_2}B^e_{\substack{e\\o}mn}[k_2a\,h_n^{(1)}(k_2a)]', \end{aligned}$$

which lead to the transcendental equation

$$\text{(19)}\qquad \frac{[N\rho\,j_n(N\rho)]'}{N^2j_n(N\rho)} = \frac{\mu_2}{\mu_1}\frac{[\rho\,h_n^{(1)}(\rho)]'}{h_n^{(1)}(\rho)}$$

[1] WATSON, *loc. cit.*, Chap. XVIII.

for the characteristic values ρ_{ns}. The natural frequencies ω_{ns} are obtained from (12).

9.23. Oscillations of a Conducting Sphere.—We shall suppose the sphere to be conducting, but the medium in which it is embedded to be a perfect dielectric.[1] Consider first the limiting case in which $\sigma_1 \to \infty$. The two determinantal equations (10) and (19) now reduce to

$$h_n^{(1)}(\rho) = 0, \qquad [\rho\, h_n^{(1)}(\rho)]' = 0, \tag{20}$$

whose roots are respectively the characteristic values of the magnetic and electric modes. However, each function $h_n^{(1)}(\rho)$ can be represented as an exponential times a polynomial, Sec. 7.4, page 405. The roots in question are the roots of these polynomials *and are, therefore, finite in number for each function of finite order n*. Take, for example, the first electric mode for which $n = 1$. By (31), page 405,

$$h_1^{(1)}(\rho) = -\frac{1}{\rho}\, e^{i\rho}\left(1 + \frac{i}{\rho}\right), \tag{21}$$

$$[\rho\, h_1^{(1)}(\rho)]' = -\frac{i}{\rho^2}\, e^{i\rho}(\rho^2 + i\rho - 1), \tag{22}$$

and the desired roots are those of $\rho^2 + i\rho - 1 = 0$, or $\rho_{11} = 0.86 - i0.5$, $\rho_{12} = -0.86 - i0.5$. Since $\sigma_2 = 0$, the natural frequencies are given by

$$\omega_{ns} = \frac{\rho_{ns}}{a}\sqrt{\epsilon_2\mu_2}; \tag{23}$$

hence, in the case of the lowest electric mode

$$e^{-i\omega t} = e^{-\frac{0.5}{a}\sqrt{\epsilon_2\mu_2}\,t \mp \frac{i0.86}{a}\sqrt{\epsilon_2\mu_2}\,t}. \tag{24}$$

The amplitude is reduced to the *e*th part of its initial value in the time required for a wave to travel a distance equal to the diameter of the sphere. The wave length in the external medium is $\lambda = \dfrac{2\pi}{0.86}a = 7.3a$. Since the conductivity is infinite, the damping must be attributed entirely to loss of energy by radiation. This loss is very rapid, even in the case of the lowest mode, and in the example just cited the amplitude is reduced to the *e*th part of its initial value in a time equal to the 0.27th part of a complete period.

The first few characteristic values for the problem of the infinitely conducting sphere are tabulated below. The minimum damping factor is that of the lowest electric mode.

[1] DEBYE, *Ann. Physik*, **30**, 57, 1909.

	Magnetic modes	Electric modes
$-i\rho_1$	-1	$-0.50 + i0.86$ $-0.50 - i0.86$
$-i\rho_2$	$-1.50 + i0.86$ $-1.50 - i0.86$	-1.60 $-0.70 + i1.81$ $-0.70 - i1.81$
$-i\rho_3$	-2.26 $-1.87 + i1.75$ $-1.87 - i1.75$	$-2.17 + i0.87$ $-2.17 - i0.87$ $-0.83 + i2.77$ $-0.83 - i2.77$

When the sphere is of metal, the conductivity is finite, but N is very large none the less. We may expect, therefore, to find roots of (10) and (19) in the neighborhood of those of (20). If $|N\rho| \gg 1$, the asymptotic representation of (10) is

$$N\rho \tan\left(N\rho - \frac{n+1}{2}\pi\right) \simeq -\frac{\mu_1}{\mu_2}\frac{[\rho\, h_n^{(1)}(\rho)]'}{h_n^{(1)}(\rho)}. \tag{25}$$

Let the deviation from any root ρ_{ns} of the equation $h_n^{(1)}(\rho) = 0$ be $\Delta\rho_{ns}$. The right side of (25) is expanded in a Taylor series about $\rho = \rho_{ns}$. Retaining only the linear term, we obtain

$$\Delta\rho_{ns} \simeq -\frac{\mu_1}{\mu_2 N}\frac{1}{\tan\left(N\rho_{ns} - \frac{n+1}{2}\pi\right)}, \tag{26}$$

which determines the corrections to be applied to the natural frequencies of the magnetic modes. Likewise for the electric modes, we replace (19) by

$$\frac{\rho}{N}\tan\left(N\rho - \frac{n+1}{2}\pi\right) \simeq -\frac{\mu_2}{\mu_1}\frac{[\rho\, h_n^{(1)}(\rho)]'}{h_n^{(1)}(\rho)}. \tag{27}$$

Let ρ_{ns} be a root of $[\rho\, h_n^{(1)}(\rho)]' = 0$ and $\Delta\rho_{ns}$ the deviation due to a finite conductivity.

$$[\rho\, h_n^{(1)}(\rho)]' = [\rho\, h_n^{(1)}(\rho)]''_{\rho=\rho_{ns}}\Delta\rho_{ns} + \cdots. \tag{28}$$

From (4), page 400,

$$\rho[\rho\, h_n^{(1)}(\rho)]'' + [\rho^2 - n(n+1)]h_n^{(1)}(\rho) = 0; \tag{29}$$

hence,

$$\Delta\rho_{ns} \simeq \frac{\mu_1}{\mu_2 N}\frac{\rho_{ns}^2}{\rho_{ns}^2 - n(n+1)}\tan\left(N\rho_{ns} - \frac{n+1}{2}\pi\right). \tag{30}$$

Thus, in the case of the fundamental electric mode, one obtains

$$-i\rho_1 \simeq -0.50 \pm i0.86\left(1 - \frac{2}{3}\frac{\mu_1}{\mu_2 N}\right), \tag{31}$$

since the tangent of a very large complex number is $+i$. In the case of a metal sphere

$$\frac{\mu_1}{\mu_2 N} \simeq \frac{1 - i}{\sqrt{2}}\frac{\omega\epsilon_2}{\sigma_1}. \tag{32}$$

It will be recalled finally that in our study of the natural modes of propagation along a cylinder, the principal wave corresponded to the condition: $\lambda_1 a$ very large inside the cylinder and $\lambda_2 a$ nearly zero outside. A complementary set of waves was shown to exist, however, for which $\lambda_1 a$ was small. The same circumstances arise in the case of the sphere. Debye has pointed out that sets of roots can be found for (10) and (19) such that $N\rho$ is a finite number even when N itself is very large. If the conductivity of the sphere is finite, the damping of the oscillations is extremely rapid, as was the case of the complementary cylindrical waves. On the other hand, if the sphere is a perfect dielectric, the characteristic values are real: *there is no damping, whence it appears that these modes do not radiate.*

9.24. Oscillations in a Spherical Cavity.—So far as it is possible to foresee at the present time, ultrahigh-frequency radio circuits will be reduced eventually to hollow-tube transmission lines and cavity resonators. The latter are electromagnetic analogues of the Helmholtz resonators used in acoustics. Such systems do not radiate; one may, therefore, ascribe to them definite values of electric and magnetic energy, and consequently effective inductance, capacitance, and resistance parameters which depend solely on the configuration. The damping results from a penetration of the field into the metal walls a distance equal approximately to the skin depth.

The character of the oscillations in a resonator can be illustrated by the example of a spherical cavity. Let us suppose now that the interior of the sphere is a perfect dielectric, so that $k_1 = \omega\sqrt{\epsilon_1\mu_1}$, while the external region has an infinite conductivity. Then $N = 0$, and the characteristic values of the magnetic and electric modes are determined respectively by

$$j_n(k_1 a) = 0, \qquad [k_1 a\, j_n(k_1 a)]' = 0. \tag{33}$$

Let u_{ns} be the roots of $j_n(k_1 a) = 0$. The lower of these roots are

$$\begin{aligned} u_{11} &= 4.50, \qquad u_{12} = 7.64, \\ u_{21} &= 5.8, \qquad \cdots . \end{aligned} \tag{34}$$

Since $k_1 a = \omega\sqrt{\epsilon_1\mu_1}\, a$, the characteristic frequencies of the magnetic modes are given by

$$\nu_{ns} = \frac{u_{ns}}{2\pi a \sqrt{\epsilon_1 \mu_1}}. \tag{35}$$

Formally this is identical with Eq. (65), page 539, for the characteristic frequencies of a hollow pipe. The free wave length, or wave length at the same frequency in an unbounded medium, is

$$\lambda_{ns} = \frac{2\pi a}{u_{ns}}. \tag{36}$$

The longest allowable wave length among the magnetic modes is $\lambda_{11} = 2\pi a/4.5 = 1.4a$. In this case the field has only the components E_ϕ, H_R, and H_θ, and the charge flows along parallels of latitude. The

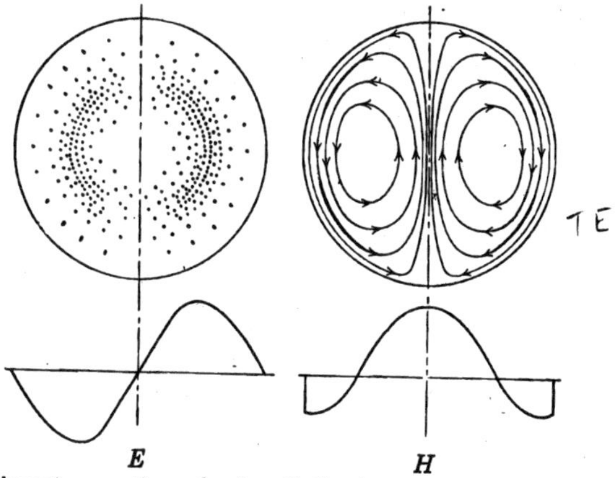

FIG. 110.—The lowest magnetic mode of oscillation in a spherical resonator, showing the field in a meridian plane.

distribution of the electric and magnetic lines of force in a meridian plane is illustrated by Fig. 110.

The roots of the second of Eqs. 33 will be denoted by u'_{ns} and the characteristic frequencies of the electric modes by ν'_{ns}. Apart from a prime, the formulas (35) and (36) remain valid and a calculation shows that the free wave length of the fundamental electric mode is

$$\lambda'_{11} = 2\pi a/2.75 = 2.28a.$$

This is the lowest of all possible modes, both electric and magnetic. The axially symmetric case corresponds to an oscillating flow of charge along meridian lines from pole to pole; the magnetic lines follow parallels of latitude. The lines of force in a meridian plane are shown in Fig. 111. At any internal point of the sphere, when $m = 0$, $n = 1$, we have from (16) and (17):

$$
\begin{aligned}
E_R &= -\frac{2Aa}{2.75R}\cos\theta\, j_1\!\left(2.75\,\frac{R}{a}\right) e^{-i\omega'_{11}t},\\
E_\theta &= \frac{Aa}{2.75R}\sin\theta\left[2.75\,\frac{R}{a}\, j_1\!\left(2.75\,\frac{R}{a}\right)\right]' e^{-i\omega'_{11}t}, \qquad (37)\\
H_\phi &= i\sqrt{\frac{\epsilon_1}{\mu_1}}\, A\sin\theta\, j_1\!\left(2.75\,\frac{R}{a}\right) e^{-i\omega'_{11}t},
\end{aligned}
$$

where $\omega'_{11} = 2.75/a\sqrt{\epsilon_1\mu_1}$.

The surface current density **K** at any point on the wall of the cavity is $\mathbf{K} = \mathbf{n} \times \mathbf{H}$, where **n** is a unit vector directed radially inwards and **H** is measured on the cavity side of the surface. Since there is only the single component H_ϕ, the current density is directed along meridian lines. The total charge crossing the equator per second is

$$
\begin{aligned}
(38)\qquad I &= 2\pi a K\\
&= 2\pi a\, i\sqrt{\frac{\epsilon_1}{\mu_1}}\, A\, j_1(2.75)\, e^{-i\omega'_{11}t}.
\end{aligned}
$$

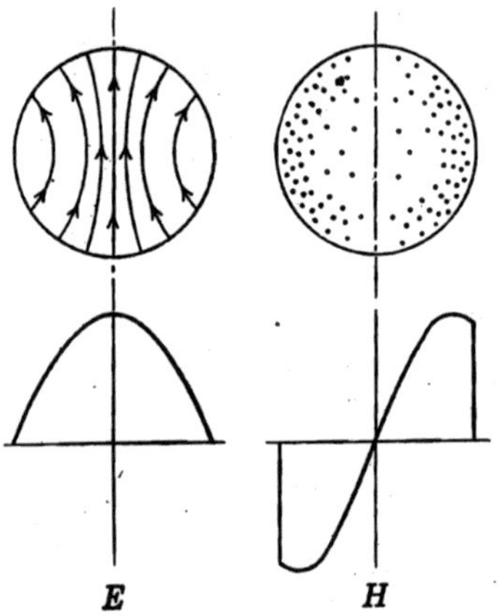

FIG. 111.—The lowest electric mode of oscillation in a spherical resonator, showing the field in a meridian plane.

An equivalent inductance and resistance for the oscillating system can be defined in terms of this current.[1] The average magnetic energy within the cavity is

$$
\begin{aligned}
(39)\quad \bar{T} = \frac{\mu_1}{4}\int \mathbf{H}\cdot\tilde{\mathbf{H}}\,dv &= \frac{2\pi\epsilon_1 A^2}{4}\int_0^a\int_0^\pi \left[j_1\!\left(2.75\,\frac{R}{a}\right)\right]^2 R^2\sin^3\theta\, d\theta\, dR\\
&= \frac{2\pi}{3}\,\epsilon_1 A^2\int_0^a \left[j_1\!\left(2.75\,\frac{R}{a}\right)\right]^2 R^2\, dR.
\end{aligned}
$$

The integral can be evaluated without difficulty either by applying general formulas for the integration of cylindrical Bessel functions, or directly by noting that

$$
(40)\qquad j_1(k_1R) = -\frac{1}{k_1R}\left(\cos k_1R - \frac{1}{k_1R}\sin k_1R\right).
$$

We find

$$
(41)\qquad \int_0^x j_1^2(x)x^2\,dx = \frac{x^3}{2}\left[j_1^2(x) - j_0(x)j_2(x)\right],
$$

whence

$$
(42)\qquad \bar{T} = \frac{2\pi}{3}\,\epsilon_1 A^2 a^3 \times 0.054.
$$

[1] HANSEN, *J. Applied Phys.*, **9**, 654, 1938.

Next we define an inductance L by the formula

$$\bar{T} = \tfrac{1}{2}L\bar{I}^2 = \tfrac{1}{4}LI\tilde{I}. \tag{43}$$

If the current I in (43) is arbitrarily taken to be that which crosses the equator, the equivalent inductance is by (38), (42), and (43)

$$L = 0.077\mu_1 a \qquad \text{henrys.} \tag{44}$$

Thus far it has been assumed that the conductivity of the walls is infinite, in which case E_θ vanishes at $R = a$ and the attenuation is zero. If, however, the walls are metallic, the tangential component of **E** at the surface may be obtained from the approximation (47), page 534. At $R = a$

$$E_\theta \simeq \sqrt{\frac{\omega\mu_2}{\sigma_2}}\, H_\phi\, e^{-\frac{i\pi}{4}} \simeq iA\sqrt{\frac{\omega'_{11}\epsilon_1\mu_2}{\sigma_2\mu_1}} \sin\theta\, j_1(2.75)\, e^{-\frac{i\pi}{4} - i\omega'_{11}t}. \tag{45}$$

There is now a radial flow of energy into the metal whose mean density is

$$\bar{S}_R = Re\, S_R^* = \frac{A^2}{2\sqrt{2}}\frac{\epsilon_1}{\mu_1}\sqrt{\frac{\omega'_{11}\mu_2}{\sigma_2}} \sin^2\theta\, j_1^2(2.75). \tag{46}$$

The energy lost per unit time is obtained by integrating (46) over the surface of the sphere.

$$W = a^2 \int_0^{2\pi}\int_0^{\pi} \bar{S}_R \sin\theta\, d\theta\, d\phi = \frac{4\pi a^2}{3\sigma_2}\frac{\epsilon_1}{\mu_1}\frac{A^2}{\delta} j_1^2(2.75), \tag{47}$$

where $\delta = \sqrt{\dfrac{2}{\omega'_{11}\mu_2\sigma_2}}$ is the skin depth defined on page 504. An equivalent series resistance $\mathcal{R}$ is expressed by

$$W = \mathcal{R}\bar{I}^2 = \tfrac{1}{2}\mathcal{R}I\tilde{I}, \tag{48}$$

whence

$$\mathcal{R} = \frac{2}{3\pi\sigma_2}\frac{1}{\delta} = \frac{0.212}{\sigma_2}\frac{1}{\delta}. \tag{49}$$

The parameter Q, defined as the ratio of inductive reactance to series resistance, is often used as a measure of circuit efficiency. The Q of the cavity oscillating at its fundamental frequency is

$$Q = \frac{\omega'_{11}L}{\mathcal{R}} = 0.725\frac{\mu_1}{\mu_2}\frac{a}{\delta}. \tag{50}$$

DIFFRACTION OF A PLANE WAVE BY A SPHERE

9.25. Expansion of the Diffracted Field.—A periodic wave incident upon a material body of any description gives rise to a forced oscillation of free and bound charges synchronous with the applied field. These constrained movements of charge set up in turn a secondary field both inside and outside the body. The resultant field at any point is then the vector sum of the primary and secondary fields. In general the forced oscillations fail to match the conditions prevailing at the instant the primary field was first established. To ensure fulfillment of the boundary conditions at all times, a transient term must be added, constructed from the natural modes with suitable amplitudes. Such transient oscillations, however, are quickly damped by absorption and radiation losses, leaving only the steady-state, synchronous term.

The simplest problem of this class and at the same time one of great practical interest is that of a plane wave falling upon a sphere. As in the preceding section, we shall suppose the sphere of radius a and propagation constant k_1 to be embedded in an infinite, homogeneous medium k_2. A plane wave, whose electric vector is linearly polarized in the x-direction, is propagated in the direction of the positive z-axis. The expansion of this incident field in vector spherical wave functions has been given in (36), page 419.

$$(1)\quad \begin{aligned} \mathbf{E}_i &= \mathbf{a}_x E_0 e^{ik_2 z - i\omega t} = E_0 e^{-i\omega t} \sum_{n=1}^{\infty} i^n \frac{2n+1}{n(n+1)} \left(\mathbf{m}_{o1n}^{(1)} - i\mathbf{n}_{e1n}^{(1)}\right), \\ \mathbf{H}_i &= \mathbf{a}_y \frac{k_2}{\mu_2 \omega} E_0 e^{ik_2 z - i\omega t} = -\frac{k_2 E_0}{\mu_2 \omega} e^{-i\omega t} \sum_{n=1}^{\infty} i^n \frac{2n+1}{n(n+1)} \left(\mathbf{m}_{e1n}^{(1)} + i\mathbf{n}_{o1n}^{(1)}\right), \end{aligned}$$

where E_0 is the amplitude and

$$(2)\quad \mathbf{m}_{\substack{o\\e}1n}^{(1)} = \pm \frac{1}{\sin\theta} j_n(k_2 R) P_n^1(\cos\theta) \begin{matrix}\cos\\ \sin\end{matrix}\,\phi\, \mathbf{i}_2 - j_n(k_2 R) \frac{\partial P_n^1}{\partial \theta} \begin{matrix}\sin\\ \cos\end{matrix}\,\phi\, \mathbf{i}_3,$$

$$(3)\quad \begin{aligned} \mathbf{n}_{\substack{o\\e}1n}^{(1)} = {} & \frac{n(n+1)}{k_2 R} j_n(k_2 R) P_n^1(\cos\theta) \begin{matrix}\sin\\ \cos\end{matrix}\,\phi\, \mathbf{i}_1 + \frac{1}{k_2 R} [k_2 R\, j_n(k_2 R)]' \\ & \frac{\partial P_n^1}{\partial \theta} \begin{matrix}\sin\\ \cos\end{matrix}\,\phi\, \mathbf{i}_2 \pm \frac{1}{k_2 R \sin\theta} [k_2 R\, j_n(k_2 R)]' P_n^1(\cos\theta) \begin{matrix}\cos\\ \sin\end{matrix}\,\phi\, \mathbf{i}_3, \end{aligned}$$

the prime denoting differentiation with respect to the argument $k_2 R$.

The induced secondary field must be constructed in two parts, the one applying to the interior of the sphere and the other valid at all external points with the necessary regularity at infinity. By analogy with the problem of plane boundaries discussed in Secs. 9.4 to 9.9, these two parts will be referred to as transmitted and reflected waves, although such terms are strictly appropriate only when the wave length is very much smaller than the radius of the sphere. Let us write

$$(4)\quad \begin{aligned} \mathbf{E}_r &= E_0 e^{-i\omega t} \sum_{n=1}^{\infty} i^n \frac{2n+1}{n(n+1)} \left(a_n^r\, \mathbf{m}_{o1n}^{(3)} - i b_n^r\, \mathbf{n}_{e1n}^{(3)}\right), \\ \mathbf{H}_r &= -\frac{k_2}{\omega \mu_2} E_0 e^{-i\omega t} \sum_{n=1}^{\infty} i^n \frac{2n+1}{n(n+1)} \left(b_n^r\, \mathbf{m}_{e1n}^{(3)} + i a_n^r\, \mathbf{n}_{o1n}^{(3)}\right), \end{aligned}$$

valid when $R > a$, and

$$\begin{aligned} \mathbf{E}_t &= E_0 e^{-i\omega t} \sum_{n=1}^{\infty} i^n \frac{2n+1}{n(n+1)} \left(a_n^t \, \mathbf{m}_{o1n}^{(1)} - i b_n^t \, \mathbf{n}_{e1n}^{(1)}\right), \\ \mathbf{H}_t &= -\frac{k_1}{\omega\mu_1} E_0 e^{-i\omega t} \sum_{n=1}^{\infty} i^n \frac{2n+1}{n(n+1)} \left(b_n^t \, \mathbf{m}_{e1n}^{(1)} + i a_n^t \, \mathbf{n}_{o1n}^{(1)}\right), \end{aligned} \tag{5}$$

which hold when $R < a$. The functions $\mathbf{m}_{^e_o 1n}^{(3)}$ and $\mathbf{n}_{^e_o 1n}^{(3)}$ are obtained by replacing $j_n(k_2 R)$ by $h_n^{(1)}(k_2 R)$ in (2) and (3). k_1 replaces k_2 in (5).

The boundary conditions at $R = a$ are

$$\begin{aligned} \mathbf{i}_1 \times (\mathbf{E}_i + \mathbf{E}_r) &= \mathbf{i}_1 \times \mathbf{E}_t, \\ \mathbf{i}_1 \times (\mathbf{H}_i + \mathbf{H}_r) &= \mathbf{i}_1 \times \mathbf{H}_t. \end{aligned} \tag{6}$$

These lead to two pairs of *inhomogeneous* equations for the expansion coefficients.

$$\begin{aligned} a_n^t j_n(N\rho) - a_n^r h_n^{(1)}(\rho) &= j_n(\rho), \\ \mu_2 a_n^t [N\rho\, j_n(N\rho)]' - \mu_1 a_n^r [\rho\, h_n^{(1)}(\rho)]' &= \mu_1 [\rho\, j_n(\rho)]', \end{aligned} \tag{7}$$

$$\begin{aligned} \mu_2 N\, b_n^t\, j_n(N\rho) - \mu_1 b_n^r\, h_n^{(1)}(\rho) &= \mu_1 j_n(\rho), \\ b_n^t [N\rho\, j_n(N\rho)]' - N b_n^r [\rho\, h_n^{(1)}(\rho)]' &= N[\rho\, j_n(\rho)]', \end{aligned} \tag{8}$$

where again

$$k_1 = N k_2, \qquad \rho = k_2 a, \qquad k_1 a = N\rho. \tag{9}$$

This system is now solved for the coefficients of the external field

$$a_n^r = -\frac{\mu_1 j_n(N\rho)[\rho\, j_n(\rho)]' - \mu_2 j_n(\rho)[N\rho\, j_n(N\rho)]'}{\mu_1 j_n(N\rho)[\rho\, h_n^{(1)}(\rho)]' - \mu_2 h_n^{(1)}(\rho)[N\rho\, j_n(N\rho)]'} \tag{10}$$

$$b_n^r = -\frac{\mu_1 j_n(\rho)[N\rho\, j_n(N\rho)]' - \mu_2 N^2 j_n(N\rho)[\rho\, j_n(\rho)]'}{\mu_1 h_n^{(1)}(\rho)[N\rho\, j_n(N\rho)]' - \mu_2 N^2 j_n(N\rho)[\rho\, h_n^{(1)}(\rho)]'}. \tag{11}$$

If the conductivity or inductive capacity of the sphere is large relative to the ambient medium, and at the same time the radius a is not too small, then $|N\rho| \gg 1$ and (10) and (11) can be greatly simplified by use of the asymptotic expressions

$$\begin{aligned} j_n(N\rho) &\simeq \frac{1}{N\rho} \cos\left(N\rho - \frac{n+1}{2}\pi\right), \\ [N\rho\, j_n(N\rho)]' &\simeq -\sin\left(N\rho - \frac{n+1}{2}\pi\right). \end{aligned} \tag{12}$$

In this case

$$a_n^r \simeq -\frac{j_n(\rho)}{h_n^{(1)}(\rho)}, \qquad b_n^r \simeq -\frac{[\rho\, j_n(\rho)]'}{[\rho\, h_n^{(1)}(\rho)]'}. \tag{13}$$

Since $h_n^{(1)}(\rho) = j_n(\rho) + i\, n_n(\rho)$, these coefficients can also be put in the form

$$a_n^r \simeq i e^{i\gamma_n} \sin \gamma_n, \qquad b_n^r \simeq i e^{i\gamma_n'} \sin \gamma_n', \tag{14}$$

where

$$\tan\gamma_n = \frac{j_n(\rho)}{n_n(\rho)}, \qquad \tan\gamma'_n = \frac{[\rho\, j_n(\rho)]'}{[\rho\, n_n(\rho)]'}. \tag{15}$$

To sum up, we have found that the primary wave excites certain partial oscillations in the sphere. These are not the natural modes discussed in Sec. 9.22, for all are synchronous with the applied field. These partial oscillations and their associated fields, however, are of electric and magnetic type for the same reasons that were set forth in Sec. 9.22. The coefficients a_n are the amplitudes of the oscillations of magnetic type, while the b_n are the amplitudes of electric oscillations. Whenever the impressed frequency ω approaches a characteristic frequency ω_{ns} of the free oscillations, resonance phenomena will occur. Now the characteristic frequencies of the magnetic oscillations have been shown to satisfy (10), Sec. 9.22. But this is just the condition which makes the denominator of a^r_n, in (10) above, vanish. Likewise the natural frequencies of the electric oscillations satisfy (19) in Sec. 9.22, which is the condition that the denominator of b^r_n shall vanish. Note, however, that the ω_{ns} are always complex, whereas the frequency ω of the forcing field is real. Consequently the denominators of a^r_n and b^r_n can be reduced to minimum values, but never quite to zero, so that the catastrophe of infinite amplitudes is safely avoided.

In Fig. 112 the electric and magnetic lines of force are shown for the first four partial waves of electric type. The drawings are reproduced from the original paper by Mie.[1]

The incident wave is linearly polarized with its electric vector parallel to the x-axis. At very great distances from the sphere the radial component of the secondary field vanishes as $1/R^2$, while the tangential components $E_{r\theta}$ and $E_{r\phi}$ diminish as $1/R$. In this radiation zone the field is transverse to the direction of propagation. Moreover, the components $E_{r\theta}$ and $E_{r\phi}$ are perpendicular to each other and differ, in general, in phase. *The secondary radiation from the sphere is elliptically polarized.* There are two exceptional directions. When $\phi = 0$, we note that $E_{r\phi} = 0$, and when $\phi = \pi/2$ we have $E_{r\theta} = 0$. Consequently, when viewed along the x- or the y-axis, the secondary radiation is linearly polarized. Inversely, if the primary wave is unpolarized, as in the case of natural light, the secondary radiation exhibits partial polarization dependent upon the direction of observation. This effect has been studied in connection with the scattering of light by suspensions of colloidal particles. The most complete numerical investigation of the problem thus far in the visible spectrum has been made by Blumer.[2]

[1] MIE, *Ann. Physik*, **25**, 377, 1908.
[2] BLUMER, *Z. Physik*, **32**, 119, 1925; **38**, 304, 1926; **38**, 920, 1926; **39**, 195, 1926.

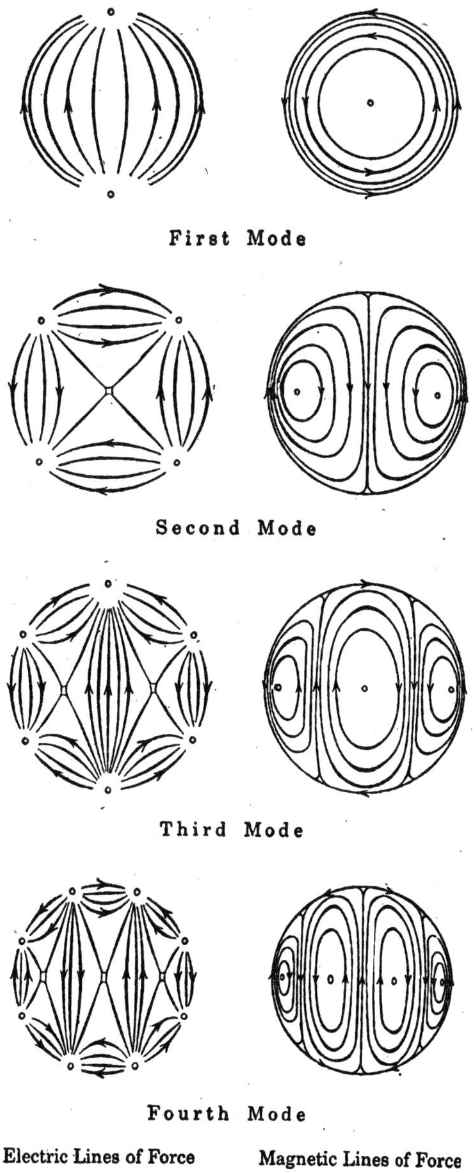

FIG. 112.—Lines of force corresponding to the first four modes of electric type.

9.26. Total Radiation.—The resultant field at any point outside the sphere is the sum of the primary and secondary fields.

$$\mathbf{E} = \mathbf{E}_i + \mathbf{E}_r, \qquad \mathbf{H} = \mathbf{H}_i + \mathbf{H}_r. \tag{16}$$

The radial component of the total complex flow vector is

$$S_R^* = \tfrac{1}{2}(E_\theta \tilde{H}_\phi - E_\phi \tilde{H}_\theta), \tag{17}$$

which can be resolved by virtue of (16) into three groups:

$$\begin{aligned} S_R^* = {} & \tfrac{1}{2}(E_{i\theta}\tilde{H}_{i\phi} - E_{i\phi}\tilde{H}_{i\theta}) + \tfrac{1}{2}(E_{r\theta}\tilde{H}_{r\phi} - E_{r\phi}\tilde{H}_{r\theta}) \\ & + \tfrac{1}{2}(E_{i\theta}\tilde{H}_{r\phi} + E_{r\theta}\tilde{H}_{i\phi} - E_{i\phi}\tilde{H}_{r\theta} - E_{r\phi}\tilde{H}_{i\theta}). \end{aligned} \tag{18}$$

Let us draw about the diffracting sphere a concentric spherical surface of radius R. The real part of S_R^* integrated over this sphere is equal to the net flow of energy across its surface. To simplify matters we shall assume the external medium (2) to be a perfect dielectric, so that $k_2 = \omega\sqrt{\epsilon_2\mu_2}$. If the diffracting sphere is also nonconducting, the net flow across any surface enclosing the sphere must be zero. If, however, energy is converted into heat within the sphere, the net flow is equal to the amount absorbed and is directed inwards. We shall call the total energy absorbed by the sphere W_a,

$$W_a = -Re \int_0^\pi \int_0^{2\pi} S_R^* R^2 \sin\theta \, d\phi \, d\theta. \tag{19}$$

The first term on the right-hand side of (18) measures the flow of energy in the incident wave. When integrated over a closed surface this gives zero so long as $\sigma_2 = 0$. The second term obviously measures the outward flow of the secondary or *scattered* energy from the diffracting sphere, and the total scattered energy is

$$W_s = \frac{1}{2} Re \int_0^\pi \int_0^{2\pi} (E_{r\theta}\tilde{H}_{r\phi} - E_{r\phi}\tilde{H}_{r\theta}) R^2 \sin\theta \, d\theta \, d\phi. \tag{20}$$

To maintain the energy balance the third term of (18) must be equal in magnitude to the sum of the absorbed and scattered energies.

$$\begin{aligned} W_t = W_a + W_s = -\frac{1}{2} Re \int_0^\pi \int_0^{2\pi} (& E_{i\theta}\tilde{H}_{r\phi} + E_{r\theta}\tilde{H}_{i\phi} - E_{i\phi}\tilde{H}_{r\theta} \\ & - E_{r\phi}\tilde{H}_{i\theta}) R^2 \sin\theta \, d\theta \, d\phi. \end{aligned} \tag{21}$$

W_t measures, therefore, the total energy derived from the primary wave and dissipated as heat and scattered radiation.

To calculate W_s and W_t we allow R to grow very large and introduce the asymptotic values

$$j_n(\rho) \simeq \frac{1}{\rho} \cos\left(\rho - \frac{n+1}{2}\pi\right), \qquad h_n^{(1)}(\rho) \simeq \frac{1}{\rho}(-i)^{n+1} e^{i\rho}, \tag{22}$$

into the field functions (2) and (3). The integrals can then be evaluated with the help of Eq. (20), page 417,

$$(23)\quad \int_0^\pi \left(\frac{dP_n^1}{d\theta}\frac{dP_m^1}{d\theta} + \frac{1}{\sin^2\theta}P_n^1P_m^1\right)\sin\theta\,d\theta \begin{cases} 0, & \text{if } n \neq m, \\ = \dfrac{2}{2n+1}\dfrac{(n+1)!}{(n-1)!}n(n+1), & \text{if } n = m, \end{cases}$$

and the relation

$$(24)\quad \int_0^\pi \left(\frac{P_m^1}{\sin\theta}\frac{dP_n^1}{d\theta} + \frac{P_n^1}{\sin\theta}\frac{dP_m^1}{d\theta}\right)\sin\theta\,d\theta = 0.$$

One finds for the scattered energy

$$(25)\quad W_s = \pi\frac{E_0^2}{k_2^2}\sqrt{\frac{\epsilon_2}{\mu_2}}\sum_{n=1}^\infty (2n+1)(|a_n^r|^2 + |b_n^r|^2);$$

for the sum of absorbed and scattered energies,

$$(26)\quad W_t = \pi\frac{E_0^2}{k_2^2}\sqrt{\frac{\epsilon_2}{\mu_2}}\,Re\sum_{n=1}^\infty (2n+1)(a_n^r + b_n^r).$$

The mean energy flow of the incident wave *per unit area* is

$$(27)\quad \bar{S}_s = \frac{1}{2}E_0^2\sqrt{\frac{\epsilon_2}{\mu_2}}.$$

The *scattering cross section* of the sphere is defined as the ratio of the total scattered energy per second to the energy density of the incident wave.

$$(28)\quad Q_s = \frac{2\pi}{k_2^2}\sum_{n=1}^\infty (2n+1)(|a_n^r|^2 + |b_n^r|^2) \qquad \text{meters}^2.$$

Likewise one may define the cross section Q_t by

$$(29)\quad Q_t = \frac{2\pi}{k_2^2}\,Re\sum_{n=1}^\infty (2n+1)(a_n^r + b_n^r) \qquad \text{meters}^2.$$

If the conductivity or inductive capacity of the sphere is so large that $|N\rho| \gg 1$, then the approximations (14) and (15) can be introduced and the cross sections reduce to

$$(30)\quad Q_s = \frac{2\pi}{k_2^2}\sum_{n=1}^\infty (2n+1)(\sin^2\gamma_n + \sin^2\gamma_n'),$$

$$(31)\quad Q_t = \frac{2\pi}{k_2^2}\,Re\sum_{n=1}^\infty i(2n+1)(e^{i\gamma_n}\sin\gamma_n + e^{i\gamma_n'}\sin\gamma_n').$$

If the sphere is nonabsorbing, γ_n and γ'_n are real, in which case it is obvious that $Q_s = -Q_t$.

9.27. Limiting Cases.—Although in appearance, the preceding formulas are exceedingly simple, the numerical calculation of the coefficients usually presents a serious task. There are two limiting cases which can be handled with ease. If $|\rho| = |k_1 a| \gg 1$, the functions $j_n(\rho)$ and $h_n^{(1)}(\rho)$ may be replaced by their asymptotic values. The quantity ρ is essentially the ratio of radius to wave length, so that this is the case of a sphere whose radius is very much larger than the wave length. Results obtained by this method should approach those deduced from the Kirchhoff-Huygens principle of Secs. 8.13 to 8.15. If, on the other hand, the wave length is very much larger than the radius of the sphere, so that $|\rho| \ll 1$, the radial functions can be expressed by the first term or two of their expansions in powers of ρ.

Case I. $|\rho| \gg 1$. If the asymptotic representations (12) and (22) are substituted into (10) and (11), one obtains for the coefficients

$$a_n^r \simeq i^n e^{-i\rho} \frac{\mu_1 \sin x - \mu_2 N \cos x \tan y}{\mu_1 - i\mu_2 N \tan y}, \tag{32}$$

$$b_n^r \simeq -i^{n+1} e^{-i\rho} \frac{\mu_1 \cos x \tan y - \mu_2 N \sin x}{\mu_1 \tan y + i\mu_2 N}, \tag{33}$$

where

$$x = \rho - \frac{n+1}{2}\pi, \qquad y = N\rho - \frac{n+1}{2}\pi. \tag{34}$$

What is apparent here is also true in the general case: the expansion coefficients are oscillating functions of ρ and the order n. Small changes in either ρ or n may give rise to wide variations in the values of the coefficients. The absolute magnitudes of the coefficients oscillate between the limits zero and one.

If in (32) we replace n by $n + 1$, it will be observed that $a_{n+1}^r \simeq b_n^r$. *The amplitude of an electric oscillation of order n is approximately of the same magnitude as the amplitude of the magnetic oscillation of next higher order.*

The asymptotic expressions used in the derivations of (32) and (33) are valid only so long as the order n is very much less than the arguments $|\rho|$ and $|N\rho|$. Asymptotic formulas for the Bessel functions which hold for all orders have been found by Debye, who has also shown that the number of terms to be retained in the series expansions is just equal to the number $|\rho|$.[1] Methods of summing the coefficients have been discussed by Jobst.[2]

Case II. $|\rho| \ll 1$. From Sec. 7.4, page 405, we take the expansions of the radial functions in powers of ρ or $N\rho$.

[1] DEBYE, *loc. cit.*
[2] JOBST, *Ann. Physik*, **76**, 863, 1925; **78**, 158, 1925.

$$(35)\quad j_n(\rho) = 2^n \frac{n!}{(2n+1)!}\rho^n\left(1 - \frac{n+1}{(2n+1)(2n+3)}\rho^2 + \cdots\right),$$

$$(36)\quad h_n^{(1)}(\rho) = \frac{i}{2^n}\frac{(2n)!}{n!}\frac{1}{\rho^{n+1}} + \cdots.$$

The coefficients can now be expanded in powers of ρ.

$$(37)\quad a_n^r \simeq -i\frac{2^{2n}n!(n+1)!}{(2n)!(2n+1)!}\frac{\rho^{2n+1}}{\mu_1 n + \mu_2(n+1)}\left\{\mu_2 - \mu_1 - \frac{\mu_2[(n+1)+(n+3)N^2] - \mu_1[(n+3)+(n+1)N^2]}{(2n+2)(2n+3)}\rho^2\right\},$$

$$(38)\quad b_n^r \simeq i\frac{2^{2n}n!(n+1)!}{(2n)!(2n+1)!}\frac{\rho^{2n+1}}{\mu_1(n+1) + \mu_2 N^2 n}\left\{\mu_1 - \mu_2 N^2 - \frac{\mu_1[(n+1)+(n+3)N^2] - \mu_2 N^2[(n+3)+(n+1)N^2]}{(2n+2)(2n+3)}\rho^2\right\}.$$

If $|\rho|$ is so small that all powers above the fifth can be neglected, only the first four coefficients need be considered.

$$a_1^r \simeq \frac{i}{3}\frac{\rho^3}{\mu_1 + 2\mu_2}\left\{\mu_1 - \mu_2 - \frac{1}{5}[\mu_1(2+N^2) - \mu_2(1+2N^2)]\rho^2\right\},$$

$$a_2^r \simeq \frac{i}{15}\frac{\mu_1 - \mu_2}{2\mu_1 + 3\mu_2}\rho^5,$$

$$(39)\quad b_1^r \simeq \frac{2i}{3}\frac{\rho^3}{2\mu_1 + \mu_2 N^2}\left\{\mu_1 - \mu_2 N^2 - \frac{1}{10}[\mu_1(1+2N^2) - \mu_2 N^2(2+N^2)]\rho^2\right\},$$

$$b_2^r \simeq \frac{i}{15}\frac{\mu_1 - \mu_2 N^2}{3\mu_1 + 2\mu_2 N^2}\rho^5.$$

Furthermore, if $\mu_1 = \mu_2$, these expressions reduce to

$$a_1^r \simeq \frac{i}{45}(N^2 - 1)\rho^5,$$

$$(40)\quad b_1^r \simeq -\frac{2i}{3}\left(\frac{N^2-1}{N^2+2}\rho^3 - \frac{1}{10}\frac{N^4-1}{N^2+2}\rho^5\right),$$

$$b_2^r \simeq -\frac{i}{15}\frac{N^2-1}{2N^2+3}\rho^5.$$

So long as the conductivity remains finite, the rule still holds that the magnetic oscillation of order $n + 1$ is of the same magnitude as the electric oscillation of order n.

When the ratio of radius to wave length is so small that ρ^5 can be neglected with respect to ρ^3, only the first-order electric oscillation need be taken into account. This case was first investigated by Lord Rayleigh. The amplitude of the oscillation is then

$$b_1^r \simeq -\frac{2i}{3}\frac{N^2 - 1}{N^2 + 2}\rho^3. \tag{41}$$

If this value is introduced in the expressions (2), (3), and (4), it will be noted on comparison with (30), page 436, that *the field of the fundamental mode is that of an electric dipole, oriented along the x-axis and with a dipole moment equal to*

$$p = 4\pi\epsilon_2 \frac{N^2 - 1}{N^2 + 2} a^3 E_0. \tag{42}$$

Reference to (32), page 206, shows this to be identical with the result obtained in the electrostatic case. The amplitude of the scattered or diffracted radiation at great distances from a small sphere increases, therefore, as the square of the frequency; and the energy, as the fourth power. This is the celebrated *Rayleigh law of scattering* which was applied by its discoverer to explain the blue of the sky. The light which reaches us from the heavens, unless the observer is looking directly at the sun, has been scattered by dust particles or the molecules of the atmosphere. Since the wave length of visible light is much longer than the radius of these submicroscopic particles, the fourth-power law is valid and the short wave lengths in the natural sunlight are strongly scattered relative to those near the red end of the spectrum. If one looks directly at the sun, the opposite effect is observed, for as the blue suffers the greatest scattering, its intensity in the direct beam is weakened relative to the red, which now predominates. This effect is accentuated when the sun is near the horizon, so that the path lies in a region heavy with dust particles.

The varied shades of mist and smoke can be accounted for in the same fashion, and it is well known that the colors of colloidal suspensions of metallic particles in liquids or gases often differ completely from the true color of the metal itself. However, as the radius of the particle increases, terms of higher order play an increasingly important part. The extremely selective character of the Rayleigh scattering disappears and it becomes impossible to predict the pattern of the diffracted radiation on any basis other than the results of a difficult computation. The higher order terms can be interpreted as electric and magnetic multipoles.[1] An experimental test of the theory for radio waves has been made by Schaefer.[2]

The diffraction of a plane wave by an ellipsoidal body can be treated in the same manner as the sphere, but the analytical properties of the

[1] GANS and HAPPEL, *Ann. Physik*, **29**, 277, 1909.

[2] SCHAEFER and WILMSEN, *Z. Physik*, **24**, 345, 1924.

wave functions are very much more complex. An approximate solution has been given by Gans[1] and a more rigorous analysis by Herzfeld.[2] The most thorough investigation to date is that of Möglich.[3] In no case, however, have the numerical calculations been carried far enough to be particularly useful, and the many practical applications of the theory make it desirable that further work be done along this line.

EFFECT OF THE EARTH ON THE PROPAGATION OF RADIO WAVES

9.28. Sommerfeld Solution.—The classical investigation of the effect of a finitely conducting plane upon the radiation of an oscillating dipole was published by Arnold Sommerfeld in 1909.[4] Since that time an enormous amount of work has appeared on the subject and it may be fairly said that no aspect of the problem of radio wave propagation has received more careful attention. It is impossible within the confines of this chapter to give a complete account of the theory, but it should prove instructive to see how the problem is set up. As in all but the simplest diffraction problems, the real difficulties arise with the necessity of reducing the formal solution to a state that permits numerical computation.

The first problem attacked by Sommerfeld was that of a vertical dipole located at the surface of a plane, finitely conducting earth. The axis of the dipole coincides with the z-axis of a rectangular coordinate system and the plane $z = 0$ represents the earth's surface. All points for which $z < 0$ lie within the earth, whose propagation constant is k_1. The propagation constant of the air in the region $z > 0$ is $k_2 = \omega/c$. A point of observation is located by the cylindrical coordinates r, ϕ, z, and its radial distance from the origin is

$$R = \sqrt{r^2 + z^2}. \tag{1}$$

These relations are illustrated in Fig. 113.

The field is cylindrically symmetrical about the z-axis, and in the neighborhood of the origin it must grow infinite like that of a dipole. Now the field of a dipole, we recall from (7), page 432, can be represented by an axial Hertzian vector which behaves as $1/R$ as $R \to 0$. We shall write, therefore, the z-component of the Hertzian vector of the *resultant* field as the sum of two parts:

[1] GANS, *Ann. Physik,* **37,** 881, 1912; **47,** 270, 1915; **61,** 465, 1920.

[2] HERZFELD, *Wiener Ber.,* **120,** 1587, 1911.

[3] MÖGLICH, *Ann. Physik,* **83,** 609, 1927.

[4] SOMMERFELD, *Ann. Physik,* **28,** 665, 1909; *Jahrbuch d. drahtl. Telegraphie,* **4,** 157, 1911. These earlier papers contain an error in algebraic sign which was corrected in the *Ann. Physik,* **81,** 1135, 1926. Also in Riemann-Weber, *loc. cit.,* p. 542. Some of the consequences of this error appear, however, to have been overlooked.

$$\Pi_1(r, z) = \frac{1}{R} e^{ik_1R} + F_1(k_1; r, z), \qquad (z < 0),$$

(2)

$$\Pi_2(r, z) = \frac{1}{R} e^{ik_2R} + F_2(k_2; r, z), \qquad (z > 0).$$

The first term accounts for the singularity at the origin. The dipole moments have arbitrarily been chosen such that the coefficient of this term is unity, and a common time factor $\exp(-i\omega t)$ has been suppressed. The functions F_1 and F_2 are finite everywhere, including the point $R = 0$, and satisfy the wave equation and the radiation condition at infinity. They represent the contribution of the diffracted wave and are to be

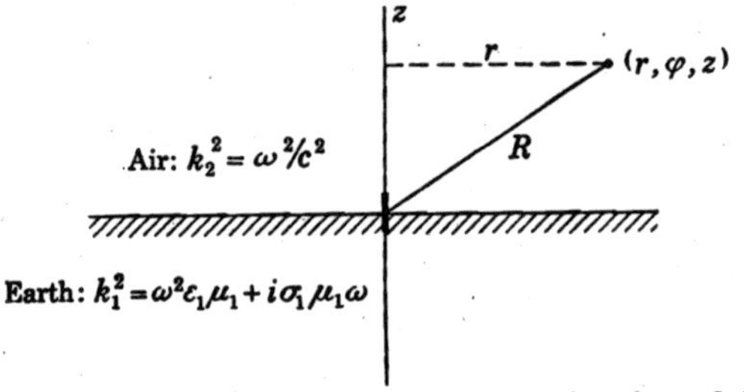

FIG. 113.—Vertical electric dipole located at the surface of a plane, finitely conducting earth.

chosen such as to ensure the fulfillment of the boundary conditions at $z = 0$. The field vectors are derived from the Hertzian vectors through

(3) $$\mathbf{E} = \nabla\nabla \cdot \mathbf{\Pi} + k^2\mathbf{\Pi}, \qquad \mathbf{H} = \frac{k^2}{i\omega\mu} \nabla \times \mathbf{\Pi},$$

whence the tangential components are

(4) $$E_r = \frac{\partial^2 \Pi}{\partial r\, \partial z}, \qquad H_\phi = -\frac{k^2}{i\omega\mu}\frac{\partial \Pi}{\partial r}.$$

It may be assumed for the earth that $\mu_1 = \mu_2 = \mu_0$. At any point on the plane $z = 0$ the functions (2) must satisfy the condition

(5) $$\frac{\partial^2 \Pi_1}{\partial r\, \partial z} = \frac{\partial^2 \Pi_2}{\partial r\, \partial z}, \qquad k_1^2 \frac{\partial \Pi_1}{\partial r} = k_2^2 \frac{\partial \Pi_2}{\partial r}, \qquad (z = 0).$$

These relations hold for all values of r and can, therefore, be integrated with respect to r. The functions and their derivatives vanish as $r \to \infty$, so that the constant of integration must be zero. Thus the boundary conditions reduce to

(6) $$\frac{\partial \Pi_1}{\partial z} = \frac{\partial \Pi_2}{\partial z}, \qquad k_1^2 \Pi_1 = k_2^2 \Pi_2, \qquad (z = 0).$$

To find functions F_1 and F_2 that satisfy these conditions, we shall expand both the "primary excitation" e^{ikr}/R and the "secondary excitation" F in terms of cylindrical wave functions. In Sec. 6.6 it was shown that every electromagnetic field within a homogeneous, isotropic medium might be represented by the superposition of elementary cylindrical wave functions of the type

$$\psi(r, \phi, z) = e^{in\phi} J_n(\sqrt{k^2 - h^2}\, r)\, e^{ihz}. \tag{7}$$

In the present instance the fields are symmetric and consequently $n = 0$. As a parameter we shall choose $\lambda^2 = k^2 - h^2$ rather than h, and multiply each wave function by an amplitude $f(\lambda)$ dependent upon λ. This parameter can assume any complex value. The sign of the root $ih = \pm\sqrt{\lambda^2 - k^2}$ must be chosen such that the fields vanish as $z \to \pm\infty$. We do this by making the real part of $\sqrt{\lambda^2 - k^2}$ always positive. Then the conditions at the origin and at infinity are fulfilled by a discontinuous Hertz vector whose radial and azimuthal components are zero and whose z-component is

$$\text{(8)}\qquad \begin{aligned} \Pi_1 &= \frac{1}{R} e^{ik_1R} + \int_0^\infty f_1(\lambda) J_0(\lambda r) e^{+\sqrt{\lambda^2 - k_1^2}\, z}\, d\lambda, \qquad (z < 0), \\ \Pi_2 &= \frac{1}{R} e^{ik_2R} + \int_0^\infty f_2(\lambda) J_0(\lambda r) e^{-\sqrt{\lambda^2 - k_2^2}\, z}\, d\lambda, \qquad (z > 0). \end{aligned}$$

The next step is to represent the known primary excitation in the same manner. To this end we make use of the Fourier-Bessel theorem (53), page 371. If $f(r)$ vanishes as $r \to \infty$ such that the integral of $f(r)\sqrt{r}$ converges absolutely, then

$$f(r) = \int_0^\infty \lambda\, d\lambda \int_0^\infty \rho\, d\rho\, f(\rho) J_0(\lambda\rho) J_0(\lambda r). \tag{9}$$

For $f(r)$ we take first e^{ikr}/r.

$$\frac{e^{ikr}}{r} = \int_0^\infty \lambda\, d\lambda\, J_0(\lambda r) \int_0^\infty d\rho\, e^{ik\rho} J_0(\lambda\rho). \tag{10}$$

Integration over ρ can be effected if the Bessel function $J_0(\lambda\rho)$ is replaced by its integral representation (37), page 367.

(11)

$$\int_0^\infty e^{ik\rho} J_0(\lambda\rho)\, d\rho = \frac{1}{2\pi} \int_{-\pi}^{\pi} d\beta \int_0^\infty d\rho\, e^{i(k + \lambda\cos\beta)\rho} = \frac{1}{2\pi i} \int_{-\pi}^{\pi} \frac{d\beta}{k + \lambda\cos\beta}.$$

It is assumed here that k has a positive imaginary part so that the integrand vanishes at the upper limit. k_2 is in fact real, but for purposes of integration the atmosphere can be assigned a very small conductivity which may be reduced to zero after the integral has been evaluated. The last integral is equivalent to an integration about a closed contour. Let

$$u = e^{i\beta}, \qquad du = iu\, d\beta. \tag{12}$$

Then

(13)
$$\frac{1}{2\pi i}\int_{-\pi}^{\pi}\frac{d\beta}{k+\lambda\cos\beta}=\frac{1}{\pi}\oint\frac{du}{2ku+\lambda(1+u^2)}=\frac{1}{\pi\lambda}\oint\frac{du}{(u-u_1)(u-u_2)},$$

where the path of integration is the unit circle about the origin in the complex u-plane and where u_1 and u_2 are the roots of

$$u^2+\frac{2ku}{\lambda}+1=0. \tag{14}$$

The product of the roots is unity, so that u_1 lies inside the circle and u_2 outside.

$$\frac{1}{\pi\lambda}\oint\frac{du}{(u-u_1)(u-u_2)}=\frac{1}{\pi\lambda(u_1-u_2)}\oint\frac{du}{u-u_1}=\frac{2i}{\lambda(u_1-u_2)}=\frac{1}{\sqrt{\lambda^2-k^2}}, \tag{15}$$

by Cauchy's theorem (133), page 315. Consequently

$$\frac{e^{ikr}}{r}=\int_0^\infty\frac{J_0(\lambda r)}{\sqrt{\lambda^2-k^2}}\lambda\,d\lambda. \tag{16}$$

This represents the primary excitation e^{ikR}/R at all points on the plane $z=0$ as a sum of elementary cylindrical waves whose amplitudes are expressed by the function $\lambda(\lambda^2-k^2)^{-\frac{1}{2}}$. A representation at any other point follows directly from (7).

(17)
$$\frac{1}{R}e^{ik_1R}=\int_0^\infty\frac{J_0(\lambda r)e^{\sqrt{\lambda^2-k_1^2}\,z}}{\sqrt{\lambda^2-k_1^2}}\lambda\,d\lambda, \qquad (z<0),$$
$$\frac{1}{R}e^{ik_2R}=\int_0^\infty\frac{J_0(\lambda r)e^{-\sqrt{\lambda^2-k_2^2}\,z}}{\sqrt{\lambda^2-k_2^2}}\lambda\,d\lambda, \qquad (z>0).$$

For the Hertzian vector of the resultant field, we now have

(18)
$$\Pi_1=\int_0^\infty\left[\frac{\lambda}{\sqrt{\lambda^2-k_1^2}}+f_1(\lambda)\right]J_0(\lambda r)e^{\sqrt{\lambda^2-k_1^2}\,z}\,d\lambda, \qquad (z<0),$$
$$\Pi_2=\int_0^\infty\left[\frac{\lambda}{\sqrt{\lambda^2-k_2^2}}+f_2(\lambda)\right]J_0(\lambda r)e^{-\sqrt{\lambda^2-k_2^2}\,z}\,d\lambda, \qquad (z>0).$$

The functions $f_1(\lambda)$ and $f_2(\lambda)$ are next determined from the boundary conditions (6). A slight difficulty is encountered with the first of these relations, for if (17) is differentiated under the sign of integration with respect to z and z then placed equal to zero, the integrals diverge. But

Sommerfeld points out that

$$\left[\frac{\partial}{\partial z}\left(\frac{e^{ikR}}{R}\right)\right]_{z=0} = \left[\frac{d}{dR}\left(\frac{e^{ikR}}{R}\right)\frac{z}{R}\right]_{z=0} = 0, \tag{19}$$

and hence in this case the primary excitation in (18) can be ignored. A simple calculation leads us then to the result:

$$\begin{aligned} f_1(\lambda) &= \frac{\lambda}{\sqrt{\lambda^2 - k_1^2}} \frac{k_2^2\sqrt{\lambda^2 - k_1^2} - k_1^2\sqrt{\lambda^2 - k_2^2}}{k_2^2\sqrt{\lambda^2 - k_1^2} + k_1^2\sqrt{\lambda^2 - k_2^2}}, \\ f_2(\lambda) &= \frac{\lambda}{\sqrt{\lambda^2 - k_2^2}} \frac{k_1^2\sqrt{\lambda^2 - k_2^2} - k_2^2\sqrt{\lambda^2 - k_1^2}}{k_1^2\sqrt{\lambda^2 - k_2^2} + k_2^2\sqrt{\lambda^2 - k_1^2}}. \end{aligned} \tag{20}$$

Upon substitution of these expressions into (18), we find at last

$$\begin{aligned} \Pi_1 &= 2k_2^2 \int_0^\infty \frac{J_0(\lambda r)}{N} e^{\sqrt{\lambda^2 - k_1^2}\,z}\,\lambda\,d\lambda, \qquad (z < 0), \\ \Pi_2 &= 2k_1^2 \int_0^\infty \frac{J_0(\lambda r)}{N} e^{-\sqrt{\lambda^2 - k_2^2}\,z}\,\lambda\,d\lambda, \qquad (z > 0), \end{aligned} \tag{21}$$

where

$$N = k_1^2\sqrt{\lambda^2 - k_2^2} + k_2^2\sqrt{\lambda^2 - k_1^2}. \tag{22}$$

A formal solution has been obtained in terms of infinite integrals expressing the Hertzian vector of the resultant field at all points both above and below the surface.

9.29. Weyl Solution.—It was recognized from the outset by Sommerfeld that his solution could be interpreted as a bundle of plane waves reflected and refracted from the earth's surface at various angles of incidence. This point of view was developed later by Weyl.[1] Its basis is an integral representation of the function $h_0^{(1)}(kR)$ demonstrated in Sec. 7.7. According to (66), page 410,

$$h_0^{(1)}(kR) = \frac{e^{ikR}}{ikR} = \int_{i\infty}^{1} e^{ikR\eta}\,d\eta. \tag{23}$$

As in Fig. 70, page 407, the wave normal of each elementary wave makes an angle α with the z-axis and an angle β with the x-axis. The point of observation at P is located by the spherical coordinates (R, θ, ϕ), while the angle between the radius vector $\mathbf{R}$ and the wave normal $\mathbf{k}$ is γ. Thus $\mathbf{k} \cdot \mathbf{R} = kR \cos \gamma$,

$$\begin{aligned} \eta &= \cos\gamma = \sin\theta\sin\alpha\cos(\phi - \beta) + \cos\theta\cos\alpha, \\ R\cos\gamma &= x\sin\alpha\cos\beta + y\sin\alpha\sin\beta + z\cos\alpha. \end{aligned} \tag{24}$$

[1] WEYL, *Ann. Physik*, **60**, 481, 1919.

If the integration in (23) be carried out with respect to γ instead of η, we have

$$\text{(25)} \qquad \frac{e^{ikR}}{ikR} = \int_0^{\frac{\pi}{2}-i\infty} e^{ikR\cos\gamma} \sin\gamma\, d\gamma$$

$$= \frac{1}{2\pi}\int^{2\pi}\int_0^{\frac{\pi}{2}-i\infty} e^{ikR\cos\gamma}\sin\gamma\, d\gamma\, d\psi,$$

where ψ is an equatorial angle in a system of polar coordinates whose axis coincides with the radius vector **R**. But $\sin\gamma\, d\gamma\, d\psi$ is an element of area on a unit sphere concentric with the origin and the value of the integral is invariant to a rotation of the reference axes. Hence,

$$\text{(26)} \qquad \frac{e^{ikR}}{R} = \frac{ik}{2\pi}\int_0^{2\pi}\int_0^{\frac{\pi}{2}-i\infty} e^{ikR\cos\gamma}\sin\alpha\, d\alpha\, d\beta.$$

This is the desired representation of the spherical wave function in terms of plane waves. Sommerfeld's representation (17) is obtained from (26) by carrying out the integration with respect to β. Weyl points out, however, that in so doing one loses the freedom to rotate the coordinate system into an orientation that facilitates evaluation of the integrals. For this purpose the axis of the radius vector **R** is more appropriate than the z-axis normal to the surface of the earth.

Let us suppose that the dipole is located a distance z_0 above the surface of the earth. The resultant field Π_2 at any point for which $z > 0$ can be expressed as the sum of the direct or primary field Π_0, and the reflected field Π_r. In the region $0 \leq z \leq z_0$ between the dipole and the earth the primary radiation is represented by

$$\text{(27)} \quad \Pi_0 = \frac{e^{ik_2R}}{R}$$

$$= \frac{ik_2}{2\pi}\int_0^{2\pi}\int_0^{\frac{\pi}{2}-i\infty} e^{ik_2[x\sin\alpha_2\cos\beta_2 + y\sin\alpha_2\sin\beta_2 + (z_0 - z)\cos\alpha_2]}\, d\Omega_2,$$

where

$$\text{(28)} \qquad d\Omega_2 = \sin\alpha_2\, d\alpha_2\, d\beta_2.$$

In this region α_2 is the angle made by the wave normal with the *negative* z-axis, as shown in Fig. 114. It is thus also the angle of incidence at which an elementary plane wave meets the earth's surface.

The Hertz potential Π_r of the reflected field can be constructed from plane waves in the same manner. Each elementary wave of the bundle will have the form

$$f_r\, e^{ik_2[x\sin\alpha_2\cos\beta_2 + y\sin\alpha_2\sin\beta_2 + (z_0 + z)\cos\alpha_2]}.$$

Likewise the refracted, or transmitted, field in the domain $z < 0$ is constructed from the elementary waves

$$f_t\, e^{ik_2z_0 \cos \alpha_2} \cdot e^{ik_1(x \sin \alpha_1 \cos \beta_1 + y \sin \alpha_1 \sin \beta_1 - z \cos \alpha_1)}.$$

The undetermined functions f_r and f_t are reflection and refraction coefficients. To simplify the notation we write

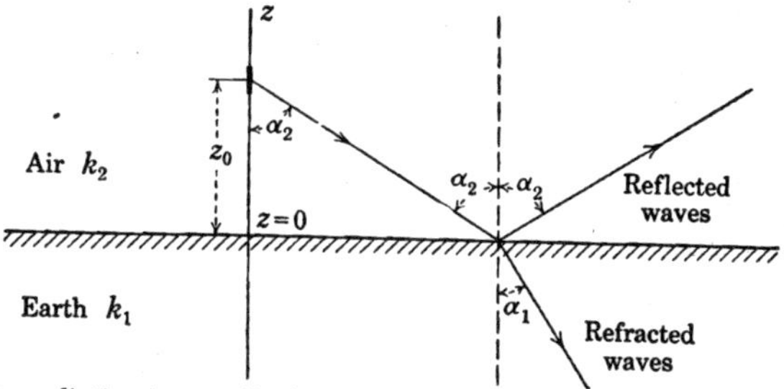

FIG. 114.—The radiation from a dipole can be represented by superposition of plane waves which are reflected and refracted at the surface of the earth.

$$\mathbf{k}_2 \cdot \mathbf{r} = k_2(x \sin \alpha_2 \cos \beta_2 + y \sin \alpha_2 \sin \beta_2), \tag{29}$$
$$\mathbf{k}_1 \cdot \mathbf{r} = k_1(x \sin \alpha_1 \cos \beta_1 + y \sin \alpha_1 \sin \beta_1).$$

Then at all points in the domain $0 \leq z \leq z_0$ the resultant field is

$$\Pi_2 = \frac{ik_2}{2\pi} \int_0^{2\pi} \int_0^{\frac{\pi}{2} - i\infty} (f_r\, e^{ik_2z \cos \alpha_2} + e^{-ik_2z \cos \alpha_2}) e^{i\mathbf{k}_2\cdot\mathbf{r} + ik_2z_0 \cos \alpha_2}\, d\Omega_2. \tag{30}$$

At all points for which $z \leq 0$,

$$\Pi_1 = \frac{ik_1}{2\pi} \int_0^{2\pi} \int_0^{\frac{\pi}{2} - i\infty} f_t\, e^{i\mathbf{k}_1\cdot\mathbf{r} - ik_1z \cos \alpha_1 + ik_2z_0 \cos \alpha_2}\, d\Omega_2. \tag{31}$$

At $z = 0$, the boundary conditions lead to

$$\frac{\partial \Pi_1}{\partial z} = \frac{\partial \Pi_2}{\partial z}, \qquad k_1^2 \Pi_1 = k_2^2 \Pi_2. \tag{32}$$

Imposing these relations on (30) and (31), one obtains first

$$k_1 \sin \alpha_1 = k_2 \sin \alpha_2, \qquad \beta_1 = \beta_2. \tag{33}$$

The elementary waves form a circular cone about the z-axis and the angles of reflection and refraction satisfy Snell's law (8), *page* 491. A further calculation gives for the coefficients:

$$
(34)\quad
\begin{aligned}
f_r &= \frac{k_1 \cos \alpha_2 - k_2 \cos \alpha_1}{k_1 \cos \alpha_2 + k_2 \cos \alpha_1} = \frac{Z_{21}^2 \cos \alpha_2 - \sqrt{Z_{21}^2 - 1 + \cos^2 \alpha_2}}{Z_{21}^2 \cos \alpha_2 + \sqrt{Z_{21}^2 - 1 + \cos^2 \alpha_2}}, \\
f_t &= \left(\frac{k_2}{k_1}\right)^3 \frac{2k_1 \cos \alpha_2}{k_1 \cos \alpha_2 + k_2 \cos \alpha_1} \\
&= \frac{1}{Z_{21}^3} \frac{2Z_{21}^2 \cos \alpha_2}{Z_{21}^2 \cos \alpha_2 + \sqrt{Z_{21}^2 - 1 + \cos^2 \alpha_2}},
\end{aligned}
$$

where $Z_{21} = \mu_2 k_1/\mu_1 k_2$. Now the lines of magnetic field intensity are circles whose centers lie on the z-axis. The meridian planes in Fig. 114 are also planes of incidence, whence it is apparent that the magnetic vector of each elementary wave is normal to its plane of incidence, while its electric vector is parallel to this plane although not transverse to the direction of propagation. Reference to (19), page 494, shows that apart from a factor in f_t, the *functions* f_r *and* f_t *are the Fresnel reflection and refraction coefficients for a plane wave.*

When $z > z_0$, Eq. (30) must be modified such as to vanish properly at infinity. We now measure α from the positive z-axis and write

$$
(35)\quad \Pi_2 = \frac{ik_2}{2\pi} \int_0^{2\pi} \int_0^{\frac{\pi}{2} - i\infty} (f_r\, e^{ik_2 z_0 \cos \alpha_2} + e^{-ik_2 z_0 \cos \alpha_2}) e^{i\mathbf{k}_2 \cdot \mathbf{r} + ik_2 \cos \alpha_2}\, d\Omega_2.
$$

If finally $z_0 = 0$, this reduces to

$$
(36)\quad \Pi_2 = \frac{ik_1^3}{2\pi k_2^2} \int_0^{2\pi} \int_0^{\frac{\pi}{2} - i\infty} f_t(\alpha_2)\, e^{ik_2[r \sin \alpha_2 \cos (\phi - \beta_2) + z \cos \alpha_2]} \sin \alpha_2\, d\alpha_2\, d\beta_2,
$$

while beneath the surface

$$
(37)\quad \Pi_1 = \frac{ik_1}{2\pi} \int_0^{2\pi} \int_0^{\frac{\pi}{2} - i\infty} f_t(\alpha_2)\, e^{ik_2 r \sin \alpha_2 \cos (\phi - \beta_2) - \sqrt{k_1^2 - k_2^2 \sin^2 \alpha_2}} \sin \alpha_2\, d\alpha_2\, d\beta_2.
$$

It is now a simple matter to effect the transition to Sommerfeld's solution. By (80), page 412, we have

$$
(38)\quad J_0(\lambda r)\, e^{ihz} = \frac{1}{2\pi} \int_0^{2\pi} e^{i\lambda r \cos (\phi - \beta) + ihz}\, d\beta.
$$

Let

$$
(39)\quad \lambda = k_2 \sin \alpha_2, \qquad h = k_2 \cos \alpha_2 = i\sqrt{\lambda^2 - k_2^2}.
$$

When $\alpha_2 = 0$, $\lambda = 0$, and when $\alpha = \frac{\pi}{2} - i\infty$ $\lambda = \infty$. In virtue of these relations, (36) transforms to

$$(40)\qquad \Pi_2 = \frac{ik_1^3}{k_2^2}\int_0^{\frac{\pi}{2}-i\infty} f_t(\alpha_2)\, J_0(k_2 r \sin\alpha_2) e^{ik_2 z\cos\alpha_2}\sin\alpha_2\, d\alpha_2$$
$$= 2k_1^2\int_0^\infty \frac{J_0(\lambda r)}{N} e^{-\sqrt{\lambda^2-k_2^2}\,z}\,\lambda\, d\lambda,$$

while the same substitutions transform (37) to

$$(41)\qquad \Pi_1 = 2k_2^2\int_0^\infty \frac{J_0(\lambda r)}{N} e^{+\sqrt{\lambda^2-k_1^2}\,z}\,\lambda\, d\lambda,$$

where N is the function defined by (22). Thus the formal solution of Weyl is identical with that of Sommerfeld in the special case $z_0 = 0$.

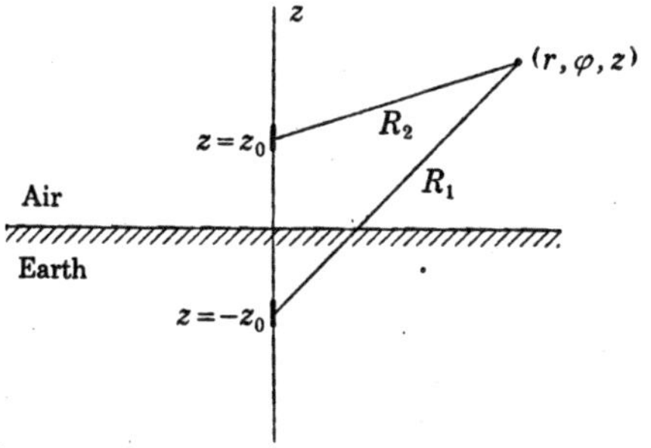

FIG. 115.—An image of the dipole is located in the earth at the point $z = -z_0$.

In the general case where $z_0 \neq 0$, we replace f_r in (35) by

$$(42)\qquad f_r = \left(\frac{k_1}{k_2}\right)^3 f_t - 1.$$

Then (35) can be resolved into three distinct parts, reducing in virtue of (27) and (40) to

$$(43)\qquad \Pi_2 = \frac{e^{ik_2R_2}}{R_2} - \frac{e^{ik_2R_1}}{R_1} + 2k_1^2\int_0^\infty \frac{J_0(\lambda r)}{N} e^{-\sqrt{\lambda^2-k_2^2}\,(z+z_0)}\,\lambda\, d\lambda,$$

where R_1 and R_2 are the distances shown in Fig. 115. This expression holds for all values of $z > 0$. The resultant field at the point of observation is thus composed of a direct contribution from the vertical electric dipole at $z = +z_0$, a contribution from its *image* at $z = -z_0$, and a supplementary term which takes into account the effect of a finite conductivity of the earth. If $k_1 \to \infty$, the last term of (43) approaches

$$(44)\qquad \lim_{k_1\to\infty} 2k_1^2\int_0^\infty \frac{J_0(\lambda r)}{N} e^{-\sqrt{\lambda^2-k_2^2}\,(z+z_0)}\,\lambda\, d\lambda = \frac{2e^{ik_2R_1}}{R_1}.$$

and

$$(45)\qquad \Pi_2 \to \frac{e^{ik_2R_2}}{R_2} + \frac{e^{ik_2R_1}}{R_1},$$

by virtue of Eq. (17).

Identical results have been obtained by Niessen[1] through an application of the Kirchhoff-Huygens method discussed in Sec. 8.13. The surface integrals are extended over the plane earth and over small spheres which exclude the singularities occurring at the source and the point of observation.

9.30. Van der Pol Solution.—An interesting and useful transformation of the Sommerfeld-Weyl expressions has been discovered by van der Pol.[2] Starting from the integral for the reflected field

$$(46)\quad \Pi_r = \int_0^\infty \frac{k_1^2\sqrt{\lambda^2-k_2^2}-k_2^2\sqrt{\lambda^2-k_1^2}}{k_1^2\sqrt{\lambda^2-k_2^2}+k_2^2\sqrt{\lambda^2-k_1^2}}\,\frac{J_0(\lambda r)e^{-\sqrt{\lambda^2-k_2^2}\,(z+z_0)}}{\sqrt{\lambda^2-k_2^2}}\,\lambda\,d\lambda,$$

obtained from (18) and (20), it is shown that the total field—primary and reflected—at any point above the earth's surface can be expressed as a volume integral extended over a certain domain in real space. The successive steps in the demonstration involve properties of Bessel functions which have been discussed in the earlier pages of this volume. Details will be omitted here. For all values of $z > 0$, van der Pol obtains the expression

$$(47)\qquad \Pi_2 = \frac{e^{ik_2R_2}}{R_2} - \frac{e^{ik_2R_1}}{R_1} + \frac{1}{\pi}\int_V \frac{\partial}{\partial\zeta}\left(\frac{e^{ik_2R_i}}{R_i}\right)\frac{\partial}{\partial\zeta}\left(\frac{e^{ik_1p}}{p}\right)dv$$
$$= \frac{e^{ik_2R_2}}{R_2} + \frac{e^{ik_2R_1}}{R_1} - \frac{1}{\pi}\int_V \frac{e^{ik_2R_i}}{R_i}\frac{\partial^2}{\partial\zeta^2}\left(\frac{e^{ik_1p}}{p}\right)dv.$$

The volume element $dv = \rho\,d\rho\,d\zeta\,d\beta$, and the domain of integration is $0 \le \rho \le \infty$, $0 \le \zeta \le \infty$, $0 \le \beta \le 2\pi$. The distance $p = \sqrt{\rho^2+\zeta^2}$, and the quantity R_i is a "complex distance" defined by

$$(48)\qquad R_i = \sqrt{\rho^2 + 2\rho r\cos\beta + r^2 + \left(z + z_0 + \frac{k_1^2}{k_2^2}\zeta\right)^2}.$$

The sign of the root is chosen such that (48) has a positive real part, and it is assumed that k_2 is real. The domain of integration can be interpreted as the half space *below* the geometrical image of the source. This is the shaded part of Fig. 116. The variables ρ, ζ, β are thus cylindrical coordinates with the image point at $z = -z_0$ taken as an origin. ζ is measured positive downwards.

When the conductivity of the earth is large, the propagation constant k_1 has a large imaginary part. Consequently the wave $\frac{\partial^2}{\partial\zeta^2}\left(\frac{e^{ik_1p}}{p}\right)$ is rapidly attenuated and contributes an appreciable amount to the integral only when p is small. In other words, *only that part of the lower medium*

[1] NIESSEN, *Ann. Physik*, **18**, 893, 1933.
[2] VAN DER POL, *Physica*, **2**, 843, 1935.

(earth) in the immediate neighborhood, but below, the geometrical image sends secondary waves to the observer. In the limit of infinite conductivity the region containing the effective sources of secondary radiation shrinks to a point coincident with the geometrical image itself.

An analogous, somewhat simpler, solution is given also by van der Pol for the case of a magnetic dipole, and the method can be extended to include horizontal dipoles. Since any source of electromagnetic radiation can be represented in terms of a distribution of electric and magnetic dipoles, the bearing of these calculations on many other problems of physical optics and electromagnetic theory is apparent. One might mention, in passing, the rather complicated phenomena associated with

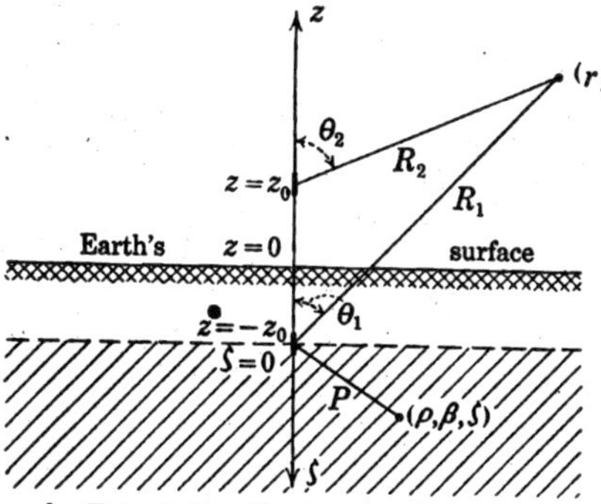

FIG. 116.—In the van der Pol solution the domain of integration is the half-space below the plane $z = -z_0$.

total reflection, which we have treated only for the idealized case of an infinite plane wave.

9.31. Approximation of the Integrals.—The several methods of approach to the problem of reflection from a plane surface all lead to integral representations of the field, or more precisely of its Hertzian vector, that are easily shown to be identical. A much more difficult task, which has occupied the attention of many investigators, is the reduction of these integrals to practical formulas for the field intensity. Sommerfeld's attack was based on a deformation of the path of integration in the complex λ plane. The function $1/N$ defined in (22) has branch points at $\lambda = \pm k_1$ and $\lambda = \pm k_2$, and a pole at

$$\lambda_0 = \pm \frac{k_1 k_2}{\sqrt{k_1^2 + k_2^2}}. \tag{49}$$

The corresponding Riemann surface has four sheets; on only one of these are the conditions fulfilled necessary for the convergence of the integral at infinity. According to Sommerfeld the path of integration can be resolved on this sheet into three parts: the first a loop from infinity about the branch point $\lambda = +k_1$, the second a similar loop about $\lambda = +k_2$, and the third any small circle about the pole at $\lambda = \lambda_0$. The contributions of the loops about the branch points are Q_1 and Q_2, while the residue at the pole gives a term P, so that $\Pi_2 = Q_1 + Q_2 + P$. The terms Q_1 and Q_2 were developed in inverse powers of the distance from the dipole, the dominant terms of the series being proportional respectively to e^{ik_1R}/R and e^{ik_2R}/R. Consequently Q_1 and Q_2 were interpreted as "space waves." On the other hand, for P Sommerfeld obtained a function of the form

$$P \sim \frac{1}{\sqrt{r}}\, e^{i\lambda_0 r - \sqrt{\lambda_0^2 - k_2^2}\, z}, \qquad (z > 0),$$

(50)

$$P \sim \frac{1}{\sqrt{r}}\, e^{i\lambda_0 r + \sqrt{\lambda_0^2 - k_1^2}\, z}, \qquad (z < 0),$$

which was interpreted as a "surface wave."

Historically these surface waves are of considerable interest. The guiding of a plane electromagnetic wave along the plane interface separating a dielectric and a good conductor seems to have been first investigated by Cohn[1] and shortly thereafter by Uller.[2] Zenneck[3] recognized the bearing of these researches on the propagation of radio waves and showed that the field equations admit a solution that can be interpreted as a surface wave guided by a plane interface separating any two media. This solution is identical with the surface wave derived in another manner in Sec. 9.13. Zenneck was quite clear as to the limitations of his results. He showed that a wave with a forward tilt, following a plane earth and attenuated in the vertical as well as the horizontal direction is compatible with Maxwell's equations, and that such a wave would explain many of the observed phenomena of radio transmission. There was no proof as yet, on the other hand, that a radio antenna does in fact generate a wave of this type. Sommerfeld's 1909 paper undertook to complete the demonstration, and the term P of Eq. (50) was interpreted as the surface wave component (in cylindrical coordinates) of the total field.

Doubt as to the validity of Sommerfeld's resolution was first raised by Weyl in his 1919 paper. Weyl obtains an asymptotic series representation of the diffracted field by applying the method of "steepest descent" (page 368) to the integral (36). In case $|k_1|/k_2 \gg 1$, Weyl's solution also reduces to a form which can be interpreted as the superposition of a space and a surface wave, but the Weyl surface wave is not identical with that of Sommerfeld and Zenneck.

[1] COHN, "Das elektromagnetische Feld," Leipzig, 1900.
[2] ULLER, Rostock Diss., 1903.
[3] ZENNECK, *Ann. Physik*, **23**, 846, 1907.

The discrepancy between the two forms of solution has been examined more closely in recent years, although it can hardly be said that the problem is closed. Burrows[1] has pointed out that numerically the transmission formulas based on Sommerfeld's results differ from those of Weyl by just the surface wave term P, and has made careful measurements which support the results of Weyl. The discrepancy is large only when the displacement current in the ground is comparable to the conduction current. Asymptotic expansions have been given by Wise[2] and by Rice[3] which show that the term P in Sommerfeld's solution is canceled when all terms of the series for Q_1, Q_2, and P are taken into account. Finally, attention was called by Norton[4] to an error in sign occurring in Sommerfeld's 1909 paper. This error does not appear in the 1926 paper, but the numerical results given there are accurate only when the conductivity of the earth is large.

From a practical standpoint the situation can be summed up as follows: The results of Sommerfeld, and the charts of Rolf[5] which are based upon them, cannot be relied upon when the displacement current is appreciable, as is the case at ultrahigh frequencies. The formulas of van der Pol, Niessen, Burrows and Norton[6] are for the most part extensions of Weyl's results. Field intensities calculated by these formulas are in accord with each other and apparently in good agreement with experiment. The same is true of the results obtained by Strutt,[7] who applies an alternative method to obtain an asymptotic solution.

To give some idea of the character of the formulas to be used for practical calculations, we present a summary of the results obtained by Norton for the vertical electric dipole. The Hertz vector of the resultant field at any point above the earth's surface is

$$\Pi_2 = \frac{e^{ik_2R_2}}{R_2} - \frac{e^{ik_2R_1}}{R_1} + V, \tag{51}$$

where V is the infinite integral expressed by the third term of (43). The approximate formula proposed for V is

$$V = [(1 - R_v)F + 1 + R_v]\frac{e^{ik_2R_1}}{R_1}, \tag{52}$$

where

$$R_v = \frac{Z_{21}^2 \cos\theta_1 - \sqrt{Z_{21}^2 - \sin^2\theta_1}}{Z_{21}^2 \cos\theta_1 + \sqrt{Z_{21}^2 - \sin^2\theta_1}}, \tag{53}$$

[1] BURROWS, *Nature*, Aug. 15, 1936; *Proc. Inst. Radio Engrs.*, **25**, 219, 1937.
[2] WISE, *Bell System Tech. J.*, **16**, 35, 1937.
[3] RICE, *Bell System Tech. J.*, **16**, 101, 1937.
[4] NORTON, *Nature*, **135**, 954, June 8, 1935; *Proc. Inst. Radio Engrs.*, **25**, 1192, 1937.
[5] ROLF, *Proc. Inst. Radio Engrs.*, **18**, 391, 1930.
[6] NORTON, *Proc. Inst. Radio Engrs.*, **24**, 1367, 1936; **25**, 1203, 1937.
[7] STRUTT, *Ann. Physik*, **1**, 721, 1929; **4**, 1, 1930; **9**, 67, 1931.

(54) $$\cos\theta_1 = \frac{z + z_0}{R_1}, \qquad Z_{21} = \frac{k_1}{k_2}.$$

R_v is the Fresnel reflection coefficient for a plane wave incident at an angle θ_1. The apparent source is in this case at a distance z_0 *below* the surface. (See Fig. 115.) The function F is defined by

(55) $$F = [1 + i\sqrt{\pi w}\, e^{-w}\, erfc(-i\sqrt{w})],$$

where

(56) $$erfc(-i\sqrt{w}) = \frac{2}{\sqrt{\pi}}\int_{-i\sqrt{w}}^{\infty} e^{-x^2}\,dx = \frac{2i}{\sqrt{\pi}}\int_{i\infty}^{\sqrt{w}} e^{x^2}\,dx,$$

(57) $$w = p_1\left(1 + \frac{Z_{21}^2\cos\theta_1}{\sqrt{Z_{21}^2 - \sin^2\theta_1}}\right)^2 = \frac{4p_1}{(1 - R_v)^2},$$

(58) $$p_1 = \frac{ik_2R_1}{2Z_{21}^4}(Z_{21}^2 - \sin^2\theta_1) = pe^{ib}.$$

Since

(59) $$Z_{21}^2 = \frac{\epsilon_1}{\epsilon_0}\left(1 + \frac{i\sigma_1}{\omega\epsilon_1}\right) = \frac{\epsilon_1}{\epsilon_0}(1 + i\eta),$$

one may also write

(60) $$p_1 = \frac{\pi R_1}{\lambda}\frac{\epsilon_0}{\epsilon_1}\left[-\eta + i\left(1 - \frac{\epsilon_0}{\epsilon_1}\sin^2\theta_1\right)\right]\frac{1}{(1 + i\eta)^2},$$

where λ is the free-space wave length. A detailed numerical discussion of the function F is given by Norton.

In order that the final formula shall express the field intensity of a current element, the Hertzian vector is multiplied by an appropriate amplitude factor. According to (44), page 431, the Hertz vector of a linear current element $I\,dz$ is

(61) $$d\Pi = \frac{i}{4\pi\omega\epsilon}\frac{e^{ikR}}{R}\,I\,dz.$$

Upon multiplication of (51) by $iI\,dz/4\pi\omega\epsilon_2$ and applying (3), Norton finds for the z-component of electric field intensity in volts per meter the expression

(62) $$dE_z = \frac{i\omega\mu_0}{4\pi}I\,dz\left[\sin^2\theta_2\frac{e^{ik_2R_2}}{R_2} + R_v\sin^2\theta_1\frac{e^{ik_2R_1}}{R_1}\right.$$
$$\left. + \frac{(1 - R_v)}{Z_{21}^4}(Z_{21}^4 - Z_{21}^2 + \sin^2\theta_1)F\frac{e^{ik_2R_1}}{R_1} + \text{higher order terms in } \frac{1}{R_2}\right].$$

The terms in $1/R_2^2$ and $1/R_2^3$ referred to represent the contributions of the "static" and "induction" fields and may be neglected at distances of ten wave lengths or so from the source. The angle θ_2 is defined by

$$\cos\theta_2 = \frac{z - z_0}{R_2}. \tag{63}$$

An analogous expression can be obtained for the component dE_r, which together with (62) determines the tilt or polarization of the field.

In the special case of a current element on or very near the earth $z_0 = 0$ and $R_1 = R_2 = R$, $\theta_1 = \theta_2 = \theta$. The total electric field intensity at any point such that $R \gg \lambda$ can then be expressed as the vector sum of two components,

$$d\mathbf{E}_{\text{space}} = -\frac{i\omega\mu_0}{4\pi} I\,dz \sin\theta\,(1 + R_v)\frac{e^{ik_2R}}{R}\mathbf{a}_\theta, \tag{64}$$

$$d\mathbf{E}_{\text{surface}} = \frac{i\omega\mu_0}{4\pi} I\,dz\,(1 - R_v)F\frac{e^{ik_2R}}{R} \times\left[\mathbf{a}_z + \mathbf{a}_r\frac{\sin\theta}{Z_{21}^2}\left(1 + \frac{\cos^2\theta}{2}\right)\sqrt{Z_{21}^2 - \sin^2\theta}\right], \tag{65}$$

where $\mathbf{a}_\theta$ is a unit vector tangent to the meridian circle through the point of observation in the direction of increasing θ and, consequently, perpendicular to the radius vector $\mathbf{R}$, while $\mathbf{a}_z$ and $\mathbf{a}_r$ are the unit vectors in a cylindrical coordinate system parallel to the z- and r-directions. As the conductivity of the ground approaches infinity, $R_v \to 1$. The component (65) then vanishes and the total field (64) at any point reduces to that of a dipole and its image. If the conductivity is finite, $R_v \to -1$ as $\theta \to \pi/2$. Near the surface of the earth, (64) now vanishes and the total field is represented by the term (65) alone. In virtue of this resolution one must agree with Norton that the terms "space" and "surface" waves are fully justified, even though the "surface wave" (65) is not identical with that predicted by Zenneck and Sommerfeld. Equation (65) is an expression for the "ground wave" which accounts for the major part of long-wave radio transmission, while (64) is the "high-angle" radiation which is reflected from the ionosphere to give rise to strong signals at great distances.

Extension of the formal solution from a plane to a spherical earth offers no serious difficulties. The field of the dipole or current element must be expressed in terms of spherical vector wave functions by such methods as were discussed in Chap. VIII. Since, however, the radius of the earth is vastly greater than the wave length, the convergence of the series representations of the field is poor and the difficulties of reducing the formal solution to practical terms have appeared truly formidable. Recently much progress has been made in overcoming these purely analytical troubles and a detailed discussion of the diffraction of a dipole field by a finitely conducting sphere for all values of the radius/wave-length ratio has been given by van der Pol and Bremmer.[1]

[1] VAN DER POL and BREMMER, *Phil. Mag.*, **24**, 141, 825, 1937; **25**, 817, 1938.

Problems

1. A plane wave reflected from a plane surface at normal incidence is expressed by

$$\psi = A(e^{ikz} + r\,e^{-ikz})e^{-i\omega t},$$

where the reflection coefficient r is the ratio of reflected to incident amplitude. Let

$$r = -e^{-2\gamma}$$

and show that

$$\psi = 2iAe^{-\gamma}\sin(kz - i\gamma).$$

Note that the relative phase of the two waves at $z = 0$ is equal to twice the imaginary part of γ, and that the ratio of amplitudes without regard to phase is $e^{-2Re(\gamma)}$.

2. Plane waves are reflected at the plane boundary separating two dielectrics. Let $n = \sqrt{\epsilon_1/\epsilon_2}$ be the relative index of refraction and assume the primary wave to be incident in medium (1) and refracted in medium (2). Let r be the ratio of the amplitude of the reflected to that of the incident electric vector. Show that when **E** is normal to the plane of incidence

$$r_\perp = \frac{n\cos\theta_0 - \sqrt{1 - n^2\sin^2\theta_0}}{n\cos\theta_0 + \sqrt{1 - n^2\sin^2\theta_0}},$$

and for **E** parallel to the plane of incidence

$$r_\parallel = \frac{\cos\theta_0 - n\sqrt{1 - n^2\sin^2\theta_0}}{\cos\theta_0 + n\sqrt{1 - n^2\sin^2\theta_0}},$$

where θ_0 is the angle of incidence.

3. Plane waves are reflected at the plane interface of two dielectrics whose inductive capacities differ by a very small amount. Following the notation of Problem 2, let $n = 1 + \Delta n$ and show that $r_\perp \simeq r_\parallel$, both equal approximately to

$$r \simeq \frac{1 - \sqrt{1 - 2\,\Delta n\tan^2\theta_0}}{1 + \sqrt{1 - 2\,\Delta n\tan^2\theta_0}} \simeq \frac{\phi - \sqrt{\phi^2 - 0.657 \times 10^{-4}\,\Delta n}}{\phi + \sqrt{\phi^2 - 0.657 \times 10^{-4}\,\Delta n}},$$

where $\phi = \frac{\pi}{2} - \theta_0$ and in the second formula is expressed in *degrees.*

Apply this result to the reflection of radio waves from an air-mass boundary in the troposphere. Assume $\Delta n = \pm 10^{-4}$ and plot r as a function of ϕ. Under what circumstances can total reflection occur?

4. A plane wave is reflected at normal incidence from an earth surface. Assume the specific inductive capacity to be 9, and the conductivity to range from 10^{-2} to 10^{-6} mho/meter due to variations in moisture content. Plot the shift in phase at the surface of reflection as a function of conductivity for a 10- and for a 500-meter wave.

5. Referring to the theory of total reflection developed in Sec. 9.7, calculate for this case the time average of the components of energy flow normal and parallel to the surface of reflection, in both the first and the second medium.

6. A 20-cm. wave from a radio altimeter strikes at normal incidence a sheet of ice which has formed over a body of salt water. Calculate the approximate thickness of ice which will lead to a minimum value of the reflection coefficient. What is this minimum value? Assume the conductivity of the ice to be negligible.

7. Surface waves are set up at the plane interface of two dielectric media in the manner described in Sec. 9.13. Discuss the field and the nature of the energy flow. Show that the angle of incidence necessary to establish such waves is in this case identical with the Brewster angle.

8. The effect of a variable transition layer on the reflection of waves passing from one medium to another is a question of considerable importance and approximate solutions have on occasion led to erroneous conclusions. The true nature of the phenomenon has been clarified through an investigation by Wallot, *Ann. Physik*, **60**, 734, 1919.

Assume that media (1), (2), and (3) are separated by plane surfaces at $z = 0$ and $z = d$ as in Fig. 96, page 511. The conductivity of all three media is zero. Regions (1) and (3) are homogeneous dielectrics with constant inductive capacities ϵ_1 and ϵ_3. The inductive capacity of (2) is a function of z of the form

$$\epsilon_2 = \epsilon_1(1 + az)^n,$$

the constant a being related to the thickness d by the relation

$$ad = \left(\frac{\epsilon_3}{\epsilon_1}\right)^{\frac{1}{n}} - 1.$$

A linearly polarized plane wave passes from medium (1) into medium (2) at normal incidence. Let

$$E_x = E(z)\, e^{-i\omega t}, \qquad H_y = H(z)\, e^{-i\omega t}.$$

Introduce a new variable w defined by

$$w - w_0 = \int_0^z \sqrt{\epsilon}\, dz$$

and show that the field equations are

$$\frac{dE}{dw} = \frac{i\omega\mu}{\sqrt{\epsilon}} H, \qquad \frac{dH}{dw} = i\omega \sqrt{\epsilon}\, E,$$

or

$$\ddot{E} + \frac{\dot{\epsilon}}{2\epsilon} \dot{E} + \omega^2\mu E = 0, \qquad \ddot{H} - \frac{\dot{\epsilon}}{2\epsilon} \dot{H} + \omega^2\mu H = 0,$$

the dot indicating differentiation with respect to w. For the assumed variation of ϵ_2 we have

$$w = \frac{2\sqrt{\epsilon_1}}{a(n+2)} (1 + az)^{\frac{n+2}{2}} = w_0(1 + az)^{\frac{n+2}{2}}$$

or

$$\epsilon_2 = \epsilon_1 \left(\frac{w}{w_0}\right)^{\frac{2n}{n+2}}.$$

Let

$$p = \frac{n+1}{n+2}, \qquad n = \frac{2p - 1}{1 - p},$$

and find for the field equations

$$\ddot{E} + \frac{2p-1}{w}\dot{E} + \omega^2\mu E = 0,$$

$$\ddot{H} + \frac{1-2p}{w}\dot{H} + \omega^2\mu H = 0,$$

and show that in the transition layer

$$E = \left(\frac{w}{w_0}\right)^{1-p}[A\,H^{(1)}_{p-1}(\omega\sqrt{\mu}\,w) + B\,H^{(2)}_{p-1}(\omega\sqrt{\mu}\,w)],$$

$$H = i\sqrt{\frac{\epsilon_1}{\mu}}\left(\frac{w}{w_0}\right)^{p}[A\,H^{(1)}_{p}(\omega\sqrt{\mu}\,w) + B\,H^{(2)}_{p}(\omega\sqrt{\mu}\,w)].$$

Determine the constants A and B from the boundary conditions and obtain expressions for the ratios of reflected and transmitted amplitudes to the amplitude of the incident wave. Show that reflection is only appreciable when the thickness of the transition layer is of the order of the wave length, and that in the limit of an infinitely thin layer one obtains the classical expression for the reflection coefficient. Note that the theory applies also to the propagation of current and voltage along a transmission line with varying parameters.

9. An infinite plane slab of thickness $2a$ is embedded in another medium. The propagation constants of the slab and external material are respectively k_1 and k_2. Show that either transverse electric or transverse magnetic waves can be propagated along the slab. In the case of transverse magnetic waves with **H** parallel to the slab show that the propagation factor h determining the phase velocity and attenuation of the allowed waves in the direction of propagation is given by the roots of

$$\tanh a\sqrt{h^2 - k_1^2} = \frac{\mu_1 k_2^2}{\mu_2 k_1^2}\sqrt{\frac{k_1^2 - h^2}{k_2^2 - h^2}}.$$

The problem is solved most quickly by applying the wave functions of Problem 1, Chap. VI, and by matching normal wave impedances at each surface.

10. Determine the pressure exerted by a plane electromagnetic wave incident on a plane surface as a function of the angle of incidence and the type of polarization.

11. A plane wave is incident at an angle θ upon an infinite slab of material bounded by parallel plane faces and of thickness d. The constants of the slab are ϵ, μ, σ and the medium on either side is free space. Calculate the pressure exerted by the wave on the slab. (Debye, *Ann. Physik*, **30**, 57, 1909; Lebedew, *Ann. Physik*, **6**, 433, 1901; Nichols and Hull, *Ann. Physik*, **12**, 225, 1903.)

12. A vertical electric dipole is located at $z = z_0$, $r = 0$, in medium (1) whose propagation constant is k_1. Medium (2) is an infinite slab with plane boundaries at $z = a < z_0$ and $z = 0$, and with a propagation constant k_2. The half space $z < 0$ is occupied by material of infinite conductivity. Assuming the primary field to be of the form $\frac{\exp(ikR)}{R}$, show that the Hertz vector of the secondary or reflected field in the first medium is

$$\Pi_{sec} = \int_0^\infty f_r J_0(\lambda r)e^{-\sqrt{\lambda^2-k_1^2}(z+z_0-2a)}\frac{\lambda\,d\lambda}{\sqrt{\lambda^2-k_1^2}},$$

$$f_r = \frac{k_2^2\sqrt{\lambda^2-k_1^2} - k_1^2\sqrt{\lambda^2-k_2^2}\tanh(\sqrt{\lambda^2-k_2^2}\,a)}{k_2^2\sqrt{\lambda^2-k_1^2} + k_1^2\sqrt{\lambda^2-k_2^2}\tanh(\sqrt{\lambda^2-k_2^2}\,a)},$$

where r is the horizontal distance from the source to the point of observation. (Van der Pol and Bremmer, *Phil. Mag.*, **24**, 825, 1937.)

13. In most practical cases the field and current distribution within a metallic conductor, even at very high radio frequencies can be determined to a satisfactory approximation from the equations

$$\mathbf{E} = i\omega\mathbf{A}, \qquad \mathbf{B} = \nabla \times \mathbf{A}, \qquad \mathbf{J} = i\omega\sigma\mathbf{A},$$
$$\nabla^2\mathbf{A} + i\omega\mu\sigma\mathbf{A} = 0, \qquad \nabla \cdot \mathbf{A} = 0.$$

(*Cf.* Problem 20, Chap. V.) This neglects entirely the effect of the external field on the propagation factor within the conductor and assumes the time to enter as a factor $\exp(-i\omega t)$. At a surface separating the conductor from an external dielectric the normal component of $\mathbf{J}$ is zero, and both $\mathbf{J}$ and $\mathbf{A}$ must be finite everywhere.

Apply these equations to find the distribution of longitudinal current over the circular cross section of a very long, straight wire. Find an expression for the longitudinal component of $\mathbf{E}$ at the surface of the wire in the form

$$E_z = RI + L_i \frac{dI}{dt},$$

where I is the total current, R the resistance, and L_i the *internal* inductance, both per unit length. Compare with the results of Sec. 9.17.

14. A tubular conductor of circular cross section has an outer radius a and inner radius b. Obtain expressions for the current density, alternating-current resistance, and internal inductance per unit length.

15. An alternating magnetic field is directed perpendicular to a long conducting cylinder of radius a. In the region considered the field is uniform and the wave length $\gg a$. Calculate the power dissipated per unit length by eddy currents induced in the conductor. (See Frenkel, "Elektrodynamik," Vol. II, p. 392.)

16. A thin, conducting, plane sheet of infinite extent is placed in a homogeneous, alternating magnetic field. The frequency is sufficiently low that the "skin depth" is very much greater than the thickness of the lamination. Find expressions for the vector potential of the induced eddy currents and the power dissipated per unit area.

17. Following the general theory of Sec. 6.2, discuss the propagation of waves along perfect cylindrical conductors. In an orthogonal system of cylindrical coordinates u^1, u^2, z, assume $E_z = H_z = 0$ and show that the components of the field can be derived from two scalar functions ϕ and ψ by the formulas

$$E_1 = -\frac{1}{h_1}\frac{\partial\phi}{\partial u^1}, \qquad E_2 = -\frac{1}{h_2}\frac{\partial\phi}{\partial u^2},$$
$$B_1 = \frac{1}{h_2}\frac{\partial\psi}{\partial u^2}, \qquad B_2 = -\frac{1}{h_1}\frac{\partial\psi}{\partial u^1},$$

and that these potential and stream functions satisfy

$$\frac{\partial\phi}{\partial z} + \frac{\partial\psi}{\partial t} = 0, \qquad \frac{1}{\mu}\frac{\partial\psi}{\partial z} + \epsilon\frac{\partial\phi}{\partial t} = 0,$$
$$\frac{\partial^2\phi}{\partial z^2} - \mu\epsilon\frac{\partial^2\phi}{\partial t^2} = 0,$$
$$\frac{\partial}{\partial u^1}\left(\frac{h_2}{h_1}\frac{\partial\phi}{\partial u^1}\right) + \frac{\partial}{\partial u^2}\left(\frac{h_1}{h_2}\frac{\partial\phi}{\partial u^2}\right) = 0,$$

$$\psi = \sqrt{\mu\epsilon}\,\phi, \qquad \phi = f(u^1, u^2)g(z - vt), \qquad v = \frac{1}{\sqrt{\mu\epsilon}},$$

where μ and ϵ pertain to the external dielectric. Let q be the charge per unit length, I the current on a particular conductor, and C a closed contour linking this conductor only and lying in a plane z = constant. Show that

$$q = -\epsilon \int_C \frac{\partial \phi}{\partial n}\,ds, \qquad I = -\frac{1}{\mu}\int_C \frac{\partial \psi}{\partial n}\,ds = \frac{1}{\sqrt{\mu\epsilon}}\,q,$$

the differentiation being in the direction of the outward normal.

The theory applies to any number of parallel conductors of constant but arbitrary cross section provided the sum of the charges q in any transverse plane is zero. Otherwise the functions ϕ and ψ are irregular (logarithmic) at infinity. In the case of two parallel conductors a and b we have $q_a = -q_b$. The capacity and inductance per unit length are defined by

$$C = \frac{q_a}{\phi_a - \phi_b}, \qquad L = \frac{\psi_a - \psi_b}{I_a},$$

exactly as in the static regime. Prove that in m.k.s. units

$$\mu\epsilon = LC$$

and show that current and charge satisfy

$$\frac{1}{C}\frac{\partial q}{\partial z} + L\frac{\partial I}{\partial t} = 0, \qquad \frac{\partial I}{\partial z} + \frac{\partial q}{\partial t} = 0,$$

$$\frac{\partial^2 q}{\partial z^2} - LC\frac{\partial^2 q}{\partial t^2} = 0, \qquad \frac{\partial^2 I}{\partial z^2} - LC\frac{\partial^2 I}{\partial t^2} = 0.$$

18. The theory of propagation along a system of parallel, cylindrical conductors discussed in Problem 17 was extended by Abraham to the case of finite conductivity. The solution is approximate but valid and useful in most cases of technical importance. One assumes (a) that the waves are transverse magnetic. We have seen in Sec. 9.15 that, apart from the case of an axially symmetric field, this is not strictly correct, but the error involved is, in general, extremely small. Assume, next, that the field is harmonic and take for the scalar potential

$$\phi = -\frac{\partial \Pi_z}{\partial z} = f(u^1, u^2)e^{i\gamma z - i\omega t}.$$

Show that at any point in the conductor or in the dielectric the components of the field are given by

$$E_1 = -\frac{1}{h_1}\frac{\partial \phi}{\partial u^1}, \qquad E_2 = -\frac{1}{h_2}\frac{\partial \phi}{\partial u^2},$$

$$H_1 = -\frac{\omega\epsilon + i\sigma}{\gamma}E_2, \qquad H_2 = \frac{\omega\epsilon + i\sigma}{\gamma}E_1, \qquad H_z = 0,$$

$$\frac{\partial}{\partial u^1}\left(\frac{h_2}{h_1}\frac{\partial \phi}{\partial u^1}\right) + \frac{\partial}{\partial u^2}\left(\frac{h_1}{h_2}\frac{\partial \phi}{\partial u^2}\right) = ih_1h_2\gamma E_z.$$

Calculate the z-component of the complex Poynting vector $\mathbf{S}^*$ and integrate this over the entire transverse xy-plane. Let the value of this integral be W^*. Show that

$$W^* = \frac{2\gamma}{\omega\mu_e\epsilon} \times \text{magnetic energy in dielectric} + \frac{2\gamma}{i\sigma\mu_i} \times \text{magnetic energy in conductor},$$

$$-\frac{\partial W^*}{\partial z} = \text{Joule heat} + 2i\omega \text{ (electric energy} - \text{magnetic energy)},$$

where the subscript e refers to the external or dielectric medium, i to the conductor, and all quantities are per unit length of line. Now make the following assumptions:

(b) In the dielectric the longitudinal component γE_z is very small relative to the transverse field, and consequently in this medium the components E_1, E_2, H_1, H_2 have the values computed in the case of perfect conductivity. Furthermore, the values of capacity C and *external* inductance L_e are unchanged, with $\mu_e\epsilon = L_eC$ as in Problem 17. The contribution of the longitudinal component to the electric energy in the dielectric is negligible.

(c) In the metal the intensity of the transverse electric components is negligible with respect to the longitudinal component, and the electric energy is negligible with respect to the magnetic energy.

Consider a pair of conductors carrying equal currents in opposite directions. Let R be twice the alternating-current resistance per unit length per conductor, L_i the internal inductance per unit length of the double circuit. Both these quantities are functions of the frequency and depend upon the proximity of the conductors as well as upon their cross sections. They are obtained approximately by the method outlined in Problem 13. Remembering that the energies involved are time averages, show that

$$W^* = \frac{\gamma}{\omega\mu_e\epsilon}\frac{1}{2}L_eI\tilde{I} + \frac{\gamma}{i\sigma\mu_i}\frac{1}{2}L_iI\tilde{I},$$

$$-\frac{\partial W^*}{\partial z} = \frac{1}{2}RI\tilde{I} + i\omega\left(\frac{\gamma\tilde{\gamma}}{\omega^2\mu_e\epsilon} - 1\right)\frac{1}{2}L_eI\tilde{I} - \frac{i\omega}{2}L_iI\tilde{I},$$

$$\frac{\partial W^*}{\partial z} = i(\gamma - \tilde{\gamma})W^*.$$

The second term in W^* can be neglected and one obtains for the propagation factor

$$\gamma^2 = \omega^2\mu_e\epsilon\left[\left(1 + \frac{L_i}{L_e}\right) + \frac{iR}{\omega L_e}\right].$$

19. According to Problem 18 the propagation factor for waves along a pair of parallel, cylindrical conductors is given by

$$\gamma^2 = \omega^2\mu_e\epsilon\left[\left(1 + \frac{L_i}{L_e}\right) + i\frac{R}{\omega L_e}\right].$$

Let $\gamma = \alpha + i\beta$ and $L = L_i + L_e$, the total inductance per unit length. Show that

$$\alpha = \omega\left\{\frac{LC}{2}\left[\sqrt{1 + \left(\frac{R}{\omega L}\right)^2} + 1\right]\right\}^{\frac{1}{2}},$$

$$\beta = \omega\left\{\frac{LC}{2}\left[\sqrt{1 + \left(\frac{R}{\omega L}\right)^2} - 1\right]\right\}^{\frac{1}{2}}.$$

If $$\left(\frac{R}{\omega L}\right)^2 \gg 1, \qquad \alpha \simeq \beta \simeq \sqrt{\frac{RC\omega}{2}}.$$

If $$\left(\frac{R}{\omega L}\right)^2 \ll 1, \qquad \alpha \simeq \omega\sqrt{LC}\left[1 + \frac{1}{8}\left(\frac{R}{\omega L}\right)^2\right], \qquad \beta \simeq \frac{R}{2}\sqrt{\frac{C}{L}}.$$

20. A Lecher wire system consists of two long parallel wires of circular cross section. The distance between centers is $2a$ and the radius of each wire is b. Show that the capacitance and external self-inductances are

$$C = \frac{\pi\epsilon}{\ln\left(\dfrac{a + \sqrt{a^2 - b^2}}{b}\right)} \qquad \text{farads/meter,}$$

$$L_e = \frac{\mu}{\pi}\ln\left(\frac{a + \sqrt{a^2 - b^2}}{b}\right) \qquad \text{henrys/meter.}$$

If $a \gg b$, the effect of one wire on the current distribution in the other can be neglected and the resistance and internal self-inductance per meter per conductor are given approximately by the formulas at the end of Sec. 9.17. To obtain an idea of the orders of magnitude assume the wires to be 2 mm. in radius (about a No. 6 gauge) and of copper, spaced 10 cm. Calculate C, L_e, L_i, R for frequencies of 10^8, 10^5, and 10^2 cycles/sec. and compute the attenuation in decibels per meter. Discuss the effect of a change in spacing or radius on the various parameters.

21. When leakage through the dielectric is considered in the preceding problems, one obtains Eq. (27), page 550, for the general equation of propagation of current along a pair of parallel conductors.

$$\frac{\partial^2 I}{\partial z^2} - LC\frac{\partial^2 I}{\partial t^2} - (LG + RC)\frac{\partial I}{\partial t} - RGI = 0.$$

This equation was first studied in detail by Lord Kelvin in connection with submarine cables. By the methods of the Fourier or Laplace transform discussed in Chap. V, obtain a general solution in terms of initial conditions imposed on current and voltage. Show that by an appropriate choice of parameters the velocity and attenuation can be made independent of frequency so that a signal is propagated without distortion.

22. The feasibility of an Atlantic cable was first demonstrated in a theoretical paper published by Lord Kelvin in 1855. For the case in question the self-inductance and leakage could be neglected with respect to the series resistance and the potential with respect to ground was governed by

$$CR\frac{\partial V}{\partial t} = \frac{\partial^2 V}{\partial z^2}.$$

According to Kelvin the time necessary to produce a given potential at a distance z from the origin is proportional to CR multiplied by the square of the distance. Verify this result.

23. The problem of eliminating "cross talk" between pairs of telephone lines by means of cylindrical shields is of great technical importance. To reduce the problem to a form which can be handled satisfactorily it is usually assumed that variation along the line can be neglected, so that the field is essentially two-dimensional. This proves justifiable as long as the wave length is very much greater than the distances between wires and between wires and shield. Show first that on this assumption the field equations resolve into two groups:

$$E_r = \frac{i\omega\mu}{k^2}\frac{1}{r}\frac{\partial H_z}{\partial \theta},$$

(I)
$$E_\theta = -\frac{i\omega\mu}{k^2}\frac{\partial H_z}{\partial r},$$

$$\frac{1}{r}\left[\frac{\partial}{\partial r}(rE_\theta) - \frac{\partial E_r}{\partial \theta}\right] = i\omega\mu H_z.$$

$$H_r = \frac{1}{i\omega\mu}\frac{1}{r}\frac{\partial E_z}{\partial \theta},$$

(II)
$$H_\theta = -\frac{1}{i\omega\mu}\frac{\partial E_z}{\partial r},$$

$$\frac{1}{r}\left[\frac{\partial}{\partial r}(rH_\theta) - \frac{\partial H_r}{\partial \theta}\right] = \frac{k^2}{i\omega\mu}E_z.$$

Interference caused by a field of type (I) is commonly called "electrostatic" cross talk; that associated with a field of type (II) is known as "electromagnetic" cross talk. The terms are unfortunate and misleading but firmly entrenched in engineering literature.

Group (II) represents the field of one or more alternating-current filaments. Show that the field due to a single filament I is

$$H_\theta = \frac{ikI}{4}H_1^{(1)}(kr)\,e^{-i\omega t}, \qquad E_z = -\frac{\omega\mu I}{4}H_0^{(1)}(kr)\,e^{-i\omega t},$$

whose radial impedance is

$$Z_r = -\frac{E_z}{H_\theta} = -\frac{i\omega\mu}{k}\frac{H_0^{(1)}(kr)}{H_1^{(1)}(kr)}.$$

A current solenoid or vortex line produces a field of type (I), but such a field is also generated by pairs of charged filaments of opposite sign, whence the term "electrostatic" interference.

24. A transmission line consists of two parallel wires of radius b whose separation between centers is $2a$. The inductance and capacitance per unit length are known from the results of Problem 20. So far as the external field is concerned, the system is equivalent to a pair of current filaments of strength I and opposite sign, and a pair of line charges of strength q per unit length and opposite sign. A study of the problem in bipolar coordinates shows that the location of these filaments does not coincide exactly with the centers of the wires, but the effect of the deviation on the field can be ignored except perhaps in the immediate neighborhood of the wire itself. Let the central line between conductors coincide with the z-axis of a cylindrical coordinate system. At distances from this axis which are very much less than the wave length the field is governed by the equations of Problem 23.

Show that at points whose distance r from the central axis $\gg a$, the field of a current pair or doublet is

$$E_z = -\frac{\omega\mu k}{2}\,aIH_1^{(1)}(kr)\cos\theta\,e^{-i\omega t},$$

$$H_r = -\frac{ik}{2}\frac{aI}{r}H_1^{(1)}(kr)\sin\theta\,e^{-i\omega t},$$

$$H_\theta = \frac{ik}{2}\frac{aI}{r}\left[H_1^{(1)}(kr) - krH_0^{(1)}(kr)\right]\cos\theta\,e^{-i\omega t},$$

and find expressions for the field of a corresponding charge doublet.

25. Extend the results of Problem 24 with the help of the addition formula (11), Sec. 6.11, and show that any pair of eccentric filaments can be replaced by a system of line sources on the central axis emitting cylindrical waves of different orders.

26. The transmission line described in Problem 24 is enclosed by a coaxial copper shield of inner radius c and thickness δ. Assume $c \gg a$ and obtain expressions for the field penetrating the shield. Discuss the results in terms of the frequency and dimensions of the system. Consider fields of both electric and magnetic type. (On this and the preceding problems see the discussion and bibliography by Schelkunoff, *Bell System Tech. Jour.*, **13**, 532, 1934.)

27. A hollow tube of rectangular cross section is bounded by thick metal walls. The inner dimensions are $x = a$, $y = b$. Assuming first perfect conductivity, obtain expressions for the allowed modes of the transverse electric field (H waves) and the corresponding transverse magnetic modes (E waves). The components of the field can be written down directly from the results of Problem 1, Chap. VI. If the tangential components of **E** are to vanish at the boundaries, the fields must be periodic in the x- and y-directions and hence $h_1 = m\pi/a$, $h_2 = n\pi/b$. Correspondingly, the various orders of E wave are designated by $E_{m,n}$; for an H wave of order m, n, one writes $H_{m,n}$.

Find expressions for the phase velocity, group velocity, wave length within the tube, critical frequency, and critical wave length. Sketch the lines of electric and magnetic field intensity for the first few orders for both E and H waves. The $H_{0,1}$ wave has the simplest structure, the lowest critical frequency, and the lowest attenuation in the case of tubes of finite conductivity. For these reasons it is the mode commonly used in practice.

Assuming the walls to be metal of finite conductivity, derive an approximate expression for the attenuation of the $H_{0,1}$ wave in decibels per meter. Show that there are optimum values of the ratio a/b and the frequency leading to a minimum attenuation.

28. Find the lowest natural frequency for electromagnetic oscillations in a cavity whose form is that of a right circular cylinder of radius a and length l. The walls are metal of conductivity σ. Calculate the equivalent inductance, resistance, and the parameter Q for this mode. Follow the procedure of Sec. 9.24.

29. Discuss the propagation of transverse electric and transverse magnetic waves in a perfectly conducting hollow tube of elliptic cross section.

30. The axis of an infinitely long, circular cylinder of radius a coincides with the z-axis of a coordinate system. The propagation factor of the cylinder is k_1, that of the external medium is k_2. A plane wave whose direction is that of the positive x-axis is incident upon the cylinder. Derive expressions for the diffracted field and the scattering cross section. Obtain approximate formulas for the case $a/\lambda \ll 1$. Discuss the two cases:

a. Electric vector of the incident wave parallel to z-axis;

b. Magnetic vector of the incident wave parallel to z-axis.

The theory has been confirmed experimentally by Schaefer, *Ann. Physik*, **31**, 455, 1910; *Zeit. Physik*, **13**, 166, 1923.

31. A plane, linearly polarized wave is scattered by a spherical body of radius a. The material of the sphere has a propagation factor k_1; that of the external medium, which is assumed to be nonconducting, is k_2. Let **R** be the radius vector from the center of the sphere to a point of observation. The plane containing **R** and the axis of propagation is called the *plane of vision*. The plane containing the direction of polarization of the incident wave and the axis of propagation is the *plane of oscillation*. Let ϕ be the angle between the plane of vision and the plane of oscillation, and θ the angle between **R** and the axis of propagation. Let λ be the wave length in the external medium. Show that, if $\lambda \gg a$, the intensity of the scattered Rayleigh radiation is given by

$$\frac{\bar{S}}{\bar{S}_0} = (2\pi)^4 \frac{a^6}{\lambda^4 R^2} \left| \frac{k_1^2 - k_2^2}{k_1^2 + 2k_2^2} \right|^2 (\cos^2\theta \cos^2\phi + \sin^2\phi),$$

where $\bar{S}$ is the mean value of the scattered intensity and $\bar{S}_0$ that of the incident wave. Discuss the polarization of the scattered radiation.

32. Linearly polarized light is scattered by very small, nonconducting, spherical particles. Show that in a direction parallel to the electric vector of the incident light the intensity of the scattered radiation varies inversely as the *eighth* power of the wave length. Polarized white light scattered in this direction appears as a purer blue than when viewed at other angles and the effect is referred to as Tyndall's "residual blue."

33. Discuss the attenuation of ultrahigh-frequency radio waves due to scattering in rain and fog from the following data, taken from Humphreys, "Physics of the Air."

	Radius, meters	Drops/meter^3
Fog	5×10^{-6}	10^7
Light rain	2×10^{-4}	4,000
Moderate rain	5×10^{-4}	500
Excessive rain	10^{-3}	450

Express the decrease in intensity in decibels per kilometer.

34. Calculate the force exerted by a plane, linearly polarized wave on a dielectric sphere, assuming the wave length to be very large relative to the radius of the sphere. (Debye, *Ann. Physik*, **30**, 57, 1909.)

35. A plane, linearly polarized electromagnetic wave is incident upon a perfectly conducting sphere. Assuming the wave length to be very large relative to the radius of the sphere, calculate the total force exerted on the sphere. (Debye, *Ann. Physik*, **30**, 57, 1909.)

Estimate the change in the result in case the sphere were of copper.

36. A spherical cavity 17.5 cm. in radius, such as was described in Sec. 9.24, is bounded by thick copper walls. Calculate the frequency and logarithmic decrement of the lowest possible mode of oscillation. Repeat for a similar sphere of half the radius.

37. An oscillating electric dipole is located at a point whose distance from the center of a sphere is b. The radius of the sphere is a, with $a < b$, and the dipole is oriented in the radial direction. The propagation factor of the sphere is k_1; that of the external medium is k_2. Show that the Hertz vector of a field satisfying the boundary conditions and behaving as e^{ik_2R}/ik_2R at the dipole is

$$\Pi = \frac{e^{ik_2R}}{ik_2R} + \sum_{n=0}^{\infty} (2n+1)R_n \frac{j_n(k_2a)}{h_n^{(1)}(k_2a)} h_n^{(1)}(k_2b)\, h_n^{(1)}(k_2r)\, P_n(\cos\theta),$$

when $r > a$, and

$$\Pi = \frac{k_2^2}{k_1^2} \sum_{n=0}^{\infty} (2n+1)(1+R_n) \frac{j_n(k_2a)}{j_n(k_1a)} h_n^{(1)}(k_2b)\, j_n(k_1r)\, P_n(\cos\theta),$$

when $r < a$, where

$$R_n = \frac{-\left[\frac{1}{x}\frac{d}{dx}\ln\,[x\,j_n(x)]\right]_{x=k_2a} + \left[\frac{1}{y}\frac{d}{dy}\ln\,[y\,j_n(y)]\right]_{y=k_1a}}{\left[\frac{1}{x}\frac{d}{dx}\ln\,[x\,h_n^{(1)}(x)]\right]_{x=k_2a} - \left[\frac{1}{y}\frac{d}{dy}\ln\,[y\,j_n(y)]\right]_{y=k_1a}}$$

and where R is the distance from the dipole to the observer, r the distance from the center to the observer, and θ the polar angle measured from the dipole axis. The Hertz vector is radial.

38. The axis of an infinite, perfectly conducting, right circular cone coincides with the negative z-axis of a coordinate system. The vertex of the cone is at the origin and its generators make an angle $\theta = \theta_0$ with the positive z-axis. The external space is dielectric. Investigate the natural modes of propagation on the cone. (Macdonald, "Electric Waves," Cambridge University Press, 1902.)

39. Discuss the radiation from the open end of a semi-infinite, straight wire carrying an oscillating current by allowing $\theta_0 \rightarrow \pi$ in the results of Problem 38. Compare this with Problem 9, Chap. VIII.

40. In Problem 9, Chap. VII, expressions were found for a field with rotational symmetry in spheroidal coordinates. Apply this theory to a perfectly conducting prolate spheroid whose eccentricity differs only slightly from unity. Calculate the frequency of the fundamental mode of oscillation and carry the approximations far enough to give an expression for the damping due to radiation. (Abraham, *Math. Ann.*, **52**, 81, 1899.)

41. A circular loop of wire 15 meters in radius carries a 60-cycle alternating current of 10 amp. The loop lies on a flat surface of earth whose constants are $\epsilon/\epsilon_0 = 4$, $\sigma = 10^{-4}$ mho/meter.

a. Calculate the electric field intensity in volts per meter at a point on the surface of the earth whose distance from the center of the loop is r.

b. What is the effective resistance at the driving point of the loop? Neglect losses in the wire and assume that all dissipation is due to the finite conductivity of the earth.

(Note: At this low frequency the loop is equivalent to a vertical magnetic dipole. The field can be calculated by the methods of Secs. 9.28 to 9.31, and the losses determined from the mean vertical energy flow at the surface of the earth.)

42. Consider a system of conductors and dielectrics characterized by the parameters ϵ, μ, σ. The media are isotropic but not necessarily homogeneous. Sharp boundary surfaces may be replaced by layers of rapid transition. There are two kinds of impressed forces, the one $\mathbf{F}_1$ causing dielectric polarization, the other $\mathbf{F}_2$ producing conduction currents. Thus

$$\mathbf{D} = \epsilon(\mathbf{E} + \mathbf{F}_1), \qquad \mathbf{J} = \sigma(\mathbf{E} + \mathbf{F}_2).$$

$\mathbf{F}_1$ and $\mathbf{F}_2$ are continuous functions of position. These forces are started during an infinitely short interval Δt, ending at the instant $t = 0$, and henceforth remain constant. Let

$$A = \int_0^t dt \int_V dv \left(\mathbf{F}_1 \cdot \frac{\partial \mathbf{D}}{\partial t} + \mathbf{F}_2 \cdot \mathbf{J}\right)$$

be the work done on the system by the impressed forces from the initial instant to the time t, and

$$W = \int_0^t dt \int_V dv\, \mathbf{F}_2 \cdot \mathbf{J}$$

the total generation of heat *at the final rate,* where **J** is the final value of current density. Then if U and T are the electric and magnetic energies of the field in the final state, show that

$$A = W + 2(U - T).$$

The theorem was stated first by Heaviside, "Electrical Papers," Vol. II, page 412, and proved by Lorentz, *Proc. Natl. Acad. Sci.*, **8**, 333, 1922. A similar theorem for electrical circuits is expressed in Pierce, "Electric Oscillations," page 40, 1st ed., 1920.

APPENDIX I

A. NUMERICAL VALUES OF FUNDAMENTAL CONSTANTS

Permittivity of free space ϵ_0,

$$\epsilon_0 = 8.854 \times 10^{-12} \simeq \frac{1}{36\pi} \times 10^{-9} \quad \text{farad/meter.}$$

Permeability of free space μ_0,

$$\mu_0 = 4\pi \times 10^{-7} = 1.257 \times 10^{-6} \quad \text{henry/meter.}$$

$$c = \frac{1}{\sqrt{\mu_0\epsilon_0}} = 2.998 \times 10^8 \simeq 3 \times 10^8 \quad \text{meters/sec.}$$

$$1/\epsilon_0 = 1.129 \times 10^{11} \quad \text{meters/farad.}$$

$$1/\mu_0 = 7.958 \times 10^5 \quad \text{meters/henry.}$$

$$\sqrt{\frac{\mu_0}{\epsilon_0}} = 376.7 \quad \text{ohms.}$$

$$\frac{1}{2\pi}\sqrt{\frac{\mu_0}{\epsilon_0}} = 59.95 \simeq 60 \quad \text{ohms.}$$

$$\sqrt{\frac{\epsilon_0}{\mu_0}} = 2.654 \times 10^{-3} \quad \text{mho.}$$

Propagation constant $k = \alpha + i\beta$,

$$k^2 = \epsilon\mu\omega^2 + i\omega\mu\sigma = \epsilon\mu\omega^2(1 + i\eta),$$

$$\eta = \frac{\sigma}{\epsilon\omega} = 1.796 \times 10^{10}\frac{\sigma}{\nu\kappa_e} \simeq 1.8 \times 10^{10}\frac{\sigma}{\nu\kappa_e},$$

where σ is the conductivity in mhos per meter; ν, the frequency; and κ_e the specific inductive capacity ϵ/ϵ_0.

If $\eta \gg 1$,

$$\alpha \simeq \beta \simeq 1.987 \times 10^{-3}\sqrt{\nu\sigma\kappa_m} \simeq 2 \times 10^{-3}\sqrt{\nu\sigma\kappa_m},$$

$$k = \alpha + i\beta \simeq \sqrt{i\omega\mu\sigma},$$

where κ_m = specific magnetic permeability μ/μ_0.

If $\eta \ll 1$,

$$\alpha = \omega\sqrt{\epsilon\mu}\,(1 + \tfrac{1}{8}\eta^2 + \cdots),$$

$$\beta \simeq \frac{\sigma}{2}\sqrt{\frac{\mu}{\epsilon}} = 188.3\,\sigma\sqrt{\frac{\kappa_m}{\kappa_e}}.$$

B. DIMENSIONS OF ELECTROMAGNETIC QUANTITIES

The quantities appearing in the table below are expressed in terms of mass M, length L, time T, and charge Q. In the Giorgi m.k.s. system used exclusively in this book mass is measured in kilograms, length in meters, time in seconds, and charge in coulombs.

Quantity	Symbol	Dimensions	M.k.s. unit
Force	—	MLT^{-2}	Newton
Energy	—	ML^2T^{-2}	Joule
Power	—	ML^2T^{-3}	Watt
Charge	q	Q	Coulomb
Current	I	$T^{-1}Q$	Ampere
Charge density	ρ	$L^{-3}Q$	Coulomb/cubic meter
Current density	$\mathbf{J}$	$L^{-2}T^{-1}Q$	Ampere/square meter
Resistance	R	$ML^2T^{-1}Q^{-2}$	Ohm
Conductivity	σ	$M^{-1}L^{-3}TQ^2$	Mho/meter
Electric potential	ϕ	$ML^2T^{-2}Q^{-1}$	Volt
Electric field intensity	$\mathbf{E}$	$MLT^{-2}Q^{-1}$	Volt/meter
Capacitance	C	$M^{-1}L^{-2}T^2Q^2$	Farad
Dielectric displacement	$\mathbf{D}$	$L^{-2}Q$	Coulomb/square meter
Inductive capacity	ϵ	$M^{-1}L^{-3}T^2Q^2$	Farad/meter
Electric dipole moment	$\mathbf{p}$	LQ	Coulomb-meter
Magnetic flux	Φ	$ML^2T^{-1}Q^{-1}$	Weber
Flux density	$\mathbf{B}$	$MT^{-1}Q^{-1}$	Weber/square meter
Magnetomotive force	m.m.f.	$T^{-1}Q$	Ampere-turn
Magnetic field intensity	$\mathbf{H}$	$L^{-1}T^{-1}Q$	Ampere-turn/meter
Inductance	—	ML^2Q^{-2}	Henry
Permeability	μ	MLQ^{-2}	Henry/meter
Magnetic dipole moment	$\mathbf{m}$	$L^2T^{-1}Q$	Ampere-square meter

C. CONVERSION TABLES

Multiply the number of m.k.s. units below	by	To obtain the number of c.g.s. electromagnetic units of
Ampere	10^{-1}	Current I
Ampere/meter²	10^{-5}	Current density $\mathbf{J}$
Coulomb	10^{-1}	Charge q
Coulomb/meter³	10^{-7}	Charge density ρ
Farad	10^{-9}	Capacitance C
Henry	10^{9}	Inductance L
Joule	10^{7}	Energy (erg) W
Mho/meter	10^{-11}	Conductivity σ
Newton	10^{5}	Force (dyne) $\mathbf{F}$
Ohm	10^{9}	Resistance R
Volt	10^{8}	Potential ϕ
Weber	10^{8}	Magnetic flux (Maxwell) Φ
Weber/meter²	10^{4}	Flux density (Gauss) $\mathbf{B}$

Multiply the number of m.k.s. units below	by	To obtain the number of c.g.s. electrostatic units of
Ampere	3×10^9	Current I
Ampere/meter^2	3×10^5	Current density **J**
Coulomb	3×10^9	Charge q
Coulomb/meter^3	3×10^3	Charge density ρ
Farad	9×10^{11}	Capacitance C
Henry	$\frac{1}{9} \times 10^{-11}$	Inductance L
Joule	10^7	Energy (erg) W
Mho/meter	9×10^9	Conductivity σ
Newton	10^5	Force (dyne) **F**
Ohm	$\frac{1}{9} \times 10^{-11}$	Resistance R
Volt	$\frac{1}{3} \times 10^{-2}$	Potential ϕ
Weber	$\frac{1}{3} \times 10^{-2}$	Magnetic flux Φ
Weber/meter^2	$\frac{1}{3} \times 10^{-6}$	Flux density **B**

The quantities listed in the preceding tables are unaffected by the question of rationalization. The m.k.s. units employed in this book are rationalized. The c.g.s. electromagnetic and electrostatic systems are ordinarily unrationalized. Conversion factors for the three most important quantities affected by rationalization are tabulated below.

Multiply the number of rationalized	by	To obtain the number of unrationalized c.g.s. electromagnetic units of
Ampere-turns	$4\pi \times 10^{-1}$	Magnetomotive force (gilbert)
Ampere-turns/meter	$4\pi \times 10^{-3}$	Magnetic field intensity **H** (oersted)
M.k.s. units of dielectric displacement	$4\pi \times 10^{-5}$	Dielectric displacement **D**

Multiply the number of rationalized	by	To obtain the number of unrationalized c.g.s. electrostatic units of
Ampere-turns	$12\pi \times 10^9$	Magnetomotive force
Ampere-turns/meter	$12\pi \times 10^7$	Magnetic field intensity **H**
M.k.s. units of dielectric displacement	$12\pi \times 10^5$	Dielectric displacement **D**

In case the c.g.s. units are also rational, the factors 4π are suppressed.

APPENDIX II

FORMULAS FROM VECTOR ANALYSIS

Scalar functions are represented in the following by Greek letters; Roman letters are used for vectors.

(1) $\mathbf{a} \cdot \mathbf{b} \times \mathbf{c} = \mathbf{b} \cdot \mathbf{c} \times \mathbf{a} = \mathbf{c} \cdot \mathbf{a} \times \mathbf{b}$

(2) $\mathbf{a} \times (\mathbf{b} \times \mathbf{c}) = (\mathbf{a} \cdot \mathbf{c})\mathbf{b} - (\mathbf{a} \cdot \mathbf{b})\mathbf{c}$

(3)
$$\begin{aligned}(\mathbf{a} \times \mathbf{b}) \cdot (\mathbf{c} \times \mathbf{d}) &= \mathbf{a} \cdot \mathbf{b} \times (\mathbf{c} \times \mathbf{d}) \\ &= \mathbf{a} \cdot (\mathbf{b} \cdot \mathbf{d}\, \mathbf{c} - \mathbf{b} \cdot \mathbf{c}\, \mathbf{d}) \\ &= (\mathbf{a} \cdot \mathbf{c})(\mathbf{b} \cdot \mathbf{d}) - (\mathbf{a} \cdot \mathbf{d})(\mathbf{b} \cdot \mathbf{c})\end{aligned}$$

(4) $(\mathbf{a} \times \mathbf{b}) \times (\mathbf{c} \times \mathbf{d}) = (\mathbf{a} \times \mathbf{b} \cdot \mathbf{d})\mathbf{c} - (\mathbf{a} \times \mathbf{b} \cdot \mathbf{c})\mathbf{d}$

(5) $\nabla(\phi + \psi) = \nabla\phi + \nabla\phi$

(6) $\nabla(\phi\psi) = \phi\nabla\psi + \psi\nabla\phi$

(7) $\nabla \cdot (\mathbf{a} + \mathbf{b}) = \nabla \cdot \mathbf{a} + \nabla \cdot \mathbf{b}$

(8) $\nabla \times (\mathbf{a} + \mathbf{b}) = \nabla \times \mathbf{a} + \nabla \times \mathbf{b}$

(9) $\nabla \cdot (\phi\mathbf{a}) = \mathbf{a} \cdot \nabla\phi + \phi\nabla \cdot \mathbf{a}$

(10) $\nabla \times (\phi\mathbf{a}) = \nabla\phi \times \mathbf{a} + \phi\nabla \times \mathbf{a}$

(11) $\nabla(\mathbf{a} \cdot \mathbf{b}) = (\mathbf{a} \cdot \nabla)\mathbf{b} + (\mathbf{b} \cdot \nabla)\mathbf{a} + \mathbf{a} \times (\nabla \times \mathbf{b}) + \mathbf{b} \times (\nabla \times \mathbf{a})$

(12) $\nabla \cdot (\mathbf{a} \times \mathbf{b}) = \mathbf{b} \cdot \nabla \times \mathbf{a} - \mathbf{a} \cdot \nabla \times \mathbf{b}$

(13) $\nabla \times (\mathbf{a} \times \mathbf{b}) = \mathbf{a}\nabla \cdot \mathbf{b} - \mathbf{b}\nabla \cdot \mathbf{a} + (\mathbf{b} \cdot \nabla)\mathbf{a} - (\mathbf{a} \cdot \nabla)\mathbf{b}$

(14) $\nabla \times \nabla \times \mathbf{a} = \nabla\nabla \cdot \mathbf{a} - \nabla^2\mathbf{a}$

(15) $\nabla \times \nabla\phi = 0$

(16) $\nabla \cdot \nabla \times \mathbf{a} = 0$

If $\mathbf{r} = \mathbf{i}x + \mathbf{j}y + \mathbf{k}z$ is the radius vector drawn from the origin to the point (x, y, z), then

(17) $\nabla \cdot \mathbf{r} = 3, \qquad \nabla \times \mathbf{r} = 0.$

In the following formulas V is a volume bounded by the closed surface S. The unit vector $\mathbf{n}$ is normal to S and directed positively outwards.

(18) $\int_V \nabla\phi \, dv = \int_S \phi \mathbf{n} \, da$

(19) $\int_V \nabla \cdot \mathbf{a} \, dv = \int_S \mathbf{a} \cdot \mathbf{n} \, da$

(20) $\int_V \nabla \times \mathbf{a} \, dv = \int_S \mathbf{n} \times \mathbf{a} \, da$

Let S be an unclosed surface bounded by the contour C.

(21) $\int_S \mathbf{n} \times \nabla\phi \, da = \int_C \phi \, d\mathbf{s}$

(22) $\int_S \nabla \times \mathbf{a} \cdot \mathbf{n} \, da = \int_C \mathbf{a} \cdot d\mathbf{s}$

APPENDIX III

CONDUCTIVITY OF VARIOUS MATERIALS

Metals and Alloys

Material	Conductivity σ, mhos/meter at 20°C.
Aluminum, commercial hard drawn	3.54×10^7
Constantin, 60%Cu, 40%Ni	0.20×10^7
Copper, annealed	5.8005×10^7
Copper, hard drawn	5.65×10^7
German silver, 18% Ni	0.30×10^7
Gold, pure drawn	4.10×10^7
Iron, 99.98% pure	1.0×10^7
Steel	$0.5\text{–}1.0 \times 10^7$
Steel, manganese	0.14×10^7
Lead	0.48×10^7
Magnesium	2.17×10^7
Manganin, 84% Cu, 12% Mn, 4% Ni	0.23×10^7
Mercury	0.1044×10^7
Monel metal	0.24×10^7
Nichrome	0.10×10^7
Nickel	1.28×10^7
Silver, 99.98% pure	6.139×10^7
Tin	0.869×10^7
Tungsten	1.81×10^7
Zinc, trace Fe	1.74×10^7

Dielectrics

Material	Approximate conductivity, mhos/meter at 20°C.
Bakelite (average range)	$10^{-8}\text{–}10^{-10}$
Celluloid	10^{-8}
Ceresin	$<2 \times 10^{-17}$
Fiber, hard	5×10^{-9}
Glass, ordinary	10^{-12}
Glyptol	10^{-14}
Hard rubber	$10^{-14}\text{–}10^{-16}$
Marble	$10^{-7}\text{–}10^{-9}$
Mica	$10^{-11}\text{–}10^{-15}$
Paraffin	$10^{-14}\text{–}10^{-16}$
Porcelain	3×10^{-13}
Quartz, crystal	
$\parallel$ to axis	10^{-12}
$\perp$ to axis	3×10^{-15}

Material	Approximate conductivity, mhos/meter at 20°C.
Quartz, fused	$< 2 \times 10^{-17}$
Rosin	2×10^{-15}
Shellac	10^{-14}
Slate	10^{-6}
Sulfur	10^{-15}
Wood, paraffined	10^{-8}–10^{-11}
Alcohol, ethyl, 15°C	3.3×10^{-4}
Alcohol, methyl	7.1×10^{-4}
Petroleum	10^{-14}
Water, distilled, 18°C	2×10^{-4}

The conductivities of geological materials vary so greatly from one locality to the next that only approximate values can be given. The following figures indicate orders of magnitude:

Sea water, 3 to 5 mhos/meter. Samples taken from the Atlantic Ocean off the coasts of Massachusetts and New Jersey showed a conductivity of 4.3 mhos/meter.

Fresh water, 10^{-3} mhos/meter. The conductivity of small fresh-water lakes may be five or ten times larger.

Wet ground, 10^{-2} to 10^{-3} mhos/meter.

Dry ground, 10^{-4} to 10^{-5} mhos/meter.

SPECIFIC INDUCTIVE CAPACITY OF DIELECTRICS

The values tabulated below are approximately independent of frequency at frequencies less than one megacycle per second. Atmospheric pressure and 20°C. unless otherwise stated.

Gases	κ_e
Air, 0°C.	1.00059
40 atmospheres	1.0218
80 atmospheres	1.0439
Carbon dioxide, 0°C.	1.000985
Hydrogen, 0°C.	1.000264
Water vapor, 145°C.	1.00705

Liquids	κ_e
Acetone, 0°C.	26.6
Air, −191°C.	1.43
Alcohol,	
amyl	16.0
ethyl	25.8
methyl	31.2
Benzene	2.29
Glycerin, 15°C.	56.2
Oils,	
castor	4.67
linseed	3.35
petroleum	2.13
Water, distilled	81.1

Solids

	κ_e
Asphalt	2.68
Diamond	16.5
Glass,	
flint, density 4.5	9.90
flint, density 2.87	6.61
lead, density 3.0–3.5	5.4–8.0
Gutta-percha	3.3–4.9
Marble	8.3
Mica	5.6–6.0
Paper (cable insulation)	2.0–2.5
Paraffin	2.1
Porcelain	5.7
Quartz,	
⊥ to axis	4.69
∥ to axis	5.06
Rubber	2.3–4.0
Shellac	3.1
Slate	6.6–7.4
Sulphur,	
amorphous	3.98
cast, fresh	4.22
Wood, dry	
red beech ⊥ to fibers	3.6–7.7
red beech ∥ to fibers	2.5–4.8
oak ⊥ to fibers	3.6–6.8
oak ∥ to fibers	2.5–4.2

The data in this appendix have been taken largely from the "Smithsonian Physical Tables" and the "Handbook of Chemistry and Physics." These references should be consulted for other materials and for the temperature coefficients.

APPENDIX IV

ASSOCIATED LEGENDRE FUNCTIONS

$$P_0(z) = 1$$
$$P_1(z) = z = \cos\theta$$
$$P_1^1(z) = (1 - z^2)^{\frac{1}{2}} = \sin\theta$$
$$P_2(z) = \tfrac{1}{2}(3z^2 - 1) = \tfrac{1}{4}(3\cos 2\theta + 1)$$
$$P_2^1(z) = 3(1 - z^2)^{\frac{1}{2}}z = \tfrac{3}{2}\sin 2\theta$$
$$P_2^2(z) = 3(1 - z^2) = \tfrac{3}{2}(1 - \cos 2\theta)$$
$$P_3(z) = \tfrac{1}{2}(5z^3 - 3z) = \tfrac{1}{8}(5\cos 3\theta + 3\cos\theta)$$
$$P_3^1(z) = \tfrac{3}{2}(1 - z^2)^{\frac{1}{2}}(5z^2 - 1) = \tfrac{3}{8}(\sin\theta + 5\sin 3\theta)$$
$$P_3^2(z) = 15(1 - z^2)z = \tfrac{15}{4}(\cos\theta - \cos 3\theta)$$
$$P_3^3(z) = 15(1 - z^2)^{\frac{3}{2}} = \tfrac{15}{4}(3\sin\theta - \sin 3\theta)$$
$$P_4(z) = \tfrac{1}{8}(35z^4 - 30z^2 + 3) = \tfrac{1}{64}(35\cos 4\theta + 20\cos 2\theta + 9)$$
$$P_4^1(z) = \tfrac{5}{2}(1 - z^2)^{\frac{1}{2}}(7z^3 - 3z) = \tfrac{5}{16}(2\sin 2\theta + 7\sin 4\theta)$$
$$P_4^2(z) = \tfrac{15}{2}(1 - z^2)(7z^2 - 1) = \tfrac{15}{16}(3 + 4\cos 2\theta - 7\cos 4\theta)$$
$$P_4^3(z) = 105(1 - z^2)^{\frac{3}{2}}z = \tfrac{105}{8}(2\sin 2\theta - \sin 4\theta)$$
$$P_4^4(z) = 105(1 - z^2)^2 = \tfrac{105}{8}(3 - 4\cos 2\theta + \cos 4\theta)$$

INDEX

D

E

CPSIA information can be obtained at www.ICGtesting.com
Printed in the USA
BVOW05s0147181214

379143BV00005B/82/P

9 781446 517390